Vladimir V. Tsukruk and Srikanth Singamaneni

Scanning Probe Microscopy of Soft Matter: Fundamentals and Practices

Further Reading

Bowker, Michael / Davies, Philip R. (eds.)

Scanning Tunneling Microscopy in Surface Science

2009

ISBN: 978-3-527-31982-4

Vickerman, J. C. / Gilmore, I. (eds.)

Surface Analysis

The Principal Techniques

2009

Hardcover

ISBN: 978-0-470-01763-0

Softcover

ISBN: 978-0-470-01764-7

Fukumura, Hiroshi / Irie, Masahiro / Iwasawa, Yasuhiro / Masuhara, Hiroshi / Uosaki, Kohei (eds.)

Molecular Nano Dynamics

Vol. I: Spectroscopic Methods and Nanostructures / Vol. II: Active Surfaces, Single Crystals and Single Biocells

2009

ISBN: 978-3-527-32017-2

van Tendeloo, Gustaaf / van Dyck, Dirk / Pennycook, Stephen J. (eds.)

Handbook of Nanoscopy

2012

ISBN: 978-3-527-31706-6

Schlücker, Sebastian (ed.)

Surface Enhanced Raman Spectroscopy

Analytical, Biophysical and Life Science Applications

2010

ISBN: 978-3-527-32567-2

Schlüter, Dieter A. / Hawker, Craig / Sakamoto, Junji (eds.)

Synthesis of Polymers

New Structures and Methods

Series: Materials Science and Technology

2012

ISBN: 978-3-527-32757-7

Vladimir V. Tsukruk and Srikanth Singamaneni

Scanning Probe Microscopy of Soft Matter: Fundamentals and Practices

WILEY-VCH Verlag GmbH & Co. KGaA

The Authors

Prof. Dr. Vladimir V. Tsukruk
Georgia Inst. of Technology
School of Mat. Science & Eng.
771, Ferst Dr. N.W.
Atlanta, GA 30332-0245
USA

Prof. Dr. Srikanth Singamaneni
Washington Univ. in St. Louis
Dept. of Mechanical Engineering
and Materials Science
One Brookings Drive
St. Louis, MO 63130
USA

Library of Congress Card No.: applied for

British Library Cataloguing-in-Publication Data
A catalogue record for this book is available from the British Library.

Bibliographic information published by the Deutsche Nationalbibliothek
The Deutsche Nationalbibliothek lists this publication in the Deutsche Nationalbibliografie; detailed bibliographic data are available on the Internet at http://dnb.d-nb.de.

Typesetting Thomson Digital, Noida, India
Printing and Binding Fabulous Printers Pte Ltd, Singapore
Cover Design Adam-Design, Weinheim

Printed in Singapore
Printed on acid-free paper

Print ISBN: 978-3-527-32743-0
ePDF ISBN: 978-3-527-63997-7
oBook ISBN: 978-3-527-63995-3
ePub ISBN: 978-3-527-63996-0
mobi ISBN: 978-3-527-63998-4

The authors dedicate this book to their parents, spouses, and children for their encourgement and continuous support of their passion, career, and life.

Contents

Preface

The invention of scanning tunneling microscopy (STM) in the early 1980s by Rohrer, Binnig, and coworkers at the IBM Zurich Laboratories led to the establishment of a new class of proximity probe microscopies, known as scanning probe microscopy (SPM) [1–5]. The milestone introduction of principles of atomic force microscopy (AFM) in 1986 was especially critical for soft material applications [6]. In the past two decades, SPM has become so widespread that modern materials' characterization cannot be imagined without a battery of SPM techniques, which provide invaluable insight into nanoscale structure and properties of various materials. Specifically, since the introduction of the first AFM instruments in 1989 and the fast commercial introduction of a range of nondestructive SPM modes in the early 1990s, the number of research groups involved in, articles published on, and research presentations delivered concerning SPM techniques grew almost exponentially. Indeed, a quick search on the publication records of both concepts of AFM and polymers shows that in contrast to only a few papers that appeared from 1989–1992, around 1500 papers were published annually in the past 5 years. And this is only a low-bound estimate with the true number of all SPM (not just AFM) studies on soft matter probably exceeding 3000 papers annually.

The current state of SPM research in soft materials (organic, polymeric, and biological) warrants the summarization of recent results in a single all-inclusive volume. This volume will serve as a comprehensive reference to current practices by providing representative examples of a wide range of soft materials, from synthetic polymers to biological molecules. Few comprehensive books and collections on the subject were published in the 1990s with most of them covering practical aspects of SPM techniques, and there are only a few prominent books that cover the important aspects of SPM studies on soft materials [7–9]. Among significant volumes, Magonov and Whangbo introduced SPM practices and discussed numerous examples of high-resolution SPM imaging of polymeric and organic materials, Vancos released a collection of SPM polymer applications, a book by Marti and Amrein included SPM studies on a variety of biological materials, a comprehensive collection of SPM applications was edited by Colton *et al.*, and the book by Ratner and Tsukruk featured a number of chapters on fundamentals and initial SPM applications to polymeric materials [10–14].

In this book, we present a broad and fairly comprehensive coverage of current and recent scanning probe studies of soft materials; starting with a brief introduction of SPM fundamentals and principles, we summarize representative examples of imaging and probing of all major classes of soft materials and touch some related applications in lithography and sensing. However, it is worth noting that this book does not focus on detailed descriptions of basic SPM principles that have already been widely presented in earlier publications. Rather, this book presents a wide range of practical examples for various classes of soft materials with discussions of the most significant contributions, limitations, artifacts, and approaches that can be valuable for understanding a particular class of materials. Here, we focus mostly on recent results published in the past decade or so, although some significant and fundamental results from the previous decade have also been included when necessary.

The book is composed of four major parts and includes 20 chapters. Part One (Chapters 1–4) is devoted to a brief introduction of SPM fundamentals and principles with separate chapters focusing on a general introduction of AFM principles (Chapter 1), an introduction to the basic instrumentation and practices (Chapter 2), a summary of the fundamentals of physical interactions and the most common artifacts and misgivings in practical scanning (Chapter 3), and, finally, an introduction to advanced imaging and probing modes that are applicable to soft materials (Chapter 4).

Part Two (Chapters 5–8) comprehensively covers selected experimental modes that are especially important for SPM imaging and probing of soft materials. Nanomechanical probing routines of elastic, viscoelastic, shearing, and plastic properties of compliant surfaces are introduced and discussed in Chapter 5. Chapter 6 discusses the measurement of microthermal properties with SPM including glass transition, melting, and thermal expansion. SPM measurements of chemical, electrochemical, and electrical properties are covered in Chapter 7. Finally, optical measurements that include near-field optical microscopy, tip-enhanced spectroscopies, and integrated optical-SPM routines are briefly summarized in Chapter 8.

Part Three, the most extensive part of this book, which includes eight chapters (Chapters 9–16), presents a comprehensive analysis of recent representative results of SPM research on a wide variety of soft materials. Chapters 9 and 10 overview SPM studies of traditional one-component, homopolymer materials, in both amorphous and crystalline microstructural states, and in the glassy, rubbery, and viscoelastic physical states. Novel macromolecular architectures, which include dendrimers, brush molecules, star block copolymers, and hybrid nanoparticles, are considered in Chapter 11. Multicomponent polymeric materials such as polymer blends, block copolymers, polymer nanocomposites, and polymer fibers are discussed in Chapter 12. Chapters 13 and 14 focus on interfacial materials in the form of self-assembled organic monolayers, brush polymer surface layers, layer-by-layer multilayered films, and Langmuir–Blodgett films. Recent and most significant SPM results on their formation, morphologies and microstructures, and physical properties are covered in these chapters. Colloidal organic and hybrid microparticles, nanoparticles, and microcapsules are discussed in Chapter 15. Finally, SPM studies of biological materials, including adsorbed biomacromolecules, cells, and cell membranes, are briefly summarized in Chapter 16.

Finally, Part Four (Chapters 17–20) focuses on modern applications of SPM techniques that go beyond traditional imaging and probing of soft materials. This part considers advanced SPM probing on working microdevices and microfabricated structures in a practical device-related environment (Chapter 17), physical and chemical nanolithography and micropatterning of different compliant and hard surfaces with controlled plastic deformation and highly localized material and molecular deposition (Chapters 18 and 19), and some chemical and biological sensing approaches based on monitoring of AFM microcantilever deflections (Chapter 20).

The results discussed in this book are based both on a comprehensive literature analysis and on, to a great extent, the experimental results collected by the authors over about 20 years of SPM studies of a wide variety of soft materials in the Surface Engineering and Molecular Assemblies (SEMA) laboratory [15]. In this book, the authors include a number of their results on SPM imaging and probing of various materials published in archival journals and provide detailed descriptions of experimental routines, guidelines for identifying common artifacts, and basic routes for the analysis of the experimental data that have been adapted, developed, and established in the SEMA lab. Considering that the authors' own research is focused mostly on polymeric and interfacial materials, a significant fraction of the results is devoted to various important classes of polymeric materials, although many literature examples from low molecular weight organic materials, colloidal microparticles, and biological materials are presented as well.

Numerous representative examples are included in all chapters from the current literature to illustrate the present status of the field. However, despite this representation and the fact that the book includes about 2000 nonunique references and hundreds of images from present research literature, the authors had to go through the painful task of selecting very few primary and representative examples from massive publication records from respectable journals and numerous research groups to make this book relatively compact and still representative. For this selection routine, we used our own judgment and past experience to make the book adequate and objective, although some elements of the authors' own preferences might be found by a rigorous reader. Overall, we believe that in the end the book presented here represents a fairly comprehensive, albeit very condensed, summary of the present research results and experimental routines in modern SPM studies of soft materials.

This book would not have been completed without contributions from a number of present and former SEMA lab members who conducted SPM research and established experimental routines. Among former members of the SEMA lab, whose significant contributions are included in this book, we would like to acknowledge Prof. V. Bliznyuk (WMU), Prof. I. Luzinov (Clemson University), Prof. V. Gorbunov (Bruker), Prof. A. Sidorenko (UPST), Prof. C. Jiang (USD), Dr. M. Lemieux (Stanford), Dr. S. Peleshanko (HP), Dr. M. Ornatska (Clarkson University), Dr. M. McConney (AFRL), Dr. D. Zimnitsky (KCI), and Prof. E. Kharlampieva

(University of Alabama). Present members of the SEMA group whose results significantly contributed to this book include K. Anderson, M. Gupta, I. Choi, D. Kulkarni, I. Drachuk, and Dr. O. Shchepelina. We owe special thanks to our students S. Young, Z. Combs, K. Anderson, R. Muhlbauer, R. Kodiyath, C. H. Lee, R. Kattumenu, L. Tian, and S. Z. Nergiz who were extremely helpful with many technical aspects of this endeavor.

The authors would like to especially thank a number of researchers who discussed some topics with the authors and provided selected images for this book: Dr. S. Magonov, Prof. S. Sheiko, Prof. I. Luzinov, Prof. A. Fery, Prof. W. Huck, Prof. S. Cheng, and Prof. D. Reneker. Also special thanks to Prof. A. Fery who hosted one of the authors (VVT) during his sabbatical and provided excellent conditions for writing this book at University of Bayreuth, Germany. This stay and book writing has also been facilitated by the generous support provided by the Humboldt Research Award. It is important to reveal that Prof. Darrell Reneker (University of Akron), one of the pioneers in SPM applications to polymers, was critical in introducing one of the authors (VVT) to SPM research during his postdoctoral years at a very early stage in his career, which ultimately predetermined the appearance of this book.

The authors acknowledge generous and continuous support of their research results of SPM studies on soft matter that are included in this book from the US National Science Foundation (DMR, CBET, and IMMI Divisions), Air Force Office for Scientific Research, Air Force Research Lab, Defense Advanced Research Project Agency, Army Research Office, Department of Energy, Semiconductor Research Corporation, Humboldt Foundation, Petroleum Research Fund, and a number of private industries. Publishing houses, which include Wiley-VCH, American Chemical Society, American Physical Society, Elsevier, Royal Society of Chemistry, American Institute of Physics, Institute of Physics, and National Academy of Sciences of the United States of America all are acknowledged for copyright permissions for a number of images presented in this book.

Summarizing our book's philosophy, we could state that this book is written by experienced SPM users for beginner, recent, and modestly experienced users as well as curious graduate students who are interested in general guidelines on the applicability of different SPM techniques and practices, from gentle imaging to harsh probing, as applied to a wide class of soft materials. We believe this book will also be useful for more experienced SPM users who intend to branch out into a new class of soft materials and need to get a quick overview of what has been done and can be done with SPM for particular soft materials. We also think about graduate students in chemistry, physics, biology, materials, and engineering who need to understand principles and applications of SPM techniques. We hope that this book can be a useful reference for both undergraduate- and graduate-level courses on soft material characterization.

Washington University, St. Louis — *Vladimir V. Tsukruk*
Georgia Institute of Technology, Atlanta — *Srikanth Singamaneni*

References

1 Binnig, G. and Rohrer, H. (1982) Scanning tunneling microscopy. *Helv. Phys. Acta*, **55** (6), 726–735.
2 Binnig, G., Rohrer, H., Gerber, C., and Weibel, E. (1982) Tunneling through a controllable vacuum gap. *Appl. Phys. Lett.*, **40** (2), 178–180.
3 Binnig, G., Rohrer, H., Gerber, C., and Weibel, E. (1982) Surface studies by scanning tunneling microscopy. *Phys. Rev. Lett.*, **49** (1), 57–61.
4 Gerber, C. and Lang, H.P. (2006) How the doors to the nanoworld were opened. *Nat. Nanotechnol.*, **1** (1), 3–5.
5 Binnig, G. and Rohrer, H. (1987) Scanning tunneling microscopy – from birth to adolescence. *Rev. Mod. Phys.*, **59** (3), 615–625.
6 Binnig, G., Quate, C.F., and Gerber, C. (1986) Atomic force microscope. *Phys. Rev. Lett.*, **56** (9), 930–933.
7 Sarid, D. (1994) *Scanning Force Microscopy with Applications to Electric, Magnetic and Atomic Forces*, Revised Edition,Oxford University Press, Oxford.
8 Chernoff, D.A. and Magonov, S. (2003) Atomic force microscopy, in *Comprehensive Desk Reference of Polymer Characterization and Analysis*, Oxford University Press, Oxford.
9 Samori, P. (ed.) (2003) STM and AFM Studies on (Bio)molecular Systems: Unravelling the Nanoworld Series: Topics in Current Chemistry, vol. 285, Springer, Berlin.
10 Magonov, S.N. and Whangbo, M. (1995) Surface Analysis with STM and AFM: Experimental and Theoretical Aspects of Image Analysis, Wiley-VCH Verlag GmbH, Weinheim.
11 Vancos, G.J. (ed.) (1994) Special Issue: Atomic Force Microscopy. A Collection of Papers presented at the 67th ACS Colloid and Surface Science Symposium Devoted to Atomic Force Microscopy, the University of Toronto, June 20–23, 1993, *Colloid Surf. A*, **87** (3).
12 Marti, O. and Amrein, M. (eds) (1993) STM and SFM in Biology, Academic Press, San Diego.
13 Colton, R.,Engel, A., Frommer, J., Gaub, H., Gewirth, A., Guckenberger, R., Heckl, W., Parkinson, B., and Rabe, J. (eds) (1998) Procedures in Scanning Probe Microscopy, John Wiley & Sons Ltd., Chichester, pp.105–109.
14 Ratner, B.D. and Tsukruk, V.V. (eds) (1998) Scanning Probe Microscopy of Polymers, ACS Symposiums Series, vol. 694, American Chemical Society, Washington D.C.
15 SEMA web site:http://polysurf.mse.gatech.edu/.

Part One
Microscopy Fundamentals

Scanning Probe Microscopy of Soft Matter: Fundamentals and Practices, First Edition.
Vladimir V. Tsukruk and Srikanth Singamaneni.

1
Introduction

The invention of the STM microscopic technique in the early 1980s by Rohrer, Binnig, and coworkers soon established a new class of proximity probe microscopies, SPM, in the 1980s, followed by an explosion of SPM studies on various mostly conductive materials in the early 1990s [1,2]. The invention of STM was soon followed by the introduction of atomic force microscopy (AFM), a milestone in the field of nanoscience and nanotechnology, especially in the case of soft materials. Fast spreading AFM applications involved all-important soft materials ranging from synthetic to biological ones. AFM utilizes intermolecular forces between the tip and the surface to obtain the topographic information on the surface and other physical properties.

In the past two decades of active SPM (both STM and AFM) studies, a number of excellent reviews have been published in this field, which discussed different aspects of SPM studies, but are naturally limited to a particular class of molecules or materials, or a particular scanning mode(s) by the limited space of articles in professional journals [3–13]. Readers might study these excellent reviews if specific aspects of SPM imaging of a particular class of materials are of interest.

The common fundamental feature among the wide range of scanning probe techniques introduced, which include a variety of specific scanning modes and probing regimes, is a sharp hard probe (usually of nanoscale dimensions) integrated with long and flexible microcantilever. The sharp probe with a radius of curvature around 10 nm interacts with a selected substrate in a gentle (imaging), modest (probing), or even damaging (lithography) manner. Monitoring of one or more physical or chemical interactions (such as van der Waals forces, electrostatic interactions, elastic and plastic resistance, electrical current, capacitance, or conductivity) is then employed to unveil the surface morphology, surface and subsurface organization, and/or physical and chemical properties of the materials under investigation with unprecedented nanoscale lateral and vertical spatial resolution.

It is worth noting that initially the traditional AFM contact mode was widely used to image soft materials in the late 1980s through the early 1990s, but excessive surface damage, frequent and prominent artifacts, and difficulties with the stable imaging of compliant materials limited its applicability and overall impact on various soft matter-related research fields. However, this instrumentation problem

Scanning Probe Microscopy of Soft Matter: Fundamentals and Practices, First Edition.
Vladimir V. Tsukruk and Srikanth Singamaneni.

was quickly realized and "fixed." A new instrumentation development led to the introduction of "noncontact" mode AFM and its practical and robust version, the so-called tapping mode version, is widely accepted [14]. The introduction of this reliable and low-damaging scanning mode that became popular very quickly was critical for the expansion of robust and near-nondamaging AFM imaging to a range of synthetic and biological soft materials and to numerous research groups besides professional AFM developers. Rapid expansion of various scanning and probing modes observed in the 1990s "converted" the SPM technique into a universal and highly versatile tool of the twenty-first century, which can be found in virtually every science and engineering department in the world. The appearance of this family of close-proximity probe microscopies dramatically affected research landscapes in many science and technology fields – an effect similar to that observed with the rapid expansion of transmission and scanning electron microscopies in the 1970s and 1980s.

The relative affordability of AFM instruments, their high (near-molecular) resolution, robust scanning procedures, fast learning curve for beginners, ambient and easily controllable (gas, liquid, or temperature) conditions for scanning, and the versatility in measuring not only surface morphology but also a wide range of important surface physical and chemical properties with a nanoscale resolution, all resulted in a fast expansion of the application of this new microscopic method toward virtually all types of soft materials and across all science disciplines in the 1990s and early 2000s.

However, this rapid spreading, especially in the very beginning of the SPM era, frequently resulted in overhyped promises and a number of high-profile artifacts in imaging of prominent molecular features published in journals across different disciplines. Many of those were later withdrawn or overturned or just forgotten. It took nearly a decade to settle the dust, sort out results and artifacts, regain trust of the non-SPM science community, and finally establish robust experimental routines in SPM operations that are accepted by the majority of the research community in science and engineering. At present, major experimental procedures and instrumentation basics are well established, and only a few imaging artifacts and overhyping scans are still published from time to time in major archival journals.

The next important round of experimental developments came with the realization that various interfacial forces and interactions such as electrical, thermal, and magnetic, apart from basic atomic interactions (short-range repulsive and van der Waals type of interactions exploited in early modes), can be readily utilized to image and probe a wide range of material properties. Corresponding developments, mostly in the end of 1980s and in the beginning of 1990s, gave rise to an extended family of new SPM techniques including scanning thermal microscopy, conductive force microscopy, electrostatic force microscopy, magnetic force microscopy, chemical force microscopy, shear force microscopy, pulse force microscopy, near-field scanning optical microscopy, and various probing methods of force spectroscopies such as surface force spectroscopy, colloidal force spectroscopy, nanoindentation, or various versions of nanolithography including dip-pin lithography and mechanical, oxidative, electrostatic, and thermomechanical nanolithography (see some related original papers [15–23]). These modes (and some other more recent modes such peak-force mode), which will be reviewed in this book, have been designed to probe surface interactions and properties as a function of the AFM tip proximity to and location on the probing surface.

Two primary physical quantities could be used to characterize the minute interactions between the AFM tip and the surface and present a broad picture of the applicability of various SPM modes to different soft materials. These physical parameters are the normal force applied to the surface by the sharp AFM tip in different modes of operation (usually expressed in pN or nN) and the characteristic lateral resolution attained using these modes (usually expressed in nanometers) (Figure 1.1). These two parameters are critical for the evaluation of SPM applicabilities because they determine if a particular soft material can be imaged or probed without ultimate physical damage and if a particular property/feature can be probed with the spatial resolution required for a given application.

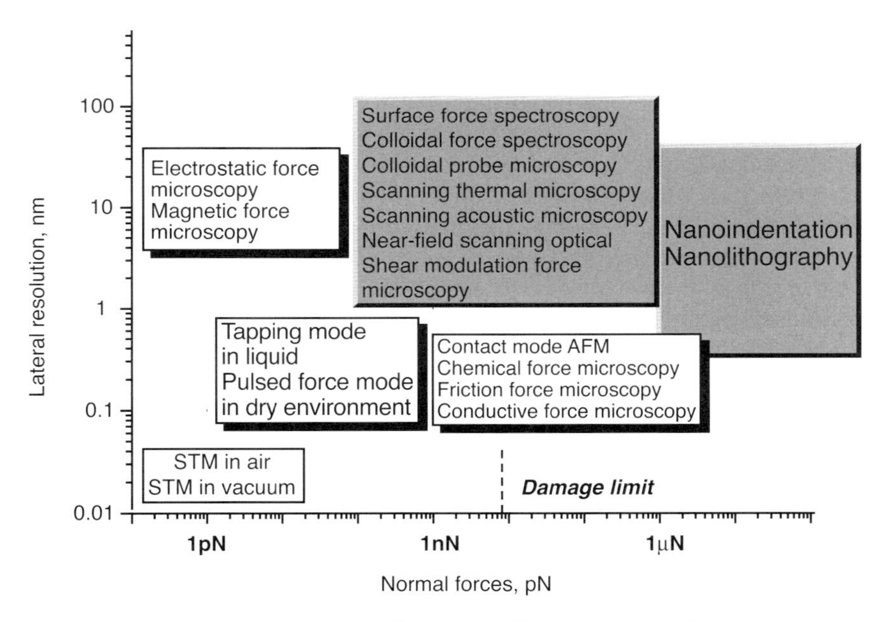

Figure 1.1 Resolution-force "diagram" of SPM modes for soft materials with common scanning parameters.

As is well known, a normal load applied to the AFM tip can vary over a wide range for a particular SPM mode of operation and can be precisely (pN) controlled by a preset microcantilever deflection, microcantilever stiffness, the tip radius of curvature, and scanning operation conditions. On the other hand, local deformation of the substrate or the gap between the SPM tip and the surface defines the achievable lateral resolution during imaging or lithographical modes as presented in Figure 1.1 for different SPM modes. Apparently, the level of surface damage (which is an especially sensitive issue for soft matter) is ultimately determined by the local mechanical deformation and shear/normal stresses and can be set to be around

10 nN assuming a regular AFM tip shape with a radius of curvature of around 10 nm. In a practical scanning regime, the actual threshold depends upon the yield strength of the soft materials and surface stiffness, and thus can be as low as less than 1 nN for delicate biological materials and hydrogels and as high as 100 nN for high-performance composites and fibers.

As can be concluded from this simplified force-resolution representation, under practical imaging conditions, a number of SPM modes are capable of resolving surface features with a spatial resolution well below 1 nm in a nondamaging regime. True (rare) and near-molecular (difficult) resolutions (below 0.5 nm) can be achieved usually only in the contact AFM mode and under special imaging conditions (e.g., scanning in liquid with very low loads). It is worth noting that in terms of the ultimate spatial resolution, STM still remains a superior mode that, however, cannot be a primary choice for soft material studies due to the mostly nonconductive nature of the polymeric materials considered here.

On the other hand, some scanning modes, if applied with high and controlled forces, are capable of inducing excessive but controlled surface deformation that might result in severe and highly localized permanent surface damage (plastic deformation, indentation), thus limiting the probing and imaging abilities of various SPM modes (Figure 1.1). However, the same ability of the AFM tip to leave permanent localized marks on soft material surface with nanoscale lateral and vertical dimensions can be used in a highly controlled manner in nanolithographical approaches (the right side of the mechanical threshold in Figure 1.1). This mode of SPM operation is widely exploited for high-resolution SPM nanolithography, as will also be discussed in this book. In Part One, we will briefly introduce the basics of SPM techniques and the fundamental principles behind SPM applications.

References

1. Binnig, G., Rohrer, H., Gerber, C., and Weibel, E. (1982) Surface studies by scanning tunneling microscopy. *Phys. Rev. Lett.*, **49** (1), 57–61.
2. Binning, G. and Rohrer, H. (1987) Scanning tunneling microscopy: from birth to adolescence. *Rev. Mod. Phys.*, **59** (3), 615–625.
3. Giessibl, F. (2003) Advances in atomic force microscopy. *Rev. Mod. Phys.*, **75** (3), 949–983.
4. Cappella, B. and Dietler, G. (1999) Force–distance curves by atomic force microscopy. *Surf. Sci. Rep.*, **34** (1), 1–104.
5. García, R. and Pérez, R. (2002) Dynamic atomic force microscopy methods. *Surf. Sci. Rep.*, **47** (6–8), 197–301.
6. Sheiko, S.S. (1999) Imaging of polymers using scanning force microscopy: from superstructures to individual molecules. *Adv. Polymer Sci.*, **151**, 61–174.
7. Magonov, S. and Reneker, D. (1997) Characterization of polymer surfaces with atomic force microscopy. *Annu. Rev. Mater. Sci.*, **27**, 175–222.
8. Tsukruk, V.V. and Reneker, D.H. (1995) Scanning probe microscopy of polymeric and organic molecular films: from self-assembled monolayers to composite multilayers. *Polymer*, **36** (9), 1791–1808.
9. Tsukruk, V.V. (1997) Scanning probe microscopy of polymer surfaces. *Rubber Chem. Techn.*, **70** (3), 430–467.
10. Sheiko, S.S. and Möller, M. (2001) Visualization of macromolecules: a first step to manipulation and controlled

response. *Chem. Rev.*, **101** (12), 4099–4124.

11 McConney, M.E., Singamaneni, S., and Tsukruk, V.V. (2010) Probing soft matter with the atomic force microscope: force-spectroscopy and beyond. *Polym. Rev.*, **50**, 235–286.

12 Hansma, P.K., Elings, V.B., Marti, O., and Bracker, C.E. (1988) Scanning tunneling microscopy and atomic force microscopy: application to biology and technology. *Science*, **242** (4876), 209–216.

13 Dürig, U. (2000) Interaction sensing in dynamic force microscopy. *New J. Phys.*, **2**, 5.1–5.12.

14 Zhong, Q., Inniss, D., Kjoller, K., and Elings, V.B. (1993) Fractured polymer/silica fiber surface studied by tapping mode atomic force microscopy. *Surf. Sci. Lett.*, **290** (1–2), 688–690.

15 Williams, C.C. and Wickramasinghe, H.K. (1986) Scanning thermal profiler. *Appl. Phys. Lett.*, **49** (23), 1587–1589.

16 Martin, Y. and Wickramasinghe, H.K. (1987) Magnetic imaging by "force microscopy" with 1000 Å resolution. *Appl. Phys. Lett.*, **50** (20), 1455–1457.

17 Burnham, N.A. and Colton, R.J. (1989) Measuring the nanomechanical properties and surface forces of materials using an atomic force microscope. *J. Vac. Sci. Technol. A*, **7** (4), 2906–2913.

18 Terris, B.D., Stern, J.E., Rugar, D., and Mamin, H.J. (1989) Contact electrification using force microsopy. *Phys. Rev. Lett.*, **63** (24), 2669–2672.

19 Hesjedal, T., Chilla, E., and Fröhlich, H.-J. (1995) Scanning acoustic force microscope measurments on grating-like electrodes. *Appl. Phys. A*, **61** (3), 237–242.

20 Frisbie, C.D., Rozsnayi, L.F., Noy, A., Wrighton, M.S., and Lieber, C.M. (1994) Functional group imaging by chemical force microscopy. *Science*, **265** (5181), 2071–2074.

21 Salmeron, M., Neubauer, G., Folch, A., Tomitori, M., Ogletree, D.F., and Sautet, P. (1993) Viscoelastic and electrical properties of self-assembled monolayers on Au(111) films. *Langmuir*, **9** (12), 3600–3611.

22 Rosa-Zeiser, A., Weilandt, E., Hild, S., and Marti, O. (1997) The simultaneous measurement of elastic, electrostatic and adhesive properties by scanning force microscopy: pulsed-force mode operation. *Meas. Sci. Technol.*, **8** (11), 1333–1338.

23 Ge, S., Pu, W., Zhang, W., Rafailovich, M., Sokolov, J., Buenviaje, C., Buckmaster, R., and Overney, R.M. (2000) Shear modulation force microscopy study of near surface glass transition temperatures. *Phys. Rev. Lett.*, **85** (11), 2340–2343.

2
Scanning Probe Microscopy Basics

2.1
Basic Principles of Scanning Probe Microscopy

Scanning probe microscopy (SPM) techniques have several common components, including an ultrasharp probe, a high-resolution piezoscanner element, one or several sensing elements, and a computer-controlled feedback loop. One key feature that sets SPM-based techniques apart from other microscopy techniques is the use of ultrasharp scanning probes in proximity (even in direct contact) with the probed surface for acquiring sensed signals [1]. Atomic force microscopy (AFM) uses a force sensor (usually a microcantilever) to control and detect the position of the probe, which is moved vertically to maintain a constant force against the surface, while it is raster scanned across the surface, thereby creating a topographical image. The feedback loop is essential for keeping the probe in a reasonable range of deflections (forces) to interact with the surface and acquire an image.

Apart from imaging the properties at nanoscale resolution, an important development in scanning probe technology is the ability to manipulate matter at the nanoscale on a surface using a finely controlled probe. Furthermore, as a natural succession to their application as force transducers in AFM, microcantilevers are being extensively investigated as a new platform for transduction in chemical, biological, and thermal sensing technologies [2–7].

Considering the 20 year history of the development of SPM techniques, it is not surprising that a number of excellent books, collections, and comprehensive reviews have already been published on SPM fundamentals [1, 8–10]. Therefore, here we will only briefly summarize major principles, fundamental causes, and practical considerations with additional detailed discussion of AFM principles and advanced imaging modes in Chapters 3 and 4.

Scanning Probe Microscopy of Soft Matter: Fundamentals and Practices, First Edition.
Vladimir V. Tsukruk and Srikanth Singamaneni.

2.2
Scanning Tunneling Microscopy

The fundamental principle of the STM was understood quickly and relatively straightforward after its discovery. In this technique, a sharp metal tip that acts as one electrode of the tunnel junction is brought in proximity (~0.3–1 nm) to the sample surface that forms the second electrode. Typical operating voltages between 10 mV and 10 V lead to tunneling currents between 1 pA and 100 nA. The tip (tunneling electrode) is scanned over the surface of the sample, while the tip–sample tunneling current is monitored. The tunneling current is highly sensitive to the distance between the electrodes due to exponential distance decay. Under typical conditions, an increase in the distance by 0.2 nm results in a decrease in the tunneling current by a factor of 2.

The STM is operated in one of two modes: constant current mode and constant height mode [11]. In the constant current mode, a feedback loop changes the tip–sample distance to maintain a constant tunneling current. The tip–sample distance is controlled by applying voltage to a piezoelectric crystal that is used to map the topography of the surface. In the constant height mode, a metal tip can be scanned across a surface at a nearly constant height and constant voltage while the current is monitored. This mode is generally used for atomic scale imaging with individual atoms, defects, clusters, and molecules being imaged. STM is one of the rare examples that do not use feedback loop to maintain the tip–sample distance. STMs have also been employed to move, deposit, and pattern atoms on surfaces [12, 13].

Unfortunately, STM mode is not practical for rough (beyond modest atomic scale) surfaces. It is important to note that the tunneling current depends on the atomic species of the sample, and in samples comprised of multiple species, the tunneling current is not a true representation of topography. While the technique is the first of its kind, a major limitation of STM is the requirement of either a conductive sample or a sample that has an atomically thin insulating surface layer. This stringent requirement limits the application of STM, making it a nonideal choice for studying polymers and biological samples, except in the case of monomolecular organic molecules adsorbed onto highly conductive substrates. Another exception is carbon nanotubes (CNTs), for which scanning tunneling microscopy has revealed the atomic chirality–electronic structure relationship of single-walled carbon nanotubes in addition to many other results that will not be discussed here. In addition, single-walled carbon nanotubes can be utilized as ultrasharp tips for high-resolution topographic imaging [14, 15].

2.3
Advent of Atomic Force Microscopy

To overcome the limitations of STM, AFM was introduced as the logical next step in SPM techniques, greatly expanding the existing capabilities [16]. While the initial developments were primarily confined to topographical imaging, a deeper insight into the probe–sample interactions offered by instrumentation improvements and theoretical developments led to a plethora of techniques for probing and imaging mechanical, chemical, thermal, optical, electrical, and magnetic properties with unprecedented spatial resolution [17, 18].

Today scanning probe microscopies are used in a variety of environments such as ambient air or various gases, liquids, vacuum, and wide temperature ranges. This outstanding capability of SPM puts it in a completely different league when compared to electron microscopy techniques, which require high vacuums for operation. Particularly, imaging in liquid enabled the study of live biological samples in their native environments with ultrahigh resolution. Soon after the invention of initial imaging techniques, the rich information provided by the tip–sample interactions led to the development of friction force microscopy, shear force microscopy, electrostatic force microscopy, scanning Kelvin probe microscopy, conductive force microscopy, scanning electrochemical microscopy, scanning thermal microscopy, scanning acoustic force microscopy, magnetic force microscopy, scanning near-field optical microscopy, and scanning thermal microscopy, to name a few (see Chapter 3).

2.4 Overview of Instrumentation

From a design viewpoint, an atomic force microscope is comprised of four key components: scanners (x, y, and z), a force sensor (detection system), the feedback system (electronics), and the coarse approach system (such as stepper motor or manual approach). Figure 2.1 shows the schematic representation of the basic components of AFM.

2.4.1 Scanners

A set of piezoelectric scanners of different designs enable the motion of the tip with respect to the surface of the sample in three mutually orthogonal directions (x, y, and z). In some cases, the sample is fixed while the tip scans across the surface and in other cases, the tip is fixed while the sample is scanned against the probe. The desired actuation is achieved by a set of piezoelements, which are integrated in tube or planar elements. The piezoelements undergo mechanical extension and contraction with the applied electrical field. The most common piezoelectric element employed in commercial AFM scanners is lead zirconate titanate (PZT) [19]. It is important to note that depolarization of piezoelectric materials can occur due to exposure to high temperatures, large reverse biases, or aging.

There are two important architectures of the piezoelectric tubes, which are the most popular designs. In the first design, three piezoelectric tubes are combined in mutually orthogonal directions that enable the motion in the three respective directions. In the second and more common design, the scanner is a single tube that enables motion in all three directions [1]. The scanner is comprised of a

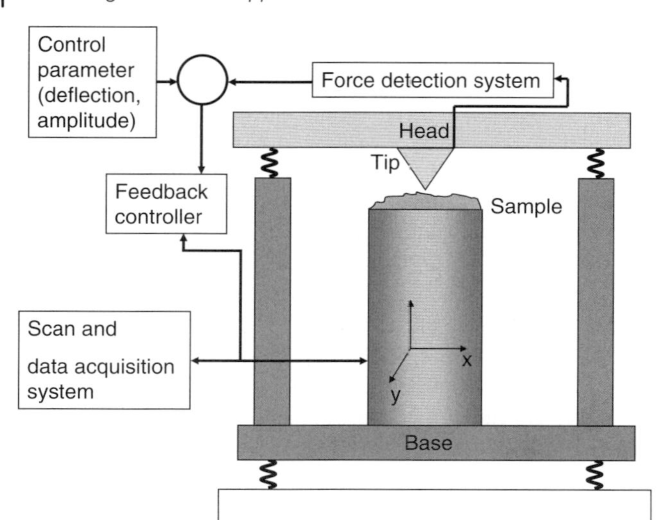

Figure 2.1 SPM system comprised of a scanner (x- and y-directions) and the feedback loop that maintains the tip–sample interaction constant in the z-direction. Adapted with permission from *Mater. Today* (2007), 11, 40–48.

piezoelectric tube with metal electrodes on the inner and outer surface. The outer electrode of the scanner is divided into four quadrants, and a voltage applied to the opposite pair quadrants results in the bending of the element. The bending of the element is translated into a lateral motion (x- and y-directions) of the scanner at the end of the tube. A voltage applied between the outer and inner electrodes results in the extension and contraction of the scanner along the length of the tube (z-direction).

2.4.2
Microcantilevers as Force Sensors

As the name suggests, atomic force microscopy involves the detection of atomic forces between the sharp probe and the sample surface. Obviously, a force sensor that can continuously monitor these forces is an important component. Microcantilevers (microscopic versions of diving boards), in which one of the ends is fixed while the other end is freely suspended, form the most common type of force sensors in AFM (Figure 2.2).

When the sharp tip is mounted on or carved at the free end, interactions between the probe and the surface cause deflection of the cantilever. Numerous methods have been developed for monitoring the minute cantilever deflections in the context of their application as force transducers in AFM.

The detection schemes can be broadly classified as optical (optical lever and interferometry) and electrical schemes (piezoresistive, piezoelectric, capacitance, and electron tunneling). The optical lever technique, in which light is reflected from

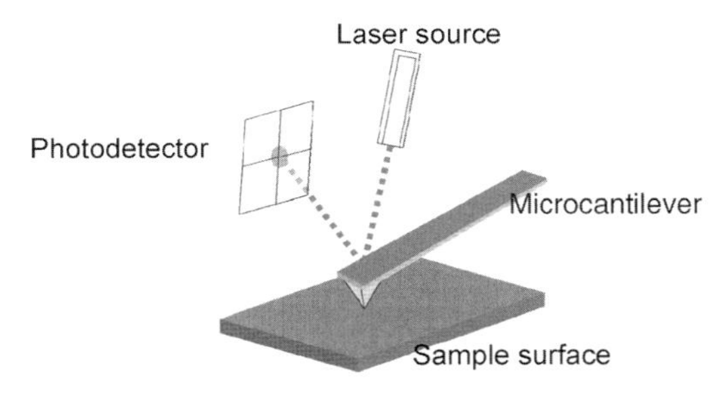

Figure 2.2 The AFM tip interacting with the sample surface and the most common optical lever technique employed to detect the deflection of the microcantilever.

the back of the cantilever onto a position-sensitive photodetector, is similar to the readout scheme widely used in most popular commercial AFM systems (schematically shown in Figure 2.2) [20]. The deflection of the cantilever is thus translated into a photodiode array output voltage that, with proper calibration, can be converted into actual z-deflection. The four-quadrant photodetector enables the detection of deflection, oscillation amplitude, and torsion of the microcantilever. This technique that offers a detection limit better than 1 Å and is mainly limited by microcantilever thermal vibrations was successfully adapted for the detection of static and dynamic signals in AFM.

While the high aspect ratio enables the imaging of deep trenches, the tips, when too long, exhibit a large vibration amplitude at the end of the tip, which limits the resolution. Some major disadvantages with the optical lever technique are the requirements of precise alignment, high reflection, a low opacity and low turbidity medium of operation, and limited bandwidth, which make it extremely difficult to extend to arrays of cantilevers and nanomechanical resonant structures.

In contrast to the optical lever technique, optical interferometry offers higher bandwidth measurement and has been introduced as a MEMS-based technique that shows great promise for the readout approach for large microcantilever arrays [21, 22]. In fact, the optical interferometry technique was employed to demonstrate parallel readout of a cantilever array simultaneously interacting with the surface. Each of the cantilevers in the array is associated with a phase-sensitive diffraction grating comprised of a fixed (reference) and a movable set of interdigitated fingers. A force on the tip causes the displacement of the movable set, which causes the intensity of the diffracted orders to change. The diffraction order intensity from each cantilever is measured, and images are acquired from multiple cantilevers (in constant height mode) simultaneously, as illustrated in Figure 2.3. The ability to monitor the deflection of an array of cantilevers simultaneously opens up the possibility of parallel imaging and manipulation of surfaces, as will be discussed in Chapter 19.

Piezoresistivity of a material (e.g., doped silicon) under external strain has been translated to monitor the deflection of the cantilevers as well [23–25]. The piezoresistive method obviates the need for a complex alignment procedure, which is often a serious problem in optic-based detection methods. It is also important to note that

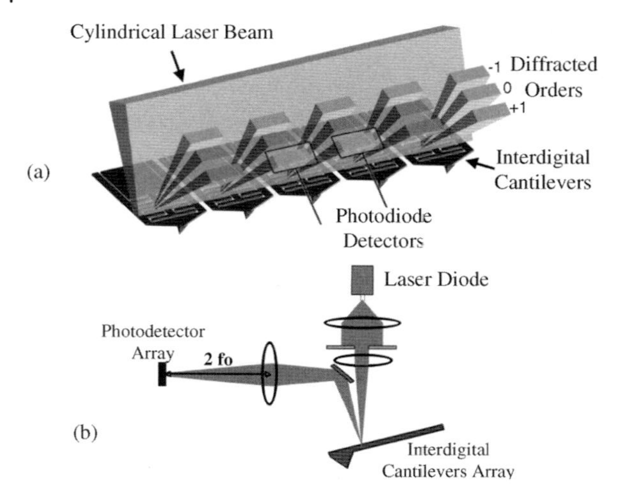

Figure 2.3 (a) The cantilever array and the diffracted orders reflected onto the photodetectors. (b) The microscope with a collimated laser beam that is focused with cylindrical optics onto the cantilever array. The reflected orders are projected onto the photodetector array with an imaging lens for concurrent monitoring of multiple cantilevers. Reprinted with permission from Ref. [21].

this method facilitates the measurement of huge deflections, while the optical detection method is limited to a smaller range (typically only a few microns). However, in addition to the lower resolution (typically 0.5–1 nm) compared to that of the optical technique, the primary disadvantage of the piezoresistive detection method is the continuous thermal drift due to the heat generated by the current flow through the piezoresistor on the cantilever that might interfere with the long-term stability.

The other important electrical method is the self-sensing piezoelectric cantilevers in which a piezoelectric material (such as ZnO) is deposited on the cantilevers [26–28]. This detection mechanism takes advantage of the piezoelectric effect, where a change in mechanical stress results in cantilever bending, causing the induction of transient charges that are translated into a voltage change. Although this approach offers freedom from the bulky optical instrumentation and inconsistencies of laser alignment, integration of a piezoelectric materials onto a cantilever requires additional steps in microfabrication, making the process significantly more expensive and cumbersome.

Another approach of cantilever deflection, the capacitance method, is based on the principle that a change in the distance between capacitor plates will effectively change the overall capacitance of the device. The deflection of the microcantilever is measured by the changes in the capacitance between a conductor electrode and the cantilever [29, 30]. Despite its simplicity, this method suffers from undesired interference effects and the changes in the dielectric medium between the capacitor plates cause changes in the capacitance and may result in a gradual discharge.

Very recently, Dravid and coworkers introduced a novel method for detecting the deflection of cantilevers by embedding a metal oxide semiconductor field effect transistor (MOSFET) in the base of the cantilever [31]. This technique was developed to monitor the deflection of microcantilevers used as transducers in chemical sensing, which will be discussed in Chapter 7. The surface stresses caused by microcantilever deflection result in an increase in the channel resistance due to the change in carrier mobility. Although the detection limit or the maximum resolution (about 5 nm) currently achieved by the technique cannot match that of the optical methods (<0.1 nm), the technique offers the unique advantage of obtaining microcantilever arrays with built-in detection elements enabling seamless monolithic integration.

2.4.3 Electronic Feedback

During lateral scanning, it is important to control the tip–sample separation distance, which should be large enough to prevent the tip or sample damage and yet small enough to sense the interaction forces between the tip and the sample. The electronic feedback system of the AFM controls the distance between the tip and the sample to maintain a constant force (specified by the user) and is comprised of four elements, namely, digital control electronics, a converter box, a high-voltage amplifier, and a laser controller and preamplifier (see Figure 2.4).

The control system depicted in Figure 2.4 assumes that the detection mechanism in the AFM is a laser beam deflection method. However, the general principles of the feedback electronics remain the same for other detection methods. The digital control electronics, most commonly a digital control card, perform all signal processing operations and calculations on the acquired raw data (which are the supplied and acquired voltages).

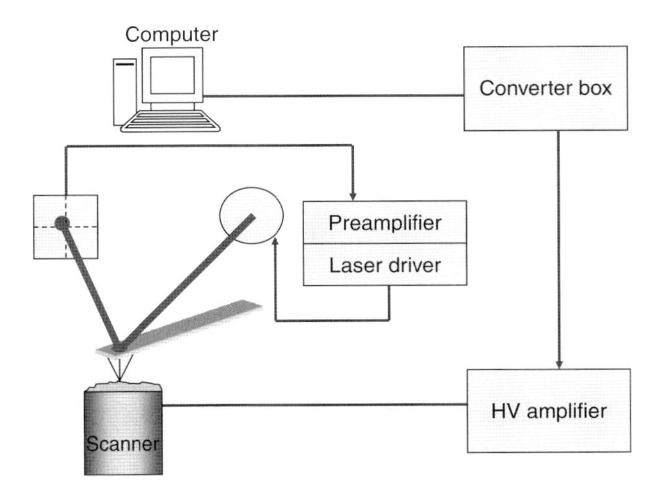

Figure 2.4 Basic components of the feedback circuit of the AFM instrument.

The converter box is comprised of an analogue to digital converter, which coverts the analogue signals from the head of the AFM to digital signals before transmitting them to the digital signal processing (DSP) card. The converter box also hosts a digital to analogue converter that converts digital signals from the digital control card to analogue signals driving the scanner. However, the analogue signals generated by the converter box are typically small voltages, which are insufficient to drive the piezo-elements of the microscope. The analogue voltage signals are amplified by a high-voltage amplifier (HV amplifier) to typically $\pm$100–150 V before feeding them to the scanner elements. The laser driver electronics, apart from supplying power to the laser LED, also ensure that the intensity of the laser remains constant over time by using a dedicated feedback loop. The preamplifier amplifies the raw signal from the photodiode before feeding it to the converter box. It is worth noting that electrical signals from power outlets and the HV amplifier are major sources of electrical noise, which can critically affect the maximum obtainable resolution.

While the piezoelements of the scanner can be moved with extremely high resolution (subangstrom), the range of motion is typically limited to several tens of micrometers. Hence, the peizoelements are complemented by a coarse approach system that enables the user to safely bring the probe to the surface (or vice versa), after which the piezoelements maintain the distance between the probe and the sample. In most modern microscopes, the coarse approach system is comprised of a stepper motor that moves the probe in discrete steps to the sample. After each step, the piezoelement is extended to check (by monitoring the deflection of the cantilever) if the sample is within the range of the piezoelement extension. If the cantilever is not within the range of the piezoelement, the cycle is repeated until the sample is finally reached. In addition to the stepper motor, a manual or motorized stage allows the user to move the scanner to within millimeters of the specimen.

2.5
Probes and Cantilevers in Scanning Probe Microscopy

In the initial stages of AFM development, cantilevers were handmade from either thin metal foils having sharp diamond fragments glued to their ends or from metal wires that were etched to form sharp ends to serve as tips. These original handmade probes suffered from numerous shortcomings such as the labor-intensive fabrication approach, lack of reproducibility, and relatively large sizes and masses, which result in large spring constants of the cantilevers. AFM probe technology has since come a long way from these initial demonstrations [1, 32, 33].

Overcoming the limitations of early fabrication processes, silicon-based micro-fabrication technology quickly replaced the handmade probes. Commercial AFM probes are commonly fabricated from Si, SiO_2, and Si_3N_4 [32]. One of the basic processes to microfabricate Si_3N_4 cantilevers involves etching a small square opening in a SiO_2 mask layer on a (100) silicon surface. KOH, which is an anisotropic etchant, terminates at the (111) planes of the silicon, resulting in a pyramidal pits in the silicon. The SiO_2 mask layer is then removed followed by the deposition of a different mask layer that defines the cantilever shape. After the creation of the etched pits in the silicon surface, the Si wafer is coated with a Si_3N_4 layer using a low-pressure chemical vapor deposition technique. Finally, the silicon is removed via etching to release the cantilever with the integrated tip.

As briefly described above, the shape of the tip is governed by the microfabrication technique employed. Three basic geometries, pyramidal, conical, and tetrahedral, are common in commercially available AFM tips [8, 34]. Si_3N_4 tips are always in the pyramidal geometry, while Si and SiO_2 tips are fabricated in all three geometries. The pyramidal tips typically have a cone angle of 20–30° and a radius of curvature of 5–20 nm for silicon, and 20–60 nm for Si_3N_4. Various techniques have been developed to improve the sharpness of these tips. For example, Si_3N_4 tips are typically oxide sharpened, which improves the aspect ratio while reducing the tip radius to as small as 5–10 nm. Silicon tip sharpness can be improved using the focused ion beam (FIB) technique, routinely reaching a radius of 5–10 nm.

Unfortunately, the microfabricated tips with high sharpness are prone to easy damage during scanning, which significantly affects the quality of both the image and the observed features. One of the important technologies that received immense attention was the implementation of carbon nanotubes as tips in AFM imaging. The nanoscale dimensions of CNTs (1–3 nm diameter for single-walled carbon nanotubes (SWNTs)), along with their high aspect ratios and extraordinary mechanical properties make them promising candidates as AFM tips [35].

The compressive strength of CNTs is approximately two orders of magnitude higher than the compressive strength of any known fiber. The mechanical properties of various types of nanotubes have been extensively studied by both experimental and computational means [36]. CNTs are characterized by exceptionally high elastic moduli [37, 38] and large elastic and fracture strain-sustaining capabilities [39, 40]. SWNTs possess a tensile strength within 100–600 GPa, which is about two orders of magnitude higher than that of specially designed high-strength carbon fibers [41, 42]. The most remarkable effect is the combination of high flexibility and strength with high stiffness, a property that is absent in graphite fibers that are brittle [38]. This property creates the possible application of CNTs as molecular mechanical springs. These springs would be very stiff for small loads, but would become soft for larger loads, accommodating large deformations without breaking [43].

Initial demonstrations of CNTs as AFM tips involved manually attaching the CNTs to the end of the conventional pyramidal tips using a micromanipulator under optical microscopy in the dark field mode [15]. The pyramidal silicon tip was initially touched to adhesive tape to form a thin layer of adhesive. The tip was then brought into contact with MWNTs to attach a single or a small bundle of CNTs, and micromanipulators were used to align the CNTs parallel to the tip axis. While this technique enabled enhanced resolution, it was limited in that the nanotubes that could be manipulated using optical microscopy were relatively large in diameter (5–20 nm). In order to overcome this limitation, micromanipulation was performed in a scanning electron microscope chamber. Apart from enabling the manipulation of CNTs with a smaller diameter, this technique provided better control over the length, size, and orientation of the nanotube tip.

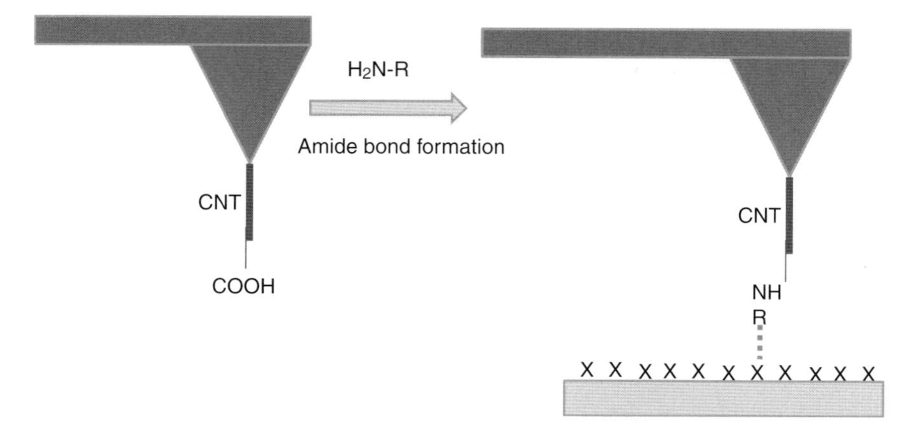

Figure 2.5 Open end of an oxidized single-walled carbon nanotube tip and the carbodiimide coupling chemistry used in one form of chemical derivatization. Schematic of a patterned self-assembled monolayer terminating with methyl and carboxylic end groups. Reprinted with permission from Ref. [45].

The high aspect ratio and excellent mechanical properties of CNTs enable probing deep crevices, and the small radius of nanotube tips (as small as 1 nm) significantly improve the lateral resolution beyond what can be achieved using microfabricated silicon tips [15, 35, 44]. Furthermore, the nanotubes' ability to buckle elastically makes them very robust while self-limiting the maximum force that is applied to delicate organic and biological samples. Apart from high-resolution topographical imaging, CNTs have been modified to create probes that can sense and manipulate matter at the molecular level. Lieber *et al.* demonstrated the capability of chemical and biological discrimination by using nanotubes with acidic functionalities and by coupling basic or hydrophobic functionalities or biomolecular probes to the carboxyl groups at the open tip ends (see Figure 2.5) (see Chapter 7 for detailed discussion) [45].

Some major problems that hinder the application of carbon nanotube tips for routine imaging are their prohibitive cost, poor reliability, and the lack of efficient and reproducible batch fabrication techniques. Further developments involving availability at reduced cost, increased tip longevity and reliability, enhanced resolution, and decreased tip–surface forces will open up CNT tips for use in structural biology, biotechnology, metrology, and nanoelectronic applications.

2.5.1
Physical Attributes of Microcantilevers

The microcantilevers are usually commercially available in beam (diving board) and V-shape geometries. The V-shaped cantilevers are preferred for contact mode imaging owing to the better mechanical stability of these cantilevers with regard to the significant torsional forces, especially when imaging in ambient air. The beam cantilevers are preferable when probing the lateral forces (LFM) and friction forces of surfaces (FFM).

Other mechanical properties (such as normal and lateral spring constants and resonance frequency) of the microcantilevers are also governed by the geometry of the cantilever, dimensions, and the material properties of the cantilever itself. The amplitude of the thermal vibrations of rectangular microcantilevers is given by

$$\sqrt{\langle \Delta z_{\mathrm{defl}}^2 \rangle} = \sqrt{\frac{k_{\mathrm{B}} T}{k}} \tag{2.1}$$

where k_{B} is the Boltzmann constant, k is the spring constant of the cantilever, and T is the absolute temperature [46].

The normal force constant and the resonance frequency of a beam cantilever are given by the equations

$$k = \frac{Ewt^3}{4l^3} \tag{2.2}$$

$$f_0 = 0.162\sqrt{\frac{E}{\varrho}} \times \frac{t}{l^2} \tag{2.3}$$

where k is the normal spring constant, E and ϱ are the elastic modulus and density, respectively, of the material from which the cantilever is fabricated, w is the width of the beam, l is the length of the cantilever, and t is the thickness of the cantilever [47]. From these relations it is clear that the spring constant of the cantilever critically depends (inverse cubed relation) on the dimensions of the microcantilever. A small increase in the length of the cantilever leads to dramatic decrease in the stiffness, which can be favorably employed to vary the stiffness of the cantilevers. It is also important to note that the length, and hence the stiffness of the cantilever, determines the amplitude of thermal vibrations of the cantilever. An increase in the length of the cantilever results in a decrease in the spring constant and an increase in the amplitude of thermal vibrations. For the V-shaped cantilevers, a simple approximation is made by considering the cantilever to be a rectangular beam with a width of $2w$ for each leg. More precise geometrical equations have been developed for V-shaped cantilevers [50, 51].

While the above expressions are sufficient for crude approximations for rectangular cantilevers, they are not the ideal choice for precise measurements that require the knowledge of probe properties. Significant differences are typically found between theoretical spring constants and experimentally measured spring constants [52]. There are several reasons for this discrepancy, the first being a lack of precision in the measurement of cantilever thickness, which can lead to significant error due to its third power relationship with the spring constant. Another source of error in the theoretical estimation of the spring constant is the large variation in the elastic modulus of the material, which can deviate from bulk values by roughly 50% depending on deposition conditions [48]. Other possible sources of error include influences of native oxide layers and reflective gold or aluminum coatings. Therefore, to ensure the accuracy of measured forces, it is important to measure the spring constant of the cantilever rather than relying on theoretical calculations.

There are several well-developed methods for directly measuring cantilever spring constants, the most common of which are the added mass method [49], geometry-based methods [50–52], the spring-on-spring method [53], and the thermal tuning method [46]. Special developments in the form of calibration plots and modified equations have been suggested for more complicated cases, such as V-shaped, gold-sputtered silicon nitride cantilevers [51, 54]. Using the cantilever resonant frequency and quality factor is a popular method that eliminates the need to know the cantilever thickness and density precisely [55]. This method produces relatively good, robust results and is very straightforward to apply.

The added mass method of calibrating cantilever spring constants involves measuring the cantilever resonant frequency shift caused by adding a known mass to the end of the cantilever [49]. For a rectangular cantilever, the shift of the resonant frequency f can be described by

$$f = \left(\frac{1}{2\pi}\right)\sqrt{\frac{k}{M_{\text{cant}}}} \tag{2.4}$$

where M_{cant} is the mass of the cantilever. After measuring the original resonant frequency, a microparticle of known size is added to the end of the cantilever and the new resonant frequency f_{M} is measured and the spring constant of the cantilever is calculated using

$$k = 4\pi^2 M\left(\frac{1}{1/f_{\text{M}}^2 - 1/f^2}\right) \tag{2.5}$$

This method can produce accurate and robust results, but unfortunately is somewhat tedious.

The spring-on-spring method can be used as an alternative means of measuring cantilever spring constants by obtaining force curves on a previously calibrated cantilever [53]. This procedure does require that the photodetector sensitivity (deflection/volt) be calibrated for the cantilever acting as the probe. The measurements should be performed in a manner so that the tip is near the end of the other cantilever, as is shown in Figure 2.6. Although it is common to use the unknown

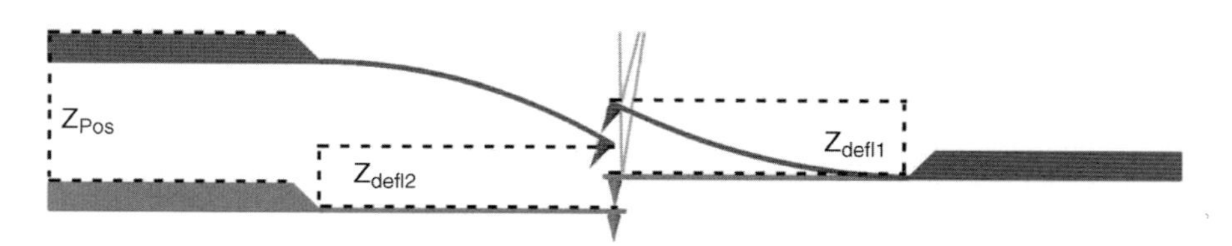

Figure 2.6 A schematic of spring-on-spring cantilever spring constant calibration. Force spectroscopy is performed on the end of a cantilever. This method utilizes a force balance, where the product of the spring constant and the deflection of each cantilever are equal. Note that the schematic was drawn assuming the convention of a sample scanning AFM (piezoelement moves sample).

cantilever as the probe, it is not critical to do so. The spring-on-spring method utilizes the following equation for force balance between the two springs:

$$k_1 \cdot z_{\text{defl1}} = k_2\left(z_{\text{pos}} - z_{\text{defl1}}\right) = k_2 \cdot P = k_2 \cdot z_{\text{defl2}} \tag{2.6}$$

where the subscript 1 refers to variables associated with the cantilever that is acting as the probe (the deflection transduced with the laser photodetector) and the subscript 2 refers to variables associated with the cantilever that is acting as the sample being probed. To calculate the spring constant, it is a simple matter of plugging in the calibrated spring constant into

$$k_{\text{unknown}} = k_{\text{known}}\left(\frac{z_{\text{known}}}{z_{\text{unknown}}}\right) \cong \left(\frac{dz_{\text{known}}}{dz_{\text{unknown}}}\right) \tag{2.7}$$

where z is the cantilever deflection and the subscripts known and unknown refer to variables associated with the calibrated cantilever and unknown cantilever, respectively [53]. This method has good accuracy (close to 10% if the ratio of spring constants is within 0.1–10) when performed with care and is relatively easy to perform and useful in the case when a thermal tuning sweep cannot cover the resonance frequency of the cantilever.

The most popular modern thermal tuning method involves measuring the mean squared deflection of the cantilever caused by thermal motion (see above). A cantilever spring constant can be described by solving Eq. (2.1) for k. However, it should be noted that the mean squared deflection in (2.1) is the sum over all harmonic modes of the cantilever, and because thermal tuning is typically performed by obtaining the data for the first harmonic mode, a correction factor must be applied. This method also requires the photodetector sensitivity to be calibrated. The thermal tuning and the spring-on-spring methods have similar accuracies of roughly 10%. Thermal tuning is generally easier to perform, but is not available on all microscopes and for a whole range of relevant frequencies (e.g., for very stiff cantilevers with very high resonant frequencies) [56, 57]. Thermal tuning also offers a novel way to estimate the photodetector sensitivity, as will be discussed later.

2.5.2
Tip Characterization

It is well known that finite tip dimensions distort the feature sizes of images because of shape convolution frequently called tip dilation. Therefore, tip shape is often the source of common scanning artifacts, such as doubled features. Furthermore, the accuracy of many quantitative force measurements strongly depends on the accurate estimation of the tip–sample contact area (discussed in detail in Chapter 3).

Several methods have been used to measure tip size and shape. One common method involves calculating tip dimensions from images obtained by scanning samples with known dimensions in tapping mode (Figure 2.7a and b) [58]. Scanning an array of standard gold nanoparticles of diameters 5–30 nm tethered to modified, atomically flat mica or silicon surfaces has proven to be quite accurate at character-

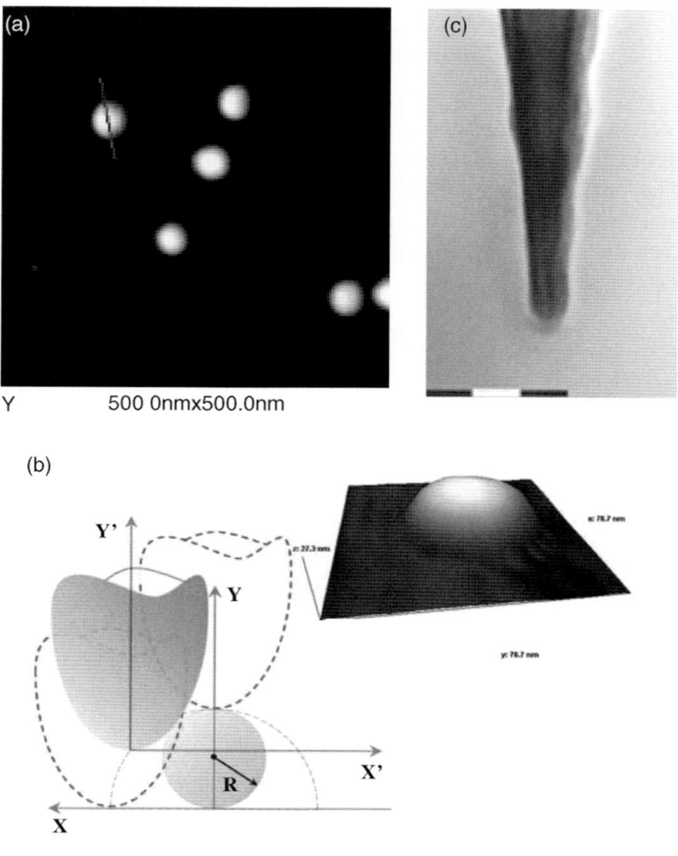

Figure 2.7 AFM tip deconvolution and dilated image of gold nanoparticle (b); AFM image of gold nanoparticles with 20 nm diameters (a) and TEM image of tip (scale bar is 20 nm) (c). NTFMT Web site, adapted with permission (c).

izing the very end portion of the tip. This method is quite appropriate for force spectroscopy measurements involving small penetration depths (several nanometers). Such reference nanoparticle samples can be prepared relatively easily from commercially available TEM kits [59]. Often the nanoparticles are tethered to a polylysine coating or attached to amine-terminated self-assembled monolayers (SAMs), which when scanned appear to help remove tip contamination and prevent nanoparticle rolling and detachment. Direct tip imaging, which involves scanning microfabricated calibration samples with sharp features (available commercially), can also be employed to calculate tip dimensions, although the lifetimes of the calibration samples are questionable. All scanning-based methods involve particular image processing algorithms to calculate the tip shape [60, 61].

Alternatively, SEM can be used relatively successfully to characterize AFM tips; however, the resolution is practically limited to 2–3 nm and this method is time-consuming. Often the imaging should be performed on conductive tips with the accelerating voltage reduced to avoid excessive charging. It should also be noted that the electron beam often cause the formation of carbon-based structures on the tip surface, thus compromising tip quality. The tip can also be characterized with much higher resolution via TEM by measuring the shadow created by the tip (Figure 2.7c) [62, 63].

2.5.3 Tip Modification

Chemical modification of probes is generally used to enhance or reduce tip–sample interactions that can be useful for a variety of applications, including chemical force microscopy and polymer chain pulling experiments (see Chapter 7) [64]. Probes are usually modified with thiol SAMs on gold precoated tips or with SAMs with silane chemistry on native silicon oxide tips. Silane modification can be done directly on silicon and silicon nitride tips after thorough cleaning [65].

Thiol modification involves noncovalent bonding that leads to a limited lifetime, especially under high stresses during contact mode AFM imaging [65]. Though thiol SAMs are important tool for surface scientists, they are poor surface modifiers for applications involving relatively high forces. Contrarily, silane-based modifications involve covalent bonds, which are quite robust and last long. Unfortunately, silane modification involves relatively stringent reaction conditions and is somewhat difficult to optimize initially to achieve a single monolayer and avoid tip fracturing. The reaction is very sensitive to the presence of water, so relative humidity must be limited to a small percentage and dry solvents must be used. Thiol modification is relatively straightforward and easy to establish. The easy preparation of thiol SAM-terminated tips has led to their widespread use, even in contact mode and friction mode AFM techniques, often resulting in characteristic artifacts.

There are numerous demonstrations in which modified tips have been employed to probe interactions between ligand–receptor pairs, antibody–antigen, functional groups, and nanomaterials [66–68]. Tsukruk and Bliznyuk probed the adhesion and friction forces between surfaces with functional terminal groups using silane-modified tips [65]. Surfaces with terminal groups of CH_3, NH_2, and SO_3H were obtained by direct chemisorption of silane-based compounds on silicon/silicon nitride surfaces. Work of adhesion, "residual forces," and friction coefficients were all obtained for four different types of modified tips and surfaces. Absolute values of the work of adhesion between various surfaces were found to be in the range of 0.5–8 mJ/m^2 and the friction forces were observed to be critically dependent upon the pH of the solution. In a recent example, the interaction between individual functional groups and groups attached to carbon nanotubes was studied by Friddle *et al.* with silane-functionalized tips in a fluidic environment, achieving single molecular group interaction conditions. These results can be used to better the design of carbon nanotube–polymer interfacial chemistry [69].

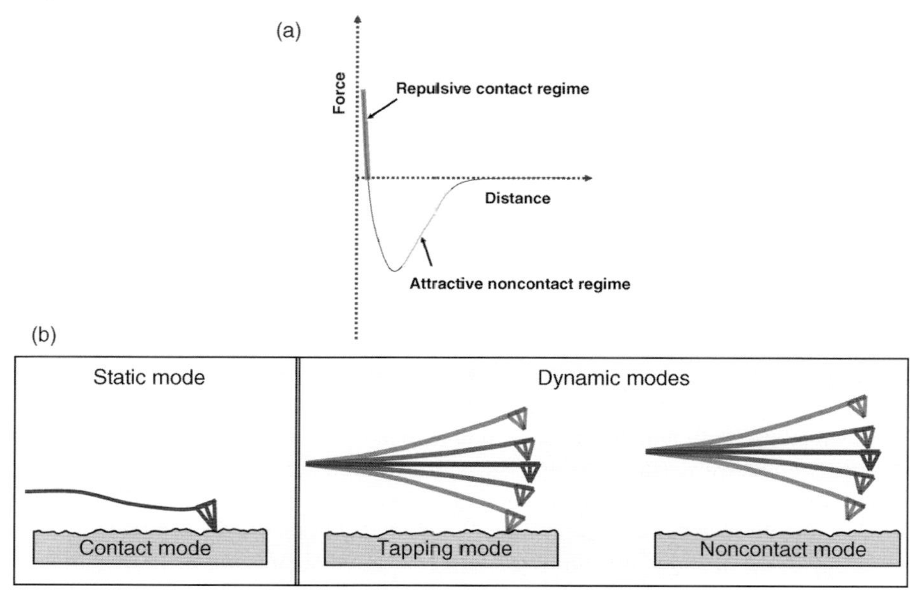

Figure 2.8 (a) Interaction force versus separation distance plot showing the long-range attractive regime (noncontact mode) and short-range repulsive regime (contact mode). (b) Schematic showing the microcantilever interaction with the sample in the three basic imaging modes of operation of AFM.

2.6 Modes of Operation

Based on scanning dynamics, AFM operation can be divided into static and dynamic modes (Figure 2.8). The readers are referred to excellent reviews for detailed information regarding discussion of AFM-based imaging and dynamic modes [70, 71].

Briefly, dynamic modes involve oscillating the cantilever (vertically or laterally), usually near its resonance frequency. Under dynamic conditions, the resonant frequency, amplitude, and phase of the oscillation change due to interactions between the tip and the sample. Dynamic modes can be performed in several variations, including amplitude modulation (AM-AFM) and frequency modulation (FM-AFM) [70].

The most common type of dynamic AFM mode, called tapping or intermittent contact mode, is a simple and robust AM-AFM approach widely employed in many AFM instruments. On the other hand, in the static mode, the tip is raster scanned across the surface and the deflection of the cantilever is kept constant by the feedback control.

AFM techniques can also be generally classified, based on tip–sample interactions, as contact and noncontact. The tip–sample interaction forces can be distinctly identified by considering interaction forces versus tip–sample separation distance, as shown in Figure 2.8a [71]. At large separation distances (several nanometers), the tip and the sample surface start experiencing weak attractive forces (van der Waals forces, capillary forces, and electrostatic interactions). These attractive forces dramatically increase with diminishing distances until the atoms on either surface are close enough (usually below 0.5 nm) for the electron clouds to interact and repel each other. In contact mode, the scanning probe operates in the repulsive regime, with the tip being in direct physical contact with the surface. In contrast, in the true noncontact mode, the probe operates in the attraction regime (Figure 2.8) [71]. Because this regime weakly depends on distance and is highly unstable, the separation distance varies around the minimum in the so-called tapping mode, which includes minute contact interactions at certain portions of an oscillating cycle.

The primary difference between the contact and noncontact modes is the total force applied to the surface during scanning, reaching 1–100 nN in the contact mode but reducing to only around 0.1 nN in the tapping mode. The forces associated with contact mode result in contact pressures exceeding the gigapascal range and frequently lead to the damage of soft surfaces and tip contamination. Thus, tapping mode is the preferential mode for scanning soft materials, such as polymers, with minimum surface damage. Moreover, tapping mode in liquid can further reduce the local forces down to the 10 pN range, thus allowing imaging of very soft, gel-like structures and cellular structures. In the following sections, we discuss the basic imaging modes in greater detail.

2.6.1 Contact Mode

As mentioned above, in contact mode imaging, the tip drags over the surface with a fixed velocity and constant normal load in a raster manner. Under these conditions, the dragged tip is in intimate contact with the sample, operating in the repulsive regime (Figure 2.8). Thus, contact mode involves a combination of localized shear forces, caused by lateral tip movement, and normal forces, caused by spring load and capillary forces. Under ambient, humid conditions, a thin film of water exists on the surface, which is responsible for the attractive capillary forces acting on hydrophilic tips. These forces can reach a level as high as several hundred nanonewtons. The shear forces depend on several variables, which are to be discussed in the section on friction force microscopy (see Chapter 4).

In contact mode, the AFM is commonly operated in the constant force regime. In other words, the normal force between the tip and the sample is maintained constant, controlled by cantilever deflection (at a user-defined value) using an electronic feedback loop. During the initial engagement of the tip onto the sample, the tip is brought in contact with the sample until the required cantilever deflection is attained. Initial engagement results in high local forces, which frequently result in tip damage and surface indentation, and thus requires special precautionary measures (e.g., lowered force for the initial engagement).

The tip is then scanned laterally over the sample and the cantilever deflection (actually, twist) is measured via a photodetector. During lateral scanning, the

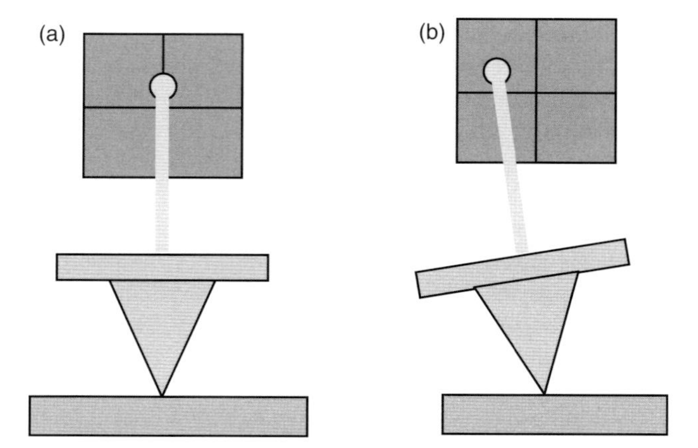

Figure 2.9 (a) Vertical and (b) torsional deflection of the cantilever and the laser spot registry on the quad photodetector.

feedback signal controls the position of the vertical piezoelectric actuator to constantly reposition the tip (or the sample) to maintain the predetermined cantilever deflection. The feedback voltage to the piezoelectric element is a measure of the surface topography and is expressed as a function of the lateral position of the sample. The microcantilever also exhibits a small torsion (or twisting) due to the lateral component of force acting on the tip when scanning in contact mode. The torsional bending of the cantilever is registered as the horizontal signal on the quad photodiode, as depicted in Figure 2.9.

This force, typically 1–100 nN, when applied to a regular tip (with a radius of curvature of several nanometers) results in a contact pressure of a few gigapascals, which is on the order of the yield stress of glassy polymers. Thus, imaging soft materials in contact mode often causes plastic deformation of the specimen. The forces can be reduced to 0.1–1 nN by performing the scanning in fluid (water, organic solvents, etc.), because capillary forces are significantly minimized. Overall, imaging in contact mode involves relatively large shear forces, frequently resulting in the damage and distortion of soft surfaces, making it unfavorable for polymeric and biological samples. Thus, it is employed only in some special cases (e.g., in friction force microscopy measurements).

2.6.2
Noncontact Mode and Tapping Mode

Noncontact and intermittent contact (tapping) modes were introduced to prevent surface damage associated with contact mode [72, 73]. Generally, in a broadly defined noncontact mode, the tip scans about 2–10 nm above the sample surface and is perturbed by the attractive van der Waals forces between the tip and the sample (Figure 2.8). In order to overcome the weak tip–sample interaction forces, the cantilever is set to oscillate at, or slightly off of, the resonance frequency of the cantilever. The small interaction force is typically detected through the dampening of the amplitude. A major drawback of this mode of imaging is that the thin surface fluid layer is often substantially thicker than the range of the van der Waals force gradient, thus limiting the resolution and stability of the scan. The lateral resolution of the dynamic mode is typically limited to 0.5 nm for topography and to around 10 nm for other properties.

The dynamic mode reduces the typical operational forces by at least one order of magnitude relative to the contact mode (usually well below 1 nN). This mode virtually eliminates the shear force associated with the lateral raster scanning and reduces the tip–sample contact duration by two orders of magnitude. The intermittent mode (tapping mode) has been used to study a wide variety of materials such as metals, semiconductors, polymers, and biological materials. This mode offers unique advantages for probing soft polymeric and biological samples when compared to contact mode AFM. Apart from tracking the surface topography using the weak van der Waals forces, the dynamic mode is also employed for probing other weak forces such as electrostatic and magnetic forces (as will be discussed later). Although in the tapping mode contact forces are considered minimal, they nonetheless can be substantial (reaching in some cases 1 nN) and may result in unintentional surface modifications and even surface damage, especially in the so-called "hard tapping" (discussed below).

In the tapping mode, the nature of the tip–sample interaction forces is complex. The force of interaction between the tip and the sample increases with the amplitude of vibration of the cantilever (drive amplitude, A_0) and the reduction of the amplitude (set by the user to image) $A_0 - A_{SP}$, where A_{SP} is the amplitude set point. A more general parameter is the set point ratio (r_{sp}), which is the ratio of the amplitude set point to the free amplitude of oscillation of the cantilever. In general, a higher value of r_{sp} represents a smaller damping of the free amplitude, and hence smaller interaction forces. While the absolute values of r_{sp} depend on many parameters, such as tip sharpness, humidity, and contamination, the relative r_{sp} values are frequently utilized to quantify the tip–sample interaction forces. Light tapping mode involves relatively larger r_{sp} values (sufficient to perform imaging) and generally the tip–sample forces can be attributed to longer–range van der Waals forces. However, in medium and hard tapping (lower r_{sp} values), the forces are mostly attributed to sample stiffness, as was demonstrated by Magonov *et al.* for two-phase polymeric materials [74]. As mentioned above, the hard tapping mode of imaging can cause modifications to the surface features or even damage to soft surface layers.

The energy loss associated with these probe–sample forces causes phase shifts in the oscillation of the probe. The shift in the phase angle of a free cantilever interacting with a surface can be expressed as

$$\Delta\phi_0 = \frac{\pi}{2} - \tan^{-1}\left(\frac{k}{Q\sigma}\right) \tag{2.8}$$

$$\sigma = \frac{\partial P}{\partial z}$$

where Q is the quality factor, P is the interaction force between the tip and the sample, and z is the tip–sample separation distance. The phase angle shift may be approximated as

$$\Delta\phi_0 \approx \frac{Q\sigma}{k} \tag{2.9}$$

for σ much smaller than the cantilever spring constant k [75]. The phase shift sign corresponds to the force derivative sign, such that repulsive forces lead to positive phase shifts and attractive forces lead to negative phase shifts.

Unfortunately, phase contrast interpretation corresponding to surface stiffness and adhesion can be somewhat confusing for complex polymeric surfaces due to poorly controlled contributions from contact area, viscoelastic response, and capillary forces [80]. In general, the phase shift is proportional to the stiffness of a material, but stiffness depends on the contact radius that is generally larger for softer materials under the same forces. These tip–surface contact area issues are less important under medium tapping forces, as compared to hard tapping forces. Furthermore, because the tip–surface contact area can significantly affect the phase shift angle, it is critically important to carefully consider the effects of topography and local deformation when interpreting phase images [74].

2.7
Advantages and Limitations

The unprecedented lateral and vertical resolution offered by SPM techniques enables the visualization of micro, nano, and molecular scale structures and morphologies of polymer and biological surfaces and interfaces with near-molecular and under some special conditions, atomic resolution [76]. Apart from advantages such as true 3D topology and minimal sample preparation, SPM techniques enable the imaging of polymer samples not only under ambient conditions but also in various fluids and finely controlled environments. The ability to probe polymer surfaces in a variety of environmental conditions provides obvious advantages when compared to traditional electron microscopies, such as SEM and TEM, which usually require a high vacuum environment and tedious sample preparation steps such as microtoming, sputtering, and staining.

One of the exceptional capabilities of AFM is its ability to measure surface topography with true atomic resolution, which can be achieved under carefully optimized conditions (thermal drift, vibrations, tip sharpness, tip–sample forces, adhesion, and artifacts). In fact, the atomic lattice of mica is frequently used as the scanner calibration standard for high-resolution AFM imaging. There has been significant skepticism regarding the possibility of atomic resolution using AFM [77, 78]; however, in the last several years, there have been numerous convincing reports in which atomic resolution was achieved via AFM [79, 80]. The readers are referred to an excellent review that provides a comprehensive and critical summary of the various original reports of atomic resolution using AFM [81].

Apart from high, near-atomic resolution imaging, various SPM techniques enable simultaneous probing of the surface morphologies and various physical properties (mechanical, electrical, thermal, or magnetic properties) with nanoscale resolution (usually within 1–50 nm), which is the subject of discussion for Part Two of this book. These measurements provide invaluable insight into the understanding of structure–property relationships of soft materials at the nanoscale.

Moreover, SPM can be used to manipulate and pattern soft matter by applying exceeding, yet well-controlled, normal and shearing forces, thus modifying surface topography by repeated scanning. Dip-pen nanolithography is another example of a soft matter patterning technology that has evolved from SPM [82]. In fact, there are many different nanolithographic techniques that evolved from SPM-based technologies, which are discussed in Part Four.

Nonetheless, most SPM-based characterization is essentially surface based, with the exception of a few techniques such as nanoindentation, scanning thermal microscopy, and acoustic force microscopy. Therefore, SPM can be strongly complemented by other surface and subsurface sensitive techniques, including transmission electron microscopy/tomography, photoelectron spectroscopy, Raman microscopy, and X-ray microscopy.

Furthermore, SPM is generally limited to resolving nanometer–submicrometer features and its image capture times are limited to the scale of seconds to minutes, with only a few cases of video-rate scanning demonstrated to date (see Chapter 4). Thus, SPM can also be greatly complemented by faster scanning, but lower resolution methods and far-field methods, including confocal microscopy, scanning electron microscopy, and optical profilometry.

References

1 Magonov, S.N. and Whangbo, M.-H. (1996) *Surface Analysis with STM and AFM: Experimental and Theoretical Aspects of Image Analysis*, Wiley-VCH Verlag GmbH, Weinheim.

2 Singamaneni, S., LeMieux, M.C., Lang, H.P., Gerber, C., Lam, Y., Zauscher, S., Datskos, P.G., Lavrik, N.V., Jiang, H., Naik, R.R., Bunning, T.J., and Tsukruk, V.V. (2008) Bimaterial microcantilevers as a hybrid sensing platform. *Adv. Mater.*, **20** (4), 653–680.

3 Raiteri, R., Nelles, G., Butt, H.-J., Knoll, W., and Skladal, P. (1999) Sensing of biological substances based on the bending of microfabricated cantilevers. *Sens. Actuator B*, **61** (1–3), 213–217.

4 Lang, H.P., Berger, R., Andreoli, C., Brugger, J., Despont, M., Vettiger, P., Gerber, C., Gimzewski, J.K., Ramseyer, J.-P., Meyer, E., and Güntherodt, H.-J. (1998) Sequential position readout from arrays of micromechanical cantilever sensors. *Appl. Phys. Lett.*, **72** (3), 383–385.

5 Dutta, P., Chapman, P., Datskos, P.G., and Sepaniak, M.J. (2005) Characterization of ligand-functionalized microcantilevers for metal ion sensing. *Anal. Chem.*, **77** (20), 6601–6608.

6 Fritz, J., Baller, M.J., Lang, H.-P., Rothuizen, H., Vettiger, P., Meyer, E., Güntherodt, H.-J., Gerber, C., and Gimzewski, J.K. (2000) Translating biomolecular recognition into nanomechanics. *Science*, **288** (5464), 316–318.

7 Singamaneni, S., McConney, M.E., LeMieux, M.C., Jiang, H., Enlow, J.O., Bunning, T.J., Naik, R.R., and Tsukruk, V.V. (2007) Polymer-silicon flexible

structures for fast chemical vapor detection. *Adv. Mater.*, **19** (23), 4248–4255.

8 Bonnell, D. (ed.) (2001) *Scanning Probe Microscopy and Spectroscopy: Theory, Techniques, and Applications*, 2nd edn, Wiley-VCH Verlag GmbH, Weinheim.

9 Sarid, D. (1992) *Scanning Force Microscopy and Its Application*, Springer, Berlin.

10 Wiesendanger, R. (1994) *Scanning Probe Microscopy and Spectroscopy*, Cambridge University Press, Cambridge.

11 Hansma, P.K. and Tersoff, J. (1987) Scanning tunneling microscopy. *J. Appl. Phys.*, **61** (2), R1–R23.

12 Lyo, I.-W. and Avouris, P. (1991) Field-induced nanometer-scale to atomic-scale manipulation of silicon surfaces with the STM. *Science*, **253** (5016), 173–176.

13 Avouris, P. (1995) Manipulation of matter at the atomic and molecular levels. *Acc. Chem. Res.*, **28** (3), 95–102.

14 Wildöer, J.W.G., Venema, L.C., Rinzler, A.G., Smalley, R.E., and Dekker, C. (1998) Electronic structure of atomically resolved carbon nanotubes. *Nature*, **391** (6662), 59–62.

15 Dai, H., Hafner, J.H., Rinzler, A.G., Colbert, D.T., and Smalley, R.E. (1996) Nanotubes as nanoprobes in scanning probe microscopy. *Nature*, **384** (6605), 147–150.

16 Binnig, G., Quate, C.F., and Gerber, C. (1986) Atomic force microscope. *Phys. Rev. Lett.*, **56** (9), 930–933.

17 Loos, J. (2005) The art of SPM: scanning probe microscopy in materials science. *Adv. Mater.*, **17** (15), 1821–1833.

18 Samori, P. (ed.) (2006) *Scanning Probe Microscopies beyond Imaging: Manipulation of Molecules and Nanostructures*, Wiley-VCH Verlag GmbH, Weinheim.

19 Bushan, B. (ed.) (2004) *Springer Handbook of Nanotechnology*, Springer, Berlin.

20 Meyer, G. and Amer, N.M. (1988) Novel optical approach to atomic force microscopy. *Appl. Phys. Lett.*, **53** (24), 1045–1047.

21 Sulchek, T., Grow, R.J., Yaralioglu, G.G., Minne, S.C., Quate, C.F., Manalis, S.R., Kiraz, A., Aydine, A., and Atalar, A. (2001) Parallel atomic force microscopy with optical interferometric detection. *Appl. Phys. Lett.*, **78** (20), 1787–1789.

22 Rugar, D., Mamin, H.J., and Guethner, P. (1989) Improved fiber-optic interferometer for atomic force microscopy. *Appl. Phys. Lett.*, **55** (25), 2588–2590.

23 Oden, P.I., Datskos, P.G., Thundat, T., and Warmack, R.J. (1996) Uncooled thermal imaging using a piezoresistive microcantilever. *Appl. Phys. Lett.*, **69** (21), 3277–3279.

24 Abedinov, N., Grabiec, P., Gotszalk, T., Ivanov, T., Voigt, J., and Rangelow, I.W. (2001) Micromachined piezoresistive cantilever array with integrated resistive microheater for calorimetry and mass detection. *J. Vac. Sci. Technol. A*, **19** (6), 2884–2888.

25 Tortonese, M., Barrett, R.C., and Quate, C.F. (1993) Atomic resolution with an atomic force microscope using piezoresistive detection. *Appl. Phys. Lett.*, **62** (8), 834–836.

26 Wang, Q.M. and Cross, L.E. (1998) Performance analysis of piezoelectric cantilever bending actuators. *Ferroelectrics*, **215** (1–4), 187–213.

27 Zurn, S., Hseih, M., Smith, G., Markus, D., Zang, M., Hughes, G., Nam, Y., Arik, M., and Polla, D. (2001) Fabrication and structural characterization of a resonant frequency PZT microcantilever. *Smart Mater. Struct.*, **10** (2), 252–263.

28 Adams, J.D., Rogers, B., Manning, L., Hu, Z., Thundat, T., Cavazos, H., and Minne, S.C. (2005) Piezoelectric self-sensing of adsorption-induced microcantilever bending. *Sens. Actuators A*, **121** (2), 457–461.

29 Britton, C.L., Jones, R.L., Oden, P.I., Hu, Z., Warmack, R.J., Smith, S.F., Bryan, W.L., and Rochelle, J.M. (2000) Multiple-input microcantilever sensors. *Ultramicroscopy*, **82** (1–4), 17–21.

30 Amantea, R., Knoedler, C.M., Pantuso, F.P., Patel, V., Sauer, D.J., and Tower, J.R.R. (1997) Uncooled IR imager with 5-mK NEDT. *Proc. SPIE*, **3061**, 210–222.

31 Shekawat, G., Tark, S.-H., and Dravid, V.P. (2006) MOSFET-embedded microcantilevers for measuring deflection in biomolecular sensors. *Science*, **311** (5767), 1592–1595.

32 Albrecht, T.R., Akamine, S., Carver, T.E., and Quate, C.F. (1990) Microfabrication of cantilever styli for the atomic force microscope. *J. Vac. Sci. Technol. A*, **8** (4), 3386–3396.

33 Nguyen, C.V., Ye, Q., and Meyyappan, M. (2005) Carbon nanotube tips for scanning probe microscopy: fabrication and high aspect ratio nanometrology. *Meas. Sci. Technol.*, **16** (11), 2138–2146.

34 Nonnenmacher, J., Wolter, O., and Kassing, R. (1991) Scanning force microscopy with micromachined silicon sensors. *J. Vac. Sci. Technol. B*, **9** (2), 1358–1362.

35 Wilson, N.R. and Macpherson, J.V. (2009) Carbon nanotube tips for atomic force microscopy. *Nat. Nanotechnol.*, **4** (8), 483–491.

36 Qian, D., Wagner, G.J., Liu, W.K., Yu, M.-F., and Ruoff, R.S. (2002) Mechanics of carbon nanotubes. *Appl. Mech. Rev.*, **55** (6), 495–553.

37 Bower, C., Rosen, R., Jin, L., Han, J., and Zhou, O. (1999) Deformation of carbon nanotubes in nanotube–polymer composites. *Appl. Phys. Lett.*, **74** (22), 3317–3319.

38 Salvetat, J.P., Bonard, J.M., Thomson, N.H., Kulik, A.J., Forro, L., Benoit, W., and Zupiroli, L. (1999) Mechanical properties of carbon nanotubes. *Appl. Phys. A*, **69** (3), 255–260.

39 Ko, F.K., Han, W.B., Khan, S., Rahman, A., and Zhou, O. (2001) Carbon nanotube reinforced nanocomposites by electrospinning process. Proceedings of the American Society for Composites 16th Annual Technical Conference, September 10–12, 2001, Blacksburg, VA.

40 Li, F., Cheng, H.M., Bai, S., Su, G., and Dresselhaus, M.S. (2000) Tensile strength of single-walled carbon nanotubes directly measured from their macroscopic ropes. *Appl. Phys. Lett.*, **77** (20), 3161–3163.

41 Cooper, C.A., Ravich, D., Lips, D., Mayer, J., and Wagner, H.D. (2002) Distribution and alignment of carbon nanotubes and nanofibrils in a polymer matrix. *Comp. Sci. Technol.*, **62** (7–8), 1105–1112.

42 Sennett, M., Welsh, E., Wright, J.B., Li, W.Z., Wen, J.G., and Ren, Z.F. (2003) Dispersion and alignment of carbon nanotubes in polycarbonate. *Appl. Phys. A*, **76** (1), 111–113.

43 Hernandez, E., Goze, C., Bernier, P., and Rubio, A. (1998) Elastic properties of C and $B_xC_yN_z$ composite nanotubes. *Phys. Rev. Lett.*, **80** (20), 4502–4505.

44 Wong, S.S., Harper, J.D., Lansbury, P.T., and Lieber, C.M. (1998) Carbon nanotube tips: high-resolution probes for imaging biological systems. *J. Am. Chem. Soc.*, **120** (3), 603–604.

45 Wong, S.S., Joselevich, E., Woolley, A.T., Cheung, C.L., and Lieber, C.M. (1998) Covalently functionalized nanotubes as nanometre-sized probes in chemistry and biology. *Nature*, **394** (6688), 52–55.

46 Hutter, J.L. and Bechhoefer, J. (1993) Calibration of atomic force microscope tips. *Rev. Sci. Instrum.*, **64** (7), 1868–1873.

47 Timoshenko, S.P. and Goodier, J. (1980) *Theory of Elasticity*, 4th edn, McGraw Hill, New York.

48 Khan, A., Philip, J., and Hess, P. (2004) Young's modulus of silicon nitride used in scanning force microscope cantilevers. *J. Appl. Phys.*, **95** (4), 1667–1672.

49 Cleveland, J.P., Manne, S., Bocek, D., and Hansma, P.K. (1993) A nondestructive method for determining the spring constant of cantilevers for scanning force microscopy. *Rev. Sci. Instrum.*, **64** (2), 403–405.

50 Sader, J.E. (1995) Parallel beam approximation for V-shaped atomic force microscope cantilevers. *Rev. Sci. Instrum.*, **66** (9), 4583–4587.

51 Hazel, J.L. and Tsukruk, V.V. (1999) Spring constants of composite ceramic/gold cantilevers for scanning probe microscopy. *Thin Solid Films*, **339** (1–2), 249–257.

52 Sader, J.E., Larson, I., Mulvaney, P., and White, L.R. (1995) Method for the calibration of atomic force microscope cantilevers. *Rev. Sci. Instrum.*, **66** (7), 3789–3798.

53 Gibson, C.T., Watson, G.S., and Myhra, S. (1996) Determination of the spring constants of probes for force microscopy/spectroscopy. *Nanotechnology*, **7** (3), 259–262.

54 Hazel, J. and Tsukruk, V.V. (1998) Friction force microscopy measurements: normal

and torsional spring constants for V-shaped cantilevers. *J. Tribol.*, **120** (4), 814–819.

55 Sader, J.E., Chon, J.W.M., and Mulvaney, P. (1999) Calibration of rectangular atomic force microscope cantilevers. *Rev. Sci. Instrum.*, **70** (10), 3967–3969.

56 Matei, G.A., Thoreson, E.J., Pratt, J.R., Newell, D.B., and Burnham, N.A. (2006) Precision and accuracy of thermal calibration of atomic force microscopy cantilevers. *Rev. Sci. Instrum.*, **77** (8), 083703.

57 Gibson, C.T., Smith, D.A., and Roberts, C.J. (2005) Calibration of silicon atomic force microscope cantilevers. *Nanotechnology*, **16** (2), 234–238.

58 Radmacher, M., Tillmann, R.W., and Gaub, H.E. (1993) Imaging viscoelasticity by force modulation with the atomic force microscope. *Biophys. J.*, **64** (3), 735–742.

59 Tsukruk, V.V. and Gorbunov, V.V. (2001) Nanomechanical probing with scanning force microscopy. *Microsc. Today*, **01-1**, 8.

60 Machleidt, T., Kästner, R., and Franke, K.-H. (2005) Reconstruction and geometric assessment of AFM tips, in *Nanoscale Calibration Standards and Methods: Dimensional and Related Measurements in Micro- and Nanometer Range* (eds G. Wilkening and L. Koenders), Wiley-VCH Verlag GmbH, Weinheim, pp. 297–310.

61 Czerkas, S., Dziomba, T., and Bosse, H. (2005) Comparison of different methods of SFM tip shape determination for various characterization structures and types of tip, in *Nanoscale Calibration Standards and Methods: Dimensional and Related Measurements in Micro- and Nanometer Range* (eds G. Wilkening and L. Koenders), Wiley-VCH Verlag GmbH, Weinheim, pp. 311–320.

62 Siedle, P., Butt, H.-J., Bamberg, E., Wang, D.N., Kühlbrandt, W., Zach, J., and Haider, M. (1992) Determining the form of atomic force microscope tips. *Inst. Phys. Conf. Ser.*, **130**, 361–364.

63 Hafner, J.H., Cheung, C.L., Oosterkamp, T.H., and Lieber, C.M. (2001) High-yield assembly of individual single-walled carbon nanotube tips for scanning probe microscopies. *J. Phys. Chem. B*, **105** (4), 743–746.

64 Shulha, H., Foo, C.W.P., Kaplan, D.L., and Tsukruk, V.V. (2006) Unfolding the multi-length scale domain structure of silk fibroin protein. *Polymer*, **47** (16), 5821–5830.

65 Tsukruk, V.V. and Bliznyuk, V.N. (1998) Adhesive and friction forces between chemically modified silicon and silicon nitride surfaces. *Langmuir*, **14** (2), 446–455.

66 Florin, E.L., Moy, V.T., and Gaub, H.E. (1994) Adhesion forces between individual ligand–receptor pairs. *Science*, **264** (5157), 415–417.

67 Frisbie, C.D., Rozsnyai, L.F., Noy, A., Wrighton, M.S., and Lieber, C.M. (1994) Functional group imaging by chemical force microscopy. *Science*, **265** (5181), 2071–2074.

68 Vezenov, D.V., Noy, A., Rozsnyai, L.F., and Lieber, C.M. (1997) Force titrations and ionization state sensitive imaging of functional groups in aqueous solutions by chemical force microscopy. *J. Am. Chem. Soc.*, **119** (8), 2006–2015.

69 Friddle, R.W., LeMieux, M.C., Cicero, G., Artyukhin, A.B., Tsukruk, V.V., Grossman, J.C., Galli, G., and Noy, A. (2007) Single functional group interactions with individual carbon nanotubes. *Nat. Nanotechnol.*, **2** (11), 692–697.

70 García, R. and Pérez, R. (2002) Dynamic atomic force microscopy methods. *Surf. Sci. Rep.*, **47** (6–8), 197–301.

71 Sheiko, S.S. (2000) Imaging of polymers using scanning force microscopy: from superstructures to individual molecules, in *New Developments in Polymer Analytics II: Advances in Polymer Science*, vol. **151** (ed. M. Schmidt), Springer, Berlin, pp. 61–174.

72 Martin, Y., Williams, C.C., and Wickramasinghe, H.K. (1987) Atomic force microscope: force mapping and profiling on a sub 100-Å scale. *J. Appl. Phys.*, **61** (10), 4723–4729.

73 Zhong, Q., Innis, D., Kjoller, K., and Elings, V.B. (1993) Fractured polymer/silica fiber surface studied by tapping mode atomic force microscopy. *Surf. Sci. Lett.*, **290** (1–2), L688–L692.

74 Magonov, S.N., Cleveland, J., Elings, V., Denley, D., and Whangbo, M.-H. (1997)

Tapping-mode atomic force microscopy study of the near-surface composition of a styrene-butadiene-styrene triblock copolymer film. *Surf. Sci.*, **389** (1–3), 201–211.

75 Magonov, S.N., Elings, V., and Whangbo, M.-H. (1997) Phase imaging and stiffness in tapping-mode atomic force microscopy. *Surf. Sci.*, **375** (2–3), L385–L391.

76 Tsukruk, V.V. and Reneker, D.H. (1995) Scanning probe microscopy of polymeric and organic molecular films: from self-assembled monolayers to composite multilayers. *Polymer*, **36** (9), 1791–1808.

77 Heuberger, M., Dietler, G. and Schlapbach, L. (1996) Elastic deformations of tip and sample during atomic force microscope measurements. *J. Vac. Sci. Technol. B*, **14** (2), 1250–1254.

78 Sokolov, I.Y., Henderson, G.S., Wicks, F.J., and Ozin, G.A. (1997) Improved atomic force microscopy resolution using an electrical double layer. *Appl. Phys. Lett.*, **70** (7), 844–846.

79 Jaschke, M., Schönherr, H., Wolf, H., Butt, H.-J., Bamberg, E., Besocke, M.K., and Ringsdorf, H. (1996) Structure of alkyl and perfluoroalkyl disulfide and azobenzenethiol monolayers on gold(111) revealed by atomic force microscopy. *J. Phys. Chem.*, **100** (6), 2290–2301.

80 Kuwahara, Y. (1999) Muscovite surface structure imaged by fluid contact mode AFM. *Phys. Chem. Miner.*, **26** (3), 198–205.

81 Gan, Y. (2009) Atomic and subnanometer resolution in ambient conditions by atomic force microscopy. *Surf. Sci. Rep.*, **64** (3), 99–121.

82 Ginger, D.S., Zhang, H., and Mirkin, C.A. (2004) The evolution of dip-pen nanolithography. *Angew Chem., Int. Ed.*, **43** (1), 30–45.

3
Basics of Atomic Force Microscopy Studies of Soft Matter

In this chapter, we briefly discuss the forces of interaction and principles that form the basis for AFM imaging and probing surface properties of soft materials in a broad range of environments. The chapter also briefly introduces the important practical considerations of AFM imaging of soft materials such as common sources of instabilities, noise, and artifacts, as well as simple methods to identify and avoid some of the most commonly generated artifacts. Only the most important aspects will be covered here and an in-depth discussion is beyond the scope of this book. Therefore, interested readers are referred to some excellent sources for a detailed discussion of intermolecular forces and tip–sample interactions already available in the literature (see Refs [1, 2] and other relevant references below).

3.1
Physical Principles: Forces of Interaction

The forces of interaction between the AFM tip and the sample can be generally classified into attractive and repulsive forces. The primary attractive forces include van der Waals (vdW) interactions and electrostatic forces (which can also be repulsive) as well as different types of "weak" interactions such as hydrogen bonding, charge–transfer interactions, and others. The attractive forces are described by different theoretical models that consider electronic structure of molecules, fluctuations of electronic clouds, redistribution of charges, and the directional aspects of short- and long-range interactions among others [1].

The primary repulsive forces usually include the hard sphere repulsion and repulsive Coulombic interactions. The repulsive forces are usually short range in nature and exhibit a rapid decay with distance, as has been discussed in detail elsewhere [1–4]. An exception is long-range Coulombic repulsion resulting from the presence of the surface bilayers or permanent surface charges caused by the presence of ionic groups in polyelectrolytes. In the following sections, we will briefly review some important aspects of these interactions relevant to AFM imaging of soft matter and interpretation of these images.

Scanning Probe Microscopy of Soft Matter: Fundamentals and Practices, First Edition.
Vladimir V. Tsukruk and Srikanth Singamaneni.

3.1.1 Long-Range Forces

The van der Waals force between macroscopic objects is due to dispersion interactions (thermal or zero-point quantum fluctuations) of electrically neutral atoms in two objects across the media [1]. The vdW potential between two atoms exhibits a strong distance dependence in the simplified form: $\sim z^{-6}$. However, for macroscopic (relatively to atoms) bodies (AFM tip and the sample) involving interaction of numerous atoms of either body, the effective force is a vector sum of the vdW forces between atoms. In this case, for a spherical tip with radius R interacting with a flat sample, the integral vdW potential force F_{vdW} is given by

$$F_{\mathrm{vdW}} = \frac{-A_{\mathrm{H}} R}{6z^2} \tag{3.1}$$

where A_{H} is the Hamaker constant, which has been calculated and well documented for different materials and media [1, 5]. Typically, A_{H} is on the order of 1 eV for common solids. The Hamaker constant, and hence the vdW forces between the tip and the sample, depends on the materials (of the AFM tip and sample), the medium in the gap, and the tip geometry [6, 7]. For a tip with a radius of 10 nm and a separation distance (z) of 1 nm, the vdW potential is nearly −1.6 eV, and the corresponding attractive force is about 0.3 nN. Thus, the force is a significant contribution to the interaction between the tip and the sample that enables topographic imaging in noncontact modes.

The effects of a liquid medium on the vdW force are shown in Figure 3.1 [8]. The plot shows that polar liquids (H_2O, H_2O_2, and glycol) significantly lower the vdW interaction between the tip and the sample. In fact, at small tip–sample separations, the vdW forces are two orders of magnitude lower than that in vacuum. Apart from the reduced vdW forces, the absence of capillary forces between the tip and the sample further lower the overall tip–sample forces, as will be discussed in Section 3.2.1. The vdW forces exerted by the molecules trapped between the tip and the surface partially cancel out the force between the tip and the sample serving as a not easily compressible and lubricating molecular layer and thus reducing significantly overall forces inserted by the AFM tip.

3.1.2 Short-Range Forces

The short-range repulsive forces dominate the tip–surface interaction for extremely small tip–sample separations (less than 0.4 nm) (Figure 3.1). As the tip approaches the sample, the electronic wave function of the tip terminal atoms and the topmost atoms of the sample will overlap, resulting in a very strong repulsion caused by Pauli exclusion (electron wave overlap). The short-range interactions at intermediate distances (0.5–1.5 nm) can be attractive in the case where the chemical species on the tip, or the tip itself, can form chemical bonds with the sample. However, these short-range attractive interactions are not beneficial for AFM imaging as they are

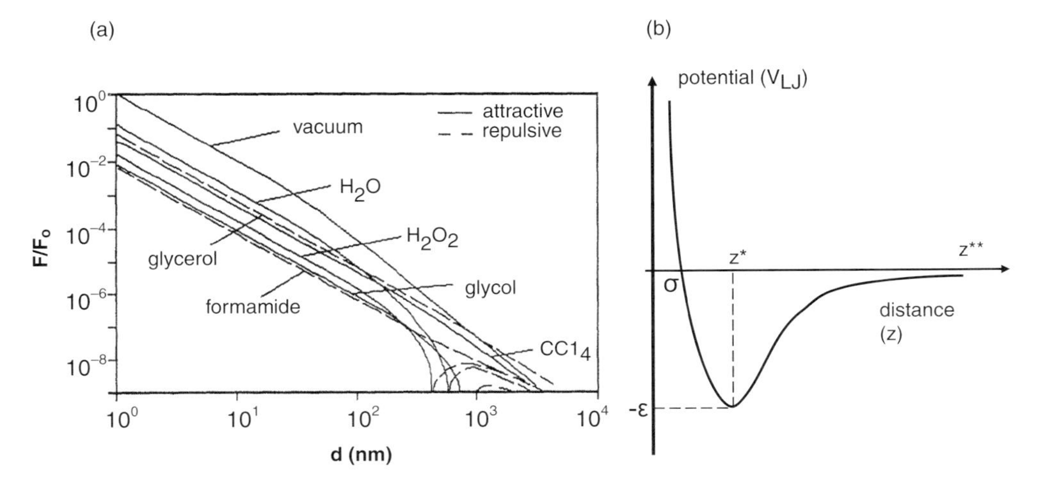

Figure 3.1 (a) Calculated van der Waals forces versus tip–surface separation in various media. The plot shows that at small tip–sample separation distances, the van der Waals forces are two orders of magnitude smaller than that in vacuum. (b) Common Lennard–Jones potential between two atoms showing the characteristic distance between the atoms with zero and minimum potential. Reprinted from Ref. [8].

more sensitive to the nature of the chemical species than to the topography of the surface.

Lennard-Jones (LJ) and Mores potentials are commonly employed to describe the balance of repulsive and attractive forces [1, 9, 10]. For instance, the LJ potential (V_{LJ}) is given by

$$V_{\text{LJ}} = -4\varepsilon\left(\left(\frac{\sigma}{z}\right)^{6} - \left(\frac{\sigma}{z}\right)^{12}\right) \tag{3.2}$$

where z is the distance between the two atoms, σ is the distance at which the potential of the two-atom system is zero, and ε is the depth of the potential well at a certain distance.

Figure 3.1b shows the plot of the generic LJ potential with the distance between the atoms. The plot identifies two important characteristic distances between the two atoms, namely, σ at which the potential is zero, Z^* at which the potential is minimum (equal to $-\varepsilon$), and Z^{**} at which overall interaction can be neglected. It is important to note that these models are typically limited to pairwise potentials between two atoms interacting with each other with additional atoms in molecular groups complicating the overall interaction potential. Generally speaking, for common hydrocarbon molecules, the repulsion dominates for distances below 0.3–0.4 nm, with minimum potential reached at 0.4–0.6 nm and near-zero interaction potential approaching for distances beyond 1.5–2 nm (Figure 3.1). From Eq. (3.2), it can be seen that the short-range repulsive force rapidly falls off with distance, which is extremely critical for achieving near-atomic resolution with AFM imaging under special conditions.

In reality, the tip–sample interaction is more complicated than this model, and it involves simultaneous interactions not only between the nearest atoms on the tip and the sample but also between the neighboring atoms of both species. However, the models mentioned above provide a qualitative understanding of short-range repulsive forces and long-range attractive forces between the tip and the sample. In practical cases, where the contact area between the tip and the sample typically involves the interaction of tens of hundreds of atoms, the repulsive force can be considered as a mechanical or elastic force involving nanoscale physical deformation of the sample. Contacts between the tip and the sample that form the basis for AFM-based nanomechanical measurements are treated using contact mechanics models such as Hertzian, Johnson–Kendall–Roberts (JKR), and Derjaguin–Muller–Toporov (DMT). These models are briefly discussed in Chapters 4 and 5.

3.1.3 Other Forces of Interaction

Apart from the short-range repulsive forces and the long-range van der Waals forces, there are other long-range forces that significantly contribute to the tip–sample interactions under practical conditions. Here, we briefly discuss capillary forces and electrostatic forces, which are the most important forces to consider for routine AFM imaging.

Capillary Forces

Water molecules adsorb to the hydrophilic (as well as hydrophobic) surface of a sample when exposed to ambient conditions, resulting in a molecular (at least few molecules even at low humidity) surface film of water present on all substrates [11, 12]. When the AFM tip (usually with hydrophilic silicon oxide surface) is brought into proximity to the surface covered with the water layer, a meniscus is immediately formed between the tip and the sample, as shown in Figure 3.2. Actual ESEM images of AFM tip meniscus are presented in Chapter 19. This water meniscus results in a strong attractive force (known as capillary force) between the tip and the sample, which is the van der Waals force between the water molecules and atoms at the surface of the tip and the sample.

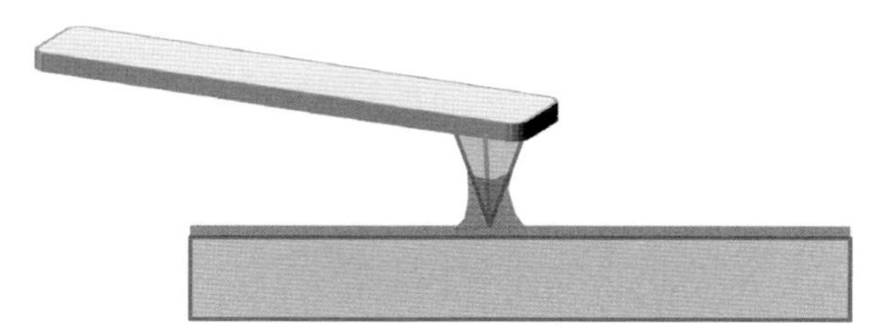

Figure 3.2 Schematic of a thin water film on the surface of a sample forming a meniscus as the tip approaches the sample, resulting in capillary forces between the tip and the sample.

This additional force experienced by the AFM tip in proximity to the surface results in an increased force required to detach the tip from the surface compared to the case when no water meniscus forms. The magnitude of the capillary force depends on the surface properties (hydrophilicity of the tip and the surface), humidity, temperature, and geometry of the tip [13, 14]. The capillary force shows longer-range dependence and is usually much stronger than the vdW forces under even moderate humidity. The capillary force between the tip and the surface is given by the Laplace equation [15]:

$$F = \frac{4\pi Rr\cos\theta}{1 + (d/D)} \tag{3.3}$$

where R is the radius of the tip, θ is the contact angle between the water vapor interface and the tip, D is the distance between the tip and the surface, and d is the length the tip extends to the water bridge. The capillary forces are typically on the order of a few nanonewton, but can easily reach tens of nanonewtons, depending on the ambient conditions (humidity) and the chemical nature of the tip and surface.

Electrostatic Forces

An electrostatic force exists between a charged or conductive tip and the sample that has a potential difference U. For a tip–sample distance, z is smaller than the tip radius R and the electrostatic force F_{el} is given by [16, 17]

$$F_{el} = -\frac{\pi\varepsilon_0 RU^2}{z} \tag{3.4}$$

where ε_0 is the dielectric constant. As F_{el} is inversely proportional to the distance between the tip and the sample, as opposed to the inverse square dependence of vdW, the electrostatic force is a relatively long-range force compared to vdW force. For a potential difference (U) of 1 V, a tip with a radius of 100 nm and a tip–sample separation (z) of 0.5 nm, the electrostatic force F_{el} experienced by the tip is approximately 6 nN [18].

Forces of Interaction in Liquid

Capillary forces are completely eliminated when both the tip and the surface are immersed in a liquid. Under these conditions, the electrostatic force forms the major component of the interaction force. To understand the electrostatic force between the tip and the sample immersed in a liquid, it is important to utilize the concept of the electrical double layer (EDL) [19]. In liquids, the surfaces of solids typically build up surface charges, which can be due to substrate atoms, impurities, adsorbates, or functional groups. The solution consists of two types of ions with opposite charges: ions with the same charge as the surfaces are called coions, whereas those with opposite charges are called counterions. The net charge neutralization process drives more counterions to the charged surface compared to coions, while the thermal motion of ions and water molecules causes counterions and coions to extend into the liquid to form an extended layer. The charged surface in conjunction with the diffused layers of the ions is often termed the electrical double layer and this concept is widely exploited in colloidal science.

In a liquid, when two surfaces (the tip and the sample) approach each other, two electrical double layers interact, resulting in an electrostatic force between the two interacting surfaces due to the osmotic pressure effect of ions between the two closely spaced EDLs. The EDL force can be attractive, repulsive, or zero, depending on the nature of participating surfaces and the media between them (solution pH and ion concentration). A description of how the forces are modulated in the liquid medium can be found in Section 3.2.1.

3.1.4
Resolution Criteria

According to the classical definition borrowed from optical microscopy research, resolution of a microscopy technique is the smallest distance between two objects that can be identified as distinct objects [20]. While the primary limitation of resolution in the case of far-field optical microscopy is from the Abbe's diffraction limit, the resolution limits in the case of AFM are from very different sources, namely, thermal noise of the force sensing system, which in most cases is a microcantilever. For the common photodiode detection system, the thermal noise of a rectangular cantilever is given by [21]

$$\Delta z = \sqrt{\frac{4k_B T}{3k}} = \frac{0.074}{\sqrt{k}} \tag{3.5}$$

where k_B is the Boltzmann constant, k is the spring constant of the cantilever, and T is the absolute temperature. Under ambient conditions ($T = 298$ K), the thermal noise is approximately 0.015 nm for a free cantilever with a spring constant of 5 N/m. This extremely small thermal vibration of the AFM at moderately stiff cantilevers allows the unprecedented high vertical resolution of the AFM technique.

It is important to recognize that the lateral resolution of the AFM technique critically depends on its vertical resolution [22]. Consider two vertical posts with same height separated by a distance d imaged by a parabolic tip with an end radius R, as shown in Figure 3.3. The posts in the resulting image will appear to be inverted parabolas due to conventional tip broadening or tip convolution effects. The small depression between the two parabolas of depth Δz aids in resolving the two posts, while the two spikes can be considered resolved if Δz is larger than the vertical resolution. In this consideration, the resolution d, the minimum separation at which the spikes are resolved, is given by

$$d = 2\sqrt{2R(\Delta z)} \tag{3.6}$$

As can be seen from this equation, the lateral resolution critically depends on the vertical resolution (Δz) and the radius of the tip. For an ultrasharp tip with a very small radius of 5 nm and a vertical resolution of 0.015 nm quoted above, a lateral resolution of about only 0.7 nm can be achieved, which effectively forbids true atomic and molecular resolution except in special cases (large molecules or special scanning conditions). The dependence of resolution on the tip radius (as well as surface roughness, see below) has been experimentally verified on numerous occasions [23].

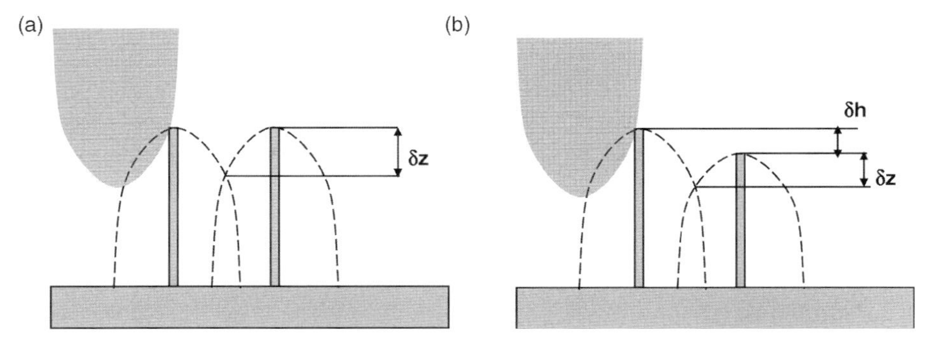

Figure 3.3 Tip-induced broadening of the features affecting the lateral resolution of features of (a) the same height and (b) with small height differences.

In the above example, the two posts were assumed to be of equal height. For the case where the two posts have a small height difference (Δh), the lateral resolution is modified and can be represented as [22]

$$d = \sqrt{2R}\left(\sqrt{\Delta z} + \sqrt{\Delta z + \Delta h}\right) \tag{3.7}$$

To put in perspective the role of such deviations from an ideal surface topography, for a pair of posts with a height difference of only 2 nm, the lateral resolution drops to 4.8 nm for a 5 nm tip and the threshold of the minimum detectable Δz is increased to 0.5 nm.

This example demonstrates the limitation imposed by practical surfaces with uneven surface features and variable roughness on the achievable spatial resolution. It is important to note that the ultimate lateral resolution of AFM imaging is also affected by factors such as deformation of the surface features. Also, the decrease of the tip radius beyond a certain point does not translate into enhancement in the resolution, but may actually result in excessive surface damage. This factor is increasingly important for soft materials with easily deformable surfaces, thus making attempts to achieve near-molecular resolution illusive in a vast majority of cases.

Furthermore, the above consideration is strictly valid for contact mode imaging in which the tip is in intimate contact with the sample surface. Contrarily, the resolution in the case of noncontact mode imaging, in which the topographic information is collected from the intermediate-range van der Waals interactions, is not completely governed by the tip radius. The resolution in noncontact mode is fundamentally limited by the distance between the tip and the sample and the surface deformations.

3.1.5 Scan Rates and Resonances

One of the limitations of the conventional AFM (in contrast to the high-speed mode, which is achieved by special techniques) is the relatively low scan rates at which imaging can be performed. Scan rate in tapping mode AFM is primarily limited by (i) inherent sluggishness of the force sensor and (ii) poor tracking of the surface at high speeds. We will briefly discuss these aspects and possible methods to overcome this limitation.

As discussed earlier, tapping mode AFM involves oscillating the cantilever close to its resonance frequency to avoid the instabilities associated with the high-Q mechanical system at this frequency. Commercial AFM microscopes offer the ability to offset the driving frequency by a fixed fraction from the resonance frequency, which is determined in the frequency sweep. The finite response time of the microcantilever, due to the slow rate at which the energy is pumped into the cantilever by the driving piezoelement, limits the ability of the cantilever to instantly respond to variations in height along the surface, resulting in the need for relatively slower scan rates (well below 10 Hz in most cases for microscopic scales).

Furthermore, the x–y piezoelements have relatively small resonance frequencies (5–10 kHz). In particular, longer piezotubes, which enable larger scan size, have even smaller resonance frequencies. In order to avoid the instabilities associated with the resonance condition, scan rates should be relatively low, even for small scan sizes. However, it is also important to pay attention to the thermal drifts associated with the piezoelements, which become significant for very small ($<$100 nm) and very large ($>$10 μm) scan sizes and under low scan rates.

3.2
Imaging in Controlled Environment

The ability of AFM to acquire surface images with near-atomic resolution under ambient conditions makes it an attractive choice for investigating the structure and properties of a wide variety of materials. However, the ability to image and probe properties of surfaces in a controlled environment is a complimentary attribute that makes it extremely versatile – broadening the gamut of applications. Apart from biological applications, liquid AFM imaging is important for minimizing the forces between the tip and the sample by virtually completely overcoming the capillary forces. The variety of fluids extends well beyond simple aqueous media into organic solvents, alcohols, pH-controlled aqueous media, and buffers.

Generally, AFM can be performed in a wide variety of conditions such as in liquid, controlled humidity, wide temperature ranges, under a desired gaseous environment or under external mechanical stresses, to name a few. In this section, we will briefly introduce AFM in controlled environments, highlighting specific considerations from an experimental standpoint.

3.2.1
AFM Imaging in Liquid

There are several important types of forces that act on the tip when immersed in a fluid with a significant difference observed for different media (Figure 3.1). For example, when the distance of separation between the tip and the sample is on the order of a few molecules, solvation forces, which are caused by the interaction between the solvent molecules and the surface, come into play. Similarly, the tip and the surface can be charged in the liquid due to the ionization or dissociation of the chemical groups, leading to the formation of electrical double layers and thus resulting in additional force between the tip and the sample.

From an experimental standpoint, AFM imaging in liquid can be as simple as confining a small volume of liquid (several microliters) between the sample surface and the glass slide attached on top of the chip hosting the cantilever to custom-made commercial fluid cell designs that allow circulation of fluid at user-defined rates. The turbidity of the liquid is an important concern as the scattering of light results in diminished intensity of the reflected laser, thus making imaging impractical. Compliant biological specimens can be mounted either directly on sample holders, which allows the continuous supply of a thin water surface layer, or by surrounding the specimens with highly porous paper in contact with the water reservoir outside the AMF stage [24]. This design prevents a specimen from drying out, while avoiding operation in a fully liquid environment.

While scanning in liquid significantly minimizes the tip–sample forces (down to $\sim$10 pN), which is extremely advantageous for soft surfaces, it comes with a premium of a lowered quality factor (Q) of the microcantilever (from $\sim$100–1000 in air to $\sim$10–50 in liquid due to its high media viscosity) and a lowered resonance frequency, resulting in lowered sensitivity and reduced scan rates. The liquid drag results in an increase of the effective mass of the cantilever by a factor of 10–40, reducing the resonance frequency and hydrodynamic interaction between the liquid and the cantilever, causing a reduction in the Q of the system mentioned above [25, 26]. There have been attempts to electronically modify the Q of the cantilever [27]. The low Q of the cantilever is increased up to three orders of magnitude by a positive feedback control. This technique also includes a phase-locked loop unit to track the resonance of the cantilever and enables scanning of soft biological samples with a force smaller than 10 pN.

Another important consideration is that the oscillation of the microcantilever by a piezoelement results in many false high-amplitude resonances and increased noise due to excitation of the liquid cell. One of the ways this problem has been addressed is by using a magnetic field to drive the magnetic cantilever [28]. However, this technique requires the cantilever to be coated with magnetic materials, which is not always acceptable.

An alternative mode called "jumping mode," which is similar to pulsed force microscopy, has been reported to improve imaging conditions [29]. Jumping mode involves the following cycle of events repeated for each point imaged: (1) lateral motion to next point when the tip is at the farthest distance from the surface, (2) approach the surface, (3) tip–sample interaction with feedback controlled using the deflection signal (since it is static mode), and (4) retraction from the surface. This technique, in theory, can be applied to perform scanning even in air. However, pulling away the tip from the sample requires large excursion of the tip from the surface to overcome adhesive forces (primarily due to capillary forces). This large pull-off distance results in painfully slow scan rates. However, when performing in liquid, the absence of capillary forces lowers the tip–sample pull-off distance to a few nanometers, thus enabling realistic scan rates. Although the technique overcomes shear forces completely, as the lateral motion of the cantilever is performed when the tip is farthest from the surface, it also lowers the tip–sample forces to below 100 pN, which is noteworthy for a static AFM mode with common forces above 1 nN [29].

3.2.2
AFM at Controlled Temperature

One of the important advantages of AFM over other microscopy techniques is the relative ease with which imaging can be performed over a wide range of temperatures. Temperature-controlled AFM imaging is a valuable tool for probing a variety of phase transitions in polymers such as the glass transition, melting, crystallization, and order–disorder transitions in block copolymers, as will be demonstrated in numerous examples in this book. A detailed discussion of the application of the temperature-controlled AFM for mechanical and physical probing of soft materials is presented in Chapter 6.

From an experimental standpoint, temperature-controlled AFM can be performed by mounting the sample on a special temperature-controlled stage (e.g., a sealed Peltier microelement). In modern commercial AFM instruments with temperature-controlled units, a wide range of temperatures are accessible (within −35 to 250 °C, usually with an accuracy of ±0.1 °C) under special cooling conditions for piezoelements and scanning stages. Without this special cooling hardware, the upper limit of temperature should be kept below 120 °C to avoid accelerating piezotube aging. Temperature-controlled AFM imaging, however, comes at the price of enhanced thermal drifts in the sample and the cantilever that affect the spatial resolution. It is important to note that the equilibration of the sample and the tip are critical to avoid artifacts associated with this significant thermal drift.

3.2.3
Imaging in Controlled Humidity

In situ variations of the relative humidity during an AFM measurement allow the user to address a number of issues such as the influence of capillary forces on AFM imaging, forces of interaction between microparticles and surfaces, conformational changes of single molecules, and changes in structure and properties of various soft materials at different humidities [30–33].

From an experimental viewpoint, there are two important techniques utilized for imaging in a controlled environment. In the first approach, which is suitable for compact systems, the AFM head is enclosed in an environmental chamber (see Figure 3.4a) in which the humidity can be controlled by using a supply of dry and humid air [34]. In the second approach, AFM designs, in which the scanning head is large and contains additional electronics, special microchambers, and custom-made setups have been reported [35].

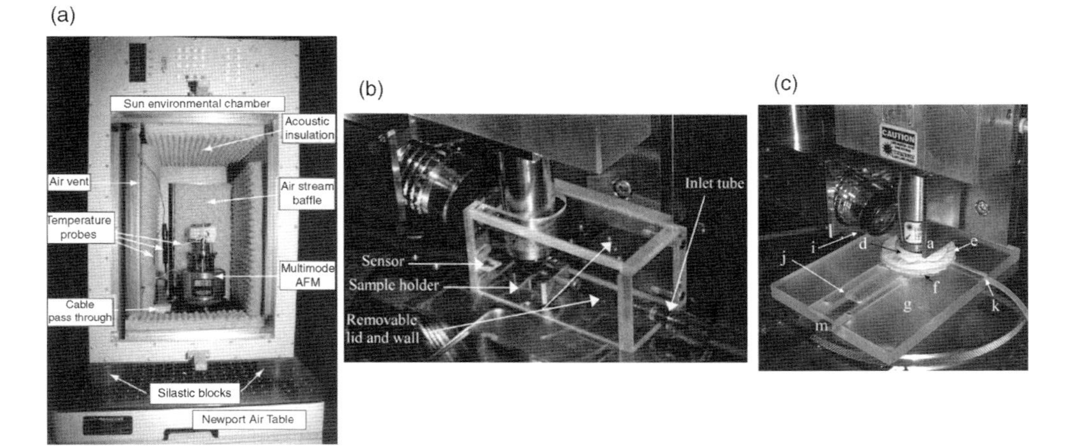

Figure 3.4 Photographs showing the three different designs for controlling the humidity while performing AFM scans. (a) Compact AFM design with the entire head enclosed in an environmental chamber. (b) Dimension series AFM with the sample and the lower portion of the scanner tube enclosed in a small chamber. (c) Dimension series AFM with the sample and the tip holder enclosed in a custom-designed chamber. Reprinted from Refs [34–36].

Maxwell and Huson have demonstrated the use of a small chamber enclosing the sample and the lower portion of the AFM head, thus creating a controlled humidity in the sample environment (Figure 3.4b) [35]. However, the design exposes the piezoelectronics to an extremely humid environment and restricts the translational motion of the stage to a certain extent. An alternative design suggested by Stukalov *et al.* overcomes this issue by completely shielding the piezoelectronics from the humid gas (Figure 3.4c) [36]. Furthermore, this design offers excellent control over ($\pm 0.2\%$) a wide range of relative humidities (RH = 5–95%).

From a tip–sample interaction standpoint, the ambient humidity significantly influences capillary forces and hence the adhesion between the tip and the sample. However, there seems to be disagreement as to whether the relative humidity enhances or lowers the adhesion. There have been several reports that the capillary force increases with RH, with several of these reports demonstrating a well-defined critical RH value above a transition from RH-independent dry adhesion to capillary adhesion [16, 37, 38]. The critical values vary from 10% to 60%, depending on the surface and tip material. Figure 3.5 shows force–distance curves acquired during retraction of a silicon tip on a freshly cleaved mica surface at different RH values obtained using the RH control chamber [36]. The plot clearly depicts the changes in capillary forces between the tip and the surface at different relative humidities with much higher pull-off forces observed for higher humidity. It is worth noting that adhesion between the tip and the surface remains insensitive to the relative humidity for a hydrophobic tip or sample, while it is found to be extremely sensitive to the relative humidity in the case of a hydrophilic tip and hydrophilic surface.

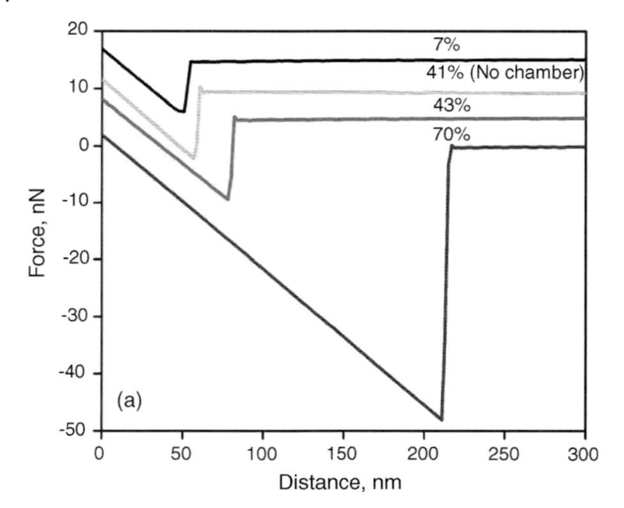

Figure 3.5 Force–distance curves obtained during the retraction of a silicon cantilever from a freshly cleaved mica surface at different relative humidities. Reprinted from Ref. [36].

AFM imaging can also be performed in the presence of specific gaseous environment to probe the morphological changes of a sample. For example, Favier *et al.* have performed AFM imaging of electrochemically deposited palladium nanowires in ambient air and in a stream of hydrogen gas [39]. They have demonstrated the formation of nanogaps in the palladium nanowires upon exposure to hydrogen owing to the formation of palladium hydride upon absorption of hydrogen. The large change in the electrical resistance of the palladium nanowires upon exposure to hydrogen was attributed to morphological changes in the palladium nanowires revealed by environmentally controlled AFM imaging.

3.3 Artifacts in AFM Imaging of Soft Materials

It is very well known that AFM imaging is prone to a wide variety of artifacts, such as deformation or damaging of the surface features, tilt or drift of the scanner, optical interference between the laser reflected from the back of the cantilever and the smooth surface of the sample, features generated by asymmetric, double, and multiple tip probes, instrumental electronic noise (poor choice of scanning parameters), and building vibrations. These rich sources of artifacts in AFM images may often elude and mislead an untrained examiner, resulting in false "breakthroughs" reported even in modern literature.

Even now, after more than two decades of availability of affordable commercial AFMs and intensive AFM studies, numerous artifacts populate scientific literature, especially if AFM imaging is conducted for soft materials and complex nanostructures, under extreme imaging conditions, and in attempts to enrich the ultimate resolution. In the following sections, we will briefly review common AFM artifacts, how they can be avoided during data acquisition, and how they can be accounted for in image analysis. Apparently, a variety of artifacts produced and still published to date cannot be covered in a single section and thus only major principles and the most commonly generated features will be underlined here.

3.3.1 Surface Damage and Deformation

While the intimate contact of the AFM tip with the surface enables unprecedented nanoscale resolution of AFM (to near-atomic scale in best cases), it also results in large deformations and, in some cases, damage of the AFM tip and, more frequently, the surface of soft materials. The large forces (commonly within 10–100 nN for contact mode) involved in AFM imaging are translated as surface stresses of tens of hundreds of megapascals to as high as 1 GPa, depending on the contact area. The surface pressure exerted by the tip on the sample is given by

$$P = \frac{\mathrm{Tip}_{\mathrm{Load}}}{A} \tag{3.8}$$

where $\mathrm{Tip}_{\mathrm{Load}}$ is the normal force acting on the tip apex and A is the area of physical contact. A normal force of 1 nN at the apex of a tip translates into a surface pressure of 1 Gpa for a contact area of 1 nm^2 (approximately 10 atoms in contact). For tips with larger radii of curvature, the normal force is still high and can easily reach hundreds of megapascals, which is much larger than the yield strength of most polymeric materials (within 5–50 MPa).

These large stresses are on the order of the ultimate stress (and even much higher) of most polymers and biological materials. Thus, local contact can easily result in large elastic deformations, plastic deformations, material wear, and irreversible surface damages. In the case of biological samples, the large forces cause perturbation of the natural state of the system, thus introducing artifacts in their structural characterization. For example, conformational changes in protein crystals have been observed due to the mechanical stress caused during scanning [40]. Reducing the normal force to around/below 1 nN in tapping mode alleviates this problem and thus allows mostly nondestructive imaging, even of the compliant surfaces, if a careful choice of parameters has been made.

3.3.2 Tip Dilation

As mentioned earlier, one of the primary limitations of the lateral resolution of the AFM is the nonzero radius of the probe used for imaging of surface features. The finite size of the tip invariably results in the contribution of the tip to the apparent lateral dimensions of the surface features, commonly known as tip dilation or geometrical convolution [41]. When the radius of the tip is comparable to the dimensions of the surface features being imaged, the convolution of tip shape with sample topography results in a significant discrepancy between the true width of the real features and that observed in an AFM image.

For a quick estimation, in the simplest example, a spherical particle with a radius R_{sphere}, when scanned with a conical tip with radius R and a half-cone angle α, will appear to have a lateral dimension (width) given by [42]

$$W = 2R_{\text{sphere}} \cos^{-1} \alpha \left[1 + \frac{R}{R_{\text{sphere}}} + \left(1 - \frac{R}{R_{\text{sphere}}} \right) \sin \alpha \right] \quad (3.9)$$

Thus, a surface feature with a diameter of 20 nm scanned with an AFM tip with a 10 nm radius and a half-cone angle of 15° will be observed as that with lateral dimensions of 52 nm, a significant overestimation of true dimensions frequently overlooked in casual AFM studies.

It is important to note that the apparent lateral dimension of particles in an AFM image not only depend on the tip radius but also on the shape of the tip (e.g., conical or parabolic). A detailed consideration of the tip dilation phenomenon associated with various geometries of the tip and surface features and under different scanning conditions may be found elsewhere [40]. A more extensive discussion of the tip dilation and the existing deconvolution approaches for better interpretation of the distorted AFM images is reserved for Section 3.3.1.

3.3.3 Damaged and Contaminated Tip or Surface

Tip contamination and damage as a result of extensive scanning are common issues resulting in numerous artifacts in AFM images. When the tip is contaminated, the size and shape of the contaminant will most likely determine the interaction between the sample surface and the tip, thus resulting in images with features convoluted and smeared by the contamination. While there is a possibility that the contamination might enhance the sharpness of the tip and result in images with higher lateral resolution compared to the original tip, most often the contamination deteriorates the lateral resolution dramatically [43].

As an example, Figure 3.6 shows the tapping mode image of spherical gold nanoparticles with each nanoparticle appearing as a "nanoheart" due to the contaminated AFM tip [44]. From the image collected with the contaminated tip, it is almost impossible to recognize true spherical shapes of the nanoparticles. Convolution of the misshaped tip with the real surface features imparts a characteristic shape to the surface features in the image to an extent that no certain conclusions can be made anymore.

Easy tip damage is yet another critical and common issue that can result in severely distorted features in the AFM images. Severe damage might be routinely caused by large tip–sample forces, especially during engagement, resulting in the fracture of the brittle tips (silicon and silicon nitride) and the excessive wear on the tip due to scanning at high normal loads. An impact of as small as 1 nN can cause a brittle fracture of a native oxide layer of a silicon tip if the feedback control is not adjusted

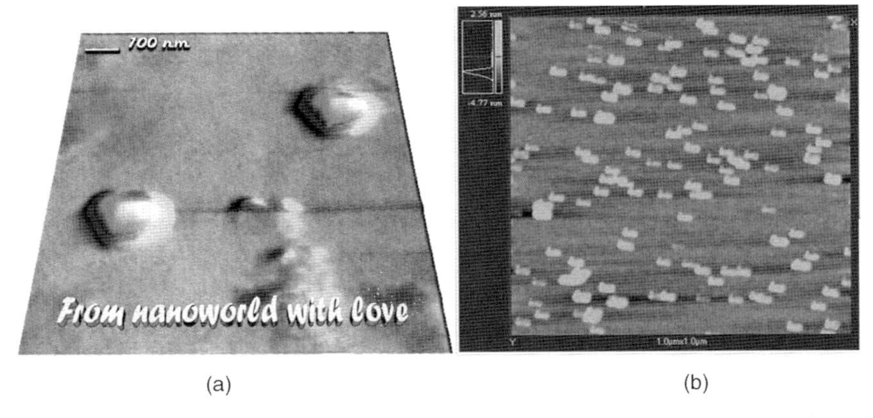

Figure 3.6 AFM image depicting (a) a contaminated tip that caused spherical nanoparticles to appear as heart shaped and (b) uniformly misshaped nanoparticles caused by a misshaped tip.

properly [45]. In contact mode especially, the tip approaches the sample with a velocity of a few microns per second until it makes contact with the surface and the cantilever deflects by a user-defined amount. This process results in rather high pressures (several gigapascals) on the silicon tip during initial contact, often causing tip fracture.

Figure 3.7 shows TEM images of a silicon tip engaged with an impact (normal force) of 0.08 nN (estimated initial impact of 5 nN) and 0.24 nN (estimated initial impact of 13 nN) confirming severe damage caused by these events even at low nominal forces. Clearly, the native oxide layer remains intact in the case of the tip engaged at smaller loads, whereas removal of the oxide layer can be observed in the case of the tip engaged at high normal forces (>0.2 nN).

Another important aspect that requires attention is the wear of the tip due to continuous scanning at relatively high forces and especially during scanning various

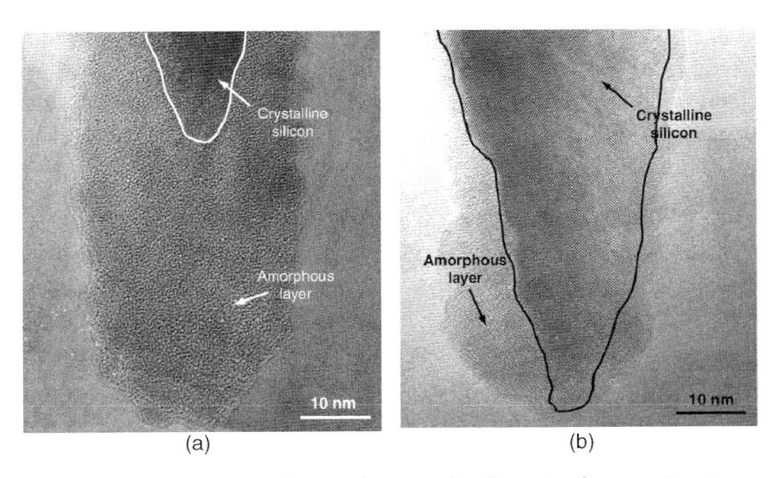

Figure 3.7 HRTEM images showing the state of a silicon tip after engaging at a normal load of (a) 5 nN and (b) 13 nN. Reprinted from Ref. [45].

standard calibration specimens made of hard materials (gratings, height standards, and tip shape arrays). It has been shown that the wear of silicon tips is higher compared to that of silicon nitride tips with higher fracture toughness, which is in turn higher compared to that of diamond-coated or carbon nanotube tips [46, 47]. In fact, carbon nanotube tips were shown to have nearly 20 times longer life compared to conventional silicon tips [48]. Both the wear and tip damage are extremely important issues, even in advanced modes of AFM imaging such as force spectroscopy. Hence, the ability to preserve and verify the shape and sharpness of the tip is of paramount importance. Some examples of verification procedures will be discussed below.

Surface contamination is a common problem for AFM imaging, especially for functionalized surfaces and for that prepared under common lab conditions. The fast adsorption of micro- and submicroparticles and organics from air happens easily during sample preparation, scanning at ambient conditions, and storage of specimens. Contaminated surfaces usually display large bumps and aggregates that can completely mask actual surface morphology and prevent quantitative measurements such as microroughness. Ideally, specimen preparation, storage, and imaging under clean room conditions should be thought, but is highly impractical and rarely utilized. Under normal lab conditions at which vast majority of AFM studies are conducted, some precautions are recommended such as specimen preparation in clean air hoods, use of highly purified solvents and filtered dry nitrogen for drying, and placement of AFM microscope in a separate location far from common sources of microparticles. A major source of contaminants is the long-term storage in nonprotected containers. Simple storage in Petri dishes is not recommended and storage in sealed vials and in vacuumed or nitrogen-filled containers is highly desirable to avoid air adsorbates and excessive surface oxidation.

3.3.4 Noises and Vibrations

Noise and vibrations are important practical limitations on the resolution of the AFM. To achieve a vertical resolution of 0.1 nm, the noise should be less than 0.01 nm, which is a stringent requirement considering the numerous sources of vibrations in ambient conditions. Even if multiple layers of isolations are used, the amplitude of noise rarely stays below 0.02 nm under conventional lab conditions.

There are numerous sources of noise such as thermal fluctuation of the cantilever, building vibrations, AC system fluctuations, vibrations of the table hosting the AFM, and electrical noise from various devices supporting the AFM operation as well as the AFM electronics itself. To achieve the highest possible resolution, it is important to take special measures to minimize the noise from these various sources and perform noise analysis of the AFM suite. It is instructive for the AFM operator and casual users to have at least a qualitative understanding of the sources of noise to avoid the common pitfalls.

Thermal fluctuations of the cantilever are probably the foremost source of noise, and thus the predominant limiting factor of the achievable vertical resolution. The amplitude of the thermal fluctuations depends on temperature, the spring constant k of the cantilever, and the method by which the cantilever deflection is detected [49]. If the deflection of the beam cantilever is measured directly, for example, with an interferometer, the thermal noise amplitude of a cantilever with a free end A is given by

$$A = \sqrt{\frac{k_B T}{k}} \tag{3.10}$$

On the other hand, when the deflection of the cantilever is measured using an optical lever in which the light from a laser diode is focused onto the back of the cantilever and the reflection angle is measured, the amplitude of the thermal fluctuation is given by

$$A = \sqrt{\frac{4k_B T}{3k}} \tag{3.11}$$

One of the important sources of intense vibrations is the building and AC system vibration, which has typical frequencies ranging between 5 and 200 Hz with maximum resonances occurring between 10 and 25 Hz. An additional problem is that low-frequency damping is less efficient with common current antivibrational measures such as air tables. The exact frequency and amplitude of the vibrations depend on the floor load, proximity to the floor support, and the distance from the ground floor. Generally higher (top) floors suffer from higher amplitude of vibrations, although examples of high-quality imaging even at 11th floor of a modern building are known (U. Akron), if proper antivibrational measures such as very long elastic cords and heavy bases that dramatically damp building vibrations are taken. Furthermore, occasional and random excitation from other sources such as opening and closing doors, people walking and talking in proximity, and elevator operation nearby all result in easily observable effects (e.g., jumped scan lines or local increases in noisy background) in the images.

3.3.5 Tip Artifacts

One of the major sources of artifacts in AFM images is the compromised (beyond contamination) probe employed for imaging, as has already been mentioned above. AFM tips, regardless of the technology or the material from which they are made, have finite radii. For practical measurements, the best AFM probes available commercially show a tip radius down to 1–2 nm for robust carbon nanotube tips, but most commonly lie within 5–10 nm for ultrasharp tips and within 10–30 nm for common robust workhorse models.

The features being imaged show apparent broadening frequently doubling or tripling true dimensions (Figure 3.8). While all the features on the surface are susceptible to this tip broadening effect, the degree of broadening depends on the nature of the features. For example, for tall features on the surface, the sidewalls of

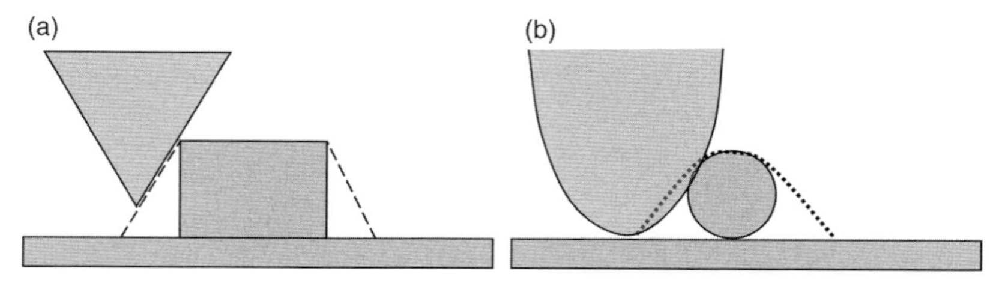

Figure 3.8 Schematics showing the tip-induced broadening of (a) pyramidal tip and (b) parabolic tip of surface features and the dependence on the shape of the probe. For surface features on the order of the tip size, the apparent shape of the feature in the image critically depends on the shape of the tip.

pyramidal tips will convolve, resulting in mirror images, as shown in Figure 3.8a. For the same reason, long carbon nanotubes (cylindrical instead of the pyramidal structures) are advantageous for imaging tall surface features. On the other hand, for relatively short structures, the tip shape can be safely assumed to be hemispherical and the broadening does not include angular cross sections (Figure 3.8b).

It is important to note that the height of features in the AFM images is not affected by the tip broadening effect if deformation is not an issue. Hence, it is better to use the height for the estimation of the size of symmetrical features (round or spherical micro- and nanoparticles). However, it is important to remember that the vertical size of the features can be affected by the deformation of soft samples caused by the tip–sample forces, which is discussed in Section 3.4.2.

Apart from the tip broadening effect, the other important sources of artifacts are the misshaped tip, multitip (double-tip), and multiprobe effects. Multitip artifacts, a very common artifact in AFM images, are due to the irregular shape of the tip often comprised of multiple particles attached at the apex [50]. There are several possible reasons for such unwanted multitip formation, the most common of them being contamination during storage, attachment of loosely bound surface features to the probe during scanning, and splitting the tip end under local stresses. A multitip causes the features on the sample to appear multiple times in a preferential direction and in most cases the two features are convoluted [50]. Figure 3.9a shows an example of a sample comprised of MWCNT on a silicon substrate scanned with a double tip. It can be clearly seen that each MWCNT appears twice in the same direction due to the double tip. This artifact can be easily recognized if scanning size is extended to cover multiple features and specimen rotation is applied.

Figure 3.9b shows an example of a tapping mode image of gold nanoparticles with a diameter of 5–50 nm deposited on a polymer surface collected with a double tip. The image clearly shows the doubled surface features with lateral dimensions much larger than the known diameter and the distortion/doubling of all nanoparticles propagates uniformly and along the same direction, two clear signs of this type of artifact.

On the other hand, multiprobe effects, which are not so common, are caused by multiprobe cantilevers and are an occasional issue for microfabricated cantilevers.

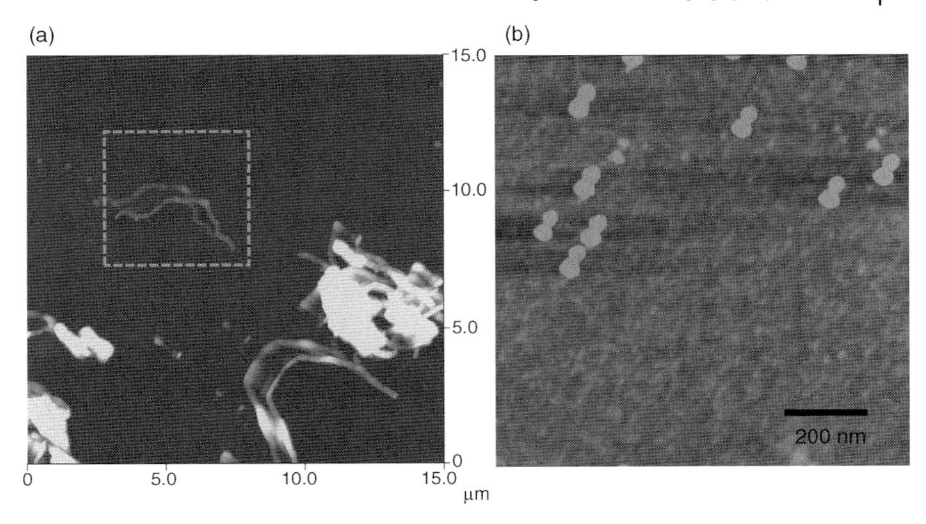

Figure 3.9 AFM images showing the doubling of multiwalled carbon nanotubes deposited on a silicon substrate (a) and gold nanoparticles deposited on a polymer film caused by the double-tip effect (b).

The cantilevers are comprised of tips with two or more distinct probes, sufficiently far apart (sometimes a few microns) such that the cantilever senses the forces of interaction associated with both the cantilevers. The multiprobe cantilevers result in images with ghost features that are far apart from the original features (generated by the primary tip), but always in a specific orientation compared to the original feature. Figure 3.10 shows two AFM images of a cell membrane scanned with a single-probe and double-probe cantilever, respectively. The ghost features identified by the arrows can be clearly identified at a specific distance (distance between the probes) and specific orientation with respect to the original features [48].

3.3.6 Thermal Drift and Piezoelement Creep

Apart from the tip-induced artifacts discussed above, there are some other types of AFM artifacts that originate from environmental conditions, scanning parameters, and the AFM instrumentation itself (e.g., laser misalignment or photodetector defocusing).

A common issue is thermal drift, which is characterized by the shift in the position of the cantilever relative to the sample surface or the shift in the position of the sample relative to the sample stage. For example, a cantilever coated with a thin metal layer (to enhance the reflection of the laser or as a conductive layer) exhibits a small drift overtime due to laser-induced localized heating of the bimorph structure with mismatched thermal expansions [51, 52]. Similarly, a sample at a certain temperature placed on a metal stage at different temperature drifts during the thermal equilibration process, thus resulting in a continuous distortion in the images. Thermal

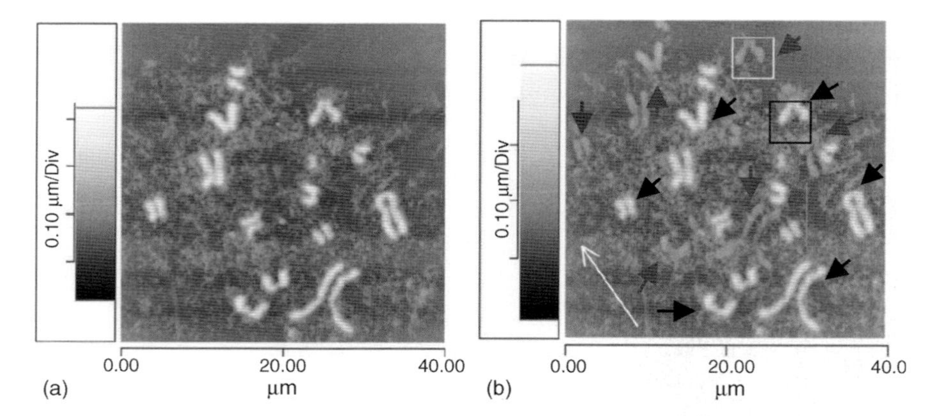

Figure 3.10 AFM images of Chinese hamster ovary (CHO) cells' genome scanned by (a) a regular probe and (b) a double-probe cantilever. The gray arrows show the real chromosomes and the black arrows show the ghost chromosomes caused by the double-probe effect. The white long arrow points in the artificial preferential direction induced by the double-probe effect. Reprinted from Ref. [50].

drifts are relatively easy to identify as they result in elongation of the surface features or loss of contact between the tip and the sample.

Sufficient equilibration time before imaging, usually by repeating scanning of the same area after zooming in or zooming out, typically avoids or minimizes the thermal drifts. However, in certain cases, especially when probing biological samples, the long waiting periods might result in a loss of valuable information. There have been significant efforts in the recent years to overcome thermal drifts associated with the cantilever. For example, Dravid and coworkers have demonstrated high-resolution imaging under a liquid using a sharp silicon tip mounted on silicon nitride cantilevers [53]. In order to enhance the reflection of the laser from the back of the cantilever, the silicon nitride tips are coated with a reflective coating (such as aluminum). Mounting the silicon cantilever on the top of silicon nitride resulted in sufficient reflection from the surface of the silicon, removing the need for a bimorph structure (metal coated on silicon nitride), thus requiring smaller waiting periods for equilibration.

Of the various factors that cause positioning errors in the AFM, piezoelement hysteresis and creep are the most significant culprits. The source of the hysteresis comes from the fact that the mechanical response of the piezoelement to the electrical voltage is not a simple linear function. Furthermore, an offset applied to the piezoelement intended to reposition it takes a finite time to complete. Imaging during this time will result in the smearing of surface features in the image, but this effect can be avoided by a repeated scan after some delay. Figure 3.11 shows the creep in the top portion of the image of a silicon grating due to the creep in the x and y piezoelement after moving to a new scanning position.

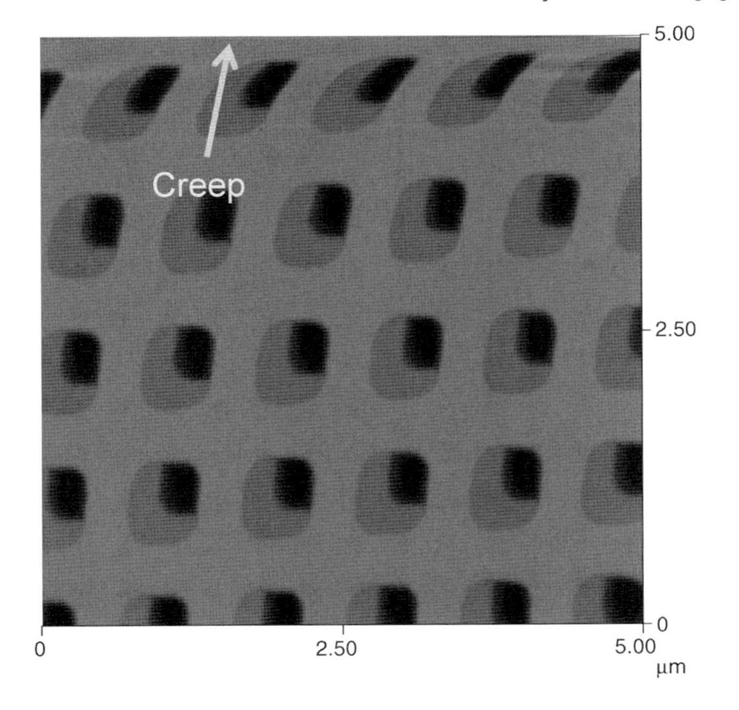

Figure 3.11 Tapping mode AFM of a silicon grating showing the distortion of the image (top portion) due to the creep in the *x*–*y* piezoelements after a change in the scanning position.

3.3.7 Oscillations and Artificial Periodicities

Oscillations and artificial periodicities in the image are serious problems and, especially for untrained operators, can be very misguiding. Oscillations appearing in AFM images can come from several sources, including a poor choice of scanning parameters (too high of feedback gains), electrical interference from numerous sources, mechanical vibrations, acoustic noise, and optical interference. There are several important pointers that can help one to identify the source of the artificial periodicities in AFM images.

For example, if the periodicity of the oscillations appears to increase with a decrease in scan rate, the source of oscillations is some fixed-frequency interference such as electrical interference or acoustic or mechanical vibrations. Knowing the scan size and the scan rate, the frequency of the oscillation can be estimated, which can point to the possible source of the spurious oscillations. Currently, commercial AFM instruments have the option to perform transformation of the image into frequency domains, which enables the identification of the spectral components of the image. In the simplest example, a frequency of 50 or 60 Hz (peak in power spectra density plot) indicates that the false oscillations are usually electrical in origin.

If changing the scan rate does not change the frequency of the oscillations, it indicates the oscillations have a constant spatial frequency. One of the most common sources of such fixed-frequency oscillations is optical interference artifact, which is produced when the stray laser light (light spilling over or even through the cantilever and then reflecting from the surface) interferes with the laser reflecting from the back of the cantilever. This is a common problem with all commercial AFM instruments based on an optical detection system, in which a laser source with coherent output is used. The interference effect is more common in contact mode on highly reflective surfaces, although they occasionally also appear in noncontact mode images, for example, tapping mode.

Another major cause of artificial periodicities is the electronic noise introduced by the feedback electronics due to the high feedback gain parameters chosen by the user. While it is important to maximize the gain to enable the cantilever to react to the asperities of the surface and avoid the damage of the probe and surface, one should note that high gains result in the electronic noise in the images. The noise is typically in the high-frequency domain and can be recognized in the AFM images as corrugations along the fast scan axis.

Before concluding this section, we would like to point out that the artificial periodicities in the AFM images introduced by various sources are routinely found, and are frequently, especially for the untrained operator, interpreted as molecular features such as molecular fragments, molecular stripes, rows of atoms, or atomic lattice steps. In fact, attributing these artificial periodicities to physical structures in literature is surprisingly not rare, although it is not as severe as during the initial stages of AFM development, which undermined the value of this technique for many regular users. Although it is expected to have such features on the surface, it is important to rigorously verify the nature of such features by performing the scanning at different scan sizes, rates, and scanning angles. Rotation of the features with changes in the scan direction (with respect to the features on the surface) should result in a corresponding change in the orientation of the features in the image. Consistency in the periodicity of the structure in the AFM image, even with a wide range of scan size variations, may further confirm that the observed features are indeed real. Reproducible imaging with several different AFM tips and different surface areas is a routine, which must be regularly undertaken to ensure that reliable data are taken from the AFM image.

3.3.8
Image Processing Artifacts

Image processing that can also generate characteristic artifacts can be broadly classified into two categories: (i) compensating for instrument contributions, and (ii) retrieving specific information from the images. It is important that the user understands the physical procedures involved in image processing to avoid some of the most common artifacts. Commercial AFMs are accompanied with image processing software containing extensive image analysis capabilities and a number of image analysis package are available for users with one of the most popular being *ImageJ* freeware [54].

From a practical standpoint, it is almost impossible to have the tip and the sample perpendicular to each other (just recall tilted arrangement of the microcantilevers in most commercial instruments). This deviation from perpendicularity as well as the pivoting motion of the x–y piezoelement results in an additional tilted plane in the raw AFM image that is compensated by subtracting a background plane calculated by a least squares fit. In some cases, the nonlinearities in the scanner (especially for larger scan sizes) demand higher-order curvature subtraction. This procedure is most often called "flattening" in the commercial AFM systems.

The commercial image processing software packages offer wide flexibility on the choice of the order of the plane and the mechanism of subtracting the plane such as whole and line by line. It is important to note that the subtraction of a background plane from an image with sharp steps or regular patterns without considering the nature of surface features, especially sharp protrusions or steep holes, results in characteristic artificial sawtoothed profiles, which can be easily found in current literature.

For instance, Figure 3.12 shows the representative AFM image along the edge of a scratch made in a thin PS film deposited on a silicon substrate. Two

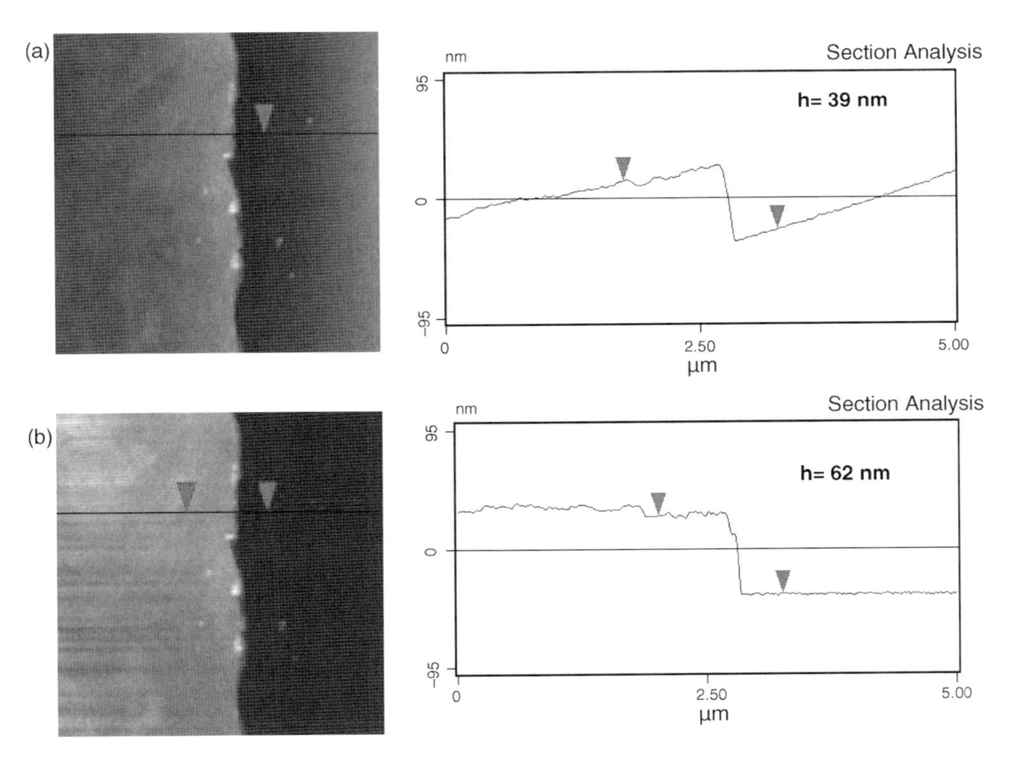

Figure 3.12 AFM image (left) and corresponding cross section (right) of the edge of a polymer film on a silicon substrate obtained by (a) flattening the entire image, resulting in a sawtooth structure at the edge and (b) flattening after omitting the polymer film portion, which resulted in the vertical step expected at the edge.

different modes of flattening (subtraction of the background plane) have been applied. In the first case (Figure 3.12a), flattening was uniformly applied over the entire image without considering the actual topography, which resulted in a sawtoothed profile at the edge, as can be seen in the cross section. In the second case (Figure 3.12b), the polymer film portion of the image was omitted to include only surface features without significant change in elevation, the correct flattening process for such features. This analysis results in a correct leveling of the image with a sharp vertical step between the polymer film and silicon, which is expected for this structure. The height difference between the polymer film and the silicon (thickness of the film) obtained from the two methods was significantly different, suggesting the possibility of large errors, let alone an overall "skewed" or "sawtooth" look to the image (Figure 3.12). This example clearly highlights that the flattening of images has to be performed carefully to avoid such artifacts in processed images. It is advisable to omit certain critical portions of images (steep edges, deep pit holes, and elevated protrusions) when flattening images.

Owing to the true 3D nature of AFM imaging, surface roughness is one of the most commonly deduced values to quantify the surface morphology of the polymer films. Surface roughness at different length scales can be computed and expressed in one of the many ways such as peak-to-peak height variation or the root mean square (RMS) of vertical heights. Evaluation of the surface roughness has been extensively discussed at the macroscale [55, 56]. The RMS surface roughness is more commonly employed in the context of AFM images because it reflects the weight-average surface characteristics.

It is important to note that surface roughness is a scale-dependent parameter. The surface roughness usually increases logarithmically with the increasing spatial scale of observation [57, 58]. Surprisingly, this fact is ignored in many AFM reports even today. To obtain, discuss, and compare any meaningful data, any values associated with surface roughness should be accompanied by the scale over which they have been calculated – a practice that many reports still do not follow. In this book, for the sake of simplicity and for comparative purposes, we will operate with microroughness values as measured within $1\,\mu m \times 1\,\mu m$ surface unless it is specially mentioned.

A more rigorous approach involves calculating the surface roughness over multiple length scales (images with different scan sizes). A plot of the surface roughness with the scan size, often termed as variational surface roughness, identifies the changes in the roughness with the characteristic length scales of the surface under investigation and allows the calculation of fractal dimension, a fundamental characteristic of the origin of the surface and the deviation from an ideal absolutely planar surface. The observed surface roughness also depends on the mode of imaging, tip size and shape, and the force employed for imaging. Although it is difficult to control all these parameters, a meaningful comparison of surface roughness values can be made by maintaining reasonably consistent conditions. A more detailed discussion on this topic may be found elsewhere [59, 60].

3.4
Some Suggestions and Hints for Avoiding Artifacts

3.4.1
Tip Testing and Deconvolution

As was discussed in Section 3.3.5, the size and shape of the AFM tip can severely compromise the size and shape of the surface features especially in the situation when these dimensions are comparable to the radius of the tip. It is worth noting that this is a nonissue for features with dimensions above 100–200 nm, but is most critical for sizes within 2–50 nm. Although it is impossible to completely eliminate tip-induced broadening, the effects can be minimized by choosing tips with a small radius of curvature (to some extent) and by preserving the shape and size of the tip by careful control of the tip–sample forces. The importance of the knowledge of tip shape and size for probing the mechanical, electrical, and thermal properties of the surface cannot be overemphasized, but are frequently ignored.

The shape and size of the tip can be independently determined by numerous imaging techniques such as electron microscopy imaging (SEM and TEM) of the tip and scanning standard samples (e.g., nanoparticles). SEM and TEM testing is useful, but can be destructive approaches. Moreover, it is labor-intensive route with a spatial resolution compromised by the poor conductivity of AFM tips.

The more practical approach includes evaluation of shape and size of the tip by analyzing AFM images with a scanning standard sample [61]. Figure 3.13a schematically shows the tip-induced dilation of a spherical particle on the surface of a flat

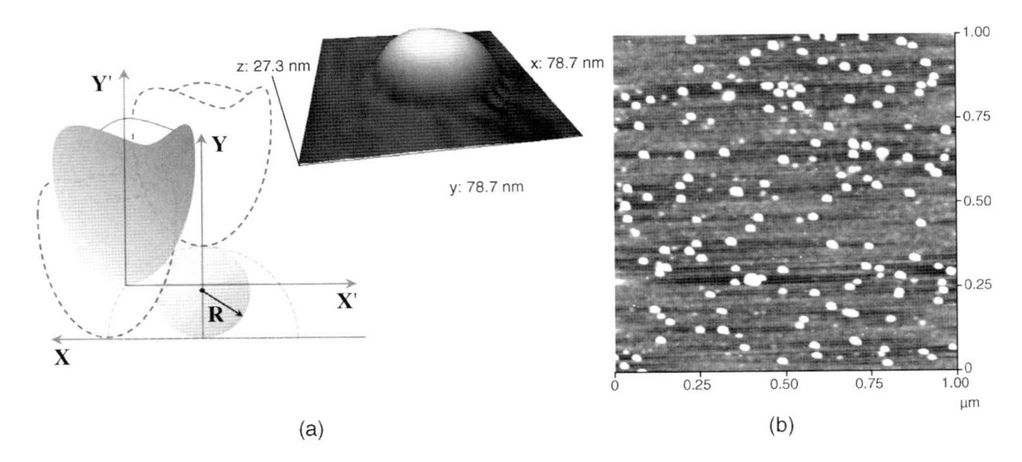

Figure 3.13 (a) Schematic showing the tip-induced broadening of a nanoparticle on a flat surface and a representative AFM image. (b) A typical nanoparticle standard sample comprised of monodisperse (5 nm) Au nanoparticles firmly attached to a silicon surface using a polylysine monolayer. The standard sample is routinely employed to estimate the tip size and shape.

substrate. Owing to the negligible elastic deformation, one of the commonly employed calibration standards is a sample with small gold nanoparticles (usually from 5–10 to 30–50 nm) on a silicon substrate firmly bound by a polymer primary layer. The AFM image (Figure 3.13b) shows the tapping mode image of gold nanoparticles (standard sample) on the surface of a silicon substrate routinely employed for estimating the radius of the tip. Direct deconvolution by applying approaches of mathematical morphology can produce the tip shape with resolution better than a few nanometers [44]. Data analysis packages that perform the deconvolution routines over the entire image can be found in the literature [62, 63].

As an alternative method, the lateral dimensions of the nanoscale surface features can be corrected by using simple analytical relations, as shown in Figure 3.14 [4]. In this approach, the "apparent" width W of spherical particles is determined by the tip radius R and height (diameter) h according to

$$W = 2\sqrt{2Rh} \tag{3.12}$$

For anisotropic particles (disks, squares) with true horizontal dimension L and height h, another simple analytical solution can be used:

$$W = L + 2\sqrt{h(2R - h)} \tag{3.13}$$

The simple analytical relations for spherical and disk-shaped structures can be employed to estimate the true lateral dimensions of surface features as well as the radius of the AFM tip with high resolution and high robustness. If the AFM tip was calibrated prior to imaging, these simple expressions would allow the fast and fair estimation of the true lateral dimensions of surface features with unknown lateral dimensions (see examples in Ref. [64]). However, "guessing" on the tip radius would result in significant error in dimension estimations.

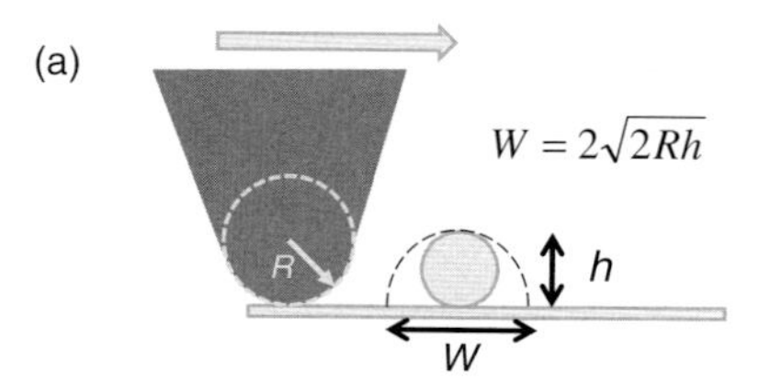

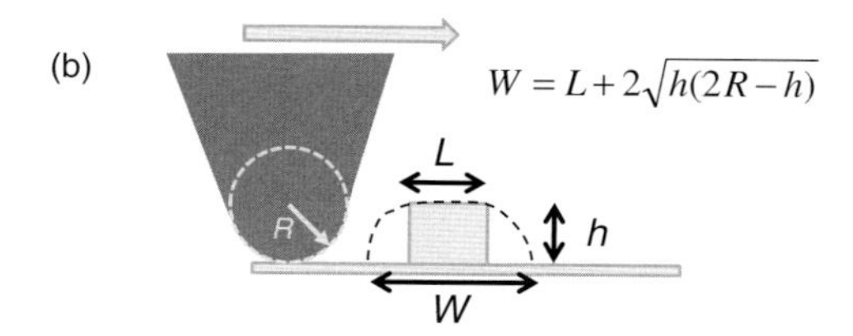

Figure 3.14 Schematic showing the tip-induced broadening of lateral dimensions and simple analytical relations to deconvolute the size of the tip by scanning spherical (a) and disk (b)-shaped structures.

With knowledge of the tip size and shape, features observed in AFM images can be better interpreted and evaluated. A major indication of a tip with a bad, damaged, or compromised shape is the presence of uniformly oriented features in the AFM images with a characteristic irregular shape, as was discussed above. Unless there is a reason for the features to show uniform orientation and dimensions over large surface areas, the user should immediately realize the possibility of a misshaped tip convoluted into the surface features. This kind of artifact can be easily confirmed by imaging the sample with a different tip and under different conditions.

3.4.2
Force Control

The force exerted by the tip results in elastic, viscoelastic, or plastic deformation of the compliant sample, such as those discussed in this book. Deformation introduces artifacts in the AFM images with the dimensions of features of the soft matter compressed relative to the actual dimensions. The force exerted by a sharp AFM tip can translate into large surface pressures. If local stresses reach the level of the yield strength of the material being probed, permanent nanoscale plastic deformation occurs.

Tapping mode AFM, which is most commonly employed for imaging soft matter, can be broadly classified into light, medium, and hard tapping based on the tip–sample interaction force, which is reflected in the set point ratio r_{sp} (ratio of the amplitude set point to the free amplitude of oscillation of the cantilever), as discussed in Section 2.6.2 with an example of variable imaging of block copolymer monolayers [65]. In general, a higher value of r_{sp} represents a smaller damping of the free amplitude, and hence smaller interaction forces. For example, a cantilever oscillating with free amplitude of 50 nm will be raised or lowered by the *z*-piezo-element to maintain a damped oscillation of 40 nm for a user-chosen r_{sp} of 0.8.

It is hard to specify the absolute range of the r_{sp} for various regimes of tapping (light, medium, and hard) that are frequently quoted in the literature since it depends on many parameters, such as tip sharpness, humidity, and contamination [66]. Generally, light tapping mode involves relatively large r_{sp} values ($r_{sp} = 0.98$–0.8), just sufficient to perform the imaging without instabilities. In this regime, the tip–sample forces can generally be attributed to longer-range van der Waals forces. In medium and hard tapping ($r_{sp} = 0.8$–0.55), the forces are mostly attributed to sample stiffness, as was demonstrated by Magonov *et al.* for two-phase polymeric materials [66]. Finally, hard tapping mode ($r_{sp} < 0.55$) imaging can cause severe deformation (even plastic deformation) of the soft surfaces or even damage to soft surface layers. The images acquired in hard tapping mode were found to be similar to that obtained in contact mode imaging where the tip senses the mechanical properties of the subsurface layers (see several examples in later chapters) [67].

As an example, Figure 3.15 shows the zoomed-out AFM imaging of a polymer film obtained in light tapping mode for a larger surface area immediately after the first scan. The image clearly reveals the damage caused by the earlier scan (rectangular region in the top) performed with much higher forces (hard tapping mode).

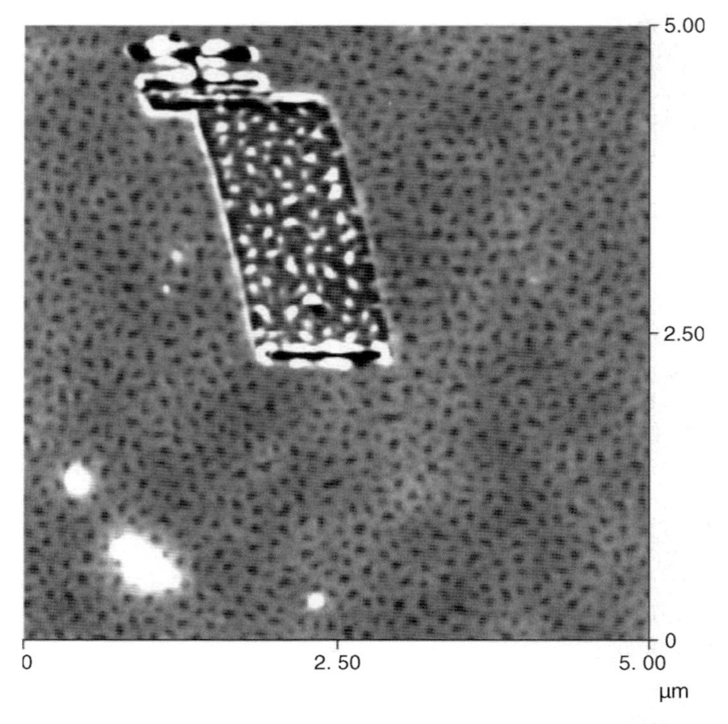

Figure 3.15 Large-scale tapping mode image (zoomed-out) of the surface of a polymer film obtained in light tapping mode. The rectangular damaged surface 9skewed shape is caused by drift) at the top of the image was caused by an earlier scan performed in the hard tapping regime, which involved high tip–sample forces.

One popular example of deformed surface feature is imaging of DNA molecules. The numerous reports in literature can be found in which the height of a DNA molecule deposited on a flat surface has been shown to be less than 1 nm, whereas the actual diameter of the DNA is known to be ~2.0 nm [68–70]. The discrepancy in the height of the DNA has been attributed to several factors such as salt concentration, strong electrostatic interaction forces between the DNA and the substrate, and dehydration of the biomacromolecule along with tip-induced deformation. In a recent example, it has been demonstrated by Yang *et al.* that the apparent height of DNA deposited on highly ordered pyrolytic graphite (HOPG) significantly depends on the imaging mode employed [71]. The height was found to be close to 2 nm in the image acquired using frequency-modulated mode, while the same was found to be close to 1 nm in the amplitude-modulated case (tapping mode). This observation clearly demonstrates the importance of careful controlling of the tip–sample interaction force in acquiring reliable quantitative surface data.

The tip–sample interaction forces are extremely important for surface force spectroscopy measurements. The force exerted by the tip on the sample has to be high enough to impart significant penetration to overcome the initial instabilities and low enough to avoid plastically deforming the surface. To measure the linear elastic modulus, it is

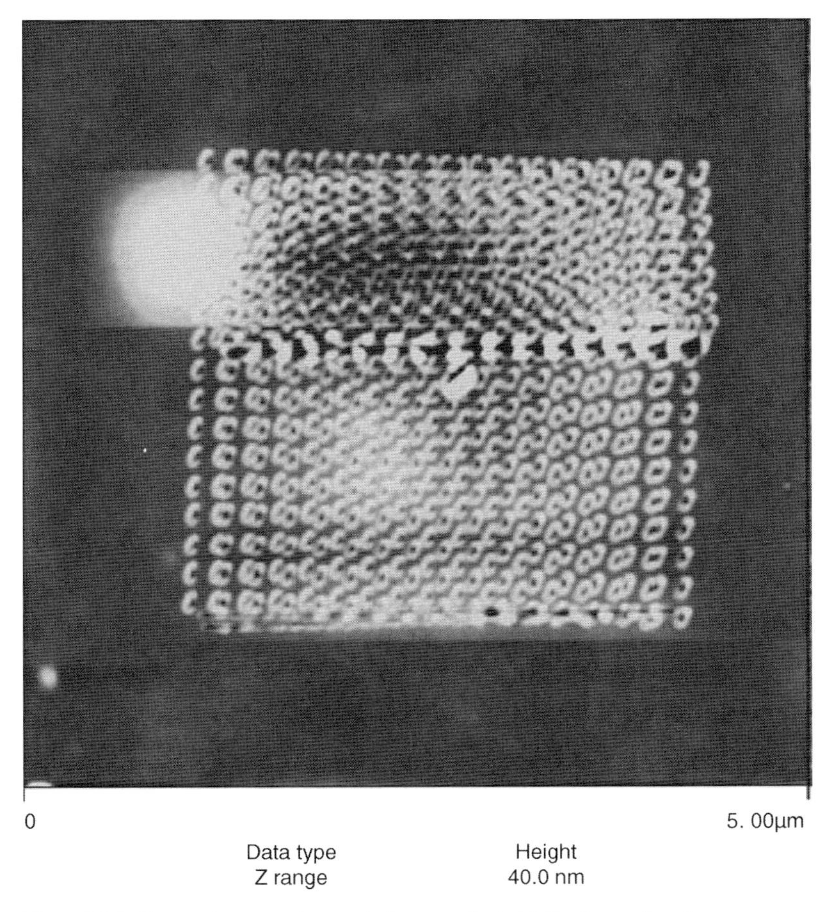

Figure 3.16 Zoomed-out tapping mode image collected following surface force spectroscopy showing the plastic deformation (appearing as holes in the film) in the polymer film due to large forces.

imperative to avoid plastically deforming the sample surface. The deflection set point chosen by the user will depend on the force constant of the cantilever and the sharpness of the tip. It is critically important that the sample be imaged in tapping mode both before (should be zoomed out prior to scanning) and after the force curves are obtained. If plastic deformation occurs, it will appear as an array of nanoindentations in the zoomed-out image, as shown in Figure 3.16. Furthermore, plastic deformation can also be recognized in the force curves as a leveling-off slope.

3.4.3 Tip Contamination and Cleaning

Tip contamination is caused by numerous sources such as contaminants in the ambient air sticking to the tip, adhesion of organic or inorganic species during scanning, and even contamination from commercial shipping gel containers. In fact, by using static time-of-flight secondary ion mass spectrometry, Lo *et al.* found that common commercial AFM cantilever shipping containers, often called PDMS gel packs, result in a thin layer of silicone oil contamination on the AFM cantilever tips [72]. Tsukruk and Bliznyuk demonstrated that simply dipping the cantilevers in piranha solution (3 : 1 mixture of concentrated sulfuric acid and hydrogen peroxide, hazardous solvent) is an easy and effective way to remove organic contamination, including silicone oils from the microcantilevers, and these clean tips can then be used for further chemical modification [73].

There are numerous other methods of tip cleaning such as plasma-, ozone-, UV-, organic solvent-, and royal solution (mixture of concentrated nitric acid and concentrated hydrochloric acid)-mediated cleaning for removing organic contaminants from the tips [74–77]. Each of these methods has distinct advantages and shortcomings, and each is more suitable than the others depending on the type of contamination. For example, organic contaminants can be effectively removed by piranha solution, while royal solution is more effective in removing metal contaminants. On the other hand, UV cleaning is effective against low-molar weight contaminants, but does not work against heavy polymeric layers. A simple method is short-time shaking of a microcantilever at a high gain between regular scannings, which can help remove large, poorly attached contaminants. However, care has to be taken during the cleaning procedures to ensure that the fragile cantilever is not broken.

Very recently, Gan and Franks have reported that both organic and inorganic contaminants can be scrubbed away by scanning the contaminated colloidal tips on calibration gratings with sharp spikes in constant force mode at reasonably high loads [78]. The authors pointed out that the technique demonstrated is highly efficient in cleaning tips with accumulated, lump-like organic/inorganic material in a nondestructive manner. Furthermore, while scrubbing the tip against the grating, one can immediately realize when the contamination is removed from the images being acquired simultaneously.

In this chapter, we briefly summarized the physical principles such as the fundamental force interactions between the tip and the sample and the resolution criteria. This discussion might serve as an introduction to advanced imaging modes and the application of AFM to a wide variety of soft materials discussed in the later chapters. One of the important take home messages of this chapter is to be aware of the artifacts that can creep into the images from numerous sources (instrumental, poor imaging, and processing). We highlighted the most common artifacts and issues and briefly suggested methods to avoid or overcome these potential pitfalls. Although initial stages of AFM applications to soft materials were plagued by numerous artifacts, currently AFM techniques are routinely employed to various soft materials and are considered to be a highly reliable form of microscopy for routine to advanced imaging if conducted with proper precautions.

References

1 Israelachvilai, J.N. (2001) *Intermolecular and Surface Forces*, Academic Press, London.
2 Rowlinson, J.S. (2002) *Cohesion: A Scientific History of Intermolecular Forces*, Cambridge University Press, Cambridge.
3 Magonov, S.N. and Whangbo, M.-H. (1996) *Surface Analysis with STM and AFM: Experimental and Theoretical Aspects of Image Analysis*, Wiley-VCH Verlag GmbH, New York.
4 Butt, H.J., Cappella, B., and Kappl, M. (2005) Force measurements with the atomic force microscope: technique, interpretation and applications. *Surf. Sci. Rep.*, **59** (1–6), 1–152.
5 Bergström, L. (1997) Hamaker constants of inorganic materials. *Adv. Colloid Interface Sci.*, **70**, 125–169.
6 Buldum, A., Ciraci, S., Fong, C.Y., and Nelson, J.S. (1999) Interpretation of long-range interatomic force. *Phys. Rev. B*, **59** (7), 5120–5125.
7 Gan, Y. (2009) Atomic and subnanometer resolution in ambient conditions by atomic force microscopy. *Surf. Sci. Rep.*, **64** (3), 99–121.
8 Hartmann, U. (1991) van der Waals interaction between sharp probes and flat surfaces. *Phys. Rev. B*, **43** (3), 2404–2407.
9 Jones, J.E. (1924) On the determination of molecular fields. II. From the equation of state of a gas. *Proc. R. Soc. Lond. A*, **106** (738), 463–477.
10 Morse, P.M. (1929) Diatomic molecules according to the wave mechanics. II. Vibrational levels. *Phys. Rev.*, **34** (1), 57–64.
11 Meyer, E. and Heinzelmann, H. (1992) *Scanning Tunneling Microscopy II* (eds R. Wiesendanger and H.J. Guntherodt), Springer, Berlin.
12 Malotky, D. L., and Chaudhury, M. K., (2001) Investigation of Capillary Forces Using Atomic Force Microscopy. Langmuir, **17**, 7823–7829.
13 Asay, D.B. and Kim, S.H. (2006) Direct force balance method for atomic force microscopy lateral force calibration. *Rev. Sci. Instrum.*, **77** (4), 043903.
14 Grobelny, J.P., Pradeep, N., Kim, D.-I., and Ying, Z.C. (2006) Quantification of the meniscus effect in adhesion force measurements. *Appl. Phys. Lett.*, **88** (9), 091906.
15 Eastman, T. and Zhu, D.-M. (1996) Adhesion forces between surface-modified AFM tips and a mica surface. *Langmuir*, **12** (11), 2859–2862.
16 Sarid, D. (1994) *Scanning Force Microscopy*, Oxford University Press, New York.
17 Olsson, L., Lin, N., Yakimov, V., and Erlandsson, R. (1998) A method for *in situ* characterization of tip shape in ac-mode using electrostatic interaction. *J. Appl. Phys.*, **84** (8), 4060–4064.
18 Giessibl, F.J. (2003) Advances in atomic force microscopy. *Rev. Mod. Phys.*, **75** (3), 949–983.
19 Lyklema, J. (1995) *Fundamentals of Interface and Colloid Science Volume 2: Solid–Liquid Interfaces*, Academic Press, London.
20 Born, M. and Wolf, E. (1993) *Principles of Optics*, Permagon Press, Oxford.
21 Butt, H.-J. and Jaschke, M. (1995) Calculation of thermal noise in atomic force microscopy. *Nanotechnology*, **6** (1), 1–7.
22 Bustamante, C. and Keller, D. (1995) Scanning force microscopy in biology. *Phys. Today*, **48** (12), 32–38.
23 Fang, S.J., Haplepete, S., Chen, W., Helms, C.R., and Edwards, H. (1997) Analyzing atomic force microscopy images using spectral methods. *J. Appl. Phys.*, **82** (12), 5891–5898.
24 Fuchigami, N., Hazel, J., Gorbunov, V.V., Stone, M., Grace, M., and Tsukruk, V.V. (2001) Biological thermal detection. I: ultra-microstructure of pit organs in infra-red imaging snakes. *Biomacromolecules*, **2** (3), 757.
25 Chen, G.Y., Warmack, R.J., Oden, P.I., and Thundat, T. (1996) Transient response of tapping scanning force microscopy in liquids. *J. Vac. Sci. Technol. B.*, **14** (2), 1313–1317.
26 Butt, H.J., Siedle, P., Seifert, K., Fendler, K., Seeger, T., Bamberg, E., Weisenhorn, A.L., Goldie, K., and Engel, A. (1993) Scan speed limit in atomic force microscopy. *J. Microsc.*, **169** (1), 75–84.
27 Tamayo, J., Humphries, A.D.L., Owen, R.J., and Miles, M.J. (2001) High-Q

dynamic force microscopy in liquid and its application to living cells. *Biophys. J.*, **81** (1), 526–537.

28 Han, W., Lindsay, S.M., and Jing, T. (1996) A magnetically driven oscillating probe microscope for operation in liquids. *Appl. Phys. Lett.*, **69** (26), 4111–4113.

29 Moreno-Herrero, F., De Pablo, P.J., Fernandez-Sanchez, R., Colchero, J., Gomez-Herrero, J., and Baro, A.M. (2002) Scanning force microscopy jumping and tapping modes in liquids. *Appl. Phys. Lett.*, **81** (14), 2620–2622.

30 Xiao, X. and Qian, L. (2000) Investigation of humidity-dependent capillary force. *Langmuir*, **16** (21), 8153–8158.

31 Zitzler, L., Herminghaus, S., and Mugele, F. (2002) Capillary forces in tapping mode atomic force microscopy. *Phys. Rev. B*, **66** (15), 155436.

32 Gallyamov, M., Khokhlov, A.R., and Möller, M. (2005) Real-time imaging of the coil–globule transition of single adsorbed poly(2-vinylpyridine) molecules. *Macromol. Rapid Commun.*, **26** (6), 456–460.

33 Maxwell, J.M. and Huson, M.G. (2005) Scanning probe microscopy examination of the surface properties of keratin fires. *Micron*, **36** (2), 127–136.

34 Tokumasu, F., Jin, A.J., and Dvorak, J.A. (2002) Lipid membrane phase behaviour elucidated in real time by controlled environment atomic force microscopy. *J. Electron. Microsc.*, **51** (1), 1–9.

35 Maxwell, J.M. and Huson, M.G. (2002) Using the scanning probe microscope sample stiffness. *Rev. Sci. Instrum.*, **73** (10), 3520–3524.

36 Stukalov, O., Murray, C.A., Jacina, A., and Dutcher, J.R. (2006) Relative humidity control for atomic force microscopes. *Rev. Sci. Instrum.*, **77** (3), 033704.

37 Qian, L., Tian, F., and Xiao, X. (2003) Tribological properties of self-assembled monolayers and their substrates under various humid environments. *Tribol. Lett.*, **15** (3), 169–176.

38 Bonaccurso, E. and Gillies, G. (2004) Revealing contamination on AFM cantilevers by microdrops and microbubbles. *Langmuir*, **20** (26), 11824–11827.

39 Favier, F., Walter, E., Zach, M., Benter, T., and Penner, R.M. (2001) Hydrogen sensors and switches from electrodeposited palladium mesowire arrays. *Science*, **293** (5538), 2227–2231.

40 Muller, D.J., Buldt, G., and Engel, A. (1995) Force-induced conformational change of bacteriorhodopsin. *J. Mol. Biol.*, **249** (2), 239–243.

41 Markiewicz, P. and Goh, M.C. (1994) Atomic force microscopy probe tip visualization and improvement of images using a simple deconvolution procedure. *Langmuir*, **10** (1), 5–7.

42 Odin, C., Aime, J.P., El Kaakour, Z., and Bouhacina, T. (1994) Tip's finite size effects on atomic force microscopy in the contact mode: simple geometrical considerations for rapid estimation of apex radius and tip angle based on the study of polystyrene latex balls. *Surf. Sci.*, **317** (3), 321–340.

43 Nie, H.-Y. and McIntyre, N.S. (2001) A simple and effective method of evaluating atomic force microscopy tip performance. *Langmuir*, **17** (2), 432–436.

44 Tsukruk, V.V. and Gorbunov, V.V. (2001) Nanomechanical probing with scanning force microscopy. *Micros. Today*, **01** (1), 8–14.

45 Koo-Hyun, C., Lee, Y.-H., and Kim, D.-E. (2005) Characteristics of fracture during the approach process and wear mechanism of a silicon AFM tip. *Ultramicroscopy*, **102** (2), 161–171.

46 Chung, K.H. and Kim, D.E. (2003) Fundamental investigation of micro wear rate using an atomic force microscope. *Tribol. Lett.*, **15** (2), 135–144.

47 Khurshudov, A.G., Kato, K., and Koide, H. (1997) Wear of the AFM diamond tip sliding against silicon. *Wear*, **203–204**, 22–27.

48 Larsen, T., Moloni, M., Flack, F., Eriksson, M.A., Lagally, M.G., and Black, C.T. (2002) Comparison of wear characteristics of etched-silicon and carbon nanotube atomic-force microscopy probes. *Appl. Phys. Lett.*, **80** (11), 1996–1998.

49 Butt, H.-J. and Jaschke, M. (1995) Calculation of thermal noise in atomic force microscopy. *Nanotechnology*, **6** (1), 1–7.

50 Chen, Y., Cai, J., Liu, M., Zeng, G., Feng, Q., and Chen, Z. (2004) Research on double-probe, double- and triple-trip effects during atomic force microscopy scanning. *Scanning*, **26** (4), 155–161.

51 Kipp, S., Lacmann, R., and Schneeweiss, M.A. (1995) Problems in temperature control performing *in situ* investigations with the scanning force microscope. *Ultramicroscopy*, **57** (4), 333–335.

52 Wenzler, L.A., Moyes, G.L., and Beebe, T.P. (1996) Improvements to AFM cantilevers for increased stability. *Rev. Sci. Instrum.*, **67** (12), 4191–4197.

53 Shekhawat, G.S., Chand, A., Sharma, S., Verawati, and Dravid, V.P. (2009) High resolution atomic force microscopy imaging of molecular self assembly in liquids using thermal drift corrected cantilevers. *Appl. Phys. Lett.*, **95** (23), 233114.

54 http://rsb.info.nih.gov/ij/ (accessed Nov. 5, 2010).

55 Russ, J.C. (1988) *The Image Processing Handbook*, 3rd edn, CRC Press, New York.

56 Thomas, T.R. (1982) *Rough Surfaces*, Longmans, London.

57 Palasantzas, G. (1993) Roughness spectrum and surface width of self-affine fractal surfaces via the K-correlation model. *Phys. Rev. B*, **48** (19), 14472–14478.

58 Mitchell, M.W. and Bonnell, D.A. (1990) Quantitative topographic analysis of fractal surfaces by scanning tunneling microscopy. *J. Mater. Res.*, **5** (10), 2244–2254.

59 Sedin, D.L. and Rowlen, K.L. (2001) Influence of tip size on AFM roughness measurements. *Appl. Surf. Sci.*, **182** (1–2), 40–48.

60 Simpson, G.J., Sedin, D.L., and Rowlen, K.L. (1999) Surface roughness by contact versus tapping mode atomic force microscopy. *Langmuir*, **15** (4), 1429–1434.

61 Markiewicz, P. and Goh, M.C. (1995) Atomic force microscope tip deconvolution using calibration arrays. *Rev. Sci. Instrum.*, **66** (5), 3186–3190.

62 Villarrubia, J.S. (1996) Scanned probe microscopy tip characterization without calibrated tip characterizers. *J. Vac. Sci. Technol. B*, **14** (2), 1518–1521.

63 Keller, D.J. and Franke, F.S. (1993) Envelope reconstruction of probe microscope images. *Surf. Sci.*, **294** (3), 409–419.

64 Ornatska, M., Jones, S.E., Naik, R.R., Stone, M., and Tsukruk, V.V. (2003) Biomolecular stress-sensitive gauges: surface-mediated immobilization of mechanosensitive membrane protein. *J. Am. Chem. Soc.*, **125** (42), 12722–12723.

65 Luzinov, I. and Tsukruk, V.V. (2002) Ultrathin triblock copolymer films on tailored polymer brushes. *Macromolecules*, **35** (15), 5963–5973.

66 Magonov, S.N., Cleveland, J., Elings, V., Denley, D., and Whangbo, M.-H. (1997) Tapping-mode atomic force microscopy study of the near-surface composition of a styrene-butadiene-styrene triblock copolymer film. *Surf. Sci.*, **389** (1–3), 201–211.

67 Luzinov, I., Julthongpiput, D., and Tsukruk, V.V. (2000) Thermoplastic elastomer monolayers grafted to a silicon substrate. *Macromolecules*, **33** (20), 7629–7638.

68 Moreno-Herrero, F., Colchero, J., and Baro, A.M. (2003) DNA height in scanning force microscopy. *Ultramicroscopy*, **96** (2), 167–174.

69 Hansma, H.G., Revenko, I., Kim, K., and Laney, D.E. (1996) Atomic force microscopy of long and short double-stranded, single-stranded and triple-stranded nucleic acids. *Nucleic Acids Res.*, **24** (4), 713–720.

70 Tang, J., Li, J., Wang, C., and Bai, C. (2000) Enhancement of resolution of DNA on silylated mica using atomic force microscopy. *J. Vac. Sci. Technol. B*, **18** (4), 1858–1860.

71 Yang, C.-W., Hwang, I.-S., Chen, Y.F., Chang, C.S., and Tsai, D.P. (2007) Imaging of soft matter with tapping-mode atomic force microscopy and non-contact-mode atomic force microscopy. *Nanotechnology*, **18** (8), 084009.

72 Lo, Y.-S., Huefner, N.D., Chan, W.S., Dryden, P., Hagenhoff, B., and Beebe, T.P., Jr. (1999) Organic and inorganic contamination on commercial AFM cantilevers. *Langmuir*, **15** (19), 6522–6526.

73 Tsukruk, V.V. and Bliznyuk, V.N. (1998) Adhesive and friction forces between chemically modified silicon and silicon nitride surfaces. *Langmuir*, **14** (2), 446–455.
74 Sirghi, L., Kylian, O., Gilliland, D., Ceccone, G., and Rossi, F. (2006) Cleaning and hydrophilization of atomic force microscopy silicon probes. *J. Phys. Chem. B*, **110** (51), 25975–25981.
75 Feiler, A., Larson, I., Jenkins, P., and Attard, P. (2000) A quantitative study of interaction forces and friction in aqueous colloidal systems. *Langmuir*, **16** (26), 10269–10277.
76 Fujihira, M., Okabe, Y., Tani, Y., Furugori, M., and Akiba, U. (2000) A novel cleaning method of gold-coated atomic force microscope tips for their chemical modification. *Ultramicroscopy*, **82** (1–4), 181–191.
77 Hinterdorfer, P., Baumgartner, W., Gruber, H.J., Schilcher, K., and Schindler, H. (1996) Detection and localization of individual antibody-antigen recognition events by atomic force microscopy. *Proc. Natl. Acad. Sci. USA*, **93** (8), 3477–3481.
78 Gan, Y. and Franks, G.V. (2009) Cleaning AFM colloidal probes by mechanically scrubbing with supersharp "brushes." *Ultramicroscopy*, **109** (8), 1061–1065.

4
Advanced Imaging Modes

In the previous chapters, we have discussed the fundamentals of scanning probe microscopy and the basic and most popular imaging modes, such as contact mode and intermittent (tapping) mode. As mentioned earlier, a better understanding of the tip–sample interactions, tip variations (e.g., conductive and magnetic), and the feedback that is generated opened up multitudes of advanced techniques, which have enabled comprehensive investigations of structure–property relationships of soft matter. In this chapter, we will briefly introduce some advanced imaging and probing modes in the atomic force microscopy (AFM) technique.

4.1
Surface Force Spectroscopy

4.1.1
Introduction to Force Spectroscopy

Surface force spectroscopy (SFS) is a powerful method to probe the nanomechanical and adhesive properties of surfaces, such as elastic modulus, adhesion, chemical binding, inter/intramolecular forces, selective interactions, chemical composition, relaxation times, and resilience. Modified SFS techniques are also extremely useful for electrical and thermal characterization of materials. SFS is a form of force spectroscopy (also known as static force spectroscopy) that is limited to probing surfaces (limited to the closest subsurfaces), as opposed to more complicated force spectroscopic measurements, such as deforming and indenting structures, or molecular chain pulling.

A so-called "pulling-off" version of SFS is widely utilized for the investigation of protein unfolding, segment flexibility, molecular weight, brush stretching, and other tensile-related mechanical properties of individual molecules or surface structures and aggregates, and it typically requires special tip modification with selectively binding groups. A brief discussion of this approach will be provided in Chapter 5 and a full discussion of this subject can be found in the literature [1–4].

Scanning Probe Microscopy of Soft Matter: Fundamentals and Practices, First Edition.
Vladimir V. Tsukruk and Srikanth Singamaneni.

An important variation of force spectroscopy is dynamic force measurement that involves the oscillation of a cantilever at a few kHz (much smaller than the resonance frequency of the cantilever) to obtain a simultaneous map of topography, relative stiffness, adhesion, and electrostatic properties much faster than the conventional, static force spectroscopy. The dynamic force spectroscopy is discussed in detail in Section 4.5.

As a surface-based technique, it is well suited to study the effect of free surfaces and confined surfaces on polymeric properties, which can be quite different from bulk properties. SFS in different modes is particularly useful in characterizing the material properties of polymeric composites and phase-separated copolymers, where the contributions of individual components and their interaction can be quantified and spatially mapped. Furthermore, SFS can be used to spatially map the mechanical properties of multicomponent materials over varying conditions, including time, temperature, light, and solvent exposure.

Force spectroscopy is a tedious multistep process that should be done with great care to ensure accurate results. Without care, force spectroscopy measurements can be very inaccurate and misleading. Therefore, it is important to fully understand the process and the sources of error. Furthermore, like many experimental methods, practice and experience with reference samples with known properties is invaluable for successful probing and analysis. Because every sample behaves somewhat differently, there is usually a learning curve associated with each new sample.

This section is focused on a brief discussion of practical SFS measurement techniques rather than the theoretical aspects and interpretation of more complex force curves in order to help introduce the field. Specifically, the main emphasis is on obtaining quantitative data, data analysis, and avoiding common pitfalls associated with this method, with coverage on basics and capabilities. A comprehensive, in-depth review of force spectroscopy including contact theories and complex force curve interpretation can be found in some reviews [5, 6].

4.1.2
Force–Distance Curves

A single force curve, commonly called a force–distance (displacement) curve, is a plot of tip–sample force versus piezoelement movement (Figure 4.1a). Figure 4.1a and b shows the plotting of deflection data with displacement and time, respectively. Figure 4.1c is a schematic explaining the particular regions of force curves in terms of piezoelement movements, cantilever deflection, and surface deformation.

In Figure 4.1a, an ideal force–distance curve is plotted in the conventional trace–retrace manner. The x-axis is the distance between the tip and the surface and a measure of the interaction between the sample and the probe. The larger the piezoelement position value, the farther the probe from the surface. First, the vertical piezoelement is moved in the extension direction, which is depicted as the solid line in Figure 4.1a and b. In the curve, line 1–2 is called the extension zero line, which corresponds to the region when the sample is not in contact with the tip but is moving toward the probe. Line 2–3 corresponds to the "jump to contact" region (also known

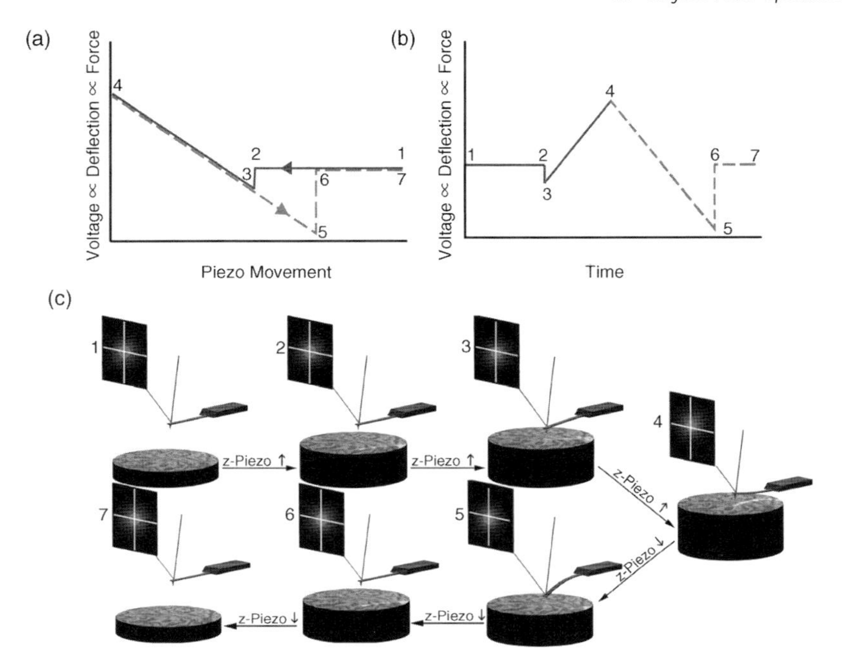

Figure 4.1 (a) An ideal force–distance curve as explained in the text. (b) The deflection data from (a) plotted with respect to time. (c) A schematic explaining the different regions of the force curve. Note that all the numbers in (a)–(c) correspond to each other. It is also important to carefully note that in the schematic, the laser spot, cantilever deflection, and sample height (piezoelement position) correspond to their positions in the force curve.

as the "snap-to" region), when the probe is initially attracted to the sample surface, thereby bending the cantilever downward. In this region, the surface might also be slightly deformed toward the tip. As the piezoelement moves the sample surface closer to the tip, the cantilever passes the zero deflection and is further bent up. This is an unstable region, like the snap-from region, which is not monitored and in the minimum point the balance of vertical forces is reached and the tip rests onto a slightly deformed local region (the contact point).

The deflection of the cantilever, when in physical contact with the sample surface, is indicated by the line 3–4. This region is linear for purely elastic deformation with a slope directly related to surface stiffness. For infinitely stiff substrates utilized for sensitivity calibrations (see below), the slope is 1 due to the fact that the cantilever deflection is exactly equal to the piezoelement displacement. In the case of time-dependent surface deformation (viscoelasticity phenomenon), nonuniform deformation, or plastic deformation, this region becomes highly nonlinear (see Chapter 5). Such an instrumentational factor as piezoelement hysteresis in this region, which can be critical for stiff materials probed with very high forces, is usually barely noticeable and not generally very critical for soft materials.

Point 4 indicates the end of the piezoelement extension sequence and the beginning of the retraction sequence. Ideally, lines 3–4 and 4–5 will partially overlap and have the same slope during extension and retraction. Generally, line 5–6 represents the force of adhesion, "pulling forces," or the "snap-from contact" region. It is vertical in ideal cases, but can display complex shapes in special cases (e.g., "saw-toothed" unfolding events, see Chapter 5). Line 6–7 is a region where the cantilever is once again free from contact with the surface and rests in the neutral position.

It is important to note that the applied force is only indirectly measured by the AFM via the cantilever deflection. The microscope itself monitors only the movement of the laser spot on the photodiode, which is related to the cantilever deflection via the photodiode sensitivity. The photodiode sensitivity, expressed in units of nm/V, must be calibrated for each set of measurements, as is described in the next section. Cantilever deflection is directly related to the applied force via the cantilever spring constant, which is expressed in units of N/m or nN/nm. The cantilever spring constant must also be calibrated for each set of force measurements, as is described in a later section. The elastic modulus of the sample can be deduced from the force–distance curves using one of the various contact mechanics models. Three popular models that are frequently employed for data analysis are the Hertzian, Sneddons, and JKR models, each more suitable than the others for different conditions as discussed in Chapter 5. In general, the elastic modulus of the sample being elastically indented is related to the indentation depth h by

$$E \propto ch^{\alpha} \tag{4.1}$$

where E is the elastic modulus, c is a coefficient that is a function of the cantilever spring constant and tip radius, and α is a constant that depends on the contact model chosen for analysis (see Chapter 5).

4.1.3 Force Mapping Mode

Force mapping (sometimes referred to as force–volume) is a spatial map of force–distance curves collected across the selected surface area by pixel-to-pixel motion. This force–distance curve matrix can be used for sampling statistics, as well as for relating surface features to mechanical and chemical (adhesion) properties. Force–distance curves are obtained by monitoring the applied force while extending the vertical piezoelement to diminish the probe–sample distance until the cantilever is deflected by a set amount (trigger), followed by retraction of the vertical piezo-element. Unless otherwise stated, for the following discussion, the convention of a sample-scanning AFM is used, where the vertical piezoactuator moves the sample with respect to the tip, as opposed to a tip-scanning AFM instruments.

4.2 Friction Force Microscopy

Friction force microscopy (FFM) is an AFM-based technique that is used to characterize the tribological properties of surfaces [7]. Common measurements

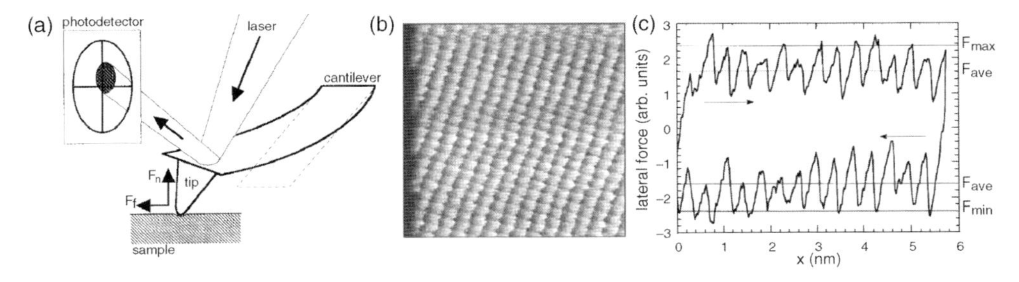

Figure 4.2 (a) A schematic showing the lateral and normal forces applied in FFM and the resulting laser spot deflection in the photodetector. (b) 6 nm × 6 nm friction force image of KF (001) imaged in UHV. (c) A friction loop indicating the average force level for each direction and the clear hysteresis between the directions. Reprinted from Ref. [7].

relate frictional forces to applied normal forces (so-called loading curves) by using an AFM in contact mode and controlling normal loads while monitoring the lateral cantilever deflection signal (see Figure 4.2a) [7]. FFM imaging can be as simple as scanning a surface in contact mode and using the lateral signal to produce a friction force image (so-called lateral force microscopy (LFM)), as is shown in Figure 4.2b.

The probe is scanned back and forth across a sample surface, and each cycle produces a "friction loop" curve, as shown in Figure 4.2c. The average friction force is defined as half of the difference of the average lateral force in each direction of the friction loop, as shown in Figure 4.2c. A sample region is scanned under several different normal loads, thus producing a friction force versus normal load curve (Figure 4.2c). Measurements are usually performed on homogeneous areas of a sample that have smooth surface. However, friction force images are often obtained for heterogeneous samples and the friction force contrast is associated with adhesive properties of the individual components, with common spikes between different areas recognized as "geometrical" friction artifacts [8].

Typically, measurements involve conventional silicon and silicon nitride probes, but diamond-coated probes are becoming popular for friction measurements because of the reduced wear associated with them. It is also possible to obtain chemical-related information from friction force images, especially when obtained with chemically functionalized probes. This can be done by simply obtaining a friction force image on a chemically heterogeneous sample [16] or by performing friction force versus normal force measurements on homogeneous samples with several different chemically functionalized tips. FFM can provide rich information on the shearing behavior of surface layers, friction coefficients, wearing dynamics, and velocity-dependent shearing.

Friction phenomena at the nanoscale are still far from being well understood, but FFM has indeed provided an avenue for a deeper understanding of this phenomenon. A notable example of the impact that FFM has had on the understanding of friction behavior is a study performed by Carpick and coworkers [9]. FFM and LFM are also very useful for polymeric studies involving reducing material wear, enhancing lubrication, stiction in MEMs, and other related tribological phenomena [10, 11].

4.3
Shear Modulation Force Microscopy

Shear modulation force microscopy (SMFM) is a probing AFM technique, which involves engaging a tip on the surface with a constant normal load (typically a few to tens of nanonewtons) followed by lateral modulation (sinusoidal oscillation) with nanoscale amplitude, which is typically on the order of a few nanometers to ensure no slip between the tip and the sample surface. The lateral amplitude response (ΔX_r) of the photodiode detector to the lateral modulation is monitored, as shown in Figure 4.3. Apparently, ΔX_r of the tip is highly sensitive to the elastic and viscoelastic properties of the surface of the sample. As with any other SFM-based technique, the observed mechanical properties depend on the tip shape, size, stiffness, surface chemistry, load, and tip modulation frequency.

SMFM has been extensively used to monitor the phase transitions in ultrathin polymer films. In a typical experiment, the AFM tip is brought into contact with the polymer surface and maintained at a constant set point (fixed normal load). The tip is laterally modulated at a constant drive amplitude and the response is monitored at varying temperatures. For example, Sokolov and coworkers have employed SMFM for monitoring the glass transition of ultrathin polymer films with varying thicknesses (17–500 nm) and surface states (substrate supported and freestanding) [12, 13]. Below the glass transition temperature, the tip penetration and contact area remain small, resulting in a small ΔX. On the other hand, above the glass transition, polymer softening results in larger penetration, larger amplitudes, and thus a higher

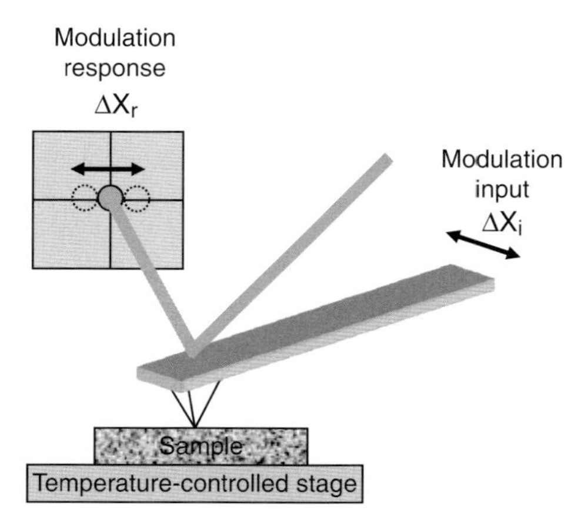

Figure 4.3 Schematic of the experimental setup of shear modulation force microscopy showing the input lateral modulation and the amplitude response (ΔX_r) as monitored by the photodiode detector.

lateral amplitude ΔX. The distinct change in the slope of the temperature versus amplitude curve is used to identify the glass transition.

4.4
Chemical Force Microscopy (CFM)

Quantifying adhesion from force spectroscopy can provide information regarding intermolecular interactions and surface energies. Usually, precise measurements aimed at understanding specific chemical interactions involve the use of chemically modified tips in what is known as chemical force spectroscopy (CFM). Chemical force spectroscopy is capable of providing rich information regarding chemical interactions with lateral resolutions down to a single molecular group. A full discussion of chemical force spectroscopy and other associated techniques is beyond the scope of this chapter and the readers are referred to elsewhere for more information [14–18].

Chemical modification of the tip in order to attain the desired functionality is an important procedure in CFM experiments. Probes are usually modified with self-assembled monolayers (SAMs) with thiol chemistry on gold precoated tips or SAMs with silane chemistry on a native silicon oxide surface. Thiol-based surface modification is a relatively straightforward method that involves coating tips with an adhesion layer followed by a gold coating. Therefore, this method utilizes noncovalent bonding, which leads to a limited lifetime [19]. Though thiol SAMs are an important tool for surface scientists, they are poor surface modifiers for applications involving relatively high forces, such as contact mode techniques. However, the ease of thiol tip modification has led to its widespread use, even in contact mode and friction mode AFM techniques, with characteristic artifacts generated widely.

Silane modification, on the other hand, can be done directly on silicon and silicon nitride tips after thorough cleaning (Figure 4.4) [19] and involves covalent bonds, which are quite robust and last long even under high stresses between mating surfaces. Unfortunately, silane modification involves relatively stringent reaction conditions and is somewhat difficult to initially optimize to achieve a single monolayer. The reaction is very sensitive to the presence of water, so the relative humidity has to be limited to a small percentage and dry solvents must be used.

Adhesion force information is collected in the retraction portion of the force curve at the snap-from region (Figure 4.1). Usually pull-off forces are considered to be representative of true interactions, although some issues relevant to instability in tip behavior should always be considered. These adhesion data are often a combination of several forces, including contributions from capillary forces, electrostatic forces, van der Waals forces, hydrogen bonding, ionic bonding, and covalent bonds. Capillary forces tend to dominate adhesive forces in ambient air, which are commonly on the order of 10–100 nN [17, 20–22]. Although capillary forces can be used to gauge hydrophobic/hydrophilic forces and thereby provide contrast, most adhesion measurements are aimed at obtaining chemical interaction information where capillary forces are considered as undesirable interference. Therefore, in order to

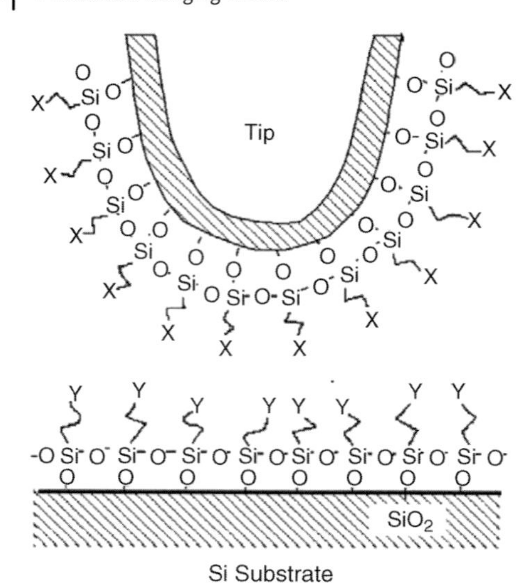

Figure 4.4 Chemically modified AFM tip probing the specific interaction with functional groups of the sample. Reprinted from Ref. [19].

avoid overwhelming contributions from capillary forces, CFM experiments are frequently performed under vacuum, dry conditions, or most commonly immersed in liquid.

In a uniform medium (vacuum, gas, or liquid), when capillary forces have been eliminated, the work of adhesion W_{SMT} can be related to the surface–tip interfacial free energy, the tip–medium interfacial free energy, and the surface–medium interfacial free energy (γ_{ST}, γ_{TM}, and γ_{SM}, respectively) by the following equation [18]:

$$W_{SMT} = \gamma_{SM} + \gamma_{TM} - \gamma_{ST} \tag{4.2}$$

The work of adhesion can be calculated from force spectroscopy data by taking the contact area into account with contact models. It should be stressed that much like elastic modulus measurements, there are errors associated with calculating the work of adhesion and surface free energies and care should be taken in both the measurements and data analysis. In this case, the Hertzian model, which assumes no adhesion between the tip and the surface, is inappropriate and, therefore, JKR and DMT contact models are typically employed instead. Under the JKR theory, the work of adhesion is defined as [23]

$$W_{SMT} = \frac{3}{2}\pi\left(\frac{R}{P_{ad}}\right) \tag{4.3}$$

where R is the tip radius and P_{ad} is the force required to separate the tip from the sample surface. The DMT theory results in a similar relationship with the 3/2 coefficient replaced with 2. Both models have shown good agreement with experimental data [24].

4.5 Pulsed Force Microscopy

Pulsed force microscopy (PFM) is another method that maps topography, relative stiffness, adhesion, and electrostatic properties at about 1 ms per pixel: about 1000 times faster than conventional static force spectroscopy [25]. Since the advent of PFM in 1997 by Marti and coworkers, the popularity of this method has gained momentum, especially since becoming commercially available [26]. Simplistically stated, PFM is essentially dynamic force curve mapping done at a much higher rate. Pulsed force mode involves driving a z-piezoelement with an amplitude of 10–500 nm at 100 Hz–10 kHz, orders of magnitude less than the cantilever resonance frequency, and thereby obtaining force curves on the millisecond timescale [27]. The z-piezoelement is usually driven with a sinusoidal profile (not necessary), as opposed to the triangular wave that drives typical force curves. A typical PFM force–distance curve can be seen in Figure 4.5, with the piezoelement driving signal plotted as the dashed line [28].

PFM uses minimized forces, avoiding plastic deformation with ease. Furthermore, PFM measures and maps adhesion force directly. This technique acquires the same information as the static force spectroscopy (described in Section 4.1.2) mapping, but the sampling is much faster. The height, stiffness, and adhesion come from a peak and trough picking routine, which is used to quickly process the data, thereby providing images. The system acts under a constant force mode, like static force spectroscopy mapping, and therefore sample stiffness can slightly affect the height

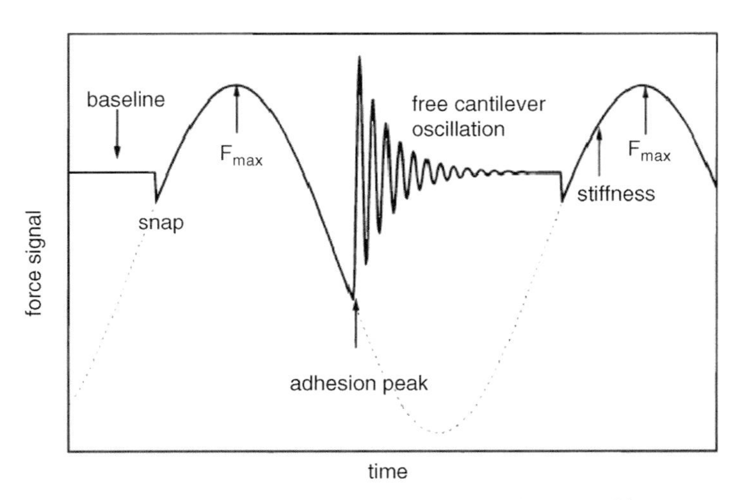

Figure 4.5 A schematic showing the resulting force data from pulsed force microscopy. The dotted line shows the relative force modulation voltage. The arrows indicate the points picked to acquire the baseline, maximum applied force, and adhesion. Reprinted from Ref. [28].

data. Softer domains would appear to be depressed compared to stiffer domains on a soft sample. The high speeds at which the stiffness and adhesion forces are probed have a significant effect on the apparent behavior of viscoelastic materials.

It is important to note that SFS (static), recently introduced peak-dice mode, and PFM (dynamic) are complementary techniques. Between SFS and PFM, one could probe polymers at frequencies 0.01 Hz–10 kHz, with SFS probing frequencies ranging from 0.01 Hz up to frequencies approaching 100 Hz and PFM probing frequencies ranging from 100 Hz to 10 kHz. These results indicate the truly complementary nature that SFS and PFM have, especially in mapping dynamics at the nanoscale. It should also be mentioned that chemical modified tips are commonly used with PFM, which adds the important ability to map chemical forces at relatively high speeds.

4.6 Colloidal Probe Microscopy

Colloidal probe microscopy is an AFM-based technique that involves the probing of surface properties of a sample (e.g., elastic modulus and adhesion) using a colloidal microparticle as an AFM tip. The technique is identical to conventional AFM and force spectroscopy discussed in Section 4.1.2. In a force measurement, the sample is moved up and down by applying a voltage to the piezoelectric translator onto which the sample is mounted while recording the cantilever deflection. In the conventional SFS experiment, the AFM tip typically has a radius of 5–50 nm, whereas the radii of colloidal probes range from 1 to 50 μm. The higher radius results in larger contact area between the probe and the sample surface, which in turn manifests as higher adhesion forces.

Colloidal probes are fabricated by carefully gluing spherical microparticles onto the end of a cantilever [29, 30]. The microparticles are available through several commercial sources and colloidal probes themselves are commercially available as well. Microparticles from silica and borosilicate glass are most commonly used and have roughnesses below 1 nm for 1 μm^2 area, which is acceptable for most measurements. Figure 4.6 shows an SEM image of a borosilicate colloidal probe attached at the end of a microcantilever.

Colloidal probes have several advantages over conventional probes for very compliant materials. A major advantage is that the applied forces per unit area are significantly lower than conventional probes, thus allowing the probing of very compliant materials such as hydrogels with an elastic modulus well below 1 MPa and down to a fraction of a kilopascal [31]. By applying less force per unit area, the total applied force can be much higher without plastically deforming the surface or damaging the probe, which provides higher resolution in force/area for each force curve. This, however, sacrifices lateral spatial resolution. It is important to note that the probing depth highly depends on the probe radius and, therefore, colloidal probes are inappropriate for characterizing the stiffness or elastic modulus of ultrathin films. Furthermore, the microparticle radius quoted by the manufacturer is generally

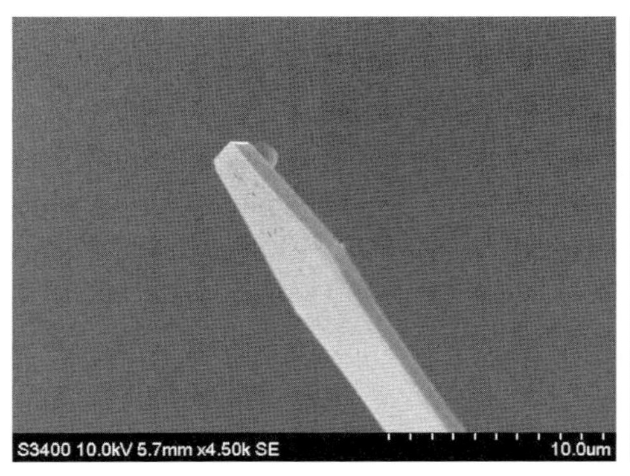

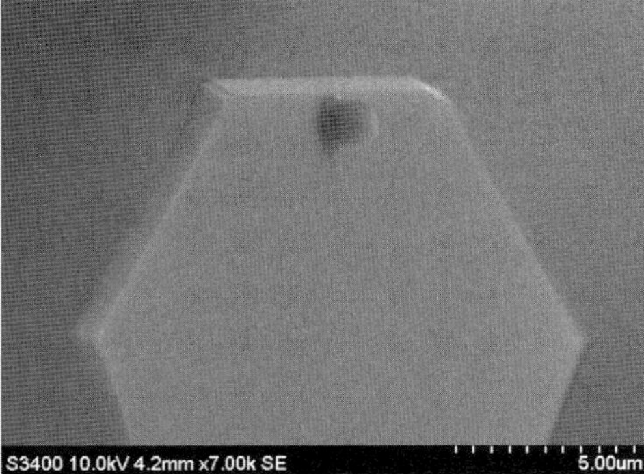

Figure 4.6 SEM image of a colloidal probe comprised of a 1 μm borosilicate sphere glued onto a microcantilever.

quite accurate as compared to the conventional probes and can be easily verified with SEM. Therefore, there is little error associated with assuming the tip size to be the quoted size, unlike conventional probes.

The preservation and well-defined tip shape allows the robust SFS analysis with contact models that assume a spherical probe shape with well-characterized dimensions. However, care should be taken in preparation to ensure good particle–cantilever contact and that the probing particle surface is not covered with glue. It is also possible that the mechanical properties of the glue between the sphere and the cantilever can be measured rather than the sample while measuring very stiff samples, such as reinforced polymers. Possible roughness of some colloidal beads should also be considered in CFS studies.

4.7
Scanning Thermal Microscopy

Studying the local thermal properties of materials is of fundamental importance in understanding a variety of phenomena, including photon–phonon interactions, electron–phonon interactions, molecular motion, and various phase transitions [32]. Although various thermal characterization techniques based on SPM have been developed, this chapter will focus on methods based on AFM techniques combined with electrical resistance thermometry, which is applicable to polymers and is relatively well developed. This includes scanning thermal microscopy (SThM) imaging of thermal conductivity and other local thermal analysis (L-TA) techniques (Figure 4.7) [33].

The term micro- or nanothermal analysis (micro/nano-TA) encompasses a variety of techniques involving characterizing localized material properties on a temperature-controlled sample. However, the term usually refers to methods that exploit a combination of AFM and one or more of the following techniques:

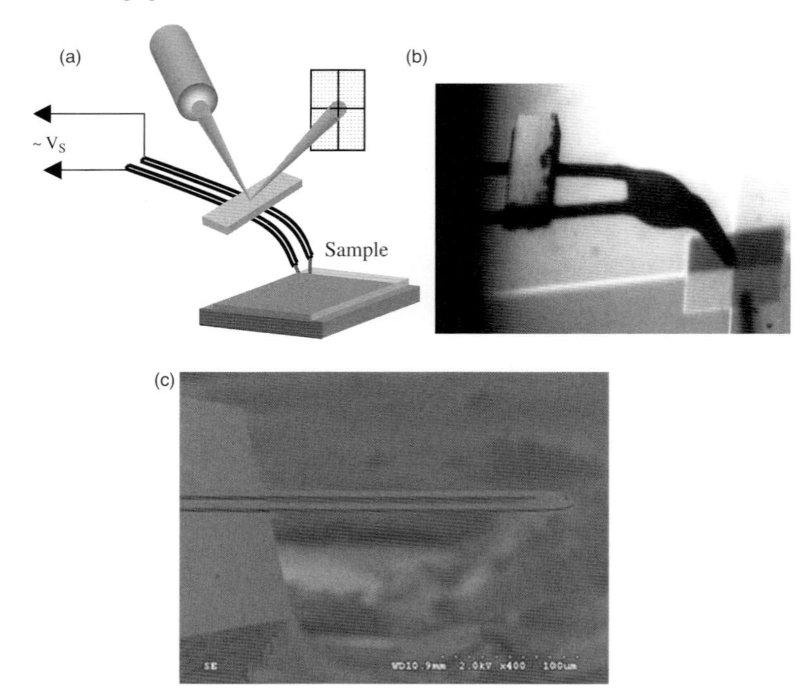

Figure 4.7 (a) Schematic and (b) optical image of a Wollaston resistive probe and corresponding micrograph. (c) A scanning electron micrograph of higher resolution (microfabricated) thermal probes. Reprinted from Ref. [33] and Anasys Instruments, Inc.

thermomechanical analysis (TMA), dynamic mechanical analysis (DMA), differential thermal analysis (DTA), and spectroscopic or analytical pyrolysis. When referring to local thermal analysis, the term differential scanning calorimetry (DSC) is not used because the mass of the sampled region that melts cannot be determined easily and, therefore, a quantitative measure of heat (J/g) is not made. These well-established material characterization methods, through the introduction of AFM, can now be carried out on preselected parts of a sample that are typically a few cubic micrometers in volume or smaller.

Control over the temperature of a sample is achieved by using a thermally active, electrically resistive probe (thermal probe) and a variable temperature microscope stage (temperature stage). If the latter is being used, practically any type of probe normally available for AFM may be mounted in the microscope. A thermal probe may function as a thermometer as well as a heat source. This enables a different type of micro/nano-TA to be carried out, in which heat is applied to the sample from an external energy source (e.g., infrared radiation) and the probe is used to sense the resulting change in temperature of the material. This enables spectroscopy to be carried out with a spatial resolution that is, in theory, better than the diffraction limit.

The development of micro/nano-TA significantly enhances the utility of these techniques by enabling variations in properties or composition to be mapped at the microscopic level. This will increase their usefulness in the field of materials science and technology, which is becoming ever more concerned with the control of material structure and hence properties at the microscale and, increasingly, at the nanoscale.

4.7.1
Thermal Resistive Probes and Spatial Resolution

Arguably, the most relevant form of SThM to polymers is carried out using near-field electrical resistance thermometry, in which the AFM probe can be used as a thermometer and a heat source. These fairly complicated probes are being actively developed to enhance thermal measurement resolution. Initially, Wollaston wire probes were the most common, which were developed by Dinwiddie *et al.* [34] and first used by Balk *et al.* [35] and Hammiche *et al.* [36].

A diagram of the construction details of this probe is shown in Figure 4.7 [37]. The Wollaston probe is relatively a massive structure compared to most inert probes used in other forms of AFM. The Wollaston probe's high and variable spring constant (5–20 N/m) and complexity render it unsuitable for all but contact mode. These probes routinely perform with spatial and thermospatial resolution on the subnanometer scale, although nanometer is occasionally reported [32, 36, 38–42].

Unfortunately, while folded probes are capable of imaging topography at the nanometer scale, these probes are not suitable for high-resolution thermal imaging. The heated area is on the top of the inverted pyramidal tip, meaning that the resistive element that is sensitive to temperature is relatively large (on the order of 10 μm) and is remote from the surface. The effect of this combination is that the heater serves very well to heat the tip, but the thermal resistance thermometer function is impaired. While the probe is poor at mapping thermal properties, such as thermal conductivity, it is excellent for local thermal–mechanical measurements.

However, the best results for robust routine local thermal analysis with high spatial resolution come from probes based on the approach adopted by King *et al.* [43]. The spatial resolution of these probes is the same as conventional AFM tips. Microfabricated bow tie probes are probes where the metal conducting layer has a bow tie shape at the tip so that electrical resistance is located at the narrow middle area. The resolution of the bow tie probe has been demonstrated on model samples of silica coated with a layer of PMMA [39]. Elongated rectangular discontinuities in the coating are detected in thermal images down to a width of 200 nm, but only when lying parallel to the raster direction. The resolution perpendicular to this is shown to be poorer roughly by a factor of 2.

Nonetheless, electrical resistance thermometry probes are an active area of development and future developments will surely lead to enhanced thermal resolution. It should be mentioned that a significant amount of information in SThM images will originate from the subsurface features of the material. Most of the other AFM modes, such as mechanical property-based imaging, are generally restricted to measuring the response of the material in the immediate vicinity of the surface. The minimal thermal mass of the probes leads to heating and cooling rates on the order of tens of degrees per second, which can be used to probe depth effects through AC temperature modulation imaging.

4.7.2
Localized Thermal Analysis

Local thermal analysis refers to a localized thermal measurement, much like force–distance curves are localized mechanical measurements. Typically, L-TA measurements do not involve mapping. These measurements are performed by contacting the surface with probe under a set force and then running a thermal measurement [36, 37, 44]. After the tip exerts a predetermined downward threshold force (utilizing the signal from the photodetector, which is proportional to the cantilever deflection and, hence, the force exerted), a temperature ramp can be applied to the sample via the probe. This is usually a linear heating program or linear heating followed by linear cooling. Heating and cooling may be set at different rates. Essentially, two signals are acquired simultaneously: the vertical deflection of the probe and the power required to ramp the probe temperature. These techniques have been extensively used in the study of thermal transitions in polymers and other materials (see Chapter 5 for several examples) [33].

Once the probe is in contact with the surface and the temperature program is initiated, the force–feedback mechanism is disabled and the fixed end of the cantilever remains at a constant height throughout the experiment. Therefore, these experiments are not constant force experiments. Thermal expansion typically results in a steady increase of the cantilever deflection (an increase in applied force) and an increase in the sample penetration. For a material that undergoes no thermal transitions over the temperature range of the experiment, the probe deflection with temperature will be essentially linear and upward as the sample beneath the probe heats and expands. Heating a sample to a phase transition results in the softening of the sample material, measured by both a dramatic increase in the sample penetration (decrease in cantilever deflection) and an increase in power required to change the probe temperature.

The heating of the probe element will itself cause some movement of the cantilever, but for the relatively massive Wollaston wire thermal probe this effect should be minimal. Providing a baseline subtraction procedure is carried out (acquired from a run with the probe in free air), the rate of power consumption of the probe over the duration of the same experiment on a sample should remain constant. When heating polymer through phase transitions (glass transition, cold crystallization, curing, melting, or degradation), the response of the micro-TA signals can be rich and distinctive [52]. There will be a large indentation at the softening temperature (around the glass transition temperature), and then a further indentation at the melting temperature. It is important to note that because the DTA signal is dominated by changes in the contact area, the glass transition and melting events are clearly detected. L-TA can also be used to analyze ultrathin films and to study differences between the surface and bulk properties by using suitable sectioning techniques [45].

As mentioned, AC heating can be applied to SThM as well as to L-TA experiments. AC heating may also be applied to the thermal probe. This produces a fixed temperature modulation in the range of $\pm 1\,^{\circ}\mathrm{C}$ to $\pm 10\,^{\circ}\mathrm{C}$, although it is usually confined to the range of $\pm 2\,^{\circ}\mathrm{C}$ to $\pm 5\,^{\circ}\mathrm{C}$. This may be seen as analogous to modulated temperature differential scanning calorimetry (MTDSC) [46]. In micro/nano-TA, the response of the sample to the modulated and underlying heat flows can be separated using a deconvolution program. The modulated regime is sensitive to the reversible changes in the heat capacity of the material associated with molecular vibrations and the latter detects changes due to kinetically controlled processes that are unable to reverse the temperature and the rate of the modulation. An obvious advantage of this technique is its ability to characterize heterogeneous samples in which different types of transitions occur over the same temperature range. Theoretically, the use of AC heating offers advantages similar to those of MTDSC over conventional DSC and it has been shown that the AC signals may be particularly sensitive to transitions that produce a relatively large change in heat capacity for a small heat input [47].

In order to measure the temperature of local transitions, a temperature–resistivity calibration of the thermal probe must first be carried out. The subject of temperature calibration has been comprehensively addressed by Blaine *et al.* [48] and Meyers *et al.* [49]. This process typically involves measuring the resistivity of the tip at the melting transition of known calibration samples. These calibration samples should be over several hundred nanometers thick to avoid substrate contributions. Furthermore, as already discussed, the thermal conductivity of calibration sample should be considerably less than the probe material thermal conductivity. The L-TA technique acquires two signals (L-TMA and L-DTA), of which one should be used to indicate the transition for calibration samples. The temperature calibration should be carried out on two or more substances whose melting temperature (T_m) is well known from literature. For this purpose, it is often more convenient to use polymer films whose melting point has previously been measured using DSC or another technique. A good correlation between bulk and local measurements has been demonstrated with the variation of the transition onset of $\pm 3\,^{\circ}\mathrm{C}$ [47, 48, 50].

4.7.3 Thermal Conductivity

Quantitative thermal conductivity measurements can be obtained from both SThM and L-TA, as described in the next two sections. Thermal measurement resolution strongly depends upon the ratio between surface thermal conductivity and tip thermal conductivity. Using the Block and Jaeger theories, the relationship between the heat dissipation through the contact area during physical contact, ΔQ, for the quasi-steady process can be presented in the following form [37, 51]:

$$\Delta Q = \frac{3}{4}\pi\lambda R_C \Delta T \tag{4.4}$$

where ΔT is the initial temperature difference between the probe and the surface, R_C is the effective contact radius of the thermal probe, and λ is the "composite" thermal conductivity in W/mK, defined as

$$\frac{2}{\lambda} = \frac{1}{\lambda_S} + \frac{1}{\lambda_P} \tag{4.5}$$

where λ_S and λ_P are the thermal conductivities of the surface of the sample and thermal probe, respectively.

Equation (4.4) shows that the ratio $\Delta Q/\Delta T \approx \lambda \cdot R_C$ should be constant for a given material if the contact radius and the thermal conductivity are unchanged. The sensitivity of the SThM method highly depends on the mechanisms of heat transfer from the thermal probe to a sample through the physical tip–surface contact. Equation (4.4) can be represented in a different form:

$$\frac{\Delta Q}{\Delta T} = \frac{3}{4} \frac{\lambda_p \pi R_c}{(1-\lambda_p/\lambda_s)} \tag{4.6}$$

which allows the direct evaluation of thermal conductivity of the surface probed by conducting a series of experiments with differently preheated SThM tips.

Figure 4.8 presents measurements of the heat dissipation for selected materials with different thermal and mechanical properties presented as a ratio of the thermal conductivities of the surface to the probe [51]. From Eq. (4.4), for the "composite" thermal conductivity coefficient, we can conclude that the method should be considerably more sensitive for materials like polymers with thermal conductivities less than the conductivity of the tip material (Figure 4.8).

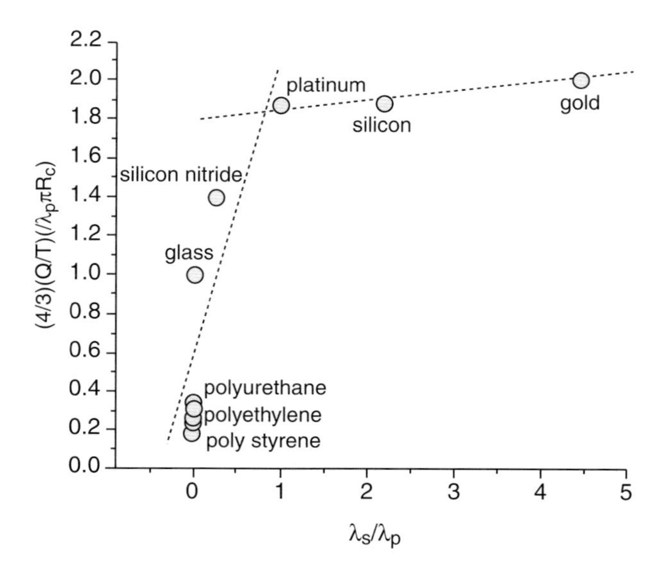

Figure 4.8 The variation of the reduced heat dissipation during physical contact versus the thermal conductivity of various materials reduced to dimensionless units. Reprinted from Ref. [52].

Materials with thermal conductivity below that of platinum (first group) behave very differently compared to high thermal conductivity materials (second group) (Figure 4.8), which can be understood considering the nature of "composite" thermal conductivity λ for the tip/surface entity, as presented by Eq. (4.4). This relationship demonstrates that for all surfaces with $\lambda_S \ll \lambda_P$ (polymers, glass, and semiconductors), the "composite" thermal conductivity for the tip/surface entity is primarily determined by the less thermally conductive component of this entity, namely, the surface. Therefore, under these conditions, the "composite" thermal conductivity that is responsible for heat dissipation at the contact point is directly proportional to the surface thermal conductivity of the materials tested; in other words, $\lambda \approx \lambda_S$. In contrast, for high thermal conductivity materials (with thermal conductivities higher than that of platinum), the surface becomes a thermal sink and the thermal tip becomes the poorly conductive counterpart. In this case, Eq. (4.5) predicts a virtually constant composite thermal conductivity $\lambda \approx \lambda_P$, with essentially no regard to the actual material tested. Therefore, a virtually constant value of measured heat dissipation should be anticipated for highly conductive materials but not for polymers.

It should also be stated that as a heated tip approaches the surface, the tip heats the surface even before the contact is made (Figure 4.9) [52]. The gradual heat increase in the heat dissipation thermal signal up to the physical contact point can be observed to start far away from the surface. The behavior of the thermal signal indicates that when the tip meets the surface, the area around the contact point is already "preheated" by the approaching thermal probe at certain distance.

The temperature of the surface when contact occurs depends on the tip temperature, the thermal properties of the test material, and the velocity of the heated tip that engages the surface. The temperature distribution of a surface has been analyzed based on heat flow theory [51]. The temperature within the thermal contact is virtually homogeneous (<3% of variation for all materials) and the temperature in the center of the heated zone, T_c, can be used for estimating the average temperature [52]. Also taking into consideration that the thermal probing of a surface can easily satisfy a

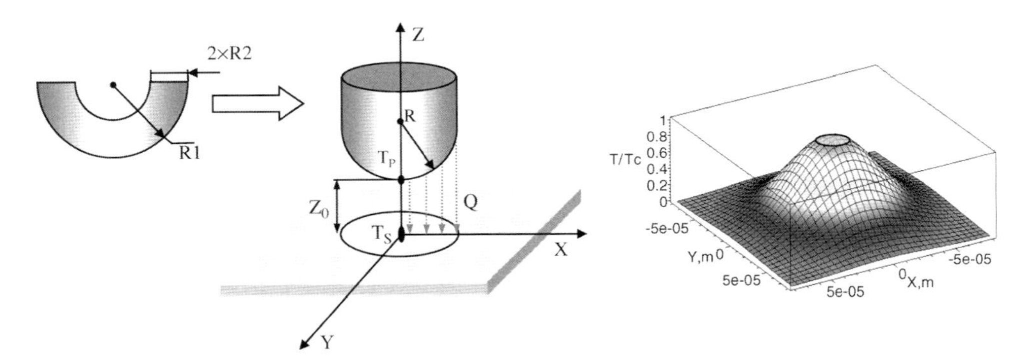

Figure 4.9 Modeling of temperature distribution in the contact area of a thermal tip. Reprinted from Ref. [52].

quasi-stationary case for heat flow from the thermal probe to a surface, the heat transfer can be described by a quasi-stationary equation to analyze the dynamics of the heat dissipation [51].

4.8
Kelvin Probe and Electrostatic Force Microscopy

Kelvin probe force microscopy (KPFM) and its simpler analogue, electrostatic force microscopy (EFM), enable the mapping of work function distributions and surface potential distributions with a lateral resolution of a few nanometers and a potential resolution of a few millivolts. The work function (Φ) is defined as the minimum energy required to remove an electron from the electronic ground state in a material [53]. It is known that Φ is a sensitive indicator of surface conditions and is affected by adsorbed layers, surface charging, imperfections, and contaminations.

Although macroscopic Kelvin probes for measuring the average work function for relatively large surfaces have been in use for a long time, KPFM was first demonstrated in 1987 using optical heterodyne to track the frequency changes in a vibrating AFM cantilever resulting from the normal electric field components of surface charges and potentials. The potential of the conductive tip to that of the surface is matched using an electronic feedback. When the potential of the tip exactly equals that of the material, electrostatic interaction between the tip and the sample is nullified. The voltage applied to nullify the electrostatic interaction is the local measure of the work function, or more directly, the contact potential difference between the tip and the surface. If the work function of the tip is known, a quantitative two-dimensional map of the surface work function can be constructed from the applied DC feedback signal. The KPFM technique is performed in the noncontact mode and does not involve the injection of any charges into the sample as in the conductive AFM.

Electrostatic force microscopy, on the other hand, is much simpler in that the technique does not involve an electronic feedback. An oscillating (at the resonance frequency) conductive tip biased at a fixed DC voltage scans over the sample, electrostatically interacting with the surface. EFM measurements are performed in both AC and DC modes. In the DC mode, a conductive tip oscillated near the resonance frequency with fixed DC bias (V_{tip}) scans over the surface at a fixed height (few tens of nanometers) above the surface of the sample. The tip interacts with the sample and the changing electrostatic force with the vertical separation distance that causes a shift in the resonance frequency and the phase ($\Delta\varphi$) of the cantilever given by

$$\Delta\phi \propto \left(d^2C/dz^2\right)\left(V_{tip} + \Phi - V_S\right)^2 \qquad (4.7)$$

where V_s is the voltage within the sample, Φ is the work function difference between the tip and the sample, V_{tip} is the tip voltage, and C is the tip sample capacitance. The observed phase shift is thus proportional to the square of the DC voltage difference between the tip and the sample. Therefore, by mapping the frequency or the phase shift as a function of the tip voltage pixel by pixel or by calibrating the phase/frequency response of the cantilever as a function of bias and maintaining a constant value of V_{tip}, EFM can be employed to measure potential profiles with high resolution.

One of the significant issues with KPFM is that the experimentally obtained potential profiles do not generally reflect the true profile in the device due to complex coupling between the tip and the sample [54]. In the initial stages, it was believed that the tip–sample separation and the tip radius were limiting factors of the resolution attained in KPFM measurements [55–59]. However, by combining experimental and finite element analysis, Charrier *et al.* quantitatively showed that the potential profiles obtained by scanning Kelvin probe microscopy do not purely reflect the electrostatic potential under the tip apex, but are strongly affected by the electrostatic coupling between the entire probe and the entire device, even for small tip–sample separations [60].

Figure 4.10a shows the experimental potential profile of the geometry shown in Figure 4.10b and the three different modeling curves [59]. The solid black line is calculated for the 3D probe consisting of an apex, a cone, and a lever. Calculations for a probe consisting of a cone and an apex (dashed line) and only a single apex (dotted line) are also shown. It can be observed that by removing the lever and leaving only the apex and cone, the full potential difference between the electrodes becomes 20% higher. Removing the cone and leaving the apex results in further deviation from the experimentally observed potential profile, clearly underscoring the importance of taking the entire probe and device geometry into account for reliable quantitative potential profiles.

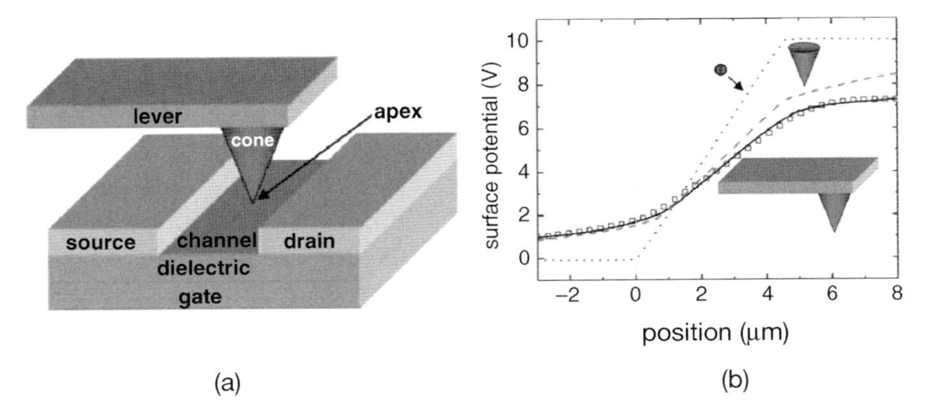

Figure 4.10 (a) Schematic showing the microcantilever orthogonal to the channel of the FET device. (b) Experimental potential profile with the lever orthogonal to the channel (squares) and modeling for a full 3D tip (black line) containing the apex, cone, and lever. The dashed line shows the simulation for a probe consisting only of the cone and an apex, and the dotted line shows the simulation for a probe consisting of only an apex. Reprinted from Ref. [60].

4.9 Conductive Force Microscopy

Conductive atomic force microscopy (c-AFM) enables the simultaneous mapping of the topographical and electrical conductivity of the sample using a conductive AFM tip. This technique involves using the electrically conducting AFM tip as one electrode and a conductive substrate or a metal electrode on the surface of the sample as the second electrode. The measurement can be performed by either applying a constant voltage between the tip and the metal electrode, while simultaneously recording images that can be used as a measure of the local conductivity, or by collecting local I–V curves by sweeping the voltage between the tip and the other electrode, which can be mapped. In a different kind of measurement, I–Z curves can be obtained by holding the voltage constant while the z-piezoelement is moved perpendicular to the sample surface, thereby changing the tip–surface separation.

Conductive AFM can be operated in two different configurations, namely, horizontal and vertical variants [61–63]. While in both the modes of operation, the conductive AFM tip acts as one electrode, in the horizontal mode, the material under investigation is deposited on an insulating surface, such as silica, and is connected to an external electrode. On the other hand, in the vertical configuration, the material is deposited on a conductive surface that acts a second electrode. Typically, c-AFM measurements are performed in the contact mode. However, this mode of operation results in the damage of soft polymeric and biological samples. To avoid the potential damage of the sample surface, some groups have adapted an alternative approach that involves acquiring the topography using a dynamic method such as tapping mode followed by point contact I–V measurements at predefined regions [64].

Apart from obtaining simultaneous topography and conductive maps, vertical configuration c-AFM has been extensively employed to obtain current density–voltage curves (J–Vcurves), which in turn were used to extract local hole mobilities using a space charge limited current (SCLC) model [65, 66]. Carrier mobility is extracted by fitting the J–V data to the Mott–Gurney law:

$$J = \frac{9}{8}\varepsilon_0 \varepsilon \mu \frac{V^2}{L^3} \tag{4.8}$$

where J is the current density, ε is the relative dielectric constant of the active layer, ε_0 is the permittivity of free space, μ is the charge carrier mobility, V is the applied voltage, and L is the thickness of the device. However, this technique usually results in higher carrier mobilities compared to those observed in a planar electrode configuration as the current spreads out under the AFM tip, enabling a larger space charge limited current density than is expected in the plane-parallel case, as shown in Figure 4.11.

Ginger and coworkers recently demonstrated that the primary cause of this observation is the fundamental difference in geometry between the two configurations, namely, planar electrodes and the vertical c-AFM [67]. The conventional Mott–Gurney law is not applicable for c-AFM measurements because SCLC measurements performed in this geometry deviate from the $J \propto L^{-3}$ dependence.

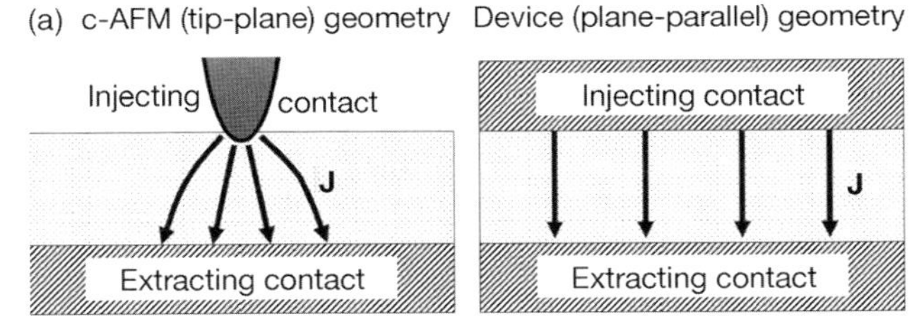

(b)

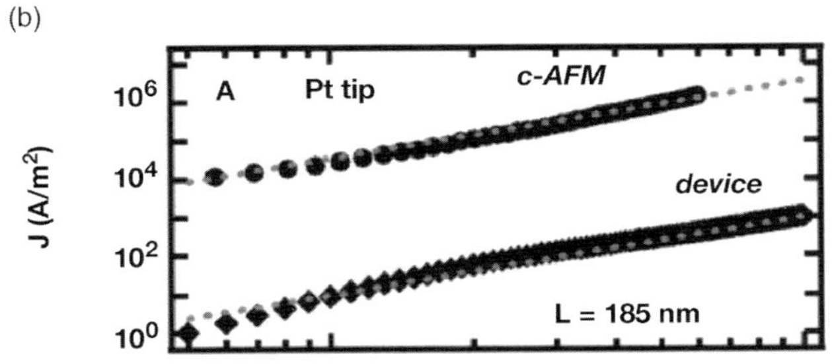

Figure 4.11 (a) Schematics showing the geometry of the c-AFM and planar macroscopic device measurement respective geometry. Current spreading laterally beneath the AFM tip results in a larger space charge limited current density than is expected in the plane-parallel case. (b) *J–V* curves measured using c-AFM (circles) and macroscopic devices (diamonds) on P3HT showing the apparently higher current density in c-AFM measurements. Dotted line shows the fit using classical Mott–Gurney law to each of the curves to extract the mobility. c-AFM measurement was made using a 50 nm diameter platinum-coated tips. Reprinted from Ref. [67].

Taking the tip sample geometry into account and using FEA simulations, they suggested a semiempirical equation for the current density as

$$J = \alpha\varepsilon\varepsilon_0\mu_0 e^{0.89\gamma(V/L)^{1/2}} \frac{V^2}{L^3}\delta\left(\frac{L}{D}\right)^{1.6\pm0.1} \tag{4.9}$$

which is valid for common tip diameters and sample thicknesses, using a scaling factor based on the ratio of tip diameter D to sample thickness L. This scaling factor enables extraction of quantitative values of charge carrier mobility from J–V curves collected by c-AFM.

4.10 Magnetic Force Microscopy

Magnetic force microscopy (MFM) is similar to EFM except that the tip interacts with the magnetic domains of the sample, as opposed to the electric. MFM operates in the noncontact mode in which a tip coated with a ferromagnetic material (such as Ni, Co, and Fe) detects the stray magnetostatic field of the magnetic dipoles of the sample. As the magnetostatic interactions are of long range (similar to the electrostatic interactions), the magnetic imaging is performed between the probe and the surface at a set distance, typically 20–50 nm, in a mode commonly referred to as lift mode. Lift mode involves a special raster scan where each line is scanned twice before the next line is scanned. In the first line scan, the topography is scanned in a conventional manner and the surface profile is stored. Then the probe is lifted by a set amount (several tens of nanometers) and the probe retraces the previous topographic line scan. During the second line scan, the cantilever deflection is monitored and used to create the MFM image. MFM has been widely employed to probe magnetic recording media and to image and record the magnetization of Co, Ni, and Fe magnetic micro- and nanostructures down to a single nanodot [68–72].

4.11 Scanning Acoustic Force Microscopy

4.11.1 Force Modulation

Force modulation microscopy (FMM) is a variant of the conventional contact mode operation of AFM, in which the sample (using an additional piezoelement) is set to oscillate at a frequency (typically tens of kilohertz) smaller than the resonance frequency of the microcantilever [73]. This oscillation causes the sample to elastically indent into the tip, resulting in a modulation of the tip, mainly at the same frequency as the sample oscillation. Under ideal conditions (an infinitely hard surface), the tip oscillates at the same frequency and amplitude as the sample. However, when the surface is soft and viscoelastic, the amplitude is reduced and the phase of oscillation might be significantly perturbed (Figure 4.12) [74]. Thus, the elastic properties of the sample can be estimated from the amplitude of the oscillation of the tip. The higher amplitude regions on the sample correspond to the stiffer regions, while the lower amplitude represents the softer regions of the sample. Force modulation provides a simultaneous map of topography and relative stiffness of the surface under investigation.

4.11.2 Ultrasonic Force Microscopy

Although conventional imaging modes offer unprecedented vertical and lateral resolution, they only provide information about the surface and, at best, shallow features buried under the surface. Conventional nondestructive imaging of subsurface features (defects, fillers, and the like) is based on acoustic microscopy in which an acoustic (ultrasonic) wave is transmitted through the sample and its amplitude and phase are monitored to image the subsurface features. However, in the far-field regime, the lateral resolution is severely limited by classical diffraction limit, which is given by

$$\delta = 0.51(\nu_0/f_{\mathrm{AW}}\,\mathrm{NA}) \tag{4.10}$$

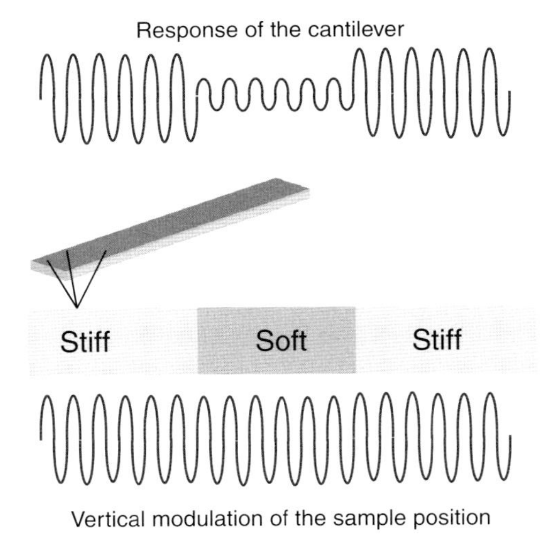

Figure 4.12 Force modulation mode of AFM showing the modulation of the *z*-position of the sample and the corresponding deflection of cantilever in response to the stiffness of the sample.

where ν_0 is the speed of sound in the coupling medium (typically water), f_{AW} is the frequency of the acoustic wave, and NA is the numerical aperture of the lens utilized [75].

Ultrasonic force microscopy (UFM) is a robust technique developed for subsurface imaging and can be considered as a modification of the standard contact mode of AFM, where the sample is oscillated at a high frequency (compared to the resonance frequency of the cantilever) by an additional piezoresonator [76, 77]. The microcantilever exhibits nearly 10^2–10^4 times higher dynamic stiffness at frequencies much higher than the primary resonance frequency. The fundamental principle involves working in the inertial regime (high dynamic stiffness) of the cantilever and sensing the nonlinearity of the tip–surface interaction.

The sample oscillating at these higher frequencies exerts a constant additional force on the apparently stiff cantilever, elastically indenting itself into the tip. The modulation of ultrasonic waves passing through the sample thickness due to the varying local stiffness and buried features is detected as modulation of the cantilever deflection. Apart from subsurface imaging, UFM has been employed to probe the local mechanical properties of thin polymer films and composites, especially high-performance materials with high elastic moduli ($>$10 GPa) [78, 79].

UFM has also been utilized to probe the morphology, local elastic properties, and internal structure of rubber-toughened poly(methyl methacrylate) (PMMA) samples [80]. AFM topography measurements combined with UFM can reveal the distribution and orientation of the rubber particles in the PMMA matrix. UFM revealed the core–shell structure of the rubber particles and their presence under the surface, which would otherwise be invisible. The filler particles were found to be either covered with a thin layer of PMMA or broken, thereby exposing parts of their internal structure. In another example, UFM has been used to study morphology and phase separation of adsorbed layers formed by two incompatible polymers, polystyrene (PS) and poly(butyl methacrylate) (PBMA) [81]. The high contrast between the two polymer components provided by UFM compared to the topographical images enabled unambiguous identification of the blend morphology.

A major issue with UFM for imaging the subsurface features is the nonlinear tip–sample interaction, which is extremely sensitive to the elastic and viscoelastic properties of the surface. Furthermore, the method is not an ideal choice for soft polymeric and biological samples due to the relatively large forces of interaction between the tip and the sample. To overcome these limitations, scanning near-field ultrasound holography (SNFUH) has been developed. This method involves setting up two ultrasonic waves, one from underneath the sample (2.1 MHz) and the other from the cantilever (2.3 MHz), forming a standing wave [82]. The phase and amplitude of the sample-scattered ultrasound wave, manifested as perturbation to the surface acoustic standing wave, are mapped to unveil the subsurface features.

Indeed, Shekawat and Dravid employed a polymer/gold nanoparticle composite thin film to demonstrate the SNFUH ability to image deeply buried features [82]. The sample was comprised of gold nanoparticles buried deep (500 nm) beneath the polymer cover layer (Figure 4.13a) [82]. Figure 4.13b shows the normal AFM topography scan with smooth featureless polymeric surface layer. On the other hand, the phase image of SNFUH clearly reveals the gold nanoparticles buried ~500 nm deep from the top surface (Figure 4.13c). The contrast in the SNFUH phase image is due to the difference in elastic modulus between the polymer and the gold nanoparticles, which results in a phase delay of the acoustic waves reaching the sample surface compared to those traversing through the polymer layer (Figure 4.13d and e).

4.12
High-Speed Scanning Probe Microscopy

Although the SPM techniques described thus far offer a wealth of information regarding structure and properties with nanoscale resolution, one of their primary limitations is that they are relatively slow imaging processes, often taking a few minutes to as much as several hours (force–volume mode of SFS). High-speed imaging (10–30 frames/s) with nanoscale resolution is extremely important to understand chemical and biological processes in real time (e.g., phase transition in polymers, motion of molecular motors, diffusion, and self-assembly). High-speed SPM is currently limited by various factors such as the dynamic behavior of the piezoscanners and the SPM probe, the limited bandwidth of the electronic components (power amplifiers that drive the piezoelectric actuators, photodetectors, and feedback control system), the availability of special tips, such as carbon nanotubes, and the speed of the data acquisition system [83].

There have been extensive efforts to overcome these limitations and enable high-speed AFM imaging. For example, the introduction of small, and hence fast, actuators for the vertical displacement significantly improved the feedback band-

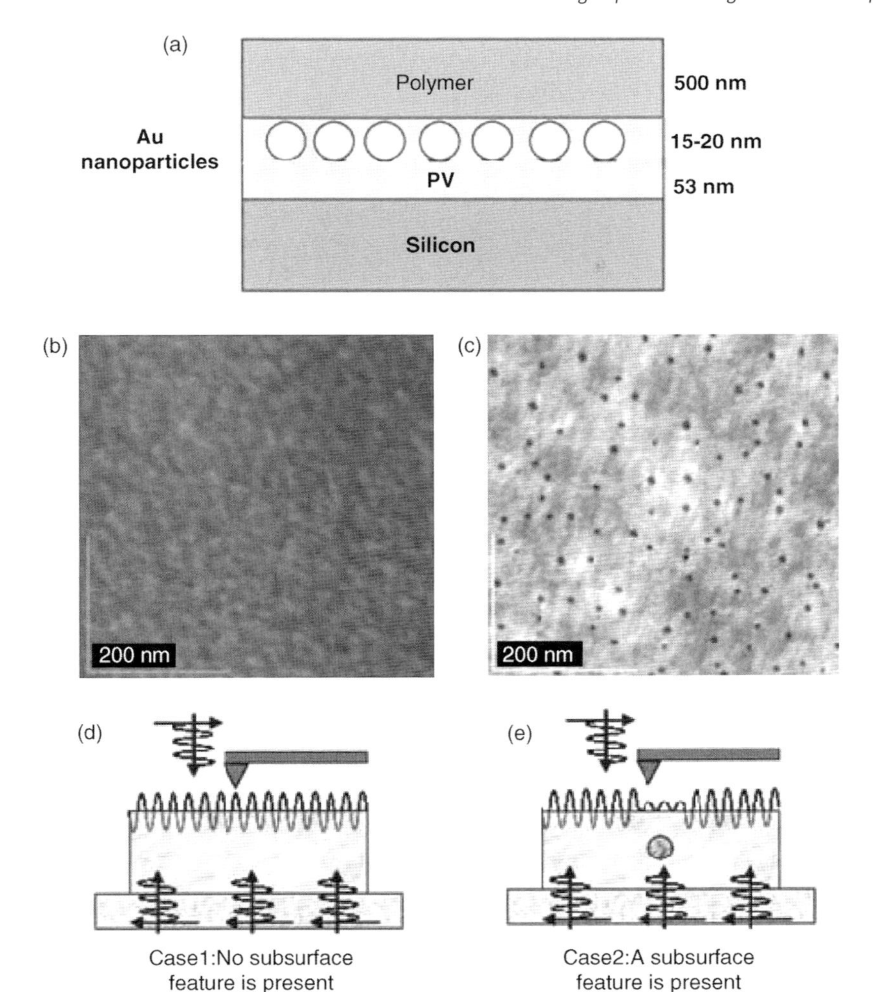

Figure 4.13 (a) Schematic of the model sample comprised of a thick polymer layer (500 nm) covering uniformly dispersed nanoparticles for demonstrating subsurface imaging using SNFUH. (b) AFM topography image showing the featureless top polymer surface. (c) The phase image of SNFUH clearly reveals the buried gold nanoparticles. (d and e) Schematics showing the interfering tip and the surface ultrasonic wave in the presence and absence of subsurface nanoparticles. Reprinted from Ref. [82].

width for tracking the tip–sample interaction using piezoelectric actuation of an AFM cantilever. Owing to the significant improvements in the bandwidth of the feedback electronics, there have been numerous reports on high-speed AFM of crystallization of polymers or biological processes, some of which will be discussed in Chapters 10 and 15. A detailed description of the technological developments concerning the high-speed AFM is beyond the scope of this chapter, and the reader is referred to several excellent reviews and original articles on the subject [83–86].

References

1 Zhang, W. and Zhang, X. (2003) Single molecule mechanochemistry of macromolecules. *Prog. Polym. Sci.*, **28** (8), 1271–1295.

2 Kühner, F. and Gaub, E.H. (2006) Modelling cantilever-based force spectroscopy with polymers. *Polymer*, **47** (7), 2555–2563.

3 Zou, S., Korczagin, I., Hempenius, M.A., Schöherr, H. and Vancso, G.J. (2006) Single molecule force spectroscopy of smart poly(ferrocenylsilane) macromolecules: towards highly controlled redox-driven single chain motors. *Polymer*, **47** (7), 2483–2892.

4 Al-Maawali, S., Bemis, J.E., Akhremitchev, B.B., Liu, H., and Walker, G.C. (2005) Single-molecule AFM study of polystyrene grafted at gold surfaces. *J. Adhes.*, **81** (10–11), 999–1016.

5 Butt, H.-J., Cappella, B., and Kappl, M. (2005) Force measurements with the atomic force microscope: technique, interpretation, and applications. *Surf. Sci. Rep.*, **59** (1–6), 1–152.

6 Cappella, B. and Dietler, G. (1999) Force–distance curves by atomic force microscopy. *Surf. Sci. Rep.*, **34** (1–3), 1–104.

7 Carpick, R.W. and Salmeron, M. (1997) Scratching the surface: fundamental investigations of tribology with atomic force microscopy. *Chem. Rev.*, **97** (4), 1163–1194.

8 Bhushan, B. (2008) *Nanotribology and Nanomechanics: An Introduction*, 2nd edn, Springer, Berlin.

9 Cannara, R.J., Brukman, M.J., Cimatu, K., Sumant, A.V., Baldelli, S., and Carpick, R.W. (2007) Nanoscale friction varied by isotopic shifting of surface vibrational frequencies. *Science*, **318** (5851), 780–783.

10 Tsukruk, V.V. (2001) Molecular lubricants and glues for micro- and nanodevices. *Adv. Mater.*, **13** (2), 95–108.

11 Tsukruk, V.V., Everson, M.P., Lander, L.M., and Brittain, W.J. (1996) Nanotribological properties of composite molecular films: C_{60} anchored to a self-assembled monolayer. *Langmuir*, **12** (16), 3905–3911.

12 Ge, S., Pu, Y., Zhang, W., Rafailovich, M., Sokolov, J., Buenviaje, C., Buckmaster, R., and Overney, R.M. (2000) Shear modulation force microscopy study of near surface glass transition temperatures. *Phys. Rev. Lett.*, **85** (11), 2340–2343.

13 Pu, Y., Ge, S., Rafailovich, M., Sokolov, J., Duan, Y., Pearce, E., Zaitsev, V., and Schwarz, S. (2001) Surface transitions by shear force microscopy. *Langmuir*, **17** (19), 5865–5871.

14 Noy, A., Vezenov, D.V., and Lieber, C.M. (1997) Chemical force spectroscopy. *Annu. Rev. Mater. Sci.*, **27**, 381–421.

15 Noy, A. (2004) Interaction at solid–fluid interfaces, in *Nanoscale Structure and Assembly at Solid–Fluid Interfaces* (eds X.Y. Liu and J.J. De Yoreo), Kluwer Academic Publishers, Norwell, pp. 57–82.

16 Takano, H., Kenseth, J.R., Wong, S.-S., O'Brien, J.C., and Porter, M.D. (1999) Chemical and biochemical analysis using scanning force microscopy. *Chem. Rev.*, **99** (10), 2845–2890.

17 Janshoff, A., Neitzert, M., Oberdörfer, Y., and Fuchs, H. (2000) Force-spectroscopy of molecular systems: single molecule spectroscopy of polymers and biomolecules. *Angew Chem., Int. Ed.*, **39** (18), 3212–3237.

18 Noy, A. (ed.) (2008) *Handbook of Molecular Force Spectroscopy*, Springer, New York.

19 Tsukruk, V.V. and Bliznyuk, V.N. (1998) Adhesive and friction forces between chemically modified silicon and silicon nitride surfaces. *Langmuir*, **14** (2), 446–455.

20 Thundat, T., Zheng, X.-Y., Chen, G.Y., and Warmack, R.J. (1993) Role of relative humidity in atomic force microscopy imaging. *Surf. Sci. Lett.*, **294** (1–2), L939–L943.

21 Binggeli, M. and Mate, C.M. (1994) Influence of capillary condensation of

water on nanotribology studied by force microscopy. *Appl. Phys. Lett.*, **65** (4), 415–417.

22 Fujihira, M., Aoki, D., Okabe, Y., Takano, H., Hokari, H., Frommer, J., Nagatani, Y., and Sakai, F. (1996) Effect of capillary force on friction force microscopy: a scanning hydrophilicity microscope. *Chem. Lett.*, **25** (7), 499–502.

23 Johnson, K.L. (1985) *Contact Mechanics*, Cambridge University Press, Cambridge.

24 Shi, X. and Zhao, Y.-P. (2004) Comparison of various adhesion contact theories and the influence of dimensionless load parameter. *J. Adhes. Sci. Technol.*, **18** (1), 55–68.

25 Marti, O. and Hild, S. (2000) Temperature-dependent surface properties of thin polystyrene films determined by scanning force microscopy, in *Microstructure and Microtribology of Polymer Surface ACS Symposium Series 741* (eds V.V. Tsukruk and K. Wahl), American Chemical Society, Washington, D.C.

26 Rosa-Zeiser, A., Weilandt, E., Hild, S., and Marti, O. (1997) The simultaneous measurement of elastic, electrostatic, and adhesive properties by scanning force microscopy: pulsed-force mode operation. *Meas. Sci. Technol.*, **8**, 1333–1338.

27 Marti, O., Stifner, T., Waschipky, H., Quintus, M., and Hild, S. (1999) Scanning probe microscopy of heterogeneous polymers. *Colloids Surf. A*, **154** (1–2), 65–73.

28 Krotil, H.-U., Stifter, T., Waschipky, H., Weishaupt, K., Hild, S., and Marti, O. (1999) Pulsed force mode: a new method for the investigation of surface properties. *Surf. Interface Anal.*, **27** (5–6), 336–340.

29 Ducker, W.A., Senden, T.J., and Pashley, R.M. (1991) Direct measurement of colloidal forces using an atomic force microscope. *Nature*, **353** (6341), 239–241.

30 Butt, H.-J. (1991) Measuring electrostatic, van der Waals, and hydration forces in electrolyte solutions with an atomic force microscope. *Biophys. J.*, **60** (6), 1438–1444.

31 McConney, M.E., Anderson, K.D., Brott, L.L., Naik, R.R., and Tsukruk, V.V. (2009) Bio-inspired material approaches to sensing. *Adv. Funct. Mater.*, **19**, 2527–2544.

32 Majumdar, A. (1999) Scanning thermal microscopy. *Annu. Rev. Mater. Sci.*, **29**, 505–585.

33 Gorbunov, V.V., Fuchigami, N., and Tsukruk, V.V. (2000) Microthermal probing of ultrathin polymer films. *High Perform. Polym.*, **12** (4), 603–610.

34 Dinwiddie, R.B., Pylkki, R.J., and West, P.E. (1994) Thermal conductivity contrast imaging with a scanning thermal microscope, in *Thermal Conductivity 22* (ed. T.W. Tong), Technomics Inc., Lancaster, pp. 668–677.

35 Balk, L.J., Maywald, M., and Pylkki, R.J. (1995) Nanoscopic detection of the thermal conductivity of compound semiconductor materials by enhanced scanning thermal microscopy. *Proceedings of the 9th Conference on Microscopy of Semiconducting Materials*, IOP Publishing Ltd., Bristol, pp. 655–658 (*Inst. Phys. Conf. Ser.*, 146).

36 Hammiche, A., Hourston, D.J., Pollock, H.M., Reading, M., and Song, M. (1996) Scanning thermal microscopy: subsurface imaging, thermal mapping of polymer blends, and localized calorimetry. *J. Vac. Sci. Technol. B*, **14** (2), 1486–1491.

37 Gorbunov, V.V., Fuchigami, N., and Tsukruk, V.V. (2000) Microthermal analysis with scanning thermal microscopy. I. Methodology and experimental. *Probe Microsc.*, **2** (1), 53–63.

38 Hammiche, A., Pollock, H.M., Song, M., and Hourston, D.J. (1996) Sub-surface imaging by scanning thermal microscopy. *Meas. Sci. Technol.*, **7**, 142–150.

39 Pollock, H.M., Hammiche, A., Song, M., Hourston, D.J., and Reading, M. (1998) Interfaces in polymeric systems as studied by C.A.S.M.: a new combination of localised calorimetric analysis with scanning microscopy. *J. Adhes.*, **67** (1–4), 217–234.

40 Price, D.M., Reading, M., Hammiche, A., and Pollock, H.M. (1999) Micro-thermal analysis: scanning thermal microscopy and localized thermal analysis. *Int. J. Pharm.*, **192** (1), 85–96.

41 Price, D.M., Reading, M., Hammiche, A., and Pollock, H.M. (2000) New adventures in thermal analysis. *J. Therm. Anal. Calorim.*, **60** (3), 723–733.

42 Pollock, H.M. and Hammiche, A. (2001) Micro-thermal analysis: techniques and applications. *J. Phys. D*, **34** (9), R23–R53.

43 King, W.P., Saxena, S., Nelson, B.A., Weeks, B.L., and Pitchimani, R. (2006) Nanoscale thermal analysis of an energetic material. *Nano Lett.*, **6** (9), 2145–2149.

44 Lawson, N.S., Ion, R.H., Pollock, H.M., Hourston, D.J., and Reading, M. (1994) Characterising polymer surfaces: nanoindentation, surface force data, calorimetric microscopy. *Phys. Scripta*, **T55**, 199–205.

45 Chui, B.W., Stowe, T.D., Ju, Y.S., Goodson, K.E., Kenny, T.W., Mamin, H.J., Terris, B.D., Ried, R.P., and Rugar, D. (1998) Low-stiffness silicon cantilevers with integrated heaters and piezoresistive sensors for high-density AFM thermomechanical data storage. *J. Microelectromech. Syst.*, **7** (1), 69–78.

46 Jones, K.J., Kinshott, I., Reading, M., Lacey, A.A., Nikolopoulos, C., and Pollock, H.M. (1997) The origin and interpretation of the signals of MTDSC. *Thermochim. Acta*, **304–305**, 187–199.

47 Häßler, R. and zur Müehlen, E. (2000) An introduction to μTA™ and its application to the study of interfaces. *Thermochim. Acta*, **361** (1–2), 113–120.

48 Blaine, R.L., Slough, C.G., and Price, D.M. (1999) Microthermal analysis calibration, repeat-ability and reproducibility. *Proceedings of the 27th NATAS, Savannah, Georgia*, Omnipress, Madison, pp. 691–696.

49 Meyers, G., Pastzor, A., and Kjoller, K. (2007) Localized thermal analysis: from the micro- to the nanoscale. *Am. Lab.*, **39** (20), 9–14.

50 Moon, I., Androsch, R., Chen, W., and Wunderlich, B. (2000) The principles of micro-thermal analysis and its application to the study of macromolecules. *J. Therm. Anal. Calorim.*, **59** (1–2), 187–203.

51 Gorbunov, V.V., Fuchigami, N., and Tsukruk, V.V. (2000) Microthermal analysis with scanning thermal microscopy. II. Calibration, modeling, and interpretation. *Probe Microsc.*, **2** (1), 65–75.

52 Tsukruk, V.V., Gorbunov, V., and Fuchigami, N. (2002) Microthermal analysis of polymeric materials. *Thermochim. Acta*, **395** (1–2), 151–158.

53 Pruton, M. (1994) *Introduction to Surface Physics*, Oxford University Press, New York.

54 Zerweck, U., Loppacher, C., Otto, T., Grafström, S., and Eng, L.M. (2005) Accuracy and resolution limits of Kelvin probe force microscopy. *Phys. Rev. B*, **71** (12), 125424.

55 Koley, G., Spencer, G., and Bhangale, H.R. (2001) Cantilever effects on the measurement of electrostatic potentials by scanning Kelvin probe microscopy. *Appl. Phys. Lett.*, **79** (4), 545–547.

56 Gil, A., Colchero Gómez-Herrero, J., and Baró, A.M. (2003) Electrostatic force gradient signal: resolution enhancement in electrostatic force microscopy and improved Kelvin probe microscopy. *Nanotechnology*, **14** (2), 332–340.

57 Takahashi, T. and Ono, S. (2004) Tip-to-sample distance dependence of an electrostatic force in KFM measurements. *Ultramicroscopy*, **100** (3–4), 287–292.

58 Sacha, G.M., Verdaguer, A., Martinez, J., Sáenz, J.J., Ogletree, D.F., and Salmeron, M. (2005) Effective tip radius in electrostatic force microscopy. *Appl. Phys. Lett.*, **86** (12), 123101.

59 Argento, C. and French, R.H. (1996) Parametric tip model and force–distance relation for Hamaker constant determination from atomic force microscopy. *J. Appl. Phys.*, **80** (11), 6081–6090.

60 Charrier, D.S.H., Kemerink, M., Smalbrugge, B.E., de Vries, T., and Janssen, R.A.J. (2008) Real versus measured surface potentials in scanning Kelvin probe microscopy. *ACS Nano*, **2** (4), 622–626.

61 Dai, H., Wong, E., and Lieber, C.M. (1996) Probing electrical transport in nanomaterials: conductivity of individual carbon nanotubes. *Science*, **272** (5261), 523–526.

62 Klein, D.L. and McEuen, P.L. (1995) Conducting atomic force microscopy of alkane layers on graphite. *Appl. Phys. Lett.*, **66** (19), 2478–2480.

63 O'Shea, S.J., Atta, R.M., Murrell, M.P., and Welland, M.E. (1995) Conducting

atomic force microscopy study of silicon dioxide breakdown. *J. Vac. Sci. Technol. B*, **13** (5), 1945–1952.

64 Kelley, T.W., Granstrom, E.L., and Frisbie, C.D. (1999) Conducting probe atomic force microscopy: a characterization tool for molecular electronics. *Adv. Mater.*, **11** (3), 261–264.

65 Douhéret, O., Lutsen, L., Swinnen, A., Breselge, M., Vandewal, K., Goris, L., and Manca, J. (2006) Nanoscale electrical characterization of organic photovoltaic blends by conductive atomic force microscopy. *Appl. Phys. Lett.*, **89** (3), 032107.

66 Frenette, M., MacLean, P.D., Barclay, L.R.C., and Scaiano, J.C. (2006) Radically different antioxidants: thermally generated carbon-centered radicals as chain-breaking antioxidants. *J. Am. Chem. Soc.*, **128** (51), 16532–16539.

67 Reid, O.G., Munechika, K., and Ginger, D.S. (2008) Space charge limited current measurements on conjugated polymer films using conductive atomic force microscopy. *Nano Lett.*, **8** (6), 1602–1609.

68 Porthun, S., Abelmann, L., and Lodder, C. (1998) Magnetic force microscopy of thin film media for high-density magnetic recording. *J. Magn. Magn. Mater.*, **182** (1–2), 238–273.

69 Folks, L. and Woodward, R.C. (1998) The use of MFM for investigating domain structures in modern permanent magnet materials. *J. Magn. Magn. Mater.*, **190** (1–2), 28–41.

70 Kleiber, M., Kümmerlen, F., Löhndorf, M., Wadas, A., and Wiesendanger, R. (1998) Magnetization switching of submicrometer Co dots induced by a magnetic force microscope tip. *Phys. Rev. B*, **58** (9), 5563–5567.

71 Gider, S., Shi, J., Awschalom, D.D., Hopkins, P.F., Campman, K.F., Gossard, A.C., Kent, A.D., and von Molnar, S. (1996) Imaging and magnetometry of switching in nanometer-scale iron particles. *Appl. Phys. Lett.*, **69** (21), 3269–3271.

72 Bliznyuk, V., Singamaneni, S., Sahoo, S., Polisetty, S., He, X., and Binek, C. (2009) Self-assembly of magnetic Ni nanoparticles into 1D arrays with antiferromagnetic order. *Nanotechnology*, **20** (10), 105606.

73 Maivald, P., Butt, H.-J., Gould, S.A.C., Prater, C.B., Drake, B., Gurley, J.A., Elings, V.B., and Hansma, P.K. (1991) Using force modulation to image surface elasticities with the atomic force microscope. *Nanotechnology*, **2** (2), 103–106.

74 Radmacher, M., Tilmann, R.W., and Gaub, H.E. (1993) Imaging viscoelasticity by force modulation with the atomic force microscope. *Biophys. J.*, **64** (3), 735–742.

75 Briggs, G.A.D. (1992) *Acoustic Microscopy*, Clarendon Press, Oxford.

76 Kolosov, O. and Yamanaka, K. (1993) Nonlinear detection of ultrasonic vibrations in an atomic force microscope. *Jpn. J. Appl. Phys.*, **32** (8A), L1095–L1098.

77 Rabe, U. and Arnold, W. (1994) Acoustic microscopy by atomic force microscopy. *Appl. Phys. Lett.*, **64** (12), 1493–1495.

78 Dinelli, F., Biswas, S.K., Briggs, G.A.D., and Kolosov, O.V. (2000) Measurements of stiff-material compliance on the nanoscale using ultrasonic force microscopy. *Phys. Rev. B.*, **61** (20), 13995–14006.

79 Rabe, U., Amelio, S., Kopycinska, M., Hirsekorn, S., Kempf, M., Göken, M., and Arnold, W. (2002) Imaging and measurement of local mechanical material properties by atomic force acoustic microscopy. *Surf. Interface Anal.*, **33** (2), 65–70.

80 Porfyrakis, K., Kolosov, O.V., and Assende, H.E. (2001) AFM and UFM surface characterization of rubber-toughened poly (methyl methacrylate) samples. *J. Appl. Polym. Sci.*, **82** (11), 2790–2798.

81 Bliznyuk, V.N., Lipatov, Y.S., Ozdemir, N., Todosijchuk, T.T., Chornaya, V.N., and Singamaneni, S. (2007) Atomic force and ultrasonic force microscopy investigation of adsorbed layers formed by two incompatible polymers: polystyrene and poly(butyl methacrylate). *Langmuir*, **23** (26), 12973–12983.

82 Shekhawat, G.S. and Dravid, V.P. (2005) Nanoscale imaging of buried structures via scanning near-field ultrasound holography. *Science*, **310** (5745), 89–92.

83 Schitter, G. and Rost, M.J. (2008) Scanning probe microscopy at video rate. *Mater. Today*, **11** (Suppl. 1), 40–48.

84 Fantner, G.E., Hegarty, P., Kindt, J.H., Schitter, G., Cidade, G.A.G., and Hansma, P.K. (2005) Data acquisition system for high speed atomic force microscopy. *Rev. Sci. Instrum.*, **76** (2), 026118.

85 Ando, T., Uchihashi, T., and Fukuma, T. (2008) High-speed atomic force microscopy for nano-visualization of biomolecular processes. *Prog. Surf. Sci.*, **83** (7–9), 337–437.

86 Ando, T., Kodera, N., Takai, E., Maruyama, D., Saito, K., and Toda, A. (2001) A high-speed atomic force microscope for studying biological macromolecules. *Proc. Natl. Acad. Sci. USA*, **98** (22), 12468–12472.

Part Two
Probing Nanoscale Physical and Chemical Properties

Scanning Probe Microscopy of Soft Matter: Fundamentals and Practices, First Edition.
Vladimir V. Tsukruk and Srikanth Singamaneni.

5
Mechanical Properties of Polymers and Macromolecules

AFM is capable of applying and detecting forces that are orders of magnitude lower than that of the chemical bonds, making it a unique tool for probing intermolecular interactions [1–4]. The photodetector has subangstrom sensitivity, resulting in the theoretical ability to measure forces down to 0.1 pN; however, noise from thermal, electronic, and optical sources limits the force sensitivity in ambient conditions to about 1 pN with practical limits closer to 5 pN. SFS, discussed in Chapter 4, is a powerful method to probe the nanomechanical and adhesive properties of surfaces, including, but not limited to, quantification of the elastic modulus, adhesion, chemical binding, inter/intra molecular forces, resilience, and elasticity. The limitation of SFS is that it can only probe surfaces (or topmost sublayers and subsurfaces) as opposed to other approaches, such as micro-/nanoindentation, which are capable of probing submicron depths. The so-called "pulling-off" version of SFS is widely utilized for investigating synthetic and protein macromolecular unfolding, brush layer stretching, and other tensile-related mechanical properties of individual molecules and usually requires special tip modification with selective binding groups [5–9].

As a surface-sensitive technique, SFS is well suited to study the effect of free and confined surfaces on physical properties, which can be quite different from conventional bulk properties. SFS is particularly useful in characterizing the local material properties of polymer-based and phase-separated polymer composites, where the contributions of individual components and their interactions can be quantified and spatially mapped. In this chapter, we outline some basic principles, issues, and assumptions of SFS probing and will demonstrate selected examples of applications for a wide range of materials. Some in-depth discussion of interfacial interactions, theoretical aspects, and interpretation of SFS approaches to complex materials can be found elsewhere [10–21]. A comprehensive in-depth review of force spectroscopy, including contact theories and complex force curve interpretation that can be found in these references, is beyond the scope of this chapter.

Scanning Probe Microscopy of Soft Matter: Fundamentals and Practices, First Edition.
Vladimir V. Tsukruk and Srikanth Singamaneni.

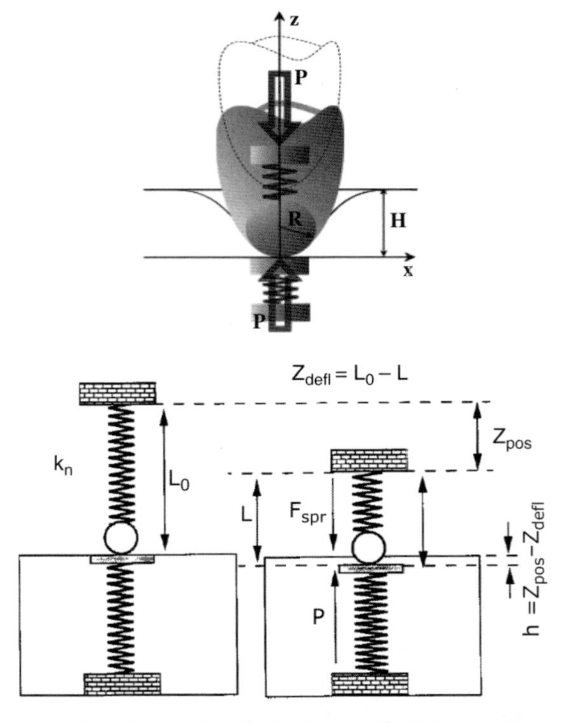

Figure 5.1 Schematics of indentation with AFM tip and two-spring model for contact mechanics analysis. Reprinted from Ref. [26].

5.1 Elements of Contact Mechanics and Elastic Modulus

5.1.1 General SFS Nanoprobing Principles

A single force curve, commonly called a force–distance curve, reflects tip–sample interaction during piezoelement movement and can be presented in terms of piezomovement, cantilever deflection, and surface deformation (see Chapter 4). For in-depth discussions of different contact mechanics models including Deryagin–Muller–Toporov (DMT) not discussed here, refer to other literatures [22–24].

In order to calculate the elastic modulus from applied force and sample penetration data, equations are derived from a quasi-static spring-on-spring or force balance approach as

$$z_{\text{defl}} k = P(h) \tag{5.1}$$

where P is the applied force and h is the sample penetration, as defined in Figure 5.1.

Assuming a spherical tip, flat surface, and no plastic deformation, one can define the elastic modulus as a derivative of the forces as

$$\left(\frac{\partial P}{\partial h}\right) = 2r\frac{E}{1-\nu^2} \tag{5.2}$$

where r is the tip–surface contact radius, E is the material elastic modulus, and ν is the material Poisson's ratio. There is currently no known way to measure the contact radius in real time at the nanoscale while the measurements are being performed. Instead, contact mechanics models are used to estimate r. These models generally differ in how they account for tip–surface interaction contributions to the contact area as well as how each model is applicable to different scenarios.

The most popular, the Hertzian contact mechanics, model is applicable for small deformation and it assumes that the adhesion forces are negligible [12, 25] and that at initial contact, the contact area is also zero. These assumptions are all far from reality for most polymeric materials. However, in the majority of practical cases, these contributions can be ignored to some extent or proper corrections can be introduced. The force as a function of penetration depth is described by the Hertzian model as

$$P = \frac{4}{3}R^{1/2}h^{3/2}E' \tag{5.3}$$

where R is the tip radius and E' is the composite modulus. The modulus associated with the probe is generally assumed to be much larger than the elastic modulus of the surface, which is true for all polymeric surfaces. Poisson's ratio is usually taken as known bulk values and possible deviations within 0.3–0.5 (0.5 for most elastic materials/scenarios) are modest considering overall contributions. For a routine estimation of the elastic modulus value for small indentation depths, the Hertzian model of a sphere–plane contact type is usually applied and generate results equivalent to more general Snedonn's model.

The Johnson–Kendal–Roberts (JKR) model of contact mechanics includes an adhesive contribution, which can be expressed in terms of a reduced load P_{JKR} [12]. The elastic modulus from the JKR can be described by the modified Hertzian relationship between the load and the contact area as

$$E = \frac{3}{4}\left(\frac{1-\nu^2}{R^{1/2}}\right)\left(\frac{dP_{JKR}}{d(h^{3/2})}\right) \tag{5.4}$$

where the reduced load P_{JKR} is found from the snap-from portion of the force curve.

On the other hand, the popular Sneddon's model suggests a specific and practical analytical relationship between the surface stiffness dP/dh and Young's modulus E' in the form

$$\frac{dP}{dh} = \frac{2\sqrt{A}}{\sqrt{\pi}}E' \tag{5.5}$$

where E' is the composite modulus and A is the contact area. Sneddon's model can also be utilized to describe tips with an elliptic paraboloid shape and for significant elastic deformations.

Although all three of the most popular models briefly discussed above are based on very different suggestions and are valid only under specific deformational conditions, testing of different approaches suggested that for all practical purposes, data analysis using Sneddon's model with a parabolic tip and without an inherent limitation on the indentation depth and the adhesive forces gives results virtually undistinguishable from the Hertzian model and generates reasonably consistent and reliable values for the elastic modulus.

Indeed, Sneddon's, Hertzian, and JKR approaches can be applied to process the force–distance data and calculate Young's modulus at different penetration depths for rubber materials (Figure 5.2) [26]. As was demonstrated for incremental analysis of force–distance curves, the Young's modulus for elastomeric materials is relatively independent of indentation depth beyond initial contact (5–20 nm for different materials). The value obtained at larger deformation is virtually constant (within ±20%) at indentation depths larger than 20 nm. However, below a 10 nm depth, unstable results are obtained that are related to the destabilizing attractive force gradient in the vicinity of surfaces.

All three approaches discussed above give convergent results and very close absolute values of Young's modulus for polyisoprene (PI) rubber at higher indentation depths (Figure 5.2). The absolute value of the rubber Young's modulus is 2.9 ± 0.6 and 2.4 ± 0.5 MPa, as determined by the Hertzian and JKR models, respectively. Both values are within the expected elastic bulk modulus for PI rubber (2 ± 1 MPa). The possibility of up to 20% overestimation of absolute values of Young's modulus comes from utilizing the Hertzian model as compared to the more complete JKR theory.

The Hertzian model was used to calculate depth dependencies of elastic moduli for a set of polymeric materials with very different elastic properties (Figure 5.2). The experimental data are shown along with bars representing the range of Young's modulus measured for different bulk materials. The most important feature of the plot in Figure 5.2 is the close correlation between the level of the elastic moduli probed by SFS and the known mechanical properties of bulk materials. SFS probing can be conducted for materials with a bulk elastic moduli differing by more than three orders of magnitude ranging from 2 MPa for PI rubber to 3 GPa for polystyrene (PS) (Figure 5.2).

As mentioned earlier, there is a discrepancy regarding the initial deformation at snap-in point, the zero contact point, in the way it is accounted in experiments and in theoretical models. This point is usually taken as either the snap-to point (the minimum deflection point in the extension curve) or the zero deflection point after the "snap-to" in the deflection curve. Therefore, it should be apparent that at initial penetrations, the modulus can be either overestimated or underestimated and as the penetration depth increases, the measured modulus will steadily decrease or increase to the "true value." If the total deformation well exceeds (two–three times) the initial contact penetration (which is possible for elastic materials), the true value of the elastic modulus can be obtained even if an initial mechanical instability is present. This is usually the case for elastic materials, where overall deformation of 10–100 nm utilized for data analysis is much higher than the initial deformation of 1–5 nm.

The expansion of the data analysis by applying a polynomial fit combined with the Snedonn's parabolic model has recently been suggested for the study of elastomeric materials [27]. This approach has been applied to very compliant glassy polymeric materials, and it has been confirmed that Sneddon's parabolic model is an excellent

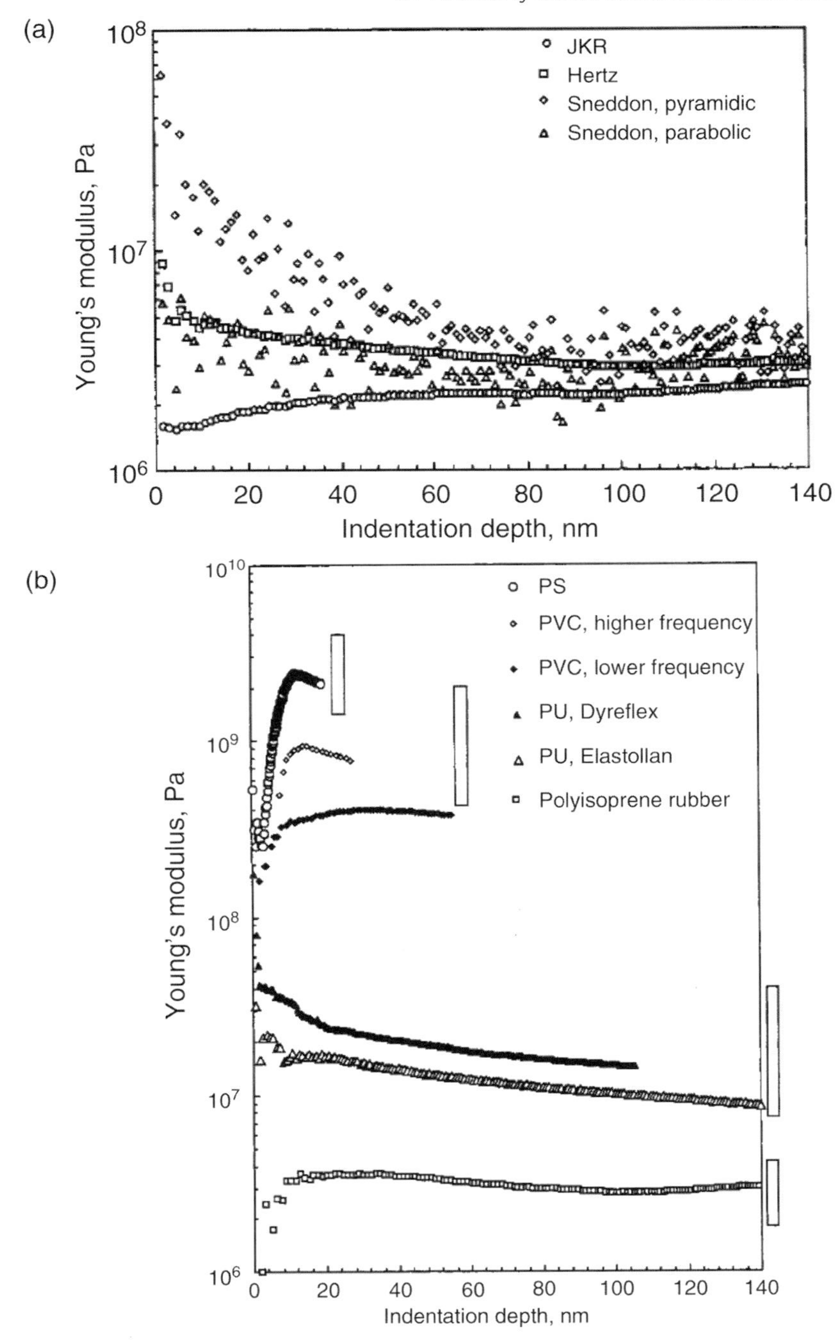

Figure 5.2 (a) Comparison of different models for evaluation of elastic modulus and modulus versus depth for the PI rubber sample. (b) Summarized data for depth variation of the elastic modulus for rubber, two different PUs, PVC at high and low frequencies, and PS. Bars demonstrate the range of elastic bulk modulus variation for a specific material (frequency, molecular weight, and composition dependence). Reprinted from Ref. [26].

choice for SFS data analysis with results coinciding with the conventional Hertzian model for polymeric materials. This approach is not a subject of the usual limitations thought for the Hertzian model and can be applied to large deformation and adhesive surfaces. Careful analysis of a series of polymeric materials demonstrated an excellent correlation between the SFS elastic modulus and an independently measured bulk elastic modulus mostly within ±10%. Overall, it has been confirmed that the deformation of polymeric materials in the course of SFS measurement is dominated by the elastic response and the onset of plastic deformation is usually not observed under regular SFS conditions.

5.1.2
Substrate Effects

SFS measurements are frequently exploited for probing nanomechanical properties of nanoscale and ultrathin organic films on solid substrates. In such cases, the presence of stiff solid substrates can severely affect the SFS measurements by affecting the overall deformation. Moreover, more complex cases of vertically layered films can no longer be considered as uniform elastic films, and data analysis must be modified to take into account nonuniform transfer of stresses and sequential deformation of layers (Figure 5.3). There are several methods for the evaluation of the elastic modulus of ultrathin films on solid substrates, which will be briefly discussed here.

By using a gelatin film with a gradient in thickness ranging from a few nanometers to micrometers, Domke and Radmacher demonstrated a clear effect of the solid substrate on the SFS measurements [28]. For thicker films, a uniform elastic response in the whole indentation range (up to 300 nm) has been observed with the average elastic modulus of about 20 kPa as was expected for this material. However, for thinner portions of the film, a significant reduction of the indentation was observed, which can be associated with the presence of the stiff substrate. A continuous "increase" in apparent elastic modulus (up to three orders of magnitude)

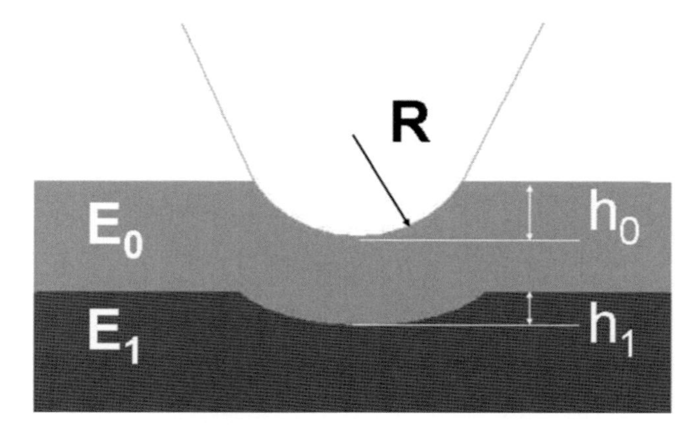

Figure 5.3 Sequential deformation in the double-layer model. Reprinted from Ref. [34].

was observed for an increasing indentation depth and exceeding 40% of total thickness.

A number of different models, which considered elastic deformation of the layered solids with a certain model of transfer of the mechanical load between adjacent layers, were proposed for the analysis of microindentation data of soft films on stiff substrates [29–33]. The models suggested in different studies considered various scenarios for interactions between films and substrates and different stress transfer models and, thus, should be applied with care considering a practical ratio of elastic moduli in different layers and without attempting to generalize a particular approach to a global case. In any case, careful examination of the applicability of multilayered models to different layered solids should be conducted (see examples below).

A generalized approach that starts with a simple definition of the depth profile as a smooth function with gradual localized changes in the elastic properties provides a means for the "visualization" of the transfer function. The interpretation of the complex layered solid with two- and three-layer architectures has been suggested by Kovalev *et al.* [34]. A key point of this approach is the representation of the composite compliance of double-layer solids (as a film–substrate system) as a superposition of individual compliances in the general form:

$$\frac{1}{E'} = \frac{1}{E_f} \cdot \left(1 - e^{-\alpha \cdot h/t}\right) + \frac{1}{E_s} \cdot \left(e^{-\alpha \cdot h/t}\right) \tag{5.6}$$

where E_f and E_s are elastic moduli of the film and the substrate, respectively, t is the total thickness of the film, h is the indentation depth, and α is a parameter defining contributions of different layers. The modulus profile in this form can be directly used for fitting experimental data for the elastic response at a variable indentation depth (Figure 5.4). The variation of the elastic modulus over a wide range creates very different profiles, including a virtually uniform distribution for a layered system with a small difference in elastic moduli (Figure 5.4). By changing the level of the elastic modulus and adding the variable transfer function, this model can be converted to the three-layer function with ascending or descending elasticity.

This approach introduces a new measure of the level of the transfer of the mechanical deformation between layers represented by a specially selected function, the transfer function, $e^{-\alpha h/t}$ (see Eq. (5.6)). This transfer function depends upon the total thickness of the layer, the indentation depth, and the properties of the interlayer interactions as reflected by the parameter α (see how the profile is affected in Figure 5.4). The transfer function for the elastic layered solid has an initial small value for very small, initial deformations ($h \ll t$) and increases for larger deformations ($h \leq t$).

For a double-layer model, two different levels of the elastic modulus are separated by a transition zone with a gradient of the elastic properties (Figure 5.4). The width of this transition zone between two layers is determined by the parameter α through the transition function. A high value of this parameter corresponds to a very sharp interfacial zone resulting in a steplike shape. Decreasing the α-value results in a

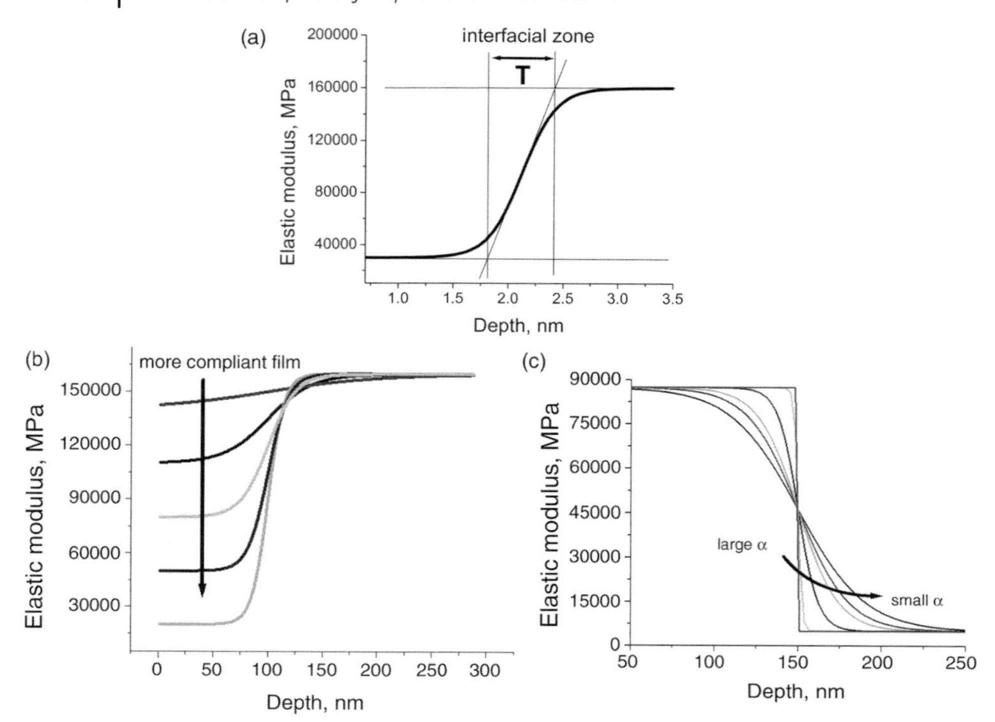

Figure 5.4 The interfacial zone *T*, as defined for double-layer model profile (a). Influence of the elastic modulus of the topmost layer (b) and the transfer function (c) on the total depth distribution of the elastic modulus of the double-layer solid. Reprinted from Ref. [34].

gradual broadening of the step function and finally forms a virtually continuous gradient for a very low value of α.

For the important case of a compliant film on a stiff substrate, assuming that the elastic properties of the substrate are known, only two unknown variables can be varied to fit experimental data (the elastic modulus of the top layer and parameter α). Usage of this proposed approach allows the analysis of different double and more complex layered structures, as will be demonstrated below.

5.1.3
Issues and Key Assumptions with Nanomechanical Probing

Here, it is important to mention several other critical contributions that could affect force–distance data and are not accounted for in the models discussed here.

A common misconception is that accurate (better than ±50%) quantitative elastic modulus data cannot be obtained from force spectroscopy. This belief comes from the inability to measure the tip–surface contact radius at the nanoscale in real time and a cumbersome, extremely time-consuming routine that is rarely followed properly without devastating shortcuts. Furthermore, the unstable nature of the "snap-to" region prevents the exact knowledge of the contact point and creates a certain discrepancy in the initial penetration, thus affecting a long chain of calculation.

Fortunately, for all practical purposes, this problem is not as critical as perceived. Overall, when performed with care, the resulting data have shown a very good agreement with known samples, and the data should be considered accurate within a ±20% deviation beyond initial engagement instabilities as demonstrated for a number of soft materials [26, 34, 35]. However, attempts to use models with limited applicability to a wide range of materials with very different elastic and adhesive properties without prior testing should be avoided. The most critical issues are to stay within limits of elastic deformation (usually 1–5 nm for glassy polymers and 100–300 nm for rubbery materials) to avoid local plastic deformation and to prevent tip contamination.

Knowledge of spring constants and the radius of curvature of tips is critical for accurate measurements. Accepting crude estimations from cantilever specs can lead to systematic and large (100–200%) deviations due to the usual overestimation of tip sharpness and significant batch-to-batch variation of cantilever thicknesses. Spring constants critically depend upon cantilever parameters (especially thickness) and geometry and must be determined individually for each cantilever to obtain the highest precision. Current methods of thermal tuning for relatively soft cantilevers (usually below 10 N/m) or direct spring-against-spring measurements for stiffer cantilevers are widely accepted and utilized [11, 12, 36].

Tip shape for individual tips must be tested before and after probing with well-established procedures. Shape can be tested with high precision (within ±10%) either with TEM/SEM (resolution is limited to 3–5 nm) or by scanning reference samples of tethered gold nanoparticles with known geometrical parameters (diameters ranging from 5 to 50 nm) [12].

Strong adhesion between the AFM tip and the polymer surface can disturb the initial portion of loading curves and might result in significant overestimation of the elastic modulus level at small indentation depths. The adhesion hysteresis and initial nonzero contact area are important for very compliant polymeric materials with high adhesion in air (e.g., polar rubbery layers with the elastic modulus below 2 MPa and surface energy much higher than common 30 mJ/m^2). For these materials, applying the Hertzian or Sneddon's models may result in a manifold overestimation of the elastic modulus for very small indentations. Proper deformation, as described earlier, must be achieved to avoid significant contributions of initial instability and removal of capillary forces by placement specimen in liquid can be excersized.. However, for compliant materials with modest adhesion and higher stiffness, this overestimation is limited to a few initial data points and, thus, the approach discussed here can be applied with care.

A complete JKR model, when applied, makes consideration significantly more complex and requires additional measurements of adhesive forces. Moreover, strong adhesion of polar rubbery materials to the hydroxyl-terminated tips might result in pulling material around the tip and compromising the assumption of the contact area and resulting in the stretching of the material during retracing and fast irreversible contamination of the AFM tip. Therefore, tip cleaning and tip radius verification procedures should be an important part of the rigorous experimental routine.

A viscous contribution (time-dependent mechanical properties) can be critical in defining the overall shape of loading curves for viscoelastic polymeric materials especially in the vicinity of the glass transition. As we discussed earlier, this phenomenon would result in a concave shape of the force–distance curves, which in fact is frequently observed for polymeric materials. The presence of the viscoelastic contribution can principally change the shape of the force–distance curve and, thus, the resulting loading curve will change from convex to concave, as will be shown below. However, for most practical cases, the viscoelastic contribution does not dominate deformation scenarios with only a few important exceptions, such as swollen gels, biological materials, or polymers in the vicinity of the glass transition.

An important practical consideration that affects the quality of the SFS measurements and puts some limits on SFS probing is directly related to the surface deformation induced by a stiff tip. If the cantilever bending stiffness is too high, then the vertical deflection of the cantilever, z_{defl}, is very small as compared to z_{pos}, the vertical displacement of the piezoelement (Figure 5.1).

In this case, the ratio T of indentation depth $h = z_{\text{pos}} - z_{\text{defl}}$ to total displacement reaches an upper measurable limit:

$$T = h/z_{\text{pos}} > 1 \tag{5.7}$$

On the other hand, if a material is much harder and its deformation is negligible compared to the cantilever deflection, then $z_{\text{pos}} \approx z_{\text{defl}}$ resulting in a lower limit of observation:

$$T = h/z_{\text{pos}} > 0 \tag{5.8}$$

Apparently, unambiguous measurements of material properties are possible only far from these limits. For practical estimations, considering the sensitivity of AFM instruments, it can be suggested that the ratio of indentation depth and cantilever deflection should not be outside the 1: 10 and 10: 1 ratio. This condition, in our opinion, takes into account the typical level of noise and nonlinearities of commercial mechanical systems and provides an acceptable signal level that can be separated from the background. Using these deflection versus indentation conditions, the range of ratios between the elastic modulus of materials and cantilever spring constants satisfying the measurable limits discussed above can be estimated, as presented in Figure 5.5 [18, 37].

As is clear from the plot presented, choosing an optimal spring constant within the range of 0.01–100 N/m should allow the measurements of elastic moduli from as low as tens of kilopascals to as high as tens of gigapascals. This range of elastic moduli spans the vast majority of polymeric materials stretching from hydrogels to hard plastics [11]. On the other hand, using the AFM cantilever with a given spring constant for surface mapping naturally limits the range of elastic modulus that can be detected within one set of probing conditions to not more than 2.5 orders of

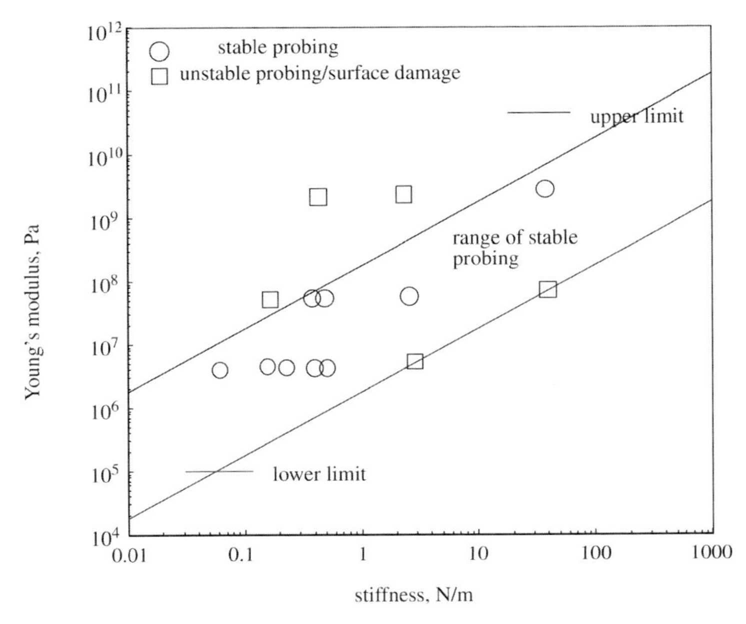

Figure 5.5 "Window" of optimal probing conditions (stiffness or spring constant of cantilevers) for surfaces with different elasticities with limits $h/z_{\mathrm{defl}} = 0.1$ (upper limit) and $h/z_{\mathrm{defl}} = 10$ (lower limit) and tested cantilevers for various materials (marks). Reprinted from Ref. [18].

magnitude (Figure 5.5). To test the validity of the proposed plot, we used a variety of commercially available cantilevers to probe the elastic properties and estimated the reliability of the limits for surface elasticities vs cantilever spring constants proposed that follow the ratios suggested by Eqs. (5.7) and (5.8) (Figure 5.5).

Another important issue to be considered in SFS probing is the stochastic nature of the highly localized measurements, as was indicated in the very early stages of these measurements [38]. The microscopic material heterogeneities can be critical in an accurate determination of the elastic modulus. The surface microroughness and topological contributions could affect the real contact area resulting in underestimation (on elevations) or overestimation (in valleys) of the local values. This effect is insignificant only for films with nanoscale roughness. For reliable SFS measurements, multiple repeated measurements over large surface areas should be conducted. The most appropriate approach is applying micromapping with collections of large arrays of force–distance curves (at least several hundreds–thousands) at multiple selected surface areas, conducting screening and averaging of these data, building histograms, and finding the average value that has been found using e.g. an MMA data analysis package such as adapted in our lab [39]. Recent experimental developments, including introduction of the so-called peak force scanning method, allow collecting the numerous array of force–distance data and thus mapping of surface distribution of mechanical properties with unprecedented lateral resolution [40].

Finally, we should mention a number of special cases for conducting and analyzing SFS experimental data, including elastic and plastic deformations of complex carbon, organic, polymeric, and biological structures [11, 12, 41] such as single-molecule deformation [42, 43], microrheology of polymer films [44], nanocomposite mapping [45, 46], freely suspended beams and plates [47], hollow microcapsules [48, 49], and deformable microparticles [50]. For these and other structures, specific models of the mechanical deformations have been adapted and applied that will not be discussed here. Some results obtained in these studies will be discussed in subsequent chapters.

5.2 Probing of Elastic Moduli for Different Materials: Selected Examples

Here, we present several selected recent examples of successful SFS applications to various polymeric materials and structures with an emphasis on approaches of data collection and analysis.

5.2.1 Bulk Materials and Blends

Elastomeric PDMS is a material of choice for SFS probing as a reference material because of its high compliance and low adhesion. As we discussed above, robust results for this material have been demonstrated on many occasions. In a recent study of Song *et al.*, PDMS with different oxidation treatments was examined by SFS probing, as well as tensile and microscopic contact measurements [51]. Direct comparison of microscopic bulk modulus and surface-derived modulus showed close correlation between all measurements (within the range of 3–6 MPa). However, after a long ozone treatment, a significant discrepancy has been observed between different measurements. Even if the bulk elastic modulus remains virtually unchanged, the elastic modulus obtained from SFS measurements increased dramatically, up to 110 MPa for the longest ozone treatment. Such a dramatic increase in the local elastic modulus was associated with both the predominant formation of a distinct surface layer with properties similar to that of stiffer silica materials and the very limited volume of compressed materials under the AFM tip during probing.

Several recent studies addressed the nanomechanical properties of poly-*N*-isopropylacrilamide (PNIPAAM) polymers with a known temperature-induced conformation transition at LCST close to 32 °C. Direct measurements of the smooth films from plasma-polymerized PNIPAAM conducted by Cheng *et al.* in water below and above LCST revealed significant change in the slope of force–distance curves caused by an increase in the surface stiffness [52]. Although the elastic modulus was constant below LCST, it increased by a factor of 10, from 200 kPa in highly swollen state to 2 MPa in a collapsed state. This behavior was similar to the expected noncross-linked PNIPAAM, thus confirming that cross-linking in the thin film does not affect the properties of the thermal transformation.

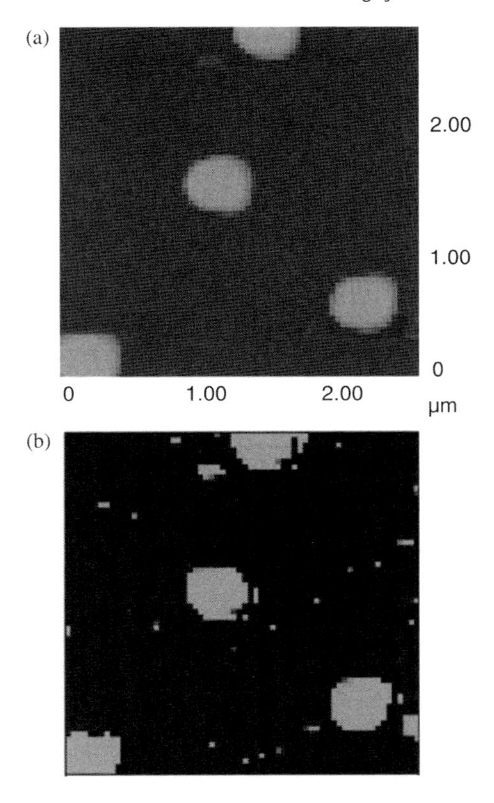

Figure 5.6 Micromapping of topography (a) and elastic response (b) for PNIPAAM microparticles. Reprinted from Ref. [54].

However, a much higher elastic modulus of 700 MPa has been obtained with the PFM mode that operated at high frequency for dry films of PNIPAAM blended with surfactants [53]. For these blends, a clear dispersed surface morphology with very different elastic responses caused by phase separation has been revealed by surface probing with high resolution.

PNIPAAM microgel particles on solid substrates have been studied with SFS micromapping by Tagit *et al.* [54]. The authors observed significant changes in particle shape in water associated with thermal transformation at varying temperature. Moreover, SFS micromapping clearly identified softer microgel particles on solid substrates (Figure 5.6). SFS probing of individual particles showed that the elastic modulus drops from 125 MPa in the dry state to 1.8 MPa in the swollen state at room temperature. However, the value of elastic modulus increases to 13 MPa after heating to temperatures above LCST. A similar increase in the elastic modulus combined with modest reduction of particle dimensions above LCST was observed for PNIPAAM by Wiedemair *et al.* [55]. This study demonstrates how a unique measurement of the mechanical properties of submicron gel microparticles can be conducted that cannot be addressed by conventional measurements.

Another type of volume phase transformation under variable solvent quality was studied for polyvinylalcohol (PVA) gels by Horkay and Lin [56]. The authors studied the surface distribution of the compressive response of polymer gels in different solvents and generated concentration dependence of the shear modulus and swelling response with SFS. The combination of macroscopic swelling measurements with micromapping of the osmotic modulus map allows the prediction of not only the elastic response but also the recovery behavior of complex viscoelastic materials with partially nonuniform distribution. Such information of the local properties of components is important for implant behavior under complex loading scenarios. Finally, the analysis of the temperature-dependent adhesive behavior of polyethyleneglycol (PEG) surface layers demonstrated reversible switching from repulsion-dominated to strongly adhesive behavior at elevated temperatures [57].

Surface distribution and their corresponding histograms of mechanical and adhesive properties for biocompatible blends based upon polylactic acid (PLA) deposited on stents have been reported by Wu *et al.* [58]. They observed a typical phase-separated morphology for PLA blends with low molar weight drags and measured the average value of the elastic modulus from SFS micromapping. The value of the elastic modulus of about 1.9 MPa for the drag PLA blend was observed to be only slightly below the elastic modulus for PLA uniform coatings. SFS measurements under high load resulted in plastic deformation of the coating with permanent marks easily visible on AFM zoom-out images. Surprisingly, the elastic modulus did not change significantly after placement of the material in liquid, but it drops by 50% if temperature rises to 36 °C. This change is interpreted as an indication of lowering of the glass transition temperature to below body temperature after swelling, which is critical for understanding drug release after stent implantation.

Another biocompatible material with soft matrix reinforced with hard domains, polyurethane (PU), was probed with SFS directly during macroscopic elastic deformation by Amitay-Sadovsky *et al.* [59]. The authors observed a significant difference in the compressive resistance and adhesion of surface layer and bulk PU materials in the course of their elongation up to 150%, with much higher elasticity response observed for the topmost polymer layer. Increasing surface elastic modulus with elongation was explained by smoothing the surface morphology and preferred segregation of highly oriented hard blocks near the surface.

Choi *et al.* demonstrated the micromapping of the elastic modulus in periodic porous polymer structures from epoxy-rich resins (bisphenol A) fabricated by multilaser beam interference lithography (Figure 5.7) [60]. The variation in the elastic modulus along periodic porous structures follows closely with the morphology. Moreover, the variation of the elastic modulus within 1–2 GPa across different elements of these microstructures (struts and nodes) was detected by conducting high-resolution micromapping of the surfaces. These measurements avoided the effects of geometry by careful control of the probed depth. Such smooth location-dependent surface variations were suggested to be due to the periodic variation in the cross-linking density resulting from the light intensity distribution in the course of

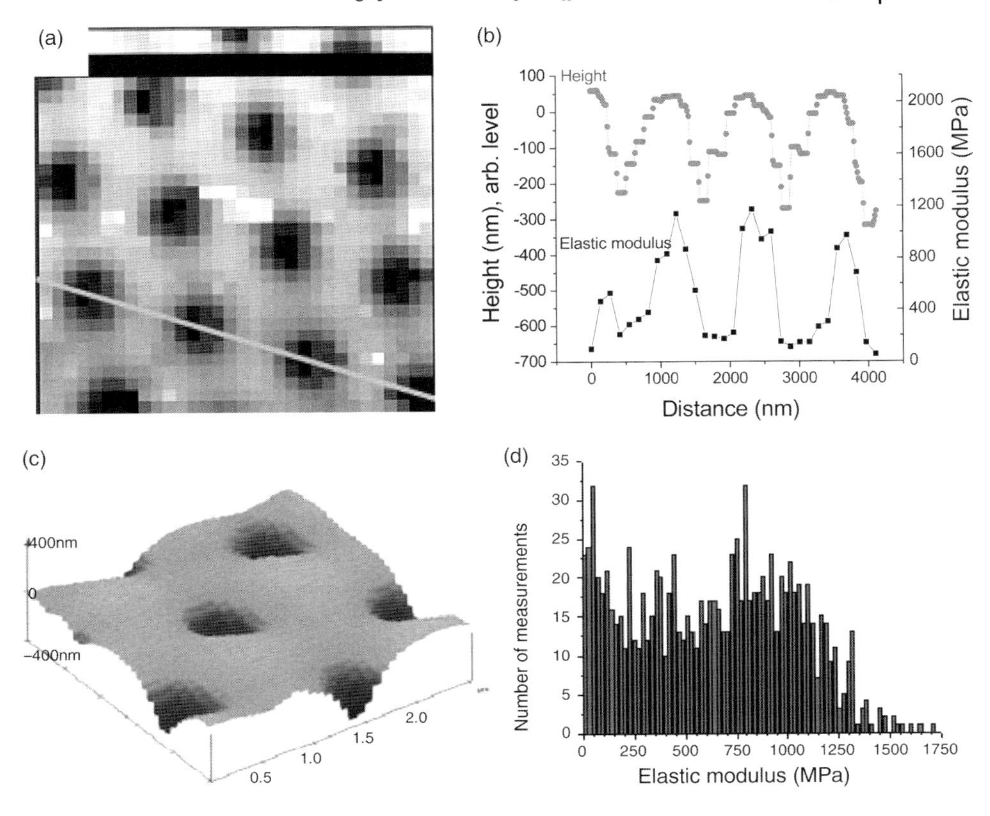

Figure 5.7 (a) Top view $4 \times 4\,mm^2$ AFM topography collected in force–volume mode with 32×32 resolution and (b) corresponding height (compiled from topographical image of this area obtained with higher resolution, left axis) and elastic modulus (right axis) cross sections along the direction (gray line in (a)) [11]. Dark line in (a) is caused by probing instabilities. (c) 3D topography of the surface areas mapped with 32×32 force spectroscopy ($2.5 \times 2.5\,mm^2$). (d) The surface distribution of the elastic modulus obtained from (c) (1024 data points). Reprinted from Ref. [59].

photopolymerization. On the other hand, macroscopic deformation measurements (tensile test and peel test) were performed to reveal the ductile failure and necking of the nanoscale struts that can be related to the reduction of the elastic modulus along the weakest elements (struts) [61, 62].

The surface distribution of the adhesive forces and elastic moduli for heterogeneous glassy–rubbery polymer films with dispersed morphology has been probed [63]. Polystyrene–polybutadiene (PS–PB) thin films probed in this study represent a classical example of highly phase-separated dispersive rubbery phases included in a glassy matrix. Micromapping has been conducted in the range of temperatures that cover the PS glass transition and flow temperatures of both phases. It has been demonstrated that for these heterogeneous films fabricated from polymer

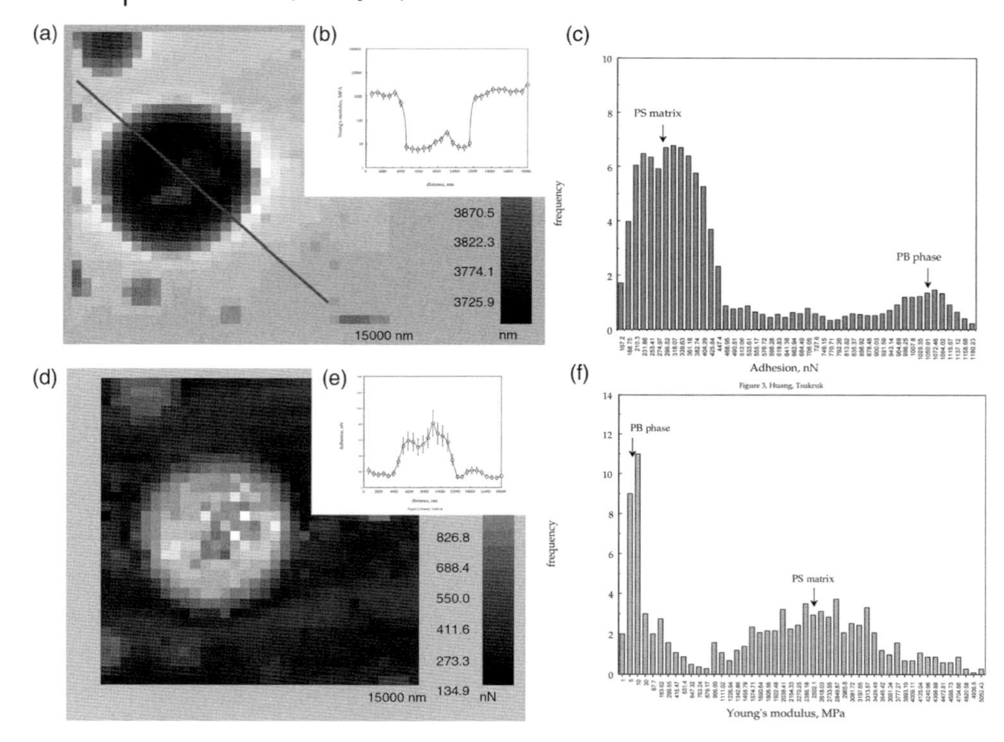

Figure 5.8 Rubber microphase inclusion in a glassy polymer matrix for PS–PB blend with examples of force mapping of topography (a) and modulus (d), corresponding cross-sections (b, e), and surface histograms of adhesion (c) and modulus (f). Reprinted from Ref. [63].

blends, the micromapping of surface properties can be obtained concurrently for glassy and rubbery phases as well as across the interface with a lateral resolution of better than 100 nm (Figure 5.8).

Histograms of the surface distribution display two very distinctive maxima for both adhesive forces and elastic moduli, which allows the concurrent measurements of micromechanical properties of glassy and rubbery phases (Figure 5.8). The elastic moduli of about 2.6 GPa for the PS matrix and about 5 MPa for the rubbery phase have been measured during *a single* micromapping session. Such concurrent probing demonstrates a unique ability of SFS measurements with properly optimized parameters to probe different phases with the difference in the elasticity of more than two orders of magnitude. On the other hand, adhesion was detected to be four times higher for the rubbery phase due to the polar nature of PB material and much higher contact area during deformation to larger depths. Glass transition temperature of the glassy matrix and the flow temperature of the rubbery phase were detected by measuring the elastic moduli over a range of temperature (see more discussion in Chapter 6 and Figure 6.2).

5.2.2
Ultrathin Polymer Films from Different Polymers

A number of different ultrathin polymer films and films with fine nanoscale morphologies have been probed with SFS and data have been analyzed taking into account component distribution in lateral and vertical directions.

The nanomechanical behavior of molecularly thick (<10 nm) compliant polymeric layers with the nanodomain microstructure from the grafted block copolymer, poly [styrene-*b*-(ethylene-*co*-butylene)-*b*-styrene] (SEBS or Kraton), was probed with micromechanical surface analysis [64, 65]. The micromapping with high lateral resolution revealed the bimodal character of the nanomechanical response caused by the different elastic moduli of the rubbery matrix and the glassy nanodomains (Figure 5.9).

High-resolution SFS probing showed a virtually constant elastic response for the compliant film compressed up to 60% of its initial thickness followed by a sharp increase of the resistance when the tip reached within 3 nm from

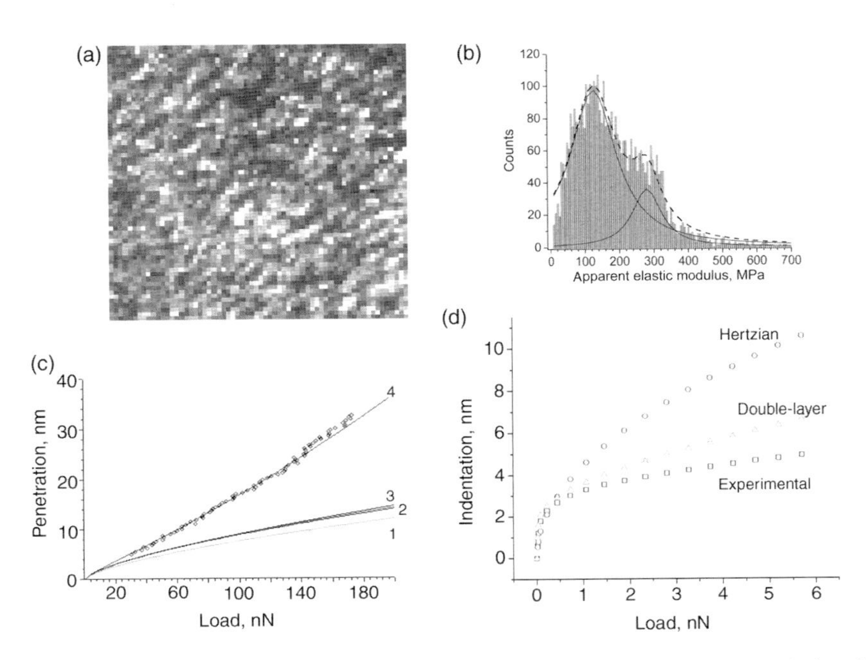

Figure 5.9 Micromapping of tethered SEBS layer, $1 \times 1\,\mu m^2$ (a) and the corresponding histogram of elastic modulus distribution (b). Penetration–load data (c) for bulk SEBS material (squares) and simulations (solid lines) – 1: Hertzian model; 2 and 3: a viscoelastic model with 50 and 100 ms probing time, 4: the viscoelastic model with the best fit to the experimental data for SEBS bulk material. An example of fitting of the loading curve for PI layer on a solid substrate with Hertzian and double-layer models (d). Reprinted from Refs [18, 64].

a stiff solid substrate making the conventional Hertzian model unacceptable and providing an example for the application of double-layer model (Figure 5.9).

Indeed, an application of the double-layer model allowed an excellent fit with experimental data and the estimation of the actual elastic moduli of different nanophases within the grafted polymer monolayer: 7 ± 3 MPa for the rubbery phase and 20 ± 7 MPa for the glassy domains. The relatively high elastic modulus of the rubbery matrix is caused by a combination of chemical cross-linking/branching and spatial confinement within a $<2R_g$ layer. On the other hand, the observed low modulus of the glassy nanodomains can be attributed to both the low molar weight of PS segments and the presence of rubber layers in the probed volume. Nanoprobing of the bulk SEBS material at various probe velocities shows velocity dependence of the measured elastic modulus due to the viscoelastic phenomenon that can be fit with the Johnson model (described in detail in Section 5.4) with different relaxation times (Figure 5.9) [64].

Careful analysis of nanoindentation experiments of several self-assembled monolayers (SAMs) was conducted by Burns *et al.* [66] within the framework of the JKR theory of elastic contact that implies purely elastic response and modest deformation and takes into account the adhesive forces. They demonstrated that this theory can be applied to a nanometer-size contact area. Elastic moduli for fluoroalkylsilane and alkylthiol SAMs were determined to be 13 and 8 MPa, respectively. Shear strength was found to be close to 13 MPa for both monolayers, which is higher than expected. Similar results were obtained by Aime and Gauthier for SAMs with shorter chain length [67]. Elastic modulus of about 20 MPa was found for crystalline monolayers and 7–10 MPa for compressed liquid monolayers [68]. These values were found for the "second" stage deformation process of indentations larger than 0.4 nm. Very gentle indentation resulted in a higher elastic modulus (>200 MPa), close to macroscopically measured values.

SFS measurements can also be very useful to probe phase transitions by performing measurements with varying probing frequencies and sample temperatures, as demonstrated in several cases of polymer brush layers [69–72]. A binary polymer brush layer was prepared by Lemieux *et al.* from rubbery poly(methyl acrylate) (PMA) and glassy poly(styrene-*co*-2,3,4,5,6-pentafluorostyrene) (PSF), both with high molecular weight, grafted on a silicon wafer by the "grafting-from" approach [72]. An example of the force–distance curves for the same binary polymer brush (PSF/PMA) placed in a good solvent for PMA (acetone) is presented in Figure 5.10 [34].

Under this solvent, polymer chains were highly swollen and possessed a very low elastic modulus as expected for polymers in a good solvent. Micromapping with very low normal forces (the spring constant of the cantilever was about 0.06 N/m) generated an array of force–distance curves with a complex shape showing three regions with different slopes. Accordingly, the loading curve with the indentation depth reaching 250 nm (the total thickness of the layer was about 300 nm) displayed a complex shape, which deviated significantly from the normal Hertzian shape expected for a uniform elastic material (Figure 5.10). Individual force–distance curves had similar shapes with a higher level of noise removed by the averaging of a significant number of experimental curves.

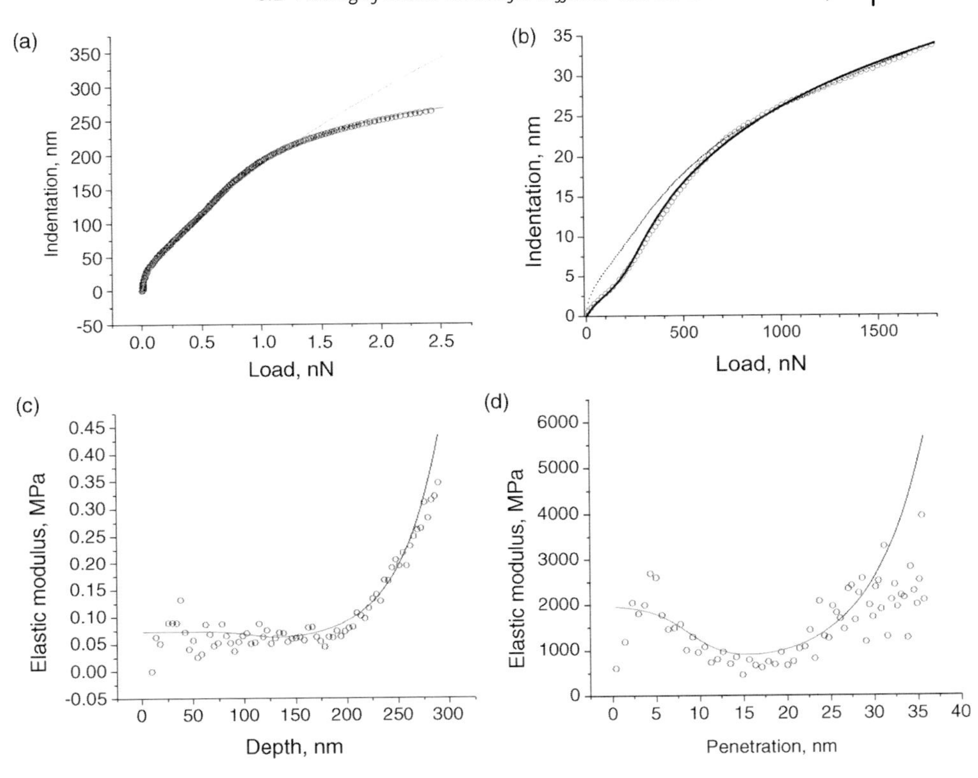

Figure 5.10 The experimental loading curve (circles), fitting with the double-layer model (solid line, almost completely buried by experimental data points) and Hertzian model (dotted line) (a); depth distribution of the elastic modulus (circles) and the best fit (solid line) for polymer brush in good solvent (c). The experimental loading curve (circles), fitting with the trilayer model (solid line, almost completely buried by experimental data points) and Hertzian model (dotted line) (b); depth distribution of the elastic modulus (circles) and the best fit (solid line) for sandwiched polymer coating (d). Reprinted from Refs [18, 74].

The best fitting of the loading curve and the depth profile can be obtained by using a very low elastic modulus value of 0.07 MPa for initial deformations not exceeding 200 nm (Figure 5.10). This very compliant region is replaced with rising elastic resistance for larger indentation depth caused by the presence of the underlying solid substrate. Low values of the elastic modulus and the large thickness of the swollen polymer layer obtained from nanomechanical testing were close to that known for these systems from independent measurements of homobrush surface layers without a second component.

In another more complex example, a polymer "sandwich" system was prepared by grafting the rubber polymer interlayer of 10 nm thickness to functionalized SAM on a silicon wafer and capping this interlayer with a photopolymerized stiff polymer layer with a thickness of 10–30 nm (Figure 5.10) [73, 74]. This system corresponds to a complex trilayer model with the elastic modulus changing from 2 GPa for the topmost layer to 5–10 MPa for the rubbery interlayer and to 1 GPa for the underlying organic layer on a silicon substrate as was independently measured for these materials. The elastic character of deformation was tested by zooming out the surface area probed and observing the absence of the indentation marks.

In fact, the force–distance curves for these sandwiched coatings demonstrated a nonmonotonic character with three different local slopes. This nonmonotonic character is visible on the loading curve, which showed a pronounced S-shape (Figure 5.10) [73]. Attempts to fit the experimental data with the Hertzian model failed – significant deviations were observed in the range of either low or high deformations depending upon the selection of the elastic modulus value. However, the experimental data could be fit with a trilayer model with a modest gradient in the transition zones, as demonstrated in Figure 5.10. The best fit was achieved with the trilayer model composed of the topmost stiff layer of 5 nm thick, a central interlayer of 20 nm thick, an apparent elastic modulus of 800 MPa, and the solid substrate with an elastic modulus of 160 GPa (Figure 5.10). The ultimate indentation of the trilayer film was about 35 nm, which was close to the total thickness of the trilayer film and indicated virtually complete compression under very high mechanical load. The thickness of the transition zone did not exceed 10 nm, indicating a modest gradient distribution between layers within the sandwiched coating.

The micromapping of topography and the elastic modulus represents a powerful tool to identify domain structure of Y-shaped amphiphilic brushes combining two dissimilar hydrophobic and hydrophilic polymer chains (polystyrene and poly-(acrylic acid) (PAA)) attached to a single focal point and grafted to a silicon surface (Figure 5.11) [75]. These molecules grafted to a solid substrate formed uniform nanoscale layers with thickness determined by the length of the arms. SFS probing revealed common loading behavior that can be described by the double-layer model for layers with different thicknesses. However, the most interesting observation is the fivefold increase in stiffness between Y1 (4.6 MPa) and Y2 (22 MPa) brushes (Figures 5.11, 5.12). The decreased compliance of Y2 brush layer can be attributed to shorter arms and higher grafting density resulting in larger space constraints for arms compressed by the AFM tip. These findings clearly demonstrate that even a subtle variation in the chemical composition has a profound impact on the mechanical and surface properties of switchable Y-shaped brushes in fluids.

The topography and elastic modulus distribution maps in Figure 5.12 result from taking 1024 FDCs over a $300 \times 300\ nm^2$ area directly in solvent with lateral resolution below 10 nm [76]. The analysis of the FDCs for Y1 and Y2 brushes in water using a Hertzian model to extract the exact modulus resulted in bimodal histograms of surface distributions (Figure 5.12). The bimodal distribution obtained in water shows one peak at around 5 MPa, coming from the swollen PAA, and another peak at around 110 MPa, which is the apparent elastic modulus of the collapsed PS domains (Figure 5.12). Modulus values from the Gaussian fits are varied ranging from 4 to 100 MPa where the individual moduli from the "crater pit" (PS) and the "rim" (PAA) were distinguished for the water state. Close investigation of the topography and modulus maps allows one to correlate the low points in the topography (PS pits) with

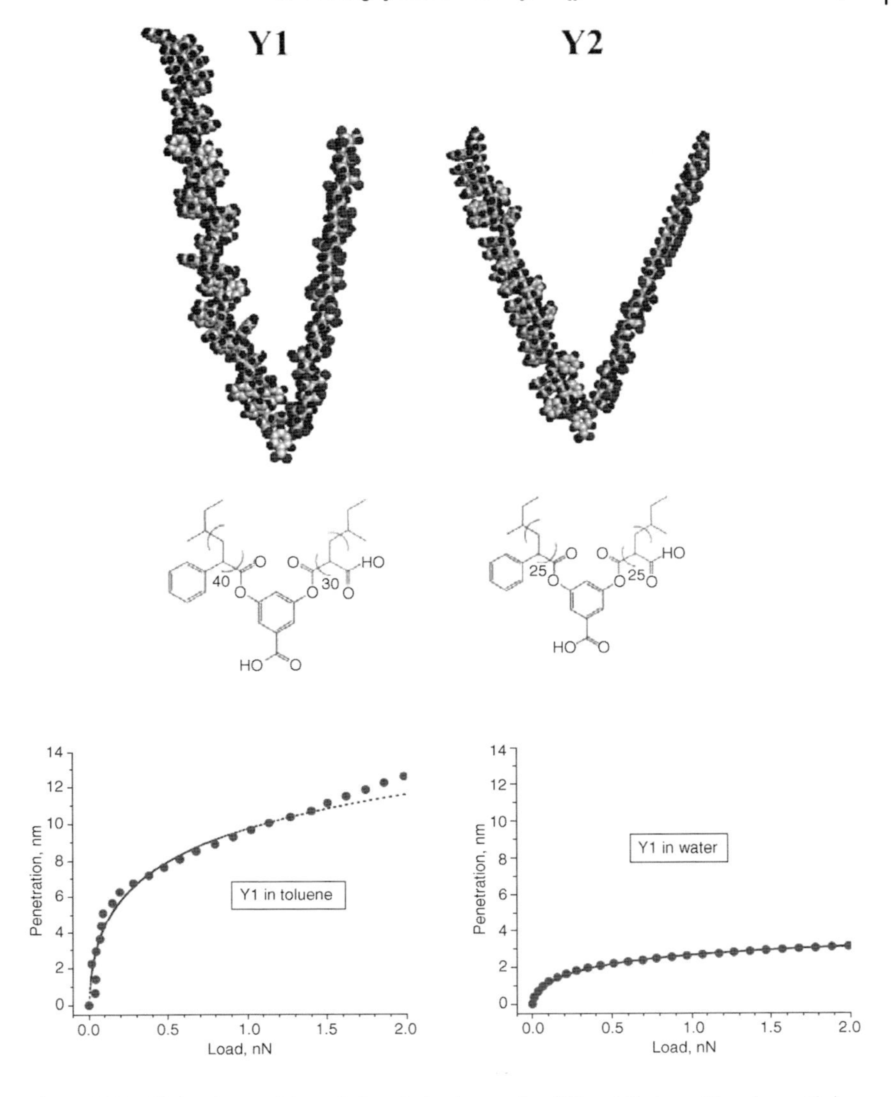

Figure 5.11 Molecular models and chemical schemes for different Y-shaped brushes with long and short arms and corresponding loading curves for long-chain brush in selective solvents (solid line is double-model fit). Reprinted from Refs [75, 76].

high points in the elastic modulus. Likewise, the high points in topography (PAA) match very closely to the low elastic modulus values in the modulus map (see circles in Figure 5.12). This force mapping confirms the presence of both arms on the surface in the water state, which leads to the higher surface roughness.

On the other hand, the surface distribution of the elastic modulus for polymer brushes in toluene is unimodal (standard deviation within 20%) with much lower values of the average apparent elastic modulus below 5 MPa (Figure 5.12).

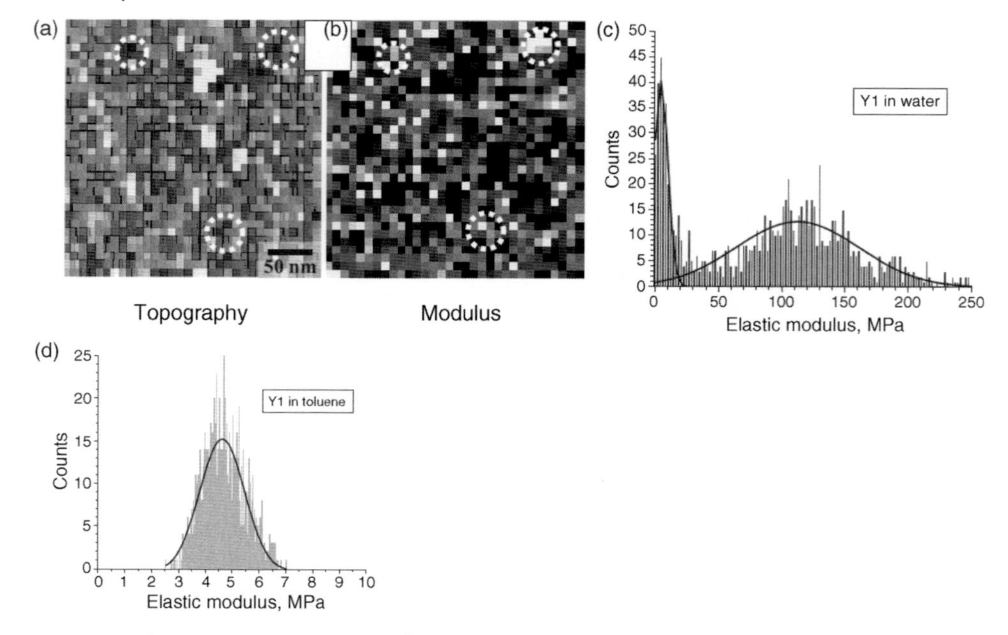

Figure 5.12 Micromapping of elastic modulus (a) and adhesive forces (b) of Y1 brush in water. Corresponding bimodal distribution of elastic modulus in water (c) and in toluene (d). Reprinted from Ref. [76].

The homogeneous character of brush layers swollen in toluene points out that the PS chains completely screen the PAA arms and form a continuous smooth surface layer over PAA chains.

5.2.3
Probing Individual Macromolecules

The elastic properties of dendritic (hyperbranched) molecules with a diameter below 3 nm have been probed with SFS allowing the micromapping of the surface stiffness with nanoscale resolution [77, 78]. To anchor dendritic molecules to hydroxyl terminal groups and reduce tip–molecule interactions, a modification of the silicon surface with an amine-terminated SAM and AFM tips with methyl-terminated SAMs was used in this study. The nanomechanical response was analyzed in terms of sequential deformation of dendritic molecules and the underlying SAM.

Figure 5.13 demonstrates a high-resolution topographical image of dendritic molecules tethered to the functionalized silicon as individual molecules or aggregates for a wide range of adsorption conditions. The number of molecules within aggregates evaluated from molecular dimensions after correction for tip dilation varied from 3 to 40 with the coexistence of individual molecules, small round

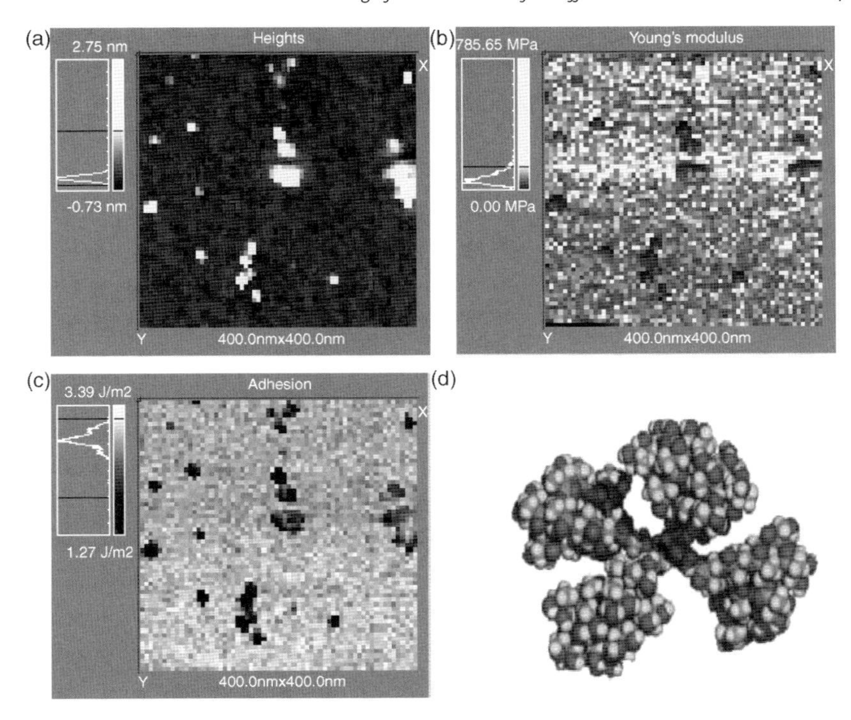

Figure 5.13 SFS micromapping of hyperbranched molecules (d), 64 × 64 array, 400 × 400 nm^2 area: topography (a) and concurrently obtained surface distribution of adhesive forces (b) and elastic modulus (c). Reprinted from Ref. [77].

aggregates, and short- and long-chain aggregates. Such surface morphologies correspond to that observed before for regular dendrimers with different levels of flattened conformations for lower- and higher-generation dendrimers and under conditions of variable molecule–surface interactions [79, 80].

The surface distribution of the nanomechanical properties was probed for randomly selected surface areas containing at least several clusters of different dimensions (Figures 5.13) [77]. Despite some random deviations due to a noise contribution, a clear correlation can be seen between locations and shapes of molecular clusters on the high-resolution AFM image and their corresponding topographical, adhesion, and elastic modulus images on SFS micromaps. As expected, the stiffness of the aggregates of dendritic molecules and individual molecules was much lower than for the surrounding SAM-terminated silicon surface. Correspondingly, adhesion is higher for the amine-terminated SAM than for the hydroxyl-terminated dendritic molecules due to stronger amine-hydroxyl interactions.

Force–distance data for the dendritic molecules clearly showed different compression behavior under low and high normal loads. An initial jump-in contact

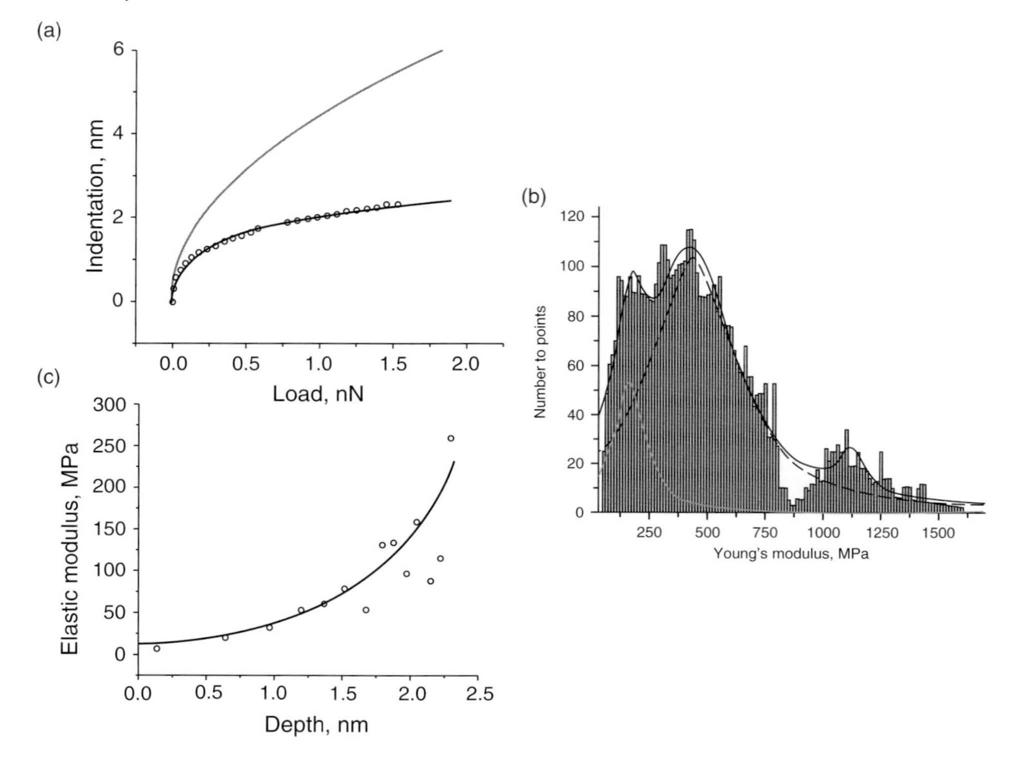

Figure 5.14 Fitting the experimental loading data and the depth distribution of the elastic moduli of hyperbranched molecules (circles) with the two-layer model (solid line) (a and c). Corresponding histogram of surface distribution of the elastic modulus with complex distribution of elastic responses (b). Reprinted from Ref. [77].

usually did not exceed 0.5 nm and, thus, did not significantly affect the deformation of the molecules. Immediately after a jump-in contact, a modest deformation was observed for the first 2–4 nm. The FDC slope changed dramatically (several times) under further compression. Correspondingly, the loading curve showed two regions of elastic response with different slopes instead of a monotonic variation expected for the elastic deformation of a homogeneous solid (Figure 5.14). Large initial deformation of 1.5 nm was detected for low loads below 1 nN. This deformation was followed by a much stiffer response with very minor, if any, deformation of the contacting surfaces at higher normal loads. All changes were completely reversible and were consistent for different surface areas and micromapping parameters. The absence of the residual plastic deformation was tested with high-resolution scanning of the surface areas immediately after the probing.

This kind of nanomechanical response is characteristic of a very compliant and thin layer on a stiff solid substrate and has been analyzed using the double-layer model. Indeed, an elastic modulus of around 20 MPa is observed for the initial 1 nm deformation (Figure 5.14). This is followed with a sharp rise in stiffness for the indentation depth higher than 1 nm caused by deformation of the SAM layer with alkyl chains with a modulus of 1–5 GPa.

Pixel-by-pixel analysis of the surface distribution of the elastic modulus clearly demonstrates the presence of three distinct levels of the surface stiffness for the dendritic molecules and their aggregates tethered to the silicon surface (Figure 5.14). These three levels corresponded to dendritic molecules, border dendritic molecules, and substrate deformations. The high elastic modulus of 1.1 GPa corresponds to the SAM-modified silicon without dendritic molecules. A broad maximum at lower values of the elastic modulus is composed of two peaks. As was identified by analyzing spatial correlation between different histogram regions and surface areas, these peaks correspond to the central areas of molecular aggregates and the aggregate borders. The absolute values of the elastic modulus for the dendritic molecules within internal aggregate regions averaged over the whole deformation range showed the apparent value of 100 MPa.

5.3 Adhesion Measurements

Quantifying adhesion forces from SFS measurements can provide information regarding the intermolecular interactions of the topmost layer and surface energies. Usually, precise measurements aimed at understanding specific chemical interactions involve the use of chemically modified tips in a mode of so-called chemical force microscopy, as was discussed in Chapter 4. SFS is capable of providing rich information regarding chemical interactions with lateral resolution down to a single molecular group. A full discussion of chemical force spectroscopy and other associated techniques is beyond the scope of this book, and readers are referred to elsewhere for more information [81–85].

Adhesion force data are collected in the retraction portion of the force curve at the snap-from region and pull-off forces are usually considered to be representative of true interactions, although some issues relevant to the spring instability in tip behavior should be considered, as has been discussed in numerous papers mentioned above. The adhesion data obtained are a combination of several forces, including contributions of capillary forces, electrostatic forces, van der Waals interactions, hydrogen bonding, ionic bonding, and covalent bonding. Capillary forces tend to dominate adhesive measurements with SFS in ambient air, which are commonly on the order of 1–100 nN [84, 86–88], and can be used to gauge hydrophobic/hydrophilic interactions and thereby provide physical contrast during AFM imaging. However, most adhesion measurements are aimed at obtaining chemical interaction information, where capillary forces are considered an interference. Therefore, in order to avoid the overwhelming contribution of capillary forces, SFS measurements are performed either in vacuum, under dry conditions (dry nitrogen), or most commonly for specimens immersed in liquid.

Monitoring adhesion forces between functionalized AFM tips and polymer surfaces with solution pH is done through the force titration measurements introduced by Vezenov *et al.* [89]. The force titration approach has the ability to map the surface energies on the nanoscale and associate any energetic contrast with nanoscale features [90]. It should be mentioned that in order to probe unknown surface pK values, it is important to use tips functionalized with hydrophilic groups that are incapable of changing ionization with pH. Force titration measurements demonstrated the important role that the surrounding fluid plays in localized adhesion measurements with SFS.

SFS measurements for specimens immersed in aqueous environments have led to more specialized measurements, namely, measuring the pK values of surfaces, which can be important for investigating surface confinement effects of the ionizability of functional groups on the surface [91]. In fact, there are significant differences in the dissociation constants from macromolecular surfaces compared to monomers in solution that can be attributed to a variety of factors, including a decreased available number of degrees of freedom associated with bonding/immobilization, the effect of the dielectric permittivity from adjacent functional groups, and the electrostatic free energy of the substrate. These differences are usually measured by quantifying the surface energy through contact angle measurements taken at different pHs.

To date, several examples of this type of modified tips with $-CH_3$, $-COOH$, $-CH_2OH$, $-CO_2CH_3$, $-CH_2Br$, and $-NH_2$ surface groups have been demonstrated [81, 91–93]. In a study by Tsukruk and Bliznyuk, analysis of surfaces with terminal groups of $-CH_3$, $-NH_2$, and $-SO_3H$ was obtained by direct chemisorption of silane-based compounds on silicon/silicon nitride surfaces (see several examples in Figure 5.15) [92]. The surface properties of the resulting SAMs in air and aqueous solutions with different pHs have also been studied in this work.

Work of adhesion, "residual forces," and friction coefficients were obtained for four different types of modified tips and surfaces, as summarized in Figure 5.16. Work of adhesion for different modified surfaces correlated with changes of solid–liquid surface energy estimated from macroscopic contact angle measurements with maximum forces observed for intermediate pH. Friction properties varied with pH and adhesive forces showing a broad maximum at intermediate pH values for a silicon nitride–silicon nitride mating pair.

These results indicated that the adhesive behavior of the modified SPM tips was controlled by the nature of the surface terminal groups and was subject to dramatic changes in the vicinity of isoelectric points (Figure 5.16) [92]. The results obtained stressed the role of electrostatic interactions between ionizable terminal groups, which are controlled by the ionic strength and pH of solution. Friction behavior followed closely the variation of the adhesive properties and can be used to identify different microphases on multicomponent surfaces with a nanoscale resolution. Similar broad maxima were observed in the acid range for different functionalized SAMs. This behavior can be understood considering changes of a surface charge state determined by the zwitterionic nature of silicon nitride surfaces with multiple isoelectric points, as suggested in Figure 5.16.

Although the contact area of these SFS measurements is usually considered to be close to 10–30 nm^2, the interaction between *individual* functional groups and

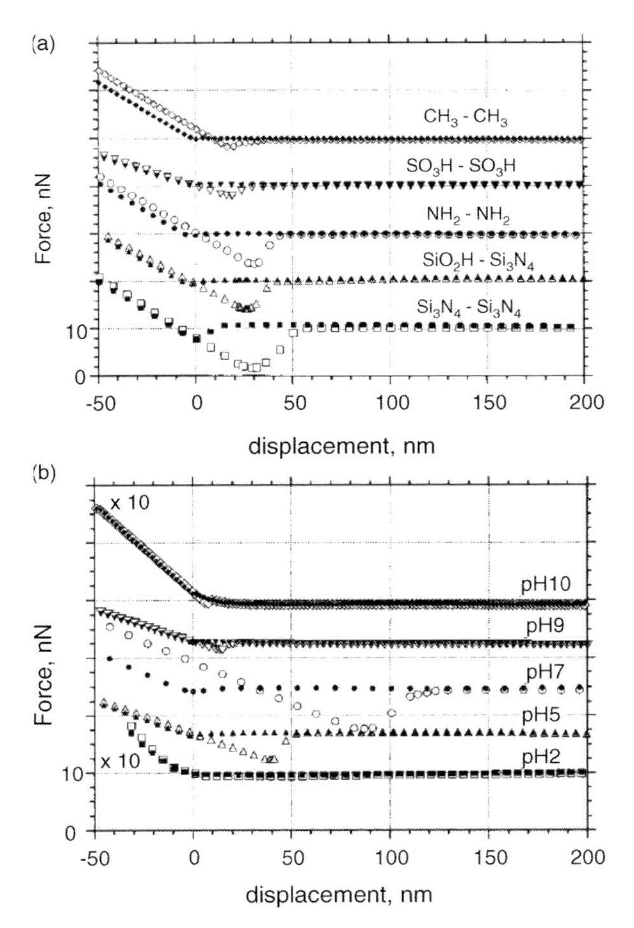

Figure 5.15 SFS measurements for modified different tips and surfaces (a) and at amine-terminated surfaces at different pH (b). Reprinted from Ref. [92].

individual groups on carbon nanotubes has been demonstrated by Friddle *et al.* [94]. These measurements in a fluidic environment achieved single molecular group interaction conditions for a variety of important functional groups. Intermolecular interactions obtained in this study were shown to follow theoretical predictions of interactions based upon the variation of the electronic state of interacting groups.

Feldman *et al.* [95] fabricated oligoethyleneglycol-terminated thiol-based SAMs to add protein resistance to gold and silver surfaces. A functionalized AFM tip in aqueous solution was used to mimic interactions between fibrinogen and SAM surfaces. It was observed that repulsive forces for this SAM deposited on gold were related to the electrostatic contribution. Switching to a silver surface resulted in long-range attractive hydrophobic interactions. The observed differences were attributed to the change of local conformation of glycol tails from helical to extended state.

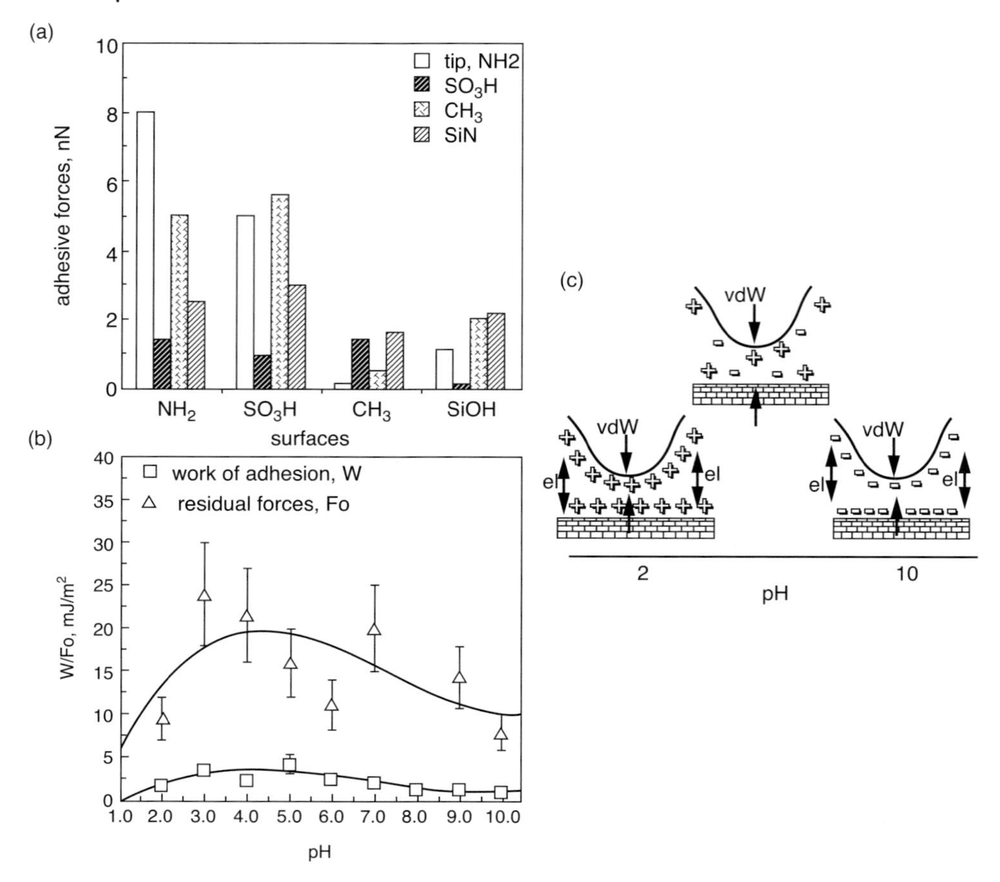

Figure 5.16 (a) Adhesive forces between different combinations of modified surfaces. (b) Titration curve derived from both SFS and FFM data for amine surfaces. (c) Overall schematics of electrostatic distribution for different pHs. Reprinted from Ref. [92].

The role of the polymer microstructure in the adhesive properties of an important piezoelectric polymer, poly-vinylidene fluorides (PVDF), has been studied by Jee *et al.* [96]. Poling and plasma treatment of PVDF resulted in changes of crystal microstructures from the initial α-phase to mixed β- and γ-phases, which in fact affected both adhesion and friction. The authors observed that increasing β-phase content results in decreasing adhesive and friction forces due to changes in electrostatic interactions.

The role of the microstructure in adhesion has been demonstrated for silk materials by Gupta *et al.* [97]. Various microscopic patterns of alternating silk I and II regions were produced, including 10 μm periodicity line patterns and a 3 μm periodicity square pattern (Figure 5.17). Swelling of the native silk I film and rearrangement of the protein chains to adapt to a β-sheet conformation during crystallization contributed to the increased roughness of the transformed regions,

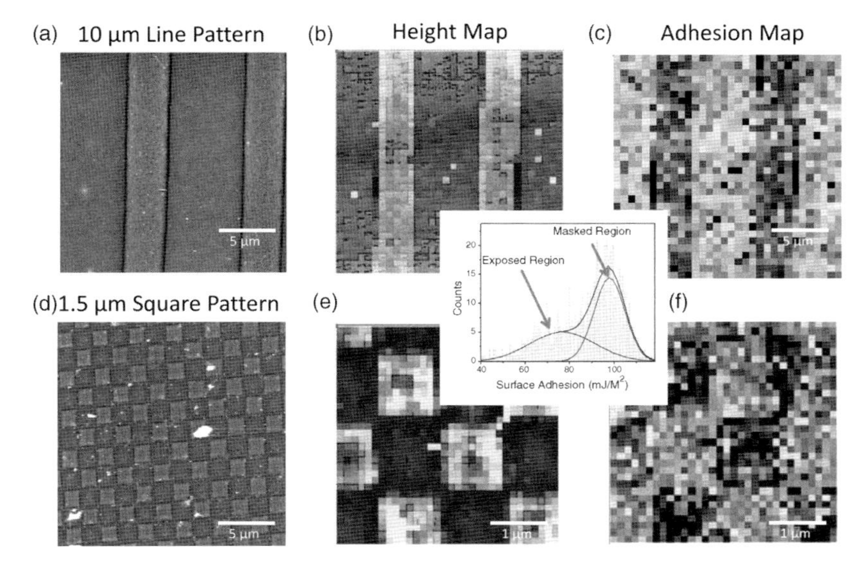

Figure 5.17 AFM topography (a and d), SFS topography (b and e), and SFS adhesion (c and f) of silk film patterned with 10 μm line pattern: 7 μm silk I and 3 μm silk II (a–c) and silk film patterned with 3 μm checkerboard pattern (d–f). *Inset*: Surface adhesion histogram showing bimodal distribution of adhesion forces. Reprinted from Ref. [97].

resulting in a characteristic grainy structure with dimensions of individual crystallites of about 50 nm common for partially crystalline silk material. The AFM topographical image shows a 10 μm line pattern with 7 μm wide silk I and 3 μm wide silk II regions with some difference in elevations. In addition, an AFM topographical image of a 3 μm periodic square pattern demonstrating similar differences in elevations of silk I and II regions each 1.5 μm wide was produced (Figure 5.17). The surface morphologies of the silk I and II regions correspond very well to the morphologies of the exposed region before and after the methanol exposure, respectively.

The SFS mapping clearly shows that silk I regions have a higher surface adhesion (pull-off forces) than silk II regions (see examples for both line and checkerboard patterns in Figure 5.17). Such a patterned response results in an overall bimodal histogram of surface distribution with a ratio of two broad contributions proportional to the fractional surface areas. The authors suggest that the difference in surface adhesion between the silk I and II regions is caused by capillary forces between the film surface and the hydrophilic silicon probe since these measurements were conducted at ambient condition. Since the silk fibroin was dissolved in water, hydrophobic interactions will force segregation of hydrophobic amino acids to the protein core. As a result, the silk I films are likely to have surfaces dominated by hydrophilic residues with a local increase in capillary-driven adhesion.

Recalling the Y-shaped amphiphilic brushes discussed earlier, the observed conformational rearrangements should affect the mechanical properties of the brushes, and their measurement in the presence of solvent gives a unique opportunity to assess these characteristics as a function of solvent–solute interactions (water or toluene) (Figure 5.11) [76]. SFS measurements show no detectable adhesive forces for both brushes in toluene. This is typical for a surface of hydrophobic PS chains, which dominate the topmost surface layer. In contrast, for Y1 in water, there was a noticeable pull-off force. Typical values of this adhesion force ranged from 200 to 400 pN. This relatively high level of adhesion is expected since the hydrophilic PAA arms extend from the surface in water and can directly interact with the hydrophilic silicon oxide AFM tip. Overall adhesives forces (normalized to the AFM tip radius) in water were twice as large for the Y2 brush layer than that for the Y1 layer. This stronger adhesion with the Y2 brush layer in water confirms the higher effective concentration of the PAA chains on the surface of the Y2 brush layer. The sensitivity of adhesive properties implies that small changes in chemical composition of the grafted Y-shaped molecules control the morphology in the presence of a solvent.

Plasma oxidized PDMS materials have been probed with CFM by Wang *et al.* [98]. Nonmonotonic force titrated data have also been observed with a peak around slightly acidic conditions (pH = 3–6), with shifts caused by the nature of the terminal groups and a balance of not only electrostatic interactions but also hydrophobic interactions and hydrogen bonding capabilities. Direct comparison of surface interactions between different functionalities for modified PDMS materials with different cross-linking densities demonstrated that unlike results on CFM measurements for SAMs on stiff substrates (e.g., silicon), mechanical contributions dominate the adhesive behavior of compliant polymeric materials [99]. A practical relationship has been suggested to estimate energy dissipation of the mechanical energy in the contact area and separate it from surface-related contributions.

Probing adhesive properties of various amorphous polymers as a function of temperature have been studied by Cappella and Stark [100]. Careful analysis of both temperature and time-dependent variations of pull-off forces resulted in the construction of master curves that can be described in terms of Williams–Landel–Ferry (WLF) equation for viscoelastic materials (see below). Moreover, by analyzing the contact behavior in elastic and plastic regimes, the authors concluded that the work of adhesion is always proportional to the contact area for the same material, and in the elastic limit, the work of adhesion does not depend upon the mechanical load. Overall, the authors concluded that the adhesion inherently depends upon both surface elasticity and plasticity and such cross-dependencies cannot be avoided in SFS measurements. Abrupt changes in the pull-off forces in the vicinity of the glass transition were analyzed as a means to determine transition temperature of the topmost polymeric layer in several publications [101, 102]. Even more complex behavior complicated by variable capillary forces has been observed for pressure-sensitive polymeric materials by modulated SFS spectroscopy by Moon and Foster [103].

Layer-by-layer (LbL) polyelectrolyte multilayers have been studied recently with SFS, and it has been demonstrated that adhesion was dominated by electrostatic interactions as governed by the topmost monolayer and the frequently observed long-range interactions were related to the pulling of polyelectrolyte chains strongly attached to the AFM tip [104, 105]. Yu and Ivanisevic [106] demonstrated measurements of the surface morphology and mechanical properties of two types of cells encapsulated in LbL polyelectrolyte shells. An increase in surface roughness was observed with the increasing number of polyelectrolyte layers. Stronger interactions between monolayers accompanied the increasing adhesion. Another example of exploitation of SFS spectroscopy to complex adhesive measurements is the evaluation of the strength of nanoparticle attachment to the modified polymer surfaces, as has been reported by Kokuoz *et al.* [107].

5.4 Viscoelasticity Measurements

Even if a majority of elastomers show pure elastic behavior under normal SFS probing conditions, a significant number of materials such as low molar weight polymers, hydrogels, polar rubbers, swollen polymers, and biological materials demonstrate well-pronounced time-dependent behavior. A few selected examples of the analysis of these data that are important from an experimental viewpoint are presented here with materials-specific results discussed in proper chapters.

The time-dependent contribution can be treated by applying Johnson's recent development [18, 108]. Johnson suggested that the relationship between contact area a, load P, and loading time t for viscoelastic solid contact can be represented in the following form:

$$a^3(\tau) = \frac{3RUT}{4E_\infty^*}[\tau-(1-k)(1-\exp(-\tau))] \tag{5.9}$$

where $\tau = t/T$ is the reduced time, $k = E_\infty^*/E_0^*$ is the reduced modulus with E_0 being the initial, instantaneous modulus and E_∞ being the "equilibrium" relaxed modulus, $U = P/t$ is the rate of loading, R is the tip radius of curvature, and T is the relaxation time of material. This relationship was derived for a three-parameter linear viscoelastic model. By varying two primary variables E_0 and T, one can fit the experimental SFS data to evaluate the elastic modulus and relaxation time.

A significant role of the viscous contribution was recognized in the very beginning of dynamic AFM measurements and continues through the current date. For instance, Friedenberg and Mate analyzed contributions of capillary and viscous forces to thin layers of low molecular weight PDMS with different thicknesses [109]. The authors concluded that viscous forces dominate the dynamic behavior at high frequencies, and the measurements of damping viscous coefficients can be used to determine the viscosity of the PDMS layer.

Gelatin films have been studied by Braithwaite and Luckham and the complex relationship between elastic and viscous responses has been found for different

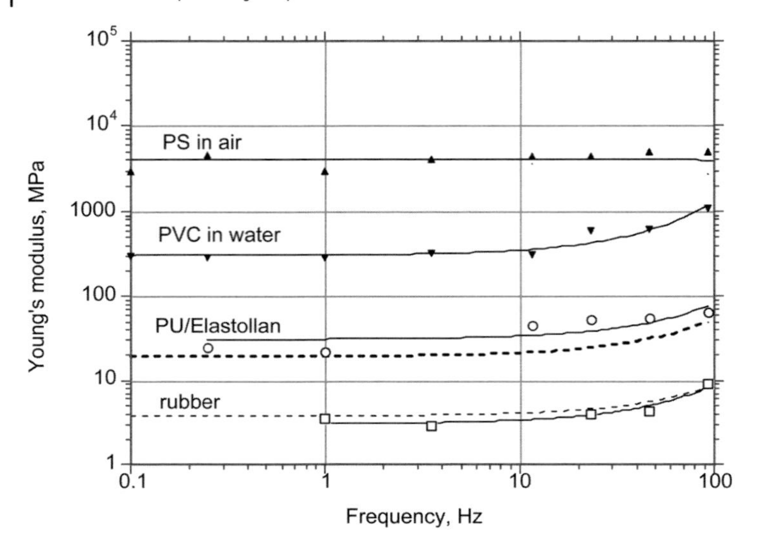

Figure 5.18 Time dependence of elastic modulus for PS, PVC, natural rubber, and Elastollan PU as derived from SFS probing in different solvents and at different frequencies. Dashed curves represent time-dependent elastic moduli calculated from the WLF equation for natural rubber and polyurethane by using known parameters for bulk materials and for PVC assuming lower glass transition temperature. Reprinted from Ref. [72].

separations of tip and substrate [110]. They concluded that careful analysis of these relationships could result in a separate calculation of loss and storage elastic moduli. Multielement spring–dashpot models have been applied to analyze force–distance curves around the glass transition. A balance of elastic, viscoelastic, and capillary contributions for polymeric materials has been discussed by Yang *et al.* [111]. The quasi-linear viscoelastic model was tested by Tripathy and Berger for SFS studies of agarose materials in order to derive complete information on the viscoelastic behavior in a swollen state [112].

In SFS studies of amorphous polymers, glassy polymers (e.g., polyvinylchloride (PVC) and PS) in air showed no time dependence of elastic modulus with probing frequency [113]. However, strong frequency dependence of the elastic modulus was observed for PVC in water and alcohol as Young's modulus increased from 200 MPa at low frequencies to 1 GPa at high frequencies (up to 100 Hz) (Figure 5.18). A similar increase of elastic modulus was observed for elastic rubber and PU materials (Figure 5.18). In this case, Young's modulus rises from 2.9 MPa for rubber and 30 MPa for Elastollan PU to 9 and 70 MPa, respectively. This time-dependent behavior is well known for elastomers and can be described within the WLF approach and by using different viscoelastic models, such as Maxwell or Voight. Indeed, the Maxwell model of viscoelastic behavior predicts exponential damping of the elastic modulus with increasing time that is close to the behavior observed for rubber and PUs (Figure 5.18). As can be concluded from direct comparison of the calculated and measured data, WLF formalism adequately describes the viscoelastic behavior of rubber and polyurethane surface layers with dynamic parameters close to that known for the bulk materials.

Significant reduction of the PVC modulus in water and strong time dependence indicate that micromechanical properties of the PVC surface layer in these solvents differ from the glassy bulk properties of solid materials. These changes can be related to the partial penetration of solvent molecules in the very uppermost surface layer (30–40 nm) without actually swelling the bulk material (surface plastification effect). As a result, a surface layer enriched with low molar mass molecules provides a higher mobility for macromolecules and the local glass transition is shifted downward. By using the WLF equation, we calculated this "apparent" glass transition temperature for the uppermost surface layer. The observed time dependence gives $T_g = 35\,°C$ for the PVC surface layer in water compared to 78 °C for the bulk material. This shift can be expected for PVC containing about 20% of low molar mass plasticizer.

Measuring force–distance curves for poly(*n*-butyl methacrylate) (PnBMA) in a range of temperatures covering both glassy and viscoelastic states allowed Cappella *et al.* to measure both plastic and viscoelastic properties in terms of WLF theory [114]. The authors obtained the parameters of the WLF equation independently and applied these values to verify SFS data. They observed a significant drop in the yield strength and the elastic modulus values as determined from Hertzian analysis in the vicinity of the glass transition temperature and constructed a master curve for a wide range of temperatures and frequencies. The possible significant contribution of material creep to longer contact times and even low mechanical loads was pointed by Moeller for a number of commercial polymeric materials [115]. Moeller applied large-radius AFM tips and multielement spring–dashpot viscoelastic models to separate a creep contribution and an elastic deformation and evaluate the elastic modulus value over a wide range, from 1 MPa to several GPas.

McConney *et al.* probed a signal filtering material under varying frequencies and related the frequency-dependent mechanical properties of the biological material to the viscous signal-filtering ability such as those found in spider hairs and fish flow receptors [116–118]. Particularly, biohydrogel cupulae of several different fish have been studied, which are essential for underwater orientation of these species (Figure 5.19) [118]. The elastic modulus of the biohydrogel cupula in the wet state was measured with colloidal probe microscopy directly in a liquid cell for folded flag-like biocupula (Figure 5.19). Force–distance curves showed significant nonlinearity that indicates time-dependent behavior and strong adhesion (Figure 5.20). Micro-mapping was conducted by collecting 16×16 arrays of force–distance curves over a $500 \times 500\,nm^2$ surface area. The loading curves, the elastic modulus, reduced adhesive forces, and surface histograms of elastic moduli and adhesive forces were obtained from experimental SFS micromapping for biological cupulae.

Similar measurements have been conducted for synthetic PEG-based hydrogels that were introduced to emulate biological material. Loading curves derived from force–distance data also showed highly nonlinear behavior associated with significant time-dependent properties of biological and synthetic hydrogel materials (Figure 5.20) [118]. The analysis of the experimental data has been conducted by utilizing a Voight viscoelastic model and Hertzian contact mechanics. This analysis

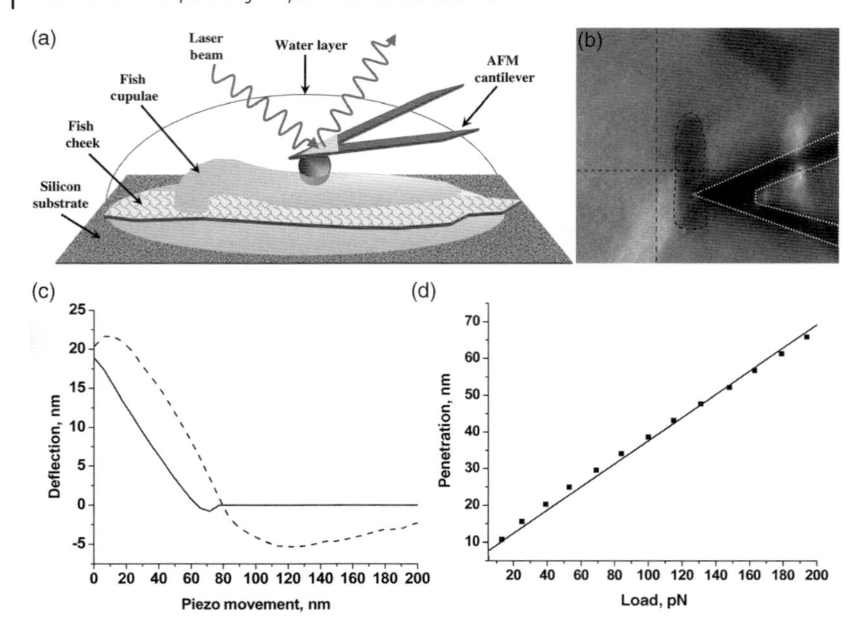

Figure 5.19 Nanomechanical properties of the cupulae of the superficial neuromasts. (a) Schematic of the experimental setup to measure nanomechanical properties of the fish cupulae. AFM cantilever with glass sphere is penetrating the fish cupulae in water. (b) Micrograph of actual AFM setup. AFM cantilever is placed over the stained fish cupula in water. (c) Typical force–distance curve of the fish cupulae. (d) Typical loading curve of the fish cupulae at low loads and higher frequencies. Reprinted from Ref. [118].

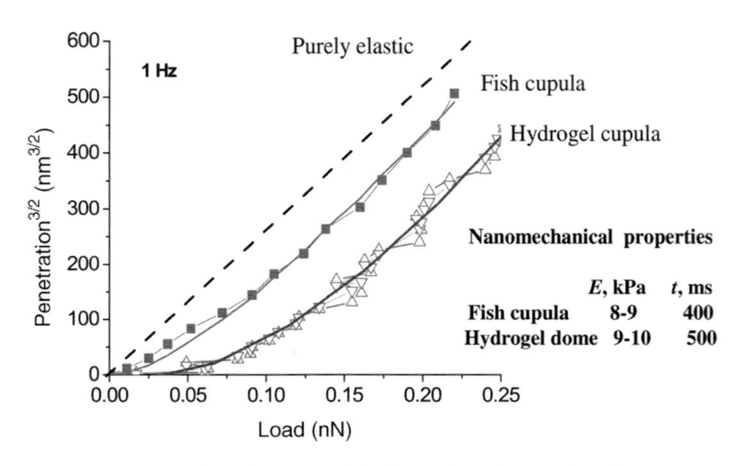

Figure 5.20 Viscoelastic behavior of the bio and synthetic hydrogels and corresponding fits with Voight model (solid lines) and mechanical parameters derived. Dashed line represents pure elastic deformation. All data were collected at scanning frequency of 1 Hz. Reprinted from Refs [118, 123].

yielded an elastic modulus of 8–10 kPa and a relaxation time close to 500 ms for the biological hydrogel. The SFS results for the synthetic hydrogel closely corresponded to that measured for natural fish cupulae. The synthetic hydrogel also showed highly viscously damping behavior similar to that observed for the biohydrogel (Figure 5.20).

Hydrogels with different surface chemistries based upon highly swellable cross-linked copoly(hydroxyethyl methacrylates) (pHEMA) at different humidities have been studied by Koffas *et al.* [119]. The elastic modulus value was estimated from Sneddon's model and viscous strain was evaluated from rate-dependent deformation at different tip velocities. In this comprehensive study, the authors observed that the relaxation times related to the rate of the dissipation of the mechanical energy as governed by the local conformation changes decreased sharply with increasing humidity. This relaxation time reached 200 ms at the highest humidity that is comparable to the loosely cross-linked PEG hydrogels discussed above. In contrast, the level of the elastic response remains unchanged at different humidities. On the other hand, the work of adhesion increased dramatically for higher humidities. This increase was interpreted as the change in overall compliance of the highly swollen hydrogel surface, not by capillary forces. Depletion of the water content on the hydrogel surface was also concluded from experimental data. Decreasing work of adhesion for copolymerized materials was discussed in terms of their role in biocompatibility and retaining properties in a protein-rich environment.

5.5 Friction

The relationship between friction forces F_f and normal load F_n can be derived by running friction loops at different cantilever deflections (see Chapter 4). These data are instrumental in deriving such a universal parameter for the characterization of lubrication properties as friction coefficient μ. The classical definition of the friction coefficient came from a generalized Amonton's law in the following form [120, 121]:

$$F_f = F_0 + \mu F_n \tag{5.10}$$

where F_0 is the adhesive force in the absence of the spring normal load. The friction coefficient for a particular mating pair is considered to be constant under various sliding conditions. As was observed in numerous studies of organic and polymer monolayers, the frictional behavior closely obeys the Amonton's law.

However, under controlled environmental conditions and for compliant organic monolayers, nonlinear loading behavior is frequently observed (Figure 5.21). Thus, the "apparent" friction coefficient derived from such nonlinear data is a variable, load-dependent parameter. Such deviations can be related to different complex variations of the contact areas under increasing mechanical load. Nonlinear behavior of this type can be interpreted within Hertzian and JKR theory, which predicts such variation

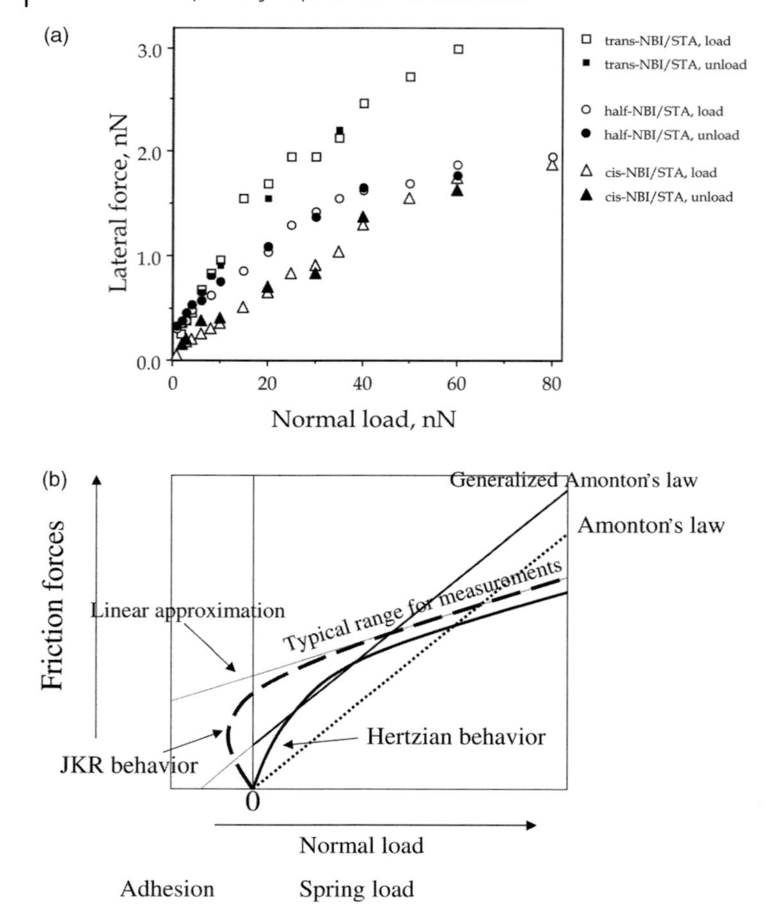

Figure 5.21 Typical friction force versus loading data for different mixed LB layers (a) and general representation of various scenarios for mating contacts (b). Reprinted from Ref. [123].

under conditions of elastic mechanical contact with the scaling relationship [122]:

$$F_{\mathrm{f}} \sim F_{\mathrm{n}}^{2/3} \tag{5.11}$$

Most practical measurements under "normal" environmental conditions include significant contribution of adhesive forces due to the hydrophilic nature of the AFM tip and the presence of a molecularly thick water layer. The downward part of the loading curve obtained at a negative spring load (adhesion or attractive regime) frequently cannot be recorded due to the inherent instability of this regime (Figure 5.21). Very small loads are usually also unavailable for measurements because of scanning instabilities. Therefore, most conventional nanotribological measurements are conducted at relatively high normal loads. Under these conditions, and within the limited range of normal forces measured and with significant data scattering for extremely small lateral and normal forces, the upper part of the loading curve can be reasonably well approximated by a linear function. This, effectively, returns us to the concept of a constant friction coefficient that can be used for nanotribological studies if applied with care.

FFM is very useful for polymeric studies involving reducing material wear, enhancing lubrication, stiction in MEMS, and other nanotribological phenomena [123, 124]. Friction at the nanoscale is still far from being well understood, but FFM has indeed provided an avenue for a deeper understanding of this phenomenon. A noticeable example of the impact FFM has had on the analysis of friction is the work performed by Carpick and coworkers [125]. FFM can be sensitive to chemical information, especially when functionalized tips are used, providing information on shearing behavior of surface layers, friction coefficients, wearing dynamics, and velocity-dependent shearing.

Application of FFM to SAMs, adsorbed molecular layers, and LB monolayers from amphiphilic molecules have been widely exploited to elucidate their morphology and applicability as molecular lubricants with a variety of regular and extremely sharp tips [126]. The very first applications of AFM and FFM techniques to study LB films brought a series of new discoveries and sharply raised the understanding of the behavior of these molecular coatings. For instance, FFM imaging of a heterogeneous, two-phase LB film from amphiphilic stearic acid presented in Figure 5.22 clearly demonstrates extremely low localized friction in selected areas coated with organic monolayer [127]. Concurrently, extremely high friction is observed on bare silicon surface areas on the same image and is associated with the high hydrophilicity of the hydroxyl-terminated solid.

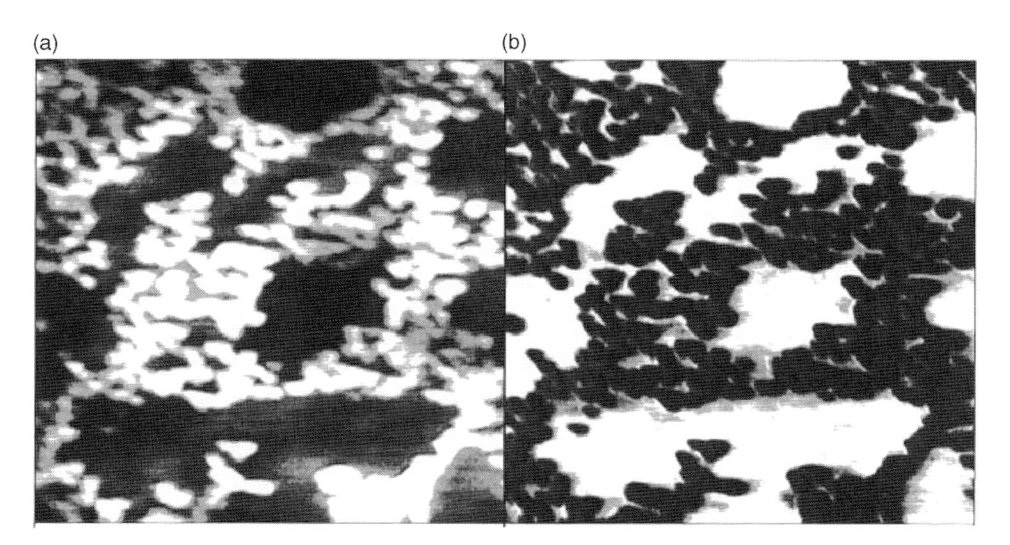

Figure 5.22 Topography (a) and friction (b) for incomplete Langmuir–Blodgett monolayer of stearic acid ($1 \times 1\,\mu m^2$) showing high friction on silicon (bright areas) and very low friction on LB monolayer (dark areas). Reprinted from Ref. [127].

Stick-slip motion is observed for certain viscoelastic layers indicating shear instabilities during transition from the static to the dynamic regime. Liu *et al.* observed that unlike bare solid substrates with a logarithmic relationship between friction forces and sliding velocity, Langmuir monolayers frequently show nonmonotonic behavior [128]. The friction forces increase initially at very low velocities (0.01–0.1 μm/s), reach a maximum at intermediate velocities, and are reduced to the initial level at velocities around 100 μm/s. The position of the maximum depends upon the normal load and is different for various monolayers.

This maximum was related to the initiation of a local phase transition similar to the one observed for liquid layers at surface force apparatus measurements. Similar studies of stearic acid LB monolayers showed that the appearance of the maximum depended upon the physical state of the monolayer and the peak was observed only for monolayers in a "liquid" state [129]. In a solid state, simple logarithmic increase was observed. Also, a monotonic increase of the friction forces with velocity was observed for alkylsilane SAMs in contrast to the sharp peak or constant level observed for composite SAMs. This phenomenon is related to a certain "match" between the rate of energy dissipation and the intralayer mobility.

Polymer brushes of PS:PVP with different compositions and grafting density have been studied with regular AFM, colloidal probe AFM, and modified colloidal probe AFM by Vyas *et al.* [130]. Adhesion and frictions were observed to be related to surface microroughness and chemical composition and can be varied dramatically by changing grafting density and PVP content. The friction coefficients of binary Y-shaped brushes (Figure 5.11) were measured against the sliding glass ball after exposure to toluene and water and they were dramatically affected by the surface composition of the layer [76]. For the Y1 brush layer treated with water, the friction coefficient remains constant with an average value of $\mu = 0.21$. In contrast, the friction of Y1 after toluene treatment shows erratic behavior and is nearly twice as high as that after water exposure. The averaged friction coefficient of Y2 after toluene exposure is $\mu = 0.55$, whereas after water treatment it is only 0.35. It is worth noting that although the absolute values of the friction coefficient for brush layers studied here are higher than the usual values for polymeric and organic layers with optimized morphology (0.1 and below), they are on a par with the friction coefficients found for many bulk glassy and rubbery polymers [72, 131].

A number of bulk polymeric materials have been studied with FFM in order to reveal distinct surface properties, analyze molecular relaxation behavior, and observe time–temperature-dependent mechanical responses. Poly(ethylene terephthalate) (PET) in different solvents showed very different frictional behaviors, both linear and nonlinear, depending upon the dielectric properties of the solvents [132]. Friction studies have been conducted for PMMA at different temperatures in both air and vacuum, and the friction coefficient has been measured below and above the glass transition temperature by Kim *et al.* [133]. Surface relaxation of PMMA has been observed in a wide range of frequencies by Tocha *et al.* [134]. The friction forces measured showed a strong nonlinear behavior with a maximum friction force revealed at intermediate frequencies. This characteristic behavior was associated with different molecular relaxations in polymer materials with much lower corresponding activation energies to be explained by greater free volume in the topmost polymer layer.

Friction, adhesive, and wetting properties of microfabricated and micropatterned PMMA and PS materials have been studied by Jung and Bhushan [135]. The authors observed superhydrophobic properties of these engineered surfaces with low contact angle hysteresis. It has been found that increasing roughness can decrease or increase the contact angle depending upon the wetting properties of materials. The friction coefficient increases with increasing environmental humidity for any surface with microscale friction to be found much higher than that measured at the nanoscale.

Dramatic stick-slip motion has been observed for a number of polymers under specific sliding conditions (loads, velocities, and environment). For instance, Wu-Bavouzet *et al.* observed periodic instabilities for soft PDMS in particular liquids [136]. Stick-slick processes observed have been associated with wet Schallamach waves known for viscoelastic materials. Correspondingly, periodic instabilities observed were explained by the dynamic formation and dissipation of microscopic adhesive bridges between mating pairs at particular velocities and loads.

5.6 Unfolding of Macromolecules

SFS measurements in the pull-off regime provide information about persistence length, number of domains, unfolding forces, molecular spring constants, and total length of the protein backbones under different strain and spatial length conditions [137–140]. Specific molecular binding and unbinding routines under variable load reveal the mechanisms of molecular recognition and site-specific interactions of biomacromolecules [141, 142].

As applied to synthetic macromolecules, unfolding measurements reveal not only specific interactions and conformational states on segmental and molecular scales but also internal mobility, interactions with external analytes, and surface adhesion and friction. Usually, the need for careful statistical analysis is emphasized in all developments due to the fact that the random nature of the tip–surface interaction and rare events of the "proper" arrangement of the chains with end-to-end attachment cause only 1–3% of meaningful force–distance curves to be collected that are appropriate for further analysis.

The stretching is controlled with a precision of around 0.1 nm that can be conducted with different levels of stretching (large and small strains) and rates while monitoring normal forces [143]. The resulting force–displacement curves with characteristic sawtooth shapes and consistent spacing were selected and used for further analysis [144, 145]. The usual criteria of periodicity, force sequence, shape of a single peak, and appropriate spatial distribution are applied for this selection. The assignment of different stretching events is conducted by using several conventional criteria: correspondence of the total stretching distance to the length of backbones calculated for different conformation; correspondence of the length of a single peak to the extended length of different segments; correspondence of the forces associated with a particular stretching event to the expected forces for unfolding, elastic deformation, or chain scission; consistency in the stretching events (random or concentrated around some mean values); quality of fit for known stretching mechanisms of flexible backbones; correspondence of the values of Kuhn segments evaluated from experimental data with expected values for given chains; and consistency of the Kuhn segment values for presumably similar stretching events (to exclude multi-chain contributions) [146].

The model of a freely jointed polymer chain is popular for the analysis of single unfolding events with such important parameters as overall contour length and Kuhn segment length to be unambiguously evaluated. More complicated models such as wormlike chain (WLC) model and others take into account kinetic barriers, multiple unfolding, and steric/torsional constraints. The interpretation of the massive arrays of complex SFS data in terms of simplistic modes of deformation, motion, confinement, or local transformation is a major challenge for the unambiguous understanding of this type of experiment. This is even more challenging considering that the experimental data are frequently collected at limits of resolution and sensitivity (subnanometer displacements and pN forces) and are superimposed by multiple potential artifacts (tip contamination, chain bundling, or chain migration).

Particularly, the role of the adhesive-controlled motion of the polymer chains at the surface during the pulling experiment has been discussed [147]. The authors considered means for the evaluation of the friction parameters and surface viscosity for different scenarios of interaction of molecular chains with the surface. Data analysis has been demonstrated to provide meaningful results for cases of low friction of DNA on mica and intermediate friction of polyallylamine on mica. A strong dependence of the pulling scenarios upon grafting density has been revealed for poly(acrylic acid) brushes grown by the grafting-from method [148]. The role of the adsorption of the polymer chain on the AFM tip that can be tuned by the environmental conditions (e.g., pH) was shown to be critical in brush response to the AFM tip intrusion and retracing.

The path to analyzing the unfolding signature of complex polymeric materials has been shown by Gunari and Walker [149]. The authors analyzed unfolding of PS–PMMA diblock copolymers considering extension profiles for individual components of the block macromolecules. They concluded that differences in the unfolding behavior caused by differences in intramolecular assembly of polymer chains in water can be used to distinguish their contributions in a chemically joined system. They demonstrated that PS chains showed three-regime extension behavior with plateau force and PMMA possessed a characteristic sawtooth pattern typical for multiple unfolding events in internally structured chains. These distinct contributions define characteristic fingerprints of the pulling behavior of PS–PMMA diblocks with different chemical compositions that allows the monitoring of internal complex unfolding of adsorbed block copolymers. In another study, chain stiffness of PNIPAAM macromolecules was related to a local conformational freedom and interactions with solute molecules during unfolding [150].

The most intensively studied silk protein to be unfolded with the SFS is fibroin from the domesticated silkworm *Bombyx mori* [151]. The structural portion of the fibers spun by the silkworm to form its cocoon consists of two proteins, a heavy-chain fibroin (~390 kDa) and light-chain fibroin (~25 kDa), linked by a single disulfide bond. The heavy-chain fibroin is dominated by large stretches of hydrophobic amino acids, particularly glycine–alanine repeats, that form β-sheet crystallites. The chemical composition of the silk protein and possible conformations of this protein and conformational transitions have been recently discussed [152–154].

Although the unfolding stretching behavior for different strains was observed for some silk proteins tested in the solid state z, the internal domain structure of *B. mori* fibroin silk protein has not been revealed [155]. For example, good force statistics were observed for spider dragline silk protein unfolding with a primary unfolding length of 14 nm assigned to specific A_nG_m-type flexible moduli. However, others observed only long-range stretching events in the range of 800 nm to several micrometers that were assigned to either stretching of a whole molecule or the complex stretching of protein aggregates [143, 154]. In all these cases, the interpretation of the SFS data is compromised because of a lack of independent information on exact and complete primary amino acid sequences for the protein backbones.

Considering the complex multidomain *B. mori* silk protein composed of 12 hydrophilic and 12 hydrophobic domains that comprise a range of possible molecular lengths, three independent types of SFS measurements were exploited to include different strategies of stretching with appropriate forces and rates by Shulha *et al.* (Figure 5.23) [155]. First, stretching was conducted with relatively high forces and extensions reaching 3 μm that exceeds the completely extended length of the molecules (1.8 μm). Second, the stretching distance was limited to below 1.5 μm with high resolution in the z-direction to collect data on sequential unfolding events for a "normal" surface layer of protein (Figure 5.23). In this case, detailed sawtooth shape analysis was conducted for the extension range within 1000 nm. Finally, gentle, low-rate, limited stretching with low stretching distances (below 300 nm) was utilized to focus on sequences of weak unfolding events related to the hydrophilic domains.

The largest and completely reversible stretching consistently collected at thicker protein layers resulted in irregular sawtooth peaks with variable heights and spacings indicating stretching of multiple and irregularly attached molecular bundles. For the largest extensions, a sharp increase in overall force occurred, reaching 3–5 nN, which is close to forces sufficient to break protein backbones and covalent bonds. In contrast, probing thin surface protein layers containing globular aggregates while keeping a limited stretching distance generated data with a regular sequence of sawtooth peaks, characteristic of single molecule stretching. Sawtooth shapes were observed with unfolding forces of 200–400 pN, and peak spacing distributed within 20–100 nm. Relaxation curves showed complete reversibility with forces gradually dropping to the initial level.

Finally, very gentle stretching to distances below 300 nm with low rates at the surface areas with globular aggregates produced sequences of regular sawtooth peaks with forces of 50–200 pN and spacing within 10–40 nm (Figure 5.23). The shape was reproducible and described by the WLC model. The shape and parameters were

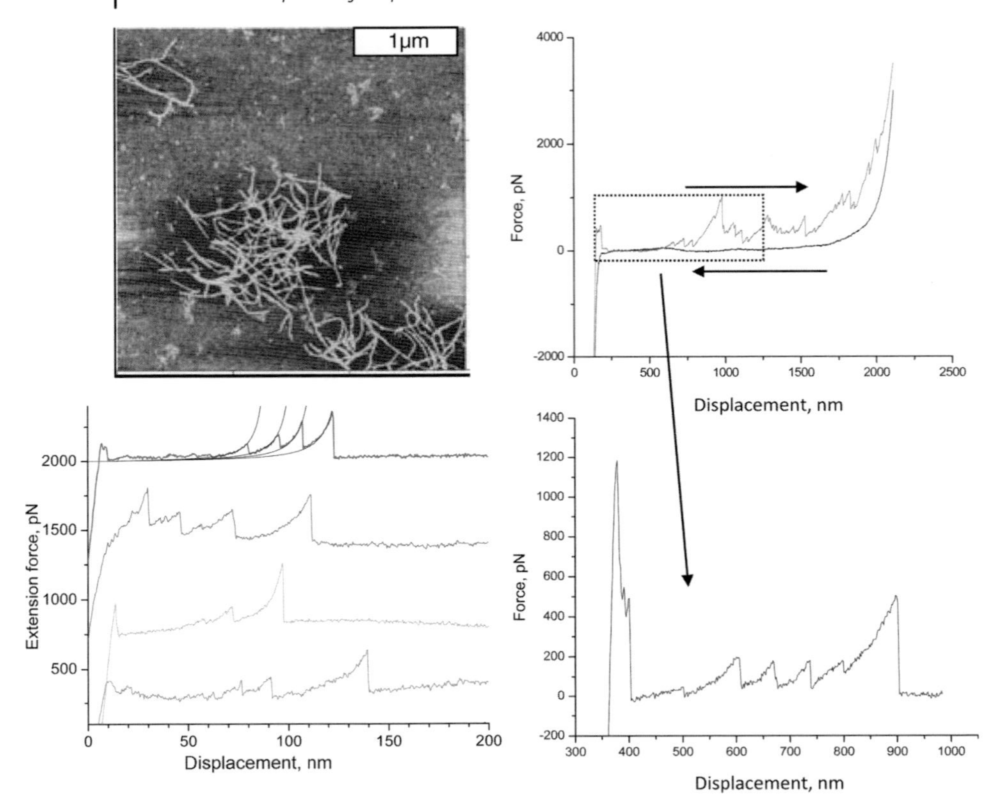

Figure 5.23 AFM images of adsorbed silk fibroin proteins and representative examples of several force–displacement curves on a small extension scale (below 300 nm) and an example of corresponding WLC fitting curves used to determine extension domain lengths and Kuhn segment. Force–distance curves of silk fibroins under different condition and including WLC model fits. Reprinted from Ref. [155].

weakly dependent upon the rate of stretching for the range used here. Finally, the overall forces required to complete unfolding event in this case increased with the extension (although remaining very scattered). Statistics of lengths and forces for different events, WLC analysis, and comparison of the experimental data with geometrical dimensions calculated from chemical composition of the molecular chains have been conducted for the final assignment of stretching events (Figure 5.24).

The total length of this stretching event was comparable to the total, fully extended length of the backbones assuming a completely unfolded structure and an extended conformation of all hydrophilic and hydrophobic domains (1.84 μm) (Figure 5.24). This behavior can be assigned to the case of complete stretching of protein backbones with random unfolding events indicating multichain attachment and concurrent stretching events under these conditions (high rate and high forces). These events are relatively rare because to complete they require forces of 1–3 nN that are nearing the

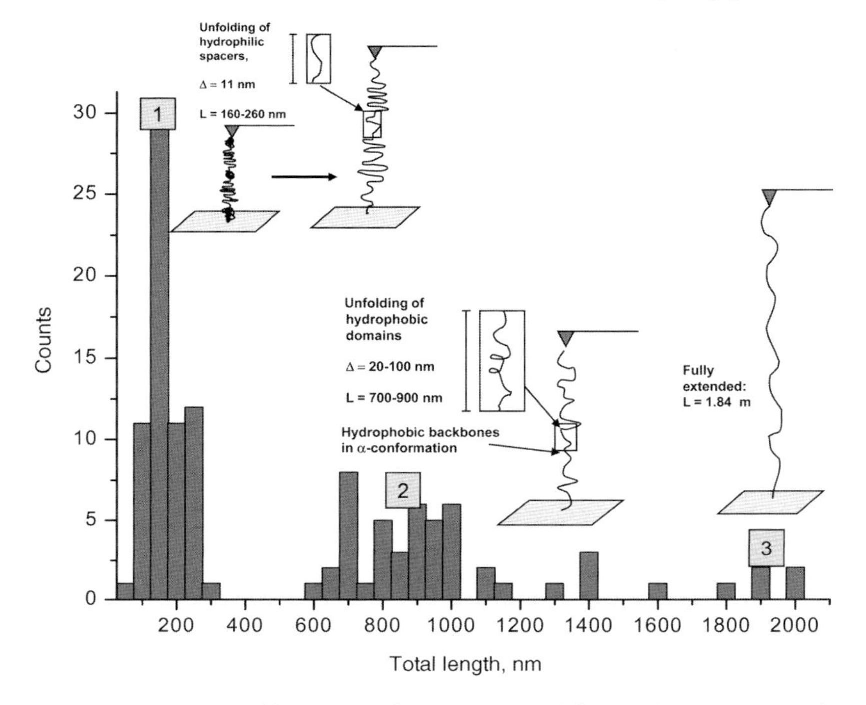

Figure 5.24 Experimental histograms of maximum stretched length (3) and the total length of the events (1 and 2) obtained in different extension scenarios (see schemes insets) and theoretical estimations (1–3 boxes) for total extended lengths as estimated from amino acid compositions of different domains. Reprinted from Ref. [155].

backbone rupture strength; thus, they can be observed only with very strong tethering of the terminal chains to the AFM tip.

In contrast, occurring at intermediate and small stretching distances were more organized events that can be associated with the regular unfolding in the protein backbone. For low stretching, the well-defined spacing of 11.3 nm dominates the unfolding pattern. Overall, 11.3, 22.5, and 32 nm spacings with a ratio close to 1 : 2 : 3 were consistently detected. The main periodicity coincided precisely with the fully extended length of the hydrophilic domains calculated from the amino acid composition (11.2 nm). Other consistent lengths can be assigned to double and triple unfolding events.

Forces for a single unfolding event were close to 60 pN, which is common for this type of backbone segment unfolding [143]. The analysis of individual domain unfolding in terms of Hooke's spring model suggested a spring constant of about 0.01 N/m, which is expected for stretching of flexible *backbones* governed by an entropic mechanism [156]. The persistent length determined from the SFS fitting with the WLC model was 0.24 nm, which is characteristic of very flexible chains [146].

Finally, the extended length of the series of events (estimated as the total length of all peaks) spread from 100 to 230 nm and are centered at 160 nm (Figure 5.24). These lengths correspond to the length of backbones with folded α-helical hydrophobic and stretched hydrophilic domains (including two terminal hydrophilic domains). The terminal domains can be stretched to the contour lengths of 53 and 18 nm, respectively. The overall stretching lengths clustered within 700–1000 nm correlated well with the end-to-end distances of the protein backbones with fully unfolded and oriented hydrophobic domains (with preserved α-helices) combined with extended hydrophilic domains.

References

1 Sarid, D. (1991) *Scanning Force Microscopy: With Applications to Electric, Magnetic, and Atomic Forces*, Oxford University Press, New York.

2 Magonov, S. and Whangbo, M.-H. (1996) *Surface Analysis with STM and AFM: Experimental and Theoretical Aspects of Image Analysis*, Wiley-VCH Verlag GmbH, New York.

3 Ratner, B. and Tsukruk, V.V. (eds) (1998) *Scanning Probe Microscopy of Polymers*, An American Chemical Society Publication, Washington, D.C.

4 Bhushan, B. (ed.) (1998) *Tribology Issues and Opportunities in MEMS*, Kluwer Academic Publishers, Dordrecht.

5 Smith, D.P.E. (1995) Limits of force microscopy. *Rev. Sci. Instrum.*, **66** (5), 3191–3195.

6 Zhang, W. and Zhang, X. (2003) Single molecule mechanochemistry of macromolecules. *Prog. Polym. Sci.*, **28** (8), 1271–1295.

7 Kühner, F. and Gaub, E.H. (2006) Modelling cantilever-based force spectroscopy with polymers. *Polymer*, **47** (7), 2555–2563.

8 Zou, S., Korczagin, I., Hempenius, M.A., Schöherr, H., and Vancso, G.J. (2006) Single molecule force spectroscopy of smart poly(ferrocenylsilane) macromolecules: towards highly controlled redox-driven single chain motors. *Polymer*, **47** (7), 2483–2892.

9 Al-Maawali, S., Bemis, J.E., Akhremitchev, B.B., Liu, H., and Walker, G.C. (2005) Single-molecule AFM study of polystyrene grafted at gold surfaces. *J. Adhes.*, **81** (10–11), 999–1016.

10 Senden, T.J. (2001) Force microscopy and surface interactions. *Curr. Opin. Colloid Interface Sci.*, **6** (2), 95–101.

11 Cappella, B. and Dietler, G. (1999) Force–distance curves by atomic force microscopy. *Surf. Sci. Rep.*, **34** (1–3), 1–104.

12 Butt, H.-J., Cappella, B., and Kappl, M. (2005) Force measurements with the atomic force microscope: technique, interpretation, and applications. *Surf. Sci. Rep.*, **59** (1–6), 1–152.

13 Bottomley, L.A. (1998) Scanning probe microscopy. *Anal. Chem.*, **70** (12), 425–476.

14 Weisenhorn, A.L., Maivald, P., Butt, H.-J., and Hansma, P.K. (1992) Measuring adhesion, attraction, and repulsion between surfaces in liquids with an atomic-force microscope. *Phys. Rev. B*, **45** (19), 11226–11232.

15 Nix, W.D. (1989) Mechanical properties of thin films. *Metall. Mater. Trans. A*, **20** (11), 2217–2245.

16 Pharr, G.M. and Oliver, W.C. (1992) Measurement of thin-film mechanical-properties using nanoindentation. *MRS Bull.*, **17** (7), 28–33.

17 Field, J.S. and Swain, M.V. (1993) A simple predictive model for spherical indentation. *J. Mater. Res.*, **8** (2), 297–306.

18 Chizhik, S.A., Gorbunov, V.V., Luzinov, I., Fuchigami, N., and Tsukruk, V.V. (2001) Surface force spectroscopy of elastomeric nanoscale films. *Macromol. Symp.*, **167** (1), 167–175.

19 Marti, O. and Hild, S. (2000) Temperature-dependent surface properties of thin polystyrene films determined by scanning force

microscopy, in *Microstructure and Microtribology of Polymer Surfaces* (eds V.V. Tsukruk and K.J. Wahl), ACS Symposium Series, American Chemical Society, Washington, D.C., pp. 212–226.

20 Giannakopoulos, A.E. and Suresh, S. (1997) Indentation of solids with gradients in elastic properties: Part I. Point force. *Int. J. Solids Struct.*, **34** (19), 2357–2392.

21 Aimé, J.P., Elkaakour, Z., Odin, C., Bouhacina, T., Michel, D., Curely, J., and Dautant, A. (1994) Comments on the use of the force mode in atomic force microscopy for polymer films. *J. Appl. Phys.*, **76** (2), 754–762.

22 Gao, H., Chiu, C.H., and Lee, J. (1992) Elastic contact versus indentation modeling of multi-layered materials. *Int. J. Solids Struct.*, **29** (20), 2471–2492.

23 Pender, D., Thompson, S., Padture, N., Giannakopoulos, A., and Suresh, S. (2001) Gradients in elastic modulus for improved contact-damage resistance. Part II: the silicon nitride–silicon carbide system. *Acta Mater.*, **49** (16), 3263–3268.

24 Suresh, S. (2001) Graded materials for resistance to contact deformation and damage. *Science*, **292** (5526), 2447–2451.

25 Johnson, K.L., Kendall, K., and Roberts, A.D. (1971) Surface energy and the contact of elastic solids. *Proc. R. Soc. Lond. A*, **324** (1558), 301–313.

26 Chizhik, S.A., Huang, Z., Gorbunov, V.V., Myshkin, N.K., and Tsukruk, V.V. (1998) Micro-mechanical properties of elastic polymeric materials as probed by scanning force microscopy. *Langmuir*, **14** (10), 2606–2609.

27 Tranchida, D., Piccarolo, S., and Soliman, M. (2006) Nanoscale mechanical characterization of polymers by AFM nanoindentations: critical approach to the elastic characterization. *Macromolecules*, **39** (13), 4547–4556.

28 Domke, J. and Radmacher, M. (1998) Measuring the elastic properties of thin polymer films with the atomic force microscope. *Langmuir*, **14** (12), 3320–3325.

29 Doerner, M.F. and Nix, W.D. (1986) A method for interpreting the data from depth-sensing indentation instruments. *J. Mater. Res.*, **1** (4), 601–609.

30 King, R.B. (1987) Elastic analysis of some punch problems for a layered medium. *Int. J. Solids Struct.*, **23** (12), 1657–1664.

31 Shulha, H., Kovalev, A., Myshkin, N., and Tsukruk, V.V. (2004) Some aspects of AFM nanomechanical probing of surface polymer films. *Eur. Polym. J.*, **40** (5), 949–956.

32 Cappella, B. and Silbernagl, D. (2008) Nanomechanical properties of polymer thin films measured by force–distance curves. *Thin Solid Films*, **516** (8), 1952–1960.

33 Oommen, B. and Van Vliet, K.J. (2006) Effects of nanoscale thickness and elastic nonlinearity on measured mechanical properties of polymeric films. *Thin Solid Films*, **513** (1–2), 235–242.

34 Kovalev, A., Shulha, H., Lemieux, M., Myshkin, N., and Tsukruk, V.V. (2004) Nanomechanical probing of layered nanoscale polymer films with atomic force microscopy. *J. Mater. Res.*, **19** (3), 716–728.

35 Tsukruk, V.V., Sidorenko, Y., Gorbunov, V.V., and Chizhik, S.A. (2001) Surface nanomechanical properties of polymer nanocomposite layers. *Langmuir*, **17** (21), 6715–6719.

36 Hazel, J.L. and Tsukruk, V.V. (1999) Spring constants of composite ceramic/gold cantilevers for scanning probe microscopy. *Thin Solid Films*, **339** (1–2), 249–257.

37 Huang, Z., Chizhik, S.A., Gorbunov, V.V., Myshkin, N.K., and Tsukruk, V.V. (2000) Scanning force microscopy probing of micromechanical properties of polymers, in *Microstructure and Microtribology of Polymer Surfaces* (eds V.V. Tsukruk and K.J. Wahl), ACS Symposium Series, American Chemical Society, Washington, D.C., pp. 117–189.

38 Vanlandingham, M.R., McKnight, S.H., Palmese, G.R., Eduljee, R.F., Gillespie, J.W., and McCulough, R.L. (1997) Relating elastic modulus to indentation response using atomic force microscopy. *J. Mater. Sci. Lett.*, **16** (2), 117–119.

39 Tsukruk, V.V. and Gorbunov, V.V. (2001) Nanomechanical probing with scanning force microscopy. *Micros. Today*, **1** (1), 8–21.

40 http://www.bruker-axs.com/dimension_icon_atomic_force_microscope.html.

41 Jagtap, R.N. and Ambre, A.H. (2005) Atomic force microscopy (AFM): basics and its important applications for polymer characterization: an overview. *J. Polym. Mater.*, **22** (1), 1–26.

42 Hugel, T., Holland, N.B., Cattani, A., Moroder, L., Seitz, M., and Gaub, H.E. (2002) Single-molecule optomechanical cycle. *Science*, **296** (5570), 1103–1106.

43 Zhang, X., Liu, C., and Wang, Z. (2008) Force spectroscopy of polymers: studying on intramolecular and intermolecular interactions in single molecular level. *Polymer*, **49** (16), 3353–3361.

44 MacKintosh, F.C. and Schmidt, C.F. (1999) Microrheology. *Curr. Opin. Colloid Interface Sci.*, **4** (4), 300–307.

45 Garcia, R., Magerle, R., and Perez, R. (2007) Nanoscale compositional mapping with gentle forces. *Nat. Mater.*, **6** (6), 405–411.

46 Dupont-Gillain, C.C., Pamula, E., Denis, F.A., and Rouxhet, P.G. (2004) Nanostructured layers of adsorbed collagen: conditions, mechanisms and applications. *Prog. Colloid Polym. Sci.*, **128**, 98–104.

47 Salvetat, J.P., Bonard, J.-M., Thomson, N.H., Kulik, A.J., Forro, L., Benoit, W., and Zuppiroli, L. (1999) Mechanical properties of carbon nanotubes. *Appl. Phys. A*, **69** (3), 255–260.

48 Fery, A., Dubreuil, F., and Mohwald, H. (2004) Mechanics of artificial microcapsules. *New J. Phys.*, **6** (18), doi: 10.1088/1367-2630/6/1/018.

49 Vinogradova, O.I., Lebedeva, O.V., and Kim, B.S. (2006) Mechanical behavior and characterization of microcapsules. *Annu. Rev. Mater. Res.*, **36**, 143–178.

50 Sukhorukov, G., Fery, A., and Mohwald, H. (2005) Intelligent micro- and nanocapsules. *Prog. Polym. Sci.*, **30** (8–9), 885–897.

51 Song, J., Tranchida, D., and Vancso, G.J. (2008) Contact mechanics of UV/ozone-treated PDMS by AFM and JKR testing: mechanical performance from nano- to micrometer length scales. *Macromolecules*, **41** (18), 6757–6762.

52 Cheng, X., Canavan, H.E., Stein, M.J., Hull, J.R., Kweskin, S.J., Wagner, M.S., Somorjai, G.A., Castner, D.G., and Ratner, B.D. (2005) Surface chemical and mechanical properties of plasma-polymerized *N*-isopropylacrylamide. *Langmuir*, **21** (17), 7833–7841.

53 Rezende, C.A., Lee, L.-T., and Galembeck, F. (2009) Surface mechanical properties of thin polymer films investigated by AFM in pulsed force mode. *Langmuir*, **25** (17), 9938–9946.

54 Tagit, O., Tomczak, N., and Vancso, G.J. (2008) Probing the morphology and nanoscale mechanics of single poly(*N*-isopropylacrylamide) microgels across the lower-critical-solution temperature by atomic force microscopy. *Small*, **4** (1), 119–126.

55 Wiedemair, J., Serpe, M.J., Kim, J., Masson, J.-F., Lyon, L.A., Mizaikoff, B., and Kranz, C. (2007) *In-situ* AFM studies of the phase-transition behavior of single thermoresponsive hydrogel particles. *Langmuir*, **23** (1), 130–137.

56 Horkay, F. and Lin, D.C. (2009) Mapping the local osmotic modulus of polymer gels. *Langmuir*, **25** (15), 8735–8741.

57 Kessel, S., Schmidt, S., Muller, R., Wischerhoff, E., Laschewsky, A., Lutz, J.-F., Uhlig, K., Lankenau, A., Duschl, C., and Fery, A. (2010) Thermoresponsive PEG-based polymer layers: surface characterization with AFM force microscopy. *Langmuir*, **26** (5), 3462–3467.

58 Wu, M., Kleiner, L., Tang, F.-W., Hossainy, S., Davies, M.C., and Roberts, C.J. (2009) Nanoscale mechanical measurement determination of the glass transition temperature of poly (lactic acid)/everolimus coated stents in air and dissolution media. *Eur. J. Pharm. Sci.*, **36** (4–5), 493–501.

59 Amitay-Sadovsky, E., Ward, B., and Somorjai, G.A. (2002) Nanomechanical properties and morphology of thick polyurethane films under contact pressure and stretching. *J. Appl. Phys.*, **91** (1), 375.

60 Choi, T., Jang, J.-H., Ullal, C.K., LeMieux, M.C., Tsukruk, V.V., and Thomas, E.L. (2006) The elastic properties and plastic behavior of two-dimensional polymer structures fabricated by laser interference lithography. *Adv. Funct. Mater.*, **16** (10), 1324–1330.

61 Jang, J.-H., Ullal, C.K., Choi, T., LeMieux, M.C., Tsukruk, V.V., and Thomas, E.L. (2006) 3D polymer microframes that exploit length-scale-dependent mechanical behavior. *Adv. Mater.*, **18** (16), 2123–2127.

62 Singamaneni, S., Chang, S., Jang, J.-H., Davis, W., Thomas, E.L., and Tsukruk, V.V. (2008) Mechanical properties of composite polymer microstructures fabricated by interference lithography. *Phys. Chem. Chem. Phys*, **10**, 4093–4105.

63 Tsukruk, V.V., and Huang, Z. (2000) Micro-thermomechanical properties of heterogeneous polymer films. *Polymer*, **41** (14), 5541–5545.

64 Luzinov, I., Julthongpiput, D., and Tsukruk, V.V. (2000) Thermoplastic elastomer monolayers grafted to a silicon substrate. *Macromolecules*, **33** (20), 7629–7638.

65 Luzinov, I., Julthongpiput, D., and Tsukruk, V.V. (2001) Stability of microdomain morphology in tethered block copolymer monolayers. *Polymer*, **42** (5), 2267–2273.

66 Burns, A.R., Houston, J.E., Carpick, R.W., and Michalske, T.A. (1999) Molecular level friction as revealed with a novel scanning probe. *Langmuir*, **15** (8), 2922–2930.

67 Aime, J.P. and Gauthier, S. (1998) Stretching a Network of Entangled Polymer Chains with a Nanotip, in *Scanning Probe Microscopy of Polymers* (eds B. Ratner and V. Tsukruk), ACS Symposium Series, An American Chemical Society Publication, Washington D.C., pp. 266–287.

68 Koike, A. and Yoneya, M. (1997) Effects of molecular structure on frictional properties of Langmuir–Blodgett monolayers. *Langmuir*, **13** (6), 1718–1722.

69 Luzinov, I., Julthongpiput, D., Liebmann-Vinson, A., Cregger, T., Foster, M.D., and Tsukruk, V.V. (2000) Epoxy-terminated self-assembled monolayers: molecular glues for polymer layers. *Langmuir*, **16**, 504–510.

70 Julthongpiput, D., LeMieux, M., and Tsukruk, V.V. (2003) Micromechanical properties of glassy and rubbery polymer brush layers as probed by atomic force microscopy. *Polymer*, **44** (16), 4557–4562.

71 Lemieux, M., Minko, S., Usov, D., Stamm, D., and Tsukruk, V.V. (2003) Direct measurement of thermoelastic properties of glassy and rubbery polymer brush nanolayers grown by “grafting-from” approach. *Langmuir*, **19** (15), 6126–6134.

72 Lemieux, M., Usov, D., Minko, S., Stamm, M., and Tsukruk, V.V. (2003) Reorganization of binary polymer brushes: switching surface microstructures and nanomechanical properties. *Macromolecules*, **36** (19), 7244–7255.

73 Sidorenko, A., Ahn, H., Kim, D., Yang, H., and Tsukruk, V.V. (2002) Wear stability of polymer nanocomposite coatings with trilayer architecture. *Wear*, **252** (11–12), 946–955.

74 Tsukruk, V.V., Ahn, H., Kim, D., and Sidorenko, A. (2002) Triplex molecular layers with nonlinear nanomechanical response. *Appl. Phys. Lett.*, **80**, 4825.

75 Julthongpiput, D., Lin, Y.-H., Teng, J., Zubarev, E.R., and Tsukruk, V.V. (2003) Y-shaped amphiphilic brushes with switchable micellar surface structures. *J. Am. Chem. Soc.*, **152** (51), 15912–15921.

76 LeMieux, M.C., Lin, Y.-H., Cuong, P.D., Ahn, H.-S., Zubarev, E.R., and Tsukruk, V.V. (2005) Microtribological and nanomechanical properties of switchable Y-shaped amphiphilic polymer brushes. *Adv. Funct. Mater.*, **15** (9), 1529–1540.

77 Shulha, H., Zhai, X., and Tsukruk, V.V. (2003) Molecular stiffness of individual

hyperbranched macromolecules at solid surfaces. *Macromolecules*, **36** (8), 2825–2831.

78 Tsukruk, V.V., Shulha, H., and Zhai, X. (2003) Nanoscale stiffness of individual dendritic molecules and their aggregates. *Appl. Phys. Lett.*, **82**, 907–909.

79 Mansfield, M.L. (1996) Surface adsorption of model dendrimers. *Polymer*, **37** (17), 3835–3841.

80 Tsukruk, V.V., Rinderspacher, F., and Bliznyuk, V.N. (1997) Self-assembled multilayer films from dendrimers. *Langmuir*, **13** (8), 2171–2176.

81 Noy, A., Vezenov, D.V., and Lieber, C.M. (1997) Chemical force microscopy. *Annu. Rev. Mater. Sci.*, **27**, 381–421.

82 Noy, A. (2004) Interaction at solid–fluid interfaces, in *Nanoscale Structure and Assembly at Solid–Fluid Interfaces* (eds X.Y. Liu and J.J. De Yoreo), Kluwer Academic Publishers, Norwell, pp. 57–82.

83 Takano, H., Kenseth, J.R., Wong, S.-S., O'Brien, J.C., and Porter, M.D. (1999) Chemical and biochemical analysis using scanning force microscopy. *Chem. Rev.*, **99** (10), 2845–2890.

84 Janshoff, A., Neitzert, M., Oberdörfer, Y., and Fuchs, H. (2009) Force spectroscopy of molecular systems: single molecule spectroscopy of polymers and biomolecules. *Angew Chem., Int. Ed.*, **39** (18), 3212–3237.

85 Noy, A. (ed.) (2007) *Handbook of Molecular Force Spectroscopy*, Springer, New York.

86 Thundat, T., Zheng, X.-Y., Chen, G.Y., and Warmack, R.J. (1993) Role of relative humidity in atomic force microscopy imaging. *Surf. Sci.*, **294** (1–2), L939–L943.

87 Binggeli, M. and Mate, C.M. (1994) Influence of capillary condensation of water on nanotribology studied by force microscopy. *Appl. Phys. Lett.*, **65**, 415–417.

88 Fujihira, M., Aoki, D., Okabe, Y., Takano, H., Hokari, H., Frommer, J., Nagatani, Y., and Sakai, F. (1996) Effect of capillary force on friction force microscopy: a scanning hydrophilicity microscope. *Chem. Lett.*, **25** (7), 499–502.

89 Vezenov, D.V., Noy, A., Rozsnyai, L.F., and Lieber, C.M. (1997) Force titrations and ionization state sensitive imaging of functional groups in aqueous solutions by chemical force microscopy. *J. Am. Chem. Soc.*, **119** (8), 2006–2015.

90 Schönherr, H., Hruska, Z., and Vancso, J. (2000) Toward high resolution mapping of functional group distributions at surface treated polymers by AFM using modified tips. *Macromolecules*, **33** (12), 4532–4537.

91 Noy, A., Frisbie, C.D., Rosznyai, L.F., Wrighton, M.S., and Lieber, C.M. (1995) Chemical force microscopy: exploiting chemically-modified tips to quantify adhesion, friction, and functional group distributions in molecular assemblies. *J. Am. Chem. Soc.*, **117** (30), 7943–7951.

92 Tsukruk, V.V. and Bliznyuk, V.N. (1998) Adhesive and friction forces between chemically modified silicon and silicon nitride surfaces. *Langmuir*, **14** (2), 446–455.

93 McDermott, M.T., Green, J.B., and Porter, M.D. (1997) Scanning force microscopic exploration of the lubrication capabilities of *n*-alkanethiolate monolayers chemisorbed at gold: structural basis of microscopic friction and wear. *Langmuir*, **13** (9), 2504–2510.

94 Friddle, R.W., LeMieux, M.C., Cicero, G., Artyukhin, A.B., Tsukruk, V.V., Grossman, J.C., Galli, G., and Noy, A. (2007) Single functional group interactions with individual carbon nanotubes. *Nat. Nanotechnol.*, **2**, 692–697.

95 Feldman, K., Hähner, G., Spencer, N.D., Harder, P., and Grunze, M. (1999) Probing resistance to protein adsorption of oligo(ethylene glycol)-terminated self-assembled monolayers by scanning force microscopy. *J. Am. Chem. Soc.*, **121** (43), 10134–10141.

96 Jee, T., Lee, H., Mika, B., and Liang, H. (2006) Effect of microstructures of PVDF on surface adhesive forces. *Tribol. Lett.*, **26** (2), 125–130.

97 Gupta, M.K., Singamaneni, S., McConney, M., Drummy, L.F., Naik, R.R., and Tsukruk, V.V. (2009) A facile fabrication strategy for patterning protein chain conformation in silk materials. *Adv. Mater.*, **22** (1), 115–119.

98 Wang, B., Oleschuk, R.D., and Horton, J.H. (2005) Chemical force titrations of amine- and sulfonic acid-modified poly (dimethylsiloxane). *Langmuir*, **21** (4), 1290–1298.

99 Noel, O., Brogly, M., Castelein, G., and Schultz, J. (2004) *In situ* estimation of the chemical and mechanical contributions in local adhesion force measurement with AFM: the specific case of polymers. *Eur. Polym. J.*, **40** (5), 965–974.

100 Cappella, B. and Stark, W. (2005) Adhesion of amorphous polymers as a function of temperature probed with AFM force–distance curves. *J. Colloid Interface Sci.*, **296** (2), 507–514.

101 Tsui, O.K.C., Wang, X.P., Ho, J.Y.L., Ng, T.K., and Xiao, X. (2000) Studying surface glass-to-rubber transition using atomic force microscopic adhesion measurements. *Macromolecules*, **33** (11), 4198–4204.

102 Bliznyuk, V.N., Assender, H.E., and Briggs, G.A.D. (2002) Surface glass transition temperature of amorphous polymers. a new insight with SFM. *Macromolecules*, **35** (17), 6613–6622.

103 Moon, S.-H. and Foster, M.D. (2002) Influence of humidity on surface behavior of pressure sensitive adhesives studied using scanning probe microscopy. *Langmuir*, **18** (21), 8108–8115.

104 Gong, H., Garcia-Turiel, J., Vasilev, K., and Vinogradova, O.I. (2005) Interaction and adhesion properties of polyelectrolyte multilayers. *Langmuir*, **21** (16), 7545–7550.

105 Mermut, O., Lefebvre, J., Gray, D.G., and Barrett, C.J. (2003) Structural and mechanical properties of polyelectrolyte multilayer films studied by AFM. *Macromolecules*, **36** (23), 8819–8824.

106 Yu, M. and Ivanisevic, A. (2004) Encapsulated cells: an atomic force microscopy study. *Biomaterials*, **25** (17), 3655–3662.

107 Kokuoz, B., Kornev, K.G., and Luzinov, I. (2009) Gluing nanoparticles with a polymer bonding layer: the strength of an adhesive bond. *ACS Appl. Mater. Interfaces*, **1** (3), 575–583.

108 Johnson, K.L. (1998) Contact mechanics and adhesion of viscoelastic spheres, in *Microstructure and Microtribology of Polymer Surfaces* (eds V.V. Tsukruk and K. Wahl), ACS Symposium Series, American Chemical Society, Washington, D.C., pp. 24–41.

109 Friedenberg, M.C. and Mate, C.M. (1996) Dynamic viscoelastic properties of liquid polymer films studied by atomic force microscopy. *Langmuir*, **12** (25), 6138–6142.

110 Braithwaite, G.J.C. and Luckham, P.F. (1999) The simultaneous determination of the forces and viscoelastic properties of adsorbed polymer layers. *J. Colloid. Interface Sci.*, **218** (1), 97–111.

111 Yang, G., Rao, N., Yin, Z., and Zhu, D.-M. (2006) Probing the viscoelastic response of glassy polymer films using atomic force microscopy. *J. Colloid Interface Sci.*, **297** (1), 104–111.

112 Tripathy, S. and Berger, E.J. (2009) Measuring viscoelasticity of soft samples using atomic force microscopy. *J. Biomech. Eng.*, **131** (9), 094507–094513.

113 Tsukruk, V.V., Gorbunov, V.V., Huang, Z., and Chizhik, S.A. (2000) Dynamic microprobing of viscoelastic polymer properties. *Polym. Int.*, **9** (5), 441–444.

114 Cappella, B., Kaliappan, S.K., and Sturm, H. (2005) Using AFM force–distance curves to study the glass-to-rubber transition of amorphous polymers and their elastic–plastic properties as a function of temperature. *Macromolecules*, **38** (5), 1874–1881.

115 Moeller, G. (2009) AFM nanoindentation of viscoelastic materials with large end-radius probes. *J. Appl. Polym. Sci. B*, **47** (16), 1573–1587.

116 McConney, M.E., Anderson, K.D., Brott, L.L., Naik, R.R., and Tsukruk, V.V. (2009) Bioinspired material approaches to sensing. *Adv. Funct. Mater.*, **19** (16), 2527–2544.

117 McConney, M.E., Schaber, C.F., Julian, M.D., Barth, F.G., and Tsukruk, V.V. (2007) Viscoelastic nanoscale properties of cuticle contribute to the high-pass properties of spider vibration receptor (*Cupoennius salei* Keys). *J. R. Soc. Interface*, **4** (17), 1135–1143.

118 Peleshanko, S., Julian, M.D., Ornatska, M., McConney, M.E., LeMieux, M.C., Chen, N., Tucker, C., Yang, Y., Liu, C., Humphrey, J.A.C., and Tsukruk, V.V. (2007) Hydrogel-encapsulated microfabricated haircells mimicking fish cupula neuromasts. *Adv. Mater.*, **19** (19), 2903–2909.

119 Koffas, T.S., Opdahl, A., Marmo, C., and Somorjai, G.A. (2003) Effect of equilibrium bulk water content on the humidity-dependent surface mechanical properties of hydrophilic contact lenses studied by atomic force microscopy. *Langmuir*, **19** (8), 3453–3460.

120 Singer, E. and Pollack, H. (eds) (1992) *Fundamentals of Friction: Macroscopic and Microscopic Processes*, Kluwer Academic Press, Dordrecht.

121 Adamson, A.W. (1990) *Physical Chemistry of Surfaces*, John Wiley & Sons, Inc., New York.

122 Persson, B.N.J. (1999) Sliding friction. *Surf. Sci. Rep.*, **33** (3), 83–119.

123 Tsukruk, V.V. (2001) Molecular lubricants and glues for micro- and nanodevices. *Adv. Mater.*, **13** (2), 95–108.

124 Tsukruk, V.V., Everson, M.P., Lander, L.M., and Brittain, W.J. (1996) Nanotribological properties of composite molecular films: C60 anchored to a self-assembled monolayer. *Langmuir*, **12** (16), 3905–3911.

125 Cannara, R.J., Brukman, M.J., Cimatu, K., Sumant, A.V., Baldelli, S., and Carpick, R.W. (2007) Nanoscale friction varied by isotopic shifting of surface vibrational frequencies. *Science*, **318** (5851), 780–783.

126 Tutein, A., Stuart, S., and Harrison, J. (2000) Role of defects in compression and friction of anchored hydrocarbon chains on diamond. *Langmuir*, **16** (2), 291–296.

127 Tsukruk, V.V., Bliznyuk, V.N., Hazel, J., Visser, D., and Everson, M.P. (1996) Organic molecular films under shear forces: fluid and solid Langmuir monolayers. *Langmuir*, **12** (20), 4840–4849.

128 Liu, Y., Wu, T., and Evans, D.F. (1994) Lateral force microscopy study on the shear properties of self-assembled monolayers of dialkylammonium surfactant on mica. *Langmuir*, **10** (7), 2241–2245.

129 Liu, Y. and Evans, D.F. (1996) Structure and frictional properties of self-assembled surfactant monolayers. *Langmuir*, **12** (5), 1235–1244.

130 Vyas, M.K., Schneider, K., Nandan, B., and Stamm, M. (2008) Switching of friction by binary polymer brushes. *Soft Matter*, **4**, 1024–1032.

131 Tsukruk, V.V. and Wahl, K.J. (eds) (2000) *Microstructure and Microtribology of Polymer Surfaces, ACS Symposium Series*, American Chemical Society, Washington, D.C.

132 Hurley, C.R. and Leggett, G.J. (2006) Influence of the solvent environment on the contact mechanics of tip–sample interactions in friction force microscopy of poly(ethylene terephthalate) films. *Langmuir*, **22** (9), 4179–4183.

133 Kim, K.S., Ando, Y., and Kim, K.-W. (2008) The effect of temperature on the nanoscale adhesion and friction behaviors of thermoplastic polymer films. *Nanotechnology*, **19** (10), 105701–105709.

134 Tocha, E., Schonherr, H., and Vancso, G.J. (2009) Surface relaxations of poly(methyl methacrylate) assessed by friction force microscopy on the nanoscale. *Soft Matter*, **5**, 1489–1495.

135 Jung, Y.C. and Bhushan, B. (2006) Contact angle, adhesion, and friction properties of micro- and nanopatterned polymers for superhydrophobicity. *Nanotechnology*, **17** (19), 4970–4980.

136 Wu-Bavouzet, F., Clain-Burckbuchler, J., Buguin, A., De Gennes, P.-G., and Brochard-Wyart, F. (2007) Stick-slip: wet versus dry. *J. Adhes.*, **83**, 761–784.

137 Zou, S., Korczagin, I., Hempenius, M.A., Schöherr, H., and Vancso, G.J. (2006) Single molecule force spectroscopy of smart poly(ferrocenylsilane) macromolecules: towards highly controlled redox-driven single chain motors. *Polymers*, **47** (7), 2483–2492.

138 Shi, W., Zhang, Y., Liu, C., Wang, Z., Zhang, X., Zhang, Y., and Chen, Y. (2006) Toward understanding the effect of substitutes and solvents on entropic and

enthalpic elasticity of single dendronized copolymers. *Polymer,* **47** (7), 2499–2504.

139 Kühner, F. and Gaub, E.H. (2006) Modeling cantilever-based force spectroscopy with polymers. *Polymer,* **47** (7), 2555–2563.

140 Zhang, Q. and Marszalek, P.E. (2006) Solvent effects on the elasticity of polysaccharide molecules in disordered and ordered states by single-molecule force spectroscopy. *Polymer,* **47** (7), 2526–2532.

141 Ludwig, M., Rief, M., Schmidt, L., Li, H., Oesterhelt, F., Gautel, M., and Gaub, H.E. (1999) AFM, a tool for single-molecule experiments. *Appl. Phys. A,* **68** (2), 173–176.

142 Zhang, X., Liu, C., and Wang, Z. (2008) Force spectroscopy of polymers: studying on intramolecular and intermolecular interactions in single molecular level. *Polymer,* **49** (16), 3353–3361.

143 Becker, N., Oroudjev, E., Mutz, S., Cleveland, J.P., Hansma, P.K., Hayashi, C.Y., Makarov, D.E., and Hansma, H.G. (2003) Molecular nanosprings in spider capture-silk threads. *Nat. Mater.,* **2**, 278–283.

144 Meadows, P.Y., Bemis, J.E., and Walker, G.C. (2003) Single-molecule force spectroscopy of isolated and aggregated fibronectin proteins on negatively charged surfaces in aqueous liquids. *Langmuir,* **19** (23), 9566–9572.

145 Carrion-Vazquez, M., Marszalek, P.E., Oberhauser, A.F., and Fernandez, J.M. (1999) Atomic force microscopy captures length phenotypes in single proteins. *Proc. Natl. Acad. Sci. USA,* **96** (20), 11288–11292.

146 Bemis, J.E., Akhremitchev, B.B., and Walker, G.C. (1999) Single polymer chain elongation by atomic force microscopy. *Langmuir,* **15** (8), 2799–2805.

147 Kuhner, F., Erdmann, M., Sonnenberg, L., Serr, A., Morfille, J., and Gaub, H.E. (2006) Friction of single polymers at surfaces. *Langmuir,* **22** (26), 11180–11186.

148 Sonnenberg, L., Parvole, J., Kuhner, F., Billon, L., and Gaub, H.E. (2007) Choose sides: differential polymer adhesion. *Langmuir,* **23** (12), 6660–6666.

149 Gunari, N. and Walker, G.C. (2008) Nanomechanical fingerprints of individual blocks of a diblock copolymer chain. *Langmuir,* **24** (10), 5197–5201.

150 Zhang, W., Zou, S., Wang, C., and Zhang, X. (2000) Single polymer chain elongation of poly(*N*-isopropylacrylamide) and poly (acrylamide) by atomic force microscopy. *J. Phys. Chem. B,* **104** (44), 10258–10264.

151 Bini, E., Knight, D.P., and Kaplan, D.L. (2004) Mapping domain structures in silks from insects and spiders related to protein assembly. *J. Mol. Biol.,* **335** (1), 27–40.

152 Tsukada, M., Gotoh, Y., Nagura, M., Minoura, N., Kasai, N., and Freddi, G. (1994) Structural changes of silk fibroin membranes induced by immersion in methanol aqueous solutions. *J. Polym. Sci. B,* **32** (5), 961–968.

153 Zhou, C.Z., Confalonieri, F., Medina, N., Zivanovic, Y., Esnault, C., Yang, T., Jacquet, M., Janin, J., Duguet, M., Perasso, R., and Li, Z.G. (2000) Fine organization of *Bombyx mori* fibroin heavy chain gene. *Nucleic Acids Res.,* **28** (12), 2413–2419.

154 Zhang, W.K., Xu, Q.B., Zou, S., Li, H.B., Xu, W.Q., Zhang, X., Shao, Z.Z., Kudera, M., and Gaub, H.E. (2000) Single-molecule force spectroscopy on *Bombyx mori* silk fibroin by atomic force microscopy. *Langmuir,* **16** (9), 4305–4308.

155 Shulha, H., Wong Po Foo, C., Kaplan, D.L., and Tsukruk, V.V. (2006) Unfolding the multi-length scale domain structure of silk fibroin protein. *Polymer,* **47** (16), 5821–5830.

156 Ikai, A., Mitsui, K., Tokuoka, H., and Xu, X.M. (1997) Mechanical measurements of a single protein molecule and human chromosomes by atomic force microscopy. *Mater. Sci. Eng. C,* **4** (4), 233–240.

6
Probing of Microthermal Properties

6.1
Introduction

Conventional studies of thermal properties using the established "*macro*" routines and standard characterization methods are typically carried out on a minimum of a few milligrams to grams of materials. In these routines, it is possible to deduce information about thermal properties (heat capacity, heat of fusion, and thermal conductivity) and transitions (glass transition, melting, or crystallization) of a variety of one-component materials. For more complex, multicomponent materials, for instance, differential scanning calorimetry (DSC) can detect that a particular sample (a polymer blend or block copolymers) is heterogeneous and can assign various thermal transitions to different components, and even under certain conditions is able to identify the constituents and their proportions in the multicomponent material. However, a major limitation of these macroscopic approaches to practical materials is that no information on the size, shape, interfacial properties, and spatial distribution of different phases can be acquired on a quantitative level.

In contrast to macroscopic approaches, micro/nanoscale imaging and probing with SPM allow the material property characterization to be conducted at the micrometer or even nanometer scale on a sample subjected to a controlled temperature regime (see Section 4.6). The development of micro/nano-TA techniques in different probing modes in recent years significantly enhanced the utility of these techniques by enabling variations in thermal and mechanical properties or composition to be reliably mapped with submicrometer resolution.

In this chapter, we will present selected examples of recent and unique SPM-based measurements of microthermal properties that are unachievable by regular thermal measurements. We selected examples that illustrate very specific and critical impacts of regular AFM measurements with thermal stage or SThM measurements as high-resolution imaging techniques. Comprehensive reviews of thermal-related SPM applications during earlier stages of the development of micro-TA methods have been published by the pioneers of these approaches: Reading *et al.*, Pollock *et al.*, and Wunderlich *et al.* [1, 2].

Scanning Probe Microscopy of Soft Matter: Fundamentals and Practices, First Edition.
Vladimir V. Tsukruk and Srikanth Singamaneni.

Finally, it is worth noting that in this topic, the overall number of quality research publications employing different SPM approaches (SThM or scanning thermal mode of AFM specimens on a hot stage) is relatively small compared to traditional AFM imaging of surface morphologies at room temperature. Such a situation is caused by not only the late arrival of these methods but also the significant technical challenges in the application of thermal TA to soft materials, including greatly increased thermally induced creep, thermal stresses and condensation, increased adhesion, decreased resistivity, and tip contaminations. Some of these technical challenges have been discussed in Section 4.6.

Moreover, significant contribution of thermal tip shape and questionable reliability of manually assembled thermal tips in the form of Wollaston microscopic wires may compromise micro-TA measurements especially if quantitative analysis is required. However, the recent arrival of durable and reproducible microfabricated thermal tips with truly nanoscale dimensions (see Section 4.6) brings new hope for achieving reliable measurements with high-resolution scanning thermal microscopy and microthermal probing in the near future.

6.2
Measurements of Glass Transition

A wide range of polymeric materials have been probed with micro-TA to identify glass transition temperature T_g and its dependence upon various specimen preparation conditions. Ultrathin polymer films of varying thicknesses, polymer blends, polymer brushes, and photodegradable polymers that were all probed with micro-TA will be discussed below.

6.2.1
Ultrathin Polymer Films

Microthermal analysis of the ultrathin polymeric films on a highly conductive substrate is a significant challenge. The huge difference in the thermal conductivity of a polymeric film and a substrate results in heat dissipation mostly to the substrate through the tested film. Only a very minor part of the heat dissipates into the ultrathin polymer film making reliable thermal measurement impossible.

In this case, micro-TA measurement procedures should be significantly modified as suggested by Gorbunov *et al.* [3]. In order to balance heat dissipation between the scanning thermal probe and a reference thermal probe in differential mode (see Section 4.6), the reference thermal probe should be gently engaged on an identical substrate without polymer film using a microscopic manipulator on a separate microstage under a stereo microscope. Moreover, it is critically important that two thermal probes with very similar thermal characteristics are utilized for measurements. These probes must be independently tested and selected prior to microthermal measurements, which adds technical difficulties in the utilization of this method. However, it is a critical and necessary step because without this modification,

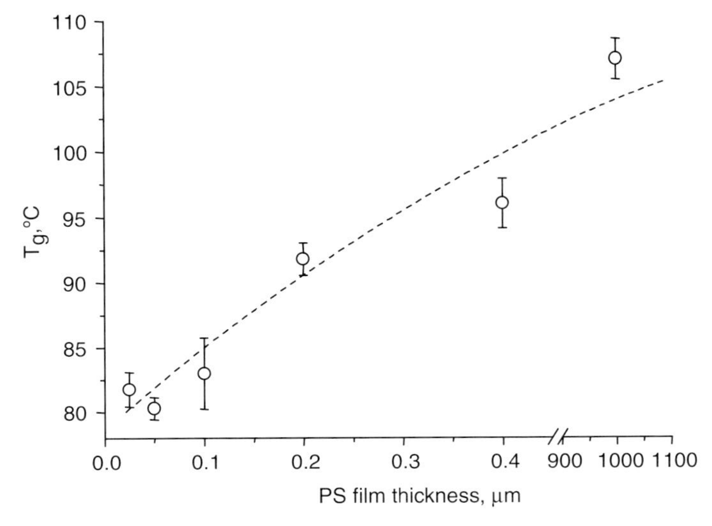

Figure 6.1 Change of glass transition temperature of PS film with the film thickness. Reprinted from Ref. [3].

a large imbalance between the heat dissipation of main and reference probes prevents any meaningful microthermal measurements of ultrathin polymer films. With the modified experimental setup, the thermal sensitivity of the microthermal probing increased dramatically, allowing the unambiguous detection of very minute heat dissipation variations (with overall effects well below 1 μW) associated with the ultrathin polymer film itself and not the highly thermally conductive substrate.

An analysis of the experimental micro-TA data for ultrathin PS films shows that T_g decreases when the film thickness decreases to less than 400 nm (Figure 6.1 compares these data with results for bulk PS film with a thickness of 1 micrometer). For the thinnest film presented in this plot (between 20 and 50 nm), T_g decreases by as much as 20 °C from its bulk value. These results confirm a general trend observed for ultrathin polymeric films deposited on solid substrates with weak film–substrate interactions derived from macroscopic thermal measurements.

6.2.2
Polymer Brushes

The microthermal and mechanical properties of ultrathin (50–90 nm) grafted-from polymer brush layers and molecularly thick (5–10 nm) grafted-to polymer brush layers have been recently studied by Tsukruk's group. Glass transitions for glassy PSF brush layers with homogeneous surface morphology were directly measured with SPM probing at different temperatures at a thermal stage [4]. At room temperature, an elastic modulus of approximately 1 GPa was determined for the PSF brush layer at below its glass transition.

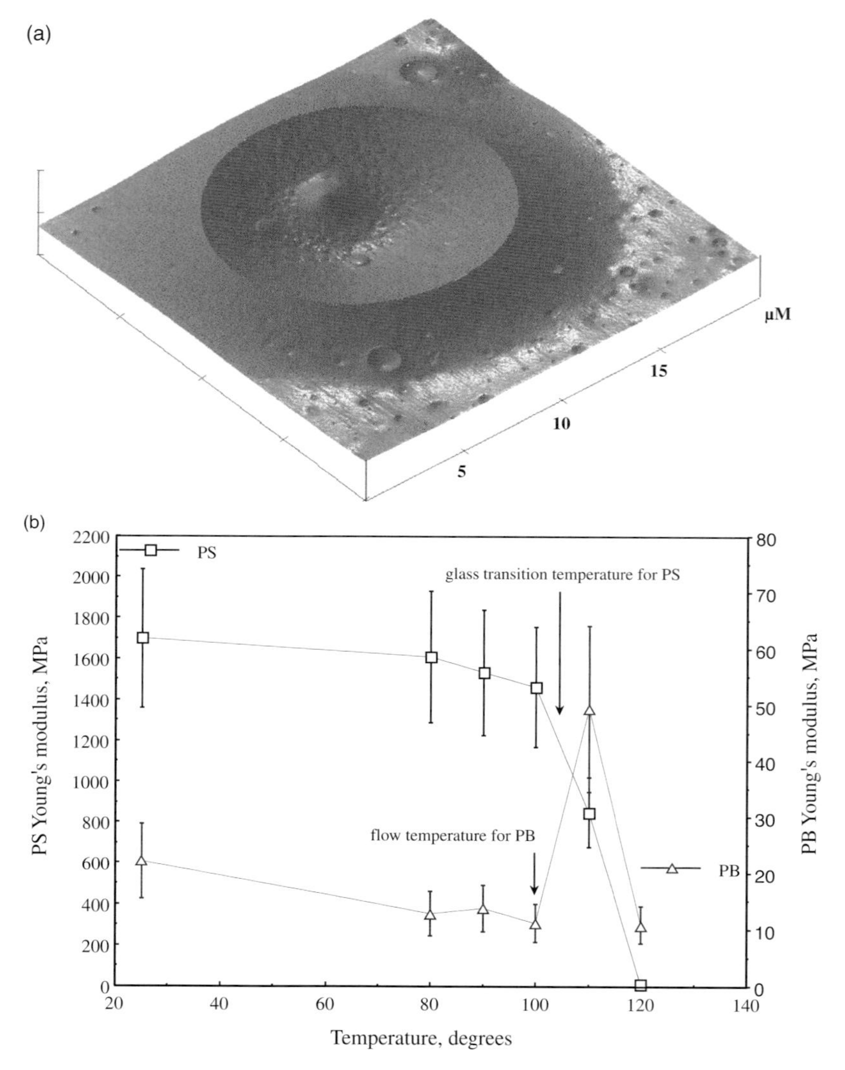

Figure 6.2 AFM topography of a PB rubber inclusion in PS matrix (a); temperature variation of elastic moduli for glassy PS and rubbery PB phases derived from surface mechanical micromapping collected at different temperatures (b). Reprinted from Ref. [5].

Heating the glassy PSF layer above room temperature resulted in a gradual decrease of the elastic modulus caused by the approaching glass transition. Finally, at highest temperatures, this transformation is completed and the brush layer is converted to the rubbery state with an elastic modulus close to 15 MPa. The fastest transformation is observed close to 100 °C indicating the glass transition behavior of a typical bulk, high molecular weight polymer with a similar T_g.

Direct measurement of heat dissipation and thermoelastic response within the PSF brush layer conducted independently with micro-TA confirmed that the glass–rubber transition occurs between 100 and 110 °C as expected for a bulk, unconfined polymer [4]. These data demonstrate the sensitivity of present micro/nano-TA designs to probe nanometer thick polymer films. The glass transition values close to those for unconfined polymers indicate an undisturbed composition and a microstructure of relatively thick polymer brush layers grown from a reactive silicon surface by the grafting-from technique.

6.2.3
Thin Films from Polymer Blends

Glassy–rubbery blends have been probed with FV micromapping in a range of temperatures by exploiting a thermal stage. To this end, Gorbunov *et al.* demonstrated that for heterogeneous thin films fabricated from mixed glassy and rubbery polymers, the micromapping of elastic properties can be obtained concurrently for both glassy and rubbery phases [5]. PS–PB films studied in detail are highly heterogeneous with well-developed phase-separated rubbery PB phase droplets of several micrometers distributed across a uniform, smooth, and glassy PS matrix (Figure 6.2).

The surface distribution of elastic and adhesive properties displays two very distinctive maxima corresponding to different phases. Such a micromapping allows the conducting of separate but concurrent measurements of micromechanical properties of glassy and rubbery phases in these composite thin films, as well as across the glassy–rubbery interface with a lateral resolution better than 100 nm (see Chapter 5). Surface micromapping including topography, elastic modulus, and adhesive force data obtained concurrently with the aforementioned spatial resolution showed that for the rubbery phase, the adhesive forces are much higher and the elastic modulus is much lower than that for the PS matrix, as has been discussed in detail in Chapter 5. At room temperature, the elastic modulus was measured to be 6 MPa and 2.5 GPa for the rubbery phase and the matrix, respectively.

The temperature dependence of both values of the elastic modulus was obtained by repeated micromapping at a series of elevated temperatures and deriving surface characteristics separately for glassy and rubbery phases. The integrated plot representing temperature variation of localized moduli are presented in Figure 6.2. In this plot, the absolute value of the elastic modulus for the PS matrix displayed a sharp drop to 11 MPa within a narrow temperature interval of 100–120 °C after being relatively constant between room temperature and 90 °C. This sharp drop indicates that glass transformation within the surface layer of the PS matrix (probed to about 5 nm depth at the largest indentation) occurs in the same temperature range as that for the bulk polymer (103 °C for bulk PS) [6].

The average elastic modulus of the rubbery PB phase was measured in the course of the same measurement cycle. The value decreases steadily for temperatures up to 100 °C and abruptly rises around 110 °C and then sharply decreases again. Because the elastic modulus decreases at the highest temperature, which is higher than the flow temperature of the PB phase (100 °C for PB with a given molecular mass), the rubbery phase loses its elastic resistance. Thus, at this temperature, the SPM tip penetrates through the viscous PB phase and detects the PS surface beneath, which is still on the verge of transformation to the elastic state, causing a sharp increase of the apparent elastic modulus value of the rubbery phase.

Alternatively, direct imaging with SThM of the PS–PB blend film revealed clear differences of rubbery and glassy phases (Figure 6.3). Thermal dissipation is much higher in the rubbery inclusions. Considering that the preheated thermal tip is in direct physical contact with the rubbery phase, high heat dissipation can be explained by a higher physical contact area under these conditions. The apparent changes in the physical contact area for heterogeneous blends are a much smaller contact area for the stiff glassy matrix and a much higher contact area when tip is in the contact with the soft rubbery phase.

The authors concluded that the differences in local heat dissipation can be easily detected with micro-TA if the difference in the surface thermal conductivities between different phases is large enough (at least two–three times). The lateral resolution can reach 20–30 nm for relatively stiff surfaces. This level of resolution has been demonstrated for a square-patterned silicon oxide with a low thermal conductivity on a single-crystal silicon substrate with a high thermal conductivity [7].

Overall, these results display that fine details of the thermal transitions of amorphous polymers (glass transitions and flow temperature) can be unambiguously probed with high spatial resolutions (below 100 nm) by the micro-TA approach. Thus, fundamentally new questions, such as the thermal behavior of selected localized

Figure 6.3 SThM scanning of rubber inclusion in glassy matrix: topography (right) and thermal dissipation (left). Taken from Tsukruk and Gorbunov, unpublished.

regions in the vicinity of interfacial zones or for surface and subsurface regions, can be addressed.

6.2.4
Depth Variation of Glass Transition in Photodegradable Polymers

It is known that chain scission can cause changes in the glass transition temperature on the polymer surface as a function of depth during photodegradation caused by exposure to light [8]. In this study, depth profiling of glass transition for these exposed surfaces has been accomplished by sectioning the sample and probing the microthermal response across the sectioned specimen. Figure 6.4 shows the T_g of a polycarbonate section as a function of depth for two UV exposure times, one before the introduction of oxygen (50 h) and one after the introduction of an oxygen atmosphere (200 h). The authors observed that at the shorter exposure time, there is a decrease in T_g at the surface relative to the bulk material, which can be probed with submicrometer resolution (Figure 6.4). It is apparent that subsurface regions experience less photodegradation as the light is predominantly absorbed in the surface and topmost regions.

On the other hand, the T_g drops for higher depths because the oxygen has not penetrated sufficiently to reverse the decline in the glass transition temperature caused by the 100 h of UV irradiation without oxygen. The T_g tends to return to the bulk value at greater depths indicating an intact inner material protected by the topmost layer (Figure 6.4). A submicrometer characterization of depth-dependent photooxidized polypropylene has also been conducted with micro-TA and a protocol was established to allow the microscopic resolution in the lateral direction [9]. These micro-TA studies can be conducted concurrently with ATR–FTIR measurements to correlate thermomechanical properties with variations in chemical composition for comprehensive depth profiling of polymeric materials.

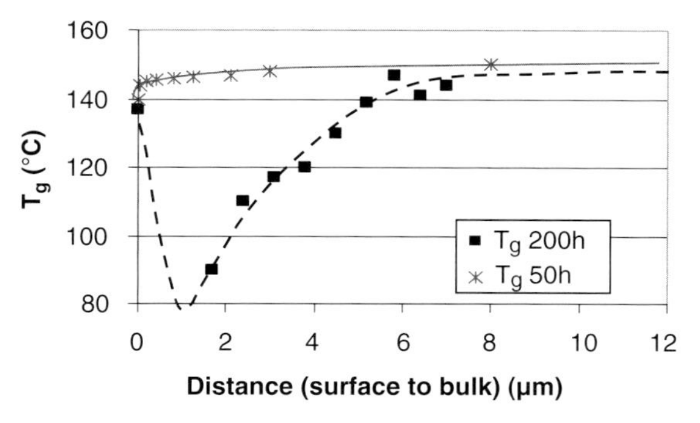

Figure 6.4 Change of glass transition temperature of polycarbonate as a function of depth with different exposure times to UV radiation and oxygen (200 and 50 h). Reprinted from Ref. [8].

6.3 Melting, Crystallization, and Liquid Crystalline Phase Transformations

During heating, a polymer may undergo one or more transitions over the temperature range in addition to a glass transition. The most commonly observed physical and chemical transformations include cold crystallization (recrystallization), curing, melting, clearing, or liquid crystalline transformations [10]. For a material that undergoes no thermal transitions, the probe deflection with temperature will be essentially linear and upward as the sample beneath the probe heats and expands in a linear manner. The rate of upward deflection will depend on the coefficient of thermal expansion, thermal conductivity, and heat capacity of the material. The heating of the probe element will itself cause some movement of the cantilever and possible thermal creep of the piezoelement, but for the relatively massive Wollaston wire thermal probe, these effects should be minimal. Provided a baseline subtraction procedure is carried out properly (acquired from a run with the probe in free air or in contact with a reference substrate), the rate of power consumption of the probe over the duration of the same experiment on a sample should remain virtually constant and can be easily removed from experimental data.

As an illustration of the microthermal behavior with multiple phase transformations at elevated temperatures, the response of the micro-TA signal of PET is presented in Figure 6.5 [10]. These data were selected to show a typical micro-TA scan for a crystallizable but initially amorphous polymer caused by fast cooling (quenching). Thus, the heating cycle covers cold crystallization, glass transition, and melting all in one run (Figure 6.5). There is a large indentation right above the glass transition temperature caused by dramatic reduction of the elastic modulus in the

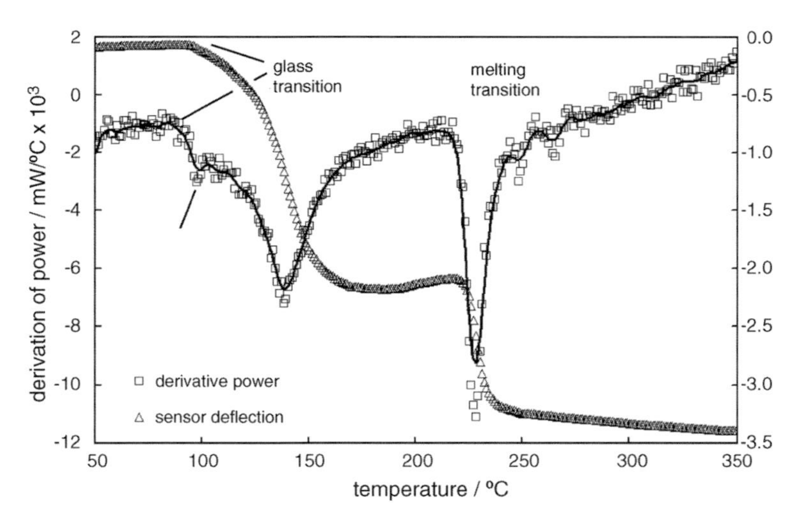

Figure 6.5 Micro-TA scan for quenched PET showing cold crystallization, glass transition, and melting. Reprinted from Ref. [8].

rubbery state. A further large and sharp indentation at the melting temperature is also associated with thermally induced reduction in the mechanical contact area during transition to the viscous melting state.

The heating rate dependence on the melting temperature for PET was evaluated independently by Gorbunov and coworkers from micro-TA measurements [11]. The authors exploited a wide range of heating rates from 10 to 1500 degree/min to verify the consistency of conventional DSC and micro-TA results. As expected, melting temperatures measured in this range shifted to higher values with the increasing heating rate for both macroscopic (DSC) and microscopic (micro-TA) measurements. The extrapolation of the apparent values of the melting temperature to zero heating rate allowed obtaining the “equilibrium” melting temperature from micro-TA data (usually obtained at much higher heating rates) that is very close to the literature value and that derived from concurrent DSC measurements at lower heating rates.

As studied with AFM with thermal stage by Luzinov *et al.*, thermoplastic elastomeric materials are capable of forming thermally reversible, physical cross-links to glassy domains, as demonstrated in Figure 6.6 for SEBS block copolymers [12]. These nanostructured materials behave as vulcanized rubber in many respects, while the PS domains soften when heated above T_g. The morphology of the SEBS grafted film of 8.4 nm thickness was investigated by the authors under room temperature to 125 °C (Figure 6.6). This temperature range covers the glass transition of PS domains (75 °C for this molecular weight), as well as the flow temperature (~100 °C) for the PS/PB matrix [13].

The high-resolution phase images of these molecularly thin films clearly show PS microdomains preserved at modestly elevated temperatures (below or close to the glass transition of PS domains) that are identical to the surface morphology observed at room temperature (Figure 6.6). Furthermore, at 105–110 °C, or above the glass transition and slightly above the flow temperature of PS microdomains, characteristic surface morphology can be still visible, although the edges of the domains become fuzzy.

When temperatures reach 115 °C, only isolated fuzzy PS domains can be found on phase images. Finally, at 125 °C, which is well above the flow temperature, the individual PS microdomains’ surface morphology is completely melted and the fine microdomain morphology is no longer detected. At this temperature regime, a SEBS film looks virtually homogeneous indicating both a complete melting (or order–disorder transition) and the forming of a uniform polymer melt still covalently grafted to the substrate. Observed changes in surface morphology are completely reversible and the microdomain structure is completely restored after cooling due to the grafted nature of these monolayers (Figure 6.6). These results reveal that at the temperature well above the T_g of PS, the PS chains are still segregated into a microdomain structure within the grafted film due to confined conditions within the grafted molecular layers. The microdomain network becomes undetectable only when the temperature of the block copolymer film reaches the order–disorder transition or flow temperature for bulk materials.

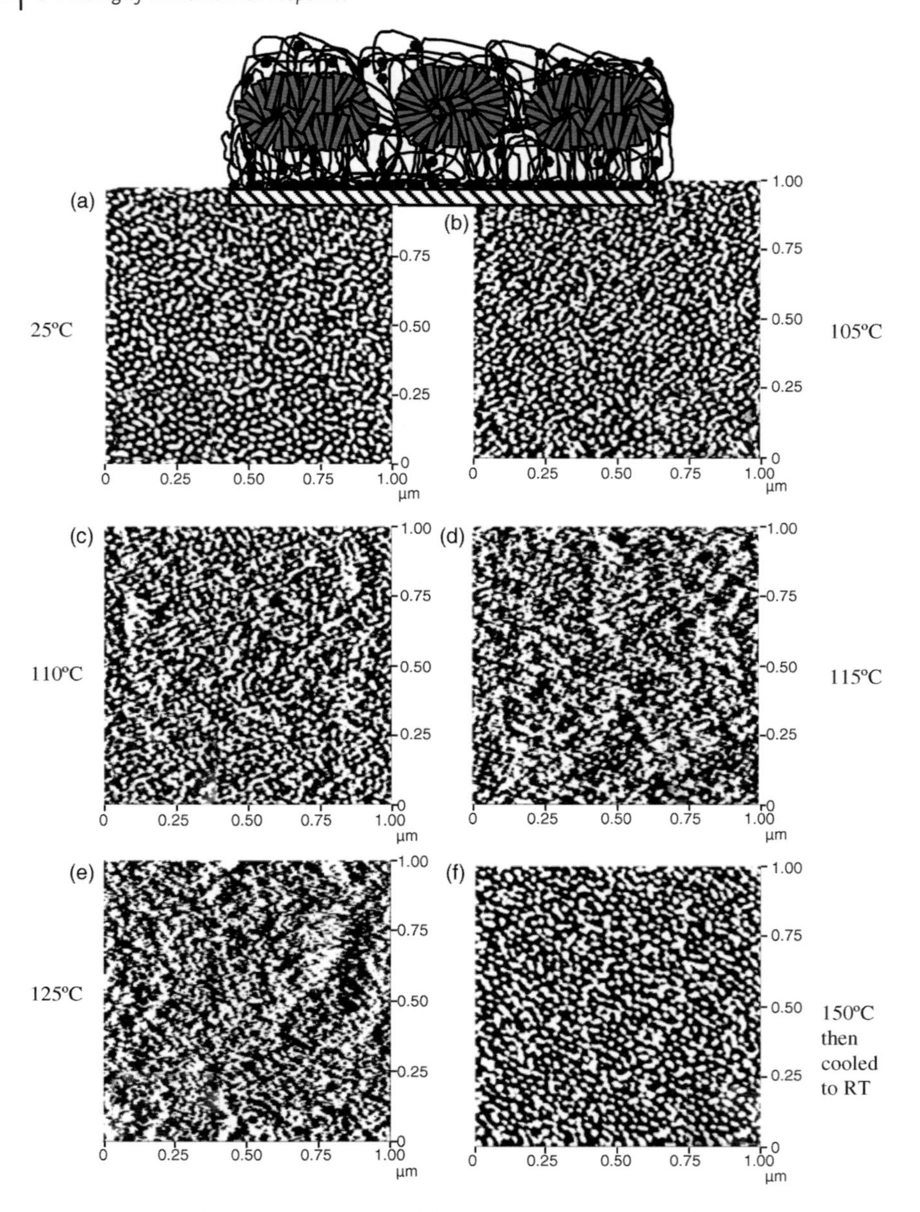

Figure 6.6 AFM phase images of SEBS grafted film (8.4 nm thick, see schematics on top) recorded at different temperatures. (a) 25 °C; (b) 105 °C; (c) 110 °C, (d) 115 °C; (e) 125 °C; and (f) SPM phase image after thermal treatment at 150 °C for 7 h. Bright parts correspond to higher phase shifts. Vertical scale is 10°. "Hard" tapping regime. Reprinted from Ref. [12].

In another recent study, triblock copolymers of PEO–PS–PE and PEO–PS–PB types with crystallizable PEO and PE blocks were investigated by Boschetti-de-Fierro *et al.* [14]. The effect of thermal treatments was considered from the point of view of the surface distribution of PEO and PS chains under different thermal treatments as controlled *in situ* on a thermal stage. Besides the direct study of thermal transformations mentioned above, it has been demonstrated that the variation of temperature regimes can be successfully exploited in a very practical manner to increase contrast and spatial resolution while observing block copolymers in a high-resolution mode [15].

A series of papers on the applications of SPM imaging at elevated temperatures to observe the variation of surface morphologies at different thermal transitions (melting, crystallization, liquid crystalline transformations) for a wide range of common polymeric materials have been published by Magonov and coworkers. In one of these studies, the crystal structure and morphology of a syndiotactic PS material under different crystallization conditions was investigated at a thermally controlled stage [16]. The authors demonstrated that the resulting surface morphology in the case of melt crystallization is determined by the annealing/preheating temperature range. Indeed, a poorly ordered α-phase is formed after preheating to 280 °C, although preheating to 320 °C leads to the thermodynamically stable β-phase. The authors suggest that low nucleation and growth rates at certain temperatures promote crystallites with 85 nm lateral dimensions. The somewhat better ordered α-crystalline phase was achieved by annealing the quenched melt at the elevated annealing temperatures. Proper annealing leads to an improvement of the β-crystallites and finally results in complete spherulitic morphology.

In situ AFM imaging on specimens at elevated temperatures has been exploited by the same authors for the visualization of the crystal morphology and nanostructure upon melting and repeated crystallization of various polymers. In this study, ultrathin (20 nm) low-density polyethylene (LDPE) films on silicon substrates have been extensively studied under different temperature regimes [17]. The original surface morphology of such films was represented by quasi-two-dimensional spherulites with clearly visible lamellar aggregates. Significant morphological changes were observed after preheating LDPE to temperatures close to the melting temperature. The original morphology disappeared upon stepwise heating and new individual lamellar aggregates have been observed. At higher annealing temperatures, the quasi-2D spherulite morphology has been transformed into the network of lamellar branches. AFM imaging also revealed that the crystallization of the ultrathin LDPE film was accompanied by the dewetting of the melted regions that causes a large number of microscopic holes.

The melt crystallization of PET, a popular practically important polymer for thermal studies, has been exploited in another study of Magonov and coworkers [18]. The direct AFM observation of the lamellar morphology in the course of secondary crystallization allowed the direct monitoring of the thickening process of crystalline lamellae, which could be directly compared with independent X-ray measurements. The melt crystallization of PET was conducted at 233 °C and AFM imaging was performed by scanning the same surface area at different annealing times to follow secondary crystallization processes at a microscale (Figure 6.7).

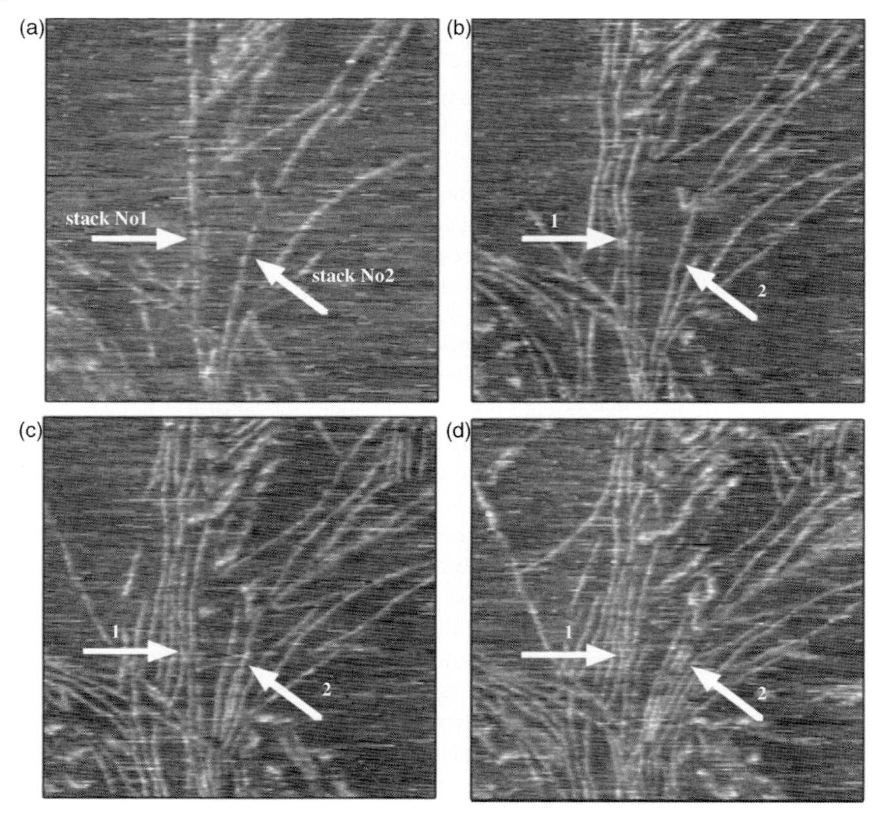

Figure 6.7 AFM phase images (1 μm²) recorded in hard tapping during crystallization of PET from the melt at 233 °C; elapsed times are (a) 72, (b) 80, (c) 96, and (d) 112 min. Thin white stripes correspond to the crystalline lamellae in almost edge-on orientation. The arrows identify the lamellar stacks and the same spots on the sample surface where the crystal occurred. Reprinted from Ref. [18].

The gradual evolution of edge-on-oriented crystalline lamellae observed on these images showed a gradual growth of new lamellar crystals in an available nonstructured, amorphous matrix. The crystalline lamellae gradually formed stacks from several individual structures, the main mode of the lamella thickening mechanism of crystallization. AFM images showed new crystallites that grow parallel to the already existing lamellar stacks, effectively increasing the average number of lamellae per stack in this surface area (compare stacks marked 1 and 2 in Figure 6.7).

The focus of another study of Magonov and coworkers was on the surface morphology and nanostructure of the elastomeric material, polydiethylsiloxane (PDES), which shows all melting, crystallization, and mesophase formations at different temperatures [19]. Morphological changes that have been recorded directly on a thermal stage with the AFM correlate with the thermal transitions determined independently by DSC. AFM studies in this temperature range discovered two different types of mesophases that possess different domain morphologies. One mesophase possesses domains of several micrometers in size, whereas the other mesophase shows much larger lamellae having a length of several tens of micrometers and a width of several hundred nanometers.

Further shearing PDES into a very thin film induced the formation of the heterogeneous mesomorphic structures, which are embedded in the amorphous material. Each lamella has a skeleton formed of 10–15 nm thick linear structures, which are separated at 40–50 nm and tapered at the ends to form a single structural element. Furthermore, crystallization of these mesomorphic lamellae at lower temperatures leads to more ordered and stiffer top layers of lamellae. A few round-shaped domains that have dark contrast in phase imaging have also crystallized with the preservation of their shape.

High-resolution AFM images of these morphologies showed that the crystalline lamellae have a complex nanostructure that manifests itself in nanoscale modulations arranged in crossed patterns. The surface of the mesomorphic lamellae exhibits some strips that are aligned perpendicular to the main direction. Cooling of the specimen to $-10\,^{\circ}\mathrm{C}$ results in the breaking of mesomorphic lamellae into multiple crystalline blocks, which is potentially a result of the nucleation processes. Decreasing the temperature to $-15\,^{\circ}\mathrm{C}$, which is well below the melting temperature, resulted in the complete crystallization of all mesomorphic structures.

6.4 Thermal Expansion of Microstructures

Direct measurement of the thermal expansion requires the precise estimation of selected morphological features' dimensions at various temperatures, which is a nontrivial task. Thus, it should be accomplished only for special cases when macroscopic methods are not suitable. Here, we present some selected recent results on AFM measurements of thermal expansion for multilayered biological receptors.

Microstructural organization of biological infrared receptors was studied to elucidate their materials properties that could be useful for prospective biomimetic design of artificial photothermal sensors from organic/polymeric materials [20]. For these studies, the authors selected *Melanophila acuminata* beetles that are capable of distant detection of forest fires via paired thoracic pit organs [21]. Studies of beetle behavior under external thermal stimuli and their physiological responses have shown that the pit organs, which serve as thermal receptors, are most sensitive to thermal radiation in the wavelength range of 2–4 μm. In behavioral experiments, a radiation intensity of only 60 $\mu W/cm^2$ at a wavelength of 3 μm was sufficient to elicit a fast response.

Unique microstructural organization of the cuticular apparatus of these beetles is represented by a massive endocuticular sphere, which is placed in an internal cavity under a thin cuticular dome (Figure 6.8a). The sphere, filled with soft matter such as water and wax, is innervated from below by the sensory dendrite of a single mechanoreceptor. At a larger scale, these receptors are arranged in two-dimensional arrays of 50–100 IR sensilla located in deep pit areas.

The thermal receptors in *M. acuminata* beetles with ultrahigh-resolution SPM in a range of temperatures have been studied by Hazel *et al.* [22]. By applying micromechanical mapping and direct visualization at a thermal stage, the micromechanical and thermomechanical properties of the cuticular apparatus of the sensillum were revealed. As was observed, the main component of the cuticular apparatus is an internal endocuticular sphere with a diameter of about 15–20 μm (Figure 6.8).

Multilayered organization of the lamellar peripheral mantle of the sphere was confirmed and characterized in great detail (Figure 6.8). High-resolution AFM imaging of the lamellated area of the sphere showed the typical structure for the endocuticle. About 5–10 curved layers are densely packed and their spatial arrangements are closely correlated. Interlayer distance varies along the perimeter with the average periodicity of 300 ± 100 nm in the vicinity of the apex. This periodicity

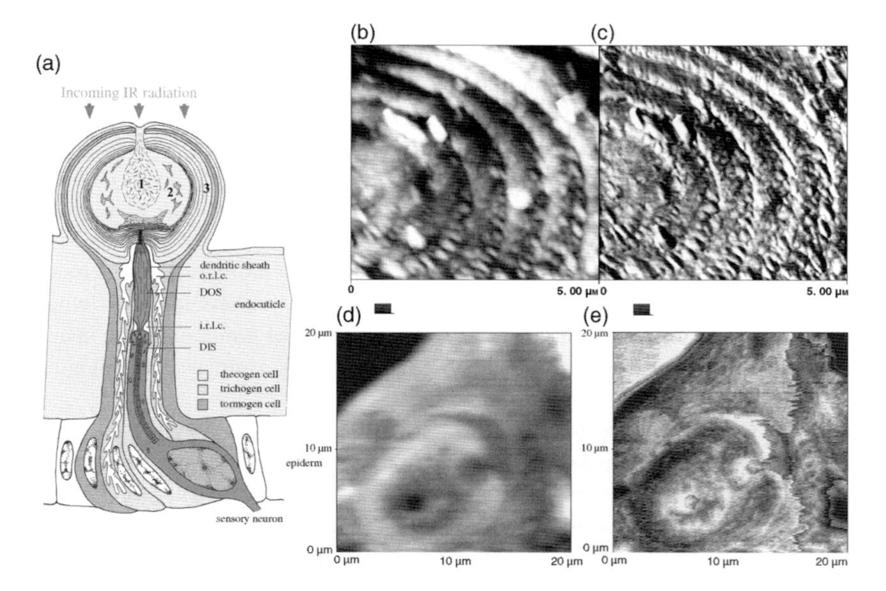

Figure 6.8 Schematic drawing of the IR sensillum (a). The internal cuticular sphere is composed of three distinguishable areas (1–3) and covered by a thin outer cuticle. Diameter of the sphere is about 12 μm. SPM topography (b) and lateral/friction force images (c) of multilayered structures in outer areas of IR receptors at intermediate magnifications. SThM topography (d) and heat dissipation (e) images of a section of the IR receptor. Reprinted from Ref. [20].

increases to 1 μm in the bottom part of thermal receptors. A multilayered structure in the areas close to the receptor tip consists of wide elevated strips separated by deep, sharp grooves.

SThM imaging of these receptors shows concurrently obtained surface distributions of topography and the local heat dissipation associated with the thermal conductivity of receptor sections (Figure 6.8). Due to the shape of the thermal probe (with an effective radius of about 5 μm), the spatial topographical resolution of this mode was limited to several tenths of a micrometer. However, even under these resolution constraints, close correlation was observed between surface topography and the thermal signal distribution (Figure 6.8). Between the areas of the receptor tip and receptor bottom, significant depletion was observed and multilayered edges were elevated. Heat dissipation detected simultaneously with topography recording was much higher than the background level for depletions in both the tip and bottom areas of receptors. The lowest heat dissipation was observed along the elevated peripheral area.

The surface distribution of heat dissipation is affected by the variation of the tip–surface contact area caused by topography. Due to obvious geometrically based contributions of the thermal tip in the depleted parts of receptors, the physical contact area increases and along the ridges, the physical contact area decreases. This variation should result in a corresponding increasing and decreasing of the heat dissipation, even if the actual local thermal conductivity is unchanged. Due to complicated tip–curved surface topography interactions, separation of these contributions cannot be done in a simple analytical and quantitative way.

Thermal expansion of the multilayered thermal receptors of these beetles was studied by scanning specimens of receptor cross sections at elevated temperatures (Figure 6.9). Samples were heated by 5–10 °C with a thermal AFM stage, equilibrated, and AFM images were carefully taken from the same surface area at identical scanning conditions. As is clear from the images obtained at different temperatures, the multilayered structure undergoes significant changes during this heating cycle. Although surface topography remained virtually unchanged during heating, the corresponding phase images changed completely at elevated temperatures; specifically, heating to intermediate temperatures (40–60 °C) led to a significant increase of dark areas on the phase images (Figure 6.9).

Further heating of receptor sections resulted in dramatic changes of the phase image: clear and contrast layering occurred with the formation of a distinctive multilayered pattern that exhibited large differences in the phase shift between adjacent layers. This change in phase contrast demonstrated that softening of the multilayered structure occurred nonuniformly, with interlayers becoming more compliant at elevated temperatures. This observation is a confirmation of the fact that the layers located in the peripheral area of the sphere are composed of alternating materials with different mechanical and thermal properties. Moreover, an intense thermal expansion of the outer mantle was observed, and the local thermal expansion coefficient of biological multilayered receptors was estimated to be about $1.5 \times 10^{-4}\,\text{grad}^{-1}$.

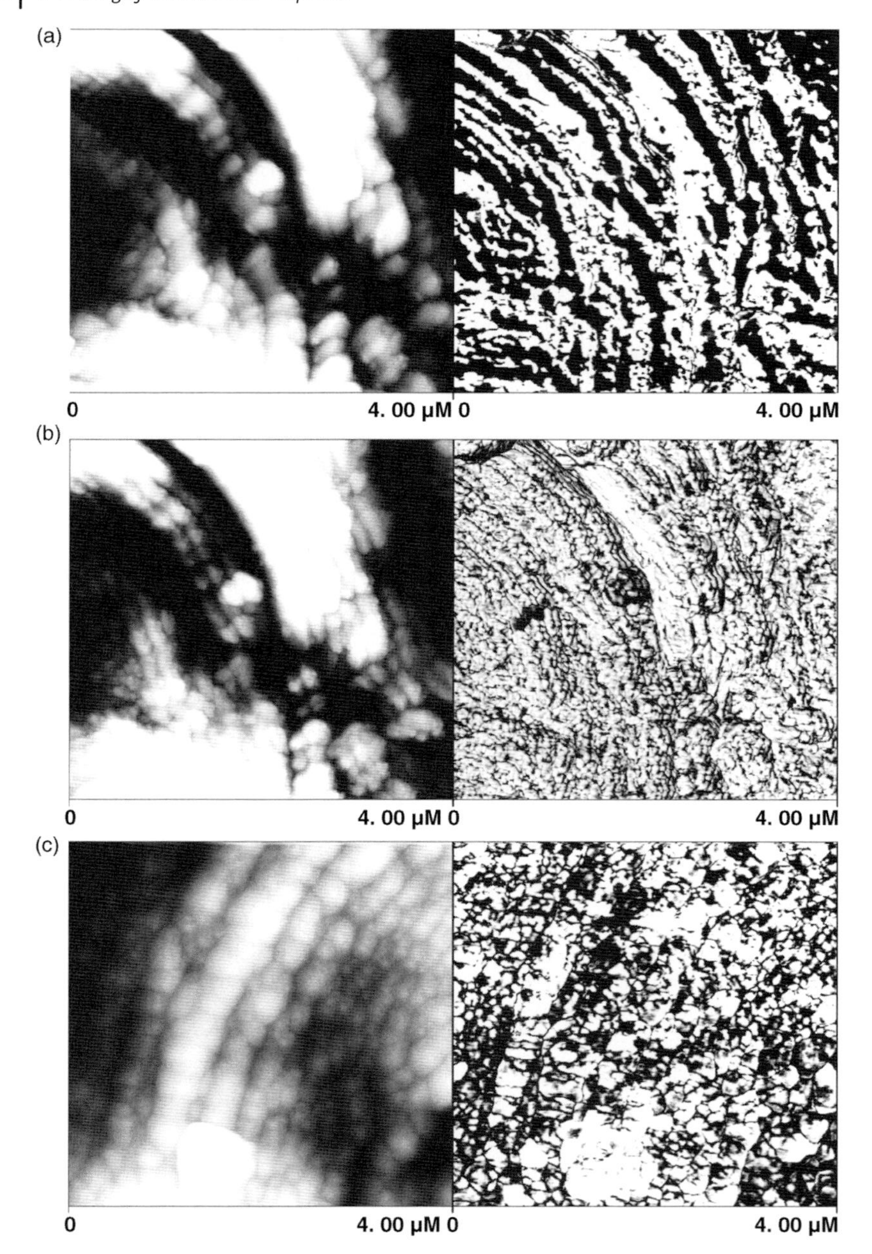

Figure 6.9 SPM topography (left) and phase (right) images of multilayered structures in outer areas of IR receptors at different temperatures utilized for the evaluation of thermal expansion: (c) 25 °C, (b) 50 °C, and (a): 70 °C. Reprinted from Ref. [22].

6.5 Surface Thermal Conductivity

Surface thermal conductivity can be derived from the differences in heat dissipation measurements for the thermal tip far from the surface and in direct physical contact after engagement, as has been discussed in Chapter 4. Here, we present an example of such an analysis conducted for a number of variously thermally conductive polymeric and metal surfaces, as well as biological tissues [20, 23, 24].

In these studies, the approaching–retracting mode of the operation was used to collect the force–distance data concurrently with the thermal signal for a number of different surfaces. The collection of force–distance data and thermal dissipation begins when the thermal tip is located several micrometers above the surface (Figure 6.10). Force–distance data were used to define the precise point of physical contact between the thermal probe and the surface. The collected thermal data for each material studied in this work ranged from 35 to 80 °C. In order to analyze the difference in heat dissipation before and after the direct physical contact between the thermal tip and the surface, ΔQ, Eq. (4.4) was exploited.

The principles developed in the course of thermally probing these materials with known thermal conductivities have been applied to biological tissues with unknown thermal properties, such as the skin and thermal receptors of snakes, by Fuchigami *et al.* [20]. Figure 6.11 displays a snake head with arrows indicating location of the thermal pit receptors that were studied in their work. Testing of the microthermal properties of a snake skin surface involved several independent experiments. First, the receptor organ surfaces were scanned in SThM mode to determine whether any singularities in the surface thermal conductivity could be observed at a microscale

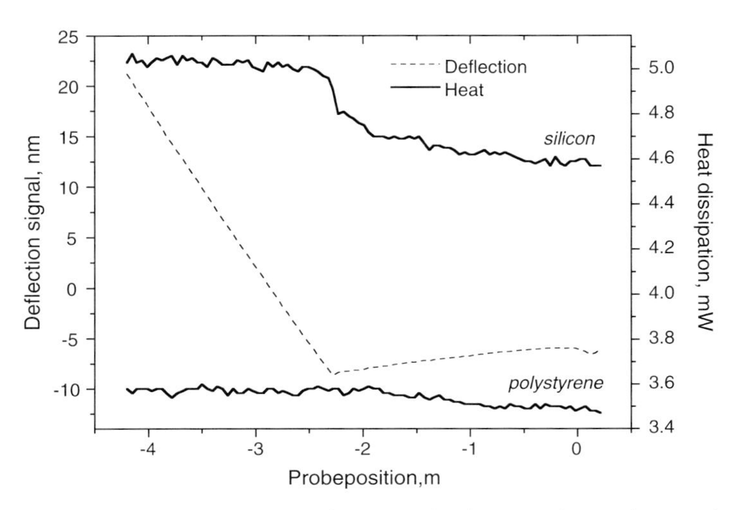

Figure 6.10 Thermal dissipation of the thermal probe versus distance between thermal probe and silicon and PS surfaces (solid lines) along with probe deflection data (dashed line). Reprinted from Ref. [3].

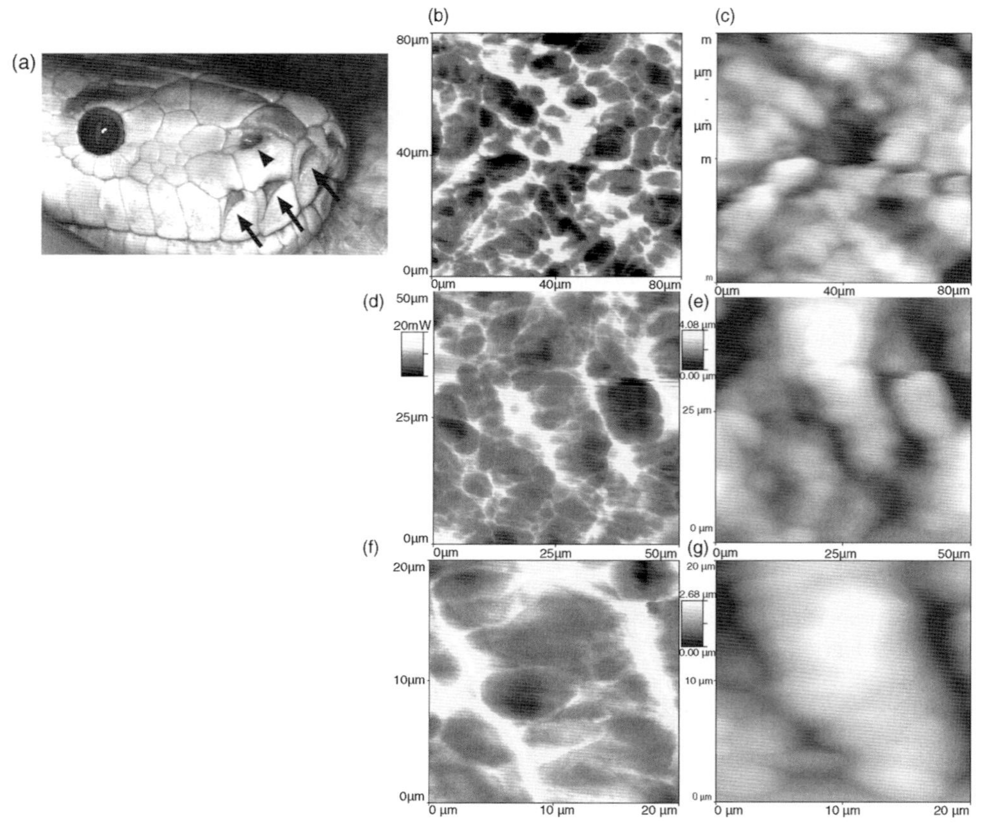

Figure 6.11 Optical image of Ball python with thermal receptors indicated by arrows (a). Topography (c, e, and g) and surface distribution of heat dissipation (b, d, and f) for formaldehyde-treated pit organ surface of Ball python at three different magnifications. Reprinted from Ref. [20].

(Figure 6.11). For all the surface skin areas studied, an uneven distribution of heat dissipation was observed with the variation of several microwatts.

The comparison of the surface topography and the heat dissipation distribution showed a close correlation between topographical features and spikes observed during heat dissipation. Shallow areas displayed higher heat dissipation, while raised areas showed much lower local heat dissipation. This difference suggests that the observed uneven surface distribution of the heat dissipation was caused mainly by the topographical contribution of uneven skin surface areas. Variable surface topography leads to a varying contact area between the thermal tip and the surface and, therefore, leads to a highly variable integrated heat transfer at the contact point.

The heat dissipation of the thermal probe in the proximity of the snake skin surface has been probed under two different scenarios. The first approach relied on the measurement of the heat dissipation ΔQ directly at the moment of the physical contact between the thermal probe preheated to temperature T_p and the snake skin surface with an initial temperature T_s (Figure 6.12). As discussed above, this

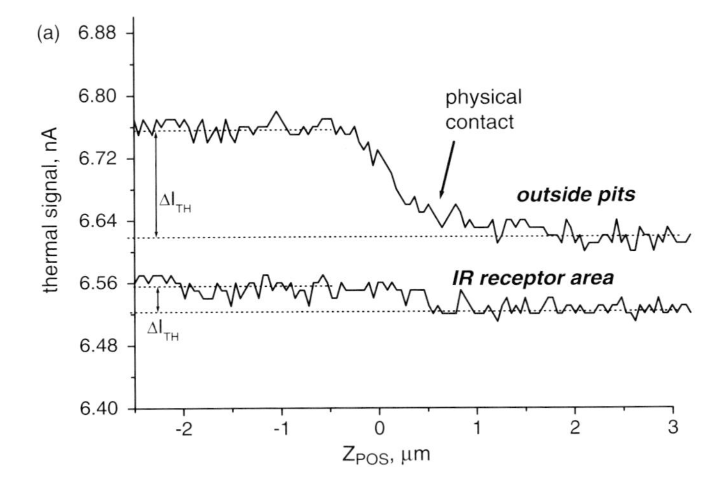

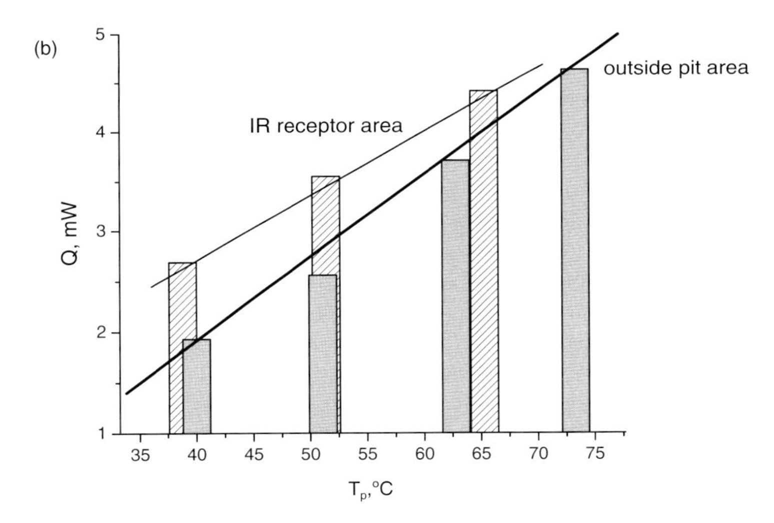

Figure 6.12 (a) Heat dissipation as a thermal probe approaches the surface of the pit organ receptor and an area outside the pit organ. (b) Heat dissipation at physical contact at different initial temperature differences between thermal probe and surface for pit organ and non-specific snake skin areas. Reprinted from Ref. [23].

approach allows the estimation of the surface thermal conductivity in accordance with Eq. (6.1).

The second approach was based upon the instant monitoring of the heat dissipation $Q(T)$ at the contact point after tip engagement during the temperature increase. The quasi-steady-state model of the heat transfer between a pointlike heat source and a planar surface predicted a linear relationship between the temperature-normalized heat dissipation and the surface thermal conductivity, which can be exploited to quantitative evaluations. Indeed, in this mode, the heat dissipation ΔQ

increased after the physical contact of the thermal tip with the snake skin surface (Figure 6.12). Repetition of such probing using different initial tip temperatures showed a virtually linear correlation of the heat dissipation and initial temperature difference ΔT. This slope provided an independent evaluation of the surface thermal conductivity. Both independent approaches gave similar results on thermal conductivity and will be discussed below.

The temperature dependence of the heat dissipation for various initial temperature differences was measured for the surface areas of pit organ and nonpit skin of snakes. These measurements revealed that the value of thermal conductivity for pit organ receptor areas was much lower than that for the surface areas outside the snake receptor organs. The absolute value of thermal conductivity was determined to be $\lambda = 0.11$ W/mK for pit organ surfaces and $\lambda = 0.34$ W/mK for nonspecific skin areas outside the pit organs. Moreover, the surface thermal conductivity of the pit organ receptor area is much lower than the typical values for the vast majority of organic and polymeric materials (0.15–0.4 W/mK).

Therefore, the major conclusion from these microthermal measurements is that snake thermal pit organs possess a surface thermal conductivity much lower than the surrounding nonspecific skin areas and common solid soft materials. To understand the implications of these results for photothermal sensing, simulations of the surface temperature variation were conducted for pit receptors and nonpit surface areas in the presence of a thermal source above the surface (Figure 6.13). For these simulations, the model of quasi-steady heat transfer adapted to SThM measurements, which

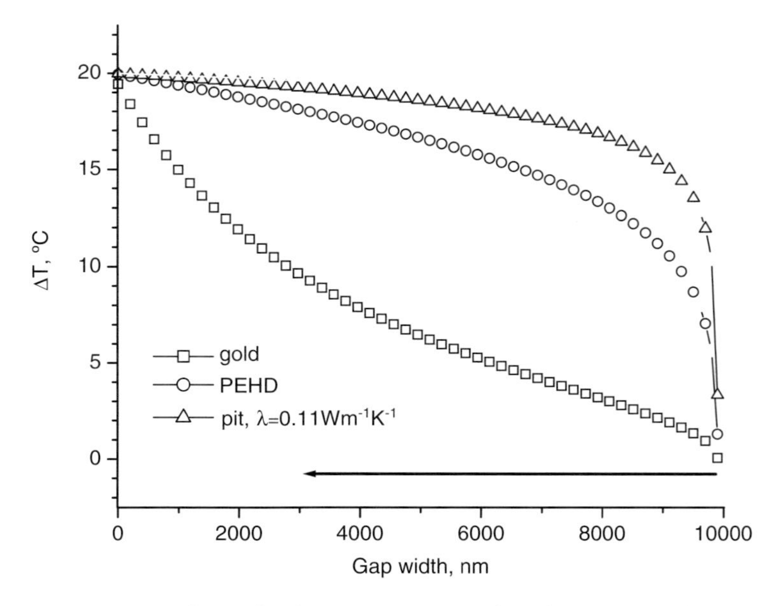

Figure 6.13 Simulation of surface temperature variation beneath the thermal tip for different materials: surface temperature for pit organ receptors, outside receptor areas, and gold surfaces for a stationary thermal tip positioned at 10 μm above the surface with the tip temperature of 40 °C "turned-on" at time zero. Reprinted from Ref. [23].

was discussed previously (Chapter 4), was used. It was assumed that the published values for the heat capacity and density of water were equivalent to that for snake tissue.

The surface temperature variation as the thermal probe approached the surface from an initial distance of 10 μm has been simulated in this study (Figure 6.13) [23, 24]. A rapid increase in the surface temperature as the thermal probe approached the low thermally conductive surface has been obtained in this simulation. The surface temperature reached the initial temperature of the thermal tip when the probe was at 5 μm above the surface for an initial difference in temperatures of 20 °C.

In contrast, the surface temperature of the areas outside the pit organ receptors with a higher thermal conductivity rose at a slower pace (Figure 6.13). The surface temperature of a highly conductive material, such as gold, simulated for comparison increased at a very slow rate under similar conditions (Figure 6.13). In addition, the authors simulated temperature variation on the surface beneath a stationary thermal probe located above the surface after its temperature was turned on.

These simulations demonstrated that the low thermal conductivity of the pit receptors prevented rapid heat dissipation from the surface area and, therefore, caused rapid local temperature rise. If the stationary thermal source is turned on far from the skin surface, a surface temperature gradient between the receptor area and the surrounding nonspecific skin area is established immediately. These simulations confirmed that the temperature difference between different surface areas (pit and regular skin) might reach 3 °C within several seconds after turning on the thermal source and this difference might persist over a long time period before thermal equilibration takes place. This behavior was very different from the observed surfaces of synthetic materials with high thermal conductivity (such as gold) where a much slower surface temperature rise is caused by high thermal diffusivity of the surface area beneath the thermal source.

References

1 Buzin, A.I., Kamasa, P., Pyda, M., and Wunderlich, B. (2002) Application of a Wollaston wire probe for quantitative thermal analysis. *Thermochim. Acta*, **381** (1), 9–18.

2 Pollock, H.M. and Hammiche, A. (2001) Micro-thermal analysis: techniques and applications. *J. Phys. D*, **34**, R23–R53.

3 Gorbunov, V.V., Fuchigami, N., Luzinov, I., and Tsukruk, V.V. (2000) Microthermal probing of ultrathin polymer films. *High Perform. Polym.*, **12** (4), 603–610.

4 Lemieux, M., Minko, S., Usov, D., Stamm, M., and Tsukruk, V.V. (2003) Direct measurement of thermoelastic properties of glassy and rubbery polymer brush nanolayers grown by "grafting-from" approach. *Langmuir*, **19**, 6126–6134.

5 Tsukruk, V.V. and Huang, Z. (2000) Micro-thermomechanical properties of heterogeneous polymer films. *Polymer*, **41** (14), 5541–5545.

6 Aklonis, J.J. and MacKnight, W.J. (1983) *Introduction to Polymer Viscoelasticity*, 2nd edn, John Wiley & Sons, Inc., New York.

7 Gorbunov, V.V., Fuchigami, N., Hazel, J.L., and Tsukruk, V.V. (1999) Probing surface microthermal properties by scanning thermal microscopy. *Langmuir*, **15** (24), 8340–8343.

8 Gorbunov, V.V., Grandy, D., Reading, M., and Tsukruk, V.V. (2009) Micro and nano

scale local thermal analysis, in *Thermal Analysis of Polymers* (eds J.D. Menzel, and R.B. Prime), John Wiley & Sons, Inc., Hoboken, pp. 615–650.

9 Grossetete, T., Gonon, L., and Vernev, V. (2002) Submicrometric characterization of the heterogeneous photooxidation of polypropylene by microthermal analysis. *Polym. Degrad. Stab.*, **78** (2), 203–210.

10 Menzel, J.D. and Prime, R.B. (2009) *Thermal Analysis of Polymers, Fundamentals and Applications*, John Wiley & Sons, Inc., Hoboken.

11 Tsukruk, V.V., Gorbunov, V.V., and Fuchigami, N. (2003) Microthermal analysis of polymeric materials. *Thermochim. Acta*, **395** (1–2), 151–158.

12 Luzinov, I., Julthongpiput, D., and Tsukruk, V.V. (2000) Thermoplastic elastomer monolayers grafted to a functionalized silicon surface. *Macromolecules*, **33** (20), 7629–7638.

13 Sperling, L.H. (2005) *Introduction to Physical Polymer Science*, John Wiley & Sons, Inc., Hoboken.

14 Boschetti-de-Fierro, A., Spindler, L., Reiter, G., Olmos, D., Magonov, S., and Abetz, V. (2007) Thin film morphology in triblock terpolymers with one and two crystallizable blocks. *Macromolecules*, **40** (15), 5487–5496.

15 Fasolka, M.J., Mayes, A.M., and Magonov, S.N. (2001) Thermal enhancement of AFM phase contrast for imaging diblock copolymer thin film morphology. *Ultramicroscopy*, **90** (1), 21–31.

16 Anokhin, D.V., Chvalun, S.N., Bessonova, N.P., Godovsky, Y.K., Nazarenko, S.I., Ivanov, D.A., Magonov, S.N., and Erina, N.A. (2005) Influence of crystallization conditions on the structure and thermal behavior of syndiotactic polystyrene. *Polym. Sci. A*, **47** (9), 1007–1021.

17 Godovsky, Y.K. and Magonov, S.N. (2001) Visualization of the morphology of ultrathin polyethylene layers and its variations over a wide temperature range by hot-stage atomic force microscopy. *Polym. Sci. A*, **43** (6), 647–657.

18 Ivanov, D.A., Amalou, Z., and Magonov, S.N. (2001) Real-time evolution of the lamellar organization of poly(ethylene terephthalate) during crystallization from the melt: high-temperature atomic force microscopy study. *Macromolecules*, **34** (26), 8944–8952.

19 Godovsky, Y.K., Papkov, V.S., and Magonov, S.N. (2001) Atomic force microscopy visualization of morphology changes resulting from the phase transitions in poly(di-*n*-alkylsiloxane)s: poly(diethylsiloxane). *Macromolecules*, **34** (4), 976–990.

20 Fuchigami, N., Hazel, J., Gorbunov, V.V., Stone, M., Grace, M., and Tsukruk, V.V. (2001) Biological thermal detection in infrared imaging snakes. 1. Ultramicrostructure of pit receptor organs. *Biomacromolecules*, **2** (3), 757–764.

21 Schmitz, H. and Bleckmann, H. (1998) The photomechanic infrared receptor for the detection of forest fires in the beetle *Melanophila acuminata* (Coleoptera: Buprestidae). *J. Comp. Physiol.*, **182** (5), 647–657.

22 Hazel, J., Fuchigami, N., Gorbunov, V.V., Schmitz, H., Stone, M., and Tsukruk, V.V. (2001) Ultramicrostructure and microthermomechanics of biological IR detectors: materials properties from a biomimetic prospective. *Biomacromolecules*, **2** (1), 304–312.

23 Gorbunov, V.V., Fuchigami, N., and Tsukruk, V.V. (2000) Microthermal analysis with scanning thermal microscopy. I. Methodology and experimental. *Probe Microsc.*, **2**, 53–65.

24 Gorbunov, V.V., Fuchigami, N., and Tsukruk, V.V. (2000) Microthermal analysis with scanning thermal microscopy. II: calibration, modeling, and interpretation. *Probe Microsc.*, **2**, 65–75.

7
Chemical and Electrical Properties

The ability to probe the topography and physical and chemical properties of the surface simultaneously with nanoscale resolution provides a deeper understanding of the structure–properties relationship of various polymeric and, more importantly, multicomponent systems. For example, the subtle variations in the topography of the structures with processing conditions of conductive polymer blends, which are being extensively investigated owing to their applications in optoelectronic devices, can result in dramatic changes in the device performance. Chemical and physical modification of AFM tips (added chemical functionality, conductive coatings, magnetic coatings) renders AFM to monitor and measure various chemical, electrical, electrochemical, conductive, and magnetic properties of the surface with unique nanoscale lateral resolution. For instance, advanced AFM techniques such as Kelvin probe force microscopy and conductive force microscopy have emerged as unique tools to bridge our understanding between the structure and the properties of these technologically important materials.

In this chapter, we discuss some of the important considerations regarding the application of AFM-based techniques for probing the above-mentioned surface properties. While the advanced AFM techniques employed to probe the surface properties have been introduced in Chapter 4, in this chapter, we focus our discussion on the most important caveats associated with quantitative probing of the surface properties and highlight some important examples from the literature.

7.1
Chemical Interactions

Some modes of AFM (force modulation, phase imaging) are sensitive to chemical contrast of the surface and the same is reflected in the images. However, this section exclusively focuses on probing chemical interactions in a quantitative manner and physically straightforward terms (adhesion force, surface energy, pull-off force). Owing to its ability to reliably probe forces significantly lower than that for an individual chemical bond, AFM is a unique tool to map the chemical interactions with unprecedented lateral resolution. Soon after the commercial AFMs were

Scanning Probe Microscopy of Soft Matter: Fundamentals and Practices, First Edition.
Vladimir V. Tsukruk and Srikanth Singamaneni.

available, chemical modification of the tip, imparting chemical sensitivity to AFM probe, was demonstrated. As discussed in Chapter 3, the technique involves in the modification of the surface of the AFM tip with desired functional groups (using SAMs), polymers, and biomacromolecules.

The introduction of chemical force microscopy opened up a possibility of experimentally probing the interaction between two mating chemical species down to single molecules in special cases. Using well-established contact models (such as JKR), a wide variety of physical parameters (e.g., adhesion force, work of adhesion, interfacial surface energy) can be quantitatively determined by force spectroscopy measurements. The general methodology developed for probing chemical interactions has been employed for probing the biological interactions (such as antibody–antigen or protein–biomaterial surface), which will not be included in this section. The readers are referred to Section 15.2 for this discussion.

7.1.1
Chemical Interactions between Molecular Assemblies

Frisbie *et al.* monitored the adhesive and frictional forces between AFM tips modified with organic monolayers with desired terminal groups and surfaces with patterned functionality [1]. Adhesion between the tip and the surface with different functionalities (like and dislike) were found to be clearly distinguishable, thus establishing the technique as a unique tool to probe chemical interactions.

Figure 7.1 presents force curves in ethanol for SAMs with different tip and surface functionalities [2]. The ability to perform AFM in aqueous environments has led to more specialized measurements, namely, measuring the pK values of surfaces, which can be important for investigating surface confinement effects of the ionization of functional groups on the surface. In fact, there are significant differences in the dissociation constants between macromolecular surfaces and monomers in solution attributed to a variety of factors, including a decreased available degrees of freedom associated with bonding/immobilization, the effect of the dielectric permittivity from adjacent functional groups, and the electrostatic free energy of the substrate [3]. These differences are usually measured by quantifying the surface energy through contact angle measurements taken at different pH values.

Similar measurements can be performed by monitoring adhesion interactions between functionalized tips and surfaces with variable solution pH through the so-called force titration measurements introduced by Vezenov *et al.* and van der Vegte *et al.* [4, 5]. In these studies, the probes and surface were functionalized to present various chemical groups (CH_3, OH, NH_2, COOH, and $CONH_2$) using self-assembly of alkanethiols. They employed JKR theory of adhesion mechanics to calculate surface free energies, the number of interacting molecules, and, hence, single-bond forces from the force–displacement curves. Here, we discuss one specific example from their study where they performed pH-dependent adhesion measurements between the tip and the substrate hosting COOH, NH_2, and OH groups under constant ionic strength.

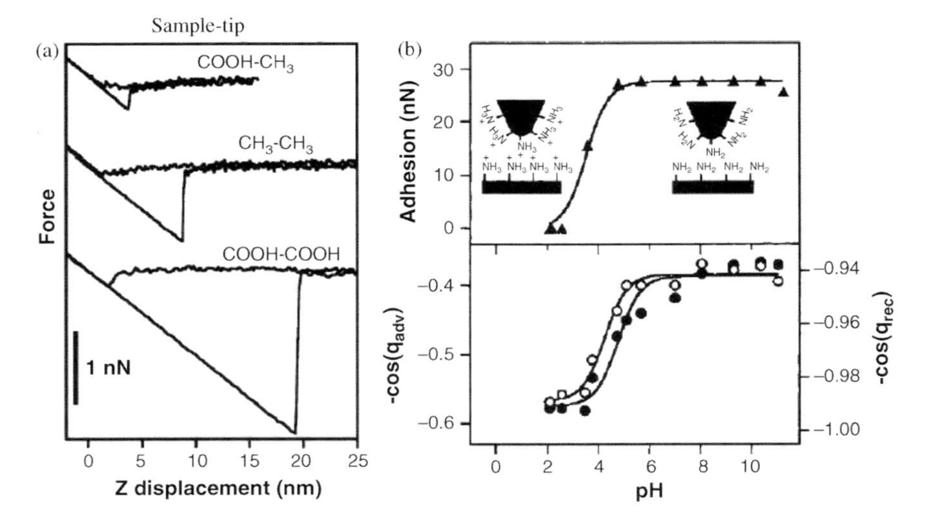

Figure 7.1 Adhesion measurements are highly affected by surface chemistry, solvent, and probe chemistry. (a) A force titration curve (top) and a corresponding contact angle titration curve (bottom). This figure demonstrates the capability of the force spectroscopy to measure interfacial energy. (b) Force–distance curves taken in different solvents and with different tip functionalities. Reprinted from Ref. [2].

Figure 7.1 shows a plot comparing results from force titrations and contact angle titrations. At low pH, the adhesion between the COOH groups, which are still in protonated state (neutral), was ascribed to the hydrogen bonding. As the pH was increased, the COOH groups deprotonated (bearing negative charges), resulting in the observed drop in the adhesion due to the like charges between the tip and the sample surface. Similar results were observed for a NH_2-functionalized tip and sample, except that the pH dependence is reversed due to the basic nature of the NH_2 groups. A nearly pH-independent adhesion force was observed for OH and CH_3 functionalizations, indicating that there is no detectable ionization of surface OH and CH_3 groups in the pH regime studied.

The force titration approach has the ability to map the surface energies on the nanoscale and associate any energetic contrast with nanoscale features [6]. It should be mentioned that, in order to probe unknown surface p*K* values, it is important to use tips functionalized with hydrophilic groups that are incapable of changing ionization with pH. Force titration measurements also demonstrate the important role that the surrounding fluid plays on adhesion measurements.

On the basis of the differences in the frictional force experienced by the AFM tip with different terminal groups, Green *et al.* mapped chemically distinct domains using frictional force microscopy [7]. FFM was performed on partially formed bilayers (formed by head-to-head hydrogen bonding of the carboxylic acid) of carboxylic acid-terminated alkanethiol ($HS(CH2)_{15}COOH$) and steric acid (CH3 $(CH_2)_{16}COOH$) on a Au(111) surface. The partial bilayer presented regions with terminal COOH and CH_3 groups. The topography images clearly revealed the thickness of the stearic acid film to be 2 nm, agreeing with the estimated length of the molecule with tilted arrangement.

FFM images revealed higher frictional forces as the unmodified (hydrophilic) tip interacted with the COOH-terminated regions and low friction in the CH_3-terminated regions. More important, the authors noted that the dependence of observed frictional force on the applied load at the contact point between the unmodified tip and the methyl-terminated component of the bilayer structure is similar to that observed for the contact point between the uncoated tip and the methyl-terminated monolayer. This observation clearly reveals that the frictional contrast observed in the FFM images arises from chemical effects and not from an elasticity difference between the CH_3- and COOH-terminated regions.

Tsukruk and Bliznyuk probed the adhesion between silicon/silicon nitride surfaces modified with robust alkylsilane SAMs with CH_3, NH_2, and SO_3H terminal groups (Figure 4.4) [8]. The chemical functionalization was performed by silanization as opposed to the thiol-based functionalization in most of the earlier work. As discussed earlier, silane-based functionalization offers distinct advantages such as high robustness without the need for additional coating (see Section 4.4). Work of adhesion and residual forces between the functionalized tips and the surfaces were studied in air and aqueous solutions at different pH values.

Figure 7.2 shows the representative force–distance curves that demonstrate significant differences in adhesive forces between modified tip–surface combinations [8]. The figure also shows unmodified silicon nitride and silicon surfaces that serve as a control highlighting the significant dependence on terminal chemistry. Strong adhesion was observed for both silicon nitride/silicon nitride and silicon nitride/silica pairs, which was comparable to that observed between NH_2-terminated/NH_2-terminated SAMs, as indicated by large pull-off forces. Both SO_3H–SO_3H and CH_3–CH_3 SAM pairs show much smaller adhesive forces. We will briefly discuss adhesive forces between surfaces, or, according to the SPM terminology, pull-off forces (point B in Figure 7.2) and friction behavior [8].

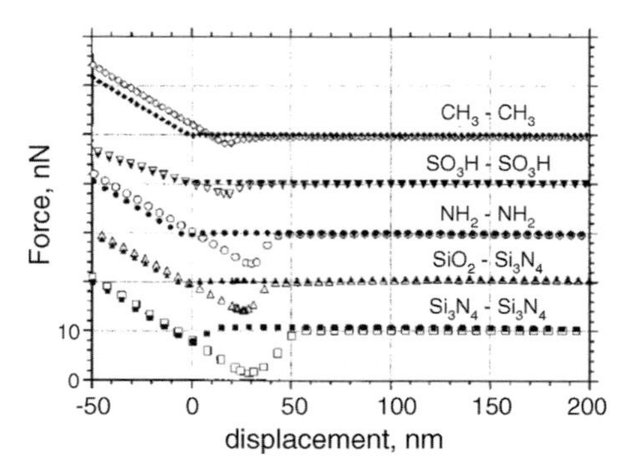

Figure 7.2 Representative force–distance curves obtained as the tip modified with different terminal groups approaches and retracts with chemically modified surface. Reprinted from Ref. [8].

In this study, adhesive forces were found to be in the range of 0.1–8 nN for different pairs of modified tips and surfaces. The highest and the lowest adhesive forces are observed for NH_2- and CH_3-terminated surfaces, respectively. While a direct comparison between the different adhesive forces is ambiguous, the authors calculated the absolute values of the work of adhesion. Work of adhesion was found to be in the range 0.5–8 mJ/m^2 and the values obtained were found to have excellent correlation with solid–liquid surface energy estimated from macroscopic contact angle measurements. Following these initial demonstrations, there have been numerous reports where the interaction between chemically modified tips and surface with a wide variety of well-defined functionalities (using SAMs or Langmuir–Blodgett layers) has been presented. The readers are referred to a recent review on the subject with comprehensive discussion of these studies [9].

7.1.2 Chemical Interactions of Polymer Surfaces

Application of chemical force microscopy to probe the chemical interactions of polymer surfaces is a significantly more challenging routine compared to probing the molecular assemblies on ideal planar stiff surfaces. The surface roughness, morphology, and elastic properties intervene with the chemical interactions making the quantitative interpretation of the data more challenging. There are many different ways to apply the capabilities of adhesion measurements and chemical force microscopy toward polymeric materials. The first approach involves using chemically modified tips to probe the distribution of the localized surface properties of the polymers [10–13]. The second approach, which is mostly applicable to polymer composites, involves probing the interaction between the filler and the matrix by pulling the filler out of the polymer surface and monitoring the pull-off force required to do so.

Schonherr and coworkers probed the lateral distribution of functional groups on oxyfluorinated films of an isotactic PP surface using OH-functionalized tips [14]. Pull-off forces were measured as a function of pH to obtain force titration curves. The pull-off forces exhibited a clear dependence on the pH of the solution. Although the forces measured show a broad distribution, possibly due to the roughness of the polymer film, the mean values could be clearly separated. These measurements revealed the force pK_a values (pH at which the pull force most rapidly changed) to be 5.5–6.0. At pH values close to or slightly higher than the pK_a, the pull-off force maps were found to be laterally inhomogeneous on a sub-50 nm scale, which was attributed to variations of local pK_a values due to the inhomogeneous distribution of the functional groups.

Yet another approach to probing chemical interactions is modifying the AFM probe with grafted polymer chains instead of SAMs to attain the desired functionality. Geissler *et al.* employed pH titration at various salt concentrations to probe the chemical interaction between polydimethylsiloxane (PDMS) substrates modified by maleic anhydride using pulsed plasma polymerization and AFM tips modified with the same [15]. The surface coating was hydrolyzed to promote the formation of dicarboxylic acid groups and the pH-dependent adhesion measurements were performed at different salt (KCl) concentrations. Owing to the dicarboxylic nature of the maleic acid groups, two distinct surface pK_a values, pK_{a1} at 3.5 and pK_{a2} at 9.5, were clearly revealed at low electrolyte concentration using adhesion measurements. The pH titration curves (shape) exhibited significant dependence on the salt concentration, which was related to the condensation of counterions on the carboxylate groups.

A major area of interest in composite material development is the interaction of functional micro/nanofillers with polymer matrices. A novel use of force spectroscopy has been in quantifying these interactions. Barber *et al.* studied the interaction of carbon nanotubes and polyethylene–butene with the use of probes with nanotubes attached to the tip [16]. The authors performed a series of pull-out experiments using AFM and estimated the force required to separate a CNT from a 300 nm thick polyethylene–butene matrix. This was accomplished by repeatedly heating the polymer matrix above the glass transition temperature, pushing the nanotubes into the polymer, cooling, and then pulling the nanotubes back out of the polymer, as shown in Figure 7.3 [16]. They measured the interfacial separation stress to be 47 MPa, indicating that CNTs are effective at reinforcing a polymer to a high degree with a great versatility of chemical and physical interactions.

Chemical interactions at the molecular level can be probed using chemical force microscopy. One of the important aspects that needs to be considered when probing such delicate interactions is the contribution of multiple interfering factors and the methods to overcome these. To capture the true molecular interactions between the desired species, it is important to exclude more dominant capillary and solvation forces between the mating species.

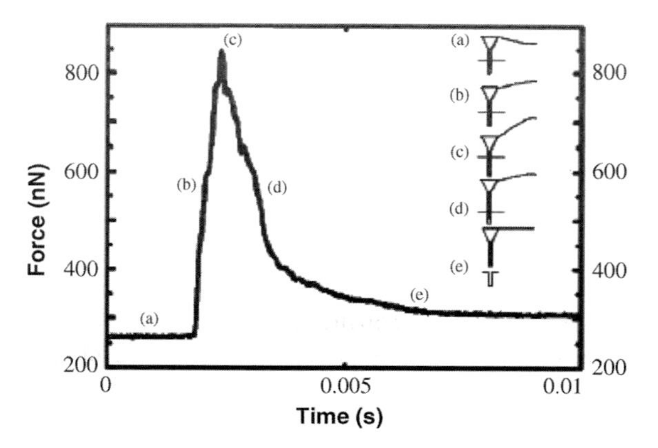

Figure 7.3 Surface force spectroscopy for measuring the interaction energy between soft matter and carbon nanotubes. The graph shows the force to pull a carbon nanotube out of a polymer matrix. The inset explains the testing approach and the various regions of the graph. Reprinted from Ref. [16].

The chemical interaction between individual functional groups and individual groups on carbon nanotubes has also been studied by Friddle *et al.* with functionalized tips in fluidic environments (toluene), achieving single molecule group interaction conditions (Figure 7.4) [17]. The high curvature of the CNTs, combined with the small tip radius of the probe (extremely small composite radius), resulted in an extremely small tip–surface contact area of 0.33 nm^2, as estimated using the Hertzian contact model. It is important to note that the contact area for conventional flat surface is an order of magnitude higher compared to the one estimated in their study. The authors noted that under ideal conditions (close-packed monolayer on the tip), this small contact area corresponds to an area occupied by a single terminal group at the very end of the modified AFM tip, enabling the probing of a single functional group with a single group on the CNT surface.

Interestingly, the chemical interaction strength did not follow conventional trends of increasing polarity or hydrophobicity, but instead reflected the complex electronic interactions between the nanotube and the functional group. *Ab initio* calculations were employed to confirm the experimental trends and predict binding force distributions for a single molecular contact that matched the experimental results.

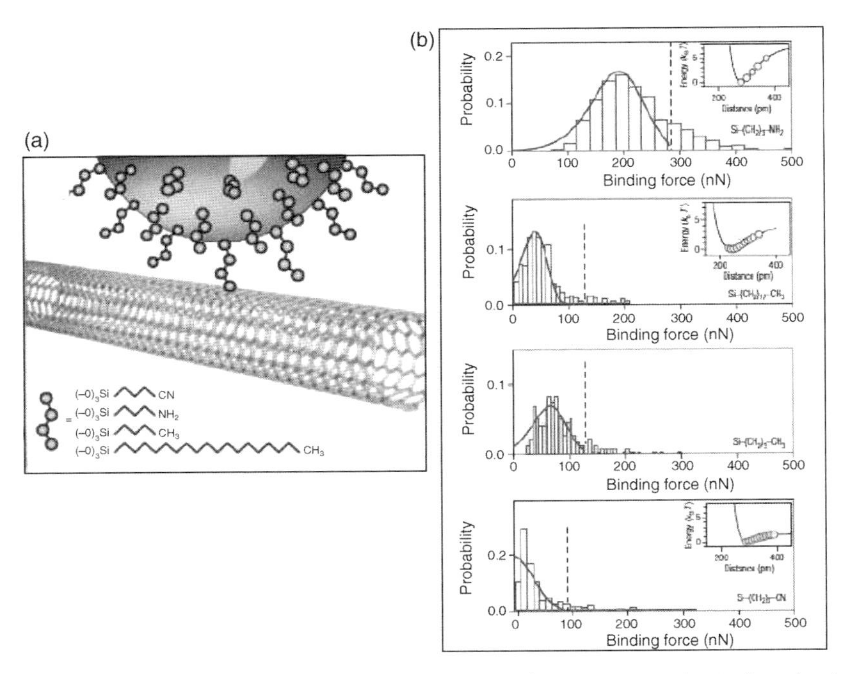

Figure 7.4 (a) The testing method of another study involving measuring the single-molecule interactions with carbon nanotubes. (b) The resulting data provide the binding force histogram of the different molecules with the wall of single-wall carbon nanotubes. Reprinted from Ref. [17].

This study clearly emphasizes the importance of molecular linkage dynamics in determining interaction strength at the single functional group level.

7.2 Electrochemical Properties

Scanning electrochemical microscopy (SECM) involves inducing chemical changes in the sample under investigation with the simultaneous collection of electrochemical information as the tip scans (approaches) the surface. It is routinely employed to probe the electrochemical phenomena such as electron, ion, and molecule transfers. SECM is operated in three primary modes: (i) feedback mode, (ii) tip generation/substrate collection mode, and (iii) substrate generation/tip collection mode [18].

The feedback mode involves immersing the tip in a solution with a redox species (RS). Applying sufficient positive potential on the tip results in the oxidation of the RS as governed by the diffusion of the RS to the tip. The current measured at the tip due to this redox process is governed by the proximity of the tip to the surface, electronic nature of the substrate, and the potential on the surface. When the tip is sufficiently far from the surface, the tip current is determined by the diffusion of the RS. However, when the tip is close to the surface (within a few tip radii), the electrically active surface causes enhancement or depletion of this current. If the surface is a conductor, the oxidized species are reduced at the surface and pumped back to the tip (for oxidation), effectively increasing flux of the redox species and hence the current observed at the tip, which is a positive feedback mechanism. Conversely, if the surface is insulating the tip, oxidized species cannot be reduced at the surface. Furthermore, the surface perturbs and reduces the diffusion of the redox species to the tip surface, a negative feedback mechanism. The tip generation/substrate collection mode involves the simultaneous measurements of both tip and substrate currents as the tip-generated electroactive species diffuse across the tip/substrate gap to react at the substrate surface.

O'Mullane *et al.* employed SECM to measure the charge transport diffusion constants in metallopolymers given by $[M(bpy)_2(Pol)_nCl]Cl$, where M is Ru or Os, bpy is 2,2′-bipyridyl, and Pol is poly(4-vinyl-4-vinylpyridine) (P4VP) [19, 20]. They used triple potential step technique in which the tip is used to electrogenerate oxidant species ($Ru(CN)_6^{3-}$) by oxidation of $Ru(CN)_6^{4-}$. The oxidant species diffuse from the tip to the underlying polymer film to cause electron transfer and local oxidation of the polymer-bound complex of Os^{II} to Os^{III}. The current–time characteristics of this process provides insight into the kinetics of the ET process between the oxidant species and the polymer-bound moiety and the concentration of redox-active species in the polymer film. The entire process is repeated after a certain waiting period, which enables the $Ru(CN)_6^{3-}$ to be converted back to $Ru(CN)_6^{4-}$ at the tip and Os^{II} to recover in concentration by electron self-exchange between Os^{III} and Os^{II} moieties. The charge transport diffusion constants on the order of 10^{-10} cm^2/s were measured in their studies. Furthermore, they showed the diffusion constants to be significantly influenced by the redox site loading and film structure (determined by AFM).

SECM was extensively employed to monitor the concentrations and fluxes of redox species generated or consumed by biomacromolecules and cells [21–24]. Zhou *et al.* investigated the catalytic behavior of horseradish peroxidase (HRP) by immobilizing HRP onto glass by cross-linking with a copolymer on a glass slide. The film was swollen in buffer solution, resulting in a hydrogel structure [25]. In an alternate approach, the copolymer and avidin were coimmobilized on a glass slide and biotinylated HRP was conjugated to the avidin in the film. The experiment was performed in solution containing H_2O_2 and benzoquinone mediator. SECM was then used to detect the presence of the bound enzyme by observing the feedback current in a solution of benzoquinone and hydrogen peroxide when hydroquinone was generated at the tip. A detection limit was found to be less than 7×10^5 HRP molecules within a 7 μm diameter area.

Very recently, Anne *et al.* introduced high-resolution atomic force electrochemical microscopy (AFM-SECM), which they called tip-attached redox mediator AFM-SECM [26]. In this technique, the redox mediator (ferrocene) was attached to the tip via flexible polyethylene glycol (PEG) chains. They demonstrated that the tip-attached ferrocene-labeled PEG chains effectively transport electrons between the tip and the substrate. The PEG chains acted as molecular sensors, probing the local electrochemical reactivity of a surface under investigation. One of distinct advantages of the approach the authors adapted compared to most of the earlier work is that the Fc-PEGylated AFM-SECM probes enabled simultaneous tapping mode topography and electrochemical feedback current images with nanometer-scale lateral and vertical resolution.

7.3 Work Function and Surface Potential

The simple definition of work function (Φ) of a material is the energy required to remove an electron from the ground state. In the case of a metal, this can be simply translated as the difference in the energy between the vacuum level and the Fermi energy. However, in the case of semiconductors and insulators (most organics), work function can be considered as the difference in the energy level between the vacuum level and the most loosely bound electrons in the material. The Kelvin probe measurement technique, which was initially introduced by Kelvin and later modified by Zisman, is widely employed for measuring the work function at macroscopic scales [27, 28]. Monitoring the current produced due to changes in the capacitance of an oscillating parallel plate capacitor (consisting of reference electrode and surface under investigation) forms physical basis of measuring the work function. Soon after the introduction of AFM, a scanning probe version of the Kelvin probe technique was demonstrated.

The readers are referred to Section 4.8 for basic operation principles of KFPM and electrostatic force microscopy. Briefly, KFPM involves matching the potential of the conductive tip (maintained at a constant distance from the surface) to that of the surface using an electronic feedback (see Figure 7.5). When the potential of the tip

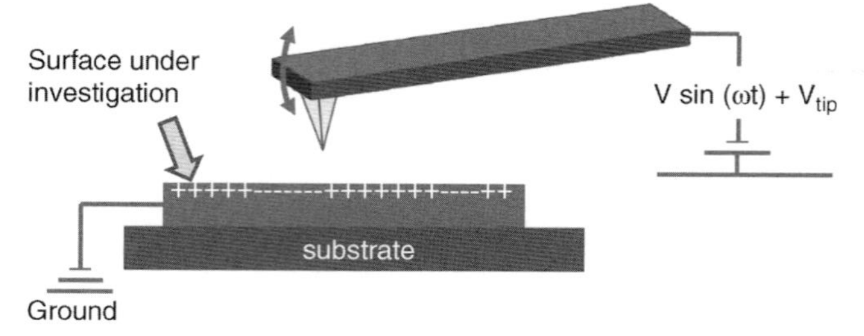

Figure 7.5 The Kelvin probe force microscopy in which an oscillating conductive tip is scanned over the sample at constant distance to probe the surface potential distribution of the sample.

exactly equals that of the material, the electrostatic interaction between the tip and the sample is nullified. The voltage applied to nullify the electrostatic interaction is the local measure of the work function of the sample. With knowledge of the work function of tip, a quantitative two-dimensional map of the surface work function can be constructed from the applied DC feedback signal.

7.3.1
Effect of Tip Shape on Surface Potential and Work Function Measurements

One of the key sources of artifacts in the KFPM comes from the finite size of the tip [29]. It is important to note that the electrostatic interaction between the tip and the sample is long range compared to the van der Waals interaction, causing the area sensed by the tip to be much wider that just below the tip. This long-range interaction results in underestimation of the measured potential. Furthermore, as discussed in Chapter 4, due to the complexity of the tip shape, simpler models such as parallel plate models are chosen to interpret the tip–sample interaction. This oversimplification often results in estimated surface potentials and work functions far from reality. Recent developments involving more complex analysis using finite element methods seem to take into account the complex nature of this interaction, providing more realistic results.

The effect of the finite size of the tip on the observed surface potential has been elegantly demonstrated by Liscio *et al.* [30]. They performed KPFM microscopy on poly(3-hexylthiophene) fibers deposited from chloroform on various substrates (HOPG, mica, and silica). The diameter of the P3HT fibers was determined by the solvent wettability on various substrates. The observed surface potential of the fibers monotonically increased with the increase in the fiber diameter. The authors noted that the observed low surface potential values of the smaller fibers were simply due to the significant contribution from the substrate, which was in the sampling surface. This example clearly demonstrates that the substrate contributions can impact the resolution that can be attained with KFPM and, more importantly, the underestimated surface potential values.

In the case of EFM, the geometry of the tip and the surface features makes quantitative analysis of the surface potential measurements nontrivial, especially when the surface features are of a complex shape. Tevaarwerk *et al.* demonstrated techniques for analysis of EFM on samples with complex surface features, using voltage modulated EFM augmented by 3D simulations [31]. The authors introduced the concept of the radius of influence, which is defined as the area of the sample responsible for 50% of the force gradient on the tip. The range of influence depends on the distance between the probe and the tip. They showed that when sample features are on the order of the radius of influence, electrostatic simulations that accurately account for sample topography are needed to extract dielectric properties. SiGe nanostructures have been employed on insulator as a model system to demonstrate the effectiveness of the technique in analyzing EFM images. Combined experimental and finite element analysis was employed to reveal the complex nature of the tip–sample interactions that need to be considered for quantitative probing of surface potential. This demonstration clearly underscores that the potential profiles obtained by scanning Kelvin probe microscopy do not purely reflect the electrostatic potential under the tip apex, but are strongly affected by the electrostatic coupling between the entire probe and the entire device, even for small tip–sample separations.

7.3.2
Surface Potential and Work Function of Molecular and Polymeric Surfaces

The surface potential difference in chemisorbed organosilane layers (octadecyltrimethoxysilane (ODS) and *p*-chloromethylphenyltrimethoxysilane (CMPhS)) have been revealed using KPFM [32]. Micropatterned photooxidation of the self-assembled monolayers was achieved by masking certain regions using a photomask during exposure to UV light. The photooxidation of the methyl end groups and the transformation into CHO and COOH groups, and the subsequent complete molecular decomposition of the SAM, resulted in changes in the surface potential that were clearly revealed in the surface potential map of the KFPM. Surface potential measurements in the different irradiated regions were performed using the non-irradiated regions as a reference. *Ab initio* calculations revealed that the change in surface potential in these molecular systems is due to the change of the molecular dipole moment [33].

Magonov *et al.* investigated the morphology and surface potential of semifluorinated alkanes self-assembled on a wide variety of substrates (silicon, mica) by spin casting of dilute solutions [34]. In the confined geometry of thin and ultrathin layers, semifluorinated alkanes self-assemble into nanoscale ribbons, spirals, toroids, and their different intermediates [35]. The self-assembled semifluorinated long-chain acids have been extensively investigated by AFM [36, 37]. Owing to the slightly tilted arrangement of the molecules with fluorinated parts closer to the air–layer interface, the height of these structures was found to be smaller than the extended length of the molecules. These structures are surrounded by thinner and self-organized layers of the semifluorinated alkanes.

Figure 7.6 shows the topography and surface potential images of $CF_3(CF_2)_{14}(CH_2)_{20}CH_3$ assembled on silicon. The authors monitored alternations during long-term studies of $CF_3(CF_2)_{14}(CH_2)_{20}CH_3$ adsorbate on Si in a humid

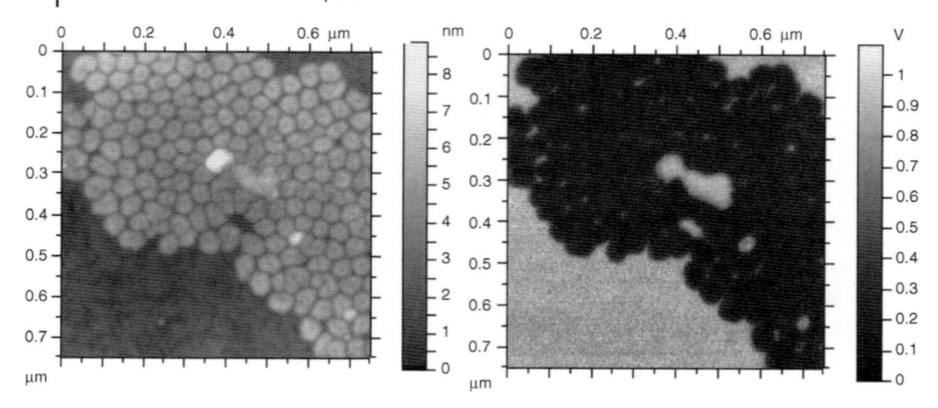

Figure 7.6 Topography and surface potential images of self-assemblies of semifluorinated alkanes $CF_3(CF_2)_{14}(CH_2)_{20}CH_3$ on Si substrate. Courtesy of Dr. S. Magonov.

atmosphere (RH > 90%). A spreading of spirals, which formed an initially compact aggregate, was noticed in the first 24 h. The authors pointed out that several factors such as a lower adhesion of the assemblies in humid air, charge interactions of the spirals, and tip-assisted material transfer might contribute to this phenomenon. At longer times, the spirals converted to toroids, which are more thermodynamically favorable structures. The heterogeneities in surface potentials are most likely related to different molecular self-assemblies that are governed by intrinsic properties of the semifluorinated molecules and their interactions with the substrates.

Photoexcited charge distributions in polymer films were monitored with a spatial resolution of 100 nm and temporal resolution of 100 μs by Coffey and Ginger [38]. They studied a model polymer blend consisting of poly-(9,9′-dioctylfluorene-*co*-benzothiadiazole) (F8BT)/poly(9,9′-dioctylfluorene-*co*-bis-*N,N*′-(4-butylphenyl)-bis-*N,N*′-phenyl-1,4-phenylenediamine (PFB) for probing the effect of morphology on photovoltaic performance. One of the important questions they addressed using the time-resolved EFM measurements was whether mesocale domain interfaces or the domain's central portions are responsible for the most photocurrent in the polymer blend. They observed slower charging near the microscale domain boundaries and related a large fraction of photocurrent with regions far from the interfaces. Moreover, the time-resolved EFM measurements were found to have excellent correlation with the independent quantum efficiency measurements. Clearly, the time-resolved EFM measurements revealed unique insight into the relation between local morphology, local optoelectronic properties, and device performance.

KPFM has been employed for the investigation of organic solar cells consisting of poly-[2-(3,7-dimethyloctyloxy)-5-methyloxy]-*para*-phenylene-vinylene/1-(3-methoxy-carbonyl)propyl-1-phenyl-[6,6]C61 (MDMO-PPV/PCBM) blends, identifying a barrier for electron transmission from the electron-rich PCBM nanoclusters to the extracting cathode [39]. Figure 7.7 shows the topography and the KPFM mapping (under light illumination) of MDMO-PPV and PCBM blend film of spin cast from chlorobenzene and toluene. The topography images clearly reveal that in the case of the films deposited from chlorobenzene, polymer nanospheres are distributed

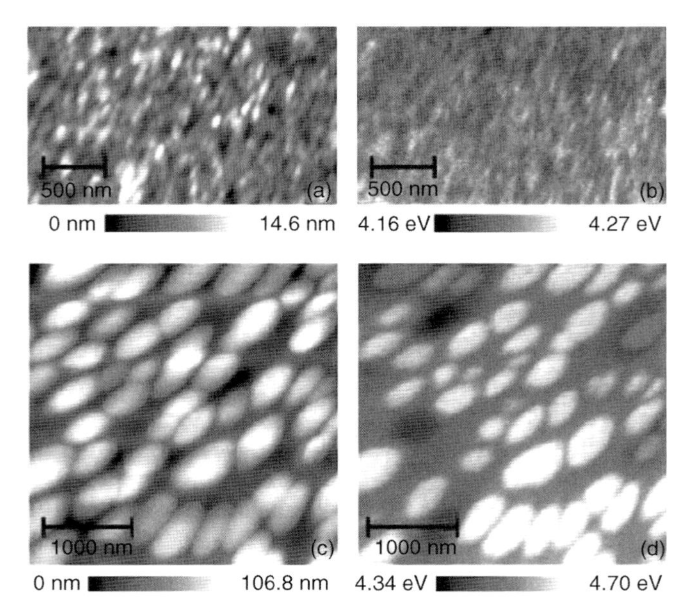

Figure 7.7 AFM and simultaneous KPFM images of the topography and work function of toluene-cast blend film of MDMO-PPV/PCBM with a mass ratio of 1 : 4, measured in the dark and under 442 nm laser illumination. Reprinted from Ref. [39].

almost evenly throughout the bulk of the film, whereas a skin layer, incorporating polymer nanospheres, surrounds the big PCBM clusters in toluene cast films. Apart from the dramatic difference in the morphology of the blend, the variation of the work function on the surface is much larger in the case of films deposited from toluene (0.1 eV) compared to that deposited from chlorobenzene (0.2 eV).

In a related study, Sirringhaus and coworkers employed KPFM to map the surface potential and the photoinduced surface photovoltage and correlated this with the topography of the polyfluorene blend-based photovoltaic devices [40]. The results clearly suggest that an optimization of an appropriate size of phase separation percolation of both electron and hole transporting phases with their respective electrodes is highly essential for improving the efficiency of energy harvesting.

7.3.3 Surface Potential and Work Function of Low-Dimensional Carbon Systems

Surface potential, surface charge distributions, and charge trapping in low-dimensional carbon systems such as carbon nanotubes and graphenes have been widely investigated, and in fact they have been employed as model systems for better understanding of the tip–sample electrostatic interactions in some cases. Cui *et al.* employed KPFM and EFM to understand the effects of coadsorption of alkanethiol, S, and O_2 on the contact potential differences and local dipoles at the carbon nanotube and metal (Au) interface [41]. They found that the coadsorbates modify the energy levels at the contacts and induce significant shifts of the CNT bands relative to the metal Fermi level. It was noted that the properties of the CNT–metal interface directly determines the characteristics of the nanotube FETs. For example, the charge injection between the metal and the nanotube is reversed in air compared to vacuum due to the change in the Au work function because of the adsorption of oxygen in ambient conditions.

It is known that the surface charge traps can significantly affect the performance of carbon nanotube-based FETs. Thus, quantitative probing of the charge traps in these nanostructures and devices is extremely important. Jespersen and Nygård employed EFM to reveal the charge pools with charge density on the order of $10^{-8}/cm^2$ that are trapped inside the nanotube loops [42]. The authors also demonstrated that the charge can be effectively dissipated from the surface by bringing the grounded AFM tip in contact with the charge trap.

Another "version" of carbon structures, graphene, an atomic sheet of carbon (ideally), has been gaining increased attention in the past years owing to its electrical, mechanical, and optical properties. Datta *et al.* employed EFM to demonstrate the increase in the surface potential in few-layer graphene with an increase in the number of layers [43]. They also found that the surface potential approaches the bulk value of graphite for graphenes with five or more layers. Figure 7.8 shows the topography of the FLG on a silicon substrate and the number of layers (n) ranging from 2 to 18 could be clearly identified from the thickness of various surface regions. EFM was performed on the FLG with two different tip voltages (-2 and 3 V) (see Figure 7.8) [43].

While the phase shift, $\Delta\Phi$, of the FLG regions with respect to the substrate is always negative, the authors noted that the FLG regions with different numbers of graphene layers exhibit different values of $\Delta\Phi$. Furthermore, the relative contrast between the FLG regions was found to be reversed when the EFM was performed with tip voltages of reverse polarity. The observed contrast flip in the images obtained with tip voltages of opposite polarity indicated the different surface potentials of the graphene layers with the different number of carbon monolayers within stacks. The data clearly revealed the variation of the surface potential of the graphenes with the number of layers.

7.4 Conductivity

As the name suggests, conductive force microscopy, also called conductive probe atomic force microscopy (C-AFM), involves using a conductive probe as an electrode to measure the conductivity of the sample with nanoscale resolution. In this section, we briefly discuss the key considerations in probing the conductivity of the sample using C-AFM and highlight several important examples where the techniques have been employed in molecular and polymeric systems. The readers are referred to several recent reviews for a comprehensive discussion of the C-AFM application to various organic systems [44–46].

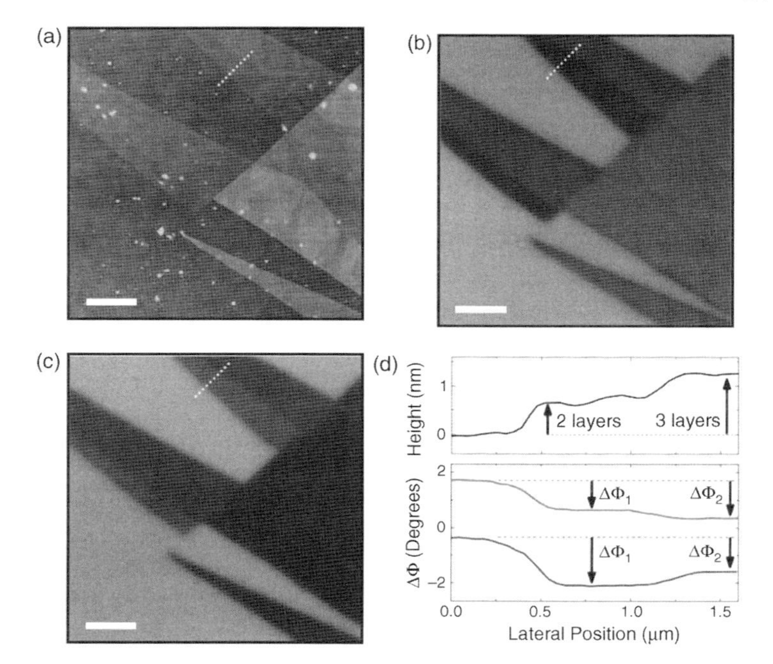

Figure 7.8 AFM and EFM images of "few-layer" graphene. (a) Topography of the graphene flake on SiO_2/Si substrate. EFM phase images of the sample, with V_{tip} of -2 V (b) and $+3$ V (c), respectively. (d) Average of 30 line scans of topography and phase centered along the dashed lines in AFM and EFM images. Black curve corresponds to (a), red curve corresponds to (B), and blue curve corresponds to (C). Scale bar in each image is 1.5 μm. Reprinted from Ref. [43].

Two basic geometries are widely employed in conductive force microscopy mode (see Figure 7.9). In the first and more common mode, the material under investigation is deposited on a conductive surface (metal film) that forms one of the electrodes. The AFM tip with a conductive coating (or heavily doped tip) is employed as the second electrode. An external bias is applied between the tip and the metal layer underneath to measure the current (translated to conductivity) of the film at the specific location along the thickness direction. In the second approach, the material is in contact with a micropatterned electrode deposited on the surface of the sample. The SPM tip forms the second electrode and the lateral conductivity of the surface structure is probed by applying an external bias.

The measurements in both of the above geometries can be performed in two different ways. The first method involves applying a constant bias, followed by scanning the sample in contact mode to acquire a conductivity map. In the second approach, static measurements are made in which the tip is engaged at specific locations and the voltage bias between the two electrodes is swept to acquire I–V curve. Alternately, the voltage bias can be maintained constant and the distance between the surface and the tip is varied to acquire I–Z curve. Finally, a more

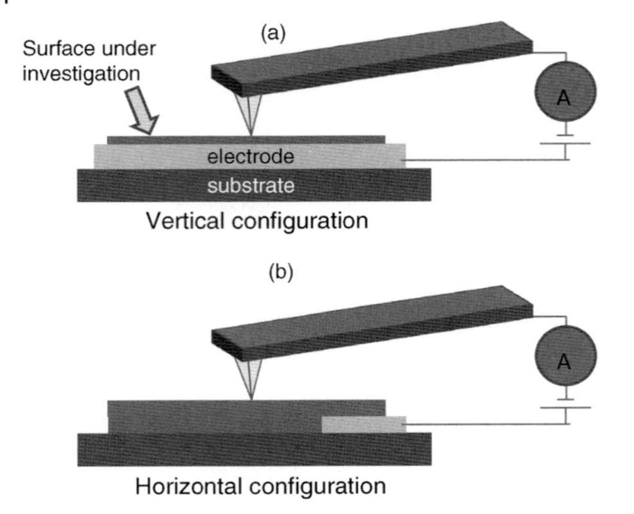

Figure 7.9 Two basic geometries of conductive force microscopy: (a) vertical configuration and (b) horizontal configuration.

exhaustive measurement mode is the combination of both the above methods involving the collection of a 2D array of I–V curves with user-defined resolution.

7.4.1
Conductive Probes

One of the key differences between the electrostatic or Kelvin probe microscopy and C-AFM is that the latter is performed in contact mode while the former is performed in tapping mode [46]. The intimate contact of the conductive probe with the sample surface obviously warrants careful consideration of the conductive coating, as will be discussed next.

Two important methods are frequently employed to fabricate a conductive probe [47]. The first method involves depositing a thin layer of metal (Au, Ag, Pt) on the conventional silicon or silicon nitride probes. An alternate approach involves heavily doping the silicon tips to enhance the conductivity of the tips. Both the techniques have their own advantages and shortcomings. Although they offer excellent electrical contact with the surface, metal-coated tips suffer from abrasion of the metal coating during scanning and compromised resolution due to the increase in the tip radius with coating. On the other hand, heavily doped tips offer high resolution (due to small contact radius), but they provide poor electrical contact.

Indeed, in related studies, Thomson and Moreland demonstrated that the electrical resistance between a metal-coated tip and a Au surface is nearly five orders of magnitude smaller than doped silicon probe [48]. They also noted that the shear forces involved in contact mode scanning of the metal-coated probes over the surface result in rapid abrasion of the coating. An alternate approach to overcome this issue is to scan the sample in tapping mode to identify the regions of interest followed by approach in these regions to acquire the electrical characteristics. However, this method cannot be easily employed to obtain a conductivity map, which is often required for understanding the final properties controlled by spatial distribution of the components, phases, and boundaries.

To acquire reliable data, one has to confirm the quality of the probe before and after measurements to ensure that the shape and the metal coating (electrical properties) of the probe are not significantly changed. Although the shape can be relatively easily verified by any one of the techniques discussed in Chapter 4, the integrity of the metal coating can be verified only by measuring the resistance as the probe is brought into contact with a smooth and uniform conductive layer on the surface.

7.4.2
Effect of Tip–Sample Interaction on Conductivity Measurements

The physical interactions (controlled by applied load and contact area) between the conductive probe and the surface of the sample critically influence the measured electrical properties in C-AFM. To achieve quantitative and reliable data, a comprehensive understanding of the tip–sample interactions from an electrical viewpoint is extremely important. Obviously, the contact resistance between the tip and the sample depends on the contact area between the tip and the sample, which in turn depends on the applied load. The variable pressure between the tip and the sample determines the degree of overlap of the electronic wave function of the mating species, responsible for injecting the charge carriers from the conductive probe to the sample.

Tivanski *et al.* investigated the influence of the adhesion force between a conducting probe and a sample surface on the measured electrical properties of conductive polymers using C-AFM [49]. The authors noted that the potential bias applied between the tip and the sample results in an additional attractive electrostatic capacitance force. The Au-coated tip–sample load critically influenced the measured I–V characteristics of the polythiophene deposited on a gold layer. Figure 7.10 shows the I–V curves obtained at various loads. It can be clearly seen from the plot that the conductivity increases with the increase in the load between the tip and the sample. Although the increase in the measured conductivity with the tip load is not unusual considering the increase in the contact area, the authors pointed out that the increase in conductivity cannot be entirely explained by this factor.

The authors also pointed out that the electronic wave function overlap of the metal (probe) and the semiconductor layer (polythiophene) determines the potential barrier created at the interface and hence the conductivity. Considering the adhesion force versus applied bias and the I–V characteristics of polythiophene, the authors concluded that the quantitative understanding of the I–V data acquired using C-AFM requires knowledge of the adhesion force.

Although briefly discussed in Section 4.9, here we would like to reemphasize the importance of understanding the geometry of the electrical measurement in C-AFM. The carrier mobility is estimated from the C-AFM experiments using the Mott–Gurney law (see Section 4.9). However, this usual approach results in higher carrier

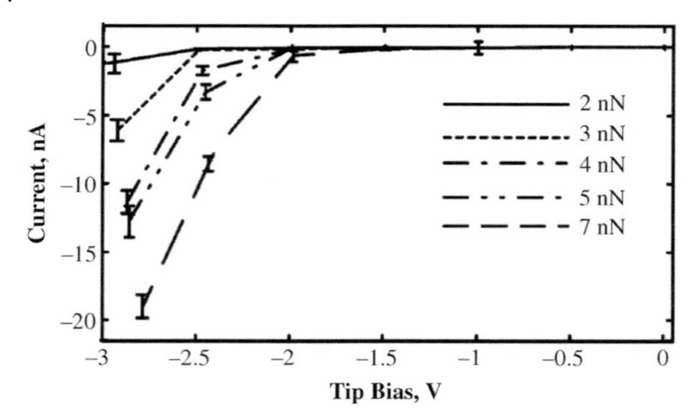

Figure 7.10 *I–V* characteristics of the gold polythiophene-gold system at five different constant interaction forces between the tip and the sample. Reprinted from Ref. [49].

mobilities compared to those observed in a planar electrode configuration as the current spreads out under the AFM tip, enabling a larger space charge-limited current density than is expected in the plane parallel case. A modified approach proposed by Ginger and coworkers accounts for this spread out and provides mobility values closer to those obtained by conventional macroscopic techniques [50].

7.4.3
C-AFM of Polymeric and Molecular Systems

C-AFM has been extensively applied to probe the nanoscale electrical characteristics of semiconducting polymers and their blends, which are employed in organic electronics and photovoltaics. C-AFM has been extensively employed to probe the morphology, conductivity, and carrier mobility of polymer thin film devices [51–55]. One of the extensively studied system is the PEDOT:PSS blend, commonly employed as a interface layer between the anode (ITO) and the organic semiconductor layer in various optoelectronic devices.

A vertical C-AFM configuration was applied to study the effect of processing conditions (such as annealing, PSS content, solvent treatment) on the vertical charge transport of PEDOT:PSS [56]. While the topography images do not show any significant change with annealing, it was observed from the conductive maps that most of the current passes through the film surface via small conductive hot spots in a relatively insulating matrix. Figure 7.11 shows the topography and the C-AFM images of PEDOT:PSS film annealed at 140 °C for different periods of time. The increase in macroscopic conductivity observed following the annealing of PEDOT:PSS films results from an increase in both the number of the conductive hot spots and current carrying capacity of the PEDOT domains observed on the film surface.

Alexeev *et al.* obtained an array of *I–V* curves to reveal the local heterogeneities of a blend of two semiconducting polymers, namely, poly[2-methoxy-5-(3,7-dimethyloctyloxy)-1,4-phenylenevinylene] (MDMO-PPV, electron donor) and poly[oxa-1,4-phenylene-(1-cyano-1,2-vinylene)-(2-methoxy-5-(3,7-dimethyloctyloxy)-1,4-phenylene)-

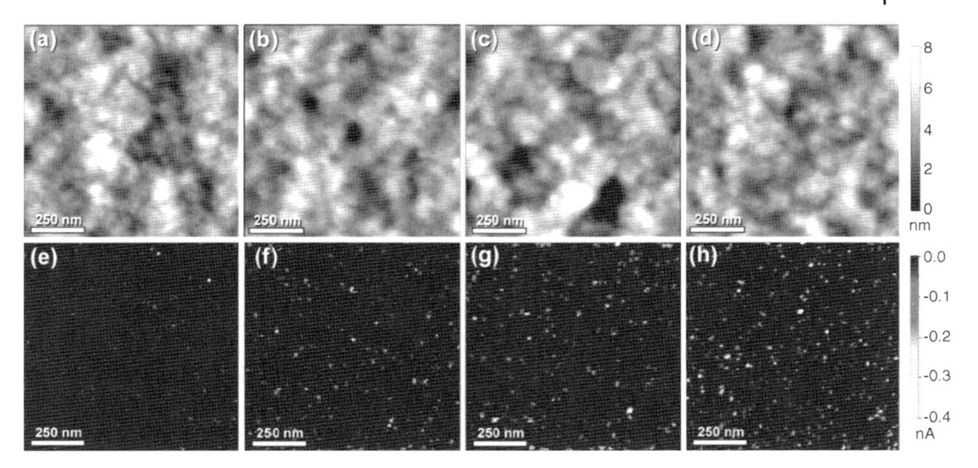

Figure 7.11 1 μm^2 area AFM topography and C-AFM images of the PEDOT:PSS films annealed at 140 °C: (a and e) 0 min, (b and f) 10 min, (c and g) 30 min, and (d and h) 100 min. The topography (top row) exhibits very little change through the course of annealing, while the c-AFM images (bottom row) clearly show that the number of conductive pathways (bright spots) increase as a function of the annealing time. Reprinted from Ref. [56].

1,2-(2-cyanovinylene)-1,4-phenylene] (PCNEPV, electron acceptor) [57]. The C-AFM measurements clearly revealed the phase separation of the components of the blend. Apart from the phase separation, nanoscale details such as the presence of additional PCNEPV domains inside the active layer and heterogeneities in the MDMO-PPV matrix were detected. *I–Z* measurements revealed a significant dependence of the measured current on the penetration of the tip into the polymer layer. For loads ranging from 10 to 30 nN, the current was found to be constant. Further increase in the load resulted in a rapid rise in the current. This example clearly highlights the importance of the applied load on the measured conductivity of the sample under C-AFM investigation.

Dante *et al.* employed C-AFM to obtain a simultaneous map of surface topography and current map of the blend of poly(3-hexylthiophene) and [6,6]-phenyl-C_{61}-butyric acid methyl ester (P3HT:PCBM), which is one of the most widely studied bulk heterojunction materials [58]. Hole and electron current images were obtained by using Pt- and Mg-coated tips, respectively. The current distribution maps revealed the composition of the phase-separated domains. Furthermore, local current–voltage curves were obtained to calculate charge carrier mobilities. Hole current images and mobilities were obtained using Pt-coated silicon probes, whereas Mg-coated silicon probes were used to measure nanoscale electron mobilities. The authors noted that the blending of P3HT and PCBM resulted in the decrease of the polymer hole mobility compared to the pristine P3HT. Hole and electron mobilities were found to increase upon annealing, which is expected considering the increase in the internal order with annealing.

Terawaki *et al.* acquired the conductivity map of a DNA network using C-AFM in horizontal mode under various humidities [59]. In dry conditions (close to 0% relative humidity), they observed no difference in the current measured on the DNA network and mica substrate. On the other hand, at a higher relative humidity (60%), the DNA network exhibited significantly higher current (20 pA) at a bias voltage of 5 V. The higher current was due to the adsorption of water molecules on the DNA network, leading to an increase in ionic current. In a different study, Fang *et al.* studied the effect of intermolecular π–π stacking on the electrical properties of monolayer film molecules containing aromatic groups using C-AFM [60]. They employed two aromatic molecules, (4-mercaptophenyl) anthrylacetylene (MPAA) and (4-mercaptophenyl)-phenylacetylene (MPPA), with different degrees of π–π stacking ability. C-AFM revealed that the conductivity of these two molecules assembled on Au(111) substrates varied by an order of magnitude, with MPAA exhibiting higher conductivity compared to MPPA.

Furthermore, the authors observed that the MPAA films exhibited distinct conductivity changes when mechanically perturbed, as opposed to the MPPA film that did not exhibit such changes. The authors reasoned that the observed difference is possibly due to the weaker π–π interactions of MPPA compared to MPAA. This example clearly demonstrates that C-AFM is a unique technique that can reveal the electronic behavior of molecules with subtle variations in the molecular structure, an extremely important capability for molecular electronics.

7.5 Magnetic Properties

Magnetic force microscopy (MFM) is an important mode for visualization of spatial distribution of magnetic microstructures on surfaces and interfaces as discussed previously. MFM has also been employed to probe the distribution of magnetic nanoparticles in polymer composites. Sun *et al.* described the polymer-mediated assembly of FePt nanoparticles using PVP and PEI polymers [61]. The assembly process involved the exchange of oleic acid/oleyl amine around the

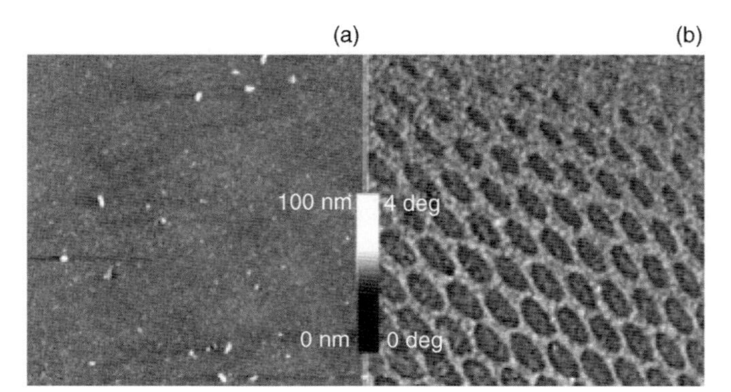

Figure 7.12 (a) AFM topography and (b) MFM image of a three-layer 4 nm $Fe_{58}Pt_{42}$ nanoparticle assembly annealed at 530 °C. The AFM reveals the smooth surface topography of the assembly, whereas the MFM image reveals the assembled particles. Reprinted from Ref. [61].

magnetic nanoparticles with a functional polymer previously deposited on a substrate. Figure 7.12 shows the topography and the corresponding MFM image of a three-layer 4 nm $Fe_{58}Pt_{42}$ assembly treated with a pulsed laser under a perpendicular magnetic field (2.5 kOe). The AFM image shows that the smooth FePt nanoparticle assembly is intact after the laser treatment. The dark spots in MFM image indicate the magnetization pointing out of the particle assembly plane.

In another study, MFM was also used for mapping the dispersion of carbon nanotubes in a polymer matrix [62]. From the MFM phase images, the carbon nanotubes were often found as agglomerates throughout the film. Furthermore, MFM also clearly revealed individual nanotube bundles, as well as surface areas with high localization of carbon nanotubes, which could not be observed in conventional topographic images. The contrast between the nanotubes and the background was strongly dependent on the distance between the tip and the surface (lift height), with lift heights greater than 15 nm exhibiting diminished contrast.

References

1 Frisbie, C.D., Rozsnyai, L.F., Noy, A., Wrighton, M.S., and Lieber, C.M. (1994) Functional group imaging by chemical force microscopy. *Science*, **265** (5181), 2071–2074.

2 Noy, A., Frisbie, C.D., Rosznyai, L.F., Wrighton, M.S., and Lieber, C.M. (1995) Chemical force microscopy: exploiting chemically modified tips to quantify adhesion, friction and functional group distributions in molecular assemblies. *J. Am. Chem. Soc.*, **117** (30), 7943–7951.

3 Zhmud, B.V. and Golub, A.A. (1994) Protolytic equilibria ligands immoblilized at rigid matrix surfaces: a theoretical study. *J. Colloid Interface Sci.*, **167** (1), 186–192.

4 Vezenov, D.V., Noy, A., Rozsnyai, L.F., and Lieber, C.M. (1997) Force titrations and ionization state sensitive imaging of functional groups in aqueous solutions by chemical force microscopy. *J. Am. Chem. Soc.*, **119** (8), 2006–2015.

5 van der Vegte, E.W. and Hadziioannou, G. (1997) Scanning force microscopy with chemical specificity: an extensive study of chemically specific tip–surface interactions and the chemical imaging of surface functional groups. *Langmuir*, **13** (16), 4357–4368.

6 Schönherr, H., Hruska, Z., and Vancso, J. (2000) Toward high resolution mapping of functional group distribution at surface treated polymers by AFM using modified tips. *Macromolecules*, **33** (12), 4532–4537.

7 Green, J.-B.D., McDermott, M.T., Porter, M.D., and Siperko, L.M. (1995) Nanometer-scale mapping of chemically distinct domains at well-defined organic interfaces using frictional force microscopy. *J. Phys. Chem.*, **99** (27), 10960–10965.

8 Tsukruk, V.V. and Bliznyuk, V.N. (1998) Adhesive and friction forces between chemically modified silicon and silicon nitride surfaces. *Langmuir*, **14** (2), 446–455.

9 Noy, A. (2006) Chemical force microscopy of chemical and biological interactions. *Surf. Interface Anal.*, **38** (11), 1429–1441.

10 Schonherr, H. and Vancso, G.J. (1998) Surface properties of oxidized LDPE by scanning force microscopy with chemically modified probes. *J. Polym. Sci. B*, **36** (14), 2483–2492.

11 Nie, H.-Y., Walzak, M.J., Berno, B., and McIntyre, N.S. (1999) Atomic force microscopy study of polypropylene surfaces treated by UV and ozone exposure: modification of morphology and adhesion force. *Appl. Surf. Sci.*, **144–145**, 627–632.

12 Eaton, P.J., Graham, P., Smith, J.R., Smart, J.D., Nevell, T.G., and Tsibouklis, J. (2000) Mapping the surface heterogeneity of a

polymer blend: an adhesion-force-distribution study using the atomic force microscope. *Langmuir*, **16** (21), 7887–7890.

13 Duwez, A.-S. and Nysten, B. (2001) Mapping aging effects on polymer surfaces: specific detection of additives by chemical force microscopy. *Langmuir*, **17** (26), 8287–8292.

14 Schonherr, H., Hruska, Z., and Vancso, G.J. (2000) Toward high resolution mapping of functional group distributions at surface-treated polymers by AFM using modified tips. *Macromolecules*, **33** (10), 4532–4537.

15 Geissler, A., Vallat, M.-F., Vidal, L., Voegel, J.-C., Hemmerle, J., Schaaf, P., and Roucoules, V. (2008) Chemical force titration of plasma polymer-modified PDMS substrates by using plasma polymer-modified AFM tips. *Langmuir*, **24** (9), 4874–4880.

16 Barber, A.H., Cohen, S.R., and Wagner, H.D. (2003) Measurement of carbon nanotube–polymer interfacial strength. *Appl. Phys. Lett.*, **82** (23), 4140–4142.

17 Friddle, R.W., LeMieux, M.C., Cicero, G., Artyukhin, A.B., Tsukruk, V.V., Grossman, J.C., Galli, G., and Noy, A. (2007) Single functional group interactions with individual carbon nanotubes. *Nat. Nanotechnol.*, **2** (11), 692–697.

18 Sun, P., Laforge, F.O., and Mirkin, M.V. (2007) Scanning electrochemical microscopy in the 21st century. *Phys. Chem. Chem. Phys.*, **9** (7), 802–823.

19 Anne, A., Demaille, C., and Goyer, C. (2009) Electrochemical atomic-force microscopy using a tip-attached redox mediator. Proof-of-concept and perspectives for functional probing of nanosystems. *ACS Nano*, **3**, 819–827.

20 O'Mullane, A.P., Macpherson, J.V., Cervera-Montesinos, J., Manzanares, J.A., Frehill, F., Vos, J.G., and Unwin, P.R. (2004) Measurement of lateral charge propagation in $[Os(bpy)_2(PVP)_nCl]Cl$ thin films: a scanning electrochemical microscopy approach. *J. Phys. Chem. B*, **108** (22), 7219–7227.

21 Zhou, J., Campbell, C., Heller, A., and Bard, A.J. (2002) Scanning electrochemical microscopy. 44. Imaging of horseradish peroxidase immobilized on insulating substrates. *Anal. Chem.*, **74** (16), 4007–4010.

22 Zhu, R.K., Macfie, S.M., and Ding, Z.F. (2005) Cadmium-induced plant stress investigated by scanning electrochemical microscopy. *J. Exp. Bot.*, **56** (421), 2831–2838.

23 Holt, K.B. and Bard, A.J. (2005) Interaction of silver(I) ions with the respiratory chain of *Escherichia coli*: an electrochemical and scanning electrochemical microscopy study of the antimicrobial mechanism of micromolar Ag^+. *Biochemistry*, **44** (39), 13214–13223.

24 Longobardi, F., Cosma, P., Milano, F., Agostiano, A., Mauzeroll, J., and Bard, A.J. (2006) Scanning electrochemical microscopy of the photosynthetic reaction center of *Rhodobacter sphaeroides* in different environmental systems. *Anal. Chem.*, **78** (14), 5046–5051.

25 Liu, B., Rotenberg, S.A., and Mirkin, M.V. (2000) Scanning electrochemical microscopy of living cells: different redox activities of nonmetastatic and metastatic human breast cells. *Proc. Natl. Acad. Sci. USA*, **97** (18), 9855–9860.

26 Anne, A., Cambril, E., Chovin, A., Demaille, C., and Goyer, C. (2009) Electrochemical atomic force microscopy using a tip-attached redox mediator for topographic and functional imaging of nanosystems. *ACS Nano*, **3** (10), 2927–2940.

27 Kelvin, L. (1898) Contact electricity of metals. *Philos. Mag.*, **46**, 82.

28 Zisman, W.A. (1932) A new method of measuring contact potential differences in metals. *Rev. Sci. Instrum.*, **3** (7), 367–471.

29 Charrier, D.S.H., Kemerink, M., Smalbrugge, B.E., de Vries, T., and Janssen, R.A.J. (2008) Real versus measured surface potentials in scanning Kelvin probe microscopy. *ACS Nano*, **2** (4), 622–626.

30 Liscio, A., Palermo, V., and Samori, P. (2008) Probing local surface potential of quasi-one-dimensional systems: a KPFM study of P3HT nanofibers. *Adv. Funct. Mater.*, **18** (6), 907–914.

31 Tevaarwerk, E., Keppel, D.G., Rugheimer, P., Lagally, M.G., and Eriksson, M.A. (2005) Quantitative analysis of electric

force microscopy: the role of sample geometry. *Rev. Sci. Instrum.*, **76** (5), 053707.

32 Sugimura, H., Saito, N., Maeda, N., Ikeda, I., Ishida, Y., Hayashi, K., Hong, L., and Takai, O. (2004) Surface potential microscopy for chemistry of organic self-assembled monolayers in small domains. *Nanotechnology*, **15** (2), S69.

33 Sugimura, H., Hayashi, K., Saito, N., Nakagiri, N., and Takai, O. (2002) Surface potential microscopy for organized molecular systems. *Appl. Surf. Sci.*, **188** (3–4), 403.

34 Magonov, S., Alexander, J., Wu, S . (2010) Advancing characterization of materials with atomic force microscopy – based electric techniques, in *Scanning Probe Microscopy of Functional Materials: Nanoscale Imaging and Spectroscopy* (eds S. Kalinin and A. Gruverman), Springer.

35 Maaloum, M., Muller, P., and Krafft, M.P. (2002) Monodisperse surface micelles of nonpolar amphiphiles in Langmuir monolayers. *Angew. Chem., Int. Ed.*, **114** (22), 4531–4534.

36 Kato, T., Kameyama, M., Eahara, M., and Iimura, K. (1998) Monodisperse two-dimensional nanometer size clusters of partially fluorinated long-chain acids. *Langmuir*, **14** (7), 1786–1798.

37 Ren, Y., Iimura, K., Ogawa, A., and Kato, T. (2001) Surface micelles of $CF_3(CF_2)_7(CH_2)_{10}COOH$ on aqueous La^{3+} subphase investigated by atomic force microscopy and infrared spectroscopy. *J. Phys. Chem. B*, **105** (19), 4305–4312.

38 Coffey, D.C. and Ginger, D.S. (2006) Time-resolved electrostatic force microscopy of polymer solar cells. *Nat. Mater.*, **5** (9), 735–740.

39 Hoppe, H., Glatzel, T., Niggemann, M., Hinsch, A., Lux-Steiner, M.C., and Sariciftci, N.S. (2005) Kelvin probe force microscopy study on conjugated polymer/fullerene bulk heterojunction organic solar cells. *Nano Lett.*, **5** (2), 269–274.

40 Chiesa, M., Bürgi, L., Kim, J.-S., Shikler, R., Friend, R.H., and Sirringhaus, H. (2005) Correlation between surface photovoltage and blend morphology in polyfluorene-based photodiodes. *Nano Lett.*, **5** (4), 559–563.

41 Cui, X., Freitag, M., Martel, R., Brus, L., and Avouris, P. (2003) Controlling energy-level alignments at carbon nanotube/Au contacts. *Nano Lett.*, **3** (6), 783–787.

42 Jespersen, T.S. and Nygård, J. (2005) Charge trapping in carbon nanotube loops demonstrated by electrostatic force microscopy. *Nano Lett.*, **5** (9), 1838–1841.

43 Datta, S.S., Strachan, D.R., Mele, E.J., and Johnson, A.T.C. (2009) Surface potentials and layer charge distributions in few-layer graphene films. *Nano Lett.*, **9** (1), 7–11.

44 Kuntze, S.B., Ban, D., Sargent, E.H., Dixon-Warren, St.J., White, J.K., and Hinzer, K. (2005) Electrical scanning probe microscopy: investigating the inner workings of electronic and optoelectronic devices. *Crit. Rev. Solid State Mater. Sci.*, **30** (2), 71–124.

45 Pingree, L.S.C., Reid, O.G., and Ginger, D.S. (2009) Electrical scanning probe microscopy on active organic electronic devices. *Adv. Mater.*, **21** (1), 19–28.

46 Berger, R., Butt, H.-J., Retschke, M.B., and Weber, S.A.L. (2009) Electrical modes in scanning probe microscopy. *Macromol. Rapid Commun.*, **30** (14), 1167–1178.

47 Trenkler, T., Hantschel, T., Stephenson, R., De Wolf, P., Vandervorst, W., Hellemans, L., Malave, A., Buchel, D., Oesterschulze, E., Kulisch, W., Niedermann, P., Sulzbach, T., and Ohlsson, O. (2000) Evaluating probes for "electrical" atomic force microscopy. *J. Vac. Sci. Technol. B*, **18** (1), 418–427.

48 Thomson, R. and Moreland, J. (1995) Development of highly conductive cantilevers for atomic force microscopy point contact. *J. Vac. Sci. Technol. B*, **13** (3), 1123–1125.

49 Tivanski, A.V., Bemis, J.E., Akhremitchev, B.B., Liu, H., and Walker, G.C. (2003) Adhesion forces in conducting probe atomic force microscopy. *Langmuir*, **19** (6), 1929–1934.

50 Reid, O.G., Munechika, K., and Ginger, D.S. (2008) Space charge limited current measurements on conjugated polymer films using conductive atomic force microscopy. *Nano Lett.*, **8** (6), 1602–1609.

51 Pingree, L.S.C., Hersam, M.C., Kern, M.M., Scott, B.J., and Marks, T.J. (2004) Spatially-resolved electroluminescence of

operating organic light-emitting diodes using conductive atomic force microscopy. *Appl. Phys. Lett.*, **85** (2), 344–346.

52 Ionescu-Zanetti, C., Mechler, A., Carter, S.A., and Lal, R. (2004) Semiconductive polymer blends: correlating structure with transport properties at the nanoscale. *Adv. Mater.*, **16** (5), 385–389.

53 Douheret, O., Lutsen, L., Swinnen, A., Breselge, M., Vandewal, K., Goris, L., and Manca, J. (2006) Nanoscale electrical characterization of organic photovoltaic blends by conductive atomic force microscopy. *Appl. Phys. Lett.*, **89** (3), 032107.

54 Coffey, D.C. and Ginger, D.S. (2005) Patterning phase separation in polymer films with dip-pen nanolithography. *J. Am. Chem. Soc.*, **127** (13), 4564–4565.

55 Coffey, D.C., Reid, O.G., Rodovsky, D.B., Bartholomew, G.P., and Ginger, D.S. (2007) Mapping local photocurrents in polymer/fullerene solar cells with photoconductive atomic force microscopy. *Nano Lett.*, **7** (3), 738–744.

56 Pingree, L.S.C., MacLeod, B.A., and Ginger, D.S. (2008) The changing face of PEDOT:PSS films: substrate, bias, and processing effects on vertical charge transport. *J. Phys. Chem. C*, **112** (21), 7922–7927.

57 Alexeev, A., Loos, J., and Koetse, M.M. (2006) Nanoscale electrical characterization of semiconducting polymer blends by conductive atomic force microscopy (C-AFM). *Ultramicroscopy*, **106** (3), 191–199.

58 Dante, M., Peet, J., and Nguyen, T.-Q. (2008) Nanoscale charge transport and internal structure of bulk heterojunction conjugated polymer/fullerene solar cells by scanning probe microscopy. *J. Phys. Chem. C*, **112** (18), 7241–7249.

59 Terawaki, A., Otsuka, Y., Lee, H.Y., Matsumoto, T., Tanaka, H., and Kawai, T. (2005) Conductance measurement of a DNA network in nanoscale by point contact current imaging atomic force microscopy. *Appl. Phys. Lett.*, **86** (11), 113901.

60 Fang, L., Park, J.Y., Ma, H., Jen, A.K.-Y., and Salmeron, M. (2007) Atomic force microscopy study of the mechanical and electrical properties of monolayer films of molecules with aromatic end groups. *Langmuir*, **23** (23), 11522–11525.

61 Sun, S., Anders, S., Hamann, H.F., Thiele, J.-U., Baglin, J.E.E., Thomson, T., Fullerton, E.E., Murray, C.B., and Terris, B.D. (2002) Polymer mediated self-assembly of magnetic nanoparticles. *J. Am. Chem. Soc.*, **124** (12), 2884–2885.

62 Lillehei, P.T., Park, C., Rouse, J.H., and Siochi, E.J. (2002) Imaging carbon nanotubes in high performance polymer composites via magnetic force microscopy. *Nano Lett.*, **2** (8), 827–829.

8
Scanning Probe Optical Techniques

8.1
Fundamental Principles

The resolution of conventional optical microscopy is limited by the diffraction-limited spot size in which a light beam can be focused using a normal optical lens. The focused light forms concentric rings known as the Airy disk, quantitatively described by Abbe [1]. The distance d from the middle of the center spot to the first node of the Airy disk is given by

$$d = 0.61\left(\frac{\lambda_0}{\mathrm{NA}}\right)$$

where λ_0 is the wavelength of the light in the free space and NA ($\mathrm{NA} = n \sin \alpha$, where n is the refractive index of the surrounding medium and α is the acceptance angle) is the numerical aperture of the objective lens. Considering that typical values of NA are 0.9–1.4 (including water and oil immersion lenses), the diffraction limit can be roughly estimated as $\lambda_0/2$. For the visible light, even under optimum conditions, the spatial resolution that can be achieved is typically limited to 250–300 nm.

As known, this diffraction limit can be overcome by performing the optical measurements with the light source or detector held closer to the sample than the wavelength of the light. In the so-called "near-field" regime, the resolution is no longer determined by the wavelength of the light but instead by the size of the source or detector [2]. The most popular way to practically realize this is by combining an optical spectroscopic technique (e.g., scattering or fluorescence) with SPM instrumentation.

8.2
Introduction to Scanning Near-Field Optical Microscopy

To overcome the fundamental, diffraction-limited resolution of conventional optical microscopy, near-field scanning optical microscopy (NSOM), a scanning probe technique comprised of an optical probe employed as a source or detector in

Scanning Probe Microscopy of Soft Matter: Fundamentals and Practices, First Edition.
Vladimir V. Tsukruk and Srikanth Singamaneni.

proximity to the sample (tens of nanometers), has been developed [2, 3]. In NSOM, confined photon flux between a local probe and the sample surface is employed to overcome the diffraction limitation. Similar to any other scanning probe technique, the probe is raster scanned over the sample surface and the optical response (depending on the desired information) is acquired as function of the lateral coordinates to create the near-field map.

The resolution of an NSOM instrument depends on the size of the probe (light source) and the separation between the probe and the sample. With both the dimensions much smaller than the wavelength of the light, the resolution attained by NSOM is typically much smaller (nearly an order of magnitude) than the wavelength of light. In the most common form of NSOM, the sample is irradiated through an aperture of the probe. Laser light is coupled to the optical fiber probe at one end using a fiber coupler. The probe is raster scanned along the sample and the distance between the tip and the sample is controlled using the lateral shear force interaction between the probe and the surface. The shear force on the probe is detected by monitoring the acoustic resonance of the probe using conventional optical techniques. In a typical experiment, the atomic force microscopy (AFM) (topography) and near-field optical image are obtained simultaneously from a selected area.

There are numerous methods to classify NSOM measurements such as the type of probe, light delivery, collection mode, and nature of the signal being collected (e.g., fluorescence signature or Raman scattering). The most common types of near-field optical imaging modes are transmission and fluorescence with visible light. In the transmission mode, the light emanating from the tip through the transparent sample is collected, whereas in the case of fluorescence imaging, the light is spectrally filtered to detect the fluorescence from the localized sample region.

We do not intend to provide a comprehensive review of all papers covering NSOM, but rather a breadth of the possible studies that could be performed using the technique, illustrated by several selected examples related to polymeric materials. The readers are referred to some excellent reviews presenting comprehensive discussions of NSOM probe technology, application to organic materials, and recent developments in different NSOM modes [4–7].

8.2.1 Aperture NSOM

Based on the type of the probe employed, NSOM can be classified into aperture and apertureless modes [6]. Here, we will briefly discuss these modes highlighting the most important similarities and differences.

In the more commonly employed aperture approach, evanescent waves resulting from passing the propagating light from an external source through a tapered tip are used to obtain surface images (see Figure 8.1). Experimentally, the light is transmitted along an optical fiber whose tip is coated with a reflective metal (usually aluminum) resulting in an evanescent field generated at the tip end. The tapered geometry of the fiber probe hosts two characteristic diameters critically affecting the

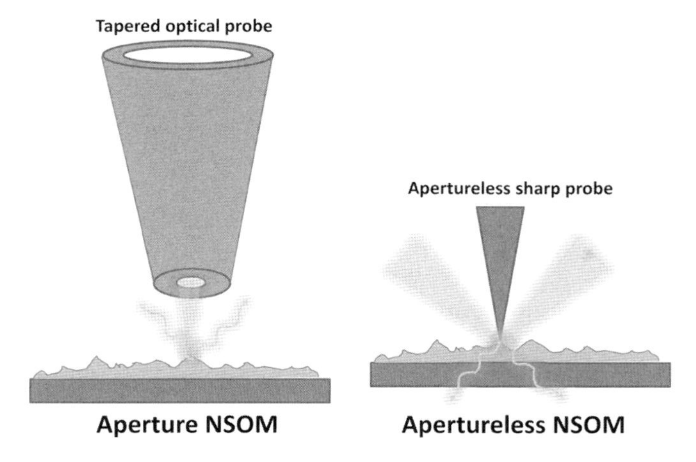

Figure 8.1 The aperture NSOM with a tapered optical probe delivering the light in the near-field regime and resulting in scattering from the sample, and the apertureless NSOM showing the sharp probe (coated or uncoated) enhancing the electromagnetic field of the incident light in the localized area of the sample.

possible electromagnetic modes that can exist. The allowable mode after the first characteristic diameter is HE_{11}, which converts into the evanescent mode after the so-called cutoff diameter [8, 9]. The field that reaches the probe end is evanescent, which is a localized field enabling imaging on the length scale of a few nanometers. It is important to note that only a tiny fraction of the light coupled to the fiber is emitted through the aperture due to the propagation cutoff of the waveguide modes, as discussed above.

8.2.2 Apertureless NSOM

In the other important variation of NSOM, the apertureless mode, the sharp probe is employed as a scatterer of evanescent waves as opposed to a waveguide in an aperture probes (Figure 8.1) [4]. The entire probe is coated with a metal (including the apex), which makes the tip blunt in comparison to conventional AFM probes and limits the spatial resolution. It is generally believed that a metallic tip is advantageous in apertureless NSOM, as the strong resonant plasmon coupling between metallic tip and sample enhances the interaction between the mating species.

However, Haefliger *et al.* demonstrated that for scattering-type NSOM operation, the image contrast and scattering efficiency of commercial Si probes after removing the native SiO_2 were significantly enhanced and are even better compared to the PtIr- or Au-coated tips [10]. The authors demonstrated that the etched tip (SiO_2-stripped) produces an immensely higher-scattering signal compared to the pristine probe, owing to the improved field enhancement capability of the Si tip. It is interesting to note that a very thin oxide layer dramatically reduces the Si tip's near-field scattering signal, highlighting the strong field confinement to the high-refractive Si tip. The authors pointed out that dielectric tips have several advantages over metal-coated tips such as (i) minimized fluorescence quenching, (ii) suppressed nonlinearity for second harmonic imaging, (iii) enhanced wear resistance, (iv) enhanced resolution due to a small tip radius, and (v) absence of corrosive decomposition that is commonly observed in metal-coated tips.

Metal-coated tips do have obvious advantages when probing the localized vibrational chemical signature, namely, Raman scattering and infrared absorption. In particular, it is well known that despite the rich information offered by regular Raman spectroscopy, it is limited by a weak signal [11]. The sharp metal tip in proximity to the surface results in electromagnetic (EM) field enhancement owing to what is known as the antenna effect [12]. When a sharp metal tip is illuminated with light polarized parallel to the tip shank, the antenna effect causes a dramatic increase in the local surface charge density at its apex, resulting in the enhancement and confinement of the EM field at the apex. EM field enhancement results in highly enhanced Raman scattering and the localization of the EM field renders the ability to investigate samples with high spatial resolution.

8.2.3
Artifacts in NSOM

As with any other scanning probe technique, it is important to be aware of potential artifacts that can creep into the NSOM images [13, 14]. In fact, this has been a significant problem in the early work related to NSOM, with some effects rippling to current studies. Since the evanescent field rapidly decays with distance, relatively small changes in the separation between the tip and the sample can lead to a large change in the intensity of the light at the sample surface. These large changes in the intensity of light (fluorescence or scattered light) can be easily misinterpreted as changes in the material properties of the sample being investigated.

Another important artifact commonly observed in the images is the slight offset in the registry between the topography and the optical map. The source of this artifact is the geometry of the fiber tip itself that has finite lateral dimensions. The feedback signal for the topography comes from the interaction between the metal coating (typically aluminum) on the outer diameter of the fiber tip, whereas the optical signal is collected at the center of the probe.

Yet another important practical consideration in NSOM studies of soft samples is the heating near the aperture of aluminum-coated optical fiber, which is commonly employed as the near-field probe. There have been numerous studies addressing this issue – using, for example, thermocouples to measure the local surface temperature [15, 16]. More recently, thermochromic polymers that exhibit distinct changes in optical properties with temperature have been employed to understand the sample heating aspects of NSOM [17, 18]. In a recent example, Erickson and Dunn elegantly addressed the question of measuring the temperature rise as a function of input and output powers using a thermochromic polymer [19]. The optical probe was positioned at a controlled distance (several nanometers) from the polymer surface using the shear force feedback method and optical spectra were recorded. The authors

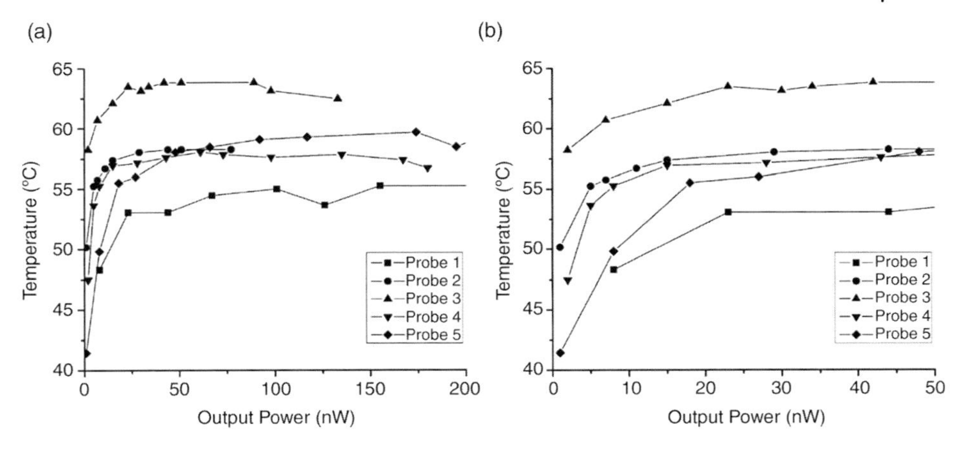

Figure 8.2 (a) Sample heating as a function of output power for five different NSOM probes. (b) An expanded view of the initial heating showing that the sample heating rises quickly with output power before leveling off between 55 and 65 °C with output powers above ~50 nW. Reprinted from Ref. [19].

observed an initial rapid rise in sample temperature as the output power is increased. For output powers of a few nanowatts, the local temperature of the polymer surface reached 45 °C, while for higher output powers (50–100 nW), slow rises in sample temperature until a maximum of ~65 °C were observed (see Figure 8.2).

8.3
Examples of NSOM Studies of Polymer and Polymer Blends

NSOM has been extensively employed to monitor the physical and chemical properties of various polymers and polymer blends with spatial resolutions on the order of single molecules. In this section, we will discuss several recent examples of NSOM utilizations for these materials.

8.3.1
NSOM for Monitoring the Composition and Physical State

Ube *et al.* monitored the conformation of a single-polymer chain in a uniaxially stretched poly(methyl methacrylate) (PMMA) film with NSOM using fluorescence intensity distribution [20]. The perylene-labeled PMMA chains embedded in the unlabeled PMMA film were observed as bright spots in the fluorescence NSOM image. Figure 8.3 shows the NSOM image of the fluorescent-labeled chains under pristine and stretched states. Since the optical near-field penetrates into the polymer film by a few hundred nanometers, the authors noted that the shape of the labeled polymer chain observed in the NSOM image corresponds to the two-dimensional projection of the chain conformation. In this study, a significant deviation has also

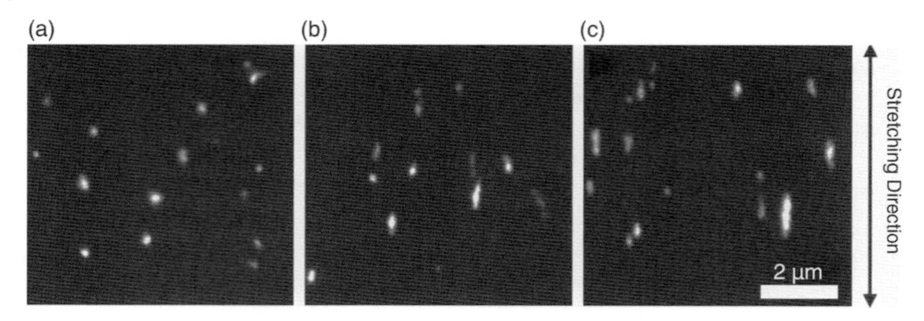

Figure 8.3 Fluorescence NSOM images of individual labeled polymer chains in the PMMA films (a) before stretching, and after stretching to the strain of (b) 1.0 and (c) 2.0. The stretching of the polymer chain seen in (c) is found to be much smaller than the strain applied on the film. Reprinted from Ref. [20].

been observed between the strain on a single chain and the macroscopic deformation applied to the film. The nonaffine deformation was more pronounced at higher deformations due to slipping of subsurface polymer chains during stretching.

Observation of birefringence caused by the local preferential orientation of molecules or functional groups in polymers using conventional polarized optical microscopy is extensively applied in monitoring crystalline and semicrystalline polymers [21]. Changes in birefringence as a function of temperature, pressure, or concentration are also utilized frequently to understand the phase transitions in liquid crystalline polymers. However, the conventional far-field polarized optical microscopy is diffraction limited. Furthermore, the conventional microscopy signal exhibits a square law dependence on the birefringence, resulting in the loss of the sign of the birefringence, which is important for identifying the relative orientation of the optical anisotropy parameters of the structures under study.

In contrast, NSOM measurements preserve the sign of the birefringence signal, so as to determine the relative orientation of the optical anisotropy parameters [22–24]. To this end, Ade *et al.* employed NSOM to monitor birefringence in partially ordered Kevlar fibers at subwavelength spatial resolution [24]. The authors successfully correlated the near-field optical observations to existing structural models. Furthermore, the simultaneous topography and optical image enables one to distinguish inherent birefringence from topographically induced and correlated birefringence, as well as to determine the origin of the observed birefringence.

In more practical studies, NSOM was employed to investigate the phase separation of dye-labeled poly(isobutyl methacrylate) (PiBMA) and poly(octadecyl methacrylate) (PODMA) at the air/water interface [25, 26]. The energy transfer efficiency between dyes tagged to the polymers and alteration of the fluorescence lifetime was mapped to reveal that the phase boundary had a finite width of a few hundred nanometers, which was significantly larger than that expected in a three-dimensional bulk state. The authors found that when the mixed polymer solution is spread onto a water surface, the PODMA chains tend to aggregate and form a solid-like domain before the phase separation occurs. This was attributed to the strong crystallinity of the long alkyl chains of PODMA. The phase separation was almost complete in a few minutes at elevated temperatures (40 °C) with further increases in the domain size with annealing time. In the case of the block copolymer PiBMA-*b*-PODMA, the polymer chains were noted to exhibit a stretched conformation and the size of the phase-separated structure was dependent on molecular weight, as observed by NSOM.

8.3.2
Optical Properties of Conjugated Polymers and Their Blends

It is well known that the molecular conformation of a conjugated polymer, which determines the interaction between the segments of the polymer chain, bears a paramount importance on its optical and electrical properties [27]. Furthermore, these interactions result in the formation of interchain species such as dimers and excimers. The conjugated polymer conformation and the effects on the optical properties have been primarily addressed using macroscopic measurements in thick films and solutions and only recently NSOM investigations were attempted for such systems.

The excellent lateral resolution of NSOM combined with the optical spectroscopic techniques places NSOM in a rather unique position to address some of these important issues for this class of materials. Conjugated polymers and their blends are being extensively investigated as the active semiconducting layers in light-emitting diodes and photovoltaic devices. NSOM and single-molecule spectroscopy have emerged as powerful tools to probe structure–property relationships, particularly in conjugated polymers [28–37]. Specifically, the polymer blend composition and morphology have been studied by means of their luminescence behavior, which can be addressed with nanoscale resolution using NSOM. Here, we will discuss several examples where NSOM has been employed to unveil the molecular structure and its influence on the macroscopic optical properties of conjugated polymers.

NSOM enabled the optical interrogation of single conjugated polymer chains. Szymanski *et al.* employed NSOM to characterize the physical nature of 5–10 nm poly [2-methoxy-5-((2-ethylhexyl)oxy)-*p*-phenylenevinylene] (MEH-PPV) nanoparticles cast on substrates from the suspension [38]. The small diameter of the nanoparticles indicates that the majority of them are comprised of a single molecule with an average molecular weight of $\sim$200 000 g/mol. The excellent lateral resolution of NSOM, down to a single nanoparticle, enabled the estimation of the optical cross sections of individual conjugated polymer molecules. The measured absorbance using NSOM versus particle diameter was explained by a simple aperture model under two assumptions: (i) NSOM aperture diameter of 50 nm and (ii) the optical cross-section scales linearly with particle volume. This demonstration was first among the studies employing NSOM to investigate single polymer chains.

NSOM has also been employed by Chappell *et al.* to study the structure of a thin film of a phase-separated blend of the conjugated polymers poly(9,9′-dioctylfluorene) (PFO) and poly(9,9′-dioctylfluorene-*alt*-benzothiadiazole) (F8BT) prepared by spin casting [39]. The simultaneous topography and UV transmission measurements enabled the unambiguous identification of the F8BT- and PFO-rich domains. The high-resolution fluorescence measurements using NSOM enabled the identification of the rather unexpected variation of fluorescence intensity within the F8BT-rich phase, where a ~300 nm wide region located at the interface of the two phases was found to be 1.6 times brighter than the center of the domain. The near-field fluorescence images with resolutions of ~70 nm confirmed the presence of a ~10 nm thick, partially crystallized PFO surface layer that preferentially wets the surface. These NSOM measurements provide a better understanding of the blend structure, critical for the development of efficient optoelectronic devices.

Dastoor and coworkers employed the NSOM method to directly map the efficiency of an organic solar cell in two dimensions, with a spatial resolution of 200 nm or less. These measurements revealed the local spatial variations in efficiency of the two-component organic blend system [40]. Near-field scanning photocurrent microscopy (NSPM) enabled the imaging of photocurrent conversion efficiency over the polymer blend samples, directly obviating the need for other techniques such as fluorescence NSOM. The technique involves local photoexcitation of the sample with simultaneous detection of the local photocurrent generated.

One of the most widely studied polymer blends for solar cell applications is based on the combination of regioregular poly(3-hexylthiophene) (P3HT) as the donor and [6,6]-phenyl C61 butyric acid methyl ester (PCBM) as the acceptor owing to its high external quantum efficiencies (above 75%) and power conversion efficiency (~4%) [41]. The high-efficiency optoelectronics based on this blend critically depends on the intrinsic properties of the two components and, more importantly, on the physical properties of the blend. The morphological evolution and the optical properties of the P3HT/PCBM blend films upon annealing were addressed using NSOM [42]. NSOM topography measurements revealed that the segregation and large-scale crystallization of PCBM take place upon thermal annealing. The component distribution (which was further confirmed by confocal Raman microscopy) and the high diffusive mobility of the PCBM chains upon annealing, even at moderate temperatures (100 °C), were revealed by the NSOM optical absorbance measurements.

Mechanochromism, a process involving color transitions induced by mechanical stress, has been studied in poly(diacetylene) (PDA) thin films at the nanometer scale using a combination of AFM and NSOM [43]. The shear forces caused by the AFM tip on PDA molecular films were monitored as blue to red chromatic transitions at the nanoscale using NSOM. The transformed regions, as small as 30 nm (also confirmed independently by AFM), were identified as red domains in the fluorescence emission signature. The irreversibly transformed domains were found to preferentially grow along the polymer backbone direction. Significant rearrangement, a characteristic feature of the mechanochromic transition, of poly-PCDA bilayer segments was also observed by AFM in transformed regions.

8.4
Multicolor NSOM Measurements

The multicolor NSOM measurement involves an acquisition of multiple color signals simultaneously, as suggested by Emonin *et al.* [44]. The authors pointed out that the main advantage of the multicolor NSOM compared to a conventional NSOM is the acquisition of three optical signals of different wavelengths at the same time, on the same area of the sample, which considerably reduces experiment time. Furthermore, sample drift, which is a common problem for successive data acquisitions in conventional NSOM experiments, is avoided with multicolor NSOM by keeping the experimental conditions exactly the same during the entire experiment. Real color images with nanoscale resolution acquired by multicolor NSOM are extremely valuable for functional biological imaging with multiple fluorescent labels.

Dual-color NSOM (at 400 and 500 nm wavelengths) was used to study thin films of poly(*p*-phenylenevinylene) and poly(dioctyloxy phenylenevinylene) copolymer and blends by Wei *et al.* [45]. Blend films were found to undergo spinodal decomposition with the domain sizes ranging from 100 to 200 nm, as revealed by NSOM, independent of the composition ratio. However, the domain size and degree of decomposition were found to depend on the free energy, surface tension, and viscosity. The correlation between the topography and the NSOM was lost when the degree of decomposition was small (300 nm) due to the nonuniform distribution of the components in the thickness direction.

Enderle *et al.* employed dual-color NSOM to perform membrane-oriented colocalization measurements with simultaneous topography mapping of parasite and host proteins in malaria-infected erythrocytes [46]. Apart from the excellent lateral resolution of the optical image and simultaneous topography image from the force feedback, dual-color NSOM offered unique advantages in studying the cell membranes in that the near-field source excites fluorescence only from the outermost layer (50–100 nm) of the cell, thereby suppressing the autofluorescence that is common problem in fluorescence measurements as the fluorescence is integrated over the entire cell thickness. Furthermore, dual-color excitation via shared aperture eliminated chromatic aberration, resulting in colocalization studies with unprecedented resolution. Furthermore, the two-color images exhibited excellent pixel-by-pixel registry since both excitation colors shared the same near-field aperture.

The multicolor NSOM application has been demonstrated by Emonin *et al.* by illuminating the sample with a single or a combined beam of different laser wavelengths [44]. They modified a conventional NSOM setup by implementing three photomultipliers for blue, green, and red light detection and color-separating dichroic filters. With this setup, the authors demonstrated the simultaneous acquisitions of three optical images (on three different channels) and the topographical image of latex beads labeled with two different dyes and transmission measurements on gold nanoparticles. The authors concluded that with future developments, the multicolor NSOM can be a powerful tool for high-resolution functional imaging, especially for biological imaging.

8.5 Tip-Enhanced Raman Spectroscopy and Microscopy

Inelastic scattering of light by a molecule, which forms the quantized vibrational signature, is the physical phenomenon behind Raman spectroscopy [47]. Despite the rich information offered by normal Raman spectroscopy (NRS), it has not been the choice as a handy analytical tool due to the inherent feebleness of its signal. The weak signal, which severely limits NRS applications, is due to the extremely small Raman scattering cross sections. For example, a benzene molecule, which is a relatively strong Raman scatterer, exhibits a scattering cross section of $2.8 \times 10^{-29}\,\mathrm{cm^2/(molecule\ sr)}$ – many orders of magnitude below elastic scattering [48, 49].

However, the discovery of surface-enhanced Raman scattering (SERS) nearly 30 years ago opened new avenues of Raman spectroscopy applications to a variety of challenging tasks, which created a huge stir in the scientific community [50, 51]. The dramatic enhancement of the intensity of the vibrational spectra from the analyte adsorbed on a rough (nanoscale) metal surface, which seems to be simple for experimental realization (working in numerous cases), was intensely investigated to understand the underlying physical phenomenon responsible for the effect [52–57].

An alternative approach to NSOM studies of Raman properties, known as tip-enhanced Raman scattering (TERS), involves the near-field probing of localized Raman scattering from the sample. Tip-enhanced Raman scattering can be considered as the very specific, scanning probe version of SERS [58]. The fundamental principle behind the tip-enhanced inelastic spectroscopic response is the strong enhancement of the electromagnetic field in the vicinity of metal nanostructures with highly curved features due to surface plasmon resonance. Electromagnetic enhancement is caused by the collective oscillations of conduction electrons (surface plasmon) in a metal particle [59]. Excitation of the surface plasmon results in the enhancement of the local field experienced by a molecule adsorbed on the surface of the particle. Although the enhancement of the local electric field is modest, the enhancement in the scattered light intensity scales to the fourth power (with intensity proportional to the square of the electric field), causing a remarkable intensity rise even for modest EM effects.

However, the enhancement effects are highly localized (within 5–10 nm) and decay rapidly as the distance of separation between the analyte and the metal particles increases, making it truly surface-sensitive technique [60]. The enhancement of the Raman scattering is very sensitive to the distance between the metal surface and the analyte. Such distance dependence has been theoretically predicted as

$$I = \left(1 + \frac{r}{a}\right)^{10}$$

where a is the average size of the field-enhancing features and r is the distance to the analyte from the surface of the metal [61].

The size of the metal nanostructures for the excitation of surface plasmons has an upper bound as the wavelength of the excitation light. While there are several aspects that affect the magnitude of the enhancement of the local electromagnetic field, and hence the inelastic response, the presence of metal nanostructures responsible for the enhanced electric field in the close (nanometer) proximity is paramount [62]. In TERS experiments, such a nanostructure (metal tip end) is brought to the vicinity of

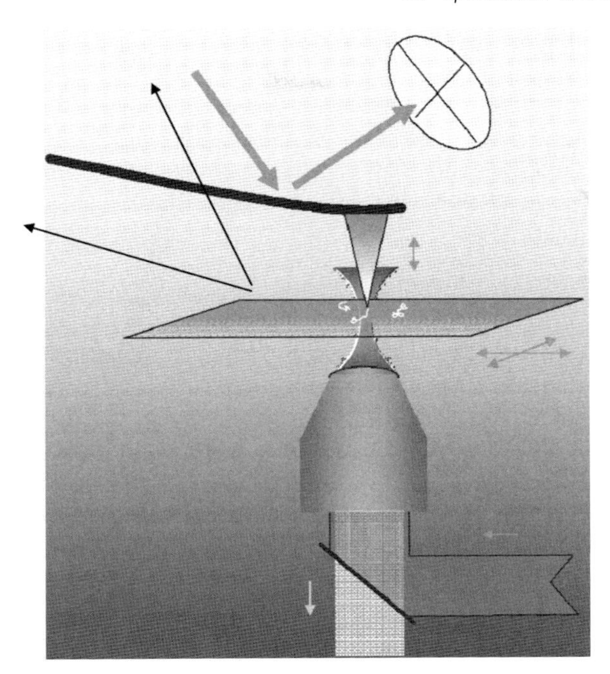

Figure 8.4 General schematics of combined TERS involving metal-coated, aperture less SPM tip interacting with the surface under shear force feedback and simultaneously collecting the topographic and spectral information.

the surface by conventional SPM means (Figure 8.4), and the sharp tip end enables highly localized enhanced Raman scattering without aperture [63]. The metal-coated tip is placed above the sample surface at a distance of about 1 nm using a sensitive shear force feedback mechanism, which provides nanoscale sensing and imaging capability. During the experiments, simultaneous acquisition of near-field Raman images and surface topography were demonstrated.

Thus, TERS provides a high degree of control over the distance between the metal nanostructures to probe the dynamic changes in the enhancement factor. The technique enables probing of the distance dependence phenomenon with a resolution as high as 0.1 nm. It is, however, important to note that this effect depends on the geometrical features of the nanostructures responsible for the enhancement. Finite element calculations based on Maxwell equations have been employed to map the electromagnetic field in the tip–surface gap [64]. These simulations provided valuable insight into the TERS mechanism, elucidating that the tip apex angle has a minimum impact on the enhancement, suggesting that significant experimental effort to make a certain shaped tip is not required. Notingher and Elfick reported an enhancement of 10^8 for AFM tips and analyte molecules adsorbed on a gold surface when excited with light in resonance [65].

Even more recently, Pettinger *et al.* investigated the dependence of the TERS intensity on the tip–metal separation distance [66]. For a relatively small tip radius

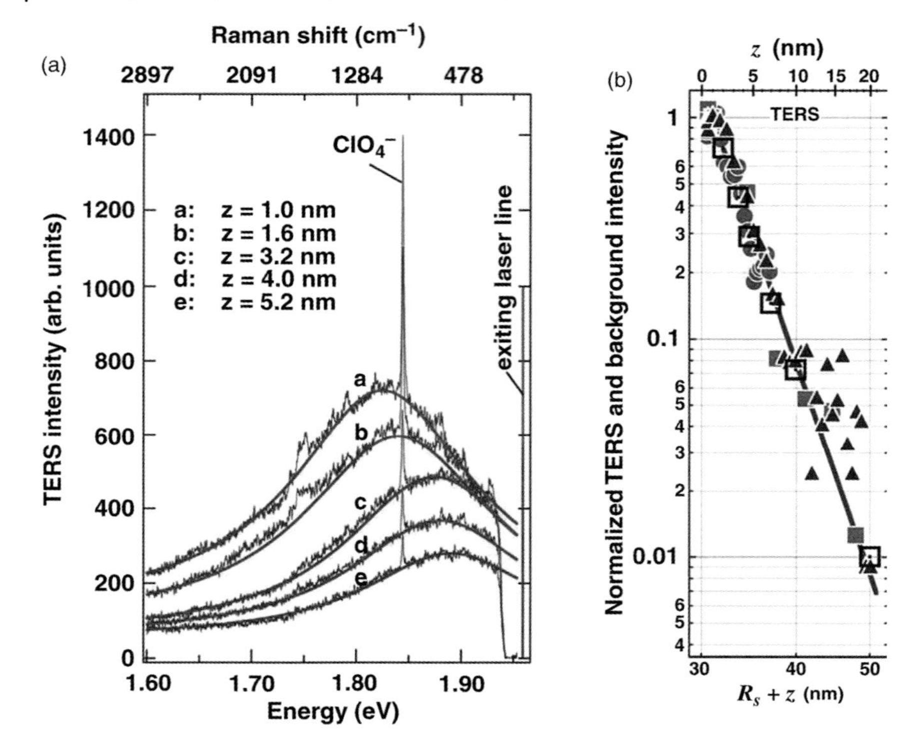

Figure 8.5 (a) Tip-enhanced Raman spectra for ClO_4 and guanine coadsorbed at Au(111) recorded at different distances (z) of the STM tip to the surface. The plot shows five representative spectra from $z = 1$ to $\sim$5.2 nm with an interval of $\sim$1 nm. Solid curves are fit to the background using a Lorentzian profile. (b) TERS intensities of varying distance parameter $R_s + z$, where R_s is the effective radius of the tip apex (30 nm) and z is the distance to the STM tip from the surface. Reprinted from Ref. [66].

($\sim$20 nm), the intensity of the Raman bands drops with increasing distance from 1 to 15 nm, providing high vertical spatial resolution of TERS. The authors attribute this dramatic drop in the intensity of the Raman bands to a rapid decrease in the electromagnetic field enhancement between the tip and the surface as the distance between them is increased. Figure 8.5 shows the TERs spectra for ClO_4 and guanine coadsorbed at Au(111) for different distances between the STM tip and the surface. The rapid decay of the intensity of spectra as the distance is increased from 1 to 5 nm can be clearly seen in this data (Figure 8.5) [66].

In any SERS experiment, resonant excitation of surface plasmons of a metal nanostructure is crucial for achieving high enhancement. TERS is not an exception to this general principle, thus requiring the resonant excitement of the surface plasmon in the metal-coated tip. However, probing the resonance of the metal-coated tip experimentally is nontrivial. This poses the important question: What is the best excitation wavelength for a given tip and experimental conditions?

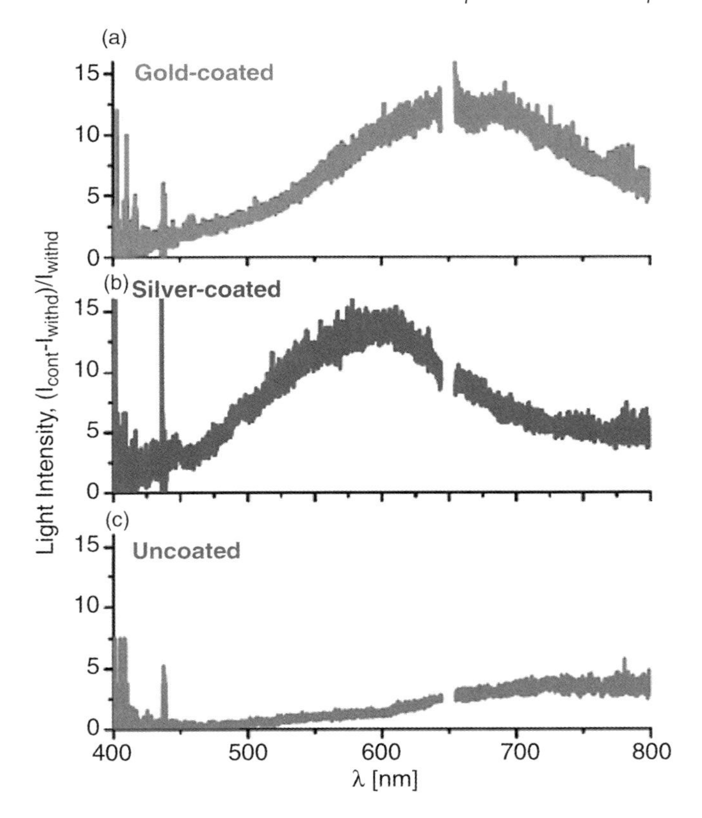

Figure 8.6 Optical scattering spectra of (a) gold-coated tungsten, (b) silver-coated tungsten, and (c) tungsten tips showing the characteristic plasmon resonance of these structures. The spectrum provides the information for the choice of the source wavelength for TERS experiment. Reprinted from Ref. [67].

Very recently, Mehtani *et al.* employed total internal reflection microscopy to measure optical properties of apertureless tips [67]. Uncoated tungsten and Si_3N_4 tips exhibited an optically flat spectrum with low signal intensity, while the metal-coated tips showed a wavelength-dependent response. The optical scattering maximum was in the range of ~650–750 nm for the gold-coated tips and ~550–650 nm for the silver-coated tips (see Figure 8.6) [67]. Using finite element simulations, optical spectra of the tips measured in the far-field were shown to correlate with the tip-enhanced near-field. TERS spectra of PEDOT/PSS collected using gold-coated tips showed the highest enhancement at a resonant excitation of 647 nm – completely agreeing with the far-field measurements. The demonstration clearly underscores the importance of the resonant excitation in TERS and a potential technique to address this.

Localized vibration modes of single-walled carbon nanotubes were probed by Anderson and coworkers using TERS method [68]. The enhanced electromagnetic field near a laser-irradiated gold tip that acts as the Raman excitation source was exploited to achieve an unprecedented spatial resolution. The authors underscored the importance of high-resolution microscopy to avoid averaging the Raman spectrum along individual carbon nanotubes.

Significant changes in the vibration modes were observed along arc-discharge-grown single-walled nanotubes, while no such changes were observed in CVD-grown nanotubes. In fact, the radial breathing mode (RBM) in the Raman spectra collected along the length of the nanotubes revealed that for arc-discharge single-walled carbon nanotubes, the peak position changes on the order of 2–3 cm^{-1}. The local changes in the vibrational modes of arc-discharge nanotubes were attributed to changes in the tube structure (n, m). Furthermore, they observed significant variations between the arc-discharge- and CVD-grown SWNTs in the ratio of the intensity of the D band to the G band (I_D/I_G), which quantifies the defects in the nanotubes. The CVD-grown SWNTs exhibited much lower value for I_D/I_G, indicating the relatively defect-free nature of such nanotubes.

In a different study, simultaneous STM imaging and TERS were performed for organic dyes in ultrahigh vacuum conditions allowing acquisition of Raman spectra of a single brilliant cresyl blue molecule adsorbed on a Au(111) surface [69]. Owing to the reduced photobleaching in UHV conditions, the acquisition time could be extended to enable TERS imaging of single molecules in dispersed state with a lateral resolution of 15 nm.

TERS is emerging as a powerful technique to probe the Raman scattering of the polymer and polymer blend samples. There are several examples of TERS of molecules on various surfaces. Stöckle *et al.* used metal-coated AFM tips with commercial scanning NSOMs as a scanning platform to obtain TERS spectra of thin brilliant cresyl blue-labeled latex spheres deposited onto a substrate [70]. They noted that although the laser spot at the sample surface was 300 nm in diameter, the enhanced Raman signal originates from an area of only 55 nm in diameter, which was in good agreement with the 50 nm tip diameter determined by SEM.

In another example, TERS has been employed to probe the structure of thin films of polystyrene/polyisoprene blends [71]. Raman-active, nonresonant PI and PS were identified at the surface and subsurface regions, respectively, using the high-quality TERS spectra and AFM topographic images. Apart from the distribution of individual components, differences in local thickness of the PI and PS layers and the subsurface nanopores of the polymer film were all identified from the TERS measurements.

8.6 AFM Tip-Enhanced Fluorescence

The near-field enhancement in the electromagnetic field around a metallic or metal-coated scanning probe has been widely employed for imaging fluorescent species down to single molecules. In fact, NSOM-based fluorescence was the first technique to spatially image single molecules (not just detect them!). Betzig and Chichester observed the fluorescent spots from single fluorescent molecules deposited on polymer films with NSOM [72]. Following this initial demonstration, NSOM has been widely employed to probe single molecules, inhomogeneities in polymer films and polymer blends, and fluorescence lifetimes (some of which were discussed in Section 8.3). In this section, we will briefly discuss the physical phenomenon behind such enhancement in the fluorescence and the most important experimental aspects to be considered.

It is important to note that in aperture NSOM-based fluorescence measurements, a sharp metallic tip in proximity to the molecules can affect the fluorescence in many ways simultaneously [73]. Apart from the enhancement in the fluorescence signal caused by the enhancement of the local electromagnetic field (see Section 8.4), the presence of a sharp metallic tip in the vicinity of a fluorescence molecule also results in the modification of both the radiative and nonradiative rates [74, 75]. The modulation of the radiative and nonradiative transfers is translated as changes in fluorescence lifetime and the emission intensity. In fact, experimentally, both the fluorescence intensity enhancement and the quenching have been reported depending on the combination and relative contribution of the effects mentioned above [76, 77].

Huang and coworkers employed a simple model to predict the relative contribution of the enhancement and the quenching of the fluorescence with respect to the distance between the tip and the fluorescent species on the surface [78]. Furthermore, they experimentally verified the same using CdSe quantum dots deposited on a surface. A gold-coated tip was modeled as a sphere and the fluorescent molecule as a dipole located at a specific distance from the surface of the sphere (see the inset of Figure 8.7) [78]. Figure 8.7 also shows the plot depicting the contributions of the expected fluorescence enhancement factor as a function of separation distance (d)

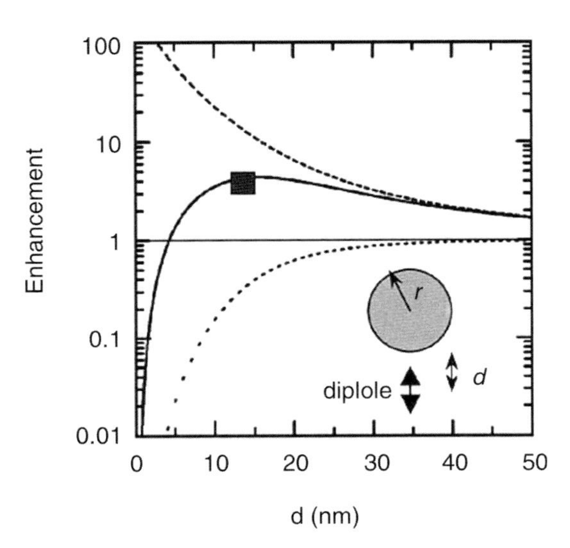

Figure 8.7 Calculated fluorescence enhancement (solid line) for distance d between the gold tip and the surface of the sample using a simple model depicted in the inset. Inset shows the contributions from both the electromagnetic field enhancement (dashed line) and the quenching (dotted line). Reprinted from Ref. [78].

and the individual contributions of the electromagnetic field enhancement and of the fluorescence quenching. A maximum signal enhancement of around 5 is predicted for the tip–fluorescent species separation distance of around 17 nm, an important result in practical considerations.

8.7 Integrating AFM with Fluorescence Optical Microscopy

Although utilization of fluorescence microscopy techniques (epifluorescence, confocal, and total internal reflection fluorescence (TIRF)) in conjunction with AFM was in vogue since the initial stages of AFM introduction, these measurements have been mainly conducted as sequential observations. However, since the beginning of the last decade, extensive efforts were dedicated to combine AFM with fluorescence imaging (simultaneous investigation) owing to the many exciting and important possibilities in the fields of biophysics, material science, and nanobiotechnology [79–81]. The two techniques are truly complementary in that fluorescence imaging offers real-time observation capabilities, functional imaging with specific labeling protocols, and sensitivity to the local physical chemistry. On the other hand, AFM provides high spatial resolution and 3D topography of the features being imaged.

Recently, Kellermayer *et al.* demonstrated synchronized TIRF and AFM that allowed the simultaneous visualization of surface topography and fluorescence of suitable cellular and biomolecular samples [82]. Furthermore, synchronized TIRF–AFM makes it possible to follow the fluorescence of mechanically manipulated molecules with high time resolution. One of the distinct features of this combined system was that the AFM was employed to mechanically manipulate the fluorescently labeled cellular and biomolecular samples. Owing to the spatial synchrony, the authors were able to image the specimen with scanning TIRF followed by the manipulation of the same into desired locations using nanolithography protocols and finally reimage the sample to detect the changes. The technique was successfully applied to ablate parts of cells and even individual actin filaments.

Integrated AFM, TIRF, and fast spinning disk (FSD) confocal microscopy was employed by Trache and Lim to probe the mechanism by which living cells sense mechanical forces and how they respond and adapt to their environment [83]. Mechanical stimulus is extremely important in vascular smooth muscle cell (VSMC) functions, which include contraction, proliferation, migration, and cell attachment. The ability to apply controlled mechanical force at a desired location using AFM was exploited to understand the effect of mechanical stimulus on the cell behavior.

Figure 8.8 shows AFM, TIRF, and FSD confocal images of the topography of VSMCs. TIRF and FSD confocal images were acquired on the same cells immediately following AFM measurements. The images show an excellent registration between the AFM image and the TIRF and FSD confocal fluorescence images (Figure 8.8d and e)) [83]. While the AFM images provided the general topography with few actin cytoskeletal fibers lying immediately beneath the membrane, the confocal image reveals a large number of actin fibers aligned along the cell axis. The limited number

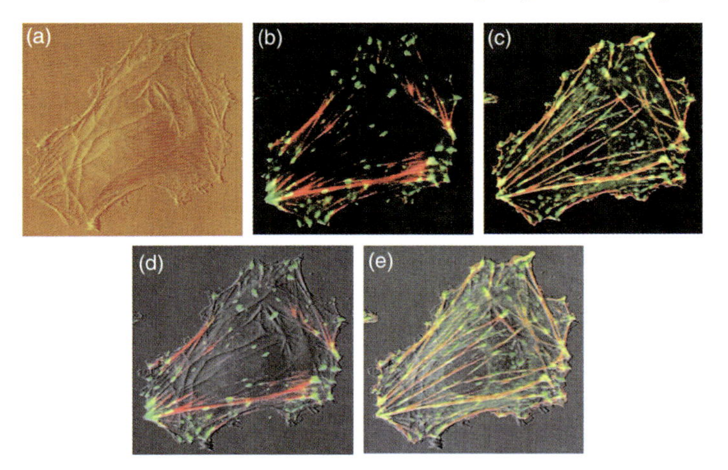

Figure 8.8 Same vascular smooth muscle cells imaged by (a) AFM at the cell surface; (b) TIRF at the basal cell surface; and (c) FSD confocal throughout the cell body shown as the maximum projection of a 3D stack. Excellent overlaps between the AFM and TIRF or FSD confocal images are shown in (d) and (e), respectively. Image size is $66 \times 64\,\mu m$. Reprinted from Ref. [83].

of actin filaments seen in the AFM images is expected as the tip senses features only immediately beneath the cell membrane. The authors observed the real-time cell shrinkage and focal adhesion rearrangement to better anchor the cell to the substrate in order to oppose the applied force by AFM mechanical stimulation.

8.8 Integrating AFM with Confocal Raman Microscopy

Raman spectroscopy provides unique information of the materials under investigation such as the chemical composition (specific chemical groups), physical state (internal stress and conformation), crystallinity, local conformation, and chain orientation [84]. Simultaneous acquisition of confocal Raman microscopy data and AFM images enables one to combine the valuable information of Raman spectra with high-resolution topographical information acquired by AFM to better understand the structure–property relationships.

Although the spatial resolution offered by the confocal Raman microscopy technique is not as high as that observed with TERS, the fact that the sample can be probed without extensive sample or tip preparation gives it additional value. The technique has been applied to study polymer blends and polymer nanostructures to better understand the structure and chemical composition of the sample under investigation [85, 86]. Of particular interest is the case of polymer blends comprised of polymers with similar stiffness. In a polymer blend, although it is relatively easy to distinguish between the different polymers when the components have a large difference in stiffness, it is extremely challenging for similar polymers.

Kim *et al.* employed an AFM in combination with a confocal Raman microscopy to simultaneously image single P3MT/Ni nanotubes and obtain correlated confocal Raman spectra [87]. They showed that the doping-related Raman bands of P3MT nanotubes were significantly affected by the existence of an outer Ni nanowall. From the Raman spectra, they concluded that the doping-related Raman band intensities of P3MT nanotubes were greatly reduced in P3MT/Ni hybrid nanotubes, indicating that the outer Ni coating plays a significant role in the dedoping of P3MT nanotubes.

The combination of a confocal Raman microscopy with an AFM was employed for the characterization of polymer blends by Schmidt *et al.* [85]. While the AFM topographical images revealed the surface morphology of the films, concurrently obtained phase images allowed the identification of two styrene–butadiene copolymers (SBR and SBS) with different chain microstructures. On the other hand, when PMMA was blended with PET, due to their similar mechanical properties (both polymers being glassy at room temperature), the assignment of the two phases just from the AFM images was difficult. Raman spectroscopy in combination with a confocal microscopy provided additional information for identifying the spatial distribution of the various phases with a resolution of 200 nm.

Ko *et al.* employed the combination of high-resolution AFM and confocal Raman microscopy to probe the structure and physical state of highly curved carbon nanotubes deposited on surfaces with micropatterned functionality [88]. Figure 8.9 shows the AFM image of highly curved nanotubes in the nestlike carbon nanotube assemblies. Corresponding confocal Raman images obtained concurrently from beneath the transparent substrate revealed that the tangential *G* mode on the Raman spectra of SWNT systematically shifts downward for the bent nanotubes at the rim. This lower-frequency shift is attributed to the tensile stress that results in the loosening of C–C bonds in the outer nanotube walls.

Singamaneni *et al.* also employed a combination of AFM and confocal Raman microscopy to monitor the internal stress distribution in periodic polymer microstructures [89, 90]. They demonstrated that mechanical instabilities in cross-linked bisphenol A Novolak epoxy microporous structures fabricated using interference lithography (IL) exhibited dramatic pattern transformation under external or internal stresses. In particular, they observed that polymerization of a rubbery component (polyacrylic acid) in square array of cylindrical pores of the IL structure resulted in the transformation of the cylindrical pores into mutually orthogonal ellipses along both (10) and (01) of the square array. Figure 8.10 shows the AFM images of the pristine and the transformed polymer structures. The AFM images clearly reveal the structural transformation at a microscopic level. The pattern transformation can be related to bending of the struts in alternate directions (along (10) and (01) directions) and the rotation of the nodes in clockwise and anticlockwise directions, as indicated on the AFM image.

Confocal Raman microscopy was employed to monitor the stress distribution at the microscale in the pristine and transformed structures to complement AFM imaging (Figure 8.10). The elastoplastic nature of the IL material deformation locked in the mechanical instabilities after the release of the external stress with internal stresses dissipated to a great extent, as was confirmed by micromapping with Raman

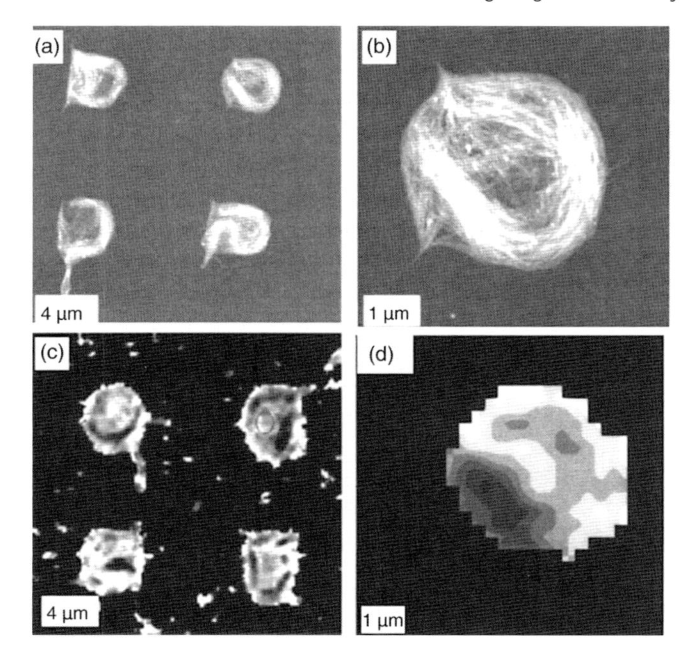

Figure 8.9 (a and b) AFM topographical images of nest-shaped patterned nanotube assembly. (c and d) Corresponding distribution of *G*-line position, varying from 1588 cm^{-1} (dark) to 1594 cm^{-1} (bright) for (c) and from 1591 cm^{-1} (bright) to 1594 cm^{-1} (dark) for (d). Reprinted from Ref. [88].

microscopy. The authors noted that this is in sharp contrast with the reversible instabilities in elastomeric solids, in which the transformed structures exhibit stress concentration in localized highly deformed elements.

In conclusion, although the idea of an optical microscopy technique with an order of magnitude smaller resolution compared to a conventional far-field microscopy

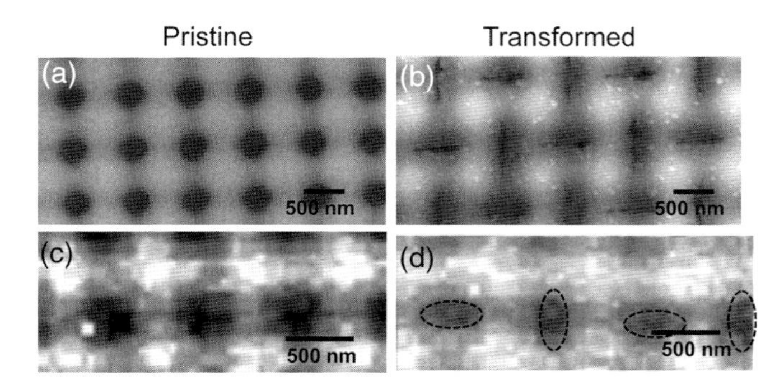

Figure 8.10 AFM topography (a and b) and corresponding Raman images (c and d) of pristine (a and c) and transformed (b and d) polymer microstructures. AFM images clearly reveal the transformation of the structure from cylindrical pores to mutually orthogonal ellipses. Reprinted from Ref. [89].

generated a lot of initial excitement, there were several issues that impaired the technique, preventing it to transform into a ubiquitous and widely exploited microscopic tool. The poor reproducibility of the images due to the poor quality and widely varying probes (especially in the aperture NSOM measurements) can be considered as the prime culprits. However, with the introduction of the microfabricated probes, the technique at present appears to be more promising than ever before. For ultimate resolutions (10–20 nm) with chemical specificity, tip-enhanced Raman and tip-enhanced fluorescence microscopy seem to be extremely promising for numerous applications if they can be utilized for different materials in a robust and reproducible fashion.

However, it is noteworthy that both these techniques are still in their infancy and require significant technical developments to ensure quantitative analysis. Most critical questions of reproducible probe fabrication and the understanding of physical factors influencing the spectral response must be fully elaborated for these techniques to become regular and user-friendly characterization tools like conventional AFM. Integration of optical microscopy with atomic force microscopy is proving to be extremely powerful, especially in studying biological systems, owing to the complementary nature of the information. In fact, recently introduced commercial AFMs with fluorescent microscopy modules are gaining popularity. These integrated systems allow acquisition of topographic data from the precise locations identified by fluorescent imaging of biological structures.

References

1 Born, M. and Wolf, E. (1997) *Principles of Optics*, Cambridge University Press, Cambridge.

2 Courjon, D. and Bainier, C. (1994) Near field microscopy and near field optics. *Rep. Prog. Phys.*, **57** (10), 989–1028.

3 Hecht, B., Sick, B., Wild, U.P., Deckert, V., Zenobi, R., Martin, O.J.F., and Pohl, D.W. (2000) Scanning near-field optical microscopy with aperture probes: fundamentals and applications. *J. Chem. Phys.*, **112** (18), 7761–7774.

4 Novotny, L. and Stranick, S.J. (2006) Near-field optical microscopy and spectroscopy with pointed probes. *Annu. Rev. Phys. Chem.*, **57**, 303–331.

5 Barbara, P.F., Adams, D.M., and O'Conner, D.B. (1999) Characterization of organic thin film materials with near-field scanning optical microscopy (NSOM). *Annu. Rev. Mater. Sci.*, **29**, 433–469.

6 Kim, J.H. and Song, B.S. (2007) Recent progress of nanotechnology with NSOM. *Micron*, **38** (4), 409–426.

7 Hayazawa, N., Tarun, A., Taguchi, A., and Kawata, S. (2009) Development of tip-enhanced near-field optical spectroscopy and microscopy. *Jpn. J. Appl. Phys.*, **48** (8), 08JA02.

8 Novotny, L. and Pohl, D.W. (1995) *Light Propagation in Scanning Near-Field Optical Microscopy*, Kluwer Academic Publishers, Dordrecht.

9 Novotny, L. and Hafner, C. (1994) Light propagation in a cylindrical waveguide with a complex, metallic, dielectric function. *Phys. Rev. E*, **50** (5), 4094–4106.

10 Haefliger, D., Plitzko, J.M., and Hillenbrand, R. (2004) Contrast and scattering efficiency of scattering-type near-field optical probes. *Appl. Phys. Lett.*, **85** (19), 4466–4468.

11 McCreery, R.L. (2000) *Raman Spectroscopy for Chemical Analysis*, John Wiley & Sons, Inc., New York.

12 Martin, Y.C., Hamann, H.F., and Wickramasinghe, H.K. (2001) Strength of the electric field in apertureless near-

field optical microscopy. *J. Appl. Phys.*, **89** (10), 5774–5778.

13 Kaupp, G., Herrmann, A., and Haak, M. (1999) Artifacts in scanning near-field optical microscopy (SNOM) due to deficient tips. *J. Phys. Org. Chem.*, **12** (11), 797–807.

14 Fenwick, O., Latini, G., and Cacialli, F. (2004) Modelling topographical artifacts in scanning near-field optical microscopy. *Synth. Met.*, **147** (1–3), 171–173.

15 Stähelin, M., Bopp, M.A., Tarrach, G., Meixner, A.J., and Zschokke-Gränacher, I. (1996) Temperature profile of fiber tips used in scanning near-field optical microscopy. *Appl. Phys. Lett.*, **68** (19), 2603–2605.

16 Thiery, L., Marini, N., Prenel, J.-P., Spajer, M., Bainier, C., and Courjon, D. (2000) Temperature profile measurements of near-field optical microscopy fiber tips by means of sub-micronic thermocouple. *Int. J. Therm. Sci.*, **39** (4), 519–525.

17 Latini, G., Downes, A., Fenwick, O., Ambrosio, A., Allegrini, M., Daniel, C., Silva, C., Gucciardi, P.G., Patané, S., Daik, R., Feast, W.J., and Cacialli, F. (2005) Optical probing of sample heating in scanning near-field experiments with apertured probes. *Appl. Phys. Lett.*, **86** (1), 011102.

18 Latini, G., Downes, A., Fenwick, O., Ambrosio, A., Allegrini, M., Gucciardi, P.G., Patané, S., Daniel, C., Silva, C., Daik, R., Feast, W.J., and Cacialli, F. (2004) Investigation of heating effects in near-field experiments with luminescent semiconductors. *Synth. Met.*, **147** (1–3), 165–169.

19 Erickson, E.S. and Dunn, R.C. (2005) Sample heating in near-field scanning optical microscopy. *Appl. Phys. Lett.*, **87** (20), 201102.

20 Ube, T., Aoki, H., Ito, S., Horinaka, J., and Takigawa, T. (2007) Conformation of single PMMA chain in uniaxially stretched film studied by scanning near-field optical microscopy. *Polymer*, **48** (21), 6221–6225.

21 Heffelfinger, C.J. and Knox, K.L. (1971) *The Science & Technology of Polymer Films*, Wiley–Interscience, New York.

22 Higgins, D.A., Liao, X., Hall, J.E., and Mei, E. (2001) Simultaneous near-field optical birefringence and fluorescence contrast applied to the study of dye-doped polymer dispersed liquid crystals. *J. Phys. Chem. B*, **105** (25), 5874–5882.

23 Vaez-Iravani, M. and Toledo-Crow, R. (1993) Pure linear polarization imaging in near field scanning optical microscopy. *Appl. Phys. Lett.*, **63** (2), 138–140.

24 Ade, H., Toledo-Crow, R., Vaez-Iravani, M., and Spontak, R.J. (1996) Observation of polymer birefringence in near-field optical microscopy. *Langmuir*, **12** (2), 231–234.

25 Aoki, H., Kunai, Y., Ito, S., Yamada, H., and Matsushige, K. (2002) Two-dimensional phase separation of block copolymer and homopolymer blend studied by scanning near-field optical microscopy. *Appl. Surf. Sci.*, **188** (1–3), 534–538.

26 Aoki, H. and Ito, S. (2001) Two-dimensional polymers investigated by scanning near-field optical microscopy: phase separation of polymer blend monolayer. *J. Phys. Chem. B*, **105** (20), 4558–4564.

27 Schwartz, B.J. (2003) Conjugated polymers as molecular materials: how chain conformation and film morphology influence energy transfer and interchain interactions. *Annu. Rev. Phys. Chem.*, **54** (3), 141–172.

28 Betzig, E. and Trautmann, J.K. (1992) Near-field optics: microscopy, spectroscopy, and surface modification beyond the diffraction limit. *Science*, **257** (5067), 189–195.

29 Straub, W., Bruder, F., Brenn, R., Krausch, G., Bielefeldt, H., Kirsch, A., Marti, O., Mlynek, J., and Marko, J.F. (1995) Transient wetting and 2D spinodal decomposition in a binary polymer blend. *Europhys. Lett.*, **29** (5), 353–358.

30 Aoki, H., Sakurai, Y., Ito, S., and Nakagawa, T. (1999) Phase-separation structure of a monolayer of binary polymer blend studied by fluorescence scanning near-field optical microscopy. *J. Phys. Chem. B*, **103** (48), 10553–10556.

31 Stevenson, R., Granström, M., and Richards, D. (1999) Fluorescence scanning near-field optical microscopy of conjugated polymer blends. *Appl. Phys. Lett.*, **75** (11), 1574–1576.

32 Credo, G.M., Lowman, G.M., DeAro, J.A., Carson, P.J., Winn, D.L., and Buratto, S.K. (2000) Probing nanoscale photo-oxidation in organic films using spatial hole burning near-field optical microscopy. *J. Chem. Phys.*, **112** (18), 7864–7872.

33 Nagahara, L.A., Nakamura, M., and Tokumoto, H. (1998) Investigation of mesoscopic domains in thin organic films using near-field optical absorption mapping. *Ultramicroscopy*, **71** (1), 281–285.

34 Tan, C.H., Inigo, A.R., Hsu, J.-H., Fann, W., and Wei, P.-K. (2001) Mesoscale structures in luminescent conjugated polymer thin films studied by near-field scanning optical microscopy. *J. Phys. Chem. Solids*, **62** (9–10), 1643–1655.

35 Arias, A.C., MacKenzie, J.D., Stevenson, R., Halls, J.J.M., Inbasekaran, M., Woo, E.P., Richards, D., and Friend, R.H. (2001) Photovoltaic performance and morphology of polyfluorene blends: a combined microscopic and photovoltaic investigation. *Macromolecules*, **34** (17), 6005–6013.

36 Teetsov, J. and Vanden Bout, D.A. (2002) Near-field scanning optical microscopy studies of nanoscale order in thermally annealed films of poly(9,9-diakylfluorene). *Langmuir*, **18** (3), 897–903.

37 Nguyen, T.Q., Schwartz, B.J., Schaller, R.D., Johnson, J.C., Lee, L.F., Haber, L.H., and Saykally, R.J. (2001) Near-field scanning optical microscopy (NSOM) studies of the relationship between interchain interactions, morphology, photodamage, and energy transport in conjugated polymer films. *J. Phys. Chem. B*, **105** (22), 5153–5160.

38 Szymanski, C., Wu, C., Hooper, J., Salazar, M.A., Perdomo, A., Dukes, A., and McNeill, J. (2005) Single molecule nanoparticles of the conjugated polymer MEH-PPV preparation and characterization by near-field scanning optical microscopy. *J. Phys. Chem. B*, **109** (18), 8543–8546.

39 Chappell, J., Lidzey, D.G., Jukes, P.C., Higgins, A.M., Thompson, R.L., O'Connor, S., Grizzi, I., Fletcher, R., O'Brien, J., Geoghegan, M., and Jones, R.A.L. (2003) Correlating structure with fluorescence emission in phase-separated conjugated-polymer blends. *Nat. Mater.*, **2** (9), 616–621.

40 McNeill, C.R., Frohne, H., Holdsworth, J.L., Furst, J.E., King, B.V., and Dastoor, P.C. (2004) Direct photocurrent mapping of organic solar cells using a near-field scanning optical microscope. *Nano Lett.*, **4** (2), 219–223.

41 Dennler, G., Scharber, M.C., and Brabec, C.J. (2009) Polymer-fullerene bulk-heterojunction solar cells. *Adv. Mater.*, **21** (13), 1323–1338.

42 Klimov, E., Li, W., Yang, X., Hoffmann, G.G., and Loos, J. (2006) Scanning near-field and confocal Raman microscopic investigation of P3HT-PCBM systems for solar cell applications. *Macromolecules*, **39** (13), 4493–4496.

43 Carpick, R.W., Sasaki, D.Y., and Burns, A.R. (2000) First observation of mechanochromism at the nanometer scale. *Langmuir*, **16** (3), 1270–1278.

44 Emonin, S., Held, T., Richard, N., Hollricher, O., and Marti, O. (2001) Multicolor images acquisition by scanning near-field optical microscopy. *J. Appl. Phys.*, **90** (9), 4820–4824.

45 Wei, P.K., Hsu, J.H., and Fann, W.S. (1999) Study of conjugated polymer blend films by a near field scanning optical microscopy. *Synth. Met.*, **102** (1–3), 1209–1210.

46 Enderle, T., Ha, T., Ogletree, D.F., Chemala, D.S., Magowan, C., and Weiss, S. (1997) Membrane specific mapping and colocalization of malarial and host skeletal proteins in the *Plasmodium falciparum* infected by dual-color near-field scanning optical microscopy. *Proc. Natl. Acad. Sci. USA*, **94** (2), 520–525.

47 Raman, C.V. and Krishnan, K.S. (1928) Polarisation of scattered light-quanta. *Nature*, **122**, 169.

48 McCreery, R.L. (2000) *Raman Spectroscopy for Chemical Analysis*, John Wiley & Sons, Inc., New York.

49 Jeanmaire, D.L. and Van Duyne, R.P. (1977) Surface Raman

spectroelectrochemistry: Part I. Heterocyclic, aromatic, and aliphatic amines adsorbed on the anodized silver electrode. *J. Electroanal. Chem.*, **84** (1), 1–20.

50 Fleischman, M., Hendra, P.J., and McQuillan, A.J. (1974) Raman spectra of pyridine adsorbed at a silver electrode. *Chem. Phys. Lett.*, **26** (2), 163–166.

51 Albrecht, M.G. and Creighton, J.A. (1977) Anomalously intense Raman spectra of pyridine at a silver electrode. *J. Am. Chem. Soc.*, **99** (15), 5215–5217.

52 Otto, A., Mrozek, I., Grabhorn, H., and Akemann, W. (1992) Surface-enhanced Raman scattering. *J. Phys. Condens. Matter*, **4** (5), 1143–1212.

53 Otto, A. (2002) What is observed in single molecule SERS, and why? *J. Raman Spectrosc.*, **33** (8), 593–598.

54 Campion, A. (1985) Raman spectroscopy of molecules adsorbed on solid surfaces. *Annu. Rev. Phys. Chem.*, **36**, 549–572.

55 Moskovits, M. (1985) Surface-enhanced spectroscopy. *Rev. Mod. Phys.*, **57** (3), 783–826.

56 Kambhampati, P., Child, C.M., Foster, M.C., and Campion, A. (1998) On the chemical mechanism of surface enhanced Raman scattering: experiment and theory. *J. Chem. Phys.*, **108** (12), 5013–5026.

57 Lewis, A., Taha, H., Strinkovski, A., Manevitch, A., Khatchatouriants, A., Dekhter, R., and Ammann, E. (2003) Near-field optics: from subwavelength illumination to nanometric shadowing. *Nat. Biotechnol.*, **21** (11), 1377–1386.

58 Anderson, M.S. (2000) Locally enhanced Raman spectroscopy with an atomic force microscope. *Appl. Phys. Lett.*, **76** (21), 3130–3132.

59 Knoll, A. (1998) Interfaces and thin films as seen by bound electromagnetic waves. *Annu. Rev. Phys. Chem.*, **49**, 569–638.

60 Moskovits, M., DiLella, D.P., and Maynard, K. (1988) Surface Raman spectroscopy of a number of cyclic aromatic molecules adsorbed on silver: selection rules and molecular reorientation. *Langmuir*, **4** (1), 67–76.

61 Kennedy, B.J., Spaeth, S., Dickey, M., and Carron, K.T. (1999) Determination of the distance dependence and experimental effects for modified SERS substrates based on self-assembled monolayers formed using alkanethiols. *J. Phys. Chem. B*, **103** (18), 3640–3646.

62 Ko, H., Singamaneni, S., and Tsukruk, V.V. (2008) Nanostructured surfaces and assemblies as SERS media. *Small*, **4** (10), 1576–1579.

63 Anderson, M.S. (2000) Locally enhanced Raman spectroscopy with an atomic force microscope. *Appl. Phys. Lett.*, **76** (21), 3130.

64 Micic, M., Klymyshyn, N., Suh, Y.D., and Lu, H.P. (2003) Finite element method simulation of the field distribution for AFM tip-enhanced surface-enhanced Raman scanning microscopy. *J. Phys. Chem. B*, **107** (7), 1574–1584.

65 Notingher, I. and Elfick, A. (2005) Effect of sample and substrate electric properties on the electric field enhancement at the apex on SPM nanotips. *J. Phys. Chem. B*, **109** (33), 15699–15706.

66 Pettinger, B., Domke, K.F., Zhang, D., Picardi, G., and Schuster, R. (2009) Tip-enhanced Raman scattering: influence of the tip–surface geometry on optical resonance and enhancement. *Surf. Sci.*, **603** (10–12), 1335–1341.

67 Mehtani, D., Lee, N., Hartschuh, R.D., Kisliuk, A., Foster, M.D., Sokolov, A.P., Čajko, F., and Tsukerman, I. (2006) Optical properties and enhancement factors of the tips for apertureless near-field optics. *J. Opt. A*, **8** (4), S183–S190.

68 Anderson, N., Hartschuh, A., Cronin, S., and Novotny, L. (2005) Nanoscale vibrational analysis of single-walled carbon nanotubes. *J. Am. Chem. Soc.*, **127** (8), 2533–2537.

69 Steidtner, J. and Pettinger, B. (2008) Tip-enhanced Raman spectroscopy and microscopy on single dye molecules with 15 nm resolution. *Phys. Rev. Lett.*, **100** (23), 236101.

70 Stöckle, R.M., Suh, Y.D., Deckert, V., and Zenobi, R. (2000) Nanoscale chemical analysis by tip-enhanced Raman spectroscopy. *Chem. Phys. Lett.*, **318** (1–3), 131–136.

71 Yeo, B.-S., Amstad, E., Schmid, T., Stadler, J., and Zenobi, R. (2009) Nanoscale probing of a polymer-blend thin film with tip-enhanced Raman spectroscopy. *Small*, **5** (8), 952–960.

72 Betzig, E. and Chichester, R.J. (1993) Single molecules observed by near-field scanning optical microscopy. *Science*, **262** (5138), 1422–1425.

73 Lakowicz, J.R. (2001) Radiative decay engineering: biophysical and biomedical applications. *Anal. Biochem.*, **298** (1), 1–24.

74 Gerton, J.M., Wade, L.A., Lessard, G.A., Ma, Z., and Quake, S.R. (2004) Tip-enhanced fluorescence microscopy at 10 nm resolution. *Phys. Rev. Lett.*, **93** (18), 180801.

75 Kramer, A., Trabesinger, W., Hecht, B., and Wild, U.P. (2002) Optical near-field enhancement at a metal tip probed by a single fluorophore. *Appl. Phys. Lett.*, **80** (9), 1652–1654.

76 Frey, H.G., Witt, S., Felderer, K., and Guckenberger, R. (2004) High-resolution imaging of single fluorescent molecules with the optical near-field of a metal-tip. *Phys. Rev. Lett.*, **93** (20), 200801.

77 Yang, T.J., Lessard, G.A., and Quake, S.R. (2000) An apertureless near-field microscope for fluorescence imaging. *Appl. Phys. Lett.*, **76** (3), 378–380.

78 Huang, F.M., Festy, F., and Richards, D. (2005) Tip-enhanced fluorescence imaging of quantum dots. *Appl. Phys. Lett.*, **87** (18), 183101.

79 Mathur, A.B., Truskey, G.A., and Reichert, W.M. (2000) Atomic force and total internal reflection fluorescence microscopy for the study of force transmission in endothelial cells. *Biophys. J.*, **78** (4), 1725–1735.

80 Nishida, S., Funabashi, Y., and Ikai, A. (2002) Combination of AFM with an objective-type total internal reflection fluorescence microscope (TIRFM) for nanomanipulation of single cells. *Ultramicroscopy*, **91** (1–4), 269–274.

81 Trache, A. and Meininger, G.A. (2005) Atomic force-multi-optical imaging integrated microscope for monitoring molecular dynamics in live cells. *J. Biomed. Opt.*, **10** (6), 064023.

82 Kellermayer, M.S.Z., Karsai, A., Kengyel, A., Nagy, A., Bianco, P., Huber, T., Kulcsár, Á., Niedetzky, C., Proksch, R., and Grama, L. (2006) Spatially and temporally synchronized atomic force and total internal reflection fluorescence microscopy for imaging and manipulating cells and biomolecules. *Biophys. J.*, **91** (7), 2665–2677.

83 Trache, A. and Lim, S.-M. (2009) Integrated microscopy for real-time imaging of mechanotransduction studies in live cells. *J. Biomed. Opt.*, **14** (3), 034024.

84 Schmidt, U., Hild, S., Ibach, W., and Hollricher, O. (2005) Characterization of thin polymer films on the nanometer scale with confocal Raman AFM. *Macromol. Symp.*, **230** (1), 133–143.

85 Schmidt, U., Hild, S., Ibach, W., and Hollricher, O. (2005) Characterization of thin polymer films on the nanometer scale with confocal Raman AFM. *Macromol. Symp.*, **230** (1), 133–143.

86 Österberg, M., Schmidt, U., and Jääskeläinen, A.-S. (2006) Combining confocal Raman spectroscopy and atomic force microscopy to study wood extractives on cellulose surfaces. *Colloids Surf. A*, **291** (1–3), 197–201.

87 Kim, D.-C., Kim, R., Kim, H.-J., Kim, J., Park, D.-H., Kim, H.-S., and Joo, J. (2007) Raman study of polymer–metal hybrid nanotubes using atomic force/confocal combined microscope. *Jpn. J. Appl. Phys.*, **46** (8B), 5556–5559.

88 Ko, H., Pikus, Y., Jiang, C., Jauss, A., Hollricher, O., and Tsukruk, V.V. (2004) High resolution Raman microscopy of curled carbon nanotubes. *Appl. Phys. Lett.*, **85** (13), 2598–2600.

89 Singamaneni, S., Bertoldi, K., Chang, S., Jang, J.-H., Young, S., Thomas, E.L., Boyce, M., and Tsukruk, V.V. (2009) Bifurcated mechanical behavior of deformed periodic porous solids. *Adv. Funct. Mater.*, **19** (9), 1426–1436.

90 Singamaneni, S., Bertoldi, K., Chang, S., Jang, J.-H., Thomas, E.L., Boyce, M., and Tsukruk, V.V. (2009) Instabilities and pattern transformation in periodic, microporous, elasto-plastic solids. *ACS Appl. Mater. Interfaces*, **1** (1), 42–47.

Part Three
Scanning Probe Techniques for Various Soft Materials

Scanning Probe Microscopy of Soft Matter: Fundamentals and Practices, First Edition.
Vladimir V. Tsukruk and Srikanth Singamaneni.

9
Amorphous and Poorly Ordered Polymers

9.1
Introduction

This chapter opens Part Three of the book that focuses on the application of scanning probe techniques to various classes of soft matter, ranging from amorphous to crystalline, from one-component to multicomponent, from interfacial materials to nanocomposites, from multilayered films to colloidal assemblies, and from branched synthetic molecules to some selected examples of biomolecules.

In this chapter, we start discussion of atomic force microscopy (AFM) data for four basic, yet important, classes of poorly ordered, noncrystalline polymers with predominantly short-range ordering that are widely exploited in many crucial applications. Namely, we consider glassy polymers, rubbers, gels, and interpenetrating polymer networks (IPNs). All these materials usually show relatively uniform surface morphology and low microroughness. From a mechanical properties point of view, the four classes of polymeric materials reviewed in this chapter represent nearly six orders of magnitude difference in their elastic moduli: with some of the hydrogels discussed here exhibiting a modulus of only a few kilopascals, while high-performance polymers in glassy state exhibiting a modulus up to tens of gigapascals [1, 2].

The ability of AFM to image surface morphology and probe the structure and properties of such a broad class of materials under wide variety of environmental conditions (temperature, pH, solvents, or humidity) is truly remarkable. The conventional glassy and rubbery polymers that have been widely studied in their bulk form mostly with scattering techniques have been routinely employed as standard samples to test and demonstrate the efficacy of AFM and related techniques in probing the structure and properties of these materials.

AFM has also served as an excellent tool to probe the nanoscale and surface behavior of these otherwise well-characterized polymers (in the bulk state) to understand the surface and interfacial properties as well as confinement effects in these systems. Polymer gels are yet another class of materials that have gained immensely from the AFM's ability to probe under various environmental conditions (in this case liquid water) enabling the imaging of these gels in their native hydrated state. There are numerous examples in which AFM or associated techniques are

Scanning Probe Microscopy of Soft Matter: Fundamentals and Practices, First Edition.
Vladimir V. Tsukruk and Srikanth Singamaneni.

employed to study the phase transition behavior of these interesting materials. With the exception of IPNs, most of the following discussion will be intentionally restricted to one-component systems, with multicomponent polymeric material discussions reserved to Chapters 12 and 15.

9.2 Glassy Amorphous Polymers

Glassy amorphous polymers are the most common class of polymer materials investigated using AFM with relatively easy operational conditions. This section is in no way a comprehensive literature review of the plethora of amorphous glassy materials that have been investigated using a wide variety of SPM techniques but rather a summary of selected examples to illustrate AFM applications [3, 4]. AFM studies of relevant materials range from simple surface morphological investigations to probe the surface uniformity and microroughness at different scales to the investigation of spatial uniformity of complex phenomena such as the mobility of macromolecules at surfaces, thus shedding new light on the elusive glass transition phenomenon.

We briefly present some of these examples in this section. Numerous other examples are discussed in different chapters in Parts One and Two of this book. Although not exclusively pointed out, many materials discussed fall into the category of glassy amorphous polymers. In is worth noting that many glassy polymers, for example, well-characterized polymers such as atactic polystyrene (PS) or poly(methyl methacrylate) (PMMA) with precise molecular weight distribution characteristics, form the benchmark for the investigation of properties with this new metrology technique owing to their well-known physical properties in the bulk state [2].

Amorphous glassy polymer thin films deposited on a solid substrate from solutions are known to exhibit highly uniform, smooth, and, in most cases, featureless surface morphologies. AFM has been routinely employed to monitor the uniformity and the surface microroughness of these films at different length scales. The true 3D imaging capability of AFM enables one to measure the thickness of these films with high precision by making an intentional scratch on the surface of these films and scanning along the edge of the film (Figure 9.1). The technique, often termed as the "scratch test," is frequently used to confirm the thickness values obtained by complementary techniques such as ellipsometry or to probe the local thickness of the films as opposed to the average thickness that is measured over large area in ellipsometry. Figure 9.1 shows a scratch made in a glassy PS film to measure the local thickness of the glassy polymer film. The readers are referred to Section 3.3.8 for details on processing the images of edges to obtain reliable values of thickness.

In earlier studies, Assender *et al.* discussed the relation between the observed surface topography and the properties of amorphous polymer films [5]. In relation to the topographic measurements of the amorphous glassy films, the most frequently measured parameter is the root mean square (RMS) microroughness

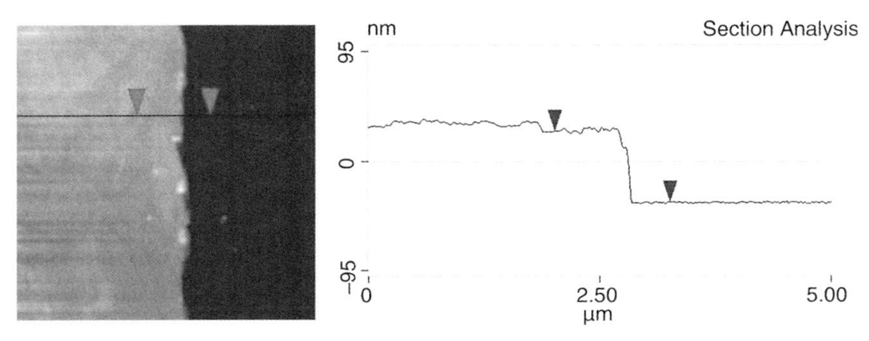

Figure 9.1 AFM image showing the scratch test performed on a thin polystyrene film on a silicon substrate. The true 3D nature of AFM imaging enables the determination of the thickness of the polymer film.

at different length scales (which must be indicated). The authors pointed out that RMS roughness does not take any account of the distance between the size of the features on the surface. For example, a surface with a few high-amplitude features (large bumps) may have the same RMS value as one with many low-lying features (small bumps).

To illustrate this point, Figure 9.2 shows the AFM images of two amorphous polymer films with significantly different surface topography and optical quality (gloss) but similar RMS microroughness. As pointed out in Chapter 3, surface roughness is a scale-dependent parameter, increasing logarithmically with the increasing spatial scale of observation. For the smooth amorphous polymer films deposited on atomically flat substrates, the surface roughness is usually on the order of 0.4–0.8 nm over a 1 μm^2 surface area (a very common size reflecting topography mostly relevant to molecular/nanoscale features that will be default surface area in

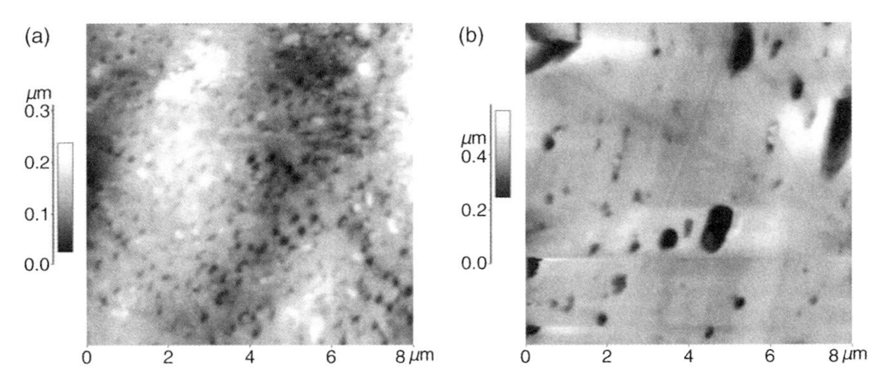

Figure 9.2 AFM images of two polymer surfaces with a similar RMS microroughness but strikingly different morphologies and optical qualities (gloss). (a) Rubber-toughened acrylic, RMS microroughness of 38 nm and gloss 18.9%. (b) PVC, RMS microroughness of 36 nm and gloss (specular reflection) 78.7%. Reprinted from Ref. [5].

this book). Larger-scale microroughness is somewhat higher and usually reflects the presence of contaminants and some microscopic defects.

These values are very common for thin films of common glassy amorphous polymers. However, the presence of more developed morphologies such as porous or grainy texture can result in a significant increase in the surface roughness under certain conditions (e.g., thermal annealing or solvent treatment).

It is worth noting that the simple RMS microroughness parameter does not reflect any anisotropy in the topography that is very important for highly oriented amorphous films. Instead, the authors pointed out a 2D autocorrelation function (ACF) that is preferred for amorphous systems due to its more comprehensive description of the surface when compared to the standard RMS microroughness value [6]. In this study, the 2D ACF is given by

$$\mathrm{ACF}(x, y) = \frac{\iint \rho(x-x', y-y') \times \rho(x', y')\,\mathrm{d}\,x'\,\mathrm{d}\,y'}{\iint \rho^2(x, y)\,\mathrm{d}\,x\,\mathrm{d}\,y} \tag{9.1}$$

where $\rho(x, y)$ is a profile function defined on the surface and the denominator is a normalization factor. Thus, it compares the height at point (x, y) with that at some second point (x', y') and maps this comparison as a function of the distance between them. The first maximum in the ACF will be at a length associated with the distance between features in the topography.

One specific example where the ACF has been employed to analyze the surface morphology of amorphous glassy polymers has been reported by Bliznyuk *et al.* [7]. Figure 9.3 shows the AFM topography image of a thin film of PMMA acquired in tapping mode. The AFM image reveals the height distribution of the free surface that

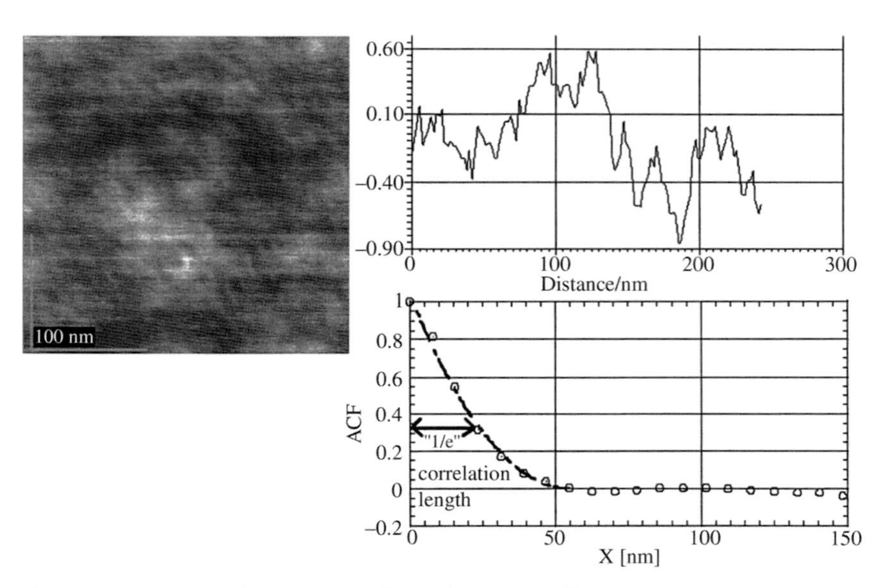

Figure 9.3 Tapping mode AFM image of amorphous PMMA film, topography cross section, and corresponding 1D ACF. Reprinted from Ref [7].

corresponds to the random distribution of surface density inhomogeneities. Two types of surface inhomogeneities of distinct length scales can be identified from the images: several nanometer-sized blobs and much smoother undulations of larger size. Figure 9.3 also shows the cross section and the corresponding ACF along the fast scanning direction.

It is worth noting that the ACF was observed to be asymmetric for fast and slow scanning directions. The correlation length (the distance at which the autocorrelation function value drops to $1/e = 0.368$) of the surface was derived from the exponential decay observed in the ACF of the topographical cross sections. Also, ACF analysis in conjunction with computer simulations performed using a lattice chain model has been applied to understand the physical origin of the microroughness of a PS film [8]. The characteristic autocorrelation length (defined above) was found to be closely related to the radius of gyration of polystyrene with molecular weights ranging from 3.9×10^3 to 9×10^6 g/mol.

In another study, Stone *et al.* employed the combination of X-ray scattering and AFM to characterize the microroughness of the free surface of a bulk amorphous polyetherimide (PEI) [9]. Both techniques show that the compression-molded PEI surface had a RMS microroughness of about 6 nm for a 1 μm^2 surface area. Although a good agreement between X-ray and AFM data was observed for the roughness exponent (Hurst coefficient, borrowed from fractal theory), the two methods resulted in different values for the correlation length. The authors noted that the observed difference in the correlation length of the amorphous surface was due in part to the approximate character of the scattering theory of X-rays for rough nonuniform surfaces.

Apart from probing surface morphologies, AFM has been employed to monitor the molecular origin of nanoscale friction between a nanoscale asperity and a glassy polymer surface. Sills and coworkers performed friction force measurements by sliding an AFM tip across the surface of a glassy PS polymer to understand the dissipation mechanism of nanoscale kinetic friction [10]. The friction forces were measured as a function of the AFM tip velocity in an isothermal manner at different temperatures. Friction–velocity relationships are found to be quasi-logarithmic that is common for boundary lubricants. Analysis of the friction force–velocity data using the method of reduced variables revealed the dissipative behavior as an activated relaxation process with a potential barrier height of 7.0 kcal/mol. The authors noted that this value is essentially identical to the activation energy of the phenyl rotation around the C—C bond to the backbone chain in atactic PS. Based on this observation, the authors opined that the primary dissipation mechanism for dry sliding of the AFM tip on glassy PS is through group mobility that manifests itself in the rotation of the side phenyl groups. It is important to note that depending on the load and the contact area of the AFM probe interacting with the surface, other relaxation processes (δ, α, β) might play a significant role in the observed friction phenomena.

In order to address thermal properties of amorphous cross-linked polymers, Singamaneni *et al.* studied the thermal expansion behavior of ultrathin glassy polymer films obtained by wet (spin-cast) and dry (plasma polymerization) approaches [11]. The plasma-polymerized films exhibited a well-developed grainy surface morphology with a grain size below 100 nm compared to the smooth surfaces of spin-cast films with no

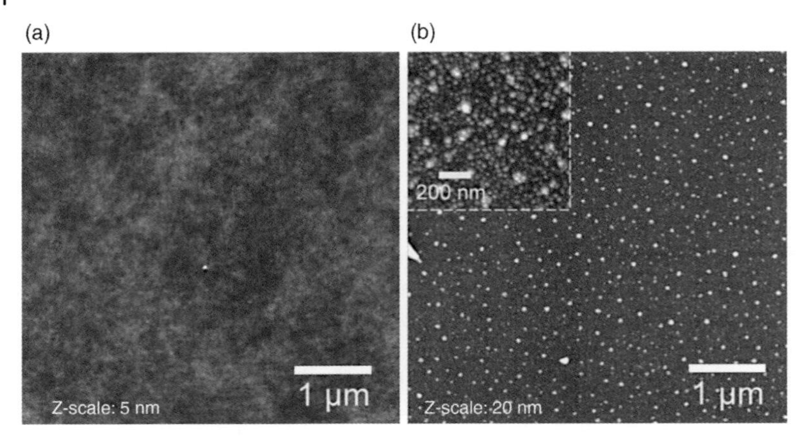

Figure 9.4 Tapping mode AFM images of a spin-coated amorphous PS film showing the smooth featureless surface morphology with low microroughness (a) and a plasma-polymerized highly cross-linked PS film showing the grainy surface morphology (b). Inset shows the higher-resolution image of the grainy surface morphology. Reprinted from Ref. [11].

specific features – possibly due to the nature of the film formation (Figure 9.4). In the case of the spin-cast films, the polymer solution subjected to higher shear forces forms uniform and smooth films with the surface microroughness defined by the spatial length scale of the polymer chains (such as radius of gyration).

In the case of plasma polymerization, subjecting the monomer to a plasma of inert gas results in the formation of energetic species (ions and radicals) of the monomer, which then polymerize on the surface of the substrate. The governing factors of the structure of the plasma polymers would be the plasma polymerization conditions (chamber geometry, monomer feed rate, and pressure in the chamber). This fundamental difference in the nature of film formation results in the dramatically different surface morphology of the films. The RMS microroughness was found to be 1.5 and 1.7 nm for plasma polymerized styrene and plasma polymerized acrylonitrile films, respectively. The surface microroughness of plasma-polymerized films is usually much higher than that observed for spin-cast films of the same linear polymers (around 0.2–0.3 nm).

They observed two important differences between the plasma-polymerized films and the spin-coated films PS and PAN films. First, both plasma-polymerized films exhibited a significantly nonequilibrium relaxation of film thickness with temperature when cooled below glass transition. The final thickness was 0.3–0.5 nm higher than the initial thickness immediately after cooling. However, the films restored to original thickness after long relaxation periods ($\sim$8 h). This behavior was observed for both plasma-polymerized films, while the spin-cast films from PS and PAN exhibited no such hysteresis behavior. Second, the thermal behavior of plasma-polymerized acrylonitrile films was absolutely uncharacteristic of conventional polymers with a reversible thermal contraction of the film in the vertical direction during heating, indicating a negative thermal expansion phenomena. The authors suggested that the unusual negative thermal expansion behavior is caused by the presence of high residual stresses in the polymer film (common for plasma-polymerized materials) and the developed grainy surface morphology, as depicted in Figure 9.4. These residual stresses arise due to the wedging of the energetic species (ions and radicals) into the existing film during polymerization. The high in-plane compressive strength common for plasma-polymerized polymers might originate from this specific, wedge-type growth mechanism [12, 13]. In fact, the authors found the in-plane compressive strength for some plasma-polymerized polymers to be as high as 50 MPa, which is close to/exceeds the yield strength of many polymeric materials.

There are numerous examples in literature where the glass transition in thin and ultrathin amorphous polymer films has been probed based on the dramatic changes in the thermomechanical and surface properties of the amorphous polymer at T_g [14–16]. One of the most common methods employed is standard surface force spectroscopy at various temperatures (and frequencies). Some of these examples were discussed in Section 5.4 to introduce SFS for probing the thermomechanical properties of various soft materials.

In a comprehensive study, the surface glass transition temperature of PS samples with molecular weights (M_n) ranging from 3900 to 1 340 000 g/mol has been measured with SFS by Bliznyuk *et al.* [17]. The values of the surface T_g for the samples with $M_n > 30\,000$ was found to be the same as the corresponding bulk values. On the other hand, for low molecular mass polymers, a depression of the glass transition at the polymer surface was observed. The magnitude of the depression increased with decreasing molecular mass, correlating with the structural and dynamical parameters of polymer chains (the presence of a virtual network of labile entanglements). The change in T_g with the molecular weight of PS was found to be steeper for the surface (as probed by the AFM tip at low loads) compared to the bulk Fox–Flory relationship.

Moreover, the authors demonstrated from observing diffusion of molecules from a LB monolayer deposited on a PS substrate that polymer chain entanglement variation rather than the end group localization on the free surface is responsible for the observed surface T_g depression effect. In the force curves obtained above T_g, the authors observed hysteresis between the loading and unloading curves that was ascribed to the viscoelasticity of the polymer surface. The hysteresis behavior for different molecular weight samples above T_g was described in terms of the contact mechanics of the tip–surface interaction by considering the relative roles of bulk viscoelasticity and peeling viscoelasticity.

Exploring a similar topic, Yang and *et al.* analyzed the viscoelastic response of the glassy polymer films near the glass transition temperature T_g [18]. The force–distance curves measured on a glassy film or a glassy surface at various temperatures were analyzed using a Burgers' model (Figure 9.5b). In particular, the authors measured the force–distance curves of PS ($T_g \sim 110\,°C$). The force–distance curves obtained at three representative temperatures: 30, 110, and 117 °C. The force–distance curves showed very small variations at different temperatures below 80 °C. All other FDC data collected at temperatures below 80 °C were found to be essentially coincident with that acquired at room temperature.

The major change that appears in the force–distance curve at temperatures above about 90 °C is a prolonged attractive region in both the approaching and retreating

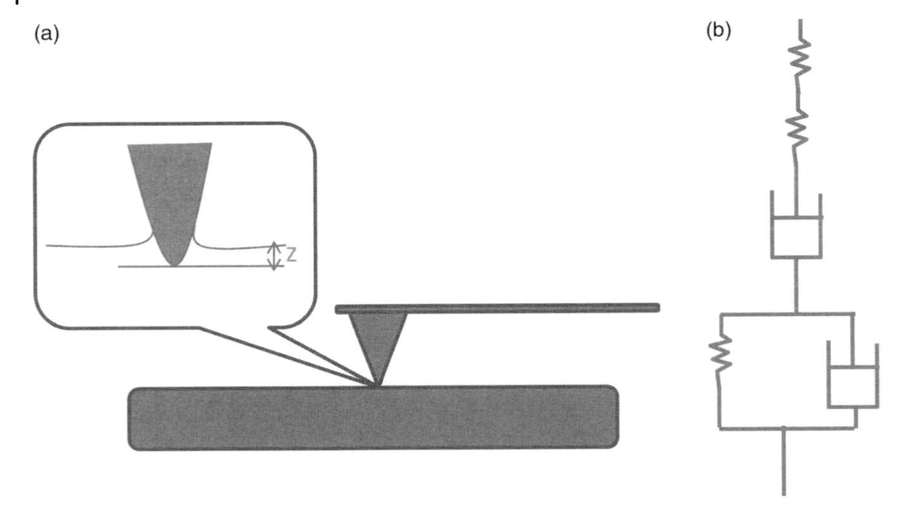

Figure 9.5 (a) Schematic configuration of an AFM tip interacting with a sample surface during the force–distance measurement. (b) Burgers' viscoelastic model.

curves. The general behaviors shown in these curves qualitatively agreed with the predicted force–distance curves using the Burgers' model. The authors found that the material's characteristics of reversible inelastic response and viscous creep can be extracted from a force–distance curve by monitoring the changes in the slope of the repulsive force-distance curve. For the constant strain rates, the inelastic response causes a shift in the repulsive FDC, while the primary viscous creep results in a decrease in the slope of the repulsive portion of the force–distance curve.

In other study, Cappella and coworkers investigated the temperature and frequency dependence of the work of adhesion on a thick film of poly(*n*-butyl methacrylate) (PBMA) and on two films of PS with different molecular weights using surface force spectroscopy measurements [19]. By applying the classical time–temperature superposition principle to the adhesion hysteresis, the authors obtained master curves of the work of adhesion E_{adh} at a fixed load. The shift factors used for building the master curves obeyed the WLF behavior with the same coefficients found for the Young's modulus. The temperature dependence of adhesion was investigated by considering the contact area both in the elastic and the plastic regimes of deformations. The work of adhesion was found to be always proportional to the contact area between the AFM tip and the surface. The authors noted that the temperature dependence of adhesion is a consequence of the temperature dependence of the stiffness and of the elastic–plastic properties of the sample, which determine the contact area between the tip and the sample for a constant load.

In a recent study, Fakhraai and Forrest employed AFM to probe the surface dynamics of amorphous PS films at various temperatures [20]. The study addressed the question of whether the polymer chain segments at the free surface (air–film interface) exhibit similar freezing (or slowing) of segmental motion as the bulk when cooled below the glass transition temperature. By partially embedding and then removing gold nanoparticles in the film, the authors produced very minute deformations on a PS surface. The time-dependent relaxation of these surface deformations was continuously monitored as a function of temperature from 4 to 96 °C using tapping mode AFM. Even at 20 °C (which is ~80 °C below the T_g of PS), as the annealing time is increased, both the depth of the nanoindents and the size of the rims were found to decrease. The authors demonstrated that time dependence of the average indentation mark depths could be described by a single-exponential function, enabling them to suggest a relationship between the time constant of annealing and the characteristic relaxation time of the system near the free surface. The study clearly revealed the enhanced surface mobility of the polymer chains on the surface of glassy polymer compared to the bulk. Furthermore, the deviation from the bulk α relaxation was found to be more pronounced as the temperature was decreased below the glass transition of the polymer.

In another recent study, Tocha *et al.* employed friction force microscopy to address the question of whether the viscoelastic properties of the surface of a glassy polymer are different compared to that of the bulk material [21]. In particular, they studied the surface relaxations of amorphous PMMA. A broad range of scanning velocities (enabled by use of a special high-velocity accessory), temperature control, and, in particular, employing AFM tips with significantly differing radii allowed the authors to probe the surface properties of PMMA over a wide range of frequencies (1–10^7 Hz). The authors exploited the fact that dynamic friction and energy dissipation at polymer surfaces encompass large contributions from internal viscoelastic dissipation (related to polymer relaxation) and thus allow the implementation of AFM-based nanotribology to address the surface behavior compared to that from the bulk material (see Section 5.5) [22, 23]. Friction data acquired at various temperatures and velocities were combined into a single master curve in which the onsets of the α- and β-relaxation processes of PMMA were identified. The substantially reduced activation energies and the significantly smaller relaxation time compared to the bulk material reiterated the notion of higher mobility of the polymer chain groups at the surface of the glassy polymer.

In an alternative approach, nanothermomechanometry (nano-TM), in which a heavily doped silicon AFM tip is resistively heated on the surface of a polymer film, has been employed to probe the softening behavior of glassy PS films by Zhou *et al.* [24]. Nano-TM measures the cantilever deflection of the AFM tip as the tip temperature is increased. A sharp increase in the penetration of the heated AFM tip into the material occurs at a specific softening temperature that is presumably related to either the glass transition or melting of the polymer being investigated (see Chapter 6 for micro-TA discussion). The apparent softening temperature of the PS film is found to depend on the logarithm of the square root of the temperature ramping rate, which enabled the authors to estimate the softening transition temperature by extrapolation to the limiting value (infinitely small heating rate). Significant shifts in the softening transitions were observed with decreasing film thickness.

Even more importantly, the authors found that the shifts (both magnitude and sign) strongly depend on the mechanical and thermal properties of the substrate. For PS films (100–500 nm thickness) on rigid substrates (silicon and glass) with large thermal conductivities, an increase in the softening transition (which was attributed to T_g) temperature with decreasing PS film thickness was observed, while a decrease in the glass transition temperature was recorded for softer and less thermally conductive substrates such as PDMS. This study reiterates that several key factors such as confinement effects and substrate and interfacial interactions are to be considered for reliable quantitative measurements of thermophysical properties of glassy polymer films at the nanoscale.

Becker *et al.* employed AFM to probe the spatial and temporal evolution of ruptured regions in ultrathin PS films deposited on silica substrates [25]. AFM images were obtained for PS films with two different thicknesses – 3.9 and 4.9 nm, which were dewetted on the silica surface. In both cases, the polymer films exhibit a long-wave instability (spinodal dewetting), but in the thicker film (4.9 nm) dewetting by heterogeneous nucleation of holes prevented the onset of the instability. The resulting differences in the dynamical evolution of film rupture are clearly discernible by comparison of corresponding AFM images. The surface of the thinner film is shown to develop a correlated pattern of indentations, while in the case of the thicker film uncorrelated, initially small holes appear to grow rapidly. Conservation of mass results in the material removed from the inner side of a hole accumulating at the boundary of the hole, with a film depression developing behind this rim.

The AFM study conducted in this work revealed a novel localized pattern formation process during the dewetting process that involves the formation of a second row of holes (satellite holes) after a certain size of the first hole and continues in a kind of hole-forming cascade. The experimental data obtained using AFM were found to have excellent quantitative agreement with their theoretical modeling using finite element analysis. Their results suggested that development of the film profiles and complex patterns can thus be monitored in both space and time using *in situ* AFM, revealing fine differences in the film dynamics caused by small changes in the amorphous glassy film conditions on different substrates.

9.3 Rubbers

Rubber materials composed of physically or chemically cross-linked long-chain molecules with free segmental motion can exhibit large and reversible mechanical deformation. These materials are arguably the least investigated class of polymer materials using AFM due to the significant difficulties in conducting nondamaging AFM studies and avoiding common artifacts as well as usually boring observations of featureless, smooth, and uniform surfaces with little to no input to be deducted on any differences in surface morphologies. There are very few studies in the literature on the morphology or other properties of pristine rubbery materials. In this section, we present some examples of the application of AFM to investigate very compliant rubbery materials.

PDMS is an excellent and very popular elastomeric material with excellent compliance and uniformity that has gained immense attention due to its common commercial applications and developments in emerging technologies such as soft

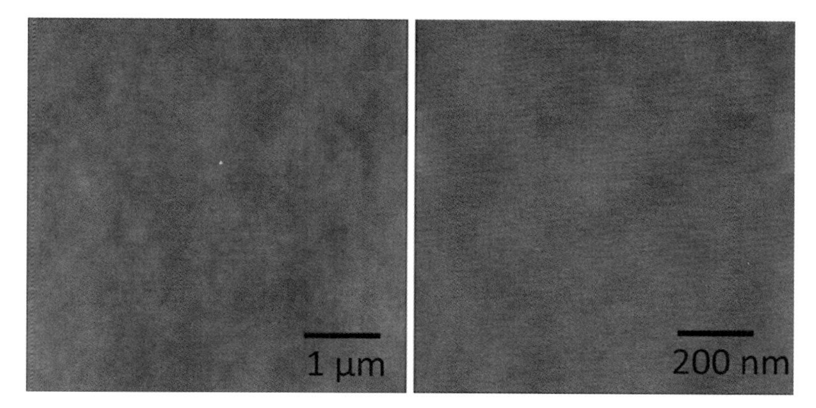

Figure 9.6 Topography images of PDMS cured on atomically flat silicon substrates showing the extremely smooth surface morphology. (Z scale: 10 nm (left) and 5 nm (right))

lithography and micro- and nanofluidics [26]. AFM has been employed to probe the structure–processing parameters relationship of this unique rubbery material. Figure 9.6 shows the topography images of a PDMS surface cured on the surface of an atomically flat silicon substrate. The surface exhibits extremely smooth morphology with a RMS microroughness of 0.25 nm over 1 μm^2 area, which is similar to that for glassy amorphous polymers. The extremely smooth surface of PDMS material results from its ability to conform, under surface tension, to the substrate upon which it is cured – silicon in the present case.

In earlier study, Tsukruk *et al.* investigated the viscoelastic effect on glassy and elastomeric polymers using surface force spectroscopy [27]. Glassy polymers (PVC and PS) in air showed no time dependence of elastic modulus within the frequency range of 0.05–100 Hz with an elastic modulus of 4.0 GPa for PS and 1.0 GPa for PVC (Figure 9.7). However, for a PVC film in water, the Young's modulus decreased significantly due to partial swelling of the surface layer. Strong frequency dependence was observed for PVC in water and alcohol: Young's modulus increased from 200 MPa at low frequencies to 1 GPa for high frequencies indicating a contribution from the viscoelastic behavior of swollen polymer films.

A significant reduction of the PVC modulus in water and its strong time dependence indicate that the micromechanical properties of the PVC surface layer in these solvents differ from the glassy bulk properties. These changes can be related to the partial penetration of solvent molecules into the very uppermost surface layer (30–40 nm) (the so-called plastification effect). As can be concluded from direct comparison of the calculated and measured data, the WLF formalism adequately describes the viscoelastic behavior of rubber and polyurethane (PU) surface layers with dynamic parameters close to the bulk values. As a result of surface segregation, the surface layer enriched with low molar mass molecules provides higher mobility for macromolecules and the local T_g is shifted downward. By using the WLF equation, the authors calculated this "apparent" glass transition temperature for the uppermost surface layer. The observed time dependence gives a T_g value of 35 °C for the PVC surface layer in water compared to 78 °C for the bulk

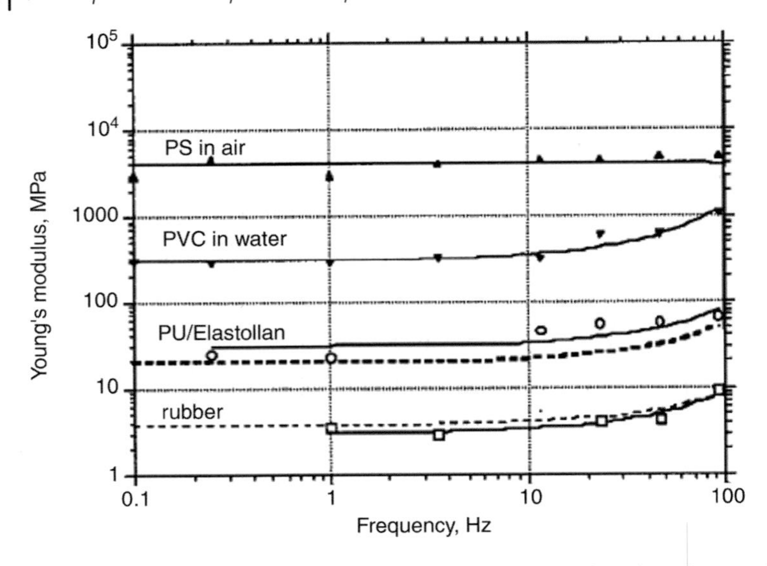

Figure 9.7 Time dependence of elastic modulus for glassy PS film, swollen PVC film, natural PI rubber, and Elastollan PU. Dashed curves represent time-dependent elastic moduli calculated from the WLF equation for natural rubber and polyurethane by using known parameters for bulk materials and for PVC assuming lowered glass transition temperature. Reprinted from Ref. [27].

material. This significant shift in glass transition can be expected for PVC containing about 20% low molar mass plasticizer.

A similar increase of elastic modulus with increasing frequency was observed for other elastomeric materials: PI rubber and polyurethane (Figure 9.8). In this case, the Young's modulus rises from 2.9 MPa for rubber and 30 MPa for Elastollan PU at low frequencies to 9 and 70 MPa at the highest frequency probed, respectively. This time-dependent viscoelastic behavior is well known for elastomers and was analyzed within the WLF approach and by using proper viscoelastic models (see Section 5.4) [2]. Indeed, the Maxwell model of viscoelastic behavior predicts exponential damping of elastic modulus with increasing time, which is close to the temporal behavior observed for both rubber and PU. It has been shown that the WLF equation with generic parameters commonly used for bulk rubbers and PUs predicted the dynamic behavior observed with AFM (Figure 9.8).

On the micrometer scale, the surfaces of all spin-cast amorphous rubbery films were homogeneous and showed very low microroughness. The PI film reported in this work possessed a microroughness (calculated within a 1 μm^2 surface, here and everywhere area) in the range of 0.2–1 nm for different samples. However, a light grainy texture caused by microphase-separated hard and soft segments was observed for all PU samples (see an example in Figure 9.8). In this case, phase imaging of polymeric materials shows a relatively even surface distribution of phase shifts, indicating a homogeneous distribution of chemical composition and viscoelastic response in amorphous rubbers. However, the PU samples generally showed a higher microroughness than PI rubber films due to partial microphase separation with peak RMS microroughness values reaching 3–4 nm or several times higher than that for traditional amorphous rubbers.

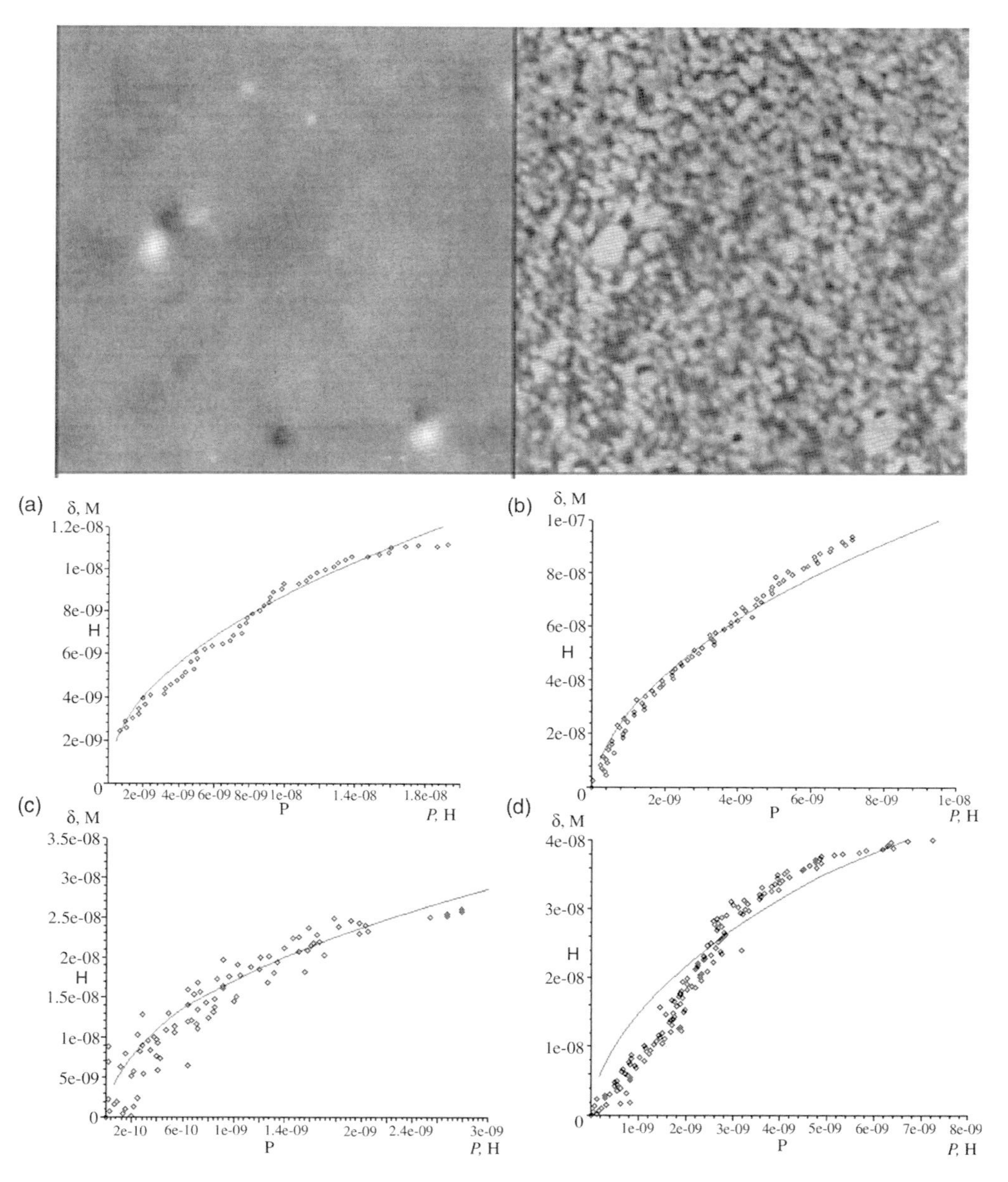

Figure 9.8 Topography (left) and phase (right) images of spin-coated films of Duraflex polyurethane film (*Top*: vertical scale for topography is 20 nm). Reprinted from Ref. [27]. *Bottom*: Loading curves for PU films with different thickness and corresponding theoretical fits: (a) 11.5 nm, $E_1 = 4.5$ MPa; (b) 20 nm, $E_1 = 0.6$ MPa; (c) 31 nm, $E_1 = 0.38$ MPa; (d) 50 nm, $E_1 = 0.75$ MPa. Reprinted from Ref. [29].

In several studies, SFS is applied to probe mechanical properties of compliant rubbers. Micromechanical probing of the elastomeric properties of ultrathin PU films with thicknesses ranging from 10 to 50 nm has been conducted by Tsukruk *et al.* [28]. The loading curves for 10 nm PU films derived from SFS data showed a characteristic non-Hertzian shape caused by the presence of the stiff substrate that can be approximated by the double-layer model with the values of elastic modulus ranging from 0.3 to 5 MPa (Figure 9.8) [29]. For thicker films, more monotonous deformational behavior was observed with clear signs of time-dependent response with purely bulk rubbery behavior observed for the film thickness more than 30 nm.

In a report on more complex rubbery materials, Achalla *et al.* employed tapping mode AFM and force spectroscopy to investigate the phase morphology of blends of incompatible elastomers (bromobutyl rubber and natural rubber) [30]. Depending on the composition of the blend, the authors observed either cocontinuous or discontinuous domain/matrix morphology. Phase images exhibited a higher contrast (compared to the topography images) between the two components of the blends. The authors noted that it was difficult to identify the domains from the phase images of a 50 : 50 polymer blend. However, a systematic investigation with varying blend composition enabled them to unambiguously identify bromobutyl rubber and natural rubber, which appeared to be bright and dark, respectively.

Indeed, Figure 9.9 shows the phase images of blends of different compositions and one can observe the gradual increase of brighter regions (with increasing bromobutyl rubber volume fraction) and concomitant decrease of dark regions. The images were collected under medium tapping conditions with a set point ratio (r_{sp}) of 0.75, ensuring that the elastic response rather than the surface adhesion was probed. Nanomechanical SFS measurements were also used to qualitatively distinguish the homopolymers based on the differences in stiffness (given by the slope of the linear portion of the force curve). From independent measurements, it was known that bromobutyl rubber is a softer material compared to natural rubber. The nanomechanical measurements confirmed the conclusions drawn from the phase images regarding the distribution of the individual rubber components.

Functionalized rubbery materials are tested in numerous AFM studies. Plasma oxidation of PDMS is a popular method to modify the surface properties (hydrophilicity and mechanical properties) of this elastomeric material [31]. Plasma oxidation is also routinely employed to bond PDMS surfaces in microfluidic devices [26, 32]. The thickness of the oxidized layer, which exhibits dramatically different properties compared to PDMS bulk, is critical in the bonding process. The thickness of this surface layer, at least in one of the mating surfaces, should be large enough to ensure uniform surface chemistry and thin enough so that the surface does not lose its ability to make conformational contact with the surface it is being bonded to.

In a recent study, Mills *et al.* employed AFM to monitor an asymmetric interface of two PDMS layers oxidized for different time periods [33]. Figure 9.10 shows the phase, topography, and the cross-sectional profiles of the phase image of the interfacial region imaged by fracturing the bonded layers. Interestingly, the phase image exhibited a high contrast (owing to the difference in the mechanical properties) between interfacial and bulk PDMS compared to the topography image. The thickness of the surface modified layer, which correlated well with the time of

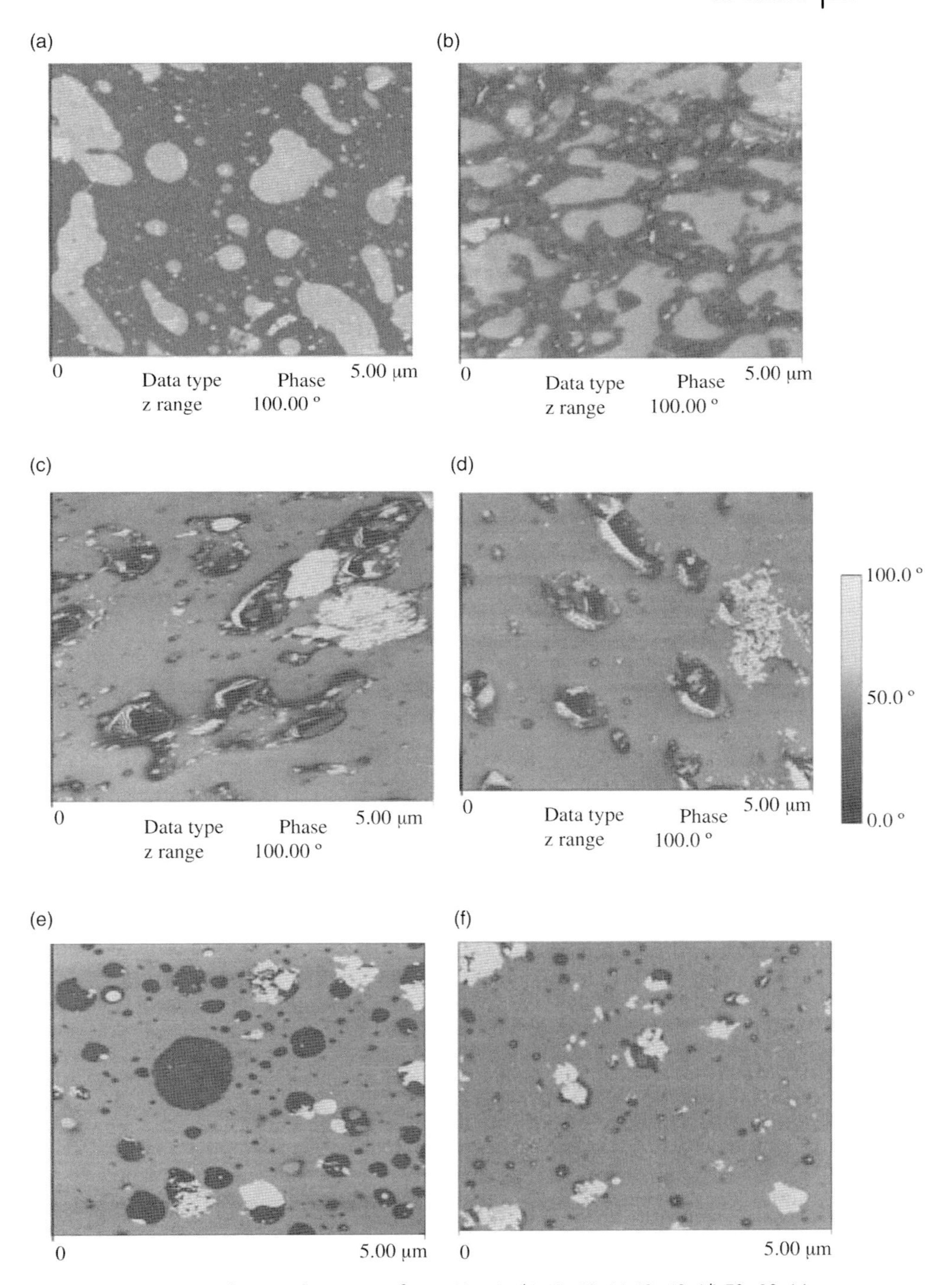

Figure 9.9 Tapping mode AFM phase image for (a) 30 : 70, (b) 40 : 60, (c) 60 : 40, (d) 70 : 30, (e) 80 : 20, and (f) 90 : 10 BIIR:NR rubber blend ($r_{sp} = 0.75$). Reprinted from Ref. [30].

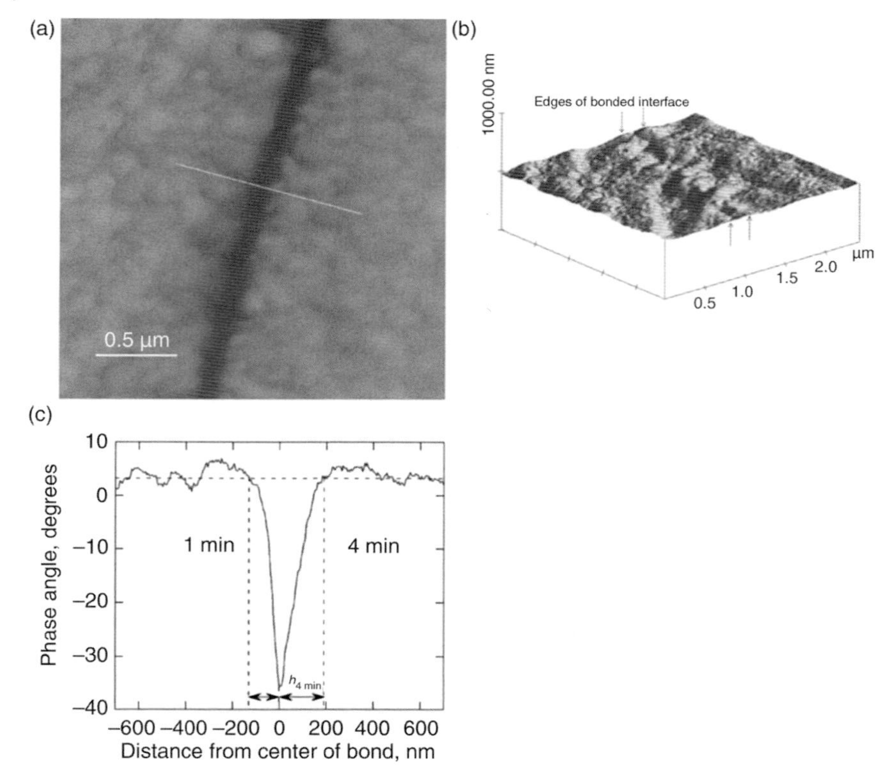

Figure 9.10 AFM (a) phase and (b) height images of the interfacial region of PDMS oxidized for 1 min (on the left) and the surface oxidized for 4 min (on the right). (c) Line trace of the phase angle across the bond at the location of the marker on the phase image in (a). Reprinted from Ref. [33].

oxidation, was measured from the lowest point of the phase angle (at the bond line) to the average value for the bulk material. From these measurements, the thickness of the plasma-oxidized layer for 1 min of treatment was found to be 100–125 nm, while that for 4 min treatment was found to be 200 nm.

AFM has also been widely employed to monitor the fidelity of PDMS stamps with complex topography to reproduce the desired features of the microfabricated master. Here, we present one example from the authors' own work where light tapping mode was employed to monitor the nanoscale details of a complex pattern replicated using a soft lithographic approach. Singamaneni *et al.* utilized microporous structures fabricated by multilaser beam interference lithography as masters for microprinting by using an intermediate PDMS structure [34]. The fabrication process involved the preparation of the negative PDMS replica, followed by the infiltration of PS solution into the receding portions of the PDMS, and, finally, transferring and depositing the PS structure onto a different substrate. Capillary transfer lithography (CTL) was chosen to reproduce a square array of air cylinders formed by interference lithography in a SU8 material [35]. A mixture of prepolymer and curing agent (10 : 1) was poured onto the 2D structure, followed by degassing and curing at 75 °C. Subsequently, the PDMS film was peeled from the SU8 to create a negative replica of the 2D organized porous template structure.

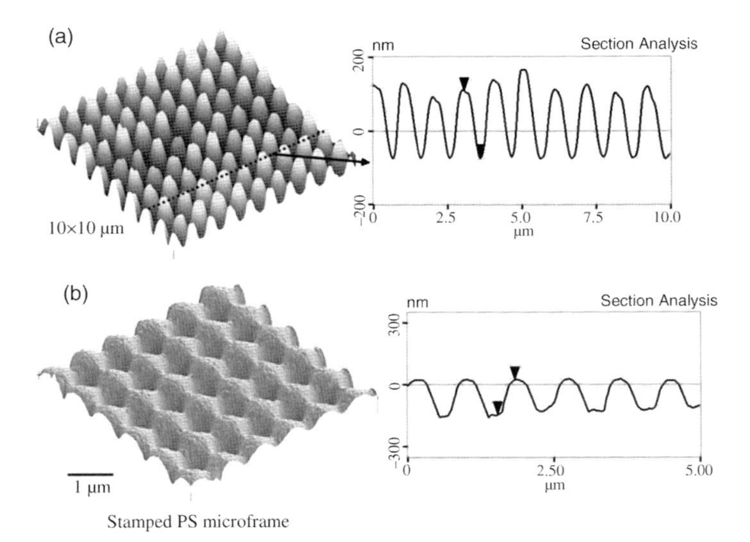

Figure 9.11 (a) AFM image depicting the vertical posts (negative replica of SU8 master) of uniform height and the cross section of AFM image showing the uniform height of the posts along the line shown in the image. (b) Stamped PS structure showing identical topographical features down to the nanoscale. Reprinted from Ref. [34].

The authors noted that due to the high viscosity of the prepolymer, only a partial infiltration (200 nm deep) into the SU8 porous structure is achieved. Complete infiltration of the viscous precursor into the pores requires external forces such as evacuation of the pores from the bottom. AFM images of the PDMS replica show uniform height of the vertical posts and the cross section of posts shows the height to be nearly 200 nm (Figure 9.11). The negative replicas obtained were then used for the fabrication of organized structures from other polymers. Figure 9.11b shows the AFM images and the corresponding cross sections of the stamped PS replica. AFM images revealed the remarkable similarity between the two structures with excellent replication of surface topology of the original SU8 IL structure. In fact, even the nanoscale grainy morphology of the SU8 surface caused by photopolymerization during the lithographical process was faithfully reproduced in the final structure [36]. It is worth mentioning that the PS stamped structures showed a significantly smaller vertical height (150 nm), measured from the AFM cross sections, compared to that of the original PDMS posts (200 nm) due to the elastic deformation of the PDMS stamp during mechanical compression.

9.4 Polymer Gels

Bio and synthetic hydrogels are unique materials composed of a loosely cross-linked, flexible polymer network and containing more than 90% water. These compliant materials show a wide range of elastic moduli from megapascals in dry state down to a few pascals in wet state and extremely high swellability [37]. Understanding the interaction of these hydrogel structures with water is critical to a vast number of fields, including biotransport phenomena, biocompatibility, and biomicrofluidics [38, 39]. Owing to their excellent biocompatibility, which is important for biomedical applications (such as tissue engineering, regenerative medicine, or drug delivery), the physical properties of bulk hydrogels have been extensively studied [40–42].

Numerous synthetic hydrogels have been developed from cross-linkable derivatives of poly(2-hydroxyethyl methacrylate) (PHEMA), poly(ethylene glycol) (PEG), and poly(vinyl alcohol) (PVA). The properties of hydrogels are widely studied for bulk, smooth, and uniform hydrogels that are dominated by volumetric phenomena [43]. Considering the extremely compliant nature of these materials, it is not surprising that the behavior of gel microstructures with variable topology is very complex and poorly understood. AFM techniques are in a rather unique position to address some of these issues (morphology, mechanical properties, or adhesion) owing to the high resolution and ability to probe in the natural wet state of these materials.

The structural features and swelling properties of responsive hydrogel films based on poly(*N*-isopropylacrylamide) (PNIPAAm) copolymers were investigated using AFM by Beines *et al.* [44]. In particular, the morphology of a PNIPAAm hydrogel containing methacrylic acid moieties with ionizable groups in the dry and swollen state has been studied. In the dry state, after spin casting followed by soft baking, the copolymer containing carboxylic groups exhibited an extremely smooth and featureless surface morphology (see Figure 9.12). Upon hydration and drying of these thin gel films, a porous morphology evolved.

The porous morphology of these films can be caused by the dissolution of the uncross-linked polymer in the hydrogel films. Furthermore, the polymer network might be partially restructured, and the more hydrophilic units may segregate to form the pores as water-release channels. On the other hand, the PNIPAAm copolymers with no ionizable groups exhibited a porous surface morphology even before hydration (see Figure 9.12). Upon hydration of these films, the pore size was found to dramatically increase with a concomitant decrease in the thickness of the film due to the dissolution of a large amount of uncross-linked polymer. From the AFM images, the authors were able to conclude that nearly 25% of the polymer dissolved with the first hydration of copolymer without ionizable groups.

Suzuki *et al.* were among the first researchers to investigate the nanoscale surface morphology of a gel surface under water using AFM [45]. The authors studied the change in the surface morphology with the variation in the network density and the effect of the temperature change of PNIPAAm gels. The gel materials prepared for the AFM imaging possessed thicknesses ranging from 10 to 50 nm, and one of the gel surfaces was chemically adhered onto a glass plate. The AFM images revealed spongelike morphology on a submicrometer scale that was found to be strongly affected by the cross-linking density, osmotic pressure, and thickness, which correspond to the network structure, environmental conditions, and constraint conditions, respectively.

To probe the effect of temperature change (as an external stimulus) on highly cross-linked PNIPAAm gel, in this study the authors performed *in situ* AFM below and above the LCST of the gel. The AFM images showed that the spongelike morphology

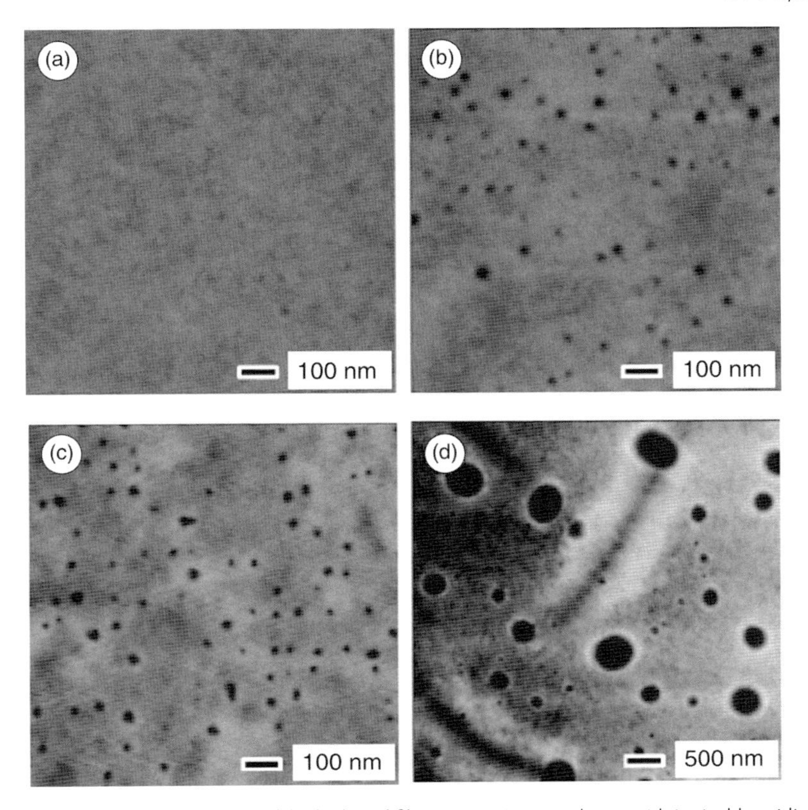

Figure 9.12 AFM images of the hydrogel films PNIPAAm copolymer with ionizable acidic groups as prepared (a) and after swelling, collapse, and drying (b) and of the polymer without ionizable groups (c) as prepared and (d) after swelling and drying. Reprinted from Ref. [44].

of the gel structure became more pronounced at 38 °C. The authors reasoned that more pronounced domain structure was caused by the collapsed state of the gel due to the attractive forces (hydrophobic interaction) at higher temperatures. The results of surface morphology were discussed in the framework of the Flory–Huggins theory for 2D gels [43].

Thermally responsive properties of gel materials from 2-(dimethylamino)ethyl methacrylate (DMAEMA) and benzophenone methacrylate (BPMA) copolymers with different compositions synthesized via atom transfer radical polymerization were probed by *in situ* contact mode AFM and lateral force microscopy by Huang *et al.* [46]. The molecular weight (M_n) of these copolymers was 30 000 g/mol, while the BPMA content varied from 2.5 to 10 mol%. The hydrogel was immobilized and patterned on a silicon wafer via UV treatment of the spin-coated polymer layer using a photomasking technique.

The contact mode AFM images acquired at different normal loads and at various temperatures revealed the presence of two lower critical solution temperature regions. One transition region was noted between 25 and 30 °C, associated with the topmost layer of the hydrogel film, and the other region, observed around 40 °C, corresponded to the bulk hydrogel response. The authors suggested that the observed differences in the thermal transitions at different depths of the hydrogel film most likely reflected the heterogeneity in cross-linking density across the film thickness. Enhanced friction in the bulk region just above the transition temperature was also observed in concurrent lateral force microscopy measurements. This particular observation suggests that employing mechanical energy dissipation probed with FFM mode might be an effective and reliable method to investigate the thermal transitions in stimulus–responsive hydrogels.

In the report on surface tethered hydrogel films, Singamaneni *et al.* demonstrated that ultrathin (20–100 nm) responsive gel films tethered to solid substrates may exhibit unusual, spontaneous, and regular self-folding under swelling-induced compressive stresses [47, 48]. P2VP materials exploited in this study exhibit strong interactions with the silicon oxide surface owing to the hydrogen bonding between the surface hydroxyl groups and nitrogen on the pyridine ring [57]. As a result of this interaction, a 90 nm thick cross-linked P2VP gel film grafted onto a silicon wafer with native oxide layer of ~1 nm shows a uniform and smooth surface morphology with a local microroughness below 1 nm (Figure 9.13).

On the other hand, in the protonated state (below pH 4.0), the electrostatic interaction between the positively charged pyridine units and negatively charged silicon oxide surface further promotes strong thin film anchoring to the surface. Surprisingly, simple exposure of the cross-linked gel films to a pH 2.0 solution resulted in the transformation of the initially smooth morphology into a network of highly anisotropic sheetlike structures with partially folded morphologies caused by stress-mediated disrupting of the initially homogeneous polymer film (Figure 9.13).

Moreover, lateral constraints imposed by microimprinting in the tethered gel film resulted in the inhibition of complete folding and the formation of individual, localized, ordered folded structures that can extend over large areas (Figure 9.14). Further exposure of the patterned gel film to acidic solution resulted in a dramatic

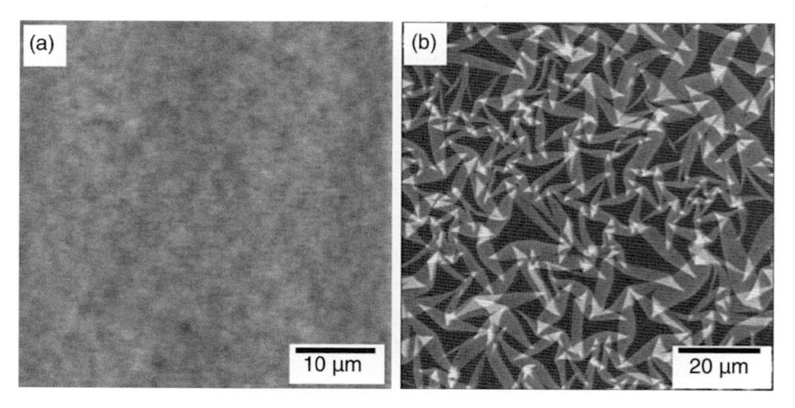

Figure 9.13 AFM images of the spin-coated P2VP gel film showing the smooth featureless surface morphology (a) and a large area of folded P2VP fractured film on exposure to acidic (pH 2) solution (b). Reprinted from Ref. [47].

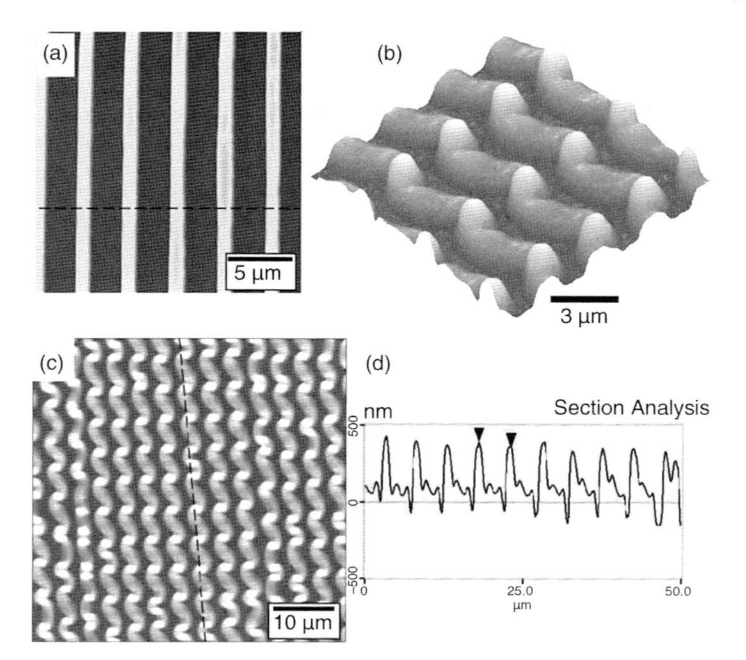

Figure 9.14 AFM images of one-dimensional P2VP gel structures with a periodicity of 3 μm (a) Three-dimensional AFM image of the folding (b). (c) Image showing the large scale uniformity of the folded structure (d) Cross section along folded structure showing the three discrete heights of the fold regions, the intermediate regions, and the substrate-grafted residual layer. Reprinted from Ref. [47].

transformation of the initial gel nanofilm to uniformly oriented, continuous "telephone cord" structures (Figure 9.14). The authors suggested that the remarkably uniform spatial alignment of the folded structures comes from the stress–strain gradient across the thinner regions in between the tall stripes. In other words, the limited supply of material between the growing structures creates long-range spatial alignment of the features through local growth competition, which causes the film to self-regulate and avoid complete separation from the substrate such as that observed for flat films. These results indicate that preexisting topological patterns acting as boundary conditions for mechanical stress–strain systems can reinforce and develop new topographically controlled self-folded patterns.

A closer observation of the tapping mode AFM images and the cross-sectional analysis clearly reveal that the 3D telephone cord structure is indeed periodic buckles transformed into twisted folds (Figure 9.14). Overall, the structure represents a chiral array with right-handed deformed helical structures that repeat themselves across the macroscopic surface areas. Moreover, wider-spaced patterned surfaces (up to 7 μm, which is more than twice the width of the folds in flat films) resulted in the misalignment and randomization of the serpentine structures observed with AFM in stark contrast to the previous case of regularly folded structures.

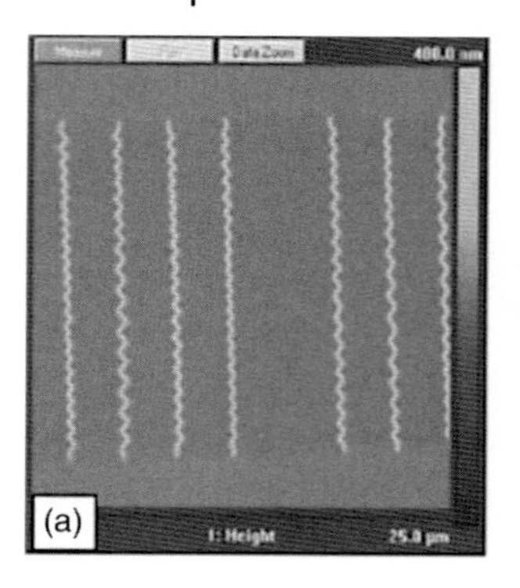

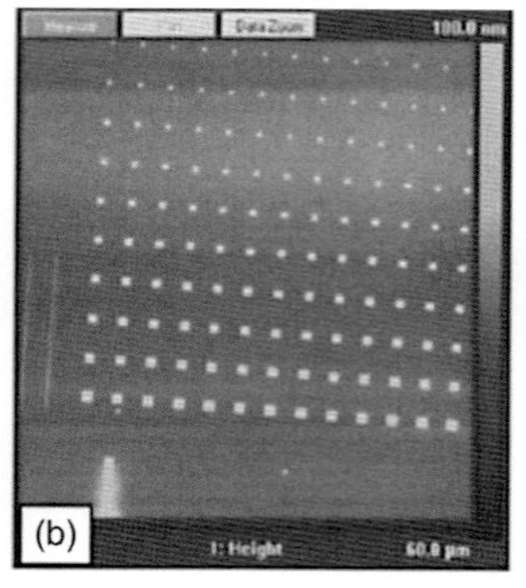

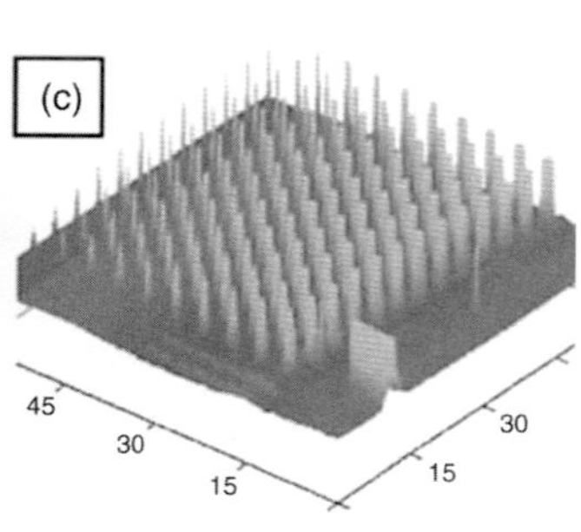

Figure 9.15 (a) AFM image of irradiated stripes (250 nm wide) of poly(vinylmethyl ether) at various doses. (b) AFM image of irradiated squares with lateral sizes ranging from 1.2×1.2 μm^2 to $250 \times 250\,nm^2$ at various radiation doses. (c) 3D plot of the AFM image of the irradiated small squares representing the lateral resolution as well as the pad height. Reprinted from Ref. [49].

In another AFM application to study hydrogel materials, Schmidt *et al.* investigated temperature-sensitive hydrogel patterns of poly(vinylmethyl ether) synthesized by irradiation with a focused low-energy electron beam (with a dose of 50–220 $\mu C/cm^2$) [49]. In this study, AFM was employed to probe the structure of the hydrogel patterns formed on the silicon surfaces. Figure 9.15 shows the tapping mode AFM image of the stripe pattern formed by e-beam lithography. The resulting patterns were found to strongly adhere to substrate without the use of any adhesion promoter. As opposed to the expected stripe patterns, one can observe that the patterns are wavy due to the confined swelling of the gel structure. The authors reasoned that the gradient swelling of the structure along the thickness resulted in the observed wavy nature.

Interestingly, higher doses of e-beam resulting in higher cross-linking suppressed the formation of the wavy patterns. The authors also fabricated small squares of hydrogels by e-beam irradiation, as shown in Figure 9.15. The lateral dimensions of these squares varied from 1.2 μm to 100 nm. Figure 9.15 also shows the 3D AFM image of the pads. The height of the square structures with different sizes was measured by AFM line analysis as a function of the applied dose after rinsing and drying. It was concluded that the smallest spatial resolution of patterned PVME films that could be achieved under the optimized experimental conditions is about 100 nm.

In a recent work, Guvendiren *et al.* demonstrated AFM images of a wide range of osmotically driven surface patterns, such as random, lamellar, peanut, and hexagonal structures, on hydrogels with gradient elastic moduli [50]. Hydrogel films were fabricated by exposing a photocurable formulation of HEMA and ethylene glycol dimethacrylate (EGDMA) to UV light open to air followed by swelling. A gradient in the elastic modulus of the films was achieved by a gradient cross-linking of the network. Competition between the saturated oxygen consumption due to curing and oxygen diffusion into the film from the top surface of the film led to high cross-linking at the surface and a monotonous decrease along the thickness of the film. The authors employed AFM imaging and colloidal force spectroscopy (CFS) to characterize the structural and mechanical properties of the cross-linked PHEMA films.

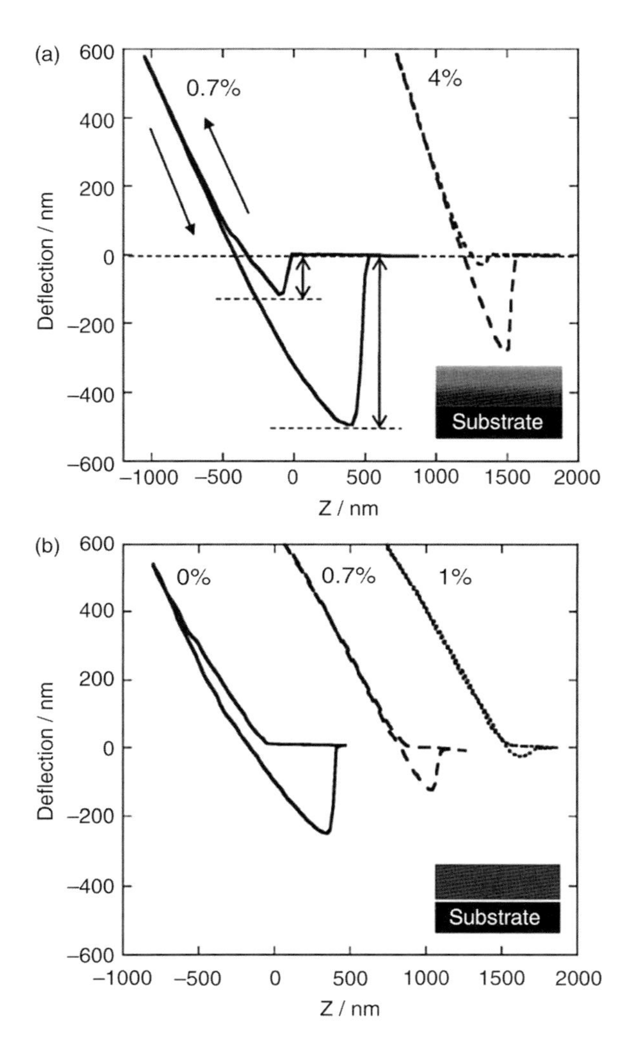

Figure 9.16 AFM force–distance curves obtained on swollen PHEMA films. Films were prepared by exposing the top surface of the precursor solution to UV light (a) open to air creating a modulus gradient with thickness and (b) covered with PDMS resulting in uniform modulus with thickness. FDC curves are also shown for various EGDMA concentrations. Reprinted from Ref. [50].

To measure nanomechanical properties, the authors used a colloidal probe with a radius of 1 μm and force constant of 0.06 N/m. As seen in Figure 9.16, for PHEMA films cross-linked in open air with 0.7 wt% EGDMA, the force curves exhibited a large "snap-to" deflection, large adhesion between the colloidal probe and the surface, and significant hysteresis between the loading and retraction curves. On the other hand, when the EGDMA content was increased to 4%, resulting in higher cross-linking, the authors observed a significant lowering of the snap-to deflection (almost absent) and very low adhesion.

For the gel films cured with PDMS protection on the top, the force curves revealed a much more uniform surface with no snap-to deflection and little adhesion. Using Hertzian analysis of CFS data, the modulus of various films was found to be between 200 KPa and 3 MPa. Long-range ordered hexagonal patterns were also observed when the osmotic pressure due to swelling was close to the critical osmotic pressure for buckling. Tuning the modulus gradient (by varying EGDMA concentration) resulted in the initially formed hexagonal pattern to collapse into peanut shapes. The lamellar and random worm surface patterns further formed with the increase of osmotic pressure relative to the critical value. The authors noted that the patterns observed with AFM were stable in both the swollen and the dry states.

In another study of responsive gels, Matzelle *et al.* investigated the swelling behavior and elastic properties of thermoresponsive PNIPAAm and PAAm hydrogels as a function of cross-linker concentration [51]. AFM nanomechanical measurements were performed in the hydrated state of the gels using a liquid cell. Elastic moduli of both PNIPAAm and PAAm hydrogels were determined at different temperatures and varying cross-linker concentrations using colloidal probes as well as regular microfabricated probes using Hertzian analysis. The authors observed a dramatic increase (nearly an order of magnitude) in the elastic modulus of the PNIPAAm surfaces when crossing the LCST phase transition at 33 °C. It has also been found that for temperatures above the LCST, the cross-linker concentration has a strong influence on the elastic modulus while only small variations were seen for temperatures below 33 °C for all cross-linking densities. On the other hand, for the PAAm hydrogels, the elastic modulus exhibited a strong dependency on the cross-linker concentration with a minimal variation with temperature. These localized measurements using AFM showed excellent agreement with the macroscopic rheological measurements conducted concurrently.

In situ AFM imaging and surface force spectroscopy were employed by Wiedemair *et al.* to reveal the volume phase transition behavior of individual thermally responsive poly(*N*-isopropylacrylamide-*co*-acrylic acid) (PNIPAAm-*co*-AAc) hydrogel microparticles [52]. AFM studies of particle deswelling were performed by varying the force applied on the particles during AFM imaging. Conventional tapping mode cantilevers (aluminum-coated silicon cantilevers) were found to significantly influence the behavior of the particles during the phase transition, leading to a significant shape change. On the other hand, the authors found that low force impact on the magnetic excitation of the AFM probe during dynamic mode measurements resulted in an undisturbed phase transition of the microparticles.

Magnetic cantilever excitation mode strongly reduces the force impact on the sample owing to smaller driving amplitudes (several nanometers) compared to that exploited in tapping mode (few tens of nanometers). The authors concluded that magnetic excitation (MAC) mode was the most suitable technique for *in situ* AFM studies on volume and shape changes of single hydrogel particles during phase transition. Furthermore, surface force spectroscopy performed on single microparticles at different temperatures (below and above the phase transition) revealed a 15-fold increase in the Young's modulus at the phase transition, indicating the transition from a soft, swollen network to a stiffer, collapsed, and deswollen state.

In another study, Nitta *et al.* measured the elastic modulus of agar gels for various concentrations using both AFM nanomechanical measurements and the tensile creep method [53]. Measurements of the local elastic modulus of agar gels obtained by force mapping were compared to values obtained by the tensile creep method. Figure 9.16 shows a representative AFM topographic and elastic modulus micromap of 1.0 wt.% agar gel. During the measurements, the temperature of the agar gels was kept constant at about 27 °C. The force–distance curves were analyzed using the Hertz contact mechanics model to deduce the elastic modulus. The AFM topography images clearly revealed the fine network structure of the agar gel that was also confirmed by SEM imaging.

Comparing the AFM topographic image with the elastic modulus map, the authors noted that higher regions in the topographic image correspond to stiffer regions of the sample. The observed spatial distributions of the local elastic modulus over the gel surface in AFM elastic images clearly corresponded to the loose network morphology of randomly oriented agar nanofibers. The peak and average values of elastic modulus distribution functions were derived from the surface histograms (Figure 9.17). It has been concluded that the average elastic modulus exhibited a monotonic increase with the increasing agar concentration. Overall, values obtained by AFM force mapping were found to be proportional to modulus values obtained by concurrent creep experiments. The authors observed that the proportionality breaks down when the agar gels become significantly stiffer than the cantilever employed in the force measurements underscoring the critical importance of the right choice of the cantilever stiffness for the compliant material being probed, as was discussed in detail in Section 5.1.3.

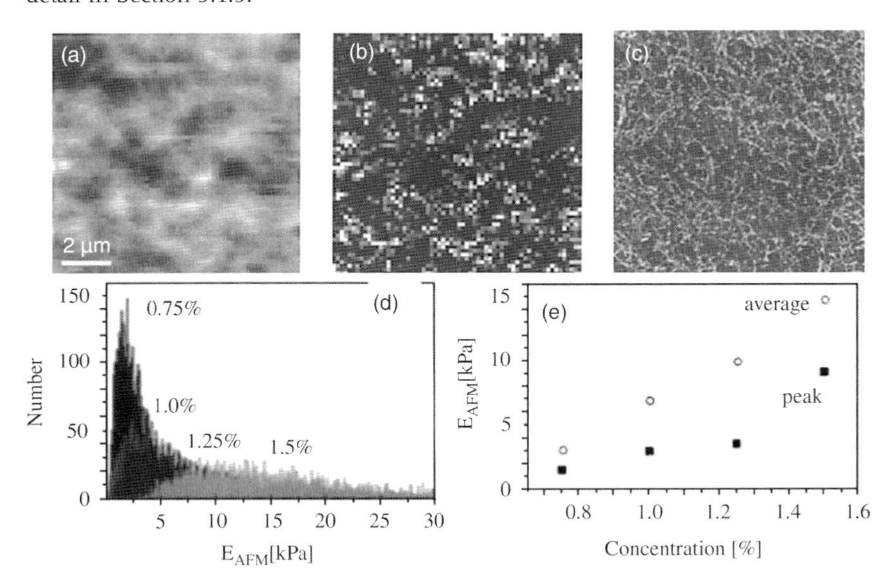

Figure 9.17 (a) AFM topography, (b) elastic modulus, and (c) SEM images of 1.0% agar gel. Z-range of the topography is 450 nm and the elastic modulus map is 0–25 kPa. (d) Histograms of elastic modulus for the various agar concentrations. (e) Average and peak values of elastic modulus as a function of the agar concentration. Reprinted from Ref. [53].

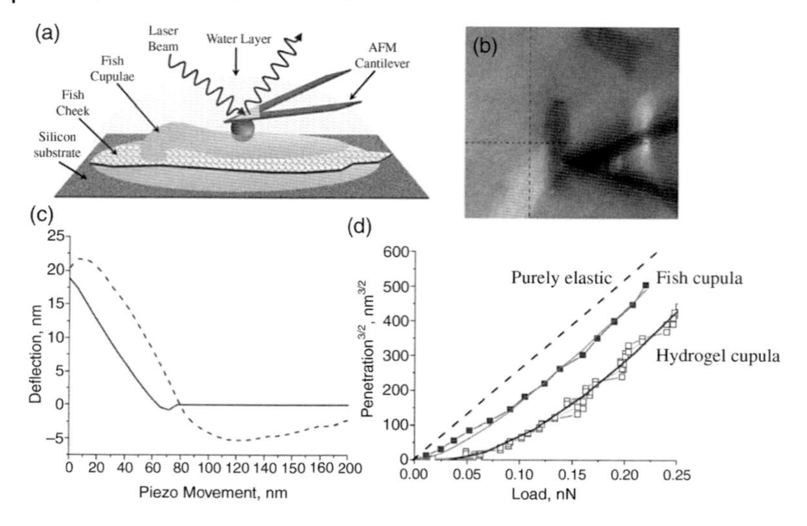

Figure 9.18 Nanomechanical measurements of the biogel cupula of the superficial neuromasts of fish. (a) Schematic of the experimental setup to measure nanomechanical properties of the fish cupula. AFM cantilever with glass sphere indents the fish cupula in water. (b) Micrograph of actual AFM setup with folded cupula. AFM cantilever is placed over the stained fish cupula in water. (c) Typical force–distance curve of the fish cupula. (d) Loading curves for biogel and synthetic gel materials. The dashed line represents pure elastic deformation. All data were collected at scanning frequency of 1 Hz.Reprinted from Ref. [55].

In one example of AFM probing of biological hydrogels, Peleshanko and coworkers studied the mechanical properties of a hydrogel-like material that makes up the cupula of a blind cave fish [54]. The cupula and support fibers are specialized structures that enhance flow-sensing properties of hair cells [55, 56]. Mechanical properties of blind cave fish cupulae were directly measured using fluid-based surface force spectroscopy with a colloidal probe. The elastic modulus of the fish cupula was measured in water using AFM in force–volume mode and a liquid cell on freshly prepared fish specimens (Figure 9.18).

The Hertzian loading plot in coordinates of penetration$^{3/2}$ versus the applied load was observed to be highly nonlinear for biohydrogel (Figure 9.18). The nonlinear response was associated with the time-dependent, viscous response of the material. The maximum applied load was extremely low, on the order of 250 pN, which is significantly less than the force needed to break a single covalent bond (on the order of several nanonewtons). Therefore, any nonlinearity of the Hertzian coordinate plot is not due to plastic deformation. The Voight viscoelastic model was combined with the Hertzian contact model to fit the nonlinear loading data for the biological cupula (Figure 9.18). This way, the biohydrogel was measured to have an elastic modulus of 9 kPa and a relaxation time of 0.42 s, which are characteristic of both compliant and viscous gel materials.

Moreover, the authors fabricated synthetic cupulae with comparable mechanical properties by photocross-linking tetra-acrylate-functionalized PEG. Photomasks were used to localize cross-linking and thereby pattern the PEG into different shapes through selective exposure of the hydrogel. The Voight viscoelastic model combined with the Hertzian contact model was used to fit the nonlinear experimental loading data obtained by colloidal probe spectroscopy on top of the synthetic cupula (Figure 9.18). This approach, when applied to the synthetic hydrogel cupula with intermediate molecular weight between cross-links, resulted in an elastic modulus of 9.5 kPa and a relaxation time of 0.5 s, fairly close to results obtained for the biological cupulae of fish (Figure 9.18).

9.5 Interpenetrating Polymer Networks

Interpenetrating networks are defined as a combination of two or more polymer networks with one or more of them polymerized/cross-linked in the presence of the other network [57–60]. These polymeric materials possess a peculiar morphology of bicontinuous microphase-separated networks due to the confined segregation of components, which has mostly been studied with light scattering and TEM techniques. In this section, we briefly highlight a few recent studies in which AFM was also employed to probe the peculiar surface morphology and mechanical properties of IPN materials.

In one of these studies, Kim and Kim investigated PU homopolymers and PU/PS IPNs, which were synthesized with varying pendant PEO chain lengths [61]. PU/PS IPNs were found to have a well-defined microphase-separated morphology with PS-rich phase domains dispersed in a PU-rich matrix. By performing AFM imaging under water, it was demonstrated that the area fraction of the hydrophilic PU-rich phase was increased by PEO grafting. The average diameter of the PS-rich domains decreased from 72 to 57 nm in air and from 66 to 41 nm in water, as the length of pendant PEO chains was increased.

From the surface force spectroscopic measurements performed in air, it was found that the hydrophilic, long pendant PEO chains readily attracted the moisture in the air as compared to the short pendant PEO chains. Due to the increase in the surface hydrophilicity and the resulting increase in the thickness of the adsorbed water layer on the surface, a dramatic increase in the tip–sample adhesion force was observed as the length of pendant PEO chains increased. On the other hand, from the force spectroscopic measurements performed in water, the authors found that the mobile and flexible pendant PEO chains increased the overall surface compliance of PU/PS IPN materials.

The morphology of IPNs combining polyisobutene (PIB) and poly(cyclohexyl methacrylate) (PCHMA) networks prepared using an *in situ* polymerization strategy was addressed by Vancaeyzeele *et al.* [62]. The authors performed tapping mode AFM imaging of three different PIB/PCHMA IPNs with different compositions: (1) 60/40, (2) 40/60, and (3) 20/80. The image contrast in these IPNs is given by the stiffness difference between a soft PIB-rich phase and a hard PCHMA phase combined with undetermined amounts of PIB. The thermomechanical properties of the individual components were determined by the authors using dynamic mechanical thermal analysis. The AFM images clearly indicate the coexistence of two phases in these materials, one phase rich in PIB and the other phase containing essentially a closely interpenetrated PIB/PCHMA network. Regardless of the observation scale, the AFM images revealed the cocontinuous morphology of the two components for this IPN material.

Highly transparent and homogeneous poly(vinylidene fluoride) (PVDF)/silica hybrids were obtained by using an *in situ* interpenetrating polymer network method and studied with AFM by Ogoshi and Chujo [63]. The simultaneous formation of a PVDF gel resulting from the physical cross-linking and a silica gel from the sol–gel process prevented the aggregation of PVDF in silica gel matrix. To form the physical cross-linking between PVDF chains, dimethylformamide (DMF) and γ-butyrolactone were used as a cosolvent system. Tapping mode AFM was employed to investigate the homogeneity of the PVDF/silica IPN hybrids. The surface microroughness of the almost transparent hybrid was found to be 25 nm that is much higher than that for homogeneous materials. On the other hand, a completely transparent polymer hybrid surface (obtained using 0.3 ml of γ-butyrolactone), was relatively smooth and flat (the microroughness was less than 10 nm). Furthermore, the AFM images of both hybrid structures with different cross-linking conditions revealed a continuous network morphology on the surface. These AFM observations suggested the nanoscale partial miscibility between PVDF and silica gel networks.

In a related study, the same authors investigated poly(urethane acrylate) (PUA)/silica hybrids formed by strong ionic interactions between the carboxylic acid groups of the PUA and amino moieties of the silica resulting from 3-aminopropyltriethoxysilane [64]. Chemical cross-linking between PUA chains in the silica gel matrix was achieved by a UV-induced cross-linking reaction. Tapping mode AFM was employed to monitor the morphological changes in the hybrid IPN structures upon cross-linking.

As has been observed, before UV irradiation, the polymer IPN hybrid material was extremely smooth (Figure 9.19). However, the nanoparticles clearly appeared on the initially smooth surface after UV irradiation. The authors noted that these nanoparticles resulted from the shrinking of PUA segments in the course of their cross-linking. The transparency of the PUA/silica IPN hybrids was not changed after UV irradiation. The AFM images also revealed the large-scale homogeneity of the hybrid

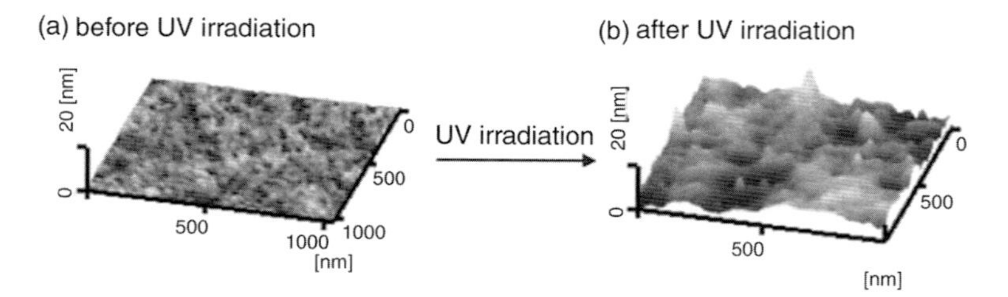

Figure 9.19 Tapping mode AFM images of the PUA/silica IPN hybrid (a) before and (b) after UV irradiation. Reprinted from Ref. [64].

structures before and after UV exposure. From a series of thermogravimetric analysis studies combined with AFM studies, the authors concluded that the IPN formation between the PUA network and the silica gel induced upon UV light irradiation gave excellent solvent resistance and surface hardness compared to the semi-IPN structures.

References

1 Ferry, J.D. (1980) *Viscoelastic Properties of Polymers*, John Wiley & Sons, Inc., New York.

2 Sperling, L.H. (2006) *Introduction to Physical Polymer Science*, John Wiley & Sons, Inc., New York.

3 Sergei, N.M. and Whangbo, M.-H. (1996) *Surface Analysis with STM and AFM: Experimental and Theoretical Aspects of Image Analysis*, Wiley-VCH Verlag GmbH, Weinheim, Germany.

4 Ratner, B.D. and Tsukruk, V.V. (1998) *Scanning Probe Microscopy of Polymers: Division of Polymer Chemistry*, American Chemical Society, Washington, D.C.

5 Assender, H., Bliznyuk, V., and Porfyrakis, K. (2002) How surface topography relates to materials' properties. *Science*, **297** (5583), 973–976.

6 Guinier, A. (1963) *X-Ray Diffraction in Crystals, Imperfect Crystals, and Amorphous Bodies*, W. H. Freeman and Co., San Francisco, London.

7 Bliznyuk, V.N., Burlakov, V.M., Assender, H.E., Briggs, G.A.D., and Tsukahara, Y. (2001) Surface structure of amorphous PMMA from SPM: auto-correlation function and fractal analysis. *Macromol. Symp.*, **167**, 89–100.

8 Goldbeck- Wood, G., Bliznyuk, V.N., Burlakov, V., Assender, H.E., Briggs, G.A.D., Tsukahara, Y., Anderson, K.L., and Windle, A.H. (2002) Surface structure of amorphous polystyrene: comparison of SFM imaging and lattice chain structures. *Macromolecules*, **35** (13), 5283–5289.

9 Stone, V.W., Jonas, A.M., Nysten, B., and Legras, R. (1999) Roughness of free surfaces of bulk amorphous polymers as studied by X-ray surface scattering and atomic force microscopy. *Phys. Rev. B*, **60** (8), 5883–5894.

10 Scott, S. and René, M.O. (2003) Creeping friction dynamics and molecular dissipation mechanisms in glassy polymers. *Phys. Rev. Lett.*, **91** (9), 095501.

11 Singamaneni, S., LeMieux, M.C., Jiang, H., Bunning, T.J., and Tsukruk, V.V. (2007) Negative thermal expansion in ultrathin plasma polymerized films. *Chem. Mater.*, **19** (2), 129–131.

12 Yasuda, H., Hirotsu, T., and Olf, H.G. (1977) Polymerization of organic compounds in an electrodeless glow discharge. X. Internal stress in plasma polymers. *J. Appl. Polym. Sci.*, **21** (11), 3179–3184.

13 Yu, Q.S. and Yasuda, H.K. (1999) Internal stress in plasma polymer films prepared by cascade arc torch polymerization. *J. Polym. Sci. A*, **37** (11), 1577–1587.

14 Tsui, O.K.C., Wang, X.P., Jacob, Y.L., Ng, T.K., and Xiao, X. (2000) Studying surface glass-to-rubber transition using atomic force microscopic adhesion measurements. *Macromolecules*, **33** (11), 4198–4204.

15 Cappella, B., Kaliappan, S.K., and Sturm, H. (2005) Using AFM force–distance curves to study the glass-to-rubber transition of amorphous polymers and their elastic–plastic properties as a function of temperature. *Macromolecules*, **38** (5), 1874–1881.

16 Overney, R.M., Buenviaje, C., Luginbühl, R., and Dinelli, F. (2000) Glass and structural transitions measured at polymer surfaces on the nanoscale. *J. Therm. Anal. Calorim.*, **59** (1–2), 205–225.

17 Bliznyuk, V.N., Assender, H.E., and Briggs, G.A.D. (2002) Surface glass transition temperature of amorphous polymers: a new insight with SFM. *Macromolecules*, **35** (17), 6613–6622.

18 Yang, G., Rao, N., Yin, Z., and Zhu, D.M. (2006) Probing the viscoelastic response of glassy polymer films using atomic force

microscopy. *J. Colloid Interface Sci.*, **297** (1), 104–111.

19 Cappella, B. and Stark, W. (2006) Adhesion of amorphous polymers as a function of temperature probed with AFM force–distance curves. *J. Colloid Interface Sci.*, **296** (2), 507–514.

20 Fakhraai, Z. and Forrest, J.A. (2008) Measuring the surface dynamics of glassy polymers. *Science*, **319** (5863), 600–604.

21 Tocha, E., Schonherr, H., and Vancso, J.G. (2009) Surface relaxations of poly(methyl methacrylate) assessed by friction force microscopy on the nanoscale. *Soft Matter*, **5** (7), 1489–1495.

22 Bhushan, B. (2002) *Introduction to Tribology*, John Wiley and Sons, Inc., New York.

23 Bhushan, B., Israelachvili, J.N., and Landman, U. (1995) Nanotribology: friction, wear and lubrication at the atomic scale. *Nature*, **374** (6523), 607–616.

24 Zhou, J., Berry, B., Douglas, J.F., Karim, A., Snyder, C.R., and Soles, C. (2008) Nanoscale thermal–mechanical probe determination of 'softening transitions' in thin polymer films. *Nanotechnology*, **19** (49), 495703.

25 Becker, J., Grün, G., Seemann, R., Mantz, H., Jacobs, K., Mecke, K.R., and Blossey, R. (2003) Complex dewetting scenarios captured by thin-film models. *Nat. Mater.*, **2** (1), 59–63.

26 McDonald, J.C. and Whitesides, G.M. (2002) Poly(dimethylsiloxane) as a material for fabricating microfluidics devices. *Acc. Chem. Res.*, **35** (7), 491–499.

27 Tsukruk, V.V., Gorbunov, V.V., Huang, Z., and Chizhik, S.A. (2000) Dynamic microprobing of viscoelastic polymer properties. *Polym. Int.*, **49** (5), 441–444.

28 Chizhik, S.A., Gorbunov, V.V., Luzinov, I., Fuchigami, N., and Tsukruk, V.V. (2001) Surface force spectroscopy of elastomeric nanoscale films. *Macromol. Symp.*, **167** (1), 167–175.

29 Chizhik, S.A., Gorbunov, V.V., Myshkin, N., Luzinov, I., Fuchigami, N., and Tsukruk, V.V. (2000) Surface force spectroscopy of polymer nanolayers on stiff substrates. Proceedings of Composite Conference, Gomel, p. 14.

30 Achalla, P., Mc Cormick, J., Hodge, T., Moreland, C., Esnault, P., Karim, A., and Raghavan, D. (2005) Characterization of elastomeric blends by atomic force microscopy. *J. Polym. Sci. B*, **44** (3), 492–503.

31 Fritz, J.L. and Owen, M.J. (1995) Hydrophobic recovery of plasma-treated polydimethylsiloxane. *J. Adhes.*, **54** (1), 33–45.

32 Hillborg, H., Ankner, J.F., Gedde, U.W., Smith, G.D., Yasuda, H.K., and Wikstrom, K. (2000) Crosslinked polydimethylsiloxane exposed to oxygen plasma studied by neutron reflectometry and other surface specific techniques. *Polymer*, **41** (18), 6851–6863.

33 Mills, K.L., Zhu, X., Takayama, S., and Thouless, M.D. (2008) The mechanical properties of a surface-modified layer on polydimethylsiloxane. *J. Mater. Res.*, **23** (1), 37–48.

34 Singamaneni, S., Chang, S., Jang, J.H., Davis, W., Thomas, E.L., and Tsukruk, V.V. (2008) Mechanical properties of composite polymer microstructures fabricated by interference lithography. *Phys. Chem. Chem. Phys.*, **10** (28), 4093–4105.

35 Ko, H., Jiang, C., and Tsukruk, V.V. (2005) Encapsulating nanoparticle arrays into layer-by-layer multilayers by capillary transfer lithography. *Chem. Mater.*, **17** (22), 5489–5497.

36 Hua, F., Sun, Y., Gaur, A., Meitl, M.A., Bilhaut, L., Rotkina, L., Wang, J., Geil, P., Shim, M., and Rogers, J.A. (2004) Polymer imprint lithography with molecular-scale resolution. *Nano Lett.*, **4** (12), 2467–2471.

37 Peppas, N.A., Hilt, J.Z., Khademhosseini, A., and Langer, R. (2006) Hydrogels in biology and medicine: from molecular principles to bionanotechnology. *Adv. Mater.*, **18** (11), 1345–1360.

38 Bellissent-Funel, M.C. (ed.) (1998) *Hydration Processes in Biology: Theoretical and Experimental Approaches*, vol. **305**, NATO Advanced Studies Institute, Series A: Life Sciences, IOS Press, Amsterdam, Netherlands.

39 Cambell, N.A., Reece, J.B., and Mitchell, L.G. (1999) *Biology*, 5th edn, Benjamin Cummings, Inc., Menlo Park, CA.

40 Ratner, B.D., Hoffman, A.S., Schoen, F.J., and Lemons, J.E. (2004) *Biomaterials Science: An Introduction to Materials in*

Medicine, 2nd edn, Elsevier Academic, Amsterdam, Netherlands.

41 Peppas, N.A., Hilt, J.Z., Khademhosseini, A., and Langer, R. (2006) Hydrogels in biology and medicine: from molecular principles to bionanotechnology. *Adv. Mater.*, **18** (11), 1345–1360.

42 Whitesides, G.M., Ostuni, E., Takayama, S., Jiang, X., and Ingber, D.E. (2001) Soft lithography in biology and biochemistry. *Annu. Rev. Biomed. Eng.*, **3**, 335–373.

43 Osada, Y. and Khokhlov, A.R. (2002) *Polymer Gels and Networks*, CRC Press.

44 Beines, P.W., Klosterkamp, I., Bernhard, M., Jonas, U., and Knoll, W. (2007) Responsive thin hydrogel layers from photo-cross-linkable poly(*N*-isopropylacrylamide) terpolymers. *Langmuir*, **23** (4), 2231–2238.

45 Suzuki, A., Yamazaki, M., and Kobiki, Y. (1996) Direct observation of polymer gel surfaces by atomic force microscopy. *J. Chem. Phys.*, **104** (4), 1751–1757.

46 Huang, J., Cusick, B., Pietrasik, J., Wang, L., Kowalewski, T., Lin, Q., and Matyjaszewski, K. (2007) Synthesis and *in situ* atomic force microscopy characterization of temperature-responsive hydrogels based on poly(2-(dimethylamino)ethyl methacrylate) prepared by atom transfer radical polymerization. *Langmuir*, **23** (1), 241–249.

47 Singamaneni, S., McConney, M.E., and Tsukruk, V.V. (2010) Swelling induced folding in confined nanoscale responsive polymer gels. *ACS Nano*, **4** (4), 2327–2337.

48 Singamaneni, S., McConney, M.E., and Tsukruk, V.V. (2010) Spontaneous self-folding in confined ultrathin polymer gels. *Adv. Mater.*, **22** (11), 1263–1268.

49 Schmidt, T., Mönch, I.J., and Arndt, K.F. (2006) Temperature-sensitive hydrogel pattern by electron-beam lithography. *Macromol. Mater. Eng.*, **291** (7), 755–761.

50 Guvendiren, M., Yang, S., and Burdick, J.A. (2009) Swelling-induced surface patterns in hydrogels with gradient crosslinking density. *Adv. Funct. Mater.*, **19** (19), 3038–3045.

51 Matzelle, T.R., Geuskens, G., and Kruse, N. (2003) Elastic properties of poly(*N*-isopropylacrylamide) and poly (acrylamide) hydrogels studied by scanning force microscopy. *Macromolecules*, **36** (8), 2926–2931.

52 Wiedemair, J., Serpe, M.J., Kim, J., Masson, J.-F., Andrew, L., Mizaikoff, B., and Kranz, C. (2007) *In-situ* AFM studies of the phase-transition behavior of single thermoresponsive hydrogel particles. *Langmuir*, **23** (1), 130–137.

53 Takahiro, N., Haga, H., Kawabata, K., Abe, K., and Sambongi, T. (2000) Comparing microscopic with macroscopic elastic properties of polymer gel. *Ultramicroscopy*, **82** (1–4), 223–226.

54 McHenry, M.J. and Van Netten, S.M. (2007) The flexural stiffness of superficial neuromasts in the zebrafish (Danio rerio) lateral line. *J. Exp. Biol.*, **210** (23), 4244–4253.

55 Peleshanko, S., Julian, M.D., Ornatska, M., McConney, M.E., LeMieux, M.C., Chen, N., Tucker, C., Yang, Y., Liu, C., Humphrey, J.A.C., and Tsukruk, V.V. (2007) Hydrogel-encapsulated microfabricated haircells mimicking fish cupula neuromasts. *Adv. Mater.*, **19** (19), 2903–2909.

56 McConney, M.M., Anderson, K.D., Brott, L.L., Naik, R.R., and Tsukruk, V.V. (2009) Bioinspired material approaches to sensing. *Adv. Funct. Mater.*, **19** (16), 2527–2544.

57 Sperling, L.H. (1979) *Interpenetrating Polymer Networks and Its Related Materials*, Plenum Press, New York.

58 Sperling, L.H. (1997) *Polymeric Multicomponent Materials: An Introduction*, John Wiley and Sons, Inc., New York.

59 Klempner, D., Sperling, L.H., and Utracki, L.A. (1994) *Interpenetrating Polymer Networks*, vol. 239, Advances in Chemistry Series, American Chemical Society, New York.

60 Lipatov, Y.S. and Alekseeva, T.T. (2007) *Phase-Separated Interpenetrating Polymer Networks*, Springer, Berlin.

61 Kim, J.H. and Kim, S.C. (2003) Effect of PEO grafts on the surface properties of PEO-grafted PU/PS IPNs: AFM study. *Macromolecules*, **36** (8), 2867–2872.

62 Vancaeyzeele, C., Fichet, O., Amana, B., Boileau, S., and Teyssie, D. (2006) Polyisobutene/polycyclohexyl

methacrylate interpenetrating polymer networks. *Polymer*, **47** (17), 6048–6056.
63 Ogoshi, T. and Chujo, Y. (2005) Synthesis of poly(vinylidene fluoride) (PVdF)/silica hybrids having interpenetrating polymer network structure by using crystallization between PVdF chains. *J. Polym. Sci. A*, **43** (16), 3543–3550.
64 Ogoshi, T., Chujo, Y. and Esaki, A. (2005) Synthesis and characterization of UV-induced interpenetrating polymer network (IPN) structure of poly(urethane acrylate) (UA polymer)/silica hybrids. *Polym. J.*, **37** (9), 686–693.

10 Organized Polymeric Materials

Almost immediately after AFM instrumentation became available for polymer researchers, the first study published on this technique presented a detailed discussion of growth and surface morphology of PE single crystals in 1990 [1]. Since then, hundreds of publications have appeared for a wide variety of polymeric materials and structures, such as low-dimensional, ordered, liquid crystalline, or mesomorphic polymers, and large-scale, periodic, patterned polymers, or periodic porous materials. While initial studies have mostly utilized the ability of AFM instruments to provide reliable information on 3D morphology, current research, if done properly, focuses more on finding ways to measure unique properties with high resolution under real-time conditions.

10.1 Crystalline Polymers

In this section, we discuss AFM-based studies of crystalline polymers with an emphasis on polymer crystallization and melting, the growth of crystal habits and crystalline polymer fibers, intralamellar and spherulitic organization, and *in situ* real-time observation of the corresponding processes. Many outstanding results on this topic have been summarized in a recent review by Hobbs *et al.*, which focuses on using the AFM to observe the crystallization behavior of polymers [2]. The authors present a number of AFM studies on crystallization phenomenon with sub-10 nm resolution (including some results on near-molecular resolution): real-time growth of individual lamellae and their bundles, screw dislocation initiation and propagation, effects of the confined state in thin polymer films, localized nucleation and growth of crystalline phases, and remelting of individual crystalline blocks. Arguably, the most important and most frequently observed feature of all polymer crystals are the individual nanoscale building blocks, which seem to be critically important for many reorganization phenomena in crystallizable, partially ordered polymers.

Although true molecular details of crystallization and melting process have not been achieved due to rapid thermal motion and the complexity of interpenetrating lamellar growth, AFM observations not only confirmed familiar trends but also

Scanning Probe Microscopy of Soft Matter: Fundamentals and Practices, First Edition.
Vladimir V. Tsukruk and Srikanth Singamaneni.

enabled an understanding of the nanoscale-level complexity of the phenomena, which should be considered in future refinement of the theories of crystallization [3]. Because of this comprehensive review and some discussion on melting and crystallization already done in Chapter 6, this chapter will be limited to the discussion of the microstructure and the surface morphology of important polymeric materials. A number of earlier AFM studies of crystalline polymers have been summarized in an excellent review [4].

10.1.1 Polyethylene Crystals

Careful study of the surface morphology of characteristic single crystals of linear PE crystallized from solution revealed their extremely smooth surface (Figure 10.1). The typical microroughness of PE single crystals does not exceed 0.2–0.5 nm, which is consistent with the dominating crystallization mechanism being regular folding of polymer chains. However, direct evidence of the lattice being composed of regularly folded chains has been observed on a rare and inconsistent basis [5]. Friction force microscopy has been exploited by Vancso *et al.* to demonstrate distinct differences in the shearing properties of the different sectors of PE single crystals with different orientations for the folded chain domains [6]. The authors suggested that the interaction of the AFM tip with the PE surface along and across folds generates a significant difference in the shear force, which can be easily detected with conventional FFM mode.

Figure 10.1 (a) AFM topographical image of a single crystal of PE on graphite Ref. [10].

The formation of shish-kebab crystalline structures of PE was studied by Hobbs *et al.* [7]. *In situ* and real-time AFM observation of the formation of shish-kebab structures reveals the critical role played by the amorphous melt phase surrounding the nanoscale building blocks in influencing the growth and mutual organization during crystallization. The factor seems to greatly affect the growth rate of individual lamellae and causes a time-variable growth rate. In this study, the authors exploited the unique ability of AFM to concurrently observe both amorphous melt and nanoscale crystalline domains in real time in the course of ongoing crystallization without employing the usual "stop-and-go" morphological studies.

The observation of lamellar branching for PE materials of different molecular weights during spherulite formation has been conducted by Toda *et al.* [8]. The authors concluded that the pressure gradient caused by the density difference between the crystalline phase and the surrounding melt is a predominant factor in branching and the formation of characteristic banded sphrulites. Confinement of polymer melt/solution to thin films during crystallization resulted in significant changes to their lamellar morphologies from banded spherulites observed for bulk materials. Such reorganization occurs if the thickness of the films decreased below 200 nm [9]. This change of organization from banded to sheaf-like morphology was attributed to surface and substrate constraints imposed on lamellar branching during 3D growth of lamellar structures causing mostly edge orientation of lamellar structures within ultrathin polymer films.

Figure 10.1 shows a representative AFM image of pyramidal lamellar single crystals of polyethylene [10]. Careful analysis of the surface morphology of PE single crystals in the course of annealing over a wide range of temperatures from well below to well above their melting point has been reported by Magonov *et al.* [11]. The authors observed smooth surface morphology of PE single crystals with some defects and fracturing caused by their precipitation from solution and flattening of solid substrate. Moreover, annealing experiments demonstrated that local melting of the single crystal begins at a temperature that is 50 °C below the melting temperature of bulk PE, and this process is concentrated mostly at the polymer–substrate interface where some residual solvent is left. High-resolution AFM imaging (down to scan sizes of several hundred nanometers) revealed that the annealed single crystals show some nanoscale strands, which aligned mostly along the crystal edges. Such undulations, which are initially found in the central location, eventually progress through the whole crystal at higher annealing temperatures.

Such reorganization eventually causes PE chains to reorient from a vertical orientation in the initial single crystal to their flattened orientation within thin surface layers. At even higher temperatures, approaching the melting temperature, the reorganized crystal surface is represented by an array of wider molecular strands. After complete melting of PE crystals at temperatures close to 170 °C (well above the bulk melting temperature of 110 °C), residual polymer ribbons can still be observed by AFM with polymer chains aligned parallel to the substrate. The authors suggest that their spatial arrangement is dictated by the chain epitaxy on the graphite substrate rather than by the initial crystal shape. The author also noted the critical role of the substrate on the reorganization of polymer crystals during annealing. The nature of a substrate (hydrophobic versus hydrophilic, amorphous versus crystalline, or smooth versus defective) determines the initiation of localized melting and final preferential orientation of the polymer chains. For instance, a characteristic saw-toothed pattern of the crystal edges can be observed for PE crystals and controlled by the prime surface decoration with alkanes, as was discussed by Loos and Tian [12].

Finally, an interesting attempt to control and tune the crystallization of PE and the formation of nanocomposites was presented by Tracz *et al.*, who included micro- and nanoparticles in substrates and studied their effect on PE crystallization [13]. The authors observed that the type of single crystals, their orientation, and their overall arrangement are controlled by the initial nanoparticle organization, which was also concurrently characterized by AFM imaging. The size, concentration, and facets of inorganic structures deposited on the substrates led to PE crystallization with very small crystal dimensions. Larger inorganic structures prevented nucleation and played a critical role as contaminants, which disturb crystallization with large crystal habits. The inorganic structures affected the initial heterogeneous nucleation only if their dimensions become comparable to PE lamellar dimensions.

10.1.2
Polypropylene Crystals and Materials

In order to gain an understanding of the morphology of single crystals of isotactic and syndiotactic polypropylenes (iPP and sPP), which are popular high-performance polymers with highly crystalline structures, iPP and sPP have been studied with AFM (usually in conjunction with TEM and SAXS). AFM studies at different length scales showed all general features of sPP single crystals known from optical microscopy, SEM, and TEM. Dimensions and overall symmetries are confirmed by AFM observations in the same range of magnifications. Various morphological habits have been observed for different crystallization processes by Tsukruk and Reneker [14]. One observed morphology included various axialites of sPP crystals grown from an isolated nucleus, which are usually composed of three to six lath-like single crystals (see example in Figure 10.2). These laths are 20–50 μm long and 5–15 μm wide. The height of a single crystal of sPP is 10.5 $\pm$ 1 nm, which correlates fairly well with independent TEM estimates. Crystals with two or three superposed lamellae with regions 22 nm and 33 nm thick are occasionally observed as well.

The surface microroughness in an area of 100 nm^2 is 0.3–0.8 nm, indicating that these surfaces are mostly flat and molecularly smooth. The roughness in the areas of 2 μm^2 is much higher, reaching 3 nm, and is caused by larger-scale surface inhomogeneities. These larger-scale features include variations in lamellar thickness from edge to center, accumulations of polymer materials along the edges of crystals, and tiny islands of undeveloped nuclei. Multilamellar structures and irregular ridges that grew along the edges of the lath crystals with different profiles of short and long edges have also been observed with high-resolution AFM imaging. It has been suggested that such a difference might reflect a different mechanism of growth and

Figure 10.2 (*Top*) AFM topographical images of the lath-like sPP single crystals grown from an isolated nucleus at low magnifications, scan size is 45 μm × 45 μm; (*Bottom*) the edge of the single crystal with accumulated ridges, surface undulations, fractures, and undeveloped nuclei, 3.5 μm × 3.5 μm. Reproduced from Ref. [14].

material supply along the a and b edges. At the molecular scale, a regular lattice was observed with the measured parameters of the unit cell of $a = 1.5$ nm and $b = 1.1$ nm. These periods are close to the lengths of the a edge and the b edge of the sPP unit cell as determined from electron diffraction for sPP single crystals.

Lotz *et al.* published an AFM study for both types of PP materials and discussed a rich variety of crystal lattices observed by molecular-resolution AFM and compared them with suggested lattices derived from X-ray analysis [15]. The study focused on the visualization of molecular packing for different crystal habits with high-resolution AFM. On the basis of the number of images collected, the authors discussed individual helical stems of PP chains and their coexistence under certain conditions for different crystal phases. The authors indicated difficulty in the interpretation of the molecular AFM images beyond the deduction of the usual average crystallographic parameters, which are well known from X-ray data.

Recent developments in AFM imaging, such as AFM nanotomography, which includes step-by-step wet etching ($KMnO_4$ in H_2SO_4) of a polymer material and AFM imaging of each slice, allows for the integration of these slices in a combined 3D image. Rehse *et al.* applied this approach and showed that the 3D PP crystal structure can be restored with 10 nm resolution [16]. The authors studied elastomeric PP, which is composed of iPP and sPP blocks, and hence, has a mixed two-phase microstructure with interconnected amorphous and crystalline phases. 3D AFM layer-by-layer reconstruction conducted to maintain exact location of the scanning area revealed a random, dendritic-like inner structure with lamellae mostly oriented perpendicular to the substrate. The analysis of the crystalline habits showed that mixed crystalline structures with different mixed crystal phases are embedded in an amorphous matrix.

Another AFM study of sPP crystals conducted by Thomann *et al.* showed bundles of needle-like and rectangular crystals with their composition controlled by the crystallization temperature [17]. They observed that sPP formed only large rectangular single crystals unlike iPP, which can form 3D spherulites. Lamellar thickness of 2D structures determined from AFM was close to that known from SAXS. AFM showed that sPP single crystals displayed transverse straight fractures along with irregular mosaic fractures that were related to their fragmented crystallization. Indeed, at a micron scale, the surface of sPP crystals possesses a clear periodic pattern that includes many periods of undulations and straight microcracks running across the total width of the single crystals (Figure 10.2) [18]. The long dimension of each undulation is transverse to the *b*-axis of the crystalline lattice and three to four cracks are observed along the length of lath crystals. The formation of such a characteristic periodic fracturing pattern was attributed to the stress relaxation of sPP crystals during their growth on a solid substrate due to thermally induced interfacial stresses caused by a significant difference in the thermal expansion of the polymer material and the substrate.

Finally, as has been observed by Schonherr *et al.* in AFM morphological studies, elastomeric PP with low crystallinity from PP copolymers showed all of the characteristic features of crystalline PP phases, including lath crystals, lamellar crystals, dendritic crystals, and various spherulites embedded within the amorphous matrix, with the noticeable absence of some other crystal morphologies observed for more highly crystalline PP materials [19]. Similarly, a variety of mixed morphological forms were observed for miscible propylene/ethylene copolymers of different compositions [20]. The presence of several crystal phases separated by characteristic defects was observed by AFM and was considered to be responsible for the multiple melting peaks observed on DSC plots.

10.1.3 Polyethylene Oxide Crystals

Polyethylene oxide (PEO) and other polymers from this class are very popular, highly crystalline materials. These materials have a low melting temperature, are water soluble, show excellent biocompatibility, and can be easily functionalized. *In situ* crystallization of PEO materials has been studied by Beekmans *et al.* with AFM and DSC techniques [21]. PEO spherulites were grown at different temperatures and over a long period of time, while fine parameters of lamellar morphology were monitored with AFM under the variable crystallization conditions. The authors concluded that the melting of the lamellar crystal is highly inhomogeneous, starting at the more defective outer or interior faces of the crystals. This process is controlled by the segregation of mobile PEO chains during melting or crystallization. The broad melting behavior of high-molecular weight PEO materials was explained by the presence of lamellae with a wide range of thicknesses, as characterized with AFM. In another study on PEO crystals, Beekmans *et al.* observed fine supramolecular organization of different spherulites [22]. They related the elongated lamellar structures to the difference in the growth rates of the {200} and {110} facets. These differences were confirmed by direct spatial measurements of crystallites using *in situ* AFM images.

Fast (video-rate) AFM imaging of the PEO crystallization process on a timescale comparable to practical processing time of this polymeric material has been reported by Hobbs *et al.* [23]. The authors claimed that this method shows a high spatial resolution similar to conventional AFM but with 1000 times faster data collection. This quick imaging allowed for the *in situ* observation of fine details of the early stages of rapid crystallization. Significant differences in PEO crystal growth in the radial and tangential directions and the corresponding spatial constraints are considered the main reason for the elongated shape of the spherulites. The authors observed screw dislocation growth in real time with nanoscale resolution of major morphological features and concluded that under high supercooling, the local lamellar growth controls the overall growth of spherulites.

Adding an amorphous PS component with a high glass transition temperature (100 °C) to the PEO phase in diblock copolymers modified the crystallization behavior of the PEO phase, as was demonstrated by Hsiao *et al.* [24]. PEO crystals, which were crystallized from solution, were grown as thin plates sandwiched between PS microphases and were monitored with AFM, TEM, and electron diffraction (ED). These trilayer PS-PEO-PS crystals showed very smooth surface morphology with elevated edges (Figure 10.3). PEO recrystallization was also conducted under these confined conditions producing crystals without crystal defects and with sectors hosting different chain orientations. The local melting temperature was measured with modulated AFM for different crystallization conditions and revealed nonuniform melting of the crystal habit with crystal thinning mostly happening in the center with materials transferred to the edges of the annealed crystal (Figure 10.3).

Adding another amorphous polymer, PMMA, in order to create amorphous crystalline polymer blends was conducted by Zhu and Wang [25]. Characteristic

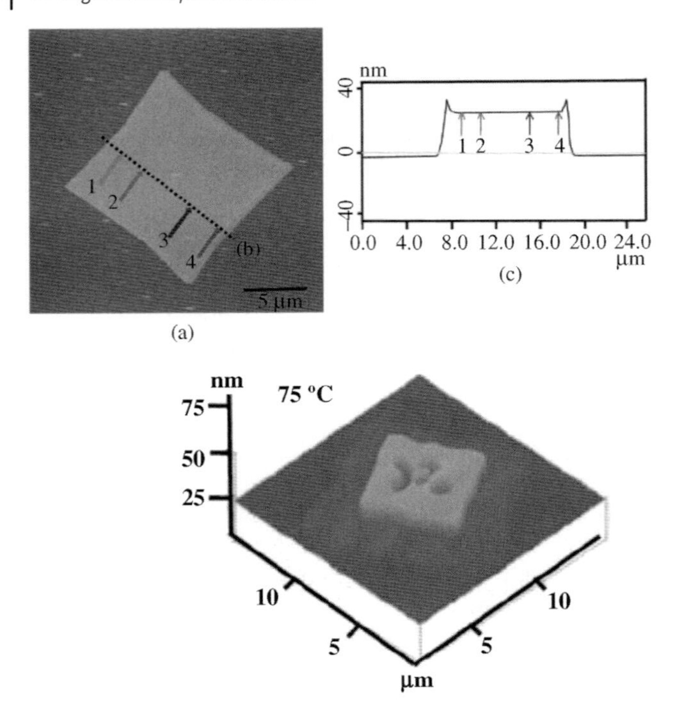

Figure 10.3 (*Top*) AFM image of a PEO-*b*-PS single crystal (a), a measurement of average thickness of the PEO-*b*-PS single crystal (b), and a height profile of the single crystal (c). (*Bottom*) *In situ* topographic change of a PEO-*b*-PS single crystal at elevated temperature close to its melting. Reproduced from Ref. [24].

dendritic growth of the PEO crystal phase included within the amorphous PMMA matrix has been observed at intermediate length scale with AFM imaging. This crystallization behavior was observed under low supercooling and was related to the anisotropic rate of growth of lamellar structures. The dendritic pattern within microscopic regions converts to a less-defined seaweed pattern at increasing undercooling conditions or under high surface coverage of PEO crystals. This reorganization was explained by the confinement of crystallizable polymer chains in ultrathin polymer films and within the PMMA dispersed PEO regions.

A similar surface morphology of dendritic type was observed for the PEO-PS star block copolymers with the high PEO content but with the lowest number of arms in the LB monolayer by Peleshanko and coworkers [26]. This block copolymer formed well-defined dendritic structures at all surface pressures similar to those observed for the ultrathin PEO phases discussed above (Figure 10.4). The height and width of the dendritic structures were very similar suggesting that a rise in the surface pressure caused uniform growth of different crystal facets with different arrangements of polymer chains covalently grafted to PS blocks. The higher-resolution AFM imaging revealed internal domain structures indicating that the dendrites are assembled by

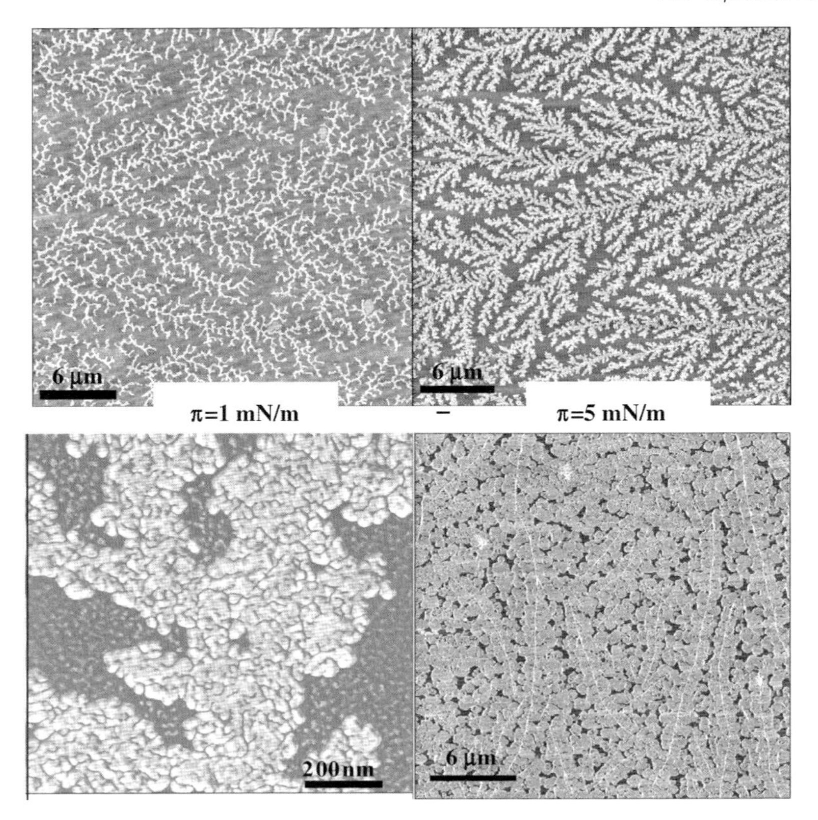

Figure 10.4 (*Top*) AFM topographical images of dendritic supramolecular nanostructures in LB monolayers from PEO-PS star block copolymer at 1 mN/m (*left*) and 5 mN/m (*right*), height is 15 nm for all images. (*Bottom*) High-resolution AFM images of dendritic structures in LB monolayers from PEO-PS star block copolymer at 5 mN/m. Height (*left*) is 15 nm, phase (*right*) is 25°. Reproduced from Ref. [26].

the aggregation of fine circular domains. These nanoscale domains merge into larger crystalline aggregates upon further compression of PEO-PS monolayers (Figure 10.4).

Furthermore, at the highest surface pressure, the dense dendritic structures were merged into a continuous fine texture (Figure 10.4). For this surface morphology, remnants of the merging and compressed branched structures are still clearly evident. The overall shapes of the compressed structures followed the initial highly branched dendritic morphology with branches squashed perpendicularly to the dipping direction. The appearance of straight ridges along the former dendrite backbones indicated folded domain boundaries. Therefore, the longer PEO arms were capable of crystallizing at the interface despite the spatial constraints imposed by the star architecture and additional vertical confinement under amorphous PS aggregates.

10.1.4 Poly-ε-Caprolactone Crystals

The morphology and microstructure of poly-ε-caprolactone (PCL) materials attracted attention due to their unique biocompatible and biodegradable properties. Moreover, it is known that the biodegradable properties of PCL are, to a large extent, controlled by the crystal structure, which can be easily characterized by AFM imaging.

Real-time PCL crystallization was observed with *in situ* AFM imaging at various temperatures with a hot stage by Beekmans and Vancso [27]. The growth of truncated lozenge single crystals from melt was observed in this study and was attributed to the anisotropic growth rate of different PCL facets with preferential growth of {110} facets. The presence of the residual solvent, which is common in thin film preparation, affected the growth mode of PCL crystals as has been observed by Mareau and Prud'homme [28]. The resulting higher molecular mobility led to dendritic growth with decreased lateral dimensions of crystals. Moreover, under certain conditions and depending upon the nature of the substrate, a severe dewetting of the PCL material has been observed on a large scale at elevated temperatures. This dewetting resulted in the formation of a characteristic spinodal decomposition pattern with long-term preservation of the amorphous phase between dendritic crystals due to spatial confinements.

Both linear and star PCL materials have been studied by Nunez *et al.* in order to observe the fine features of their crystallization behavior under different conditions [29]. Striated fold surface morphology was observed with AFM for both solution- and melt-grown crystals with a smoother morphology being observed for star PCL molecules. Sectorial dependence of the melting process with constant growth rate under different supercooling conditions was observed for crystals from linear PCL molecules. The authors observed that the anisotropic crystal growth and the formation of screw dislocations under surface confinements caused a sharp transition from edge-on to flat-on orientation of lamellar structures for ultrathin PCL films.

The crystallization morphology of PCL materials was modified by cocrystallization in physical blends with a PEO phase, as was observed by Hou *et al.* [30]. The characteristic morphology of incompatible polymer blends with nanoscale-size phase-separated domains has been observed for different compositions. In the course of crystallization, the individual morphology develops in microscopic phase-separated regions. Nonuniform melting of different crystal morphologies has been observed for these blends with the shortest lamellae melting at the lowest temperature. The authors observed that highly localized melting of the PCL phase started at crystal defects that were induced by confined spatial conditions.

Oriented crystallization of PCL thin films with the formation of the extended PCL crystals and complex shish-kebab structures has been studied by Fujita *et al.* [31]. Local mechanical stresses were introduced by the AFM tip, and a high localized load (with rsp values ranging from 0.03 to 0.7 corresponding to normal loads of $\sim$0.15 nN to 0.036 nN) was exploited to nucleate crystallization and initiate a highly oriented, crystallized PCL phase. The authors observed that the rate of growth of mechanically induced PCL crystals is comparable to the rate of growth of regular PCL spherulites.

The authors also demonstrated the ability to grow highly oriented PCL crystal morphologies of different types by applying local stresses. Moreover, the tip-induced crystallization favors extended PCL chains to become anisotropic nuclei only if the stress exceeds a certain threshold value.

10.1.5
Polylactic Acid Crystals

Another important biomedical polymer, polylactic acid (PLA), possesses biocompatible properties, rapid biodegradability, and good mechanical strength. In these materials, amorphous regions are predominantly hydrolizable, and thus, the overall behavior critically depends upon crystallization processes and the resulting phase structure. A recent AFM study done by Fujita and Doi focused on the *in situ* observation of crystallization and annealing behavior of PLA crystals. They observed both hexagonal and truncated lozenge shapes, folded chain lamellae thicknesses of nearly 12 nm, and the occasional spiral overgrowths caused by inner stresses due to adhesion to mica substrates [32]. Morphological changes during thermal treatment were observed with *in situ* imaging. Annealing was initiated at the defects on crystal structures, which are observed mostly along the edges and can serve as centers for enzyme degradation initiation. Local thickening of the lamellar structure was observed, and these regions remained more stable during crystal melting.

Yuryev *et al.* studied several different types of PLAs (L and D) and their blends and measured localized crystalline phase growth under different supercooling conditions [33]. Spherulitic growth was observed to be very different for PLAs of different chiralities with the characteristic maximum crystallization rate observed at intermediate supercooling. Various crystal shapes were observed that include either flat-on or edge-on stacks of lamellar structures, as resolved by high-resolution AFM. The formation of stereocomplexes on L/D mixtures of PLA was observed to be the favorable α-crystal structure under certain crystallization conditions. The formation of rod-like and spherical aggregates was observed for annealed spin-cast films of PLA by Li *et al.* with nucleation sites mostly located along the edges of the holes in partially dewetted polymer films [34]. Dominance of rod-like crystal aggregates during annealing was detected and explained by the ongoing dewetting contribution on unfavorable surfaces. The orientation of PLA lamellae and overall roughness of partially-crystalline films was observed to be strongly affected by both the evaporation rate and the quality of the solvent used for spin casting.

10.1.6
Crystalline Block Copolymers

A significant portion of block copolymer materials exhibiting unique mechanical properties are comprised of immiscible blocks with one or more of them being crystallizable. Here, we will summarize a few examples of AFM studies on crystallizable block copolymers with more discussion on block copolymeric materials presented in Chapter 11.

Real-time crystallization of two different diblock copolymers containing hydrogenated polybutadiene (PB) and PS-*r*-PB blocks (E-SEB) have been studied with AFM by Hobbs and Register [35]. They observed that the high contrast in AFM phase imaging is mostly caused by the difference in responses of the melt of crystallizable regions to the localized pressure applied by the AFM tip. Moreover, the characteristic mesophase morphology was completely disturbed by the crystallization of E blocks with the formation of highly branched structures observed mostly on the polymer surface. The rate of crystal growth differs significantly among different crystalline domains with the highest rate of growth observed for E-rich domains. *In situ* AFM observations during gradual heating of this block copolymer material resulted in highly localized melting processes, which lead to significant fragmentation of the initially uniform crystalline structures.

Block copolymers based upon popular crystallizable blocks, PEO, PLA, and PCL (discussed above), show interesting surface morphologies, which are caused by interference between different crystalline morphologies of two different blocks. The growth of different characteristic dendritic morphologies of PEO and PLA phases was observed for thin and ultrathin films from PEO-PLA diblock copolymers by Huang *et al.* [36]. They observed that the formation of the dendritic structures occurs in three-step processes with the final morphologies controlled by the total thickness of the films. AFM observations showed surface morphology, which is represented by hierarchical dendritic structures for PEO phase and hexagonal or lozenge-shaped single crystals of predominantly the PLA phase. The decrease in the growth rate of the PLA phase for ultrathin films was considered to be a key factor in causing the significant change in the overall morphology for these block copolymers.

Another type of block copolymers, triblocks, of PCL-PEO-PCL type were studied by Kang and Beers [37]. The authors observed that the initial cylindrical morphology of cast films reorganized to lamellar structures both during the course of heating and crystallization of PEO blocks and after increasing PEO chain mobility by treatment in water vapor. PEO block crystallization within LB films of PEO-polyisobutylene (PIB) block copolymers resulted in a fine dendritic morphology of the PEO phase, which is similar to that observed for spin-cast films and/or rounded multilayered stacks of the amorphous (mesophase) PIB phase [38]. AFM revealed that iron oxide nanoparticles added to the subphase were finely dispersed within the film without significant aggregation at modest surface pressure. However, these nanoparticles mostly occupy selected surface regions at higher compressions.

Synthesis of block copolymers of P3HT-PS allowed for the observation of how the fine morphology of block copolymers can be utilized to control conductive properties of polymeric materials [39]. The authors demonstrated that AFM imaging under hard tapping conditions of cast films reveals very fine and unique nanofibrillar morphology with interwoven nanofibers (Figure 10.5). This peculiar morphology is revealed only under high loading forces and not under light tapping, which probes mostly surface properties. On the other hand, the topmost surface layer showed more uniform morphology that was related to a PS "coating" of the nanofibers (Figure 10.5). This difference in appearance of the nanofibrillar morphology was related to the characteristic core–shell nanostructure caused by initial composition that provides

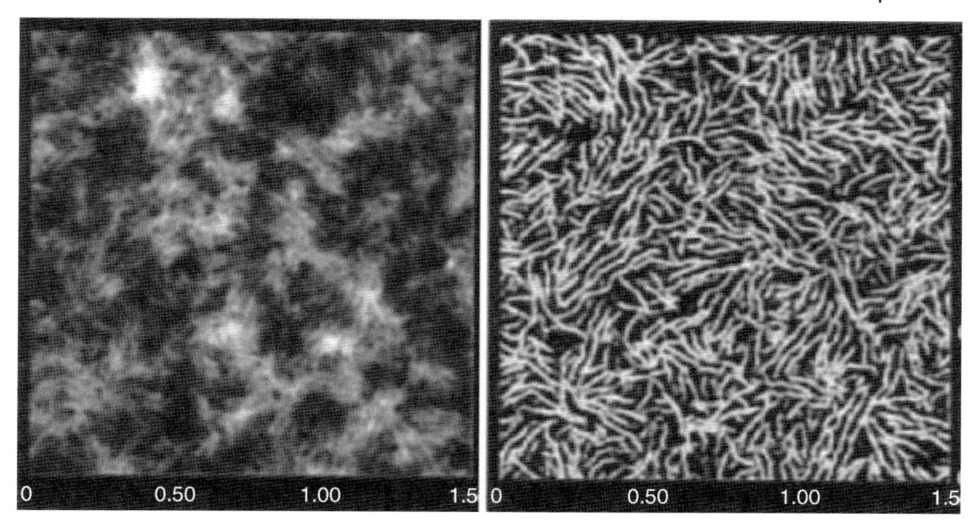

Figure 10.5 Nanofibrillar morphology in P3HT-PS copolymers. (*left*) Height image; (*right*) phase image at intermediate tapping. Reproduced from Ref. [39].

the possibility for having shelled conducting cores with π-stacking of thiophene segments.

10.1.7 Other Polymer Crystals

AFM studies of the lamellar structure of isotactic PS crystals showed sectioning of hexagonal crystals with clear cracks separating different sections and surface stripes perpendicular to the growth facet, which were related to buckling instabilities caused by the internal stresses developed during crystallization [40]. Characteristic surface undulations, which were developed perpendicular to the facets' orientation, were observed with AFM and their periodicity was measured at different supercooling conditions. Finally, on the basis of these measurements, conditions for the development of critical stresses required to lead to the generation of buckling instabilities were estimated by the authors.

Isotactic polymethamethacryles (itPMMAs) capable of crystallization (in contrast to regular amorphous PMMA) under certain crystallization conditions were studied with high-resolution AFM in their monolayer form after transfer to a mica surface from a Langmuir trough by Kumaki *et al.* [41]. The AFM imaging revealed individual random-coil macromolecular chains (amorphous phase) at low compression of the monolayers. However, at higher compression at the air–water interface, the random coils formed folded crystals with itPMMA chains lying flat-on solid substrate and the average thickness of lamellae of 17 nm. In addition to regularly folded chains, tie segments, loose loops, and chain-end defects were observed for different 2D lamellar structures. Individual chain conformation was not resolved, but the spacing of chains in lamellar regions of 1.29 nm was assigned to the exposed *b*-facets of dense crystal packing of itPMMA chains in their double-stranded helical state, the state of the bulk crystalline material [42].

Block copolymers with conjugated blocks are of great interest due to their ability to form large-scale morphologies with potentially well-controlled conductive properties [43, 44]. One recent example of the formation of nanostructures was demonstrated for diblock copolymers P3HT-ethylenhexylthiphenes (3EHT) by Zhang *et al.* [45]. A well-defined worm-like microphase separation pattern was observed for thin polymer films unlike random copolymers with featureless morphology. Annealing at high temperatures resulted in the crystallization of different blocks in different crystal unit cells within fiber-like morphologies of two blocks as confirmed by X-ray diffraction.

Rigid, rod-like polymers with conjugated backbones are capable of crystallization and their crystal structure critically affects their conductive properties. LB films from either conjugated ladder polymers, complexes of poly(naphthoylene benzimidazole) (PNBI) precursor, and stearic acid, or thermocyclized LB films of PNBI possessed a relatively flat surface morphology [46]. The microroughness (in an area of $1\,\mu m^2$) decreases steadily from 1.5 nm for the first layer to 0.5 nm for the LB film containing five layers. The major defects are pinholes 0.7 nm deep and 80 ± 20 nm in diameter. This depth corresponds quite well with the average thickness of a single molecular layer derived from X-ray data.

Needle-like single crystals of poly-3-octylthiophene (P3OT), a popular conducting polymer widely exploited for organic electronic devices, have been characterized with AFM in combination with optical microscopy, SEM, TEM, and X-ray diffraction by Xiao *et al.* [47]. The thickness of these crystals exceeded 20 nm with a gradual decrease observed toward the tips of the crystals. Electron diffraction revealed that the side alkyl chains of thiophene backbones are oriented along the normal to the substrate, and thus, backbones are arranged in the flat-on manner on the surface. Chains are oriented along the long axis of the crystals, which, in the end, control the measured conductivity of these crystals. As observed with AFM, annealing and solvent treatments significantly changed the crystal morphology with the formation of more uniform and larger single crystals as a result of dissipation of small crystals and high mobility of polymer chains in the presence of the residual solvent, which promoted intense Ostwald ripening processes.

Partially crystalline polyurethanes (PU) are very important and widely utilized commercial polymeric materials with a unique soft–hard domain morphology and excellent elastomeric mechanical properties such as strength and elongation to break combined with good biocompatibility and stability under physiological conditions [48]. Depending upon the nature and length of soft and hard segments, many PU materials show a significant crystallization ability that dramatically affects their surface and mechanical properties [49, 50]. In one of the earlier AFM-X-ray combination studies, regular spherulitic surface morphology was observed for polyester segmented PUs with this morphology becoming more refined after thermal annealing [51].

Segmented PU materials, composed of different soft segments and identical hard segments, have been synthesized and studied by Hernandez *et al.* [52]. For PDMSpolyhexamethyleneoxide (PHMO) PU materials, a complex three-phase, core–shell morphology has been observed in contrast to a single soft component PU with the conventional two-phase structure. The authors applied both light and hard tapping to show that for these materials soft domains are usually located at the surface of films and hard domains form dispersed subsurface structures with nanoscale dimensions. Various segmented PUs based upon PCL and polytetrahydrofuranes (PTHF) with biodegradable properties showed a variety of surface morphologies with an interconnected network of hard domains [53]. Nanoscale dimensions of the hard domains depend upon the chemical composition, with surface coverage and interconnectivity significantly increasing with the increase in their volume fraction.

Nanodomain and nanoporous morphologies of important ionic materials from ionomeric polymers have recently been studied [54, 55]. The authors conducted AFM studies of widely exploited commercial materials: Nafion membranes, Surlyn membranes, and other ionomers. High-resolution imaging revealed nanoscale ion clusters with concurrently visible crystal regions and aggregated domains with crystalline lamellae oriented along the normal to the surface. Moreover, a fluoro-containing barrier surface layer was detected with AFM, which extends across the entire surface of these materials and defines the barrier properties of these membranes along with a nanoporous morphology.

10.2 Liquid Crystalline Polymeric Materials

Liquid crystalline polymers (LCP) with side-chain or backbone mesogenic groups are important materials with a wide use for structural components with high strength. LCPs have been widely studied with AFM to elucidate their ordered topology, defect distribution, supramolecular organization, and their highly oriented morphologies under electromagnetic fields or during their fiber processing [56–58].

High-performance fibers from LCPs spun from both solutions and melt have been studied by Gould *et al.* [59]. In this study of LCP materials with AFM, films from Vectra polymers were imaged at different spatial scales revealing the presence of fibrillar morphologies of several microns in diameter with stacking defects and disordered regions. Higher-resolution AFM images showed that these microfibrils of LCPs were formed with platelet-like or band-like structures with submicron dimensions. The presence of large internal stresses during film formation is considered the key factor in the formation of disperse fibrillar and microfibrillar morphologies and their bundles.

Very common band structures for LCPs observed in earlier studies have been discussed in detail in a number of recent publications demonstrating high-quality AFM images [60–62]. Ge *et al.* considered main-chain LCPs with flexible side chains, abbreviated as PEFB, and their behavior under shear strain and during crystallization from an LC state [60]. A primary band structure of about 3 μm wide was observed after shearing at high temperatures with polymer chains oriented along the shearing direction. The formation of this band structure, common for oriented LCPs, was related to the periodic modulation of birefringence due to the variation of film thickness. Additional annealing at 105 °C resulted in some crystallization and formation of secondary fibrillar bands of 250 nm wide with a zigzag arrangement, as visualized by AFM. These bands are composed from aggregated lamellae with a thickness of 20 nm in an edge-on orientation. It has been suggested that this structure represents periodic buckling instability pattern with correlated tilt of polymer chains that originates from the thermally induced expansion of the lattice dimensions along the chain direction during the transition from a nematic to a crystalline phase.

A similar microband structure was observed for polyaryletherketones (PEK) copolymers after high-temperature annealing, which led to the formation of uniform, monodomain smectic LC phases [61]. Even more complex helical morphologies were observed for different chiral phases of main-chain LCPs with backbone chiral centers by Weng *et al.* [63]. Twisted aggregates of lamellar structures with twisting either within a single lamella or between two lamellae were observed not only in the LC state but also after crystallization of LCPs. Spontaneous flat-on and edge-orientation of lamellar structures was observed with combined AFM, X-ray, and electron diffraction for drop-cast films of main-chain columnar LCs of polypropylsiloxanes (PDPS) by Defaux *et al.* [64].

Supramolecular mesoscopic structural defects of LCPs were studied with AFM for side-chain LCP films by Zhang *et al.* [65]. AFM imaging shows that the transition to the smectic LC phase of thin films leads to the formation of 2D nanostripes and the appearance of a variety of topological defects such as disclinations and inversed walls due to confined film conditions. Nanostripes are caused by a mechanical instability governed by molecular alignment controlled by substrate and the film–air interface. The cores of all microstructures, including radial, spiral, and circular positive and negative disclinations, were observed to show depletion in topological images that indicates the reorientation of the director. These defects were suggested to act as stabilizers for the surface topology. The analysis of the director distribution in the regions of these defects allows for the estimation of bend and splay elastic constants of LCPs from AFM imaging.

The formation of lamellar morphologies and disc-like structures has been observed during thermotropic transitions in flexible polysiloxanes within thin films by Magonov *et al.* [66]. Polymerized diskotic LCs showed homeotropic orientation of hexagonal columnar mesophases with the AFM-determined lattice parameters similar to those known for the bulk materials from X-ray diffraction [67]. In another study, AFM imaging revealed significant changes in the homeotropic orientation of ferroelectric LC polymers at variable annealing temperatures with residual layered structures remaining even after the transition to the isotropic state at elevated temperatures [68].

The attachment of the mesogenic groups to flexible dendritic cores creates novel LC structures with unusual morphologies. The attachment of mesogenic groups to PAMAM dendritic cores forces the molecules to adopt a cylindrical shape or lamellae structures depending upon the generation number of the dendritic cores [69, 70]. The inclusion of mesogenic groups at the periphery forced a transition to a disk shape, facilitating a columnar ordering [69]. Similarly, the grafting of mesogenic groups to a PAMAM core produced a transition from smectic ordering to columnar ordering as the molecular cross section increased as observed by near-molecular AFM resolution [71]. A fifth-generation carbosilane dendrimer with 128 cyanobiphenyl groups was shown to have significant peculiarities in the LC behavior upon heating unlike lower-generation dendrimers of the same type [72, 73]. The first four generations were observed to form conventional lamellar (smectic A and C) mesophases, while the fifth generation transitioned from lamellar ordering at 40 °C to hexagonal ordering of rounded columns at 130 °C with the formation of elliptical columns at intermediate temperatures. Surface studies of the fifth-generation dendrimers with LC phases showed that the film thickness and the substrate controlled the molecular packing with mixed edge-on and flat-on orientations of smectic layers [74].

The molecular packing in the LC carbosilane dendrimers with 128 polar cyanbiphenyl groups and polar butoxyphenylbenzoate groups has been studied with AFM by Genson *et al.* (Figure 10.6) [75]. In this study, it was observed that dendrimers with

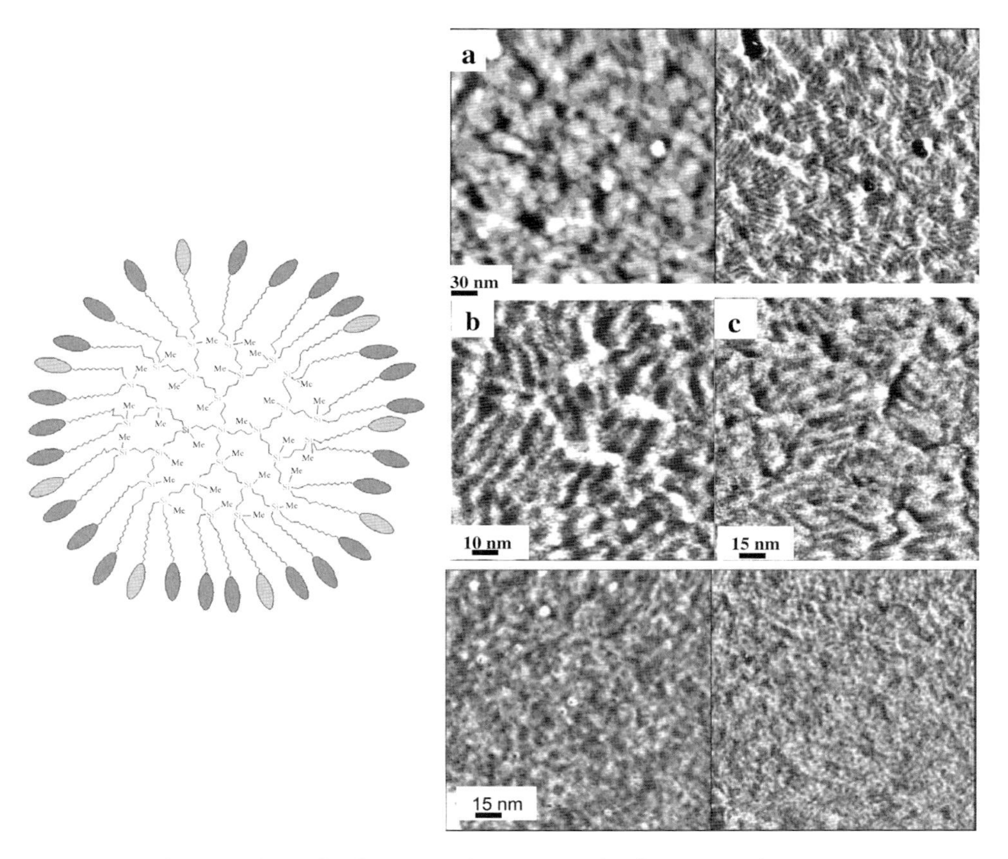

Figure 10.6 Schematic of LC carbosilane dendrimer. The grain nanostructure with internal lamellae of 2.2 nm surface films is discerned from the high-resolution AFM image (a) within condensed area, (b) the uniform monolayer, (c) and (d) phase images demonstrating the short-range ordering observed at highest resolution. (*Bottom*) Surface films on a hydrophobic surface: (b) finer grainy texture with circular molecular shapes. Topography image (*left*) and phase image (*right*). Reproduced from Ref. [75].

128 butoxyphenylbenzoate terminal groups and hydrophobic tails form organized undulated layering caused by phase separation of the flexible cores and mesogenic groups and the strong trend of the latter to form bilayer packing. Complete disappearance of the regular layer ordering in molecular surface layers was observed if butoxyphenylbenzoate terminal groups are replaced with shorter and more polar cyanbiphenyl groups despite the fact that these molecules tend to form smectic phases in the bulk state. The replacement of the hydrophilic silicon surface with hydrophobic substrates prevents the formation of dense surface films for all dendrimers studied due to antagonistic interactions between polar terminal groups and the methyl-terminated surface of the hydrophobic substrate.

A peculiar microscopic surface texture with regularly spaced circular surface areas with elevated heights (0.3 nm above the surrounding film) and an average diameter of 350 nm was observed on the monolayer surface of LC dendrimers (Figure 10.6). Closer AFM examination of the surface film surrounding these areas revealed an internal lamellae structure with poor, but still visible, periodicity limited only to a very few adjacent layers (Figure 10.6). The interlamellar spacings calculated from cross sections of the AFM images fell between 5.4 and 5.9 nm for ultrathin films prepared under different conditions. High-resolution AFM imaging revealed that the surface areas of condensed packing had a similar nanostructure with internal lamellae as the more loosely packed monolayer surrounding them. The grain nanostructures are packed with 3–6 stacks of correlated lamellae with a spacing of 5.4–5.9 nm. An abrupt change in orientation was observed for lamellar stacks with correlated defects propagating across multiple lamellae (Figure 10.6). The in-plane lamellar structure observed with d-spacing equal to approximately half the molecular diameter can be formed if the laterally compressed molecules are staggered in an alternating manner. The layered structure with these molecular dimensions suggests a highly compressed dendrimer core and layered packing of polar terminal groups in close contact with the hydrophilic silicon surface. Very flat arrangement of dendrimer molecules makes them incommensurate with the lamellar spacing and structure typically observed in the LC bulk state.

Recently, it has been demonstrated that carbon nanotubes, like other anisotropic and one-dimensional molecules with certain rigidity, might form a lyotropic LC phase [76]. Above a critical concentration, carbon nanotubes have shown a phase transition to the nematic LC phase. The full exploitation of these phenomena in the high-performance thin film transistors requires long-range order with low misaligned defects within the surface layer. A tilted drop casting of carbon nanotube solution on functionalized micropatterned geometries provided a means for the carbon nanotube trapping at the liquid–solid–air interface, causing the concentrated solution to form a nematic phase (Figure 10.7) [77, 78]. The confined geometry of the micropatterned surface induces a uniform long-range orientation of densely packed carbon nanotubes during the deposition between two electrodes. The AFM image of the resulting monolayer film revealed that the microstructure is similar to nematic-type ordering with typical topological defects observed for LC carbon nanotube solution (Figure 10.7) [76, 79].

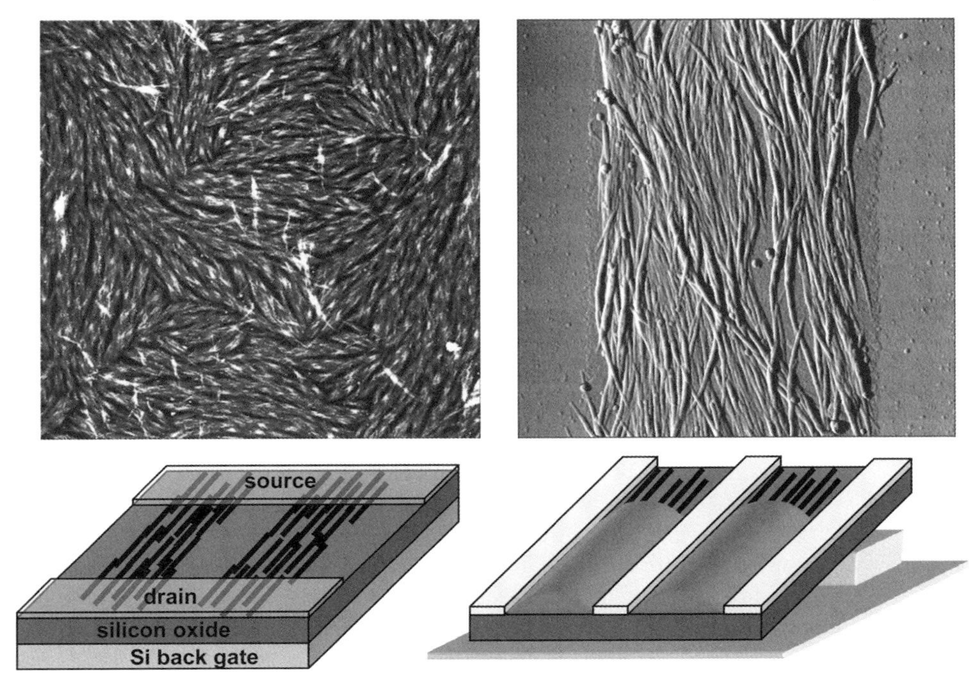

Figure 10.7 (*Bottom*) Schematics of formation of oriented stripes of carbon nanoptubes on patterned surfaces. (*Top*) AFM phase images of characteristic LC defects in oriented carbon nanotubes (*left*) and nematic LC structures of carbon nanotubes (*right*). Reproduced from Ref. [78].

10.3
Periodic Polymeric Structures

In this section, we will consider periodic polymeric structures with nanoscale and microscopic spacing that are either microfabricated or assembled from various polymeric biological and inorganic materials. Periodic materials with two- and three-dimensional periodic organization are achieved by employing various technologies such as classical microprinting, microparticle self-organization, photolithography, holographic lithography, selective chemical etching, ink printing, or laser-based polymerization [80–86]. Soft lithography, such as microcontact printing has been widely employed for the fabrication of polymer structures in few steps, with a submicron resolution, and at a low cost [87–89]. Examples of micropatterned assemblies of nanoparticles, microchannels, or microparticles have already been reported [90–93]. A variety of functional materials have been utilized for prospective applications such as antiwetting polymer coatings, sensitive films for solar cells, fuel cells, ultrastrong nanomaterials, microcapsules, and membranes for controlled drug release [94–97]. Microstructural arrays based upon the principles of either diffraction or refraction with imprinted microscopic modulations are extensively used as optical components such as gratings, beam splitters, microlenses, displays, and mirrors [98–100]. Recent interesting results have been published for interference lithographical 3D structures with complex topology, replicated metallic nanostructures in contact with biological structures [101–103].

AFM is the most popular tool for quick and reliable characterization of periodic materials with numerous publications providing regular topographical and phase images, 3D topography, cross sections, surface histograms of heights, or Fourier-transforms for dimension evaluation. Generally, AFM imaging of periodic structures is relatively easy and reliable considering the modest requirements for the resolution and robustness of the structures studied. Common problems are accurate imaging of periodic structures with large variations in elevations, the presence of the deep holes in 3D structures that interfere with stable scanning, and imaging large areas composed of alternating stiff and compliant materials. Here, we do not aim on reviewing numerous routine AFM images but rather present a few representative examples from the authors' own studies, and some typical AFM utilization for the imaging and probing complex periodic structures.

Highly elastomeric and compliant PDMS materials have proven to be an outstanding material for micro- and nanotechnology as a virtually universal microstamp for microprinting technology [104]. For example, PDMS has been employed to make microfluidic devices such as pumps, valves, channels, and cell culture systems [105, 106]. PDMS can be deformed reversibly and repeatedly without residual distortion and is thermally stable, inexpensive, nontoxic, and commercially available. Although PDMS can be potentially cross-linked using light [107], it is not generally used as a photoresist due to the inconvenient processing conditions [108, 109]. 2D periodic structures patterned by photolithography and then subsequently replicated in PDMS have been demonstrated as deformable optical and acoustic components such as lenses, waveguides, and couplers [110, 111]. The relatively low modulus of PDMS ($\sim$2 MPa) can lead to distortions such as feature–feature pairing and feature sagging in conventional microcontact printing of surface features with high aspect ratio [112–114]. In 3D structures fabricated by microcontact printing or replica molding, which can possess residual physical stresses from the molding procedure, collapse of anisotropic structures begins to occur at an aspect ratio of around 2.

Complex PDMS structures can be made by a layer-by-layer approach, which is time consuming and requires registration of subsequent layers [115]. The 3D PDMS structure can be obtained easily as a negative replica of ITL structures as shown by Jang *et al.* [116]. The basic motif is comprised of a vertical post 1100 nm in length and 500 nm in diameter with three shorter struts directed outward from the post that follows the theoretical light intensity distribution (Figure 10.8). Despite aspect ratios of about 2, the structures do not collapse in the dry state during AFM scanning. This stability can be attributed to both the interconnected nature of the structure and a lower residual stress. The PDMS elastomeric structure shows the expected periodicity in the (0001) plane based on the ITL parameters of 980 nm that agrees with the SEM and AFM images of the (0001) plane of the experimental structure, confirming that the transfer of the light intensity pattern into PDMS occurs with high fidelity.

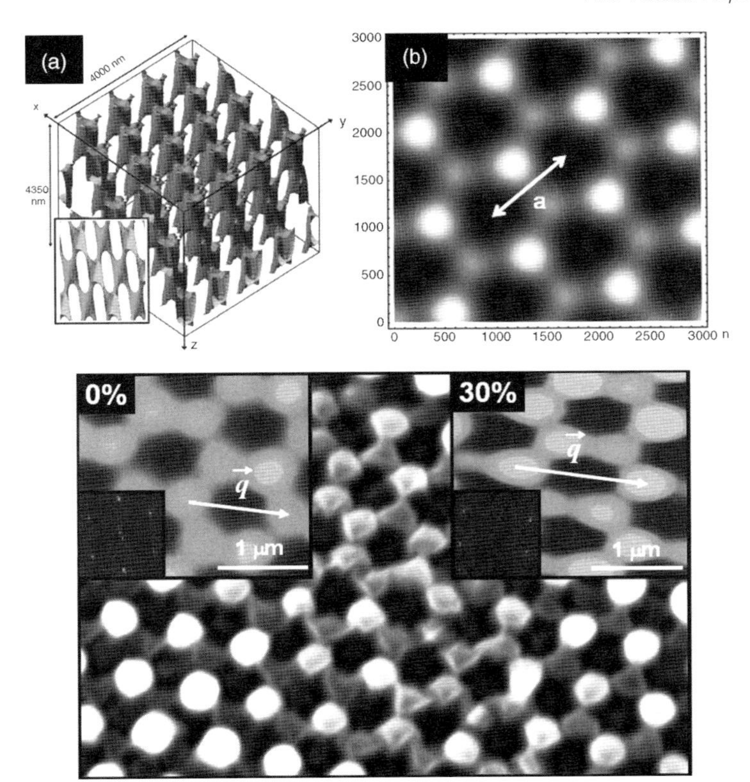

Figure 10.8 Theoretical structure of 3D PDMS ITL structures (a) and theoretical light intensity model profile interference in the (0001) plane (b). (*Bottom*) SEM image of 3D templated PDMS network/air structure having the complementary structure to original ITL structure. AFM images with 0% and 30% tensile strain along the same direction clearly show the change in lattice parameter and symmetry upon deformation. Insets are FFT of the AFM images. Reproduced from Ref. [116].

In situ monitoring of the PDMS structure stretching under tensile in-plane deformation was conducted by securing the PDMS sample in a microstretcher mounted on an AFM microscope. *In situ* AFM images obtained in the course of PDMS compression and tensile measurements show the details of the change in the lattice parameter and symmetry of the unit cells due to the 30% unidirectional strain applied along the $[10\bar{1}0]$ direction (Figure 10.8). As is clear from this image, the unit cell size along the tensile strain direction increases by 30% accompanied by the reduction of spacing in the transverse direction, thus demonstrating affine deformation of PDMS structure at microscale.

In recent study by Jiang *et al.*, LbL films of encapsulated gold nanoparticles in a poly(allylamine hydrochloride) (PAH) and poly(sodium 4-styrenesulfonate) (PSS) matrix were fabricated with spin-assisted LbL assembly combined with microprinting [117, 118]. The AFM image shows the two-phase microstructure of the film with

Figure 10.9 AFM topographical image of patterned LbL film showing the vertical regions of encapsulated gold nanoparticles (brighter stripes) alternating with all-polymeric regions along with the side view of the patterned LbL film. Reproduced from Ref. [118].

alternating regions of purely polyelectrolyte multilayers without (54 nm thick) and with (60 nm thick) gold nanoparticles (Figure 10.9). The stripes (3 μm wide) reinforced with gold nanoparticles (~4 vol%) are separated by 7 μm wide regions making the overall periodicity to be 10 μm as predetermined by the original silicon master and resulting PDMS stamp utilized for gold nanoparticle transfer. The hypothesis in this study was that the presence/absence of gold nanoparticles in the selected regions resulting in different elastic properties could significantly affect their mechanical properties leading to complex buckling instability pattern under compressive stresses.

In fact, a gentle compression (~0.2% strain) of the LbL film along stripes resulted in an immediate appearance of peculiar transversal periodic buckling patterns strictly confined to stiffer gold-containing regions with very few of them extending across all-polymeric regions, as observed in both optical micrographs and AFM images (Figure 10.10) [118]. AFM images demonstrate that these wrinkles indeed represent modulated surface topography with amplitude of 80 nm (Figure 10.10). The spacing of these wrinkles determined from optical micrographs was 2.9 ± 0.3 μm, which is consistent with the spacing obtained from AFM images. The spacing quoted here was obtained from 1D and 2D FFT data that provided the parameters averaged over a

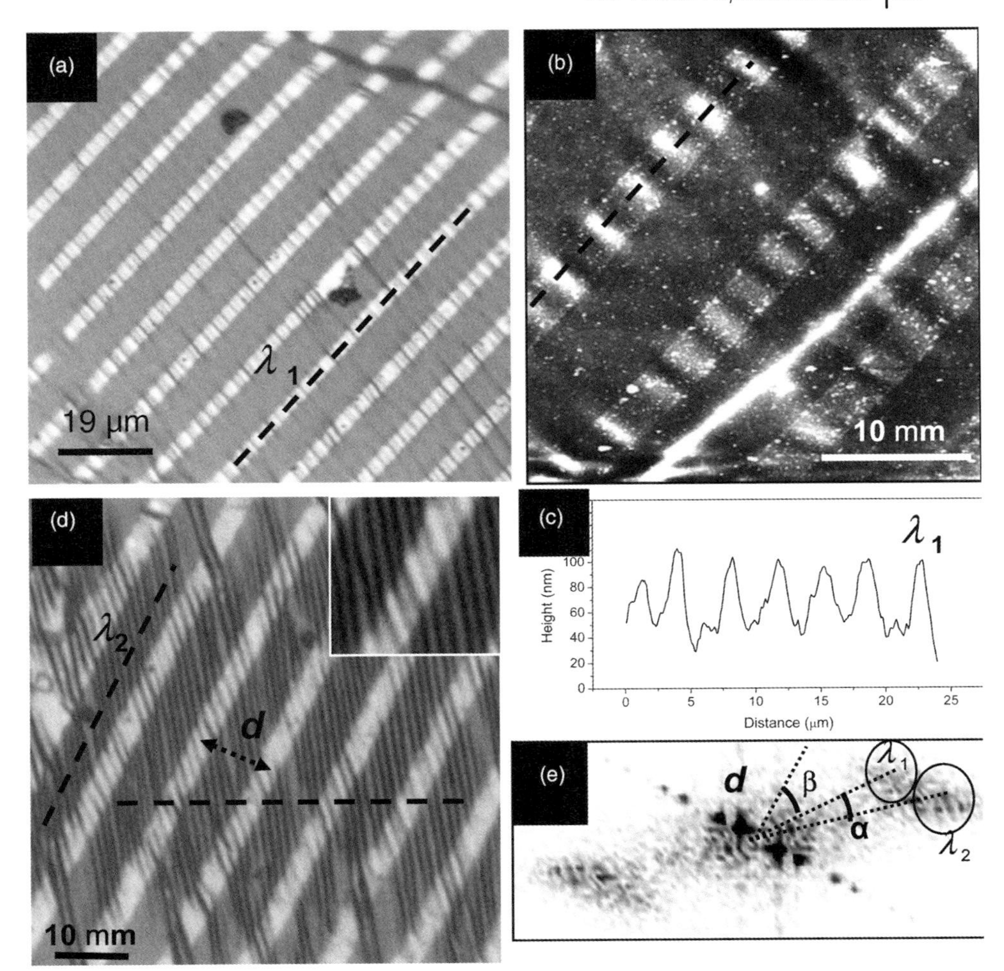

Figure 10.10 Buckling patterns of patterned LbL film: (a,b) one-dimensional buckling mode at low strain (about 0.2%), optical (a) and AFM (b) images. (c) Height profile of buckling mode (dashed line in (b)). (d) Binary buckling pattern at higher compression strain (>0.3%); the inset is a 14 μm × 14 μm image giving a close look at the boundary line. (e) 2D FT of the buckling pattern (a) with three distinct Fourier components. Reproduced from Ref. [118].

whole image that typically includes about 100 wrinkles (Figure 10.10). Therefore, the overall analysis performed here included about 1000 wrinkles making the AFM results reliable and comparable to bulk behavior.

In another study, Lin *et al.* reported an example of 3D LbL grating structures with the ability to diffract light due to the modulation of local LbL film shape with microscopic periodicity and nanoscale vertical modulations [119]. In these freely suspended 3D LbL films, the effective modulation of the refractive properties is caused by the topological variation in the local film shape that represents a negative

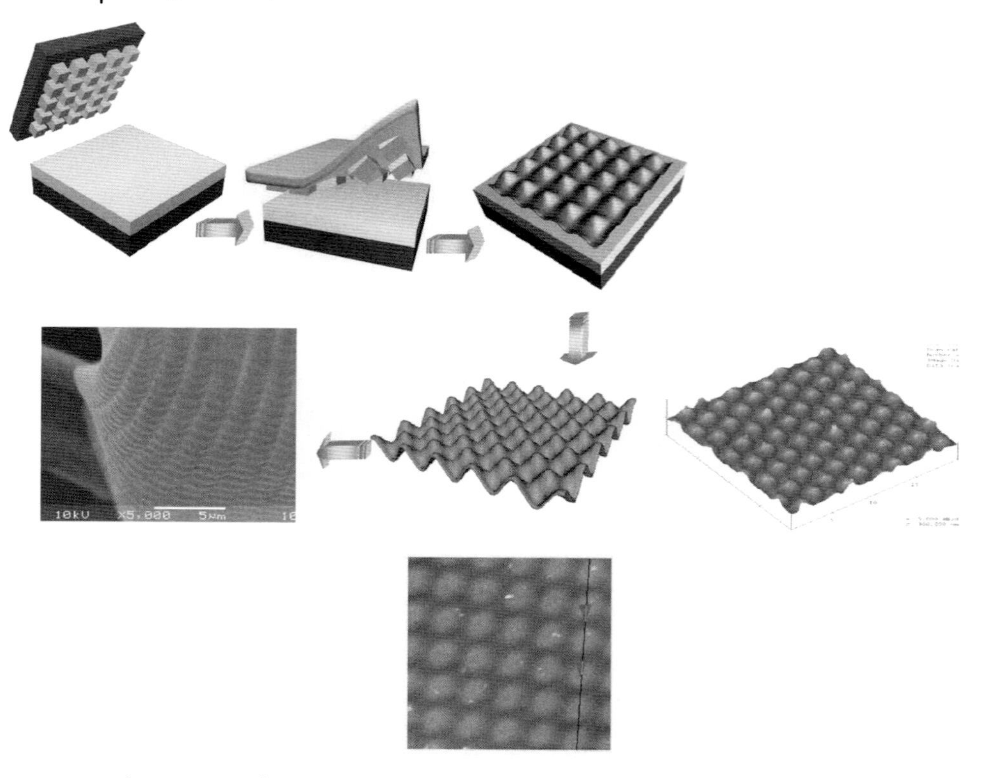

Figure 10.11 Schematic illustration of the route for the fabrication of sculptured LbL films and SEM image (*left*) of the freestanding film with a square pattern along with AFM topographical images of top (*right*) and bottom (*bottom*) side of the LbL films. Reproduced from Ref. [119].

replica of the periodic master, thus representing a purely structural color effect. In order to fabricate sculptured LbL films, a modified experimental procedure was employed, which includes the use of a micropatterned modulated sacrificial substrate (Figure 10.11). A sacrificial PS micropatterned template was first obtained through capillary transfer microprinting. The patterned LbL film was fabricated on this substrate and subsequently released to form a freestanding sculptured film with dimensions about 0.3 mm^2 and with a visible square micropattern extending over a large surface area (Figure 10.11). A simple and economical spin-assisted LbL assembly of conjugated polyelectrolytes on a sacrificial microimprinted modulated substrate was employed to generate the robust, freestanding sculptured LbL structure with an effective thickness of 60 nm.

AFM topographical imaging demonstrated a square lattice on a PDMS stamp exploited to form a micropatterned sacrificial substrate (Figure 10.11). The spacing in this sacrificial micropattern was 2.4 μm with a diagonal distance of 3.4 μm and a 400 nm difference in elevations. The AFM imaging of the freely sculptured LbL film released from substrate, conducted in light tapping mode on both sides of the film, revealed a topographical 3D replica of the original micropatterned substrate with the periodicity of the lateral modulation identical (within experimental error) to that in the original microstamp (see topography of both sides of the film in Figure 10.11). The 3D structure of the sacrificial micropatterned PS template was successfully imprinted into the 3D topology of the nanoscale LbL film with the original square shape of the template somewhat smeared during assembly (Figure 10.11). These films demonstrated efficient optical grating properties and bright structural colors in a reflective mode controlled by the in-plane spacing and the angle of incidence.

The overall "thickness" (peak-to-peak value) of the sculptured 3D LbL film reached 160 nm, which is smaller than the elevation difference in the original template, indicating some contraction in the process of drying and transfer. Although AFM images of top and bottom surfaces display the modulated shape of the LbL film with "imprinted" periodicity, the difference in the cross-sectional shape suggests an asymmetric morphology. The bottom side of the LbL film, which was in contact with the micropatterned sacrificial layer, has a difference in elevation of 125 nm, while the top side of film shows a smaller difference of 55 nm. The difference between the two sides of the freestanding film is due to the relative small channel length of micropatterned PS sacrificial layer. In addition, it might be related to the features of the array processing, such as uneven water access during LbL assembly.

Choi *et al.* studied the porous, periodic, cross-linked solid polymer structures fabricated with multiple laser beam ITL from a negative photoresist epoxy derivative of a bisphenol-A Novolac photocurable resin, SU8, which is widely used in photolithography technology [120–125]. Figure 10.12 shows the AFM image of the pristine ITL periodic structure that represents the square array of cylindrical pores. The periodicity of the square lattice was 830 nm, the radius of the cylindrical pores was 190 nm, and the porosity was 20%. The thickness of the microframe structures was 3 μm, making the aspect ratio of corresponding structures nearly 8.

SFS micromapping was applied to measure surface distribution of the elastic modulus [126]. The surface distribution of the elastic moduli collected for larger surface areas of the patterned polymer film with lower resolution (130 nm^2 per pixel) demonstrates the expected symmetry known from AFM imaging (Figure 10.12). A regular variation in the "apparent" elastic modulus on this spatial scale is determined by the surface topography due to the presence of the solid polymer and deep holes, which results in characteristic bimodal histograms of the surface distribution of the elastic modulus (Figure 10.12).

To address the question of the elastic modulus distribution on polymer surfaces on/between nodes and far from the holes, the higher-resolution micromapping (pixel size down to 60 nm^2) was implemented. The selected surface areas that included a complete set of nodes and beams were used to calculate elastic moduli separately for node and beam areas averaged over six locations (Figure 10.13). A statistically significant difference in the average elastic modulus between nodes and beams was derived from this analysis. This difference was consistently observed for multiple nodes and beams probed independently (see marks in Figure 10.13). For selected surface areas representing nodes and beams, the statistical distribution of the elastic moduli showed a bimodular character with the value of 1.5 GPa for the nodes and 1.1 GPa for the beam areas. The observed location-dependent variation in the elastic

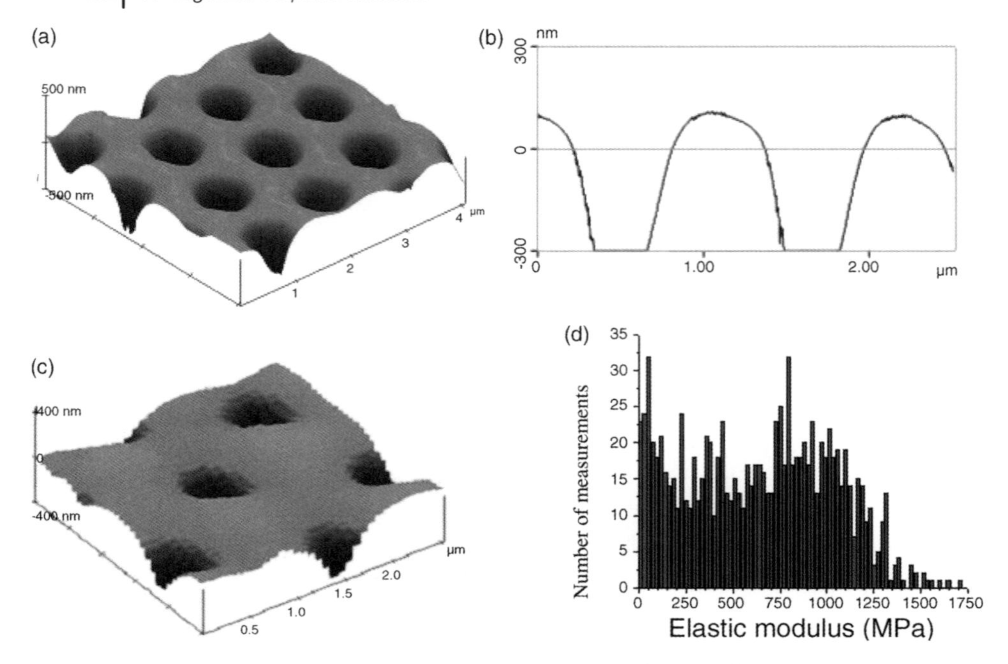

Figure 10.12 AFM topography (a) cross section (b) for periodic porous SU8 ITL film; (c) 3D topography of the surface areas mapped with 32 × 32 force spectroscopy ($2.5\,\mu m^2 \times 2.5\,\mu m^2$); (d) the bimodal surface distribution of the elastic modulus obtained from (c) (1024 data points). Reproduced from Ref. [120].

moduli is caused by the spatial variation in the material properties "templated" by the light distribution within the interference pattern [127].

Singamaneni *et al.* described the transformation of the periodic IL microporous structures caused by internal compressive stresses that can undergo sudden structural transformation at a critical strain (Figure 10.14) [128, 129]. The pattern transformation of collapsed pores is caused by the stresses originated during the localized polymerization of acrylic acid (rubbery component) inside of cylindrical pores and the subsequent solvent evaporation in the periodic microporous structure. Indeed, *in situ* solution photopolymerization of acrylic acid monomer performed directly in the cylindrical pores resulted in fundamental transformation of the initial porous structure with the cylindrical micropores transformed into periodic, mutually orthogonal, and highly collapsed elliptical pores (Figure 10.14).

A dramatically different lattice with perfectly regular, nontrivial geometry of alternating anisotropic slit-like pores expands across the macroscopic area with the original periodicity being constant (within 1%) despite the shape transformation. Transformed regions are extremely uniform and extend to surface areas to a fraction of a millimeter. AFM shows fine details of the transformed pattern morphology (Figure 10.14). The pattern of alternating elliptical micropores is identical along (10) and (01) directions with a characteristic, double-bump shape reflecting the alternating depth of the AFM tip penetration along short and long axes of the collapsed pores

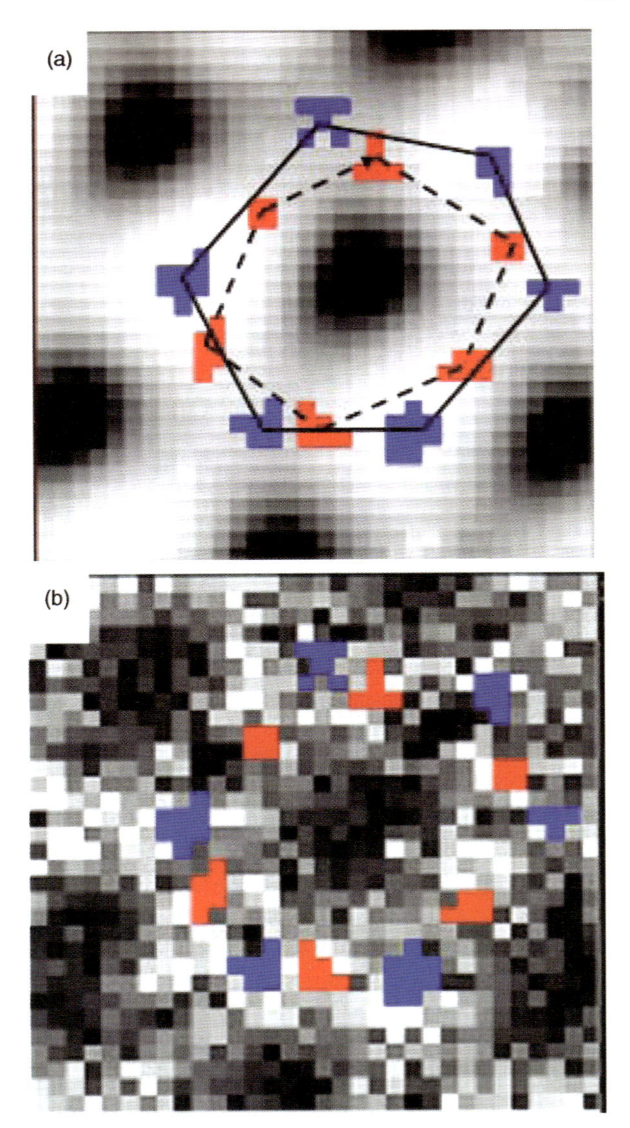

Figure 10.13 SFS micromapping of the periodic SU8 ITL film: (a) 32 × 32 mapping of topography and (b) elastic modulus collected for the 2.5 μm^2 × 2.5 μm^2 surface area (two designated areas are marked by squares of pixels for nodes and for beams. Reproduced from Ref. [120].

(Figure 10.14). The AFM image at higher magnification shows the nodes (elevated round areas) and struts (ridges connecting nodes) of the square lattice. At a microscopic level, the pattern transformation was related to the bending of the struts in alternate directions (along (10) and (01)) and the rotation of the nodes in clockwise and counterclockwise directions as indicated in the AFM image (Figure 10.14).

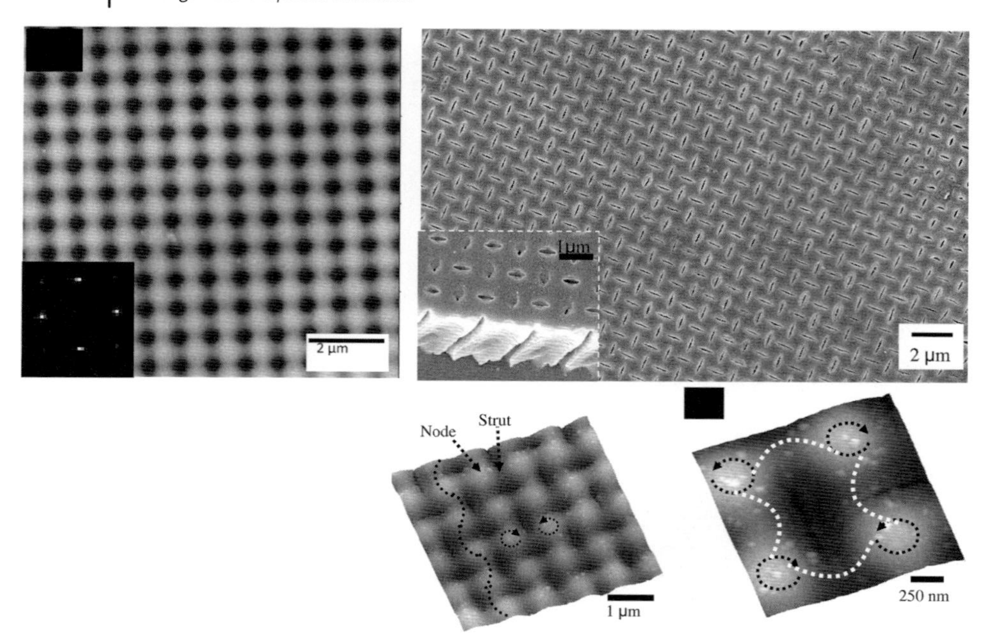

Figure 10.14 AFM image and FFT of the pristine periodic SU8 ITL structure with square lattice (*left*) and large-scale SEM images of transformed pattern of square lattice (*right*). Inset shows the SEM image of the elliptical pores. (*Bottom*) AFM topographical image of the transformed pattern showing the deformation modes of the struts (bending) and the nodes (rotation) of square lattice (*left*). Higher magnification AFM image showing rotation of the nodes resulting in the bending of the struts (*right*). Reproduced from Ref. [128].

Pattern transformation was localized by preventing PAA penetration into selected regions [129]. Figure 10.15 shows the AFM surface morphology of the confined pattern transformation resulting in a hybrid (coexisting pristine and transformed regions) porous structure when the confining PS pattern was parallel to the (10) direction of the original square lattice (Figure 10.15). The AFM image reveals narrow regions with high deformation running at $\sim 45^\circ$ to the lattice direction, releasing the stress caused by overall shrinkage of the volume due to the lateral confinement of the transformation to certain localized regions. The cross section of the patterned structure depicts the alternating single-bump (pristine regions) and double-bump morphology (transformed regions) (Figure 10.15).

One important feature common to both the confined samples was the compression of the transformed regions along the normal direction compared to the pristine regions that is apparent from the 3D topography of the micropatterned porous structure and the corresponding cross section (Figure 10.15). In fact, the cross-section analysis reveals that the transformed regions were compressed by ~ 150 nm (corresponding to 5% strain) in the vertical direction compared to the pristine porous regions (lower regions in Figure 10.15). The vertical compression is due to the vertical component of the stress exerted during the isotropic collapse of the PAA network

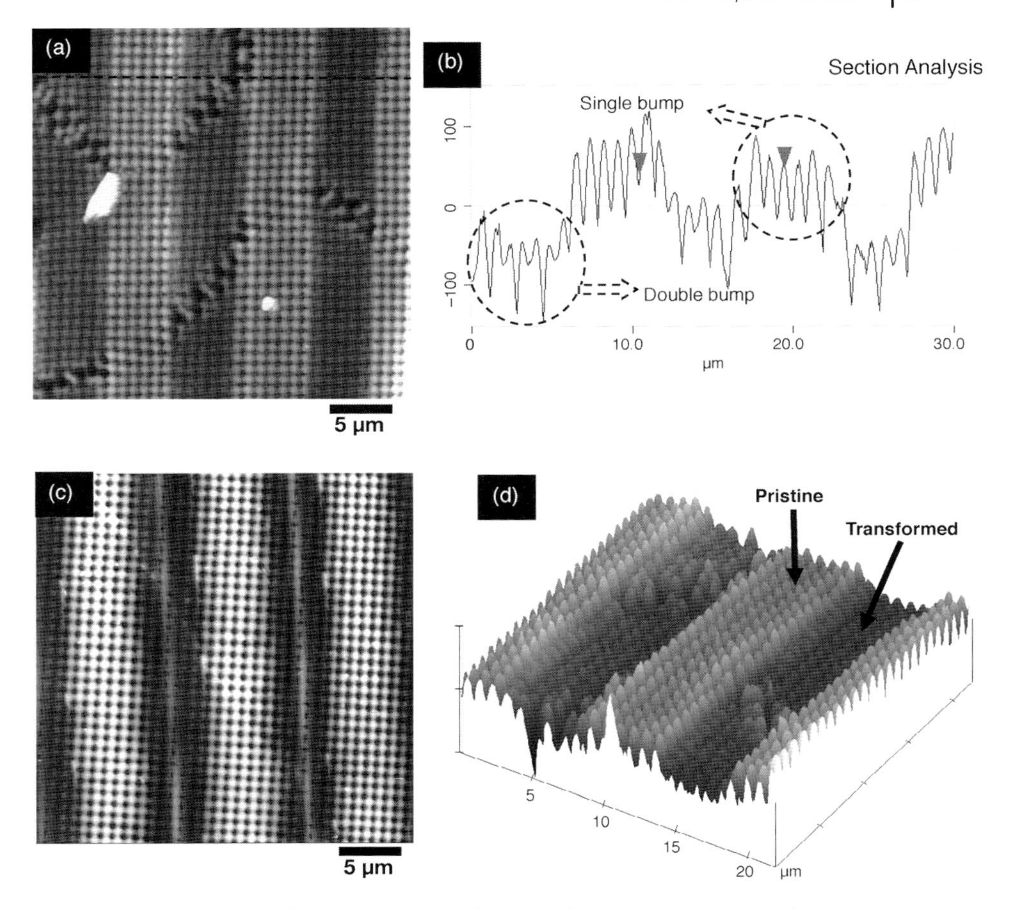

Figure 10.15 AFM image of the periodic transformation with array of circular holes interleaved with array of mutually orthogonal pores. (a) PS pattern aligned with the (10) direction of the microtruss pattern; (b) cross section along the horizontal direction showing the alternating single-bump and double-bump structure. Reproduced from Ref. [129].

during the solvent evaporation. The patterning of instabilities achieved in this study clearly revealed such significant vertical compression in the transformed areas, confirming that the pore collapse not only occurs in the (x, y) plane of the microstructure but also significantly perturbs the porous structure in the z-direction.

Singamaneni *et al.* demonstrated biometallized periodic, porous polymer microstructures composed of polymer IL templates decorated with metal nanoparticles (Figure 10.16) [130]. The direct growth on both the outer and the inner surfaces of these 3D structures was enabled by the biofunctionalization with uniform amino-acid coatings. The diameter of gold nanoparticles was varied from 10 to 90 nm by the variation of parameters of biofunctionalization. The size and the spatial distribution

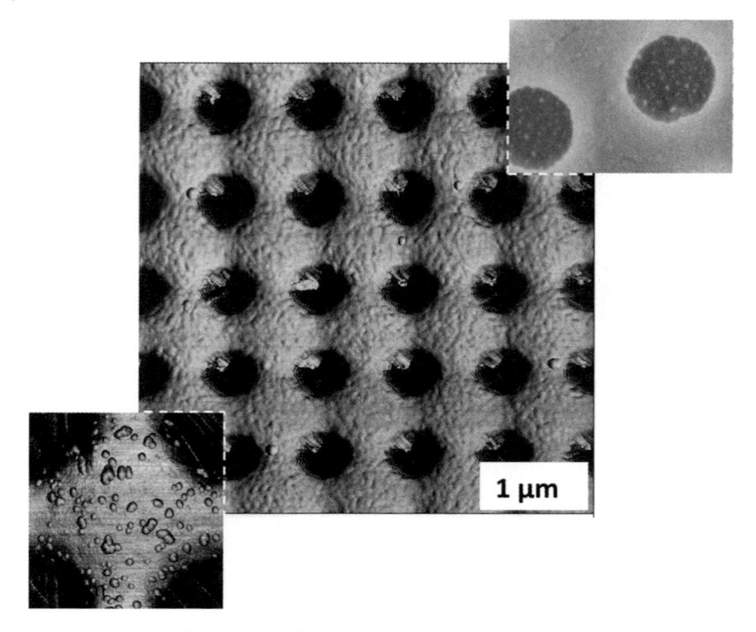

Figure 10.16 Gold nanoparticles grown on SU8 structure modified with poly-L-tyrosine coating. Insets show these gold nanoparticles uniformly grown on the surface (AFM) and inside pores (SEM). Reproduced from Ref. [130].

of the gold nanoparticles on and inside ITL structures were tailored by varying the nature of the polyamino acid-nanocoating.

To extend the range of sizes and surface coverage of metal nanoparticles, Anderson *et al.* exploited an alternative vapor-based deposition approach [131]. The tyrosine monomer was sublimated and plasma polymerized on the periodic structure resulting in cross-linked, pinhole-free films with preserved surface biofunctionality and vapor-based biometallization of gold nanoparticles (Figure 10.16).

Furthermore, patterning of the tyrosine during plasma deposition was done by securing a TEM grid with 10 μm^2 square openings to the substrate to act as a mask during the PECVD deposition of L-tyrosine. This method allowed regular patterns to be created on the silicon surface across entire masked region (Figure 10.17). AFM analysis of the patterned area shows that the peak height of the squared tyrosine regions was 80 nm (Figure 10.17). Since the TEM grid was secured only around the edges to the substrate, there was some tyrosine that permeated under the masked area leaving a thin residual film in the valleys between the raised areas with a thickness of less than 10 nm.

After exposure of the patterned tyrosine substrate to the gold chloride solution, it was seen that these patterns are capable of selectively reducing gold nanoparticles on the squared tyrosine-coated regions, while leaving the areas of exposed silicon relatively clean and free of nanoparticles, creating density-controlled regular regions of gold nanoparticles selectively reduced with the squared surface regions. In these

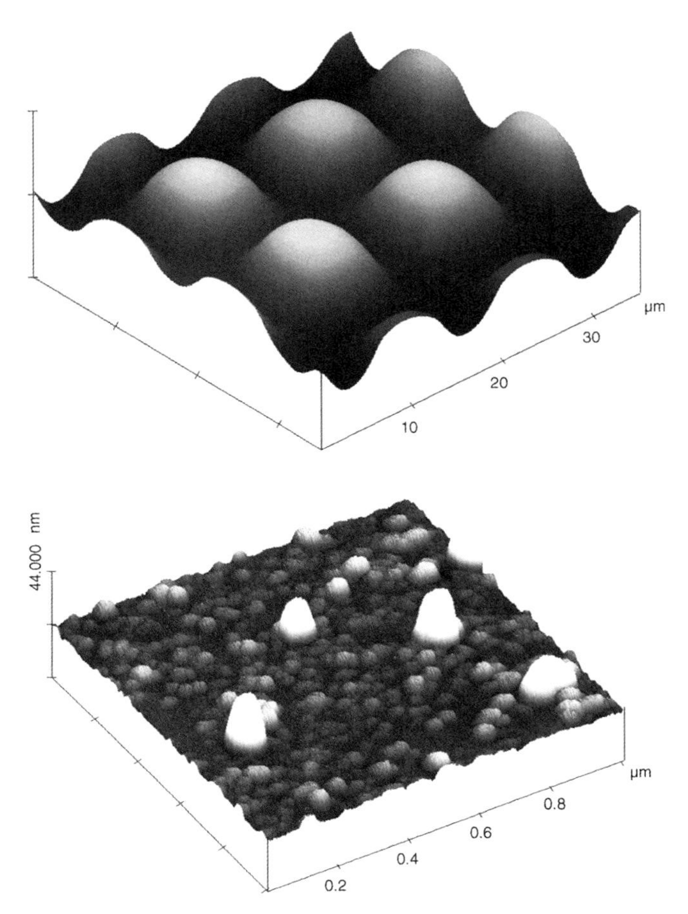

Figure 10.17 (*Top*) Micropatterned PECVD amino acid films, 35 × 35 µm scan, *z*-range 230 nm. (*Bottom*) Gold nanoparticles after reduction on selected PECVD tyrosine film, 1 µm × 1 µm, *z*-range 15 nm. Reproduced from Ref. [131].

regions, 2–3 nm nanoparticles covered the surface, matching the results seen on the uniform tyrosine films, which also had larger gold aggregates present occasionally (Figure 10.17).

References

1 Reneker, D.H. and Mazur, J. (1988) Small defects in crystalline polyethylene. *Polymer*, **29** (1), 3–13.

2 Hobbs, J.K., Farrance, O.E., and Kailas, L. (2009) How atomic force microscopy has contributed to our understanding of polymer crystallization. *Polymer*, **50** (18), 4281–4292.

3 Wunderlich, B. (1976) Macromolecular physics, in *Crystal Nucleation, Growth, Annealing*, vol. 2, Academic Press, New York.

4 Magonov, S.N. and Reneker, D.H. (1997) Characterization of polymer surfaces with atomic force microscopy. *Annu. Rev. Mater. Sci.*, **27**, 175–222.

5 Patil, R., Kim, S.J., Smith, E., Reneker, D.H., and Weisenhorn, A.L. (1990) Atomic force microscopy of dendritic crystals of polyethylene. *Poly. Comm.*, **31** (12), 455–457.

6 Vancso, G.J., Beekmans, L.G.M., Pearce, R., Trifonova, D., and Varga, J. (1999) From microns to nanometers: morphology development in semicrystalline polymers by scanning force microscopy. *J. Macromol. Sci. Phys.*, **38** (5–6), 491–503.

7 Hobbs, J.K., Humphris, A.D.L., and Miles, M.J. (2001) *In-situ* atomic force microscopy of polyethylene crystallization. 1. Crystallization from an oriented backbone. *Macromolecules*, **34** (16), 5508–5519.

8 Toda, A., Taguchi, K., and Kajioka, H. (2008) Instability-driven branching of lamellar crystals in polyethylene spherulites. *Macromolecules*, **41** (20), 7505–7512.

9 Jeon, K. and Krishnamoorti, R. (2008) Morphological behavior of thin linear low-density polyethylene films. *Macromolecules*, **41** (19), 7131–7140.

10 Patil, R. and Reneker, D.H. (1994) Molecular folds in polyethylene observed by atomic force microscopy. *Polymer*, **35** (9), 1909–1914.

11 Magonov, S.N., Yerina, N.A., and Godovsky, Y.K. (2006) Annealing and recrystallization of single crystals of polyethylene on graphite: an atomic force microscopy study. *J. Macromol. Sci. B*, **45** (2), 169–194.

12 Loos, J. and Tian, M. (2006) Annealing behavior of solution grown polyethylene singe crystals. *Polymer*, **47** (15), 5574–5581.

13 Tracz, A., Kucinska, I., Wostek-Wojchiechowska, D., and Jeszka, J. (2005) The influence of micro- and nanoparticles on model atomically flat surfaces on crystallization of polyethylene. *Eur. Polym. J.*, **41** (3), 501–209.

14 Tsukruk, V.V. and Reneker, D.H. (1995) Periodic surface instabilities in stressed polymer solids. *Phys. Rev. B*, **51** (9), 6089–6092.

15 Lotz, B., Wittman, J.C., and Lovinger, A.J. (1996) Structure and morphology of poly (propylenes): a molecular analysis. *Polymer*, **37** (22), 4979–4992.

16 Rehse, N., Marr, S., Scherdel, S., and Magerle, R. (2005) Three-dimensional imaging of semicrystalline polypropylene with 10nm resolution. *Adv. Mater.*, **17** (18), 2203–2206.

17 Thomann, R., Wang, C., Kressler, J., Jungling, S., and Mulhaupt, R. (1995) Morphology of syndiotactic polypropylene. *Polymer*, **36** (20), 3795–3801.

18 Tsukruk, V.V. and Reneker, D.H. (1995) Surface morphology of syndiotactic polypropylene single crystals observed by atomic force microscopy. *Macromolecules*, **28** (5), 1370–1376.

19 Schonherr, H., Wiyatno, W., Pople, J., Frank, C.W., Fuller, G.G., Gast, A.P., and Waymouth, R.M. (2002) Morphology of thermoplastic elastomers: elastomeric polypropylene. *Macromolecules*, **35** (7), 2654–2666.

20 Hu, Y.S., Kamdar, A.R., Ansems, P., Chum, S.P., Hiltner, A., and Baer, E. (2006) Crystallization of a miscible propylene/ethylene copolymer blend. *Polymer*, **47** (18), 6387–6397.

21 Beekmans, L.G.M., van der Meer, D.W., and Vancso, G.J. (2002) Crystal melting and its kinetics on poly(ethylene oxide) by *in situ* atomic force microscopy. *Polymer*, **43** (6), 1887–1895.

22 Beekmans, L.G.M., Hampenius, M.A., and Vancso, G.J. (2004) Morphological development of melt crystallized poly (propylene oxide) by *in situ* AFM: formation of banded spherulites. *Eur. Polym. J.*, **40** (5), 893–903.

23 Hobbs, J.K., Vasilev, C., and Humphris, A.D.L. (2005) Real time observation of crystallization in polyethylene oxide with video rate atomic force microscopy. *Polymer*, **46** (23), 10226–10236.

24 Hsiao, M.-S., Chen, W.Y., Zheng, J.X., Van Horn, R.M., Quirk, R.P., Ivanov,

D.A., Thomas, E.L., Lotz, B., and Cheng, S.Z.D. (2008) Poly(ethylene oxide) crystallization within a one-dimensional defect-free confinement on the nanoscale. *Macromolecules*, **41** (13), 4794–4801.

25 Zhu, J. and Wang, M. (2008) Temperature-induced pattern transition in crystallizing ultrathin polymer films. *J. Macromol. Sci. B*, **47** (2), 401–108.

26 Gunawidjaja, R., Peleshanko, S., Genson, K.L., Tsitsilianis, C., and Tsukruk, V.V. (2006) Surface morphologies of Langmuir–Blodgett monolayers of PEO_nPS_n multiarm star copolymers. *Langmuir*, **22** (14), 6168–6176.

27 Beekmans, L.G.M. and Vancso, G.J. (2000) Real-time crystallization study of poly(e-caprolactone) by hot-stage atomic force microscopy. *Polymer*, **41** (25), 8975–8981.

28 Mareau, V.H. and Prud'homme, R.E. (2005) Crystallization of ultrathin poly(e-caprolactone) films in the presence of residual solvent, an *in situ* atomic force microscopy study. *Polymer*, **46** (18), 7255–7265.

29 Nunez, E., Vancso, G.J., and Gedde, U.W. (2008) Morphology, crystallization, and melting of single crystals and thin films of star-branched polyesters with poly(e-caprolactone) arms as revealed by atomic force microscopy. *J. Macromol. Sci. B*, **47** (3), 589–607.

30 Hou, W.-M., Zhou, J.-J., Gan, Z.-H., Shi, A.-C., Chan, C.-M., and Li, L. (2007) The crystallization morphology and melting behavior of polymer crystals in nano-sized domains. *Polym. Commun.*, **48** (17), 4926–4931.

31 Fujita, M., Takikawa, Y., Sakuma, H., Teramachi, S., Kikkawa, Y., and Doi, Y. (2007) Real-time observations of oriented crystallization of poly(e-caprolactone) thin film, induced by an AFM tip. *Macromol. Chem. Phys.*, **208** (17), 1862–1870.

32 Fujita, M. and Doi, Y. (2003) Annealing and melting behavior of poly(L-lactic acid) single crystals as revealed by *in situ* atomic force microscopy. *Biomacromolecules*, **4** (5), 1301–1307.

33 Yuryev, Y., Wood-Adams, P., Heuzey, M.-C., Dubois, C., and Brisson, J. (2008) Crystallization of polylactide films: an atomic force microscopy study of the effects of temperature and blending. *Polymer*, **49** (9), 2306–2320.

34 Li, H., Nie, W., Deng, C., Chen, X., and Ji, X. (2009) Crystalline morphology of poly (L-lactic acid) thin films. *Eur. Polym. J.*, **45** (1), 123–130.

35 Hobbs, J.K. and Register, R.A. (2006) Imaging block copolymer crystallization in real time with the atomic force microscope. *Macromolecules*, **39** (2), 703–710.

36 Huang, S., Jiang, S., Chen, X., and An, L. (2009) Dendritic superstructures and structure transitions of asymmetric poly (L-lactide-*b*-ethylene oxide) diblock copolymer thin films. *Langmuir*, **25** (22), 13125–13132.

37 Kang, J. and Beers, K.J. (2009) Effect of temperature and water on microphase separation of PCL-PEO-PCL triblock copolymers. *Polym. Bull.*, **63** (5), 723–724.

38 Li, H., Sachsenhofer, R., Binder, W.H., Henze, T., Thurn-Albrecht, T., Busse, K., and Kressler, J. (2009) Hierarchical organization of poly(ethylene oxide)-*block*-poly(isobutylene) and hydrophobically modified Fe_2O_3 nanoparticles at the air/water interface and on solid supports. *Langmuir*, **25** (14), 8320–8329.

39 Liu, J., Sheina, E., Kowalewski, T., and McCullough, R.D. (2002) Tuning the electrical conductivity and self-assembly of regioregular polythiophene by block copolymerization: nanowire morphologies in new di- and triblock copolymers. *Angew. Chem. Int. Ed.*, **41** (2), 329–332.

40 Taguchi, K., Miyamoto, Y., Miyaji, H., and Izumi, K. (2003) Undulation of lamellar crystals of polymers by surface stresses. *Macromolecules*, **36** (14), 5208–5213.

41 Kumaki, J., Kawauchi, T., and Yashima, E. (2005) Two-dimensional folded chain crystals of a synthetic polymer in a Langmuir–Blodgett film. *J. Am. Chem. Soc.*, **127** (16), 5788–5789.

42 Kusanagi, H., Chatani, Y., and Tadokoro, H. (1994) The crystal structure of isotactic poly(methyl methacrylate): packing-mode of double stranded helices. *Polymer*, **35** (10), 2028–2039.

43 Wang, H., Wang, H.H., Urban, V.S., Littrell, K.C., Thiyagarajan, P., and Yu, L. (2000) Synthesis of amphiphilic diblock copolymers containing a conjugated block and their self-assembling properties. *J. Am. Chem. Soc.*, **122** (29), 6855–6861.

44 Iovu, M.C., Craley, C.R., Jeffries-EL, M., Krankowski, A.B., Zhang, R., Kowalweski, T., and McCullough, R.D. (2007) Conducting regioregular polythiophene block copolymer nanofibrils synthesized by reversible addition fragmentation chain transfer polymerization (RAFT) and nitroxide mediated polymerization (NMP). *Macromolecules*, **40** (14), 4733–4735.

45 Zhang, Y., Tajima, K., and Hashimoto, K. (2009) Nanostructure formation in poly (3-hexylthiophene-*block*-3-(2-ethylhexyl) thiophene)s. *Macromolecules*, **42** (18), 7008–7015.

46 Tsukruk, V.V., Bliznyuk, V.N., and Reneker, D.H. (1994) Morphology and molecular ordering in Langmuir–Blodgett films from ladder polyheteroarylenes. *Thin Solid Films*, **244** (1–2), 745–749.

47 Xiao, X., Hu, Z., Wang, Z., and He, T. (2009) Study on the single crystals of poly (3-octylthiophene) induced by solvent-vapor annealing. *J. Phys. Chem. B*, **113** (44), 14604–14610.

48 Korley, L.T.J., Pate, B.D., Thomas, E.L., and Hammond, P.T. (2006) Effect of the degree of soft and hard segment ordering on the morphology and mechanical behavior of semicrystalline segmented polyurethanes. *Polymer*, **47** (9), 3073–3082.

49 Revenko, I., Tang, Y., and Santerre, J.P. (2001) Surface structure of polycarbonate urethanes visualized by atomic force microscopy. *Surf. Sci.*, **491** (3), 346–354.

50 Garrett, J.T., Siedlecki, C.A., and Runt, J. (2001) Microdomain morphology of poly (urethane urea) multiblock copolymers. *Macromolecules*, **34** (20), 7066–7070.

51 Xu, M.X., Liu, W.G., Wang, C.L., Gao, Z.X., and Yao, K.D. (1996) Surface crystalline characteristics of polyurethane investigated by atomic force microscopy. *J. Appl. Polym. Sci.*, **61** (12), 2225–2228.

52 Hernandez, R., Weksler, J., Padsalgikar, A., Choi, T., Angelo, E., Lin, J.S., Xu, L.-C., Siedlecki, C.A., and Runt, J. (2008) A comparison of phase organization of model segmented polyurethanes with different intersegment compatibilities. *Macromolecules*, **41** (24), 9767–9776.

53 Rueda-Larraz, L., d'Arlas, B.F., Tercjak, A., Ribes, A., Mondragon, I., and Eceiza, A. (2009) Synthesis and microstructure–mechanical property relationships of segmented polyurethanes based on a PCL–PTHF–PCL block copolymer as soft segment. *Eur. Polym. J.*, **45** (7), 2096–2109.

54 Mauritz, K.A. (1988) Review and critical analysis of theories of aggregation in ionomers. *J. Macromol. Sci. Rev. C*, **28** (1), 65–98.

55 McLean, R.S., Doyle, M., and Sauer, B.B. (2000) High-resolution imaging of ionic domains and crystal morphology in ionomers using AFM techniques. *Macromolecules*, **33** (17), 6541–6550.

56 Donald, A.M., Windle, A.H., and Hanna, S. (1992) *Liquid Crystalline Polymers*, Cambridge University Press, New York.

57 Shibaev, V.P. and Lam, L. (eds) (1994) *Liquid Crystalline and Mesomorphic Polymers*, Springer, New York.

58 Tsukruk, V.V. and Bliznyuk, V.N. (1997) Side chain liquid crystalline polymers at interfaces. *Prog. Polym. Sci.*, **22** (5), 1089–1132.

59 Gould, S.A.C., Shulman, J.B., Schiraldi, D.A., and Occelli, M.L. (1999) Atomic force microscopy (AFM) studies of liquid crystalline polymer (LCP) surfaces. *J. Appl. Polym. Sci.*, **74** (9), 2243–2254.

60 Ge, J.J., Zhang, J.Z., Zhou, W., Li, C.Y., Jin, S., Calhoun, B.H., Wang, S.-Y., Harris, F.W., and Cheng, S.Z.D. (2000) Phase structures, transition behavior and surface alignment in polymers containing rigid-rod backbones with flexible side chains. Part VI. Novel band structures in a combined main-chain/side-chain liquid crystalline polyester: from liquid crystal to crystalline states. *J. Mater. Sci.*, **35** (20), 5215–5223.

61 Zhang, S., Fu, L., Zhang, J., Liu, J., Yang, D., Ge, J., Li, C., and Cheng, S. (2004) Ordering-induced micro-bands in thin films of a main-chain liquid crystalline chloro-poly(aryl ether ketone). *Polymer*, **45** (11), 3967–3972.

62 O'Rourke, M.J.E., Ding, D.-K., and Thomas, E.L. (2001) Morphologies and energies of Neel inversion wall defects in a liquid crystal polyether. *Macromolecules*, **34** (19), 6658–6669.

63 Weng, X., Li, C.Y., Jin, S., Zhang, D., Zhang, J.Z., Harris, F.W., and Cheng, S.Z.D. (2002) Helical twist senses, liquid crystalline behavior, crystal microtwins, and rotation twins in a polyester containing main-chain molecular asymmetry and effects of the number of methylene units in the backbones on the phase structures and morphologies of its homologues. *Macromolecules*, **35** (26), 9678–9686.

64 Defaux, M., Vidal, L., Moller, M., Vearba, R.I., Dimasi, E., and Ivanov, D.A. (2009) Thin films of a main-chain columnar liquid crystal: studies of structure, phase transitions, and alignment. *Macromolecules*, **42** (10), 3500–3509.

65 Zhang, S., Terentjev, E.M., and Donald, A.M. (2005) Atomic force microscopy study for supramolecular microstructures in side-chain liquid crystalline polymer films. *Langmuir*, **21** (8), 3539–3543.

66 Magonov, S.N., Elings, V., and Papkov, V.S. (1997) AFM study of thermotropic structural transitions in poly (diethylsiloxane). *Polymer*, **38** (2), 297–307.

67 Schonherr, H., Manickam, M., and Kumar, S. (2002) Surface morphology and molecular ordering in thin films of polymerizable triphenylene discotic liquid crystals on HOPG revealed by atomic force microscopy. *Langmuir*, **18** (18), 7082–7085.

68 Brodowsky, H.M., Boehnke, U.-C., and Kremer, F. (1997) Temperature dependent AFM on ferroelectric liquid crystalline polymer and elastomer films. *Langmuir*, **13** (20), 5378–5382.

69 Pastor, L., Barbera, J., McKenna, M., Marcos, M., Martin-Rapun, R., Serrano, J.L., Luckhurst, G.R., and Mainal, A. (2004) End-on and side-on nematic liquid crystal dendrimers. *Macromolecules*, **37**, 9386.

70 Donnio, B., Barbera, J., Gimenez, R., Guillion, D., Marcos, M., and Serrano, J.L. (2002) Controlled molecular conformation and morphology in poly (amidoamine) (PAMAM) and poly (propyleneimine) (DAB) dendrimers. *Macromolecules*, **35** (2), 370–381.

71 Rueff, J.M., Barbera, J., Donnio, B., Guillion, D., Marcos, M., and Serrano, J.L. (2003) Lamellar to columnar mesophase evolution in a series of PAMAM liquid-crystal codendrimers. *Macromolecules*, **36** (22), 8368–8375.

72 Ponomarenko, S.A., Boiko, N.I., Shibaev, V.P., Richardson, R.M., Whitehouse, I.J., Rebrov, E.A., and Muzafarov, A.M. (2000) Carbosilane liquid crystalline dendrimers: from molecular architecture to supramolecular nanostructures. *Macromolecules*, **33** (15), 5549–5558.

73 Agina, E.V., Ponomarenko, S.A., Boiko, N.I., Rebrov, E.A., Muzafarov, A.M., and Shibaev, V.P. (2001) Synthesis and phase behavior of carbosilane LC dendrimers with terminal mesogenic groups based on anisic acid derivative. *Poly. Sci. Series A*, **43** (10), 1757–1765.

74 Ponomarenko, S.A., Boiko, N.I., Shibaev, V.P., and Magonov, S.N. (2000) Atomic force microscopy study of structural organization of carbosilane liquid crystalline dendrimer. *Langmuir*, **16** (12), 5487–5493.

75 Genson, K.L., Holzmueller, J., Leshchiner, I., Agina, E., Boiko, N., Shibaev, V.P., and Tsukruk, V.V. (2005) Organized monolayers of carbosilane dendrimers with mesogenic terminal groups. *Macromolecules*, **38** (19), 8028–8035.

76 Zhang, S.J., Li, Q.W., Kinloch, I.A., and Windle, A.H. (2010) Ordering in a droplet of an aqueous suspension of single-wall carbon nanotubes on a solid substrate. *Langmuir*, **26** (3), 2107–2112.

77 Ko, H., Peleshanko, S., and Tsukruk, V.V. (2004) Combing and bending of carbon nanotube arrays with confined

microfluidic flow on patterned surfaces. *J. Phys. Chem. B*, **108** (14), 4385–4393.

78 Ko, H. and Tsukruk, V.V. (2006) Liquid-crystalline processing of highly-oriented carbon nanotube arrays for thin film transistors. *Nano Lett.*, **6** (7), 1443–1448.

79 Pujari, S., Rahatekar, S.S., and Gilman, J.W. (2009) Orientation dynamics in multiwalled carbon nanotube dispersions under shear flow. *J. Chem. Phys.*, **130** (21), 214903-1–214903-9.

80 Kim, E., Xia, Y., and Whitesides, G.M. (1995) Polymer microstructures formed by moulding in capillaries. *Nature*, **376**, 581–584.

81 Lu, M.H. and Zhang, Y. (2006) Microbead patterning on porous films with ordered arrays of pores. *Adv. Mater.*, **18** (23), 3094–2098.

82 Lin, S.Y., Fleming, J.G., Hetherington, D.L., Smith, B.K., Biswas, R., Ho, K.M., Sigalas, M.M., Zubrzycki, W., Kurtz, S.R., and Bur, J. (1998) A three-dimensional photonic crystal operating at infrared wavelengths. *Nature*, **394**, 251–253.

83 Jang, J.-H., Ullal, C.K., Choi, T., Lemieux, M.C., Tsukruk, V.V., and Thomas, E.J. (2006) 3D polymer microframes that exploit length scale-dependent mechanical behavior. *Adv. Mater.*, **18** (16), 2123–2127.

84 Noda, S., Tomoda, K., Yamamoto, N., and Chutinan, A. (2000) Full three-dimensional photonic bandgap crystals at near-infrared wavelengths. *Science*, **289** (5479), 604–606.

85 Gratson, G.M., Xu, M., and Lewis, J.A. (2004) Microperiodic structures: direct writing of three-dimensional webs. *Nature*, **428**, 386.

86 Campbell, M., Sharp, D.N., Harrison, M.T., Denning, R.G., and Turberfield, A.J. (2000) Fabrication of photonic crystals for the visible spectrum by holographic lithography. *Nature*, **404**, 53–56.

87 Zaumseil, J., Meitl, M.A., Hsu, J.W.P., Acharya, B.R., Baldwin, K.W., Loo, Y.-L., and Rogers, J.A. (2003) Three-dimensional and multilayer nanostructures formed by nanotransfer printing. *Nano Lett.*, **3** (9), 1223–1227.

88 Whitesides, G.M., Ostuni, E., Takayama, S., Jiang, X.Y., and Ingber, D.E. (2001) Soft lithography in biology and biochemistry. *Annu. Rev. Biomed. Eng.*, **3**, 335–373.

89 Suh, K.Y., Kim, Y.S., and Lee, H.H. (2001) Capillary force lithography. *Adv. Mater.*, **13** (18), 1386–1389.

90 Park, J., Fouche, L.D., and Hammond, P.T. (2005) Multicomponent patterning of layer-by-layer assembled polyelectrolyte/nanoparticle composite thin films with controlled alignment. *Adv. Mater.*, **17** (21), 2575–2579.

91 Hiller, J.A., Mendelsohn, J.D., and Rubner, M.F. (2002) Reversibly erasable nanoporous anti-reflection coatings from polyelectrolyte multilayers. *Nat. Mater.*, **1**, 59–63.

92 Hua, F., Cui, T., and Lvov, Y.M. (2004) Ultrathin cantilevers based on polymer-ceramic nanocomposite assembled through layer-by-layer adsorption. *Nano Lett.*, **4** (5), 823–825.

93 Jiang, C. and Tsukruk, V.V. (2006) Freestanding nanostructures via layer-by-layer assembly. *Adv. Mater.*, **18** (7), 829–840.

94 Zhai, L., Berg, M.C., Cebeci, F.C., Kim, Y., Milwid, J.M., Rubner, M.F., and Cohen, R.E. (2006) Patterned superhydrophobic surfaces: toward a synthetic mimic of the namib desert beetle. *Nano Lett.*, **6** (6), 1213–1217.

95 Jiang, C., Markutsya, S., Pikus, Y., and Tsukruk, V.V. (2004) Freely suspended nanocomposite membranes as highly sensitive sensors. *Nat. Mater.*, **3**, 721–728.

96 Tang, Z., Kotov, N.A., Magonov, S., and Ozturk, B. (2003) Nanostructured artificial nacre. *Nat. Mater.*, **2**, 413–418.

97 Lynn, D.N. (2006) Layers of opportunity: nanostructure polymer assemblies for the delivery of macromolecular therapeutics. *Soft Matter*, **2** (3), 269–273.

98 Kim, J., Serpe, M.J., and Lyon, L.A. (2005) Photoswitchable microlens arrays. *Angew. Chem. Int. Ed.*, **44** (9), 1333–1336.

99 Wu, M., Odom, T.W., and Whitesides, G.M. (2002) Generation of chrome masks with micrometer-scale features using microlens lithography. *Adv. Mater.*, **14** (17), 1213–1216.

100 Jeong, K.-H., Kim, J., and Lee, L.P. (2006) Biologically inspired artificial compound eyes. *Science*, **312** (5773), 557–561.

101 Wu, C.Y., Chiang, T.H., Lai, N.D., Do, D.B., and Hsu, C.C. (2009) Fabrication of microlens arrays based on the mass transport effect of SU-8 photoresist using a multiexposure two-beam interference technique. *Appl. Optics*, **48** (13), 2473–2479.

102 Zhang, Y., Reed, J.C., and Yang, S. (2009) Creating a library of complex metallic nanostructures via harnessing pattern transformation of single PDMS membrane. *ACS Nano*, **3** (8), 2412–2418.

103 Yu, F., Mucklich, F., Li, P., Shen, H., Mathur, S., Lehr, C.-M., and Bakowsky, U. (2005) *In vitro* cell response to a polymer surface micropatterned by laser interference lithography. *Biomacromolecules*, **6** (3), 1160–1167.

104 Xia, Y. and Whitesides, G.M. (1998) Soft lithography. *Angew. Chem. Int. Ed.*, **37** (5), 550–575.

105 Chiu, D.T., Jeon, N.L., Huang, S., Kene, R.S., Wargo, C.J., Choi, I.S., Ingber, D.E., and Whitesides, G.M. (2000) Patterned deposition of cells and proteins onto surfaces by using three-dimensional microfluidic systems. *Proc. Natl. Acad. Sci.*, **97** (6), 2408–2413.

106 McDonald, J.C. and Whitesides, G.M. (2002) Poly(dimethylsiloxane) as a material for fabricating microfluidic devices. *Acc. Chem. Res.*, **35** (7), 491–499.

107 Choi, K.M. and Rogers, J.A. (2003) A photocurable poly(dimethylsiloxane) chemistry designed for soft lithographic molding and printing in the nanometer regime. *J. Am. Chem. Soc.*, **125** (14), 4060–4061.

108 Ryu, K.S., Wang, X., Shaikh, K., and Liu, C. (2004) A method for precision patterning of silicone elastomer and its applications. *J. Microelectromech. Syst.*, **13** (4), 568–575.

109 Chiang, W.-Y. and Shu, W.-J. (1988) Preparation and properties of UV-curable poly(dimethylsiloxane) urethane acrylate. II. Property–structure/molecular weight relationships. *J. Appl. Polym. Sci.*, **36** (8), 1889–1907.

110 Rogers, J.A., Schueller, O.J.A., Marzolin, C., and Whitesides, G.M. (1997) Wave-front engineering by use of transparent elastomeric optical elements. *Appl. Optics*, **36** (23), 5792–5795.

111 Wilbur, J.L., Jackman, R.J., Whitesides, G.M., Cheung, E.L., Lee, L.K., and Prentiss, M.G. (1996) Elastomeric optics. *Chem. Mater.*, **8** (7), 1380–1385.

112 Tanaka, T., Morigami, M., and Atoda, N. (1993) Mechanism of resist pattern collapse during development process. *Jpn. J. Appl. Phys.*, **32**, 6059–6064.

113 Delamarche, E., Schmid, H., Biebuyck, H.A., and Michel, B. (1997) Stability of molded polydimethylsiloxane microstructures. *Adv. Mater.*, **9** (9), 741–746.

114 Lee, T.-W., Mitrofanov, O., and Hsu, J.W.P. (2005) Pattern-transfer fidelity in soft lithography: the role of pattern density and aspect ratio. *Adv. Funct. Mater.*, **15** (10), 1683–1688.

115 Jo, B.-H., van Lerberghe, L.M., Motsegood, K.M., and Beebe, D.J. (2000) Three-dimensional micro-channel fabrication in polydimethylsiloxane (PDMS) elastomer. *Microelectromech. Syst.*, **9** (1), 76–81.

116 Jang, J.-H., Ullal, C.K., Gorishnyy, T., Tsukruk, V.V., and Thomas, E.L. (2006) Mechanically tunable three-dimensional elastomeric network/air structures via interference lithography. *Nano Lett.*, **6**, 740–743.

117 Jiang, C., Markutsya, S., Shulha, H., and Tsukruk, V.V. (2005) Freely suspended gold nanoparticle arrays. *Adv. Mater.*, **17** (13), 1669–1673.

118 Jiang, C., Singamaneni, S., Merrick, E., and Tsukruk, V.V. (2006) Complex buckling instability patterns of nanomembranes with encapsulated gold nanoparticle arrays. *Nano Lett.*, **6** (10), 2254–2259.

119 Lin, Y.H., Jiang, C., Xu, J., Lin, Z., and Tsukruk, V.V. (2007) Sculptured layer-by-layer films. *Adv. Mater.*, **19** (22), 3827–3832.

120 Choi, T., Jang, J.-H., Ullal, C.K., Lemieux, M.C., Tsukruk, V.V., and Thomas, E.L. (2006) The elastic properties and plastic behavior of two-dimensional polymer structures fabricated with laser

interference lithography. *Adv. Funct. Mater.*, **16** (10), 1324–1330.

121 Jang, J.-H., Ullal, C.K., Maldovan, M., Gorishnyy, T., Kooi, S., Koh, C., and Thomas, E.L. (2007) 3D micro- and nanostructures via interference lithography. *Adv. Funct. Mater.*, **17** (16), 3027–3041.

122 Moon, J.H., Ford, J., and Yang, S. (2006) Fabricating three-dimensional polymeric photonic structures by multi-beam interference lithography. *Polym. Adv. Technol.*, **17** (2), 83–93.

123 Maldovan, M., Ullal, C.K., Carter, W.C., and Thomas, E.L. (2003) Exploring for 3D photonic bandgap structures in the 11 f.c.c. space groups. *Nat. Mater.*, **2**, 664–667.

124 Ullal, C.K., Maldovan, M., Thomas, E.L., Chen, G., Han, Y.-J., and Yang, S. (2004) Photonic crystals through holographic lithography: simple cubic, diamond-like, and gyroid-like structures. *Appl. Phys. Lett.*, **84**, 5434–5436.

125 Choi, T., Jang, J.-H., Ullal, C.K., Lemieux, M.C., Tsukruk, V.V., and Thomas, E.L. (2006) The elastic properties and plastic behavior of two-dimensional polymer structures fabricated by laser interference lithography. *Adv. Funct. Mater.*, **16** (10), 1324–1330.

126 Singamaneni, S., Chang, S., Jang, J.-H., Davis, W., Thomas, E.L., and Tsukruk, V.V. (2008) Mechanical properties of 2D polymer microstructures via interference lithography. *Phys. Chem. Chem. Phys.*, **10** (28), 4093–4105.

127 Decker, C., Viet, T.N.T., Decker, D., and Weber-Koehl, E. (2001) UV-radiation curing of acrylate/epoxide systems. *Polymer*, **42** (13), 5531–5541.

128 Singamaneni, S., Bertoldi, K., Chang, S., Jang, J.-H., Thomas, E.L., Boyce, M.C., and Tsukruk, V.V. (2009) Instabilities and pattern transformation in periodic, porous elastoplastic solid coatings. *ACS Appl. Mater. Interfaces*, **1** (1), 42–47.

129 Singamaneni, S., Bertoldi, K., Chang, S., Jang, J.-H., Young, S.L., Thomas, E.L., Boyce, M.C., and Tsukruk, V.V. (2009) Bifurcated mechanical behavior of deformed periodic porous solids. *Adv. Funct. Mater.*, **19** (9), 1426–1436.

130 Singamaneni, S., Kharlampieva, E., Jang, J.-H., McConney, M.E., Jiang, H., Bunning, T.J., Thomas, E.L., and Tsukruk, V.V. (2010) Metallized porous interference lithographic microstructures via biofunctionalization. *Adv. Mater.*, **22** (12), 1369–1373.

131 Anderson, K.D., Slocik, J.M., McConney, M.E., Enlow, J.O., Jakubiak, R., Bunning, T.J., Naik, R.R., and Tsukruk, V.V. (2009) Facile plasma enhanced deposition of ultrathin crosslinked amino acid films for conformal biometallization. *Small*, **5** (6), 741–749.

11
Highly Branched Macromolecules

A steady and growing interest in research directed toward molecules and macromolecules with highly branched structures is quite noticeable (Figure 11.1) [1]. This growing interest is primarily related to the fact that the multiple branches provide high concentration of functional terminal groups combined with a compact shape. This combination, along with architectural constraints, leads to significant changes in chemical and physical properties of highly branched molecules compared to their linear counterparts [2]. For example, highly branched polymers generally exhibit much lower solution and melt viscosities compared to linear polymers of the same molar mass, a characteristic that may help facilitate coating, extrusion, or other manufacturing processes [3].

Dendrimers, the most popular class of highly branched molecules, have been intensively studied for two decades as discussed in a number of excellent reviews and books [4–8]. A lot of exciting results related to chemical architectures, synthetic routines, encapsulating properties, aggregation behavior, assembly in solution and surfaces in conjunction with prospective applications in drug delivery, nanocomposite materials, and catalytic systems have been reported [9–11]. Other types of related highly branched structures with potentially interesting properties and microstructures have been introduced and intensively studied, and some selected AFM studies will be presented in this chapter.

11.1
Dendrimers and Dendritic Molecules

Classical dendrimers with all branches growing from a single center and dendritic molecules with dendritic branches being combined with other (rod-like or coils) blocks are the most popular class of highly branched molecules. These unique molecules can potentially be used as nanocapsules for drug delivery, molecular cages for inorganic nanoparticles, synthetic nanoparticles with multifunctional surfaces, and intriguing asymmetric block copolymers for unique morphologies. Surface and interfacial behavior of these molecules is of critical importance for their functioning and thus AFM studies are widely utilized to elucidate the shape of molecules after

Scanning Probe Microscopy of Soft Matter: Fundamentals and Practices, First Edition.
Vladimir V. Tsukruk and Srikanth Singamaneni.

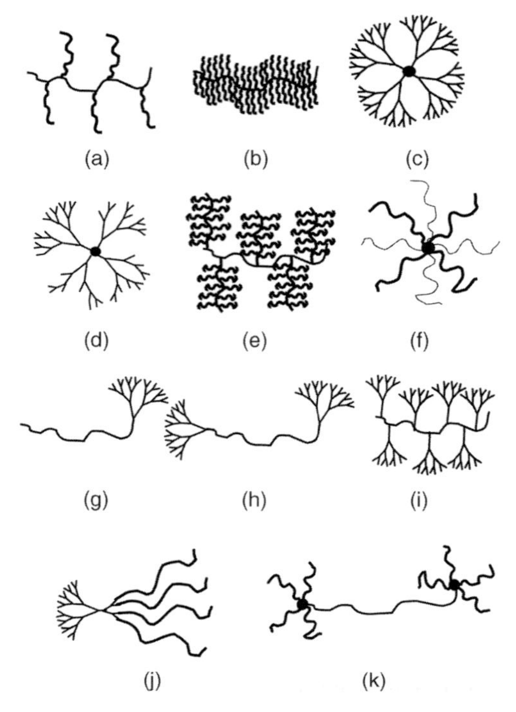

Figure 11.1 Selected common types of chain architectures for highly branched molecules of (a) graft, (b) brush, (c) dendrimer, (d) hyperbranched, (e) arborescent, and (f) stars, (g) dendron-linear, (h) barbell-like, (i) dendronized, (j) "bow-tie," and (k) dumbbell types. Reproduced from Ref. [1].

adsorption or grafting, grafting morphology of dendrimers, encapsulation of inorganic nanoparticles, functionality of dendrimer terminal groups, or physical properties of individual compacted molecules.

The variation of hydrophobic and hydrophilic interactions by changing the composition of terminal branches and cores of these tree-like molecules is considered to be a powerful tool for inducing organized supramolecular structures at surfaces and interfaces [12–18]. Because of architectural symmetry, a vast majority of dendritic molecules studied possess globular or near-globular shapes in solution, bulk, and at weakly interacting interfaces [11, 13, 19, 20]. Only specially designed dendritic molecules with peculiar architectures usually related to sterically asymmetric fragments were shown to be capable of forming self-assembled supramolecular structures such as rods, fibers, ribbons, and helices, all of which are shapes with a special interest for nanotechnology. The chemical architectures successfully used for these arrangements included shape-persistent planar dendrimers, hairy rod and diskotic polymers, rod-coils, and tapered molecules [21–26]. A proper combination of steric constraints, stacking interactions, internal rigidity, and hydrogen bonding is postulated to be critical for precise assembly of these molecules into large-scale, uniform, supramolecular structures.

Careful AFM analysis of the dendrimer molecules has been conducted by a number of groups on poly(amidoamine) (PAMAM) dendrimers. Imaging of dendrimers on different surfaces demonstrated that their conformations depend upon interfacial interactions as controlled by the nature of both their terminal groups and the surface groups [11]. In the case of strong interactions and low generation molecules, dendrimers are capable of forming uniform or aggregated monolayers with an overall thickness well below the expected diameter of spherical dendrimers in a bad solvent [13, 27, 28]. Even under weak surface interactions, the preservation of the round shapes of dendrimers of higher generation is challenging. Adsorbed dendrimers in the dry state are usually presented by an oblate shape in the slightly squashed state regardless of whether they are adsorbed as individual molecules or as a monolayer of ordered individual molecules [29, 30]. The direct AFM "counting" approach has been exploited for the estimation of the molecular volume of adsorbed PAMAM dendrimers by Li and coworkers [24].

The elasticity of the PAMAM dendrimers has been studied using AFM by several research groups. For example, Mecke *et al.* utilized direct comparison with the position of the center of gravity of dendrimers at air–solid and liquid–solid interfaces with the results of the computer simulations to conclude that the flexible nature of dendrimers is responsible for their significant flattening with the energy penalty for vertical deformation increasing significantly for higher generations [31].

In another study, Tomczak and Vancso conducted direct CFM measurements of the elastic response of PAMAM dendrimers and applied the Hertzian model for data analysis [32]. High-resolution AFM imaging revealed well-dispersed individual dendrimer molecules and their molecular aggregates with vertical layering (Figure 11.2). The elastic modulus value for the fifth generation of PAMAM was measured to be relatively high, 700 MPa, and this level of resistance was related mainly to changes in configurational entropy of tree-like molecules. The ability of

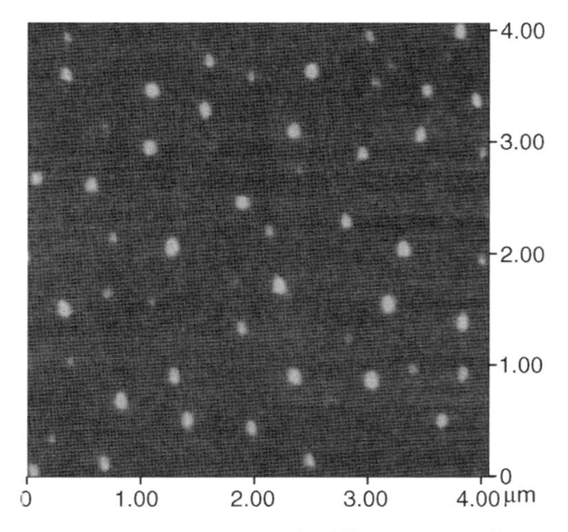

Figure 11.2 AFM image of the fifth generation of PAMAM dendrimers spin cast on the silicon substrate. Reproduced from Ref. [27].

making micropatterned surfaces from dendrimers has been also demonstrated by this group [33]. CFM studies conducted by Kim *et al.* demonstrated that the incorporation of dendrimers into LbL multilayers is an efficient way to soften the multilayers by decreasing the elastic modulus to values as low as 150 MPa due, mostly, to the change in the permeability of the multilayer shells [34].

Direct force measurements with CFM (negatively charged silica beads) have been applied to measure surface charge density of dendrimers [35]. The authors suggested that weak attractive forces at small separations are caused by the heterogeneous distribution of positively charged dendrimers on negatively charged surfaces with repulsive forces dominating larger separations. The charge density was observed to be significantly lower than that expected for charged spherical molecules. This deviation was suggested to be related to a small available volume for the diffuse layer of adsorbed dendrimers with smeared charges over a larger surface area and partial charge neutralization.

Surface adsorption of azobenzene-containing poly(phenylene) dendrimers on graphite was studied by Ding *et al.* [36]. The formation of well-ordered nanoscale stripes was observed with high-resolution AFM and ascribed to skewed packing of dendritic cores separated by alkyl tails. Similar ordered layered structures have been observed for LB monolayers of fan-shaped dendrimers by Yoon and Jung [37]. Liu *et al.* demonstrated that individual straight nanofibers can be formed on various surfaces from this type of dendrimer [38]. Strong fluorescence has been observed due to the grafting of dye molecules to the peripheral groups. Azobenzene-containing amphiphilic monodendrons were studied by Genson *et al.* for their ability to form flat LB monolayers on silicon surfaces at different surface pressures [39]. In this study, four generations of monodendrons with multiple dodecyl alkyl tails, an azobenzene spacer group, and a carboxylic acid polar head were studied at the air–water and air–solid interface using AFM and X-ray reflectivity. The increase in the cross-sectional mistmatch caused by the presence of the multiple chains for the higher generations disrupted the long-range ordering and forced the alkyl tails to adopt a quasi-hexagonal intralayer structure. The higher generations formed a kinked structure with the alkyl tails oriented perpendicular to the surface with the azobenzene group tilted at a large degree toward the surface. Chemical grafting of several different monodendrons has been exploited to "freeze" different level of dynamic dewetting, as was studied with AFM by Xiao *et al.* and Genson *et al.* [40, 41].

The introduction of rigid rod-like segments into dendritic molecules (monodendrons) proved instrumental in obtaining unusual one-dimensional surface assemblies well known for regular dendrimers. Dendritic amphiphilic rods formed peculiar one-dimensional structures such as microfibrils, ribbons, tubules, and twisted tubules depending upon chemical composition. Using a novel approach, the authors designed four-armed, star-shaped molecules with rigid thiophene segments linked to a carbosilane core [42]. They observed that these molecules form smooth and uniform monolayers, which are more stable than those formed from linear molecules of similar composition. Moreover, the variable surface morphologies of these molecules, including nanowires, were found to improve the charge transfer parameters of thin-film transistors with a noticeable increase in charge mobility and threshold voltage, making them promising materials for organic electronic devices.

Peculiar web structures have been formed by rod-dendritic molecules after slow evaporation of solvent on graphite and glass surfaces, as was demonstrated in a recent study by Genson *et al.* (Figure 11.3) [43]. The authors reported a novel mechanism of

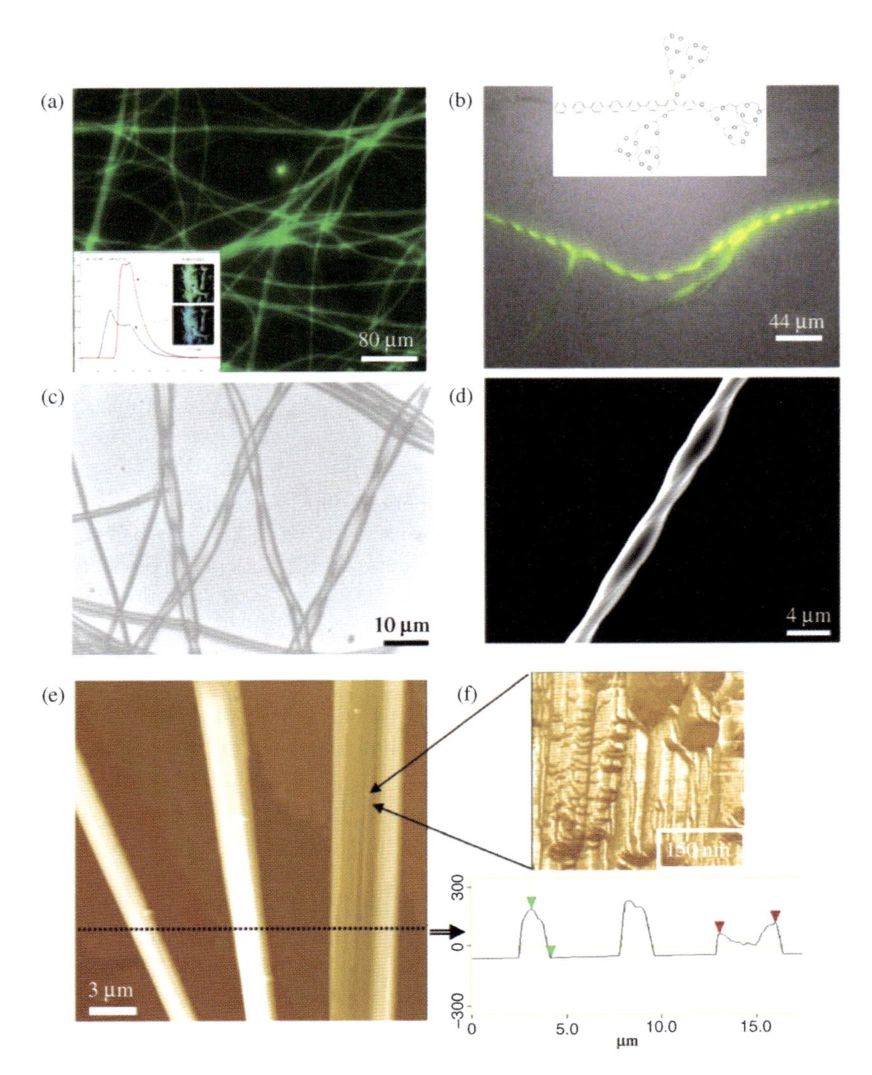

Figure 11.3 Dendronized rods: (a) high magnification fluorescent micrograph of the network; the inset shows fluorescence spectra under two different excitations. (b) Fluorescent micrograph of a twisted arm of the star-shaped aggregate; the inset shows chemical formula. (c) Optical micrograph of twisted tubular arms. (d) Higher-resolution SEM image of a twisted arm. (e and f) AFM topography image of fibrillar and collapsed tubular arms; cross section of different arms along with a high-resolution AFM image of the top area. Reproduced from Ref. [38].

assembly of dendronized rod molecules into a dense supramolecular fluorescent web featuring self-propelled motion of star-shaped aggregates within a solution droplet in the course of one-dimensional growth of the pinned arms. An intriguing assembly mechanism discovered in this work involves microscopic (hundred micrometers) directional motion of the aggregates driven by one-dimensional molecular assembly, resulting in a dense mesoscopic supramolecular fluorescent web composed of micro- and nanofibrils radiating from central regions. AFM analysis of these ribbons showed lamellar morphology with the thickness of a single step controlled by molecular dimensions (Figure 11.3).

Another type of amphiphilic, dendron-rod molecule with three hydrophilic PEO branches attached to a hydrophobic octa-*p*-phenylene rod stem was investigated due to its ability to form two-dimensional, micellar structures on a solid surface during LB deposition (Figure 11.4) [44]. The tree-like shape of the molecules was reported to be a major factor in the formation of nonplanar micellar structures in solution and in the bulk (cylindrical and spherical). However, within LB monolayers, these amphiphilic dendritic molecules assembled themselves into layered or circular micellar structures (Figure 11.4). The authors suggested that the formation of the planar ribbon-like structures within the loosely packed monolayers is caused by interdigitated linear molecular fragments. However, reduced compression stress and weaker interfacial interactions could result in the formation of circular, ring-like structures (2D circular aggregates) within the second layer, formed on top of the monolayer in the precollapsed state, as revealed by high-resolution AFM images (Figure 11.4).

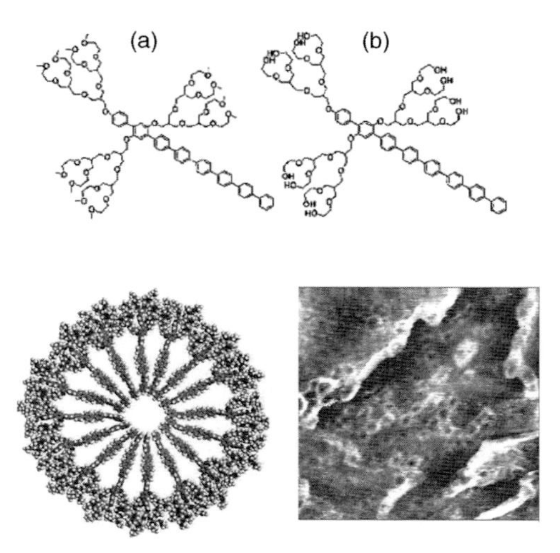

Figure 11.4 Chemical structures of dendritic rod molecules with methyl-terminated (a) and hydroxyl-terminated (b) branches and a molecular model of molecule A packing showing a circular planar structure (*bottom*). AFM phase image ($300\,nm^2 \times 300\,nm^2$) of LB monolayer from molecule A at molecular area $1.2\,nm^2$. Reproduced from Ref. [39].

11.2
Brush Molecules

Brush molecules with long grafted chains show intriguing differences in their behaviors caused by different conformations of backbones and side groups and their competition in overall chain conformation, surface interactions, and intramolecular phase separation. AFM imaging is critical for direct visualization of individual molecules adsorbed onto surfaces in static and dynamic regimes, evaluation of fine molecular characteristics, or measuring chain flexibility.

Brush molecules of different types (Figure 11.1) have been intensively studied in a series of publications that resulted from collaboration between the groups of Sheiko, Moller, and Matyjaszewski as summarized in an excellent review [45]. Initial efforts of these groups were focused on the application of high-resolution AFM for the characterization of individual brush molecules at different surfaces in order to evaluate contour molecular lengths, molecular weight characteristics, persistence lengths, adsorption kinetics, conformational states in different solvents, and different constrained conditions, surface-mediated segregation of various molecular fragments, Brownian surface motion of synthetic and biological macromolecules, and cluster formation on planar and defective surfaces. The authors took advantage of the much larger effective diameter of brush molecules that makes their visualization with AFM and quantitative analysis of the morphological data much easier and less ambiguous. The initial results reported were direct AFM-based restoration of the fine details of molecular weight distribution that cannot be obtained with routine analytical techniques for these complex branched macromolecules.

Among the recent developments summarized in the recent review from Sheiko and Matyjaszewski, there are two interesting results for a variety of cylindrical brush molecules with other studies also to be highlighted below [46]. First, visualization of individual brush molecules with AFM for evaluation of their molecular characteristics was extended to complex star brush molecules of different compositions (Figure 11.5) [47, 48]. Simple brush stars with only four arms as well as stars with a large number of arms can be visualized with fine details of side-chain conformations. Amphiphilic multiarm star block copolymers with a large number of arms demonstrated in this image were prepared by consecutive ATRP of *n*-butyl methacrylate (BMA) and poly(ethylene glycol) methyl ether methacrylate (PEGMA), using a 2-bromoisobutyric acid-modified hyperbranched polyester (Boltorn® H40) as macroinitiator. The use of such a multifunctional macroinitiator allows for the preparation of star polymers that contain significantly more arms compared to most systems that have been reported to date, which is expected to be beneficial to explore the unimicellar characteristics of these polymers for encapsulation and release applications.

This approach extended the AFM-based principles of direct "counting" of molecular lengths and gyration radii of whole molecules or individual fragments (arms) toward such complex brush-star molecules and even in the course of a controlled chemical reaction and growth [49, 50]. Sheiko *et al.* demonstrated that combining mass per unit area from LB measurements with direct counting of molecular density

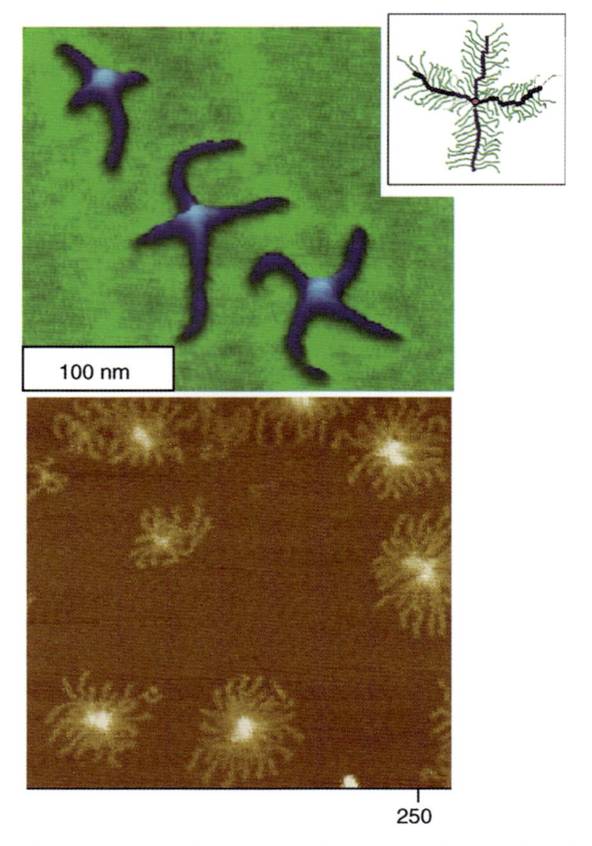

Figure 11.5 Visualization of star-like brush molecules. (*Top*) Four-arm brush star. Reproduced from Ref. [46]. (*Bottom*) AFM micrograph of multiarm H40-PBMA58-b-PPEGMA40 stars. Courtesy of Prof. S. Sheiko.

from AFM images allows for straightforward evaluation of the average molecular weight and polydispersity index [51].

Second of the interesting series of results is the elucidation of the role of 2D confinement on conformational states of flexible backbones as assisted by variable grafting density of the side chains [46]. AFM imaging demonstrated how the selection of grafted side chains can control transitions between bent coiled states or extended compacted states. In addition, restriction in available surface area per molecule was observed to result in both compact disk-like conformation and the formation of ordered hexagonal packing in the case of four-arm molecules [52]. In another study, the variations of counterions have been exploited to induce collapse transitions in polyelectrolyte brushes and were monitored by AFM [53, 54].

Amphiphilic cylindrical brush-coil block copolymers consisting of a hydrophobic PS coil and a cylindrical brush block with hydrophilic PAA side chains synthesized by ATRP have been studied with AFM [55]. The formation of micelles in aqueous solution upon hydrolysis of the PtBA was confirmed by AFM imaging of the deposited macromolecules onto a bare silicon oxide surface. Tadpole-shaped (or rod-coil) block-graft copolymers, consisting of a pentafluorostyrene polymer (PFS) block and a glycidyl methacrylate polymer (PGMA) block with grafted PtBA side chains, or PFS-*b*-(PGMA-g-PtBA) copolymers, were reported [56]. Hydrolysis of the PtBA side chains in the block-graft copolymer gave rise to a strong amphiphilicity of PFS-*b*-(PGMA-g-PAAC) macromolecules and their domain texture with phase-separated appearance upon adsorption. A peculiar shape of the adsorbed macromolecules with a brush-shaped hydrophilic rod-like head and a collapsed hydrophobic tail (coil) was suggested to be controlled by the block composition. Another investigation of macromolecular brushes with a gradient of side-chain spacing along the backbone demonstrated the characteristic anisotropy of the molecular structures adsorbed onto mica with a similar bulky head and a thin tail.

In another example, cylindrical brush-like molecules with poly(L-lysine) and poly (L-glutamate) side chains displayed a uniform worm-like chain conformation when deposited on a mica [57]. New dendritic macromolecules of polypyrrole-graft-poly (ε-caprolactone) (Ppy-g-PCL) copolymers exhibited microphase separation between the polypyrrole and the polycaprolactone segments after deposition onto mica leading to the formation of an ultrathin film with domain morphology [58]. In addition, the statistical, asymmetric P2VP-PMMA cylindrical brush macromolecules after their spin casting onto the mica surface from different solvents displayed a variety of structures: worm-like, horseshoe, and meander-like [59].

The question of real-time dynamics of surface spreading, wetting behavior, and surface restructuring remains a crucial issue for the branched molecules. To date, very few studies have addressed this issue. Sheiko and coworkers have demonstrated a real-time spreading behavior of a grafted polymer dropped on a solid substrate, with molecular resolution, by using a high-resolution *in situ* AFM monitoring technique (Figure 11.6) [60]. Individual and collective motions of brush molecular chains confined within molecular layers have been tracked with excellent quality on a flat graphite surface, and pinning and molecular reorientation around the surface defects

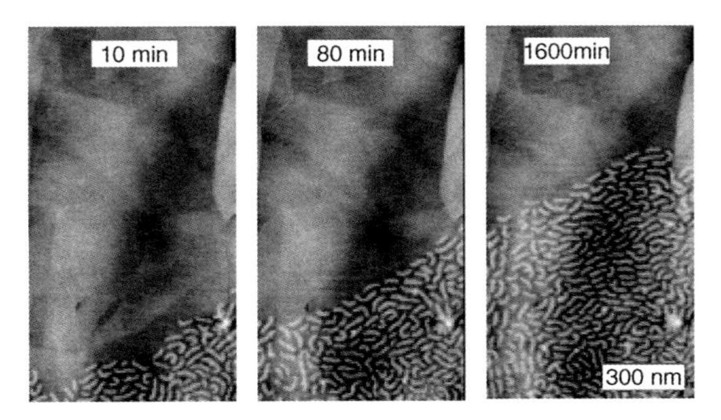

Figure 11.6 Wetting behavior of the monolayer of PBA brushes on the HOPG surface. The images were captured at different spreading times following deposition: 10, 80, and 160 min. Reproduced from Ref. [60].

were visualized in real time. From the analysis of the independently AFM measured microscopic spreading rate, the flow-induced diffusion rate of molecules, and the thermal diffusion coefficient of single molecules, the authors concluded that the plug flow of polymer chains was the main mass transport mechanism of molecular spreading of brush molecules. A very insignificant contribution from the actual molecular diffusion in the spreading process was suggested, as well.

As the next step, Sheiko and coworkers studied the real-time conformational transition of brush macromolecules in the process of their adsorption on a mica surface [61, 62]. They visualized the spreading behavior of individual branched molecules and interpreted the observed variations in terms of interplay between enthalpic and entropic terms of individual molecules and accompanying variations of surface energies. The authors suggested that the coadsorption of small organic molecules from solution controls the adherence and spreading of the macromolecules, tuning the surface expansion and the collapse of the molecules in a predictable manner. Coadsorption of macromolecules with different surfactants was also considered as an efficient way to control conformation and the mobility of macromolecular chains [63].

In a recent study, the same group demonstrated that a physical adsorption of brush-like macromolecules with very long side chains can induce not only conformational reorganization and deformations but also spontaneous rupture of covalent bonds in the macromolecule backbone, as was monitored with continuous molecular-level observation with AFM over a long period of time following initial adsorption [64]. This unexpected surface behavior for dendritic polymers with very high molecular mass was attributed to the fact that the attractive interactions between the side chains and the substrate is maximized by the spreading of the side chains, which, in turn, induces significant tension along the polymer backbone. The 2D spreading confined to the monomolecular surface layer conflicts with the inherited 3D radial architecture of highly branched molecules bringing significant intramolecular stresses into the overcrowded molecular interior on the flat surface. The authors suggested that similar adsorption-induced backbone scission of bonds located at the center might occur for sufficiently long macromolecules with highly branched architectures.

The cylindrical molecular brushes *p*[(MMA-*grad*-BPEA)-*graft*-nBA] with a gradient of grafting density along the backbone displayed a transition from rod-like to tadpole conformation upon compression. AFM revealed the shapes of individual molecules and two distinct microphases composed of different blocks coexisting within individual molecules adsorbed onto mica [65]. Even more complex 3D morphologies have been observed within LB films of dendronized block copolymers and were attributed to the coil-rod architecture of the molecules [66].

Hammond and Iyer presented a very different type of branched block copolymers, namely, a series of PEO-PAMAM linear-dendritic diblock copolymers with various terminal functionalities [67]. These hybrid, amphiphilic, stearate-terminated diblocks were found to give stable Langmuir monolayers with the surface molecular area in the condensed state controlled by the architecture of molecules and their chemical composition. LB films of these diblocks transferred onto hydrophobically functionalized surfaces at high surface pressure displayed continuous topography with a relatively smooth surface. Another recent example of highly branched hybrid molecules with a hydrophobic, helical polypeptide comb block and a hydrophilic polyester dendron block modified with PEG was presented by the same group [68]. The authors demonstrated that the cone-shaped molecules formed spherical micelles at very low solution concentration. These micelles were robust enough to preserve virtually spherical morphology after adsorption onto a solid substrate and complete drying.

11.3 Hyperbranched Polymers

Due to relatively simple and accessible synthetic routines, many hyperbranched materials became available in relatively large quantities (grams to kilograms) at a reasonable cost (Figure 11.1). Several successful examples of commercial materials are already available (hyperbranched polyesters introduced by Polyols) [69, 70]. AFM observations of hyperbranched molecules are widely utilized to address similar issues for typical dendrimers – the shape of molecules and their grafting abilities or molecular functionalities at different surfaces.

From the viewpoint of general characterization, Voit and coworkers investigated the surface properties of ultrathin surface films of hyperbranched polyesters with different end groups by zeta-potential and contact angle measurements [71]. They observed that the differences in the molecular structure and the surface properties for different hyperbranched molecules with carboxylic, hydroxyl, and acetoxyl end groups caused significant differences in their swelling behavior in a humid atmosphere. Similar studies have been conducted in order to test different types of hyperbranched molecules with various chemical compositions. The hyperbranched poly(urea urethane)s with different terminal group substitutions have been analyzed [72]. It has been shown that the surface properties were mainly controlled by the strong interactions of the urea and urethane terminal groups within the internal branches.

Recent studies by Nishide *et al.* investigated surface morphology of the hyperbranched poly[(4-(3′,5′-di-*tert*-butyl-4′-yloxyphenyl)-1,2,(6)-phenylenevinylene)] from a dilute solution onto mica and graphite surfaces [73]. Although the authors observed that these molecules collapsed on a graphite surface due to unfavorable surface interactions, regular globular structures were detected on a mica surface. In another study, the surface microstructure of highly branched sulfonated polydiphenylamine, H-PSDA, was found to be sensitive to pH [74]. In the dedoped state ($pH > 7$), AFM imaging identified nanoscale particles of uniform size with diameters of about 40 nm, although larger and irregular particles (>200 nm) were formed at the same pH. Furthermore, when the pH was decreased below 5, the surface aggregates became dispersed due to the external doping interaction caused by the HCl molecules. Cylindrical hyperbranched poly(chloroethyl vinyl ether)-PS molecules at different surface coverages have been observed with high-resolution AFM by Viville *et al.* [75]. The authors observed either lamellar organization at high surface coverage or individual ring-shaped molecules for sparse distribution of molecules. The latest images have been successfully utilized for calculation of the molecular weight of hyperbranched molecules.

Modified hyperbranched polyesters formed ordered and dense surface layers because of strong adsorption onto properly functionalized surfaces [76]. The surface behavior of third and fourth generations of hyperbranched polyesters (Boltorn®) with an average of 32 and 64 hydroxyl terminal groups was studied on a surface of silicon oxide with a high concentration of hydroxyl groups (Figure 11.7). The spontaneous adsorption of individual hyperbranched polyester molecules and their coalescence into uniform monomolecular surface layers have been observed under different conditions. It has been shown that the molecular adsorption on a bare silicon oxide surface followed a typical Langmuir adsorption isotherm. However, the resulting surface structures were very different for low and high generations of hyperbranched molecules. The higher-generation molecules kept their close-to-spherical shape in a highly dispersed state (Figure 11.7). The globular molecular shape of individual

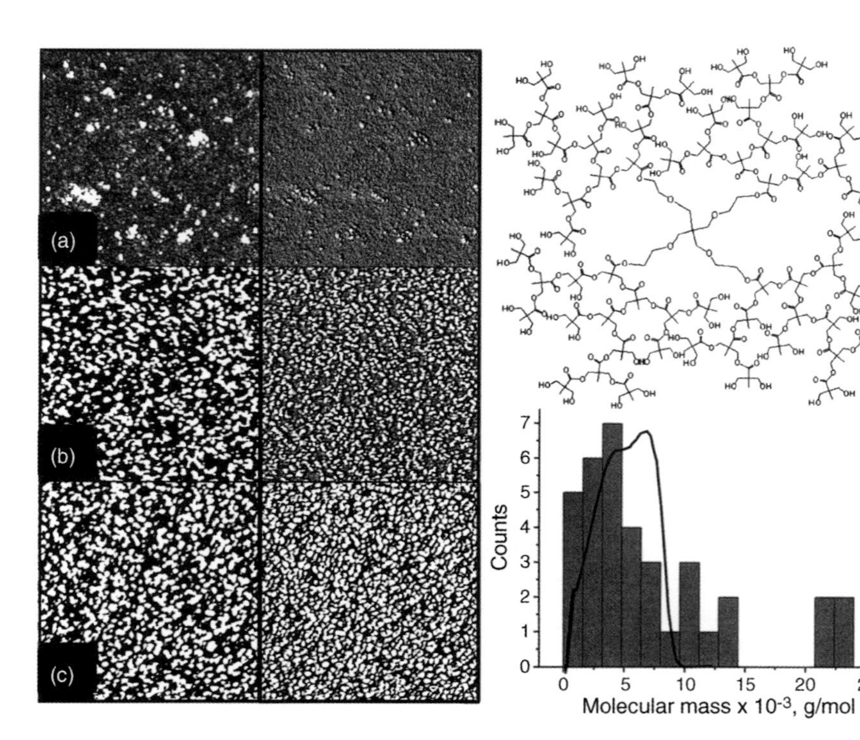

Figure 11.7 AFM images of third (*left*) and fourth (*right*) generations of hyperbranched polyesters (chemical schematics in the inset) adsorbed from different solution concentrations: 0.5 g/L (a), 1.5 g/L (b), and 5.0 g/L (c). Scan size is 2 μm × 2 μm; height scale is 5 nm. Molecular weight distribution of fourth generation is based on the diameter distribution data (histogram) and GPC data (solid line). Reproduced from Ref. [76].

molecules and their aggregates was also preserved within the densely packed adsorbed molecular layers. The average thickness of the molecular layer was close to the molecules' diameters in the compacted state.

In contrast, lower-generation hyperbranched polyester molecules, with a more compliant architecture and lower terminal group densities, went through significant conformation changes after adsorption on a silicon surface [77]. This restructuring resulted in the highly flattened shape of the molecules at a low surface coverage, which was replaced with patch-like structures at intermediate coverage. At higher surface coverage, these hyperbranched molecules formed a densely packed monolayer with a uniform surface morphology. Worm-like nanoscale structures were observed, indicating coalescence of the individual molecules into uniform matter that can be related to the lower charge density of the terminal groups and higher compliance of the structure.

The data collected for hyperbranched molecules allowed for the evaluation of the average molecular weight and molecular weight distribution of the molecules on the surface (Figure 11.7). The calculation assumed disk- or ellipsoidal shape and the bulk density of polyesters. The average volume of adsorbed fourth-generation hyperbranched molecules as calculated from this distribution was 8.5 nm^3, and the averaged molecular weight was estimated to be 7800 g/mol. These values were in good agreement with the GPC data, and the shape of the histogram was fairly close to the molecular weight distribution from GPC data (Figure 11.7).

The nanoscale and elastic properties of these hyperbranched polyester molecules were probed using SFS of individual molecules [78, 79]. A double-layer elastic deformational model was applied to calculate nanomechanical properties of the monomolecular aggregates from hyperbranched polyesters by the application of the modified Hertzian approach (for details, see Chapter 5). Higher nanoscale stiffness was also detected for molecules closely packed within long-chain molecular aggregates as compared to individual molecules dispersed all over the surface or packed within small molecular aggregates.

Moore and coworkers reported the strong effect of molecular architecture on the surface properties of hyperbranched polyether imides [80]. It has been shown that the surface properties of the spin-cast films obtained from these hyperbranched polymers strongly depend both on the functionality of end groups and on the fine details of molecular architecture. They observed that the increasing number of branches in the molecules limited the mobility of polymeric segments, which inhibited the migration of interior end groups to the film surfaces. As a result, the surface energy of randomly branched polymers scaled closely with the terminal segment composition. The variation in the functionality of end groups of these molecules extensively altered the surface properties of the resulting films. In a different approach, Hult and coworkers implemented the light-controlled switchability of molecular surface layers by using photosensitive hyperbranched polymers with azobenzene end groups [81]. Similar results have been obtained for the covalently grafted azo-containing photosensitive monodendrons [82, 83]. AFM images demonstrated surface relief on the order of 1 nm in films undergoing a photoisomerization process [84].

The kinetics of adsorption of hyperbranched polyesteramides on silicon oxide surfaces as a function of pH and salt concentration has been studied by Ondaral *et al.* [85]. The authors observed that these hyperbranched polymers formed thicker and more rigid adsorbed layer compared to similar linear polymers with very minor dissipation of the mechanical energy and a reduced viscous response. The difference in properties observed in this study was related to the absence of mobile loops and tails in adsorbed hyperbranched molecules. Recently, Haag and coworkers presented nanostructured hyperbranched PMMA gradient surfaces obtained via self-assembly of semifluorinated hyperbranched polyglycerol amphiphiles (FPG) at the PMMA surface during free radical bulk polymerization [86]. The surface exhibited unusual nanolayered superstructures composed of alternating fluorine-containing hydrophobic and fluorine-free hydrophilic nanometer-scaled regions.

In a recent publication, Wang and Wooley investigated the temporal evolution during the process of degradation of the cast hyperbranched poly(silyl ester) films in liquid environment by monitoring the changes in the surface topography [87]. The authors concluded that the unique two-stage surface degradation process observed in their studies was controlled by the composition and the architecture of the tethered branched molecules. In a related study, Wang and Wooley also studied the surface properties of hyperbranched polyfluorinated benzyl ether polymers [88]. Analysis of the AFM data suggested a phase separation within surface films, which caused a significant decrease in friction and adhesive forces. In a later study, these hyperbranched fluoropolymers with a large number of pentafluorophenyl ends were cross-linked with diamino-terminated PEG or diamino-terminated poly(dimethylsiloxane) (PDMS) to form a much more stable hyperbranched network of different compositions with variable surface wettability on glass substrates, prefunctionalized by APTS SAM [89]. The AFM data revealed well-developed phase segregation of HBFP contained in PEG or PDMS matrices with the domain dimensions dependent upon the stoichiometry and the thickness of these films. As a result, the increase in PEG content leads to an increasing hydrophilicity of the multicomponent coatings and their increasing resistance toward the adsorption of bimolecules (antibiofouling).

Intriguing fibrillar structures with helical morphologies have recently been observed for hyperbranched polymers crystallized from slow evaporating solution [90]. Yan and coworkers presented the macroscopic molecular self-assembly of an amphiphilic hyperbranched copolymer, poly(3-ethyl-3-oxetanemethanol) core and hydrophilic arms (HBPO-PEO), cast from acetone. These tubular aggregates have a wall thickness of 400 nm, are several millimeters in diameter, and are a few centimeters in length. It has been reported that this hyperbranched copolymer can also self-assemble from water to form giant polymer vesicles [91].

The formation of long and straight microfibers in the course of crystallization from solution was observed for modified hyperbranched polyester molecules with bulky anthracene end groups [92]. In this case, straight nanoribbons with a sheet-like lamellar surface morphology have been observed with AFM at a higher magnification. The authors speculated that the multiple intermolecular hydrogen bonding, polar interactions between flexible cores, and strong π–π interactions between anthracene terminal groups stabilize these nanofibers and make them well-defined, shape-persistent, and near-perfect structures spanning over a range of spatial scales from nano- to macroscale. The authors suggested that the directional crystallization of multiple peripheral fragments attached to irregular cores could be responsible for assembling near-perfect, straight, and uniform supramolecular one-dimensional structures.

The authors suggested that in order to form nanofibrillar structures of this type, the amphiphilic, hyperbranched molecules adapt a highly asymmetric conformation, which is very different from the symmetrical extended conformation observed for unmodified molecules. Ornatska *et al.* investigated the surface structures of the modified hyperbranched polyester cores with varying compositions of the terminal groups [93, 94]. The esterification of hyperbranched cores generated modified branched molecules with 50 hydrophobic palmitic (C_{16}) alkyl tails, 14 amine-, and 1–2 hydroxyl terminal groups, capable of forming nanofibrillar structures. The amine groups were introduced to enhance polar interactions with the silicon oxide surface in order to stabilize surface nanostructures and were proven to be critical for the formation of nanofibrillar structures (Figure 11.8) [93]. The core–shell architecture of the amphiphilic dendritic molecules provided exceptional stability for these one-dimensional nanofibrillar structures after complete drying.

The proposed model suggests semicylindrical conformation in which the hydrophilic cores are squashed against the solid surface, and the hydrophobic terminal

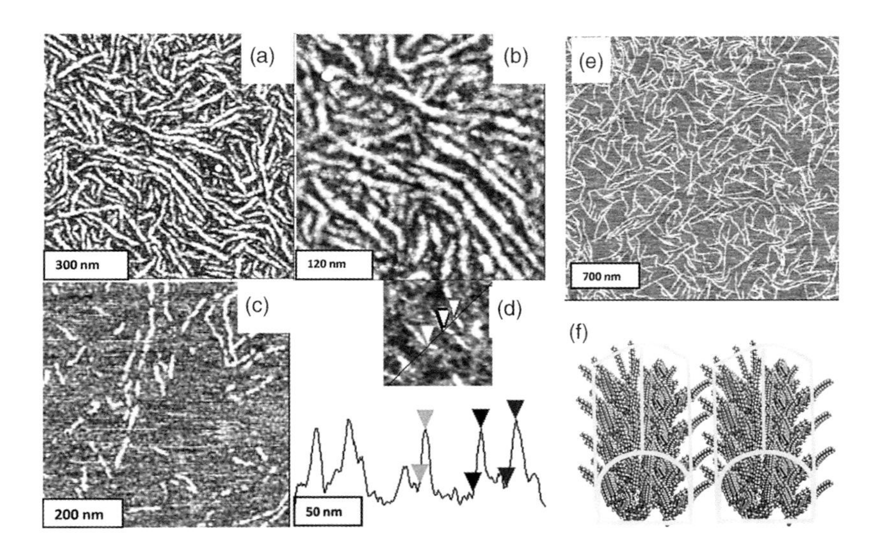

Figure 11.8 High-resolution AFM phase (a,b) and topographical (c,d; z-scale is 3 nm) images of nanofibrillar structures assembled from hyperbranched molecule at 20 mN/m (a,b) and 0.2 mN/m (c,d). An example of a cross section across different nanofibers (see insert, 200 nm × 200 nm) collected with the carbon nanotube tip and used for true nanofiber diameter calculations (d). AFM topographical images of nanofibers (e). Molecular models for assembling in the semicylindrical, one-dimensional structures (f). Reproduced from Ref. [87].

branches are concentrated in the topmost layer (Figure 11.8). This shape is appropriate for the face-on packing at the hydrophilic surface. The critical condition for the chemical composition resulting in the formation of the nanofibrillar structures is the presence of both alkyl tails in the outer shell and amine groups in the core–inner shell that provide for the multiple intermolecular hydrogen bonding and polar interactions between flexible cores.

The significant role of the amount of alkyl-terminated groups on surface behavior was also studied using a series of pseudo second-generation modified hyperbranched polyesters [95]. For these molecules, the critical number of alkyl-terminated groups was determined to be two chains per hydrophilic core in order to assure formation of the stable monolayers. For larger numbers of alkyl tails, the formation of well-developed domain surface morphology with round and dendritic shapes of individual monolayer regions with smooth surfaces was observed with AFM. Careful AFM cross-sectional measurements of the height of these monolayer domains in conjunction with molecular modeling revealed that the hydrophilic cores show prolate shape and are significantly submerged into water subphase. A larger number of alkyl tails and a higher compression led to the transformation of the core into an oblate, flattened shape with the preservation of the standing-off orientation of the alkyl tails.

Partially epoxy-terminated hyperbranched molecules were firmly tethered to the silicon oxide surface to form robust elastic surface layers of 4–6 nm thickness [96]. Grafted layers formed from these molecules were homogeneous on the nanoscale without any signs of the microphase separation usually observed for mixed chemical composition. The authors suggested that architectural constraints caused by the attachment of the dissimilar branches with epoxy and alkyl terminal groups to a single core effectively suppress their phase separation. Finally, it was demonstrated that the efficient grafting of these surface layers leaves a fraction of epoxy groups localized at the film surface that are readily available for further hydrolyzation and grafting. To test the micromechanical properties, SFS micromapping of the hyperbranched layer was performed as described in Chapter 5 (Figure 11.9) [96].

The matrix of 32×32 probing pixels was used to evaluate surface distribution of compression elastic modulus and adhesive forces with a lateral resolution close to 30 nm. It was observed that this grafted layer sustained very significant reversible deformations with compressions as high as 80%. As is made clear from elastic and adhesive micromappings, the surface distribution of both properties was statistically uniform (Figure 11.9). Surface histograms were relatively narrow with a standard deviation below 5% for adhesive forces and 25% for elastic modulus. The apparent elastic modulus of the layer was about 300 MPa due to the contribution of a stiff silicon substrate, and an elastic modulus of 11 MPa was obtained for the layer itself by exploiting the double-layer model (Figure 11.9). This value was close to the typical moduli for highly cross-linked rubbery materials and demonstrated superior elastic response due to the presence of both internal cross-linkings and multiple grafting to the surface.

Sheiko *et al.* investigated the surface behavior of hyperbranched polymers containing trimethylsilyl or hydroxyethyl end groups at the air–water interface [97]. Comparing the spreading behavior of the OH-terminated molecules with that of the

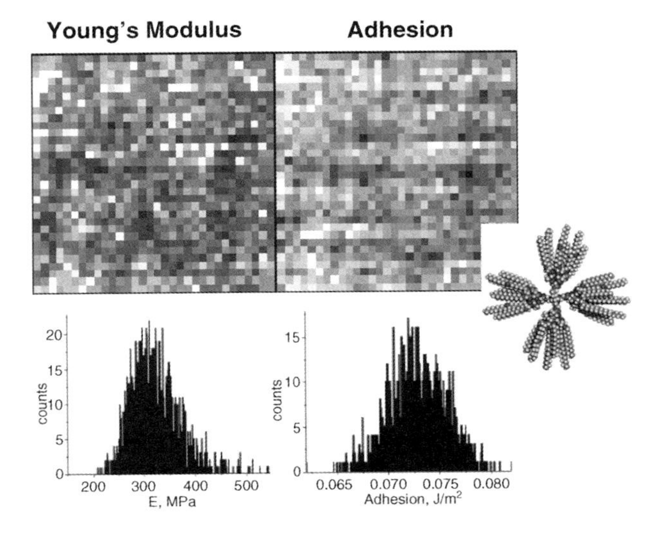

Figure 11.9 Surface distribution of the elastic modulus and adhesive forces for the epoxy-terminated hyperbranched molecules (inset) grafted to silicon obtained with 32 × 32 probing within the 1 μm × 1 μm surface area (*top*); corresponding histograms of surface distribution of elastic and adhesive properties (*bottom*). Reproduced from Ref. [96].

hyperbranched polymer with identical chemical composition revealed that the hyperbranched polymer did not show any well-defined intramonolayer transitions in the course of the monolayer compression and, instead, exhibited a more smeared surface behavior, similar to that of isotropic amphiphilic compounds.

An alternative routine for the formation of 3D structured films with nanoscale thickness is electrostatically driven LbL assembly for dendrimers and has been recently expanded to include charged hyperbranched molecules (polyelectrolytes) [98]. Qiu *et al.* used carboxyl-terminated aryl-alkyl hyperbranched polyester poly(5-hydroxyethoxyisophthalic acid) (PHEIA) for LbL assembly in combination with poly(diallyldimethylammonium chloride) (PDAC) [99]. AFM studies revealed sphere-like particles with sizes in the range of 100–225 nm, which were attributed to the aggregation of hyperbranched PHEIA within LbL films. Another study of anionic hyperbranched polymers with the same backbone but different end groups in the presence of PDAC confirmed the critical role of the terminal groups in the surface organization of resulting LbL films.

An amphiphilic polymer composed of a hydrophobic divinylbenzene-styrene microgel core, PMMA branches, and some pendent vinyl groups attached to the core has been presented by Cha *et al.* [100]. The surface morphology of LB monolayers and multilayers showed that these polymers spread on the water surface, forming surface aggregates, which become finer with increasing hydrophilic content. The authors demonstrated that LB multilayers can be polymerized further to form a robust, cross-linked structure. Janus micelles from asymmetric star block copolymers with the cross-linked PB block of PS-*b*-PB-*b*-PMMA triblock copolymers were demonstrated to form stable monolayers on a water surface [101]. It is worth noting that another study of Langmuir monolayers fabricated from oxazoline-functionalized PMMA microgels and hyperbranched PVE and PS showed nanoscale polymer hyperbranched particulates with long-range ordered packing [102].

11.4
Star Molecules

Star molecules, with a large number of long arems that are dissimilar in chemical composition, show very different behavior from that discussed above due to the length of their arms, which result in a very strong segregation phenomenon that controls their surface behavior. In more than one sense, they behave as peculiar block copolymers. AFM studies of this class of molecules mostly focuses on characterization of microphase separated structures formed by star molecules and the role of environmental conditions on the resulting aggregated surface structures.

Single molecular conformations and the associated surface morphologies for heteroarm star copolymers, which are deposited on hydrophilic mica or silicon oxide surfaces and exposed to a controlled environment, have been studied using AFM in a number of recent studies [45, 46, 103]. Similarity between the surface behavior of rubbery star polymers and soft colloids was indicated by Glynos *et al.* [104]. Conformational rearrangements of star molecules in the course of aggregation was monitored by AFM and related to a changing balance of surface and intermolecular interactions. Significant restructuring of the surface anchored cationic polyelectrolyte stars in the presence of counterions under UV light has been observed by Plamper *et al.* [105].

Very few publications considered the morphology of bulk star-shaped materials with respect to their crystallization behavior. In one such study, Nunez *et al.* considered solution- and bulk-grown partially crystalline materials and star polyesters with poly (e-caprolactone) (PCL) arms with different thermal histories [106]. The authors observed similar crystalline morphologies in star-shaped and linear polymers with trends toward surface-segregated branched cores. Moreover, covalent attachment of PCL arms to a single core resulted in different thermal stability of different crystal facets and, hence, more irregular growth of crystals from star polymers.

The PS_7-$P2VP_7$ molecules were observed to form unimolecular or multimolecular micelles in acidic conditions with the aggregation type depending on the concentration and pH of the solution (Figure 11.10) [107]. The micelles deposited on mica from acidic water were trapped with P2VP extended arms caused by a strong interaction with the mica surface. In contrast, the hydrophobic PS arms collapsed and formed elevated cores of star-like micellar aggregates (Figure 11.10). Upon the treatment of these surface structures with toluene, the PS cores become highly swollen with the PS arms gradually adapting an extended conformation. Eventually, their spreading results in the very thin (down to a single chain) "squashed" structures that could not be obtained from a simple, single-step adsorption procedure. The authors demonstrated that the star block copolymers, weakly anchored onto the solid

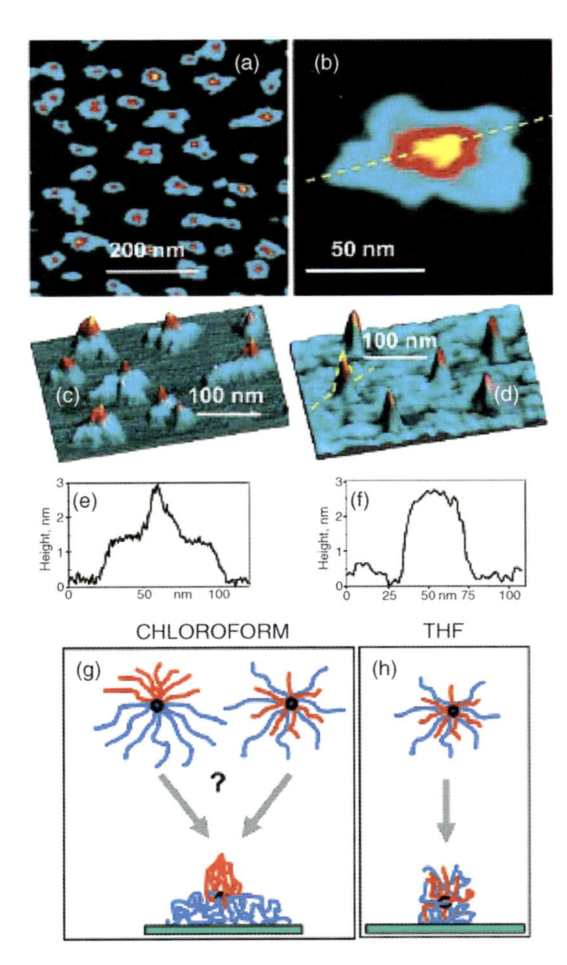

Figure 11.10 AFM topographic images (a–c) and cross sections (e,f) of the PS_7-$P2VP_7$ molecules adsorbed from chloroform on mica. 3D-AFM image (d) and cross section (f) of unimers adsorbed onto mica from THF. Suggested conformations for the PS_7-$P2VP_7$ adsorbed from chloroform (g) and THF (h) (red color indicates PS arms). Reproduced from Ref. [107].

substrate, could form environmentally responsive nanoscale structures with symmetry controlled by the chemical architecture. Moreover, they observed that the cooperative reorganization of multiple arms and cores of molecular aggregates upon solvent treatment resulted in switchable "dimple" and "ripple" surface structures within dense molecular layers. This example illustrates how the dramatic response of the molecular shape upon alternating external stimuli is facilitated by core–shell transitions of a single star molecule, an unprecedented behavior of star-like molecules, which cannot be observed for linear block copolymers.

At the next step, the authors fabricated unimolecular hybrid nanostructures by decorating the star molecules deposited on the surface in various conformational states with hexacyanoferrate anions or negatively charged clusters of cyanide-bridged complexes [108]. The decoration process, guided by selective interactions between ions and molecular backbones, initially allowed for an increasing contrast of AFM imaging. Hence, the fine molecular morphology with multiple arms became easily observable. However, the authors further advanced this approach toward a molecular template-guided metallization on surface-pinned PS_7-$P2VP_7$ molecules. To accomplish this, they exploited highly selective Pd cluster deposition along the extended P2VP chains and reduced metal ions. Metallization of ion-containing P2VP arms of these star-shaped molecules led to the peculiar organic–inorganic hybrid single-molecule structures with nanoscale palladium clusters predominantly localized along the extended polymer arms [109]. Molecules with different numbers of metallized extended arms have been obtained and readily visualized with AFM.

Excellent AFM images with single molecule resolution of four-arm star-like and comb-like single molecules deposited on different surfaces have been obtained [110]. Different initial configurations were observed as controlled by initiation and propagation reactions with predominant four-arm shape easily visible on AFM images (Figure 11.11). These molecules adopted various shapes with either widely extended arms or closely packed arms depending upon deposition conditions demonstrating conformational flexibility on weakly interacting surfaces.

In another study, six-armed star molecules with PS and PMMA arms and a triphenylene core show different self-assembling behavior on a surface [111]. For these star polymers, isolated PS cylinders on mica have been observed in dense states. In contrast, highly ordered cylindrical pores appeared within the films on a silicon

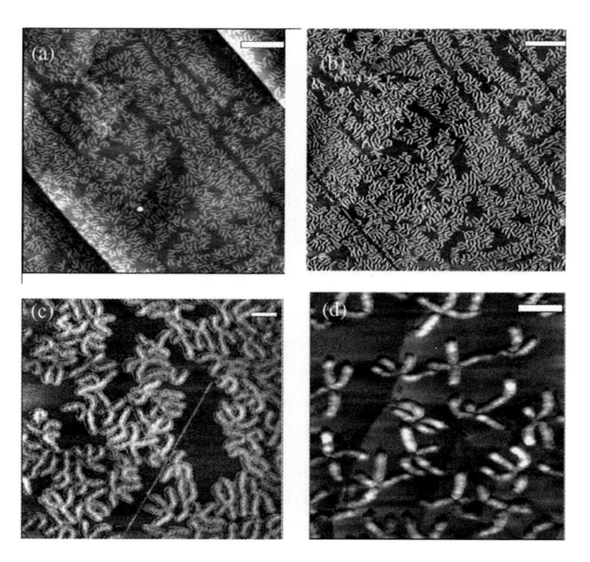

Figure 11.11 Tetra-arm comb-stars (PS1-*b*-PCEVE-*g*-PS2)4 molecules deposited on HOPG surface. (a,b) Higher-resolution phase and topographical images; scale bar is 300 nm. (c,d) Zoomed images; scale bar is 100 nm. Reproduced from Ref. [110].

oxide surface after drying at an ambient temperature. Finally, slow evaporation of the solvent and the lower molecular weight of arms both favored the formation of periodic surface patterns not observed previously.

Viville *et al.* demonstrated that highly branched molecules with a comb-like architecture, poly(chloroethyl vinyl ether)-*g*-PS (PCEVE-*g*-PS), form an ordered, layered organization within very thin film deposits [66]. Within these surface films, the molecules adopt an extended conformation with parameters controlled by the chemical composition favorable for lamellar organization [112]. The authors demonstrated that the surface organization of the PCEVE-*g*-PS molecules is controlled by the orientation of the side branches with respect to the surface. In contrast, molecules with a PS core and PS-*b*-poly(methyl vinyl ether) branches (PS-*b*-PMVE) were observed to form egg-like or long, cylindrical surface structures after deposition on a solid substrate [113]. These structures can self-assemble into segregated domains, forming flower-like morphologies under certain conditions, such as adsorption from specific solvents. In yet another example, dendronized polymers based on PS-functionalized dendritic L-lysine macromonomers showed that all rod, short rod, or globular surface structures can be formed on mica depending upon the degree of the polymerization of the backbone [114].

AFM studies of surface films fabricated from novel arborescent block copolymers comprised of rubbery polyisobutylene (PIB) and PS blocks on a silicon wafer by drop casting showed mixed spherical/cylindrical/lamellar PS domains irregularly distributed within the continuous PIB phase [115]. In a related example, hydrogen bonding for polyurea-malonamide dendron molecules controlled a wide range of surface structures with soft domains reinforced by the rigid dendritic side chains via strong hydrogen bonding interactions [116]. For these surface films, AFM imaging showed microphase-separated surface structures with random morphology on a scale of 100 nm.

A wide variety of LB monolayers have been fabricated from a series of amphiphilic PS-PEO and PS-PAA star molecules of different compositions, the most popular choice of blocks. In studies by the Duran group, three-arm star PS-PEO block copolymers of various architectures have been analyzed [117, 118]. Within LB monolayers, circular domains representing 2D micelle-like aggregated molecules were observed at low surface pressures. Upon further compression, these domains underwent additional aggregation in a systematic manner, exhibiting micellar chaining into long aggregates. Longer PEO arms led to greater intermolecular separation and enhanced the trend to reduce domain aggregation. The surface properties of LB monolayers of a new set of $(PB\text{-}b\text{-}PEO)_4$ and $(PS\text{-}b\text{-}PEO)_4$ amphiphilic four-arm star block copolymers have also been presented [119]. Surface pressure isotherms at the air–water interface possessed three characteristic regions, which correspond to a compact brush region, a pseudoplateau, and a pancake region, where the observed surface area is mainly controlled by the PEO content. Functionalized amphiphilic $PS_n\text{-}b\text{-}PEO_n$ star block copolymers have also formed stable LB films [120].

Different types of star diblock copolymers containing PS and PEO segments were investigated at the air–water interface by Duran and coworkers [121]. Both conventional and dendritic-like architectures of star molecules were studied. Each of these materials contained either a PS core or a PEO corona. These polymers displayed reproducible surface pressure area isotherms with little hysteresis. For star molecules containing 20% or more PEO segments, three distinct regions appeared on the surface pressure isotherms with the PEO blocks absorbing into pancake-like structures and PS globules forming at low surface pressure. Comparing the surface behavior to that of conventional linear molecules showed that the star molecules with the PS core spread much more on the water surface. On the contrary, the star molecules with PEO cores were much more compact.

Star polymers composed of an equal number of PEO and PS arms (up to 38 arms total), variable lengths of arms, or a large number of arms have been examined for their ability to form domain nanostructures at the air–water and air–solid interfaces [122]. All amphiphilic star polymers presented formed stable Langmuir monolayers, readily transferable to a solid substrate. A variety of nanoscale surface morphologies were observed for various molecules, ranging from cylindrical or circular 2D domains to well-developed bicontinuous structures as the fraction of the PEO block varied from 19% to 88% and the number of arms increased from 8 to 19. Furthermore, for the PS-rich stars at elevated surface pressure, a two-dimensional, supramolecular, net-like nanostructure has been observed.

The asymmetric heteroarm PEO_n-PS_m amphiphilic star polymers with different numbers of hydrophobic arms differed by their architecture have been studied by Peleshanko *et al.* [123]. AFM revealed well-developed circular domain morphology composed of PS arms for LB monolayers at low surface pressures (Figure 11.12). At higher surface pressures, the packing of circular domains became denser, but no clear transition to cylindrical structures was observed in condensed monolayers, as is

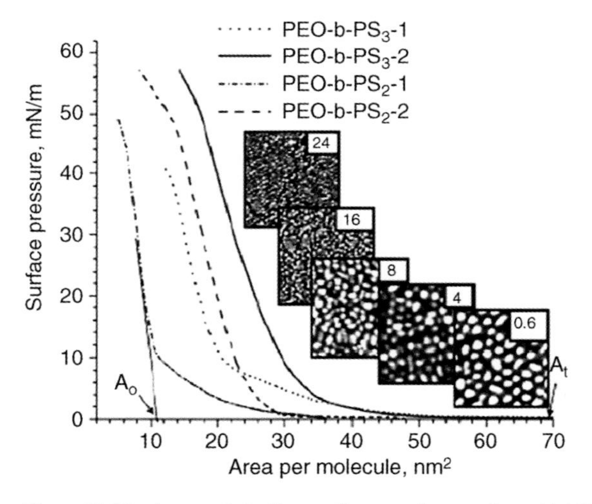

Figure 11.12 Langmuir isotherms for monolayers of amphiphilic heteroarm PEO-PS star block copolymers and corresponding AFM images of surface morphologies at different surface pressures. Reproduced from Ref. [123].

usually observed for linear block copolymers of similar composition [124]. The authors suggested that star architecture favors the formation of highly packed surface structures on highly curved interfaces and circular domains.

The role of functional terminal groups for $(X\text{-PS})_2\text{-}(PEO\text{-}Y)_2$ heteroarm star copolymers with respect to their interfacial behavior and surface morphology has been addressed [125]. A series of star copolymers with different combinations of bromine, amine, TBDPS, hydroxyl, and carboxylic terminal groups were considered in this investigation. The study concluded that hydrophilic functional groups attached to hydrophobic chains and hydrophobic functional groups attached to hydrophilic chains result in the stabilization of the spherical domain morphology, rather than the cylindrical morphology predicted for the given chemical composition of star copolymers. The replacement of functional groups of hydrophobic polymer chains was found to be even more effective in promoting stable and fine circular domain morphology. In addition, the authors demonstrated that the ionization of carboxylic terminal groups at higher pH led to greater solubility of PEO chains in the water subphase. This phenomenon along with the deionization of amine terminal group prevents the lateral aggregation of PS domains, further promoting the formation of the nanoscale circular morphology.

A combination of hydrophobic and hydrophilic PS-PAA blocks is another popular choice in many studies. The synthesis of well-defined branched PS_nPtBA_{2n} copolymers (a PS core with 2, 4, 6, or 8 arms and a corona of PtBA with 4, 8, 12, or 16 arms) has been presented [126, 127]. PS_nPAA_{2n} amphiphilic molecules subsequently generated from these molecules by the hydrolysis of the *tert*-Bu ester groups showed stable Langmuir films with surface morphology significantly changing with increasing surface pressure. It has also been demonstrated that high-resolution AFM can be exploited as an independent and powerful tool for the characterization of molecular parameters similar to that discussed above for brush molecules [128]. These studies enabled quantitative analysis of the molecular length distribution, molecular lengths, and side-chain distribution for some multiarmed brushes with high molecular weight.

A new amphiphilic heteroarm star polymer containing 24 alternating hydrophobic and hydrophilic arms of PS and PAA connected to a well-defined rigid aromatic core was investigated by Genson *et al.* (Figure 11.13) [129]. The authors reported that at the air–water interface, the molecules spontaneously formed pancake-like micellar aggregates, which measure up to several microns in diameter and 5 nm in thickness. Upon the reduction of the surface area per molecule to 7 nm^2, the two-dimensional micelles merged into a dense and uniform monolayer (Figure 11.13). The authors suggested that the confined phase separation of dissimilar polymer arms occurred upon their segregation on the opposite sides of the rigid disk-like aromatic core, forcing the rigid cores to adopt a face-on orientation at the interface with dissimilar arms placed on opposite sides. Upon transfer of these molecules to the hydrophilic solid support, the PS chains localized at the air–film interface makes the surface completely hydrophobic. In contrast, the PAA chains collapsed beneath the PS-dominated topmost layer and formed a thin flattened underlayer.

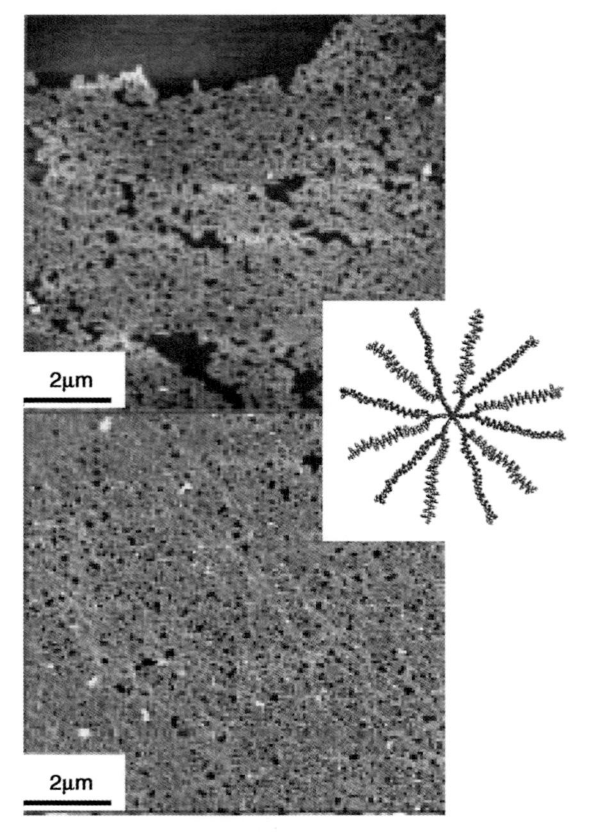

Figure 11.13 Molecular model of $(PAA_{25})_6$-s-$(PS_{25})_6$ (inset) and AFM images (topography) of two-dimensional micellar domains at moderate surface pressure (10 mN/m, *top*) and high surface pressure (30 mN/m, *bottom*). Reproduced from Ref. [129].

11.5 Highly Branched Nanoparticles

Highly branched molecules have been used as templates in the synthesis of various inorganic nanoparticles. For example, branched polyglycerols were found useful in the synthesis of gold and palladium nanoparticles [130–132]. PAMAM hyperbranched polymers have been exploited for gold nanoparticle synthesis [133]. Similarly, amphiphilic hyperbranched PEI amide and acrylate copolymers have been used for silver nanoparticle growth [134, 135]. Other types of hyperbranched polymers have also been found to be very efficient in the stabilization of nanoparticles in solution. For example, the PPV-based hyperbranched conjugated polymer has been used to facilitate gold nanoparticle growth [136]. Hyperbranched poly(amine-ester)s helped to reduce the gold and silver nanoparticles, and hyperbranched aramids have been used for the formation of palladium nanoclusters [137, 138].

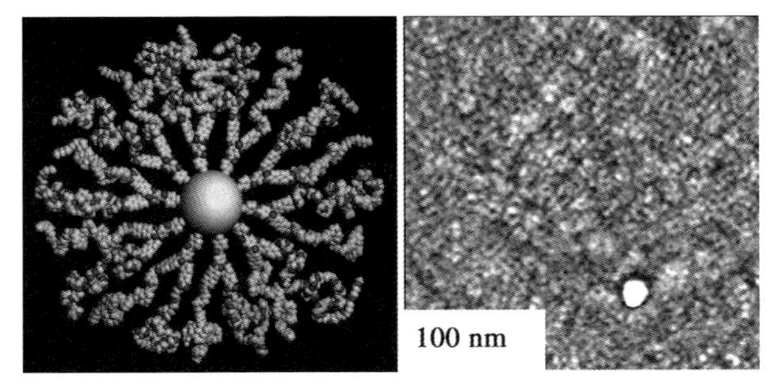

Figure 11.14 Molecular model of gold nanoparticle with amphiphilic binary shell from V-shaped polymeric arms (*left*) and AFM phase image for the LB monolayer with ring-like structures (*right*). Reproduced from Ref. [140].

Gold nanoparticle growth within the core region of a star copolymer with grafting of thiol-functionalized arms was reported by Fustin *et al.* [139]. This synthetic routine led to hairy nanoparticles with mixed hydrophilic and hydrophobic arms such as the one presented in Figure 11.14 [140]. Uniform gold nanoparticles with an average diameter of 3.7 nm were grown using this approach. AFM and TEM revealed the presence of 2 nm gold cores surrounded by the polymer shell with a diameter of 11 nm in extended state of arms. The authors suggested that the amphiphilic, mixed shell drives the spontaneous organization of these hybrid nanoparticles into discrete 2D pancake-like structures with a high density of gold–polymer clusters.

Nanofibrillar micellar structures formed by amphiphilic hyperbranched molecules within a Langmuir monolayer were utilized as a template for silver nanoparticle formation from the ion-containing water subphase [141]. The authors observed that uniform silver nanoparticles were formed from the subphase within the nanofibrillar surface structures of the multifunctional amphiphilic hyperbranched molecules at the water surface. The diameter of these nanoparticles varied from 2 to 4 nm and was controlled by the core dimensions. Furthermore, adding potassium nitrate to the subphase allowed for nanoparticle formation along the nanofibrillar structures rather than in a random manner. An example of the formation of helical assemblies from inorganic nanoparticles by using assembled dendritic molecules has been demonstrated by Stupp and coworkers [142, 143]. Dendron-rod-coil (DRC) macromolecules, which formed supramolecular ribbons, were successfully applied as ordered organic templates for the formation of single and double helices of cadmium sulfide.

Even though the star polymers' utilization is more complicated and less straightforward compared to hyperbranched polymers due to complexity of their microstructures, they have been successfully employed in the synthesis of gold, silver, and platinum nanoparticles [144–147]. Recent results showed that by using multifunctional star polymers, the surface properties of the hybrid gold nanoparticles could be altered from hydrophobic to hydrophilic by varying hydrophilic branched ligands [145].

Recently, examples of highly branched nanoscale silica inorganic cores, polyhedral oligomeric silsesquioxanes (POSS) and their hyperbranched analogous (POSS-M) have been synthesized as prospective optically transparent blocks [148–150]. Stable LB

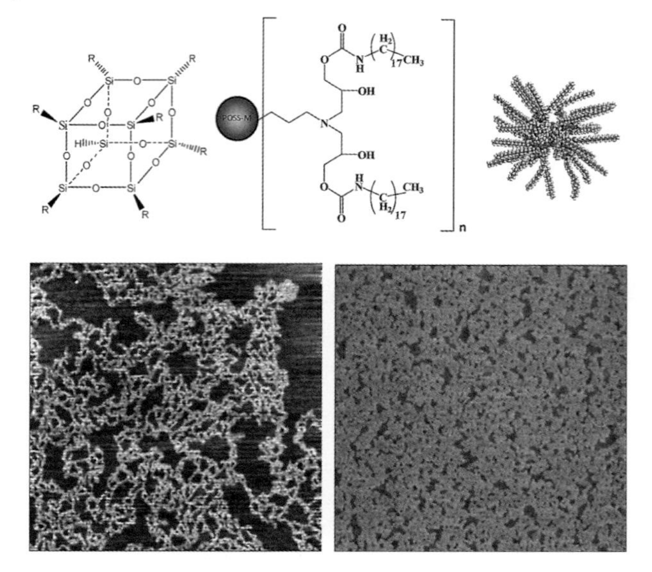

Figure 11.15 POSS-M composition with the 3D cage-like POSS structure and model (*top*). AFM images (5 μm × 5 μm) of LB monolayers at intermediate ($p = 10$ mN/m) and high surface pressures (40 mN/m) for POSS-M with 50 : 50 composition of hydrophilic and hydrophobic terminal groups. Reproduced from Ref. [151].

monolayers of the amphiphilic POSS-M compounds were observed for a range of terminal group compositions (Figure 11.15) [151]. The absence of hydrophobic groups in a fully hydroxylated POSS-M compromised the ability to form a stable Langmuir monolayer. The variation in terminal group composition led to diverse morphologies that ranged from needle-like domains for low hydrophobic content to merged aggregates for high alkyl arms content (Figure 11.15). The surface morphologies, especially for 25% and 50% alkyl content, resembled the two-phase solid–liquid state that is common for alkyl-containing hyperbranched systems. At high surface pressure, all of the POSS-M molecules eventually aggregated into a uniform monolayer.

References

1 Peleshanko, S. and Tsukruk, V.V. (2008) The architecture and surface behavior of highly branched molecules. *Prog. Polym. Sci.*, **33** (5), 523–580.

2 Kroschwitz, J.I. (ed.) (1985) Encyclopedia of Polymer Science and Engineering, *Anionic Polymerization to Cationic Polymerization*, vol. 2, Wiley-Interscience, New York.

3 Mishra, M.K. and Kobayashi, S. (eds) (1999) *Star and Hyperbranched Polymers*, Marcel Dekker, Inc., New York.

4 Zeng, F. and Zimmerman, S.C. (1997) Dendrimers in supramolecular chemistry: from molecular recognition to self-assembly. *Chem. Rev.*, **97** (5), 1681–1712.

5 Bosman, A.W., Janssen, H.M., and Meijer, E.W. (1999) About dendrimers: structure, physical properties, and applications. *Chem. Rev.*, **99** (7), 1665–1688.

6 Inoue, K. (2000) Functional dendrimers, hyperbranched and star polymers. *Prog. Polym. Sci.*, **25** (4), 453–571.
7 Vögtle, F., Gestermann, S., Hesse, R., Schwierz, H., and Windisch, B. (2000) Functional dendrimers. *Prog. Polym. Sci.*, **25** (7), 987–1041.
8 Frauenrath, H. (2005) Dendronized polymers: building a new bridge from molecules to nanoscopic objects. *Prog. Polym. Sci.*, **30** (3–4), 294–384.
9 Fréchet, J.M.J. and Tomalia, D.A. (eds) (2001) *Dendrimers and Other Dendritic Polymers*, John Wiley & Sons, Inc., New York.
10 Newkome, G.R., Moorefield, C.N., and Vögtle, F. (2001) *Dendrimers and Dendrons: Concepts, Syntheses, Applications*, John Wiley & Sons, Inc., New York.
11 Tsukruk, V.V. (1998) Dendritic macromolecules at interfaces. *Adv. Mater.*, **10** (3), 253–257.
12 Brunsveld, L., Folmer, B.J.B., Meijer, E.W., and Sibesma, R. (2001) Supramolecular polymers. *Chem. Rev.*, **101** (12), 4071–4097.
13 Tsukruk, V.V., Rinderspacher, F., and Bliznyuk, V.N. (1997) Self-assembled multilayer films from dendrimers. *Langmuir*, **13** (8), 2171–2176.
14 Weener, J.-W. and Meijer, E.W. (2000) Photoresponsive dendritic monolayers. *Adv. Mater.*, **12** (10), 741–746.
15 Jonkheim, P., Hoeben, F.J.M., Kleppinger, R., van Herrikhuyzen, J., Schnenning, A.P.H.J., and Meijer, E.W. (2003) Transfer of π-conjugated columnar stacks from solution to surfaces. *J. Am. Chem. Soc.*, **125** (51), 15941–15949.
16 Sui, G., Micic, M., Huo, Q., and Leblanc, R.M. (2000) Synthesis and surface chemistry study of a new amphiphilic PAMAM dendrimer. *Langmuir*, **16** (20), 7847–7851.
17 Lee, M., Kim, J.-W., Peleshanko, S., Larson, K., Yoo, Y., Vaknin, D., Markutsya, S., and Tsukruk, V.V. (2002) Amphiphilic hairy disks with branched hydrophilic tails and a hexa-peri-hexabenzocoronene core. *J. Am. Chem. Soc.*, **124** (31), 9121–9128.
18 Tsukruk, V.V., Luzinov, I., Larson, K., Li, S., and McGrath, D.V. (2001) Intralayer reorganization of photochromic molecular films. *J. Mater. Sci. Lett.*, **20** (9), 873–876.
19 Tsukruk, V.V., Shulha, H., and Zhai, X. (2003) Nanoscale stiffness of individual dendritic molecules and their aggregates. *Appl. Phys. Lett.*, **82** (6), 907–909.
20 Shulha, H., Zhai, X., and Tsukruk, V.V. (2003) Molecular stiffness of individual hyperbranched macromolecules at solid surfaces. *Macromolecules*, **36** (8), 2825–2831.
21 Zubarev, E.R., Pralle, M.U., Sone, E.D., and Stupp, S.I. (2001) Self-assembly of dendron rodcoil molecules into nanoribbons. *J. Am. Chem. Soc.*, **123** (17), 4105–4106.
22 Zubarev, E.R., Pralle, M.U., Li, L.M., and Stupp, S.I. (1999) Conversion of supramolecular clusters to macromolecular objects. *Science*, **283** (5401), 523–526.
23 Won, Y.Y., Davis, H.T., and Bates, F.S. (1999) Giant wormlike rubber micelles. *Science*, **283** (5404), 960–963.
24 Djalali, R., Li, S.Y., and Schmidt, M. (2002) Amphipolar core–shell cylindrical brushes as templates for the formation of gold clusters and nanowires. *Macromolecules*, **35** (11), 4282–4288.
25 Loi, S., Butt, H.J., Wiesler, U.-W., and Mullen, K. (2000) Formation of nanorods by self-assembly of alkyl-substituted polyphenylene dendrimers on graphite. *Chem. Commun.*, **13**, 1169–1170.
26 Liu, D., Zhang, H., Grim, P.C.M., De Feyter, S., Wiesler, U.-M., Berresheim, A.J., Muellen, K., and De Schryver, F.C. (2002) Self-assembly of polyphenylene dendrimers into micrometer long nanofibers: an atomic force microscopy study. *Langmuir*, **18** (6), 2385–2391.
27 Bliznyuk, V.N., Rinderspacher, F., and Tsukruk, V.V. (1998) On the structure of polyamidoamine dendrimer monolayers. *Polymer*, **39** (21), 5249–5252.
28 Hierlemann, A., Campbell, J.K., Baker, L.A., Crooks, R.M., and Ricco, A.J. (1998) Structural distortion of dendrimers on

gold surfaces: a tapping-mode AFM investigation. *J. Am. Chem. Soc.*, **120** (21), 5323–5324.

29 Li, J., Swanson, D.R., Qin, D., Brothers, H.M., Piehler, L.T., Tomalia, D., and Meier, D.J. (1999) Characterizations of core–shell tecto-(dendrimer) molecules by tapping mode atomic force microscopy. *Langmuir*, **15** (21), 7347–7350.

30 Pericet-Camara, R., Papastavrou, G., and Borkovec, M. (2004) Atomic force microscopy study of the adsorption and electrostatic self-organization of poly (amidoamine) dendrimers on mica. *Langmuir*, **20** (8), 3264–3270.

31 Mecke, A., Lee, I., Baker, J.R., Jr., Banaszak Holl, M.M., and Orr, B.G. (2004) Deformability of poly (amidoamide) dendrimers. *Eur. Phys. J. E*, **14** (1), 7–16.

32 Tomczak, N. and Vancso, G.J. (2007) Elasticity of single poly(amido amine) dendrimers. *Macromol. Rapid Commun.*, **28** (16), 1640–1644.

33 Tomczak, N. and Vancso, G.J. (2007) Microcontact printed poly(amidoamine) dendrimer monolayers on silicon oxide surface. *Eur. Polym. J.*, **43** (5), 1595–1601.

34 Kim, B.-Y., Lebedeva, O.V., Dong, H.K., Caminade, A.-M., Majoral, J.-P., Knoll, W., and Vinogradova, O.I. (2005) Assembly and mechanical properties of phosphorus dendrimer/polyelectrolyte multilayer capsules. *Langmuir*, **21** (16), 7200–7206.

35 Pericet-Camara, R., Papastavrou, G., and Borkovec, M. (2009) Effective charge of adsorbed poly(amidoamine) dendrimers from direct force measurements. *Macromolecules*, **42** (5), 1749–1758.

36 Ding, K., Grebel-Koehler, D., Berger, R., Mullen, K., and Butt, H.-J. (2005) Structure of self-assembled *n*-dodecyl substituted azobenzene poly(phenylene) dendrimers on graphite. *J. Mater. Chem.*, **15** (33), 3431–3436.

37 Yoon, D.K. and Jung, H.-T. (2003) Self-organization of a fan-shaped dendrimer at the air–water interface. *Langmuir*, **19** (4), 1154–1158.

38 Liu, D., De Feyter, S., Cotlet, M., Wiesler, U.-M., Weil, T., Herrmann, A., Mullen, K., and De Schryver, F.C. (2003) Fluorescent self-assembled polyphenylene dendrimer nanofibers. *Macromolecules*, **36** (22), 8489–8498.

39 Genson, K.L., Holzmuller, J., Villacencio, O.F., McGrath, D.V., Vaknin, D., and Tsukruk, V.V. (2005) Monolayers of photochromic amphiphilic monodendrons: molecular aspects of light switching at liquid and solid surfaces. *J. Phys. Chem. B*, **109** (43), 20393–20402.

40 Xiao, Z., Cai, C., Mayeux, A., and Milenkovic, A. (2002) The first organosiloxane thin films derived from SiCl3-terminated dendrons. Thickness-dependent nano- and mesoscopic structures of the films deposited on mica by spin-coating. *Langmuir*, **18** (20), 7728–7739.

41 Genson, K.L., Vaknin, D., Villavicencio, O.F., Holzmueller, J., McGrath, D.V., and Tsukruk, V.V. (2005) Langmuir monolayers of functionalized amphiphiles with epoxy terminal groups. *Thin Solid Films*, **493** (1–2), 237–248.

42 Ponomarenko, S.A., Tatarinova, E.A., Muzafarov, A.M., Kirchmeyer, S., Brassat, L., Mourran, A., Moeller, M., Setayesh, S., and de Leeuw, D. (2006) Star-shaped oligothiophenes for solution-processible organic electronics: flexible aliphatic spacers approach. *Chem. Mater.*, **18** (17), 4101–4108.

43 Genson, K.L., Holzmueller, J., Ornatska, M., Yoo, Y.S., Par, M.H., Lee, M., and Tsukruk, V.V. (2006) Assembling of dense fluorescent supramolecular webs via self-propelled star-shaped aggregates. *Nano Lett.*, **6** (3), 435–440.

44 Holzmueller, J., Genson, K.L., Park, Y., Yoo, Y.S., Park, M.H., Lee, M., and Tsukruk, V.V. (2005) Amphiphilic treelike rods at interfaces: layered stems and circular aggregation. *Langmuir*, **21** (14), 6392–6398.

45 Sheiko, S.S. and Moller, M. (2001) Visualization of macromolecules: a first step to manipulation and controlled response. *Chem. Rev.*, **101** (12), 4099–4124.

46 Sheiko, S.S., Sumerlin, B.S., and Matyjaszewski, K. (2008) Cylindrical molecular brushes: synthesis, characterization, and properties. *Prog. Polym. Sci.*, **33** (7), 759–785.

47 Qin, S., Matyjaszewski, K., Xu, H., and Sheiko, S.S. (2003) Synthesis and visualization of densely grafted molecular brushes with crystallizable poly(octadecyl methacrylate) block segments. *Macromolecules*, **36** (3), 605–612.

48 Kreutzer, G., Ternat, C., Nguyen, T.Q., Plummer, C.J.G., Månson, J.-A.E., Castelletto, V., Hamley, I.W., Sun, F., Sheiko, S.S., and Klok, H.-A. (2006) Water-soluble, unimolecular containers based on amphiphilic multiarm star block copolymers. *Macromolecules*, **39**, 4507–4516.

49 Matyjaszewski, K. and Qin, S. (2003) Effect of initiation conditions on the uniformity of three-arm star molecular brushes. *Macromolecules*, **36** (6), 1843–1849.

50 Schappacher, M. and Deffieux, A. (2005) AFM image analysis applied to the investigation of elementary reactions in the synthesis of comb star copolymers. *Macromolecules*, **38** (11), 4942–4946.

51 Sheiko, S.S., da Silva, M., Shirvaniants, D., LaRue, I., Prokhorova, S., Moeller, M., Beers, K., and Matyjaszewski, K. (2003) Measuring molecular weight by atomic force microscopy. *J. Am. Chem. Soc.*, **125** (22), 6725–6728.

52 Boyce, J.R., Shirvanyants, D., Sheiko, S.S., Ivanov, D.A., Qin, S., Borner, H., and Matyjaszewski, K. (2004) Multiarm molecular brushes: effect of the number of arms on the molecular weight polydispersity and surface ordering. *Lanmguir*, **20** (14), 6005–6011.

53 Xu, Y., Bolisetty, S., Drechsler, M., Fang, B., Yuan, J., Harnau, L., Ballauff, M., and Muller, A.H.E. (2009) Manipulating cylindrical polyelectrolyte brushes on the nanoscale by counterions: collapse transition to helical structures. *Soft Matter*, **5**, 379–384.

54 Duschner, S., Storkle, D., Schmidt, M., and Maskos, M. (2008) Topologically controlled interpolyelectrolyte complexes. *Macromolecules*, **41** (23), 9067–9071.

55 Khelfallah, N., Gunari, N., Fischer, K., Gkogkas, G., Hadjichristidis, N., and Schmidt, M. (2005) Micelles formed by cylindrical brush-coil block copolymers. *Macromol. Rapid. Commun.*, **26** (21), 1693–1697.

56 Fu, G.D., Phua, S.J., Kang, E.T., and Neoh, K.G. (2005) Tadpole-shaped amphiphilic block-graft copolymers prepared via consecutive atom transfer radical polymerizations. *Macromolecules*, **38** (7), 2612–2619.

57 Zhang, B., Fischer, K., and Schmidt, M. (2005) Cylindrical polypeptide brushes. *Macromol. Chem. Physic*, **206** (1), 157–162.

58 Mecerreyes, D., Stevens, R., Nguyen, C., Pomposo, J.A., Bengoetxea, M., and Grande, H. (2002) Synthesis and characterization of polypyrrole-graft-poly(ε-caprolactone) copolymers: new electrically conductive nanocomposites. *Synthetic Met.*, **126** (2–3), 173–178.

59 Stephan, T., Muth, S., and Schmidt, M. (2002) Shape changes of statistical copolymacromonomers: from wormlike cylinders to horseshoe- and meander-like structures. *Macromolecules*, **35** (27), 9857–9860.

60 Xu, H., Shirvanyants, D., Beers, K., Matyjaszewski, K., Rubinstein, M., and Sheiko, S.S. (2004) Molecular motion in a spreading precursor film. *Phys. Rev. Lett.*, **93** (20), 206103–206106.

61 Gallyamov, M.O., Tartsch, B., Khokhlov, A.R., Sheiko, S.S., Boerner, H.G., Matyjaszewski, K., and Moeller, M. (2004) Real-time scanning force microscopy of macromolecular conformational transitions. *Macromol. Rapid. Commun.*, **25** (19), 1703–1707.

62 Gallyamov, M.O., Tartsch, B., Khokhlov, A.R., Sheiko, S.S., Boerner, H.G., Matyjaszewski, K., and Moeller, M. (2004) Conformational dynamics of single molecules visualized in real time by scanning force microscopy: macromolecular mobility on a substrate surface in different vapours. *J. Microscopy*, **215** (3), 245–256.

63 Gallyamov, M.O., Tartsch, B., Mela, P., Borner, H., Matyjaszewski, K., Sheiko, S.S., Khokhlov, A., and Moller, M. (2007) A scanning force microscopy study on the motion of single brush-like macromolecules on a silicon substrate induced by coadsorption of small molecules. *Phys. Chem. Chem. Phys.*, **9**, 346–352.

64 Sheiko, S.S., Sun, F.C., Randall, A., Shirvanyants, D., Rubinstein, M., Lee, H., and Matyjaszewski, K. (2006) Adsorption-induced scission of carbon–carbon bonds. *Nature*, **440**, 191–194.

65 Sheiko, S.S., da Silva, M., Shirvaniants, D., LaRue, I., Prokhorova, S., Moeller, M., Beers, K., and Matyjaszewski, K. (2003) Measuring molecular weight by atomic force microscopy. *J. Am. Chem. Soc.*, **125** (22), 6725–6728.

66 Cheng, C.X., Jiao, T.F., Tang, R.P., Chen, E.Q., Liu, M.H., and Xi, F. (2006) Compression-induced hierarchical nanostructures of a poly(ethylene oxide)-block-dendronized polymethacrylate copolymer at the air/water interface. *Macromolecules*, **39** (19), 6327–6330.

67 Iyer, J. and Hammond, P.T. (1999) Langmuir behavior and ultrathin films of new linear-dendritic diblock copolymers. *Langmuir*, **15** (4), 1299–1306.

68 Tian, L. and Hammondm, P.T. (2006) Comb-dendritic block copolymers as tree-shaped macromolecular amphiphiles for nanoparticle self-assembly. *Chem. Mater.*, **18** (17), 3976–3984.

69 Malmstrom, E. and Hult, A. (1996) Kinetics of formation of hyperbranched polyesters based on 2,2-bis(methylol) propionic acid. *Macromolecules*, **29** (4), 1222–1228.

70 Dodiuk H., Gold Z., Kenig S (2004) Tailoring new architectures for polyurethanes using dendritic and hyper-branched polymers and their adhesion behavior. J. Adhes. Sci. Tech. 18 (3), 301–311.

71 Beyerlein, D., Belge, G., Eichhorn, K.J., Gauglitz, G., Grundke, K., and Voit, B. (2001) Preparation and properties of thin films of hyperbranched polyesters with different end groups. *Macromol. Symp.*, **164** (1), 117–131.

72 Elrehim, M.A., Voit, B., Bruchmann, B., Eichhorn, K.J., Grundke, K., and Bellmann, C. (2005) Structural and end-group effects on bulk and surface properties of hyperbranched poly(urea urethane)s. *J. Polym. Sci. A Polym. Chem.*, **43** (15), 3376–3393.

73 Nishide, H., Nambo, M., and Miyasaka, M. (2002) Hyperbranched poly (phenylenevinylene) bearing pendant phenoxys for a high-spin alignment. *J. Mater. Chem.*, **12** (12), 3578–3584.

74 Hua, F. and Ruckenstein, E. (2005) Hyperbranched sulfonated polydiphenylamine as a novel self-doped conducting polymer and its pH response. *Macromolecules*, **38** (3), 888–898.

75 Viville, P., Deffieux, A., Schappacher, M., Bredas, J.L., and Lazzaroni, R. (2001) Surface organization of single hyperbranched polymer molecules, as studied by atomic force microscopy. *Mater. Sci. Eng. C*, **15** (1–2), 311–314.

76 Sidorenko, A., Zhai, X.W., Peleshanko, S., Greco, A., Shevchenko, V.V., and Tsukruk, V.V. (2001) Hyperbranched polyesters on solid surfaces. *Langmuir*, **17** (19), 5924–5931.

77 Barriau, E., Frey, H., Kiry, A., Stamm, M., and Groehn, F. (2006) Negatively charged hyperbranched polyether-based polyelectrolytes. *Colloid & Polym. Sci.*, **284** (11), 1293–1301.

78 Shulha, H., Zhai, X., and Tsukruk, V.V. (2003) Molecular stiffness of individual dendritic macromolecules and their aggregates. *Macromolecules*, **36** (8), 2825.

79 Tsukruk, V.V., Shulha, H., and Zhai, X. (2003) Nanoscale stiffness of individual dendritic molecules and their aggregates. *Appl. Phys. Lett.*, **82** (6), 907–909.

80 Orlicki, J.A., Viernes, N.O.L., Moore, J.S., Sendijarevic, I., and McHugh, A.J. (2002) Roles of molecular architecture and end-group functionality on the surface properties of branched polymers. *Langmuir*, **18** (25), 9990–9995.

81 Helgert, M., Wenke, L., Hvilsted, S., Ramanujam, P.S. (2001) Surface relief measurements in side-chain azobenzene

polyesters with different substituents Appl. Phys. B Lasers O., 72(4), 429–433.

82 Genson, K., Vaknin, D., Villacencio, O., McGrath, D.V., and Tsukruk, V.V. (2002) Microstructure of amphiphilic monodendrons at the air–water interface. *J. Phys. Chem. B*, **106** (43), 11277–11284.

83 Sidorenko, A., Houphouet-Boigny, C., Villavicencio, O., McGrath, D.V., and Tsukruk, V.V. (2002) Low generation photochromic monodendrons on a solid surface. *Thin Solid Films*, **410** (1–2), 147–158.

84 Tsukruk, V.V., Luzinov, I., Larson, K., Li, S., and McGrath, D.V. (2001) Intralayer reorganization of photochromic molecular films. *J. Mater. Sci. Lett.*, **20** (9), 873–876.

85 Ondaral, S., Warberg, L., and Enarsson, L.E. (2006) The adsorption of hyperbranched polymers on silicon oxide surfaces. *J. Colloid Interface Sci.*, **301** (1), 32–39.

86 Thomann, Y., Haag, R., Brenn, R., Delto, R., Weickman, H., Thomann, R., and Muelhaupt, R. (2005) PMMA gradient materials and *in situ* nanocoating via self-assembly of semifluorinated hyperbranched amphiphiles. *Macromol. Chem. Phys.*, **206** (1), 135–141.

87 Wang, M., Gan, D., and Wooley, K.L. (2001) Linear and hyperbranched poly (silyl ester)s: synthesis via cross-dehydrocoupling-based polymerization, hydrolytic degradation properties and morphological analysis by atomic force microscopy. *Macromolecules*, **34** (10), 3215–3223.

88 Mueller, A., Kowalewski, T., and Wooley, K.L. (1998) Synthesis, characterization and derivatization of hyperbranched polyfluorinated polymers. *Macromolecules*, **31** (3), 776–786.

89 Gan, D., Mueller, A., and Wooley, K.L. (2003) Amphiphilic and hydrophobic surface patterns generated from hyperbranched fluoropolymer/linear polymer networks: minimally adhesive coatings via the crosslinking of hyperbranched fluoropolymers. *J. Polym. Sci. A Polym. Chem.*, **41** (22), 3531–3540.

90 Yan, D., Zhou, Y., and Hou, J. (2004) Supramolecular self-assembly of macroscopic tubes. *Science*, **303** (5654), 65–67.

91 Zhou, Y. and Yan, D. (2004) Supramolecular self-assembly of giant polymer vesicles with controlled sizes. *Angew. Chem. Int. Ed.*, **43** (37), 4896–4899.

92 Ornatska, M., Peleshanko, S., Rybak, B., Holzmueller, J., and Tsukruk, V.V. (2004) Supramolecular multiscale fibers through one-dimensional assembly of dendritic molecules. *Adv. Mater.*, **16** (23), 2206–2212.

93 Ornatska, M., Bergman, K.N., Rybak, B., Peleshanko, S., and Tsukruk, V.V. (2004) Nanofibers from functionalized dendritic molecules. *Angew. Chem. Int. Ed.*, **43** (39), 5246–5249.

94 Ornatska, M., Peleshanko, S., Genson, K.L., Rybak, B., Bergman, K.N., and Tsukruk, V.V. (2004) Assembling of amphiphilic highly branched molecules in supramolecular nanofibers. *J. Am. Chem. Soc.*, **126** (31), 9675–9684.

95 Zhai, X., Peleshanko, S., Klimenko, N.S., Genson, K.L., Vaknin, D., Vortman, M.Y., Shevchenko, V.V., and Tsukruk, V.V. (2003) Amphiphilic dendritic molecules: hyperbranched polyesters with alkyl-terminated branches. *Macromolecules*, **36** (9), 3101–3110.

96 Sidorenko, A., Zhai, X.W., Simon, F., Pleul, D., Greco, A., and Tsukruk, V.V. (2002) Hyperbranched molecules with epoxy-functionalized terminal branches: grafting to a solid surface. *Macromolecules*, **35** (13), 5131–5139.

97 Sheiko, S.S., Buzin, A.I., Muzafarov, A.M., Rebrov, E.A., and Getmanova, E.V. (1998) Spreading of carbosilane dendrimers at the air/water interface. *Langmuir*, **14** (26), 7468–7474.

98 Bliznyuk, V.N., Rinderspacher, F., and Tsukruk, V.V. (1998) On the structure of polyamidoamine dendrimer monolayers. *Polymer*, **39** (21), 5249–5252.

99 Qiu, T., Tang, L., Tuo, X., Zhang, X., and Liu, D. (2001) Study on self-assembly properties of aryl-alkyl hyperbranched polyesters with carboxylic end groups. *Polym. Bull.*, **47** (3–4), 337–342.

100 Cha, X., Yin, R., Zhang, X., and Shen, J. (1991) Investigation into monolayers and multilayers of star-shaped graft copolymer based on styrene-divinylbenzene microgel cores. *Macromolecules*, **24** (18), 4985–4989.

101 Xu, H., Erhardt, R., Abetz, V., Mueller, A.H.E., and Goedel, W.A. (2001) Janus micelles at the air/water interface. *Langmuir*, **17** (22), 6787–6793.

102 Wolert, E., Setz, S.M., Underhill, R.S., Duran, R.S., Schappacher, M., Deffieux, A., Hoelderle, M., and Muelhaupt, R. (2001) Meso- and microscopic behavior of spherical polymer particles assembling at the air–water interface. *Langmuir*, **17** (18), 5671–5677.

103 Stepanek, M., Uchman, M., and Prochazka, K. (2009) Self-assemblies formed by four-arm star copolymers with amphiphilic diblock arms in aqueous solutions. *Polymer*, **50** (15), 3638–3644.

104 Glynos, E., Chremos, A., Petekidis, G., Camp, P.J., and Koutsos, V. (2007) Polymer-like to soft colloid-like behavior of regular star polymers adsorbed on surfaces. *Macromolecules*, **40** (19), 6947–6958.

105 Plamper, F.A., Walther, A., Muller, A.H.E., and Ballauff, M. (2007) Nanoblossoms: light-induced conformational changes of cationic polyelectrolyte stars in the presence of multivalent counterions. *Nano Lett.*, **7** (1), 167–171.

106 Nunez, J., Vancso, G.J., and Gedde, U.W. (2008) Morphology, crystallization, and melting of single crystals and thin films of star-branched polyesters with poly (ε-caprolactone) arms as revealed by atomic force microscopy. *Macromol. Sci. B Physics*, **47** (3), 589–607.

107 Kiriy, A., Gorodyska, G., Minko, S., Stamm, M., and Tsitsilianis, C. (2003) Single molecules and associates of heteroarm star copolymer visualized by atomic force microscopy. *Macromolecules*, **36** (23), 8704–8711.

108 Kiriy, A., Gorodyska, G., Minko, S., Tsitsilianis, C., Jaeger, W., and Stamm, M. (2003) Chemical contrasting in a single polymer molecule: AFM experiment. *J. Am. Chem. Soc.*, **125** (37), 11202–11203.

109 Gorodyska, G., Kiriy, A., Minko, S., Tsitsilianis, C., and Stamm, M. (2003) Reconformation and metallization of unimolecular micelles in controlled environment. *Nano Lett.*, **3** (3), 365–368.

110 Schappacher, M. and Deffieux, A. (2005) AFM image analysis applied to the investigation of elementary reactions in the synthesis of comb star copolymers. *Macromolecules*, **38** (11), 4942–4946.

111 Yu, X., Fu, J., Han, Y., and Pan, C. (2003) AFM study of the self-assembly behavior of hexa-armed star polymers with a discotic triphenylene core. *Macromol. Rapid. Commun.*, **24**, 742–747.

112 Viville, P., Leclere, P., Deffieux, A., Schappacher, M., Bernard, J., Borsali, R., Bredas, J.L., and Lazzaroni, R. (2004) Atomic force microscopy study of comb-like vs arborescent graft copolymers in thin films. *Polymer*, **45** (6), 1833–1843.

113 Schappacher, M., Putaux, J.L., Lefebvre, C., and Deffieux, A. (2005) Molecular containers based on amphiphilic PS-*b*-PMVE dendrigraft copolymers: topology organization and aqueous solution properties. *J. Am. Chem. Soc.*, **127** (9), 2990–2998.

114 Luebbert, A., Nguyen, T.Q., Sun, F., Sheiko, S.S., and Klok, H.A. (2005) L-Lysine dendronized polystyrene. *Macromolecules*, **38** (6), 2064–2071.

115 Puskas, J.E., Kwon, Y., Antony, P., and Bhowmick, A.K. (2005) Synthesis and characterization of novel dendritic (arborescent hyperbranched) polyisobutylene-polystyrene block copolymers. *J. Polym. Sci. A Polym. Chem.*, **43** (9), 1811–1826.

116 Dai, S.A., Chen, C.P., Lin, C.C., Chang, C.C., Wu, T.M., Su, W.C., Chang, H.L., and Jeng, R.J. (2006) Novel side-chain dendritic polyurethanes based on hydrogen bonding rich polyurea/malonamide dendrons. *Macromol. Mater. Eng.*, **291** (4), 395–404.

117 Logan, J.L., Masse, P., Dorvel, B., Skolnik, A.M., Sheiko, S.S., Francis, R., Taton, D., Gnanou, Y., and Duran, R.S. (2005) AFM study of micelle chaining in surface films of polystyrene-block-poly(ethylene oxide) stars at the air/water interface. *Langmuir*, **21** (8), 3424–3431.

118 Francis, R., Skolnik, A.M., Carino, S.R., Logan, J.L., Underhill, R.S., Angot, S., Taton, D., Gnanou, Y., and Duran, R.S. (2002) Aggregation and surface morphology of a poly(ethylene oxide)-*block*-polystyrene. Three-arm star polymer at the air/water interface: studied by AFM. *Macromolecules*, **35** (17), 6483–6485.

119 Matmour, R., Francis, R., Duran, R.S., and Gnanou, Y. (2005) Interfacial behavior of anionically synthesized amphiphilic star block copolymers based on polybutadiene and poly(ethylene oxide) at the air/water interface. *Macromolecules*, **38** (18), 7754–7767.

120 Francis, R., Taton, D., Logan, J.L., Masse, P., Gnanou, Y., and Duran, R.S. (2003) Synthesis and surface properties of amphiphilic star-shaped and dendrimer-like copolymers based on polystyrene core and poly(ethylene oxide) corona. *Macromolecules*, **36** (22), 8253–8259.

121 Logan, J.L., Masse, P., Gnanou, Y., Taton, D., and Duran, R.S. (2005) Polystyrene-*block*-poly(ethylene oxide) stars as surface films at the air/water interface. *Langmuir*, **21** (16), 7380–7389.

122 Gunawidjaja, R., Peleshanko, S., Genson, K.L., Tsitsilianis, C., and Tsukruk, V.V. (2006) Surface morphologies of Langmuir–Blodgett monolayers of PEO_nPS_n multiarm star copolymers. *Langmuir*, **22** (14), 6168–6176.

123 Peleshanko, S., Jeong, J., Gunawidjaja, R., and Tsukruk, V.V. (2004) Amphiphilic heteroarm PEO-*b*-PS_m star polymers at the air–water interface: aggregation and surface morphology. *Macromolecules*, **37** (17), 6511–6522.

124 Peleshanko, S., Gunawidjaja, R., Jeong, J., Shevchenko, V.V., and Tsukruk, V.V. (2004) Surface behavior of amphiphilic heteroarm star-block copolymers with asymmetric architecture. *Langmuir*, **20** (22), 9423–9427.

125 Gunawidjaja, R., Peleshanko, S., and Tsukruk, V.V. (2005) Functionalized (X-PEO)2-(PS-Y)2 star block copolymers at the interfaces: role of terminal groups in surface behavior and morphology. *Macromolecules*, **38** (21), 8765–8774.

126 Matmour, R., Lepoittevin, B., Joncheray, T.J., El-khouri, R.J., Taton, D., Duran, R.S., and Gnanou, Y. (2005) Synthesis and investigation of surface properties of dendrimer-like copolymers based on polystyrene and poly(*tert*-butyl acrylate). *Macromolecules*, **38** (13), 5459–5467.

127 Joncheray, T.J., Bernard, S.A., Matmour, R., Lepoittevin, B., El-Khouri, R.J., Taton, D., Gnanou, Y., and Duran, R.S. (2007) Polystyrene-*b*-poly (*tert*-butyl acrylate) and polystyrene-*b*-poly (acrylic acid) dendrimer-like copolymers: two-dimensional self-assembly at the air–water interface. *Langmuir*, **23** (5), 2531–2538.

128 Matyjaszewski, K., Qin, S., Boyce, J.R., Shirvanyants, D., and Sheiko, S.S. (2003) Effect of initiation conditions on the uniformity of three-arm star molecular brushes. *Macromolecules*, **36** (6), 1843–1849.

129 Genson, K.L., Hoffman, J., Teng, J., Zubarev, E.R., Vaknin, D., and Tsukruk, V.V. (2004) Interfacial micellar structures from novel amphiphilic star polymers. *Langmuir*, **20** (21), 9044–9052.

130 Wan, D., Fu, Q., and Huang, J. (2006) Synthesis of amphiphilic hyperbranched polyglycerol polymers and their application as template for size control of gold nanoparticles. *J. Appl. Polym. Sci.*, **101** (1), 509–514.

131 Chen, Y., Frey, H., Toman, R., and Stiriba, S.E. (2006) Optically active amphiphilic hyperbranched polyglycerols as templates for palladium nanoparticles. *Inorganica Chim. Acta*, **359** (6), 1837–1844.

132 Mecking, S., Thomann, R., Frey, H., and Sunder, A. (2000) Preparation of catalytically active palladium nanoclusters in compartments of amphiphilic hyperbranched polyglycerols. *Macromolecules*, **33** (11), 3958–3960.

133 Perignon, N., Mingotaud, A.F., Marty, J.D., Rico-Lattes, I., and Mingotaud, C. (2004) Formation and stabilization in water of metal nanoparticles by a hyperbranched polymer chemically analogous to PAMAM dendrimers. *Chem. Mater.*, **16** (24), 4856–4858.

134 Garamus, V.M., Maksimova, T., Richtering, W., Aymonier, C., Thomann, R., Antonietti, L., and Mecking, S. (2004) Solution structure of metal particles prepared in unimolecular reactors of amphiphilic hyperbranched macromolecules. *Macromolecules*, **37** (21), 7893–7900.

135 Sato, T., Nobutane, H., Hirano, T., and Seno, M. (2006) Hyperbranched acrylate copolymer via initiator-fragment incorporation radical copolymerization of divinylbenzene and ethyl acrylate: synthesis, characterization, hydrolysis, dye-solubilization, Ag particle-stabilization, and porous film formation. *Macromol. Mater. Eng.*, **291** (2), 162–172.

136 Lin, H., He, Q., Wang, W., and Bai, F. (2004) Preparation and photophysical properties of a hyperbranched conjugated polymer-bound gold nanoassembly. *Res. Chem. Intermed.*, **30** (4), 527–536.

137 Bao, C., Jin, M., Lu, R., Xue, P., Zhang, T., Tan, C., and Zhao, Y. (2003) Synthesis of hyperbranched poly(amine-ester)-protected noble metal nanoparticles in aqueous solution. *J. Mater. Res.*, **18** (6), 1392–1398.

138 Tabuani, D., Monticelli, O., Komber, H., and Russo, S. (2003) Preparation and characterization of Pd nanoclusters in hyperbranched aramid templates to be used in homogeneous catalysis. *Macromol. Chem. Phys.*, **204** (12), 1576–1583.

139 Fustin, C.A., Colard, C., Filali, M., Guillet, P., Duwez, A.S., Meier, M.A.R., Schubert, U.S., and Gohy, J.F. (2006) Tuning the hydrophilicity of gold nanoparticles templated in star block copolymers. *Langmuir*, **22** (15), 6690–6695.

140 Genson, K.L., Holzmueller, J., Jiang, C., Xu, J., Gibson, J.D., Zubarev, E.R., and Tsukruk, V.V. (2006) Langmuir–Blodgett monolayers of gold nanoparticles with amphiphilic shells from V-shaped binary polymer arms. *Langmuir*, **22** (16), 7011–7015.

141 Rybak, B., Ornatska, M., Bergman, K.N., Genson, K.L., and Tsukruk, V.V. (2006) Formation of silver nanoparticles at the air–water interface mediated by a monolayer of functionalized hyperbranched molecules. *Langmuir*, **22** (3), 1027–1037.

142 Sone, E.D., Zubarev, E.R., and Stupp, S.I. (2002) Semiconductor nanohelices templated by supramolecular ribbons. *Angew. Chem. Int. Ed.*, **41** (10), 1705–1709.

143 Sone, E.D., Zubarev, E.R., and Stupp, S.I. (2005) Supramolecular templating of single and double nanohelices of cadmium sulfide. *Small*, **1** (7), 694–697.

144 Filali, M., Meier, M.A.R., Schubert, U.S., and Gohy, J.F. (2005) Star-block copolymers as templates for the preparation of stable gold nanoparticles. *Langmuir*, **21** (17), 7995–8000.

145 Youk, J.H., Park, M.K., Locklin, J., Advincula, R., Yang, J., and Mays, J. (2002) Preparation of aggregation stable gold nanoparticles using star-block copolymers. *Langmuir*, **18** (7), 2455–2458.

146 Ishizu, K., Furukawa, T., and Yamada, H. (2005) Silver nanoparticles dispersed within amphiphilic star-block copolymers as templates for plasmon band materials. *Eur. Polym. J.*, **41** (12), 2853–2860.

147 Zhang, L., Niu, H., Chen, Y., Liu, H., and Gao, M. (2006) Preparation of platinum nanoparticles using star-block copolymer with a carboxylic core. *J. Colloid Interface Sci.*, **298** (1), 177–182.

148 Mitsuishi, M., Zhao, F., Kim, Y., Watanabe, A., and Miyashita, T. (2008) Preparation of ultrathin silsesquioxane nanofilms via polymer Langmuir–Blodgett films. *Chem. Mater.*, **20** (13), 4310.

149 Carroll, J.B., Frankamp, B.L., Srivastava, S., and Rotell, V.M. (2004) Electrostatic self-assembly of structured gold nanoparticle/polyhedral oligomeric silsesquioxane (POSS) nanocomposites. *J. Mater. Chem.*, **14** (4), 690–694.

150 Mori, H., Lanzendörfer, M.G., Müller, A.H.E., and Klee, J.E. (2004) Silsesquioxane-based nanoparticles formed via hydrolytic condensation of organotriethoxysilane containing hydroxy groups. *Macromolecules*, **37** (14), 5228.

151 Gunawidjaja, R., Huang, F., Gumenna, M., Klimenko, N., Nunnery, G.A., Shevchenko, V., Tannenbaum, R., and Tsukruk, V.V. (2009) Ordering and behavior of branched amphiphilic polyhedral silsesquioxane POSS-M compounds. *Langmuir*, **25** (2), 1196–1209.

12
Multicomponent Polymer Systems and Fibers

As the name suggests, multicomponent polymer systems can be broadly defined as composite materials obtained by mixing two or more distinct polymers or polymeric and nonpolymeric materials, such as inorganic particles. In this chapter, referring to specific examples, we discuss how AFM has enabled better understanding of structure–property relationships of this broad class of materials and contributed to the rapid progress witnessed over the past two decades.

Apart from high-resolution morphological investigations to identify the distribution and the orientation of individual components, which became a routine task long ago, AFM has been proven to be extremely valuable to probe local physical and chemical properties of the multicomponent systems with true nanoscale resolution and frequently on a quantitative level. The AFM imaging techniques clearly reveal the micro- and nanophase separation, corresponding morphologies, and interfacial aspects of these systems. Interfacial interactions and morphologies of the dissimilar polymer chains, or polymer chains with the filler, and their effect on the macroscopic properties (such as strength, toughness, hydrophilicity, electrical conductivity, or optical appearance) can be addressed with AFM techniques. For example, surface force spectroscopy measurements with AFM enable unambiguous identification of the individual phases in nanocomposites and block copolymers based on the elastic and adhesive properties of the individual components. Electrical AFM methods enable quantitative understanding of the conductivity, surface potential, and work function distribution in conjugated polymer composites in device-ready configurations.

On numerous occasions in previous chapters in Part One and Two of this book, we have discussed polymer blends as model systems to describe the fundamental operation of various AFM modes, characteristic results, and approaches for probing various properties. For example, conductive and optical properties of conductive polymer blends have been discussed in Chapter 7, composite molecular layers will be discussed in Chapter 13, and composite polymer films will be presented in Chapter 14. The readers are referred to these and other relevant chapters for fundamental modes of operation and for some examples of the specific multicomponent systems. In this chapter, we will not reiterate these aspects, but expand the discussion to more specific examples of multicomponent systems where AFM forms

Scanning Probe Microscopy of Soft Matter: Fundamentals and Practices, First Edition.
Vladimir V. Tsukruk and Srikanth Singamaneni.

a unique tool to understand the structure and properties of these polymeric materials at different spatial scales and under different external conditions.

12.1 Polymer Blends

Polymer blends can be simply defined as a mixture of two or more distinct polymers often with complementary properties such as brittle–ductile or glassy–rubbery. These dual systems are used to improve the properties (e.g., toughness, ultimate strain, elastic modulus, etc.) and performance of the system or to create novel properties [1]. It is well known that most of the polymers phase separate upon mixing with the individual incompatible phases forming distinct regions on the microscopic level. The composition of these individual phases depends upon the nature of the individual polymer components, the degree of incompatibility, the ratio of viscosities, and the processing and post-treatment conditions. The phase separation typically occurs on the spatial scale of a few microns, except in cases of partially compatible systems and block copolymers where intermixing at various spatial scales can be observed.

Conventional tapping and contact AFM imaging and its numerous probing modes have been extensively applied to characterize and understand the phase-separated polymer blends. Obviously, the key advantage of the AFM technique in the context of studying polymer blends is the ability to reveal fine morphological features of phase separation down to the nanoscale level under ambient conditions and in controlled environments (air, solvents, and variable temperatures). Various probing modes allow the measure of physical properties of the individual phases (mechanical, electrical, magnetic, and optical) that may not be achievable by conventional methods. This helps provide a better understanding of the structure–property relationships of these materials.

The heterogeneous composition and morphology of immiscible polymer blends form an excellent model system for numerous AFM studies aimed at a better understanding of the tip–sample interaction and its influence on the observed contrast in topography and phase images obtained in tapping mode. We have discussed some of these studies while introducing the concept of set-point ratio and tapping hardness in Section 3.4. Many of these earlier AFM studies provided the general guidelines for choosing optimal scanning conditions and, more importantly, the interpretation of the AFM images well beyond the simple visualization of heterogeneous morphologies of polymer blends.

In one comprehensive study, Bar *et al.* considered contrast variation phenomenon to determine the optimal scanning conditions for polymer blends [2]. They demonstrated the contrast reversal in the tapping mode with drastic changes in the topographical and phase images of poly(ethene-*co*-styrene) (PES) and poly(2,6-dimethyl-1,4-phenylene oxide) (PPO) scanned using various values of the driving amplitude (A_0) and set-point amplitude ratio ($r_{sp} = A_{sp}/A_0$). In topography and phase images of PPO/PES blend samples, the relative contrast of different regions was shown to critically depend on the r_{sp} and A_0 values. Indeed, the authors observed that as the tip–sample force is increased, both phase and topography images of the PPO/PES blend samples can undergo a contrast reversal *twice*. The authors concluded that "true" topography of a sample surface can be obtained by using sufficiently high amplitude ($\sim$45 nm) and as high r_{sp} values as possible – that is, the lightest tapping forces possible in stable imaging. On other hand, the use of high amplitudes and moderate r_{sp} values results in an image contrast determined mainly by the stiffness of a sample surface.

In a different example, Raghavan *et al.* have employed AFM to study the heterogeneous structure of thin film blends of glassy and rubbery polymers: PS and polybutadiene (PB) [3]. AFM topographic and phase imaging in tapping mode were performed on the blended films with thicknesses of approximately 250 nm, prepared by spin casting from solutions onto silicon substrates. The authors found that the contrast observed between the two phases was affected by the hardness of the tapping. Figure 12.1 shows the AFM topography and phase images of this PS/PB physical blend collected under three different tapping regimes: hard, medium, and light. There are several notable features in these images obtained at different regimes. First, the phase contrast between the two polymer phases of the film with very different compliance is found to be higher in the hard tapping regime compared to the medium tapping regime, which, in turn, is higher than in the light tapping regime. As the authors noted, phase imaging provides good contrast between the phase-separated PS and PB regions, primarily because of the large compliance difference between the two materials revealed during hard tapping.

The second interesting feature of these images is the absence of contrast reversal in the phase images (a common phenomenon for multicomponent materials with very different components) as the tapping regime was changed from hard to medium to light. As discussed in Section 2.6.2, the change in the tapping regime, as characterized by the ratio of amplitudes, might result in a switch in the tip–sample interaction from attractive to repulsive, reversing the phase contrast (also see above). The absence of such phase contrast reversal in this case is possibly due to readjustment of the drive frequency to the lower side of the resonance frequency after the stable tip–sample contact that maintained repulsive interaction conditions.

The authors further observed that phase contrast for phase AFM images of PS/PB blend decreased with increasing annealing time because of intense thermal oxidation, which resulted in the crosslinking in PB, and thus, the increase in the elastic modulus of the PB regions, reducing compliance differences. In addition to AFM imaging, static force spectroscopy was employed to identify the PS- and PB-rich domains and to better understand the influence of the relative surface stiffness observed on the phase images. The force–displacement curves clearly revealed the much higher penetration of the tip into the PB-rich phase enabling unambiguous identification of the rubbery PB domains.

In another study of glassy–rubbery blends, Tsukruk and Huang studied the surface distribution of the adhesive forces and elastic moduli of the PS/PB thin films over a range of temperatures (from room temperature to above the glass transition of the PS matrix) [4]. PS/PB blend films are highly heterogeneous with a well-developed phase separation of the rubber phase forming circular microscopic regions distributed into

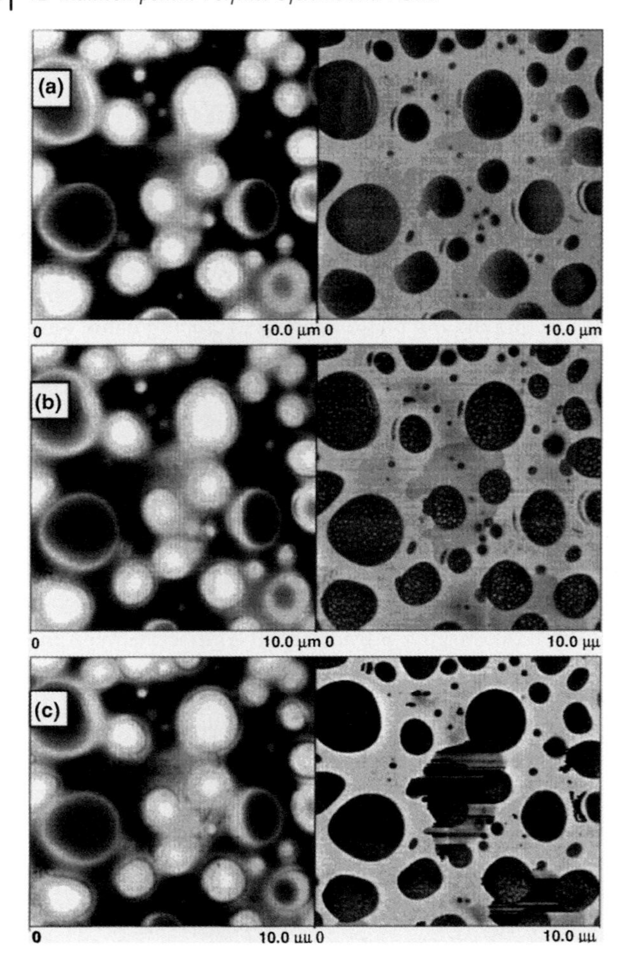

Figure 12.1 Tapping mode topography (*left*) and phase images (*right*) of PS/PB (50/50 w/w) blend film obtained under (a) light, (b) moderate, and (c) hard tapping conditions. Color contrast from black to white represents a total range of 75 nm in the topography image and 90° in the phase image. Reproduced from Ref. [3].

glassy matrix (Figure 12.2). The lateral size of the dispersed rubber phase varies from 20 μm to less than 1 μm and a different morphology (elevated or shallow droplets) is observed at various locations. Such complex surface topographies with different components forming either holes or bumps are controlled by the ratio of surface tensions and the processing conditions. Within the larger PB phase, submicron-sized nanoinclusions of the PS phase are frequently observed due to the phase trapping effect.

The authors demonstrated that for these heterogeneous composite films, micromapping of surface properties can be obtained concurrently for glassy and rubbery phases as well as across the interface with a lateral resolution better than 100 nm by applying SFS mode. The surface mapping (topography, elastic modulus, and

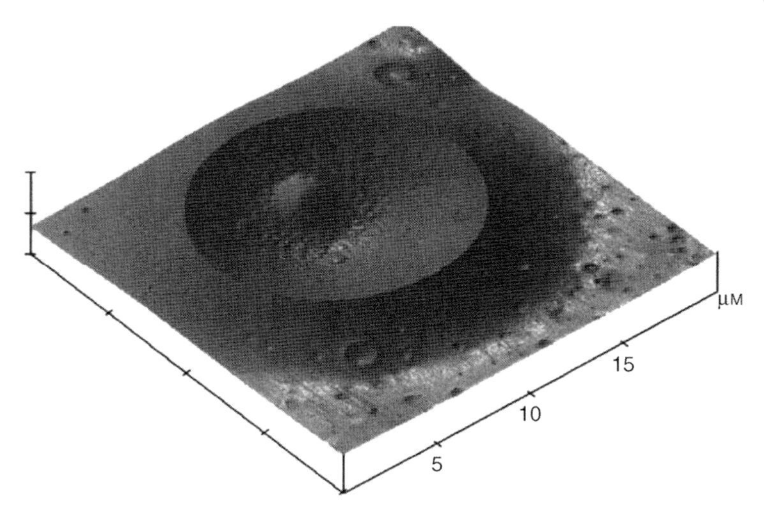

Figure 12.2 (a) Topography of the PS/PB film with PB as the dispersed phase (20 µm × 20 µm; z-scale is 1.5 µm). Note the dimension variation in the PB phase and the presence of the PS microphase inclusions inside the larger PB droplet. Reproduced from Ref. [4].

adhesive forces) were obtained concurrently for the PS/PB blends and utilized to concurrently measure the elastic moduli of different phases in a wide range of temperatures (see Chapter 6 for a detailed discussion of these results). Histograms of the surface distribution displayed two very distinctive maxima, which allowed concurrent measurements of micromechanical properties of glassy and rubbery phases under identical conditions within one mapping.

There are numerous studies where high-resolution surface morphology at different stages of phase separation enabled better understanding of the phase evolution of polymer blends. In one such study, three distinct stages in the phase evolution of a critical blend thin film of deuterated poly(methyl methacrylate) (dPMMA) and poly (styrene-ran-acrylonitrile) (SAN) were revealed by Wanga and Composto [5]. For the early stages of phase separation, the authors observed dPMMA-rich wetting layers rapidly formed at the air/polymer and polymer/substrate interfaces by hydrodynamic flow during blend formation.

In the next stage of phase separation and dewetting caused by annealing, the dPMMA phase from the wetting layers (at both of the interfaces) flows back into dPMMA-rich domains that span the middle layer accompanied by capillary wave fluctuations, which cause thickness undulations. When these fluctuations become large enough, they rupture the middle layer, which transforms into an interconnected 2D network. In the final stage, this 2D network coarsens and forms isolated droplets of the SAN-rich phase covered by a thick dPMMA-rich layer. AFM was employed to reveal the blend morphology evolution, which is the result of dynamic interplay among wetting, phase separation, capillary fluctuation, hydrodynamic flow, and confinement.

Very recently, the temperature dependence of the surface composition and morphology of the same blend (PMMA-SAN) has been addressed by You *et al.* in an *in situ* temperature AFM study [6]. The authors found that in addition to phase separation, the blend component preferentially diffuses to the air–film interface leading to the variation in surface composition with elevated temperature. At 185 °C (above the lower critical solution temperature), the amounts of PMMA and SAN phases were found to be comparable while at lower temperatures PMMA migrated to the surface, leading to a much higher PMMA surface content than in the bulk.

In another study, Virgilio *et al.* employed a combination of the focused ion beam technique and AFM (FIB-AFM) to reveal the partial wetting morphology of ternary and quaternary immiscible polymer blends prepared by melt processing [7, 8]. The interfacial tension of these polymer blend systems were computed by a Neumann triangle method that involves calculation of the interfacial tension between immiscible components from their morphological features [7]. The high-resolution morphological details offered by AFM in conjunction with FIB makes this approach a powerful (although very labor-intensive) tool to reveal internal morphology of polymer blends. It was shown that blends of PS/polypropylene (PP)/high-density polyethylene (HDPE), PMMA/PS/PP, and a quaternary blend system comprised of HDPE/PP/PS/PMMA all display a partial wetting morphology with a three-phase line of contact. Figure 12.3 shows an example of the geometric constructions on the AFM image, used to measure the contact angles for the PS/PP/HDPE ternary blend [8]. The φ angles were measured manually from the AFM image and the contact angles were subsequently calculated. In their investigation of the effect of five different SEB, SB, and SEBS interfacial modifiers on the position of the PS droplet at the PE/PP interface, the authors found that except for the SEBS, addition of even 1% copolymer modifiers resulted in a significant morphology transition with the PS droplets exclusively relocating at the PP/HDPE interface. Furthermore, it was shown that the position of the PS droplet at the interface provides an indication of the ability of this particular copolymer to reduce interfacial tension.

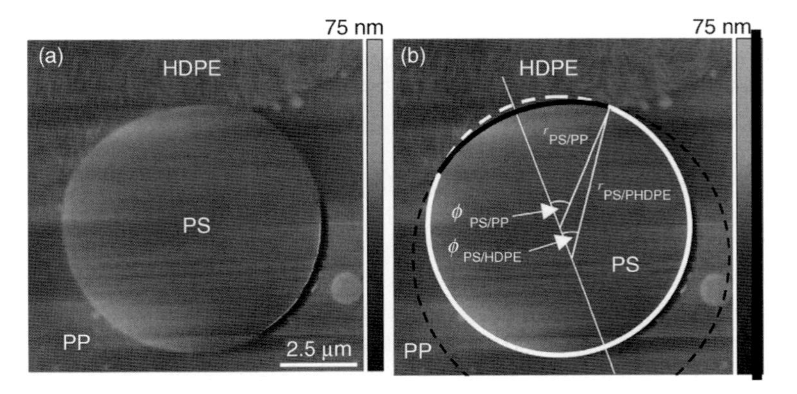

Figure 12.3 AFM image and the image with geometrical construction displaying the fitted circles and the Φ angles used to calculate the contact angles and the radius of curvature of the PS/HDPE interface (in black) and PS/PP interface (in white) of PS/PP/HDPE (10/45/45) blend after 30 min of quiescent annealing time. Reproduced from Ref. [8].

In a comprehensive publication, Galuska *et al.* have exploited force modulation mode for probing the morphology and physical properties of a number of industrially important polymer blends [9]. In particular, when combined with cryogenic sectioning, force modulation AFM mapping is shown to be a very valuable characterization technique to probe the structure–property relationships in these polymer blends with a resolution lower than 10 nm. The authors concluded that one particular advantage of force modulation mode in the context of blends of polymers with different elastic properties is the ability to distinguish polymer phases and related morphology with ease and with good specificity. This capability is demonstrated in this study on a variety of filled and unfilled elastomer/plastic and elastomer/elastomer blends.

In the same study, it has also been demonstrated that the force modulation maps are sensitive to the degree of cross-linking in polymer blends with rubbery components. For instance, in a polymer blend of PB with 95% isobutylene and 5% brominated p-methyl styrene copolymer (BIMS), the authors showed that the phase contrast observed between the PB-rich and BIMS-rich domains deteriorates with higher cross-linking of PB rubber component (achieved by curing at higher temperatures and for longer times) due to reducing compliance of this phase.

Force modulation can also enhance the contrast between polymers with very small differences in elastic modulus, thus enhancing the contrast in polymer blends of two glassy polymers. For example, Lipatov and coworkers employed force modulation mode to study the simultaneous adsorption on the surface of silica of two incompatible glassy polymers: PS and poly(butyl methacrylate) (PBMA) [10, 11]. Although PS and PBMA components are both glassy at room temperature, the small difference in the stiffness between the two phases resulted in an enhanced contrast of dissimilar domains in the force modulation images compared to smaller or no contrast difference observed in the conventional AFM topography images. These results clearly highlight that the force modulation mode of AFM is a useful technique to qualitatively probe polymer blends with low mechanical contrast between components with high lateral resolution.

In the examples discussed so far, the polymer blends exhibited phase separation with a highly polydisperse size and distribution of different phases caused by nonequilibrium processing conditions. However, an ordered phase separation in polymer blends can be induced by fine control over the processing conditions, such as additional long-term annealing, solvent evaporation rate, or addition of surfactants to modify the interfacial tension between the components [12, 13]. To this end, Cui *et al.* studied the ordered phase separation phenomenon for PS/poly(2-vinylpyridine) (P2VP) blend films cast on a mica substrate from an ethylbenzene solution by controlling different weight ratios and solvent evaporation rates. An almost honeycomb-like surface morphology of the PS/P2VP blend film formed under controlled solvent evaporation due to the known Marangoni–Benard effect [14].

Tapping mode AFM imaging was employed to reveal the ordered phase separation of these novel binary films. The authors found that solvent evaporation rates and PS/PVP weight ratio critically determined the surface morphologies of the films formed on the mica substrates under different conditions. At a very low solvent evaporation rate, disordered holes distributed on the surfaces of the blend films were observed since the PS-rich phase layer covered the P2VP due to lower surface free energy of PS phase compared to that of P2VP phase. On the other hand, when the solvent evaporation rate was increased in a controlled manner, a quasihexagonal long-range arrangement of holes was formed on the surfaces of the PS/P2VP blend films reflecting the underlying convection process.

In a different approach, Singamaneni *et al.* demonstrated ordered binary polymer blends by infiltration of the rubbery component into a preformed porous glassy matrix [15]. The periodic porous matrices were fabricated using interference lithography (IL) allowing the creation of periodically patterned solids and predetermined highly porous matrices with predetermined topology [16–19]. IL is promising for fast micro-fabrication of complex periodic structures unachievable by other techniques, and more importantly, it facilitates top down, one-shot synthesis of bicontinuous open structures that can serve as an ordered matrix for fabrication of ordered composite materials [20–22].

The organized polymer composites studied with AFM by Singamaneni *et al.* were fabricated by infiltration of PB (rubbery polymer) using a capillary infiltration approach (Figure 12.4). AFM topography and phase images of the binary infiltrated

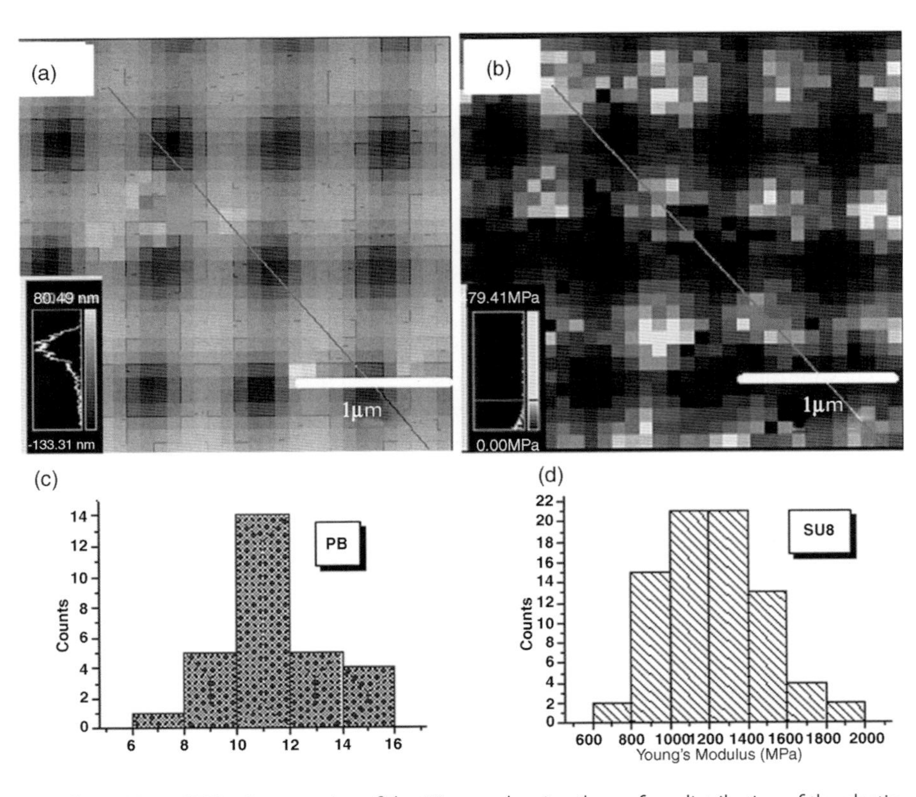

Figure 12.4 AFM micromapping of the PB infiltrated SU8 square microframe: (a) 32 × 32 topography and (b) elastic modulus collected during force micromapping, and (c) histograms showing the surface distribution of the elastic modulus in the SU8 and PB phases of the bicomponent structure. Reproduced from Ref. [15].

structure clearly depict the distinct SU8-based original glassy structure (square matrix) and the lattice of rubbery PB phase. The authors employed surface force spectroscopy to elucidate the mechanical properties of the composite structures. Figure 12.4 shows the topography and the corresponding elastic modulus distribution of the bicomponent ordered structure with the initially porous glassy matrix filled with a rubbery phase. The elastic modulus distribution shows that the glassy regions are much stiffer, with an elastic modulus of 1.2 GPa, similar to that measured for the pristine SU8 structure (see histograms of modulus distribution for different phases in Figure 12.4). Conversely, the patterned PB-filled regions exhibit much lower elastic resistances with an elastic modulus of 11 MPa, close to that expected for conventional rubbery bulk PB material.

Figure 12.5a shows the load versus penetration data obtained from three different regions on the SU8-PB microcomposites: SU8, PB, and the interface between SU8 and PB phases. From the loading data, it is clear that the rubbery PB regions (PB phase in the small pores) exhibit much higher elastic deformations compared to the interfacial regions, which in turn undergo higher deformations than those of the glassy material (Figure 12.5a). It is interesting to note that the initial portion of the SU8/PB indentation curve (about 5 nm deformation) closely matches that for the pure PB phase at low indentations. However, a drastic increase in elastic modulus is observed at larger indentations (Figure 12.5b). This drastic change in the apparent elastic modulus value for high indentation depths indicates that the AFM probe encounters underlying glassy material with higher stiffness during the indentation process at higher deformation.

12.2 Block Copolymers

Block copolymers, one of the most intriguing and widely exploited classes of nanocomposite materials, are comprised of two or more chemically distinct, and most often immiscible, polymer chains self-organized in nanodomain morphology via confined microphase separation of different blocks. The microphase-separated domains of the block copolymers organized in lattices with different symmetries are on the order of a few to tens of nanometers, depending on the molecular weight of the block copolymer as well documented in numerous publications and textbooks (e.g., see Ref. [1]).

AFM imaging in tapping regime is routinely used to establish the microphase-separated nanodomain structures in thin block copolymer films. Different aspects of block copolymer morphologies have been systematically studied with AFM usually in combination with TEM [23–31].

Several key aspects of tapping-mode AFM imaging of block copolymers, such as unambiguous phase attribution, the relationship between topography and interior structure, and contrast reversal artifacts, have been addressed by Wang *et al.* using styrene-ethylene/butylene-styrene (SEBS) as a model system [32]. Combining AFM imaging with TEM investigations enabled them to shed light on several aspects critical for the understanding of tapping mode AFM images of block copolymers. The

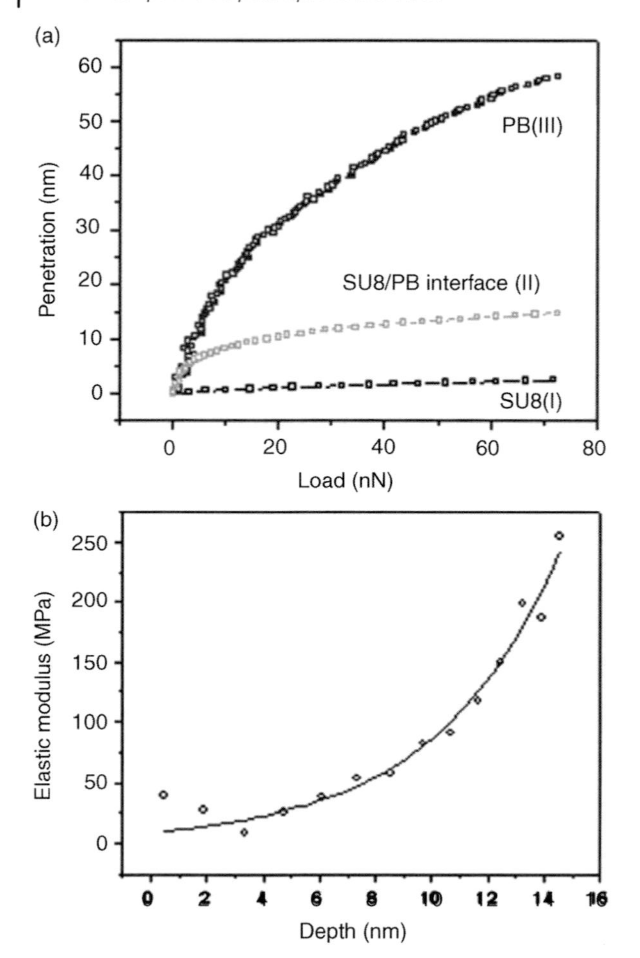

Figure 12.5 (a) Representative loading curves from three different regions (SU8, PB, and interface of the bicomponent microcomposite); (b) depth profile of the elastic modulus at the interface of SU8 and PB (along the rim of the pore) showing the increasing elastic modulus with indentation depth. Reproduced from Ref. [15].

authors noted that lateral tip indentation differences between different phase domains rather than real surface topography plays a key role in providing contrast in the AFM topography image. On the basis of the comparison of AFM results with TEM images, the authors concluded that under medium tapping higher spots in the AFM topography image and brighter domains in the phase image correspond to the hard PS phase. However, when just employing AFM images to understand phase-separated structures of soft block copolymers, one should bear in mind the common occurrence of contrast reversal artifacts.

The authors observed imaging artifacts of contrast reversal first in the phase image and then in the topography image, which were attributed not only to the changes in the tip–sample interaction from attractive forces to repulsive forces but also to the difference of changing scanning velocity on soft and hard phase domains. A systematic error in the form of larger PS domains compared to the TEM images was also observed and discussed. The authors reasoned that this systematic error is due to the presence of the top rubber layer on the hard PS spheres and due to the possible deformation of the PS domains by the AFM tip. Together with accurate AFM measurements of the absolute film thicknesses, it is possible to relate the particular morphology to the local film thickness and in this way the thickness-dependent phase behavior of block copolymers.

In another study of SEBS block copolymers, Luzinov *et al.* demonstrated robust and uniform ultrathin elastomeric films with complete surface coverage [33]. These films were formed from SEBS material functionalized with 2% maleic anhydride by thermally grafting to a chemically modified silicon surface via an epoxy terminated SAM. The thickness of the SEBS film was varied from 1.4 to 8.5 nm to test the limits of the stability of microphase-separated nanodomain structures under confined conditions.

The authors of this study observed that the in-plane cylindrical/spherical nanodomain morphology is similar to the bulk microstructure of SEBS material with the composition studied, but it is compressed in the vertical direction due to film–air and film–substrate interfacial constraints (Figure 12.6). Such a microstructure is formed at thicknesses ranging from 2.6 to 9 nm and is perfectly defined at $L/L_0 = 0.3$ (SEBS interdomain spacing, $L_0 = 28$ nm). However, microphase separation is completely suppressed only for extremely thin films with $L/L_0 < 0.08$. Unlike physically adsorbed SEBS monolayers, which dewet the silicon surface during annealing, tethered block copolymer monolayers obtained under identical conditions are very stable even under high shear stresses and at elevated temperatures.

Figure 12.6a and b present topographical and phase images of the SEBS films (thickness of 8.5 nm) recorded at the highest set point ($r_{sp} = 0.9$–0.95), or the lowest tapping forces applied while still maintaining the contact. For comparison, Figure 12.6c and d show AFM images from the same SEBS films recorded with the low set point ($r_{sp} = 0.45$–0.5) or high tapping forces applied. As discussed above, under light tapping conditions, the topographical images reflect the morphology of the topmost layer and phase imaging is mostly controlled by the distribution of the surface adhesion. In contrast, the AFM images under hard tapping conditions are recorded in the repulsive mode and a major contribution comes from the elastic response of the glassy and rubbery regions. The authors suggested that under these scanning conditions, the AFM tip squeezes the topmost compliant rubbery layer and interacts directly with underlying hard PS phase. The images clearly reveal the contrast reversal with variable tapping conditions of the soft–hard block copolymer.

In continuation of this work, ultrathin poly[styrene-*b*-butadiene-*b*-styrene] copolymer (SBS) films deposited on PS brushes were studied with AFM by Luzinov *et al.* [34]. The thickness of the films was kept constant while the grafting density and molar mass of the grafted polymer layers were varied to reveal the influence of the brush interface on the structure of the topmost SBS films. The development of a zeroth layer (a layer without an internal microphase-separated structure) of SBS block copolymer on the top of the grafted layer was observed. Both PS and PB blocks were

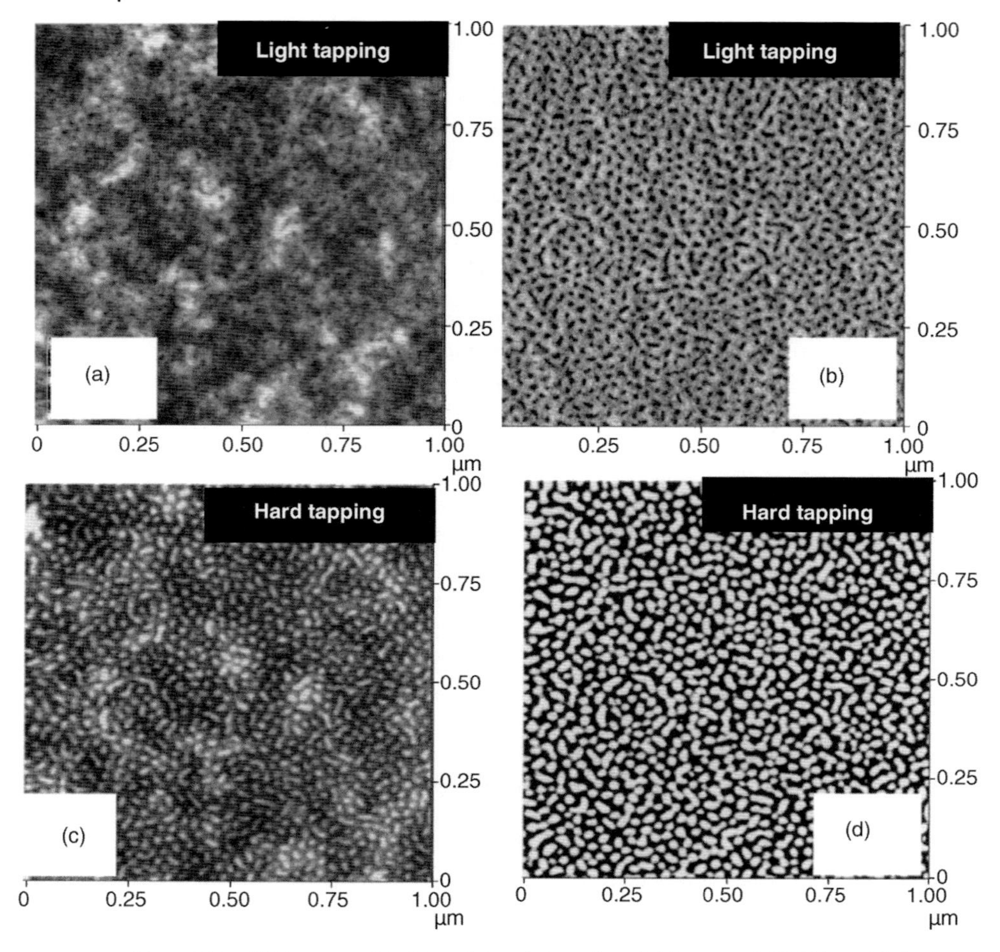

Figure 12.6 SPM topographical (a and c) and phase (b and d) images of SEBS films with thickness 8.5 nm obtained in light (a and b) and hard (c and d). Vertical scale is 7.0 nm and 20° for (a) and (b) and 7.0 nm and 10° for (c) and (d). Reproduced from Ref. [33].

present inside the zeroth layer in the intermixed state. The first truly block copolymer layer with nanodomain morphology formed on top of the zeroth layer possessed the surface microstructure typical of the SBS block copolymer in the bulk state. The polymer brushes were actively involved in the formation of the zeroth layer, and the structure of the block copolymer films was influenced by the grafting density and the degree of polymerization of the underlying grafted polymer layer. With high-resolution AFM studies, the authors have observed significantly distinct morphologies for the block copolymer films of the same thickness deposited on the different polymer brushes as discussed below.

For the SBS film deposited on the top of a grafted polymer layer with a M_n of 143 000 and a grafting density of 0.003, the formation of only the zeroth layer was found (Figure 12.7). This layer was smooth and had a uniform thickness and nondeveloped morphology that did not change much after annealing. For all other samples, surface morphology with isolated islands, holes of different size, and intermediate structures between islands and holes have been developed during annealing (Figure 12.7). A variety of morphologies of isolated islands, holes of

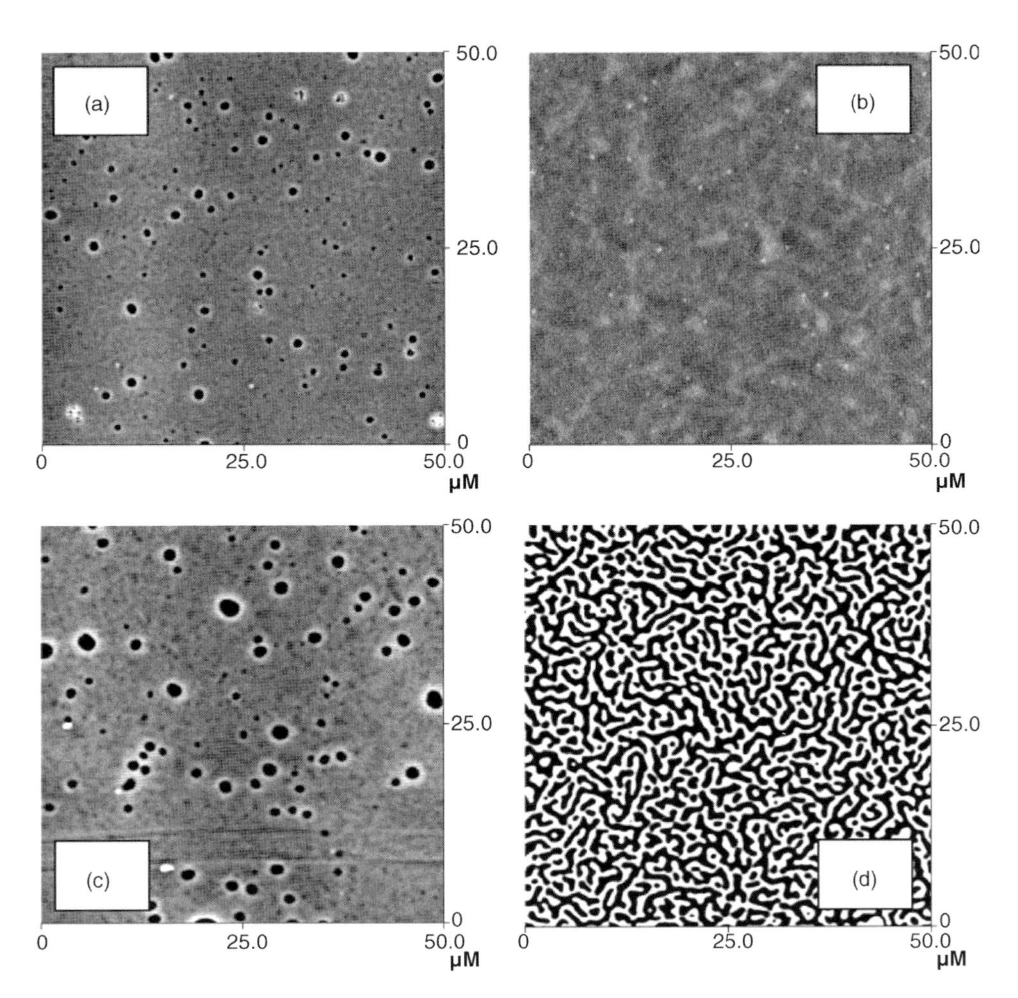

Figure 12.7 SPM topographical images of the SBS films after annealing. The films were deposited over grafted PS layers with molar mass (g/mol)/grafting density (chains/nm^2) of (a) 143 000/0.004, (b) 143 000/0.012, (c) 672 000/0.001, and (d) 672 000/0.004, which represent a change in grafting density and molar mass of the underlying polymer brush. Morphology of holes of different size (a and b) and intermediate between islands and holes (d) developed during annealing. For the film deposited on the top of grafted polymer layer with M_n of 143 000 and low grafting density (b), the formation of only zeroth layer was found. Bright parts correspond to higher features. The vertical scale is 70 nm. Scanning at high r_{sp}. Reproduced from Ref. [34].

different sizes, and intermediates between islands and holes as well as a uniform zeroth layer were all observed with AFM for SBS films in the course of their annealing.

AFM imaging has been successfully used to classify and analyze topological defects in the local microdomain structures since defects are known to compromise the long-range periodicity and limit the technological performance of block copolymer materials [35–39]. In comparison to space-averaged scattering or rheological methods, *in situ* AFM probing has proved to be a useful real-space and real-time imaging technique in determining the precise microscopic mechanisms at different length scales by which overall organization evolves in nanostructured polymer materials.

Common features in block copolymer nanostructures formed by hard and soft domains that can be relatively easily observed with AFM imaging are point defects, line defects, grain boundaries, metastable phases, and distortions in microdomain orientation. In a recent comprehensive study, Tsarkova *et al.* collected AFM images of several examples of classical and grain boundary defect configurations that have been considered in detail by Horvat *et al.* [40] as is illustrated in Figure 12.8.

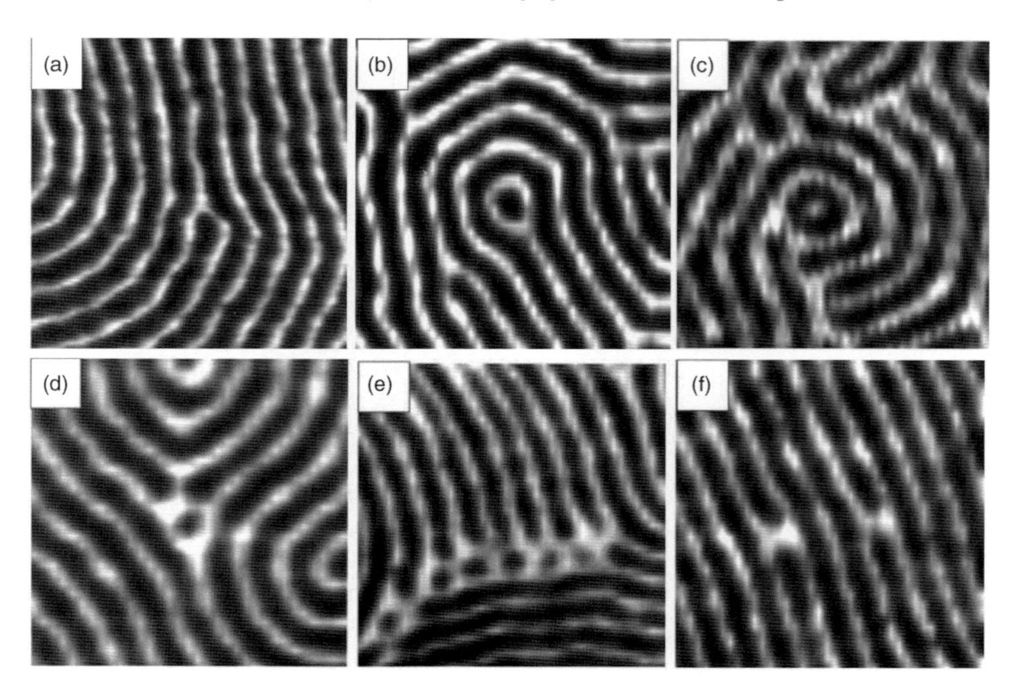

Figure 12.8 SFM phase images (300 × 300 nm) presenting topological defects in a striped pattern of cylinder forming PS-*block*-PB diblock copolymer film. White stripes correspond to the A minority phase (glassy PS cylinders); dark stripes correspond to B majority phase (soft PB matrix): classical topological defect, which is identified as a *cylinder* dislocation (a); modified classical defects: + 1/2 disclination (b) and − 1/2 disclination (d) with an incorporated PL fragment, + 1/2 dot-disclination (c); grain boundary defect: T-junction with incorporated PL rings (e); specific neck defects (f). Courtesy of L. Tsarkova.

Figure 12.8a displays an edge dislocation that is formed by white stripes and is identified as a *cylinder* dislocation. It is topologically identical to a *matrix* dislocation, which is formed by black stripes; however, the thermal history of these defects crucially differs, with a cylinder dislocation being significantly more stable [40]. Pure topological arguments are sufficient to describe classical dislocations and disclinations in films with upstanding lamella domains as a topological defect always implies the abruption of one component. In contrast, in a cylinder phase, the majority matrix (dark-looking PB phase in Figure 12.8) is always interconnected. The important conclusion made was that the conventional 2D representation of topological defects on AFM images might conceal this important property of block copolymer films.

Early studies on the dynamics of individual defects were focused on thin films of asymmetric block copolymers and employed cyclic annealing of hexagonal structures followed by post-snapshot imaging of single defects and their evolution [41–43]. The high-order periodicity of microdomains observed here enables one to draw similarities with liquid crystals systems (where single defects are well characterized) in aspects of defect interactions and annihilation. Only recently have the dynamics (origination, development, and healing) of individual defects within a block copolymer surface structures been visualized with direct AFM imaging while conducting prolonged annealing [38, 39, 44, 45]. For instance, with *in situ* AFM observation, the phase transition from the cylinder to the perforated lamella phase was monitored and characterized in a thin film obtained from a concentrated block copolymer solution [46].

In these studies, it has been observed that due to the fast evaporation of the solvent during the process of phase separation, the dissimilar blocks mostly remain in the disordered state. Numerous treatment methods, such as temperature annealing, solvent annealing, surface premodification, and a combination of these techniques, have been applied to induce ordered (especially long-range) domain morphology caused by phase separation of block copolymer films. It has been concluded that solvent annealing, which involves prolonged exposure of the polymer film to saturated solvent vapor for different durations, results in highly ordered morphologies with almost defect free lattices of ordered nanoscale domains.

In a recent intriguing study, Tsarkova *et al.* captured fast dynamics of individual defects and associated interfacial morphological changes in a thin film of a cylinder-forming melt (Figure 12.9). Their lifetime and spatial–temporal correlations were monitored on a length scale that covers several adjacent domains [39, 45]. Fast and repetitive transitions between distinct defect configurations can be captured by *in situ* AFM imaging at elevated temperatures. The structure marked with white circles fluctuates mainly between the configuration with three "open ends" (frame 158) and two "open ends" (frame 160). The upper limit of a characteristic transition timescale is given by the measurement time between two frames, which is about 46 s. On the basis of the time-resolved imaging data, the authors evaluated both the typical activation energy of the break up of a cylinder ($0.4k_BT$) and the diffusion coefficient of $D \sim 10^{-13}\,cm^2/s$. Comparison of typical transition times with characteristic diffusion coefficients of polymer chains suggests that the motion proceeds via correlated movement of clusters of chains rather individual chains.

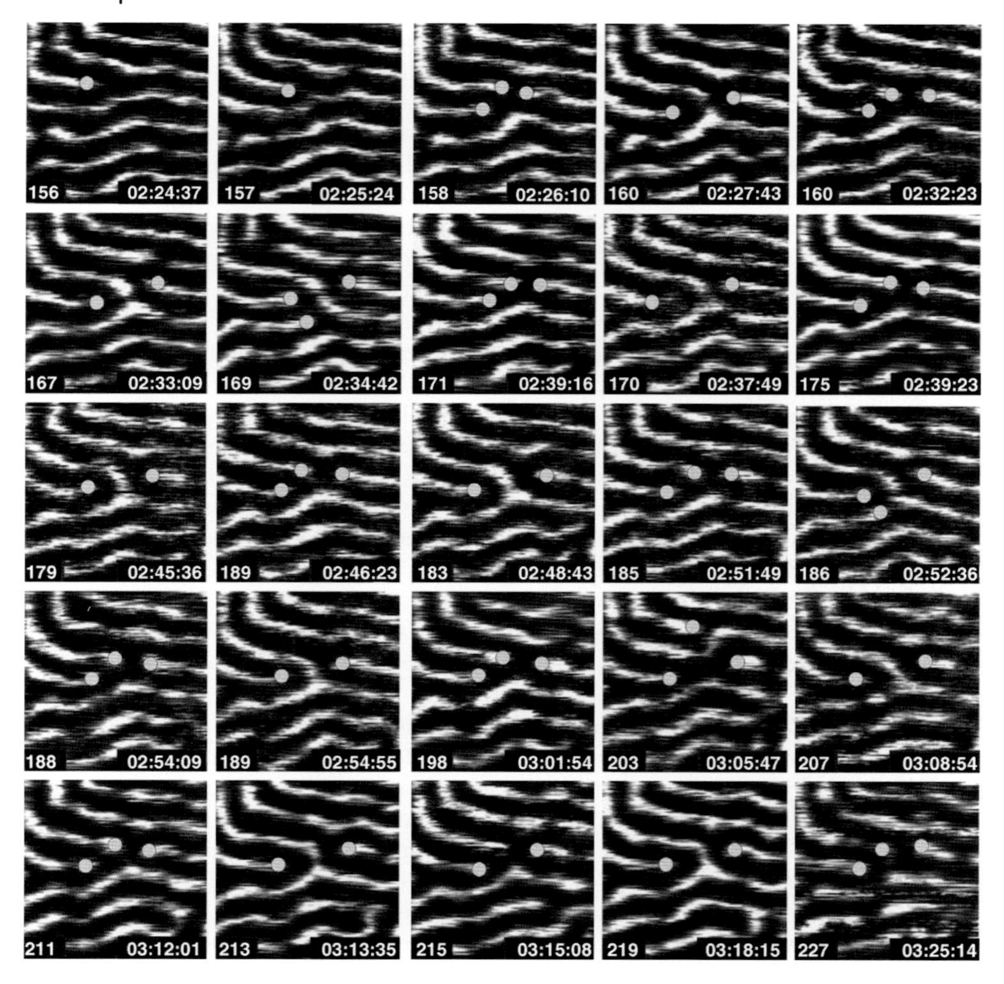

Figure 12.9 Crops (250 nm × 250 nm) from selected consequently saved frames showing oscillations between distinct defect configurations. White dots mark the open ends of cylinders. The structure fluctuates mainly between the configuration with three "open ends" (Frame 158) and the configuration with two "open ends" (Frame 160). The labels indicate the frame number and the elapsed time of the measurement. Courtesy of L. Tsarkova.

In another study, Wang *et al.* employed AFM to monitor the morphology evolution of poly(styrene-ethylene/butylene-styrene) thin films upon annealing in cyclohexane vapor, a selective solvent for the majority poly(ethylene/butylene) block [47]. The morphology evolution pathway from disordered short cylinders to aligned long cylinders and then to hexagonal spheres was tracked by obtaining a series of AFM images in the same marked location. From the series of AFM images, the authors concluded that the whole ordering process in block copolymers consists of (i) the cyclic transitions between poorly ordered cylinders and a semidisordered phase via poorly ordered spheres, during which the ordering of the cylinders gradually improves, and (ii) the pinching off from partially ordered cylinders into hex spheres.

Very recently, Yufa *et al.* employed AFM to probe *in situ* the self-healing capabilities of PS-PMMA diblock copolymer films on a silicon substrate by observing the evolution of the patterned substrate [48]. The authors have employed contact-mode AFM to generate a patterned damaged surface area (a bull's eye pattern) of the polymer film by using high local forces and controlled tip lateral movement (see Figure 12.10). The outer circle of the bull's eye pattern was created by using

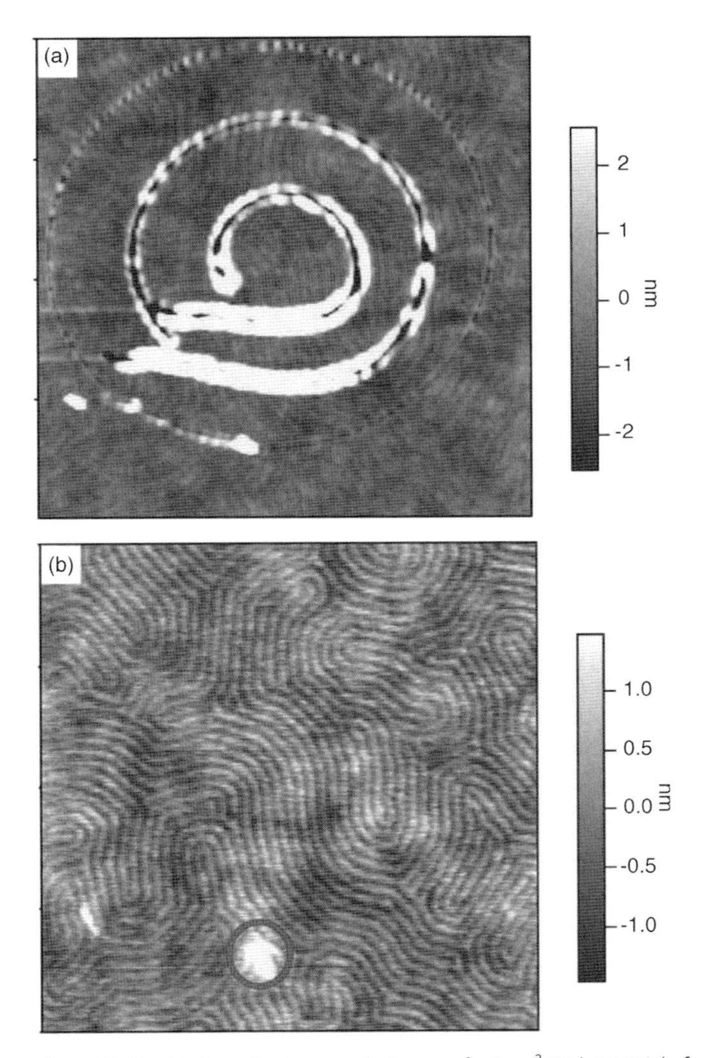

Figure 12.10 AFM tapping topography image of a 2 μm^2 PS-*b*-PMMA before (a) and after (b) healing a scratch. The pattern was created in contact mode AFM. Reproduced from Ref. [48].

a force of 70 nN and inner circle was formed by applying a force of 14 nN, thus creating grooves with different depths.

Tapping mode AFM (utilized to avoid any perturbation of the surface features) was further employed to image the patterned surface area and the healing of the block copolymer surface (Figure 12.10). The authors observed changes in the polymer morphology as it was subsequently heated *in situ* and monitored the "scars" healing. The authors noted that the polymer heals rapidly at temperatures above the glass transition temperature owing to the enhanced polymer mobility under these conditions. Furthermore, the classical fingerprint patterns associated with microphase separation of the block copolymer, which were destroyed by the tip-induced surface damage, reformed completely upon extended thermal annealing. These *in situ* AFM observations at elevated temperatures provided a fundamental insight into the healing of the domain morphology at the nanoscale.

12.3 Polymer Nanocomposites

Incorporation of functional nanostructures (e.g., metal nanoparticles, quantum dots, clay, or carbon nanotubes) into polymers has received intense attention in the past two decades owing to the synergistic enhancement in the properties of the composite structures and, in some cases, novel properties absent in both components [1]. AFM has proved to be a valuable tool to understand the morphology and microstructure of these nanomaterials. The powerful combination of high-resolution morphological AFM probing under a wide variety of conditions in conjunction with its ability to probe the localized properties (mechanical, electrical, or thermal) with nanoscale resolution makes AFM a unique tool for the interrogation of polymer nanocomposites. In this section, we will present several examples on how AFM has been employed to provide critical insight in studying some traditional nanocomposites with many examples relevant to various particular classes of nanocomposites discussed in other chapters.

In a recent study, Ramesh *et al.* employed AFM for *in situ* monitoring of the formation and growth of silver nanoparticles inside spin-coated thin films of poly (vinylpyrrolidone) (PVP) containing silver nitrate under ambient conditions [49]. The growth of the nanoparticles was also monitored with surface plasmon resonance absorption extending over periods of several hours of nanoparticle formation. While the thicker polymer films exhibited growth of high density of nanoparticles, thinner films (~25 nm) fabricated from more dilute solutions enabled formation of relatively large and well-separated silver nanostructures (see Figure 12.11). Larger dimensions allowed for tracking the growth of individual structures within the first hour itself. The profile of each of the nanostructures observed above the film surface indicated that nanoparticles are nearly spherical. The authors noted that the nanostructures result from local aggregation of silver atoms inside the film.

In a seminal study, Tang *et al.* employed LbL assembly of polyelectrolytes and clay to mimic the brick and mortar arrangement of organic and inorganic layers, which is

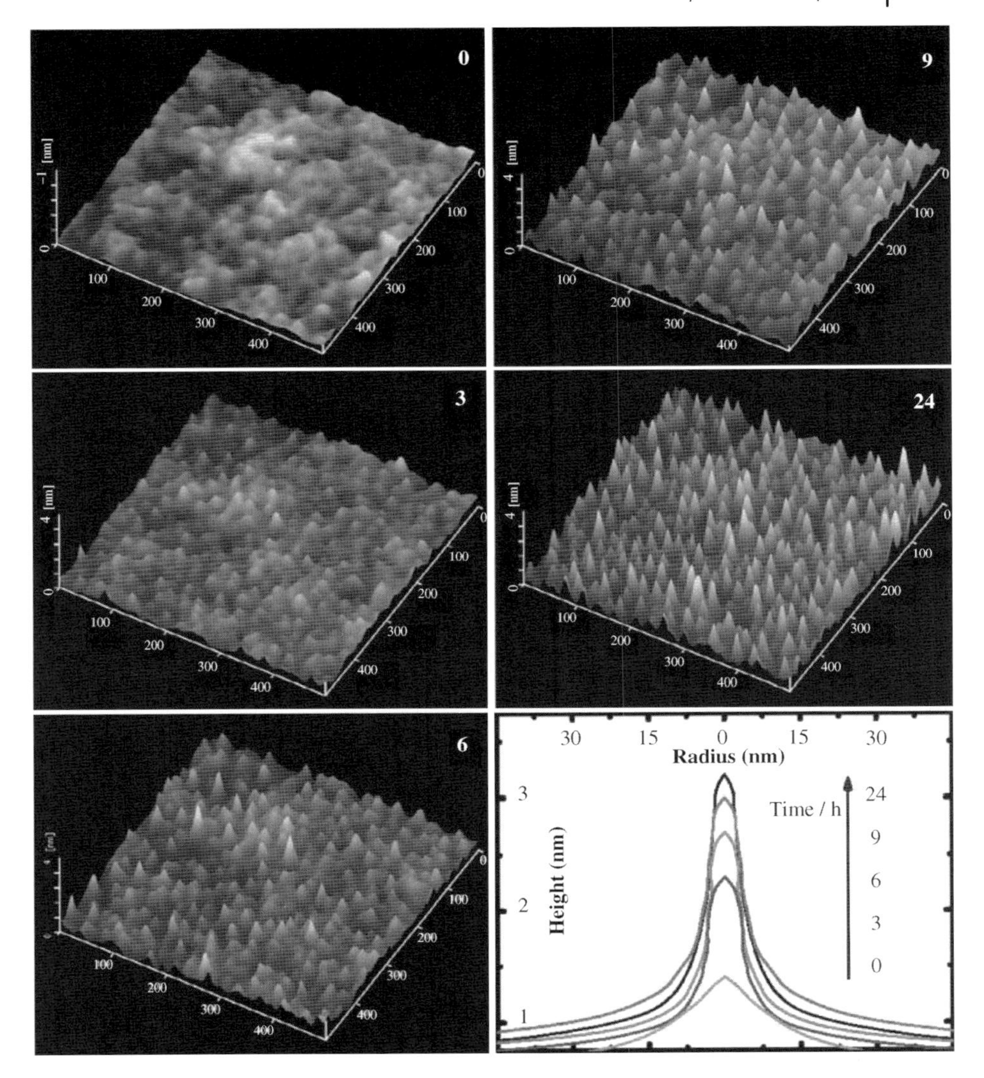

Figure 12.11 Real-time AFM images (0.5 μm × 0.5 μm × 4 nm) showing the growth of silver nanoparticles in PVP film after different number of hours (shown in the image); first image was collected 15 min after the fabrication of the sample. The last panel is a schematic representation of the growth, using average height and radius. Reproduced from Ref. [49].

believed to be the most important structural feature of nacre [50]. They successfully replicated the morphological and mechanical properties of nacre and other biocomposites using a simple multilayer composite obtained by sequential layering. In their work, high-resolution AFM was employed to monitor the actual conformation of polyelectrolyte macromolecules (poly(diallydimethylammonium) chloride (PDDA)) at the organic–inorganic interface. The intentionally rarified submonolayers of polyelectrolyte were made from 2×10^{-7} wt% solutions to resolve single chains. These single polymer chains were found to be intricately entangled to form a continuous film at higher concentrations.

AFM imaging revealed polyelectrolyte molecules adsorbed onto mica in different conformational states: extended, partially coiled, and most of them (>75%) tightly coiled. In the extended state, the length of the chain was found to be in agreement with the expected length based on the molecular weight of the polymer. The dominance of the coiled conformation was also observed for PDDA adsorbed directly onto clay platelets, the filler employed to build artificial nacre. Although a substantial portion of chains were found to be stretched between two different platelets forming coils on the clay surface, the authors noted that quantitative evaluation of the percentage of coiled and stretched conformation is difficult due to a higher substrate roughness in the presence of clay platelets.

In a recent study, Kulkarni *et al.* employed LbL assembly to fabricate highly ordered, freestanding, layered nanocomposites with embedded graphene oxide sheets [51]. Graphene oxide sheets were uniformly incorporated inside the LbL PAH/PSS polyelectrolyte matrix resulting in a highly stratified composite. AFM images revealed uniform morphology with microroughness (within $1 \times 1\,\mu m^2$) below 0.5 nm, common for LbL films [52–54]. By controlling the surface pressure in a Langmuir–Blodgett approach, the surface coverage of the graphene oxide sheets was manipulated to give a uniform deposition with a density reaching 90% while showing only occasional wrinkles and overlaps. The thickness of the graphene oxide was found to be ~0.9 nm, which corresponds to a bilayer structure (Figure 12.12). Following their deposition on the polyelectrolyte multilayers, high-resolution AFM imaging showed that the graphene oxide sheets followed the morphology of the polyelectrolyte layers. The microroughness of graphene oxide sheets of 0.38 nm indicates atomic smoothness. High contrast in phase images obtained at higher resolutions showed large differences in the surface properties of PEMs and graphene oxide sheets caused by their very different surface functionalities and stiffness (Figure 12.12). These freestanding composites exhibited excellent toughness and

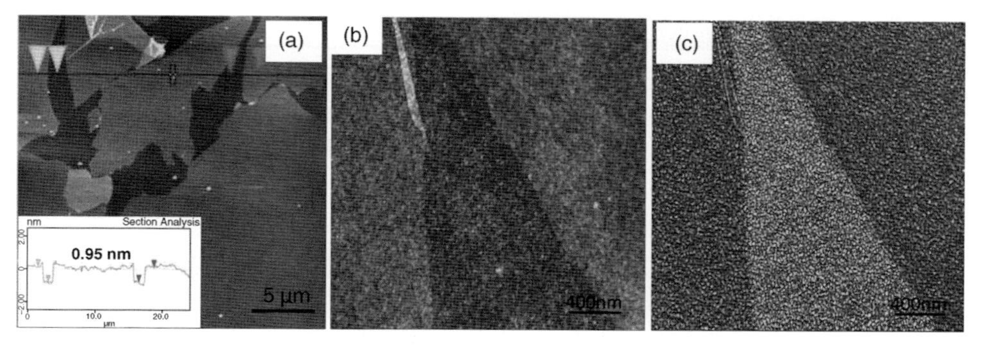

Figure 12.12 AFM showing (a) graphene oxide sheets deposited on silicon (inset showing the sectional image). Graphene oxide on top of polyelectrolyte bilayers (b) topography (z-scale: 5 nm) and (c) phase variation (z-scale: 5°). Reproduced from Ref. [51].

improved elastic modulus, reaching 1.9 MJ/m^3 and 20 GPa for a modest content of the graphene oxide component (about 3%).

In an attempt to fabricate *in situ* inorganic–organic nanocomposites, Kharlampieva *et al.* have demonstrated that rSilC protein deposited on a properly charged surface of polyelectrolyte LbL films initiates nucleation and growth of titania nanoparticles in a uniform and dispersed manner under ambient conditions [55]. The authors suggested that the confinement of titania nanoparticles to nanoscale protein surface domains controls nanoparticle growth beyond certain dimensions, thus preventing formation of larger microscopic aggregates. The nanoparticles grown on surface-bound proteins are composed of 4 nm titania cores (composed of mixed amorphous and anatase phases) surrounded by 1 nm protein shells as concluded from comparison of AFM and TEM results as well as XPS and SERS results.

In a follow-up study, aimed to further understand the mechanism of bioenabled and surface-mediated titania nanoparticle synthesis, surface force spectroscopy was used to analyze the distribution of protein domains on a polyelectrolyte LbL surface [56]. The authors observed a uniform adhesion over the 1 μm^2 area with a unimodal distribution of the adhesive forces of the adsorbed protein despite the domain morphology being visible in high-resolution AFM images (Figure 12.13). Figure 12.13 also shows the AFM image of the silafin protein deposited on polyelectrolyte bilayers. One can clearly observe the fine domain morphology in the AFM image. The silicon oxide surface of the AFM tip is moderately negatively charged, and in the case of a nonuniform distribution (only local domains without protein molecules in between domains), a bimodal distribution of the adhesion of the positively charged rSilC protein across the negatively charged PSS surface would be expected.

Therefore, the unimodal distribution of the tip–surface interactions suggests that the entire surface is covered with rSilC molecules (Figure 12.13). As the highest resolution of SFS achievable in their study (30 nm per pixel) is not sufficient to resolve the individual rSilC domains, the result indicates that the rSilC domains are surrounded by a molecular layer of rSilC molecules. Using a combination of high-resolution AFM, SFS, and Fourier-transformed infrared spectroscopy, it was found that protein deposition on PSS-terminated polyelectrolyte surfaces results in its reorganization from a random-coil to a mixture of random-coil and β-sheet secondary structures.

In the same series of studies, redox-active nanoscale LbL films with polyaminoacid-decorated surfaces were demonstrated to enable both nucleation and growth of uniformly distributed gold nanoparticles at ambient conditions by Kharlampieva *et al.* [57]. In particular, the authors found that poly-L-tyrosine (pTyr), a synthetic polyaminoacid, was able to direct nanoparticle formation to solid, flexible, and patterned surfaces preventing particle agglomeration. The gold particles were 8 nm in diameter, surrounded by a 3–6 nm polyaminoacid shell, and confined to the topmost polyaminoacid layer. AFM served as a key technique to monitor the growth of the nanoparticles on the surface with high resolution. In particular, comparing the size of the nanoparticles before and after annealing at high temperatures to ablate the organic molecules enabled them to estimate the thickness of the organic shells.

Singamaneni *et al.* have also employed colloidal probe surface force spectroscopy to investigate the micromechanical properties of periodic porous polymer interference

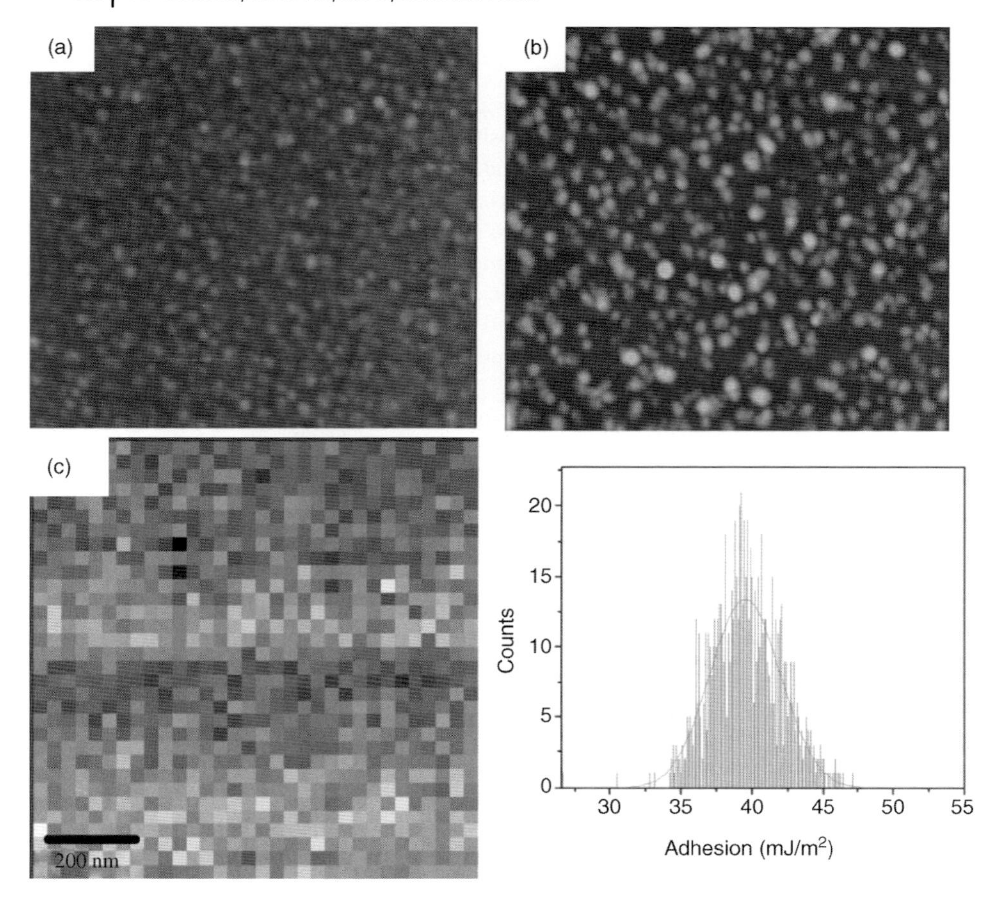

Figure 12.13 AFM images of (a) (PAH/PSS)$_2$–rSilC film and (b) after exposure to titania precursor showing the uniformly grown titania nanoparticles. (c) Force–volume measurements of a (PAH/PSS)2–rSilC film with an adhesion map showing the smooth surface morphology and uniform adhesion. Histogram of the adhesion and the corresponding Gaussian fits. Reproduced from Ref. [56].

lithographic structures with *in situ* mineralized gold nanoparticles using a biomediated approach [58]. The authors have achieved gentle, wet chemistry metallization without destroying the highly porous and collapse-prone nanoporous polymer microstructure using conformal synthetic aminoacid nanocoatings capable of metal ion binding and further reduction to metal nanoparticles. For mechanical measurements, the biomineralized microframe structure selected was infiltrated with PS, forming a bicomponent polymeric structure. Figure 12.14 shows the pristine and metallized bicomponent (SU8/PS) microframe structures and micromapping of these structures by collecting an array of force–distance curves with a pixel size of 60 nm. These images display well-resolved periodic topography and higher elastic modulus along nodes, consistent with binary composition.

Figure 12.14f shows the average load versus penetration plots collected for nodes of pristine and biomineralized structures. It is worth noting that the penetration achieved even at high normal loads (up to 120 nN) is about 1.5 nm, indicating that probing is occurring only on the topmost surface layer. Using Sneddon's analysis, the elastic modulus of the polytyrosine-coated microframe was found to be 1.7 GPa, while that of the IL structure after the biometallization procedure was much higher, 2.9 GPa [59, 60]. The authors suggested that the increase in the elastic modulus of the surface of composite structures is due to the presence of the extremely small gold nanoparticles and the gold ions bound to the topmost polytyrosine layer.

In one of the studies of nanocomposites with carbon nanotubes, Bliznyuk and coworkers employed force modulation mode AFM to probe the dispersion and phase separation of single-walled carbon nanotubes (SWNT) in thin and ultrathin polymer films [61, 62]. Composite films of SWNTs dispersed in polymer matrices of PS and polyurethane elastomers with the thickness ranging from 100 nm to 3 μm were formed by dip coating. The authors found that the phase separation in the composite system is controlled by the surface free energy of the components and the interfacial interaction with the substrate. Vertical phase separation was revealed for nanoscale-

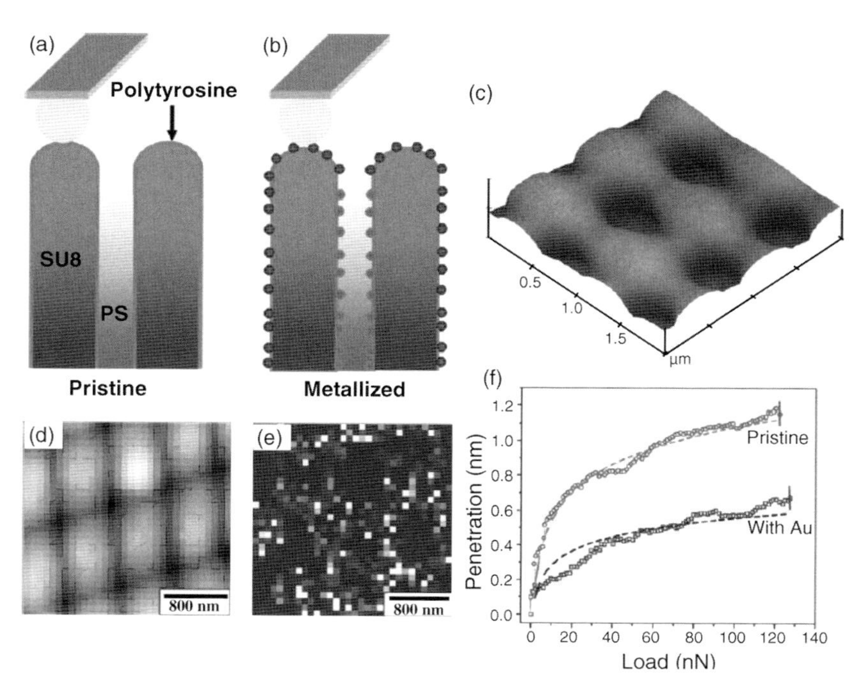

Figure 12.14 Schematic image showing the colloidal probe force spectroscopy performed on microframes: original (a) and mineralized (b). (c) 3D AFM image showing the SU8 structure with infiltrated PS (z-scale: 30 nm). (d,e) Force–volume measurements: height (d) and elastic modulus micromapping of the biometallized structure (e). (f) Load versus penetration data showing the increasing surface stiffness of the surfaces exposed to Au ions (dashed lines represent the fits using Hertzian approximation). Reproduced from Ref. [58].

thick films (100 nm to 1 μm thickness) with the location of SWNTs near the free surface of the film (in contact with air) for PU-SWNT composites. In the case of PS-SWNT composites, the filler was found to segregate to the substrate interface. In addition, the authors have employed Raman spectroscopy to confirm the presence of the SWNTs in the PS matrix, which were invisible in the AFM images.

Electrical conductivity in the plane of the film was measured with the application of silver electrodes deposited through shadow mask techniques at polymer–air and polymer–substrate interfaces. The preferential segregation of the SWNT to the film/air interface in the case of PU/SWNT composite and to the film/substrate interface in the case of PS/SWNT resulted in higher lateral electrical conductivity at these interfaces compared to the other interface (film/substrate in PU and film/air in PS) in each of these films. Peculiarities of the surface electrical conductivity in the nanocomposite films have been related to the surface free energy of the components and the strength of polymer–substrate interfacial interactions, which promote a nonuniform distribution of the conductive nanotube filler across the film thickness (vertical phase separation).

Mueggenburg *et al.* have probed the mechanical properties of hybrid nanoparticle superlattices composed of close-packed gold nanoparticles (~6 nm in diameter) separated by short organic spacers (dodecanethiol) [63]. Two-dimensional arrays of close-packed nanoparticles have been stretched across micrometer-sized holes resulting in freestanding monolayer membranes that extend over hundreds of nanoparticle diameters without fracturing. Surface force spectroscopy was employed to measure the mechanical properties of these unique monolayer structures. The elastic modulus of the superlattices was found to be relatively high, nearly 4 GPa, which is quite remarkable considering the absence of chemical cross-linking or long-chain physical entanglements (absence of a polymeric component). The authors suggested that the large elastic modulus resulted from the confinement of the short organic ligand between close-packed nanostructures. The high elastic modulus of these structures is accompanied with high flexibility, enabling the freestanding structures to bend easily while draping over the edges.

Another remarkable feature of these hybrid membranes is the resilience at elevated temperatures, which was tested using a high-temperature AFM stage. The authors found that the membrane can withstand repeated indentation up to at least 50 °C in air and typically show significant damage only above 100 °C. AFM images of the membrane were acquired following a force–displacement curve at room temperature. The image showed intact membrane following the deformation at elevated temperatures. At about 100 °C, initial damage of the membrane showed up as local rips or tears at pre-existing weak spots and finally the failure of the membrane in the form of ripping was observed at nearly 140 °C.

12.4 Porous Membranes

Porous polymer structures, especially with highly defined, controlled, and organized morphologies are being intensely studied considering their applications in controlled drug release delivery systems, as active sensing layers to maximize surface area, in separation and gating of small molecules, biomolecules, scaffolds for tissue engineering, substrates for fuel cells, organized multifunctional coatings, and elements of ultrafiltration devices. The true 3D surface morphology of the porous structures, which can be continually monitored under a wide variety of environmental conditions (temperature, pH, and light) makes AFM an ideal choice to probe the fine porous structures.

Lee *et al.* have demonstrated *in situ* AFM as a valuable tool to monitor the pH-induced hysteretic gating of track-etched polycarbonate membranes modified with LbL polyelectrolyte multilayers [64]. These multilayer-modified membranes exhibited reversible gating properties as the pH of feed solution was alternated between 2.5 and 10.5. AFM images provided insight into the time-dependent swelling (and hence closing of the pores) behavior of multilayers and gating properties of the multilayer-modified membranes. The authors noted that the degree of swelling of the multilayers in the confined geometry is smaller compared to the same multilayers on planar substrates under the same conditions.

An alternative approach to achieve responsive porous structures involves the fabrication of the porous structure using a responsive polymer as opposed to the modification of a static porous structure with a responsive polymer. Minko and coworkers have extensively investigated P2VP, a weak cationic polymer owing to its excellent pH responsive properties [65, 66]. The polymer exhibits a swelling–shrinking transition at pH 3.8. The authors showed that the spin casting of quaternized P2VP results in a porous structure. Response of the porous P2VP thin films to changes in the external pH has been investigated by monitoring the swelling properties of P2VP gel structures using *in situ* AFM experiments in a liquid cell. The AFM topography images revealed a decrease in pore size upon an increase in acidity of the aqueous solution due to significant swelling of the polyelectrolyte matrix.

Figure 12.15 shows the AFM image of a porous P2VP structure at pH 5 and pH 2 [67]. The AFM images clearly reveal the changes in the closing of the pores at pH 2. The authors found that the changes in pore size due to swelling/shrinking of the gel were fully reversible. The change in pore size of these responsive porous membranes was accompanied with a change in the permeability of the membranes with the highest permeability for neutral water, and decreased two- and fivefold for aqueous solutions with pH 3 and 2, respectively. The dramatic change in the porous structures has been employed to demonstrate tunable gel materials with strong plasmonic responses that can be employed as electrochemical gates for cholesterol [68, 69].

Another important class of porous polymer structures are fabricated using multi-laser beam interference lithography [70, 71]. Interference lithography enables the fabrication of submicron-sized elements (nodes and beams) with complex 2D and 3D lattices with micron to submicron spacings [20, 72, 73]. Jang *et al.* have demonstrated that an IL template can be employed as a facile mold for fabricating three-dimensional bicontinuous PDMS elastomeric structures and demonstrated the use of such a structure as a mechanically tunable PDMS/air phononic crystal [74]. A positive

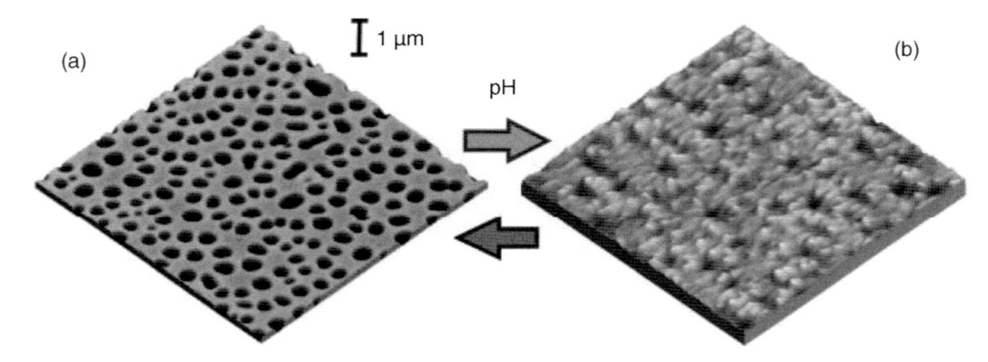

Figure 12.15 AFM topography images (7.5 μm × 7.5 μm) of P2VP gel membrane under water at (a) pH 5.5 (b) at pH 2 clearly revealing the open and closed pores at the two different pH. Reproduced from Ref. [66].

photoresist was used to make the IL templates, and after infiltration with PDMS, the resist was removed in a water-based basic solution that avoided PDMS swelling or pattern collapse occurring during the template removal process. *In situ* AFM monitoring of the PDMS material under tensile in-plane deformation was conducted by securing the PDMS sample in a microstretcher mounted on an AFM stage.

AFM images show the details of the change in the sample lattice parameter and symmetry due to the 30% unidirectional strain applied along the $[10\bar{1}0]$ direction (Figure 12.16). As is clear from this image, the unit cell size along the tensile strain direction increases by 30% accompanied by the reduction of spacing in the transverse direction, thus demonstrating affine deformation of the PDMS structure. Since the period of the structure is approximately 1 μm, the density of states of gigahertz

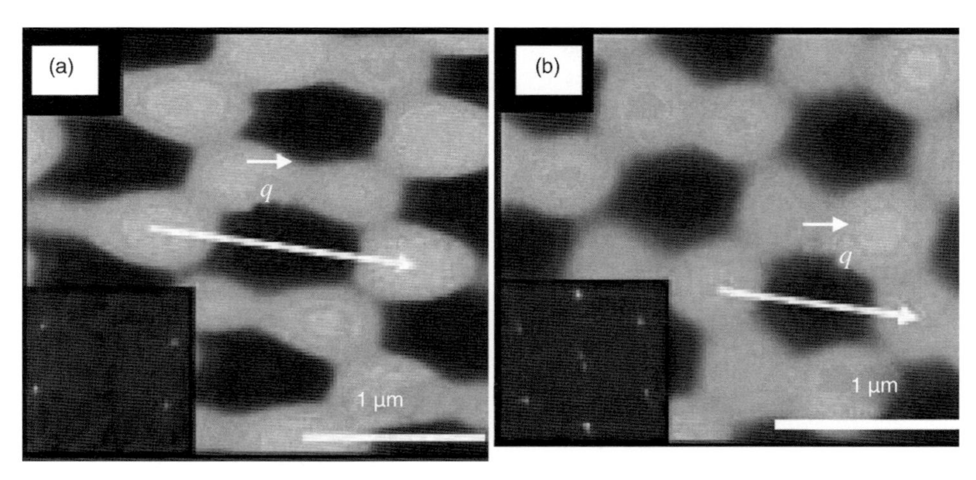

Figure 12.16 AFM images with 0% (a) and 30% tensile strain (b) along the same direction clearly show the change in lattice parameter and symmetry upon deformation. Insets are FFT of the AFM images. Reproduced from Ref. [74].

phonons is altered by the phononic porous PDMS structure. Brillouin light scattering was employed to measure phononic modes of the structure as a function of mechanical strain in conjunction with *in situ* AFM deformational studies.

In another study from the same group, Singamaneni *et al.* investigated the buckling instabilities in porous IL polymer microstructures (with square and hexagonal lattice) [75, 76]. The elastoplastic nature of the Bisphenol A Novolak epoxy (SU8) enabled the freezing of the buckling instabilities, providing a better system for understanding the details of the microstructure in transformed structures. The periodicity of the square lattice in this study was 830 nm, the diameter of the cylindrical pores was 380 nm, and the porosity was 20%. The corresponding oblique lattice fabricated in a similar manner had a periodicity of 1 μm, a radius of 200 nm, and a porosity of 35%. The thickness of the microframe structures was 3 μm, making the aspect ratio to be nearly 8, which is a high value.

In order to induce mechanical instabilities in these porous structures, photopolymerization of acrylic acid monomer was performed directly in the cylindrical pores. The slow evaporation of water from the inside of open cylindrical micropores caused the swollen PAA network grafted to the pore walls to shrink, resulting in high compressive stresses inside pores. These stresses resulted in a dramatic transformation of the periodic circular holes to alternating ellipses in the case of the square lattice and sheared ellipses in the case of the oblique lattice (see Figure 12.17 for AFM images of square lattice transformations under mechanical stresses). Transformed regions of porous morphology were extremely uniform, extending to surface areas up to few square millimeters.

The AFM image clearly reveals the structural transformation at a microscopic level (Figure 12.17b). The pattern transformation can be related to bending of the struts in

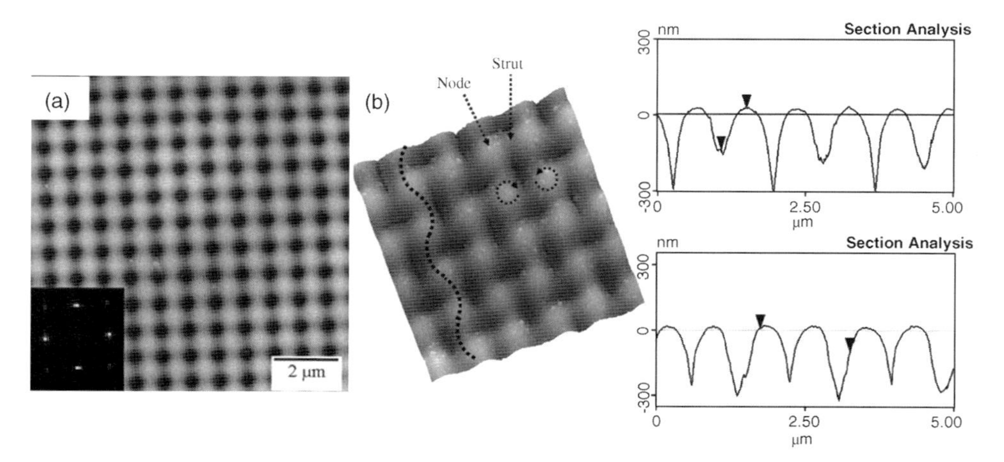

Figure 12.17 (a) AFM image of the pristine IL structure showing the square lattice of circular pores with a periodicity of 830 nm. (b) AFM topographical image (5 × 5 μm^2) of the transformed pattern showing the deformation modes of the struts (bending) and the nodes (rotation) on square lattice and the cross sections along the (01) and (10) directions showing the double bump morphology. Reproduced from Ref. [80].

alternate directions (along (10) and (01) directions) and the rotation of the nodes in clockwise and anticlockwise directions, as indicated in the AFM image. The pattern of alternating elliptical micropores is identical along (01) and (10) directions with the characteristic, double-bump shape reflecting alternating depths of AFM tip penetration along short and long axes of the collapsed pores. This is caused by its interaction with narrowing, slit-like pores (Figure 12.17, see cross sections in different directions). The elastoplastic nature of the material locks in the mechanical instabilities after the release of the external stress with internal stresses dissipated to a great extent, as was confirmed by micromapping with confocal Raman spectroscopy. This observation is in sharp contrast with the reversible instabilities in elastomeric solids, in which the transformed structures exhibit stress concentration in localized highly deformed elements [77].

In a related study, Zhang *et al.* have employed the square array of vertical posts as a master for the fabrication of a microporous elastomeric PDMS structure [78]. The structure was comprised of a square array of circular pores with a pore diameter ranging from 350 nm to 2 μm and an aspect ratio (pore depth/diameter) ranging from 2 to 20. They have triggered similar pattern transformations in these structures by swelling-mediated stresses caused by exposing the PDMS microporous structure to toluene. The instability was completely elastic in nature and structures reversed to their original state upon the evaporation of toluene. The elastic nature of the transformation and the evaporation of the toluene during high-resolution microscopy imposed severe limitations on the structural characterization. SEM images of the transformed structures were found to be partially relaxed due to the gradual evaporation of the toluene and material shrinkage.

12.5 Micro- and Nanofibers

Polymer and biological fibers (natural and manmade) are probably the most important class of polymer structures, without which it is hard to imagine the modern world. The applications of polymer fibers span across broad disciplines ranging from textiles to life sciences. In numerous studies, AFM has been extensively employed to characterize the size, true 3D morphology, sectional composition, and orientation of polymer fibers to understand their processing–structure–property relationships and resulting morphologies. In particular, in the case of nanofibers with diameters below 100 nm where conventional methods (such as SEM for imaging and tensile testing of mechanical properties) cannot be readily applied, AFM imaging and force spectroscopy measurements prove to be extremely valuable. In this section, we will discuss some of aspects of AFM research of small-diameter fibers by pointing out specific examples from recent literature.

In earlier studies, Magonov *et al.* studied structure and deformation mechanisms of gel-drawn ultrahigh molecular weight polyethylene using AFM [79]. High-resolution AFM images allowed the discrimination of different, well-defined levels of the fibrillar morphology, such as bundles of microfibrils with a diameter between 4 and

7 μm strongly depending on the elongation, nanofibrils that form the elementary fibrillar building blocks, and regular chain patterns on the molecular level that correspond to the crystalline packing of linear PE chains at the surface of the nanofibrils.

AFM images collected in this study clearly revealed the heterogeneity of the deformation process at the submicron scale. For example, after necking, some microfibrils appear to contain remnants of the original lamellae, while other microfibrils consist of nanofibrils. On the basis of the AFM images, the authors concluded that once the fibrous structure is formed, the diameters of the smallest fibrils remain constant, whereas the diameters of the microfibrils and the initially visible bundles of microfibrils are strongly affected by the draw ratio. The deformation process is essentially based on two interdependent modes of plastic deformation: the sliding motion of microfibrils and shear deformation of microfibrils due to the sliding of the nanofibrils.

In another study devoted to nanofibers, Casper *et al.* employed AFM to probe the effect of humidity on electrospun PS nanofibers [80]. Porous polymer fibers were formed when electrospinning was performed in an atmosphere with more than 30% relative humidity. The authors noted that the increase in the ambient humidity causes an increase in the number, diameter, shape, and distribution of the pores. For example, AFM images revealed that the pores formed in the PS fibers (with a molecular weight of 170 000 g/mol) electrospun at relative humidity of 60–72% had an average depth of about 50 nm, while those electrospun at 50–59% had pores too small for the AFM tip to penetrate. The authors also observed that an increase in molecular weight of the PS results in the formation of larger pores and less uniform distribution of the material.

Influence of the crystalline morphology on the mechanical properties of electrospun polymeric nanofibers was investigated by Lim *et al.* [81]. The authors found that PCL nanofibers spun from dilute polymer solutions exhibited higher degrees of molecular orientation, material crystallinity, stiffness, and strength, but lower ductility. More importantly, using high-resolution AFM the authors revealed that crystalline morphology is influenced by whether complete crystallization of polymer chains takes place before or after the electrospinning jet has reached the collector.

Figure 12.18 shows the highly textured surface morphology of PCL nanofibers investigated in this study. AFM images are accompanied by the corresponding schematic diagrams of the suggested crystalline morphology. The fibers with smaller diameters exhibited both densely packed aligned lamellae and fibrillar structures with amorphous regions most likely consisting of extended tie molecules. The fibers with larger diameters were found, with high-resolution AFM images, to host misaligned lamellae and an absence of fibrillar structures (Figure 12.18). The authors suggested that the amorphous regions most likely consist of relaxed tie molecules in this case.

In another study, Ge *et al.* demonstrated composite nanofiber sheets comprised of polyacrylonitrile (PAN) and highly oriented surface-oxidized multiwalled carbon nanotubes (MWNTs) using electrospinning [82]. AFM was employed to reveal the surface morphology and height profile of the composite PAN-MWNT nanofibers.

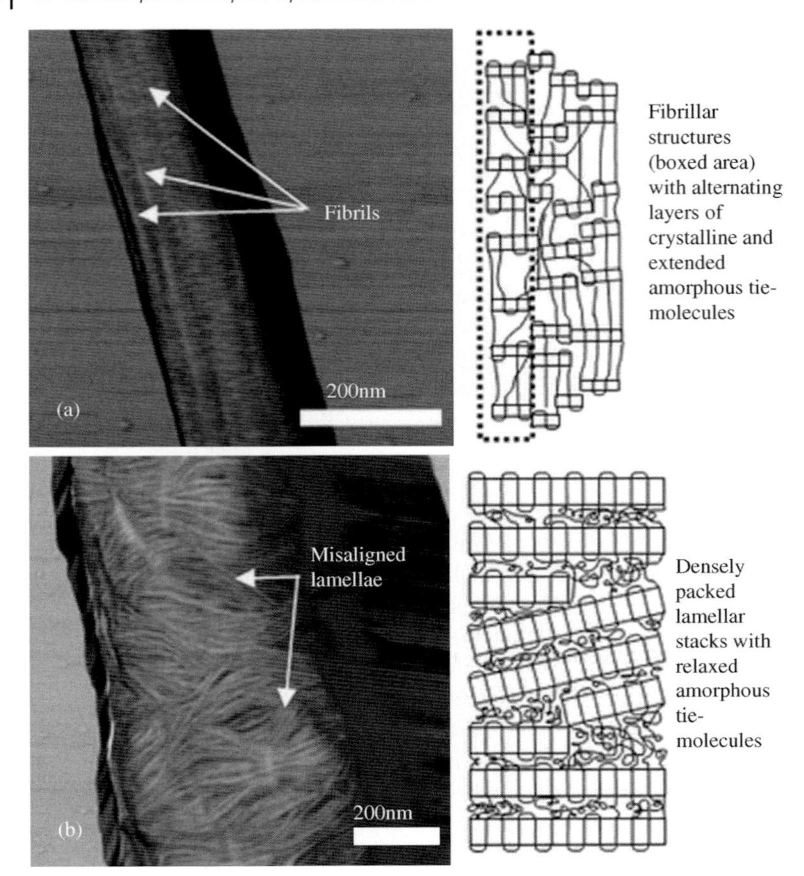

Figure 12.18 AFM phase images of PCL nanofibers with diameter of (a) 150 and (b) 450 nm and the corresponding schematic diagrams of the suggested nanostructure of the crystallites in the polymer fibers. The nanofibers were fabricated from 10 wt% PCL solution (scale bars represent 200 nm). Reproduced from Ref. [81].

The smooth surface morphology of pure PAN nanofibers was also observed in both the AFM topography and the phase images using tapping mode. PAN/MWNT (90/10) composite fiber (110 nm diameter) exhibited nonuniform, streaked surface irregularities along the fiber axis. The authors noted that the irregularities observed in the composite fibers with AFM are associated with the arrangement of the nanotube ends and with changes in the diameter of the nanofibers. Furthermore, in some places, the nanotubes were found to "poke through" the polymer fiber surface.

Mechanical properties of very stiff microscopic carbon fibers embedded in an epoxy matrix followed by transverse sectioning have been determined using surface force spectroscopy by Tsukruk and coworkers. Figure 12.19 shows the topography and friction force images depicting the axial and transverse-sectioned carbon fibers. The friction signal was much lower on stiff and nonpolar sections of carbon fibers as

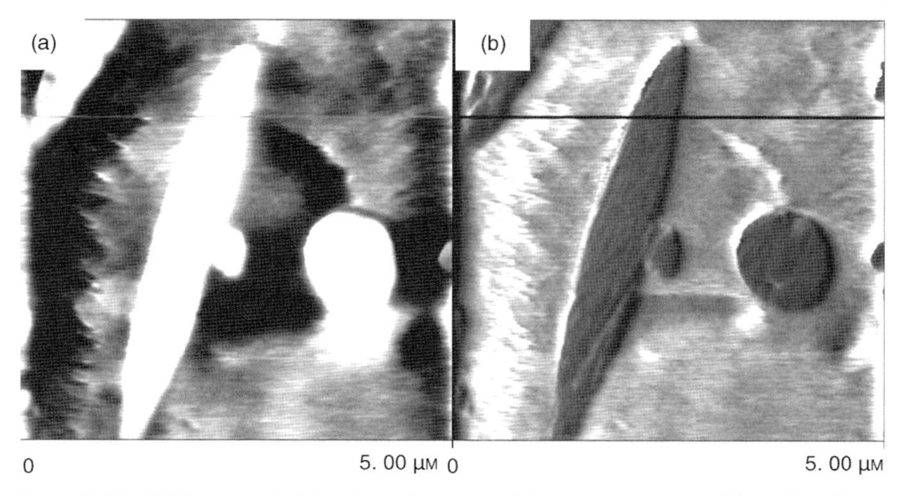

Figure 12.19 AFM topography (a) and FFM (b) images of the transverse section of the carbon fiber embedded in epoxy matrix, $5 \times 5\,\mu m^2$. Tsukruk and Dzenis, unpublished.

expected in FFM imaging. The unidirectional striations observed in the carbon fibers (clearly visible in the FFM image) are due to the shear forces during the microtoming and micropolishing processes and are common imaging artifacts.

Figure 12.20 shows the simultaneously obtained maps of topography, elastic modulus, and adhesion using surface force spectroscopic measurements on the same carbon fibers embedded in an epoxy matrix. The high contrast of elastic modulus and adhesion maps clearly reveals the much higher stiffness and much lower adhesion of the carbon fiber surfaces compared to the polar epoxy polymer matrix, thus confirming conclusions based upon contact mode AFM and FFM discussed above. The sharp interface between these two regions underscores the high-resolution nanomechanical probing that can be achieved using surface force spectroscopic measurements for these composite fiber materials.

In another study of polymer fibers, Wang and Barber employed AFM (topography and nanomechanical measurements) to directly probe the melting temperature of individual electrospun PEO nanofibers [83]. Figure 12.21 shows the AFM images of the PEO fibers imaged under 60 and at 63 °C showing the dramatic changes in fiber morphology of the thin fiber (II) at 63 °C, while the fiber with the larger diameter remains intact. An important conclusion derived from these temperature AFM observations was that the melting temperature of the PEO nanofibers decreases with decreasing diameter.

In this study, the onset of local melting was marked by both geometric changes (changes in the diameter) of the nanofibers, as obtained from AFM image cross sections, and a decrease in the elastic modulus, as probed by SFS. The plot in Figure 12.21 shows the height of the polymer fibers (equivalent to different diameters) with temperature, showing that the diameters remain unchanged at temperatures close to 60 °C. The diameter of finer nanofibers measured from

Figure 12.20 Surface force spectroscopy of the carbon fibers embedded in epoxy matrix showing the map of the (a) topography (b) adhesion (c) elastic modulus, $2 \times 2\,\mu m^2$. Tsukruk and Dzenis, unpublished.

topographical cross sections at different temperatures showed a rapid drop that corresponds to their melting, followed by the larger fiber diameters, as the temperature increases.

Various AFM nanomechanical measurement approaches have been extensively applied to probe the mechanical properties of individual polymer nanofibers as reported in numerous publications [84–87]. There are two important techniques that are commonly employed for this purpose. In the first method, conventional static SFS measurements (routinely employed for flat surfaces, as discussed in Chapter 4) with additional corrections to take the curvature of the fiber surface into account are employed [88].

The second method involves freely suspending the polymer fibers across a trench followed by bending the fiber with an AFM tip along the trench and obtaining force–displacement curves [84, 87]. The obtained force–displacement data is then modeled using classical beam bending theory to obtain the mechanical properties of the polymer fibers. However, this method has resulted in some controversy as to

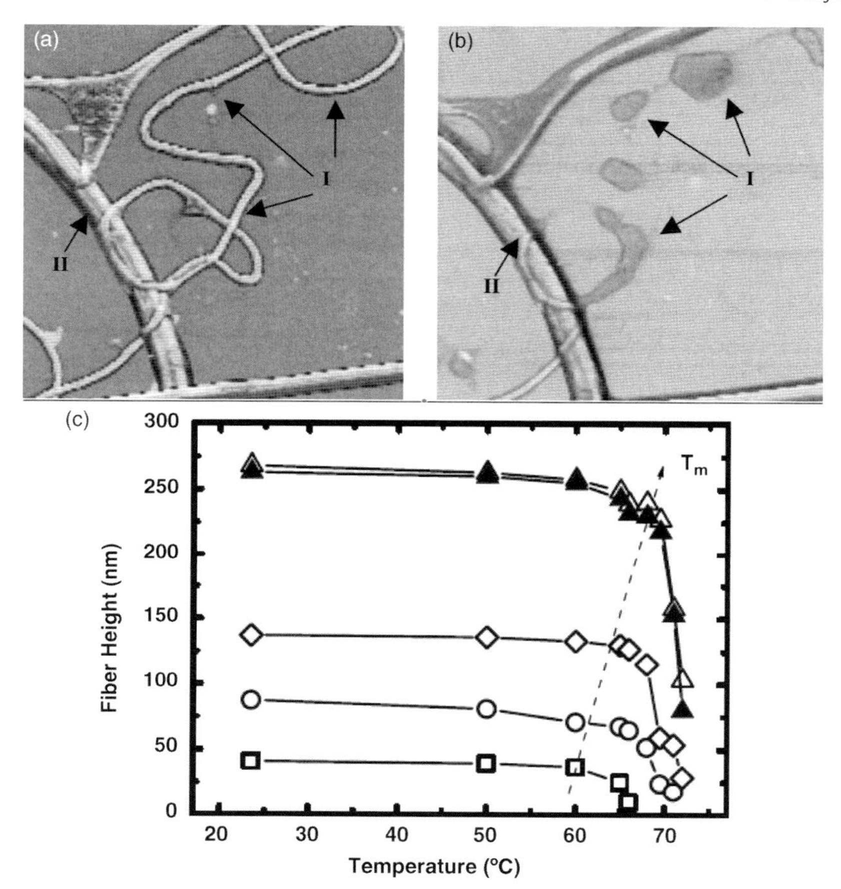

Figure 12.21 AFM topographic images for several individual electrospun PEO fibres with different diameters. The arrow indicates the change in geometry from a fibrous cylinder to droplets for a small fiber (I) as the temperature increases from below 60 °C (a) to 63 °C (b), whereas the larger fiber diameter (II) retains its geometry. (c) Plot of the measured height of electrospun PEO fibres of various diameters, taken from AFM images obtained at various temperatures. Reproduced from Ref. [83].

whether the measurements comply with all the presumptions of the beam bending theory related to bending mechanisms and boundary conditions.

For example, the theory can be applied to practical cases only when the deformation of the polymer fiber by the AFM tip is purely bending with negligible or no shear deformation and local indentation. Failure of this condition in few cases led to questioning the validity of the mechanical properties derived using this method [89]. Apart from the above-mentioned two commonly employed methods, shear force modulation mode and spring-on-spring approaches have also been demonstrated to probe the mechanical properties of the nanofibers, which will be discussed below.

Within this approach, AFM micromechanical bending of the freely suspended (across microchannels in a glass substrate) electrospun collagen nanofibers with diameters ranging from 100 to 600 nm was employed to probe the mechanical properties by Yang *et al.* [84]. Mechanical measurements were performed on pristine and glutaraldehyde cross-linked nanofibers. The bending moduli of the nanofibers were derived from the slopes of the force–displacement curves obtained along the suspended nanofiber. The bending modulus, E_{bending}, is calculated by fitting the experimentally derived slope of the force–displacement curves obtained at various distances from the edge (x) of suspended fiber according to [84]

$$\frac{dF}{dZ} = \frac{3L^3 E_{\text{bending}} I}{(L-x)^3 x^3} \tag{12.1}$$

where dF/dZ is the slope of the force displacement curve, L is the width of the channel, and I is the moment of inertia.

Figure 12.22 shows three representative force–displacement curves obtained at the edge, between the edge and the middle, and the middle of the suspended electrospun nanofiber. The corresponding dF/dZ values at various distances from the edge of the channel can be fit accurately according to Eq. (12.1) (Figure 12.22). The bending moduli of the electrospun fibers were found to be in the range of 1.3 to 7.8 GPa under ambient conditions and reduced dramatically to 0.07 and 0.26 MPa after placement in a buffer solution. Furthermore, the authors noted that an increase in the diameter of the fibrils resulted in a decrease in the bending modulus, clearly indicating the mechanical anisotropy of the fiber.

In an alternative approach, Li and coworkers have employed shear force modulation AFM to probe the nanomechanical properties of electrospun PS/clay nanocomposite fibers with varying diameters (4 μm to 150 nm) and at different temperatures [85]. As described in detail in Section 4.3, this method involves engaging the AFM tip on the surface with a constant normal load (25 nN in this case) followed by lateral modulation (sinusoidal oscillation) with nanoscale amplitude (3 nm in this case) to ensure no slip between the tip and the sample surface. The authors developed a scaling relationship using a thin pristine PS film as a reference to relate the amplitude response of the shear modulation to the actual shear modulus of the PS/clay fibers. The authors have found that the shear modulus of the fiber decreases from 3.7 GPa to 1.3 GPa as the fiber diameter increases from 150 nm to 4 μm. The authors reasoned that the decrease in the elastic modulus is caused by higher polymer chain alignment in the case of fibers with smaller diameters. Furthermore, the authors have found that the shear modulus values restored to the bulk values as the polymer fibers were treated with temperatures above T_g due to the relaxation of the aligned polymer chains in the fibers.

In another study, Gu *et al.* employed a spring-on-spring approach for determining the mechanical properties of the polymer fibers [86]. The Young's modulus of a single-electrospun PAN fiber was measured by the bending of a single fiber attached to an AFM cantilever. The spring-on-spring method is similar to the tip-on-tip approach, which was described in Section 2.5 for determining the unknown spring constant of a cantilever using a cantilever with a known spring constant. A single

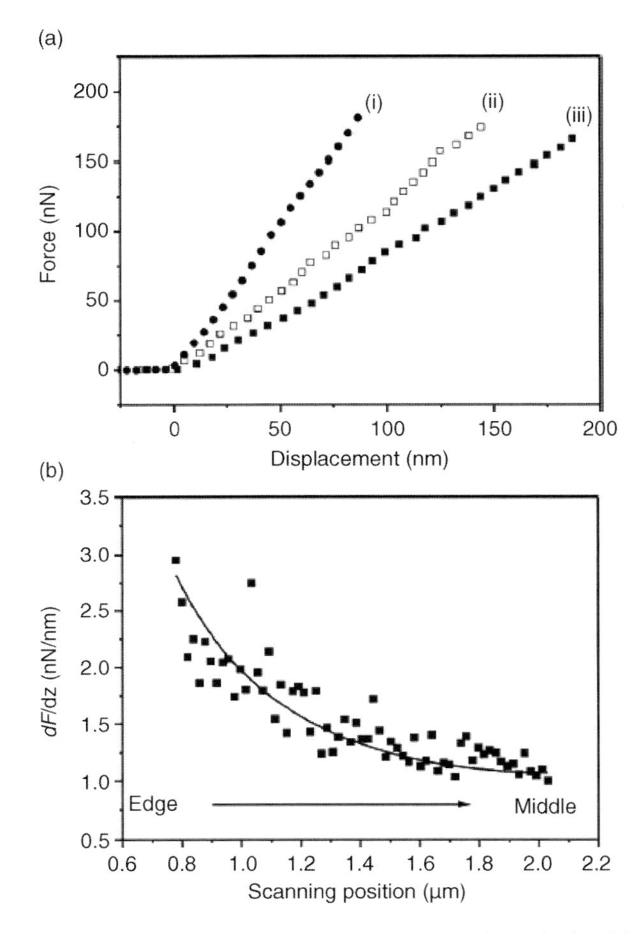

Figure 12.22 (a) Three representative force–displacement curves obtained from bending the electrospun collagen fiber at (i) the edge (ii) between the edge and the middle, and (iii) middle of the channel. (b) Plot showing the slope (dF/dz) of the force–displacement curves as a function of the scanning position. A curve fit using Eq. (12.1) is shown as a solid line. Reproduced from Ref. [84].

electrospun fiber was attached to the end of a cantilever with a small amount of epoxy glue, which was then used in AFM bending experiments. The force needed to bend the fiber attached to a cantilever is monitored by force displacement curves.

For analysis of the elastic modulus, the fiber was considered to be an ideal spring with an equal spring constant over its full length, which implicates a constant Young's modulus over the whole length of the fiber. Furthermore, the Young's modulus of the fiber is considered the same for both compressive and tensile deformations, and the fiber cross section is assumed to be circular with a constant fiber diameter over the whole length. Although the elastic modulus values showed a broad distribution, the fibers obtained by electrospinning at 18 kV have an average modulus of 5.7 GPa, while the average modulus of fibers obtained at 22 kV is 26.6 GPa. The authors concluded that the Young's moduli of most electrospun nanofibers obtained at 22 kV have higher values than those obtained at 18 kV possibly because of higher polymer chain orientation in the fibers electrospun at high voltages.

In conclusion, it is worth noting that, as demonstrated by numerous examples in this chapter, the field of multicomponent systems witnessed tremendous leaps especially in understanding structure–property relationships, with AFM playing an important role in these developments over the past decade. Blends of conducting and semiconducting polymers are being extensively investigated owing to their potential application in electrooptical devices such as solar cells and light emitting devices. The discussion of these systems is intentionally omitted in this chapter and the readers are referred to Chapter 17 for this discussion. In some cases of multicomponent systems (polymer blends, block copolymers, and porous membranes), interpretation of even tapping mode images becomes nontrivial and is severely affected by the tapping conditions, as discussed in Section 12.1. Advanced modes such as nanomechanical measurements require additional care (in both acquisition and interpretation of the data) when applied to multicomponent systems and surfaces with complex topographic features (porous structures and fibers) compared to smooth uniform surfaces.

References

1 Sperling, L.H. (1997) *Polymeric Multicomponent Materials: An Introduction*, John Wiley & Sons, Inc., New York.

2 Bar, G., Thomann, Y., Brandsch, R., and Cantow, H.J. (1997) Factors affecting the height and phase images in tapping mode atomic force microscopy. Study of phase-separated polymer blends of poly(ethene-*co*-styrene) and poly(2,6-dimethyl-1,4-phenylene oxide). *Langmuir*, **13** (14), 3807–3812.

3 Raghavan, D., Gu, X., Nguyen, T., VanLandingham, M., and Karim, A. (2000) Mapping polymer heterogeneity using atomic force microscopy phase imaging and nanoscale indentation. *Macromolecules*, **33** (7), 2573–2583.

4 Tsukruk, V.V. and Huang, Z. (2000) Micro-thermomechanical properties of heterogeneous polymer films. *Polymer*, **41** (14), 5541–5545.

5 Wanga, H. and Composto, R.J. (2000) Thin film polymer blends undergoing phase separation and wetting: identification of early, intermediate, and late stages. *J. Phys. Chem.*, **113** (228), 10386–10397.

6 You, J., Shi, T., Liao, Y., Li, X., Su, Z., and An, L. (2008) Temperature dependence of surface composition and morphology in polymer blend film. *Polymer*, **49** (20), 4456–4461.

7 Virgilio, N., Desjardins, P., L'Esperance, G., and Favis, B.D. (2009) *In situ* measure of interfacial tensions in ternary and quaternary immiscible polymer blends demonstrating partial wetting. *Macromolecules*, **42** (19), 7518–7529.

8 Virgilio, N., Desjardins, P., L'Espérance, G., and Favis, B.D. (2010) Modified interfacial tensions measured *in situ* in ternary polymer blends demonstrating partial wetting. *Polymer*, **51** (6), 1472–1484.

9 Galuska, A.A., Poulter, R.R., and McElrath, K.O. (1997) Force modulation AFM of elastomer blends: morphology, fillers and cross-linking. *Surf. Inter. Anal.*, **25** (6), 418–429.

10 Lipatov, Y.S., Bliznyuk, V.N., Todosiychuk, T.T., Chornaya, V.N., Katumenu, R.K., and Konovalyuk, V.D. (2006) On the structure

of polymer layers formed by adsorption from binary and ternary solutions on a solid SiO_2: a combined adsorption and atomic force microscopy (AFM) study. *Colloid Poly. Sci.*, **284** (8), 893–899.

11 Bliznyuk, V.N., Lipatov, Y.S., Ozdemir, N., Todosijchuk, T.T., Chornaya, V.N., and Singamaneni, S. (2007) Atomic force and ultrasonic force microscopy investigation of adsorbed layers formed by two incompatible polymers polystyrene and poly(butyl methacrylate). *Langmuir,* **23** (26), 12973–12983.

12 Cui, L. and Han, Y. (2005) Honeycomb pattern formation via polystyrene/poly(2-vinylpyridine) phase separation. *Langmuir,* **21** (24), 11085–11091.

13 Cui, L., Li, B., and Han, Y. (2007) Transformation from ordered islands to holes in phase-separating P2VP/PS blend films by adding triton X-100. *Langmuir,* **23** (6), 3349–3354.

14 Mitov, Z. and Kumacheva, E. (1998) Convection-induced patterns in phase-separating polymeric fluids. *Phys. Rev. Lett.*, **81** (27), 3427–3430.

15 Singamaneni, S., Chang, S., Jang, J.-H., Davis, W., Thomas, E.L., and Tsukruk, V.V. (2008) Mechanical properties of composite polymer microstructures fabricated by interference lithography. *Phys. Chem. Chem. Phys.*, **10** (28), 4093–4105.

16 Berger, V., Gauthier-Lafaye, O., and Costard, E. (1997) Photonic band gaps and holography. *J. Appl. Phys.*, **82** (1), 60–64.

17 Moon, J.H. and Yang, S. (2010) Chemical aspects of three-dimensional photonic crystals. *Chem. Rev.*, **110** (1), 547–574.

18 Maldovan, M. and Thomas, E.L. (2009) *Periodic Materials and Interference Lithography: For Photonics, Phononics, and Mechanics*, Wiley-VCH Verlag GmbH, Weinheim, Germany.

19 Jang, J.-H., Ullal, C.K., Maldovan, M., Gorishnyy, T., Kooi, S., Koh, C.Y., and Thomas, E.L. (2007) 3D micro- and nanostructures via interference lithography. *Adv. Funct. Mater.*, **17** (16), 3027–3041.

20 Meisel, D.C., Deubel, M., Hermatschweiler, M., Busch, K., Koch, W., von Freymann, G., Blanco, A., Enkrich, C., and Wegener, M. (2003) Functional Nanomaterials for Optoelectronics and other Applications. *Sol. State Phen.*, **55**, 99–100.

21 Maldovan, M. and Thomas, E.L. (2004) Structured photonic crystals. *Nat. Mater.*, **3**, 593–600.

22 Jang, J.-H., Ullal, C.K., Choi, T., Lemieux, M.C., Tsukruk, V.V., and Thomas, E.L. (2006) 3D polymer microframes that exploit length-scale-dependent mechanical behavior. *Adv. Mater.*, **18** (16), 2123–2127.

23 van Dijk, M.A. and van den Berg, R. (1995) Ordering phenomena in thin block copolymer films studied using atomic force microscopy. *Macromolecules,* **28** (20), 6773–6778.

24 Kim, H.-C. and Russell, T.P. (2001) Ordering in thin films of asymmetric diblock copolymers. *J. Polym. Sci. Polym. Phys.*, **39** (6), 663–668.

25 Knoll, A., Lyakhova, K.S., Krausch, G., Sevink, G.J.A., Zvelindovsky, A.V., and Magerle, R. (2002) Phase behavior in thin films of cylinder-forming block copolymers. *Phys. Rev. Lett.*, **89** (3), 035501–035501/4.

26 Tsarkova, L., Knoll, A., Krausch, G., and Magerle, R. (2006) Substrate-induced phase transitions in thin films of cylinder-forming diblock copolymer melts. *Macromolecules,* **39** (10), 3608–3615.

27 van Zoelen, W., Polushkin, E., and ten Brinke, G. (2008) Hierarchical terrace formation in PS-*b*-P4VP(PDP) supramolecular thin films. *Macromolecules,* **41** (22), 8807–8814.

28 Sohn, K.E., Kojio, K., Berry, B.C., Karim, A., Coffin, R.C., Bazan, G.C., Kramer, E.J., Sprung, M., and Wang, J. (2010) Surface effects on the thin film morphology of block copolymers with bulk order–order transitions. *Macromolecules.* doi: 10.1021/ma1001194.

29 Knoll, A., Magerle, R., and Krausch, G. (2004) Phase behavior in thin films of cylinder-forming ABA block copolymers: experiments. *J. Chem. Phys.*, **120** (2), 1105–1116.

30 Ludwigs, S., Schmidt, K., Stafford, C.M., Amis, E.J., Fasolka, M.J., Karim, A., Magerie, R., and Krausch, G. (2005)

Combinatorial mapping of the phase behavior of ABC triblock terpolymers in thin films: experiments. *Macromolecules*, **38** (5), 1850–1858.
31 Tsarkova, L., Sevink, G., and Krausch, G. (2010) Nanopattern evolution in block copolymer films: experiment, simulations and challenges. *Adv. Poly. Sci.*, **227**, 1–41.
32 Wang, Y., Song, R., Li, Y., and Shen, J. (2003) Understanding tapping-mode atomic force microscopy data on the surface of soft block copolymers. *Surf. Sci.*, **530** (3), 136–148.
33 Luzinov, I., Julthongpiput, D., and Tsukruk, V.V. (2001) Stability of microdomain morphology in tethered block copolymer monolayers. *Polymer*, **42** (5), 2267–2273.
34 Luzinov, I. and Tsukruk, V.V. (2002) Ultrathin triblock copolymer films on tailored polymer brushes. *Macromolecules*, **35** (15), 5963–5973.
35 Hahm, J., Lopes, W.A., Jaeger, H.M., and Sibener, S.J. (1998) Defect evolution in ultrathin films of polystyrene-block–poly (methyl methacrylate) diblock copolymers observed by atomic force microscopy. *J. Chem. Phys.*, **109** (23), 10111–10114.
36 Hammond, M.R., Cochran, E., Fredrickson, G.H., and Kramer, E.J. (2005) Temperature dependence of order, disorder, and defects in laterally confined diblock copolymer cylinder monolayers. *Macromolecules*, **38** (15), 6575–6585.
37 Darling, S.B. (2007) Directing the self-assembly of block copolymers. *Prog. Polym. Sci.* doi: 10.1016/j.progpolymsci.2007.05.004.
38 Yufa, N.A., Li, J., and Sibener, S.J. (2009) *In-situ* high-temperature studies of diblock copolymer structural evolution. *Macromolecules*, **42** (7), 2667–2671.
39 Tsarkova, L., Horvat, A., Krausch, G., Zvelindovsky, A.V., Agur Sevink, G.J., and Magerie, R. (2006) Defect evolution in block copolymer thin films via temporal phase transitions. *Langmuir*, **22** (19), 8089–8095.
40 Horvat, A., Agur Sevink, G.J., Zvelindovsky, A.V., Krekhov, A., and Tsarkova, L. (2008) Specific features of defect structure and dynamics in the cylinder phase of block copolymers. *ACS Nano*, **2** (6), 1143–1152.
41 Hahm, J. and Sibener, S.J. (2001) Time-resolved atomic force microscopy imaging studies of asymmetric PS-*b*-PMMA ultrathin films: dislocation and disclination transformations, defect mobility, and evolution of nanoscale morphology. *J. Chem. Phys.*, **114** (10), 4730–4740.
42 Harrison, C., Adamson, D.H., Cheng, Z., Sebastian, J.M., Sethuraman, S., Huse, D.A., and Register, P.M. (2000) Mechanism of ordering in striped patterns of diblock copolymers. *Science*, **290** (5496), 1558–1560.
43 Segalman, R.A., Hexemer, A., Hayward, R.C., and Kramer, E.J. (2003) Ordering and melting of block copolymer spherical domains in 2 and 3 dimensions. *Macromolecules*, **36** (9), 3272–3288.
44 Ivanov, D.A. and Magonov, S.N. (2003) Atomic force microscopy studies of semicrystalline polymers at variable temperature. *Lecture Notes in Physics*, **606** (Polymer Crystallization), 98–130.
45 Tsarkova, L., Knoll, A., and Magerle, R. (2006) Rapid transitions between defect configurations in a block copolymer melt. *Nano Lett.*, **6** (7), 1574–1577.
46 Knoll, A., Lyakhova, K.S., Horvat, A., Krausch, G., Sevink, J.A., Zvelindovsky, A.V., and Magerie, R. (2004) Direct imaging and mesoscale modelling of phase transitions in a nanostructured fluid. *Nat. Mater.*, **3**, 886–891.
47 Wang, Y., Hong, X., Liu, B., Ma, C., and Zhang, C. (2008) Two-dimensional ordering in block copolymer monolayer thin films upon selective solvent annealing. *Macromolecules*, **41** (15), 5799–5808.
48 Yufa, N.A., Li, J., and Sibener, S.J. (2009) Diblock copolymer healing. *Polymer*, **50** (12), 2630–2634.
49 Ramesh, G.V., Sreedhar, B., and Radhakrishnan, T.P. (2009) Real time monitoring of the *in situ* growth of silver nanoparticles in a polymer film under ambient conditions. *Phys. Chem. Chem. Phys.*, **11**, 10059–10063.

50 Tang, Z., Kotov, N.A., Magonov, S., and Ozturk, B. (2003) Nanostructured artificial nacre. *Nat. Mater.*, **2**, 413–418.

51 Kulkarni, D., Choi, I., Singamaneni, S., and Tsukruk, V.V. (2010) Graphene oxide-polyelectrolyte nanomembranes. *ACS Nano*. doi: 10.1021/nn101204d.

52 Jiang, C. and Tsukruk, V.V. (2006) Freestanding nanostructures via layer-by-layer assembly. *Adv. Mater.*, **18** (7), 829–840.

53 Jiang, C. and Tsukruk, V.V. (2005) Organized arrays of nanostructures in freely suspended nanomembranes. *Soft Matter*, **1**, 334–337.

54 Jiang, C., Markutsya, S., and Tsukruk, V.V. (2004) Compliant, robust, and truly nanoscale free-standing multilayer films fabricated using spin-assisted layer-by-layer assembly. *Adv. Mater.*, **16** (2), 157–161.

55 Kharlampieva, E., Tsukruk, T., Slocik, J.M., Ko, H., Poulsen, N., Naik, R.R., Kröger, N., and Tsukruk, V.V. (2008) Bio-enabled surface-mediated growth of titania nanoparticles. *Adv. Mater.*, **20** (17), 3274–3279.

56 Kharlampieva, E., Slocik, J.M., Singamaneni, S., Poulsen, N., Kröger, N., Naik, R.R., and Tsukruk., V.V. (2009) Protein-enabled synthesis of monodisperse titania nanoparticles on and within polyelectrolyte matrices. *Adv. Funct. Mater.*, **19** (14), 2303–2311.

57 Kharlampieva, E., Slocik, J.M., Tsukruk, T., Naik, R.R., and Tsukruk, V.V. (2008) Polyaminoacid-induced growth of metal nanoparticles on layer-by-layer templates. *Chem. Mater.*, **20** (18), 5822–5831.

58 Singamaneni, S., Kharlampieva, E., McConney, M.E., Jiang, H., Jang, J.-H., Thomas, E.L., Bunning, T.J., and Tsukruk, V.V. (2010) Metallized porous microframes via biofunctionalization. *Adv. Mater.*, **22** (12), 1369–1373.

59 Shulha, H., Kovalev, A., Myshkin, N., and Tsukruk, V.V. (2004) Some aspects of AFM nanomechanical probing of surface polymer films. *Eur. Polymer. J.*, **40** (5), 949–956.

60 Kovalev, A., Shulha, H., Lemieux, M., Myshkin, N., and Tsukruk, V.V. (2004) Nanomechanical probing of layered nanoscale polymer films with atomic force microscopy. *J. Mater. Res.*, **19**, 716–728.

61 Foster, J., Singamaneni, S., Kattumenu, R., and Bliznyuk, V. (2005) Dispersion and phase separation of carbon nanotubes in ultra thin polymer films. *J. Coll. Interface Sci.*, **287** (1), 167–172.

62 Bliznyuk, V.N., Singamaneni, S., Kattumenu, R., and Atashbar, M.Z. (2006) Surface electrical conductivity in ultrathin single-wall carbon nanotube/polymer nanocomposite films. *Appl. Phys. Lett.*, **88**, 164101.

63 Mueggenburg, K.E., Lin, X., Goldsmith, R.H., and Jaeger, H.M. (2007) Elastic membranes of close-packed nanoparticle arrays. *Nat. Mater.*, **6**, 656–660.

64 Lee, D., Nolte, A.J., Kunz, A.L., Rubner, M.F., and Cohen, R.E. (2006) pH-induced hysteretic gating of track-etched polycarbonate membranes: swelling/deswelling behavior of polyelectrolyte multilayers in confined geometry. *J. Am. Chem. Soc.*, **128** (26), 8521–8529.

65 Tokarev, I., Orlov, M., and Minko, S. (2006) Responsive polyelectrolyte gel membranes. *Adv. Mater.*, **18** (18), 2458–2460.

66 Tokarev, I. and Minko, S. (2010) Stimuli-responsive porous hydrogels at interfaces for molecular filtration, separation, controlled release, and gating in capsules and membranes. *Adv. Mater.* doi: 10.1002/adma.201000165.

67 Okarev, I. and Minko, S. (2009) Multiresponsive, hierarchically structured membranes: new, challenging biomimetic materials for biosensors, controlled release, biochemical gates, and nanoreactors. *Adv. Mater.*, **21** (2), 241–247.

68 Tokarev, I., Tokareva, I., and Minko, S. (2008) Gold-nanoparticle-enhanced plasmonic effects in a responsive polymer gel. *Adv. Mater.*, **20** (14), 2730–2734.

69 Tokarev, I., Tokareva, I., Gopishetty, V., Katz, E., and Minko, S. (2010) Specific biochemical-to-optical signal transduction by responsive thin hydrogel films loaded with noble metal nanoparticles. *Adv. Mater.*, **22** (12), 1412–1416.

70 Berger, V., Gauthier-Lafaye, O., and Costard, E. (1997) Photonic band gaps and holography. *J. Appl. Phys.*, **82**, 60.

71 Moon, J.H. and Yang, S. (2010) Chemical aspects of three-dimensional photonic crystals. *Chem. Rev.*, **110** (1), 547–574.

72 Maldovan, M. and Thomas, E.L. (2004) Diamond-structured photonic crystals. *Nat. Mater.*, **3**, 593–600.

73 Jang, J.-H., Ullal, C.K., Choi, T., Lemieux, M.C., Tsukruk, V.V., and Thomas, E.L. (2006) 3D polymer microframes that exploit length-scale-dependent mechanical behavior. *Adv. Mater.*, **18** (16), 2123–2127.

74 Jang, J.-H., Ullal, C.K., Gorishnyy, T., Tsukruk, V.V., and Thomas, E.L. (2006) Mechanically tunable three-dimensional elastomeric network/air structures via interference lithography. *Nano Lett.*, **6** (4), 740–743.

75 Singamaneni, S., Bertoldi, K., Chang, S., Jang, J.-H., Thomas, E.L., Boyce, M., and Tsukruk, V.V. (2009) Instabilities and pattern transformation in periodic, porous elastoplastic solid coatings. *ACS Appl. Mater. Interfaces*, **1** (1), 42–47.

76 Singamaneni, S., Bertoldi, K., Chang, S., Jang, J.-H., Young, S., Thomas, E.L., Boyce, M., and Tsukruk, V.V. (2009) Bifurcated mechanical behavior of deformed periodic porous solids. *Adv. Funct. Mater.*, **19** (9), 1426–1436.

77 Mullin, T., Deschanel, S., Bertoldi, K., and Boyce, M.C. (2007) Pattern transformation triggered by deformation. *Phys. Rev. Lett.*, **99**, 084301.

78 Zhang, Y., Matsumoto, E.A., Peter, A., Lin, P., Kamien, R.D., and Yang, S. (2008) One-step nanoscale assembly of complex structures via harnessing of an elastic instability. *Nano Lett.*, **8** (4), 1192–1196.

79 Magonov, S.N., Sheiko, S.S., Deblieck, R.A.C., and Möller, M. (1993) AFM of gel-drawn ultrahigh molecular weight polethylene. *Macromolecules*, **26** (6), 1380–1386.

80 Casper, C.L., Stephens, J.S., Tassi, N.G., Chase, D.B., and Rabolt, J.F. (2004) Controlling surface morphology of electrospun polystyrene fibers: effect of humidity and molecular weight in the electrospinning process. *Macromolecules*, **37** (2), 573–578.

81 Lim, C.T., Tan, E.P.S., and Ng, S.Y. (2008) Effects of crystalline morphology on the tensile properties of electrospun polymer nanofibers. *Appl. Phys. Lett.*, **92**, 141908.

82 Ge, J.J., Hou, H., Li, Q., Graham, M.J., Greiner, A., Reneker, D.H., Harris, F.W., and Cheng, S.Z.D. (2004) Assembly of well-aligned multiwalled carbon nanotubes in confined polyacrylonitrile environments: electrospun composite nanofiber sheets. *J. Am. Chem. Soc.*, **126** (48), 15754–15761.

83 Wang, W. and Barber, A.H. (2010) Diameter-dependent melting behaviour in electrospun polymer fibres. *Nanotechnology*, **21** (22), 225701.

84 Yang, L., Fitié, C.F.C., van der Werf, K.O., Bennink, M.L., Dijkstra, P.J., and Feijen, J. (2008) Mechanical properties of single electrospun collagen type I fibers. *Biomaterials*, **29** (8), 955–962.

85 Yuan, J., Li, B., Ge, S., Sokolov, J.C., and Rafailovich, M.H. (2006) Structure and nanomechanical characterization of electrospun PS/clay nanocomposite fibers. *Langmuir*, **22** (3), 1321–1328.

86 Gu, S., Wu, Q., Ren, J., and Vancso, G.J. (2005) Mechanical properties of a single electrospun fiber and its structures. *Macromol. Rapid Commun.*, **26** (9), 716–720.

87 Almecija, D., Blond, D., Sader, J.E., Coleman, J.N., and Boland, J.J. (2009) Mechanical properties of individual electrospun polymer-nanotube composite nanofibers. *Carbon*, **47** (9), 2253–2258.

88 Tan, E.P.S. and Lim, C.T. (2005) Nanoindentation study of nanofibers. *Appl. Phys. Lett.*, **87**, 123106.

89 Chaudhry, H. and Findley, T. (2008) Comment on "Physical properties of a single polymeric nanofiber." *Appl. Phys. Lett.*, **93**, 166101.

13
Engineered Surface and Interfacial Materials

A variety of modern applications, such as those relevant to microelectromechanical systems (MEMS) (micromotors, actuators, valves), require robust ultrathin coatings with finely controllable surface properties and topography. Such surface modification can be achieved by tethering various organic and polymer molecules to a substrate with the resulting surface properties to be controlled by composition, morphology, and chemical functionalities and thereby masking its original properties. The most popular choices for surface modification to date include self-assembled monolayers (SAMs), adsorbed layers, Langmuir–Blodgett monolayers, brush layers, and layer-by-layer (LbL) assemblies, all of which will be considered in this and the following chapters. AFM studies of these surface films with nanoscale thickness, grainy nanoscale morphology, dense molecular packing, and nanometer-scale microroughness are critical for understanding their behavior and properties.

These interfacial nanomaterials also might be capable of responding to very subtle changes in the surrounding environment such as pH, surface pressure, temperature, light, and solvent quality – all of which can be directly studied with AFM imaging under proper external conditions [1–10]. The macroscopic responses are caused by the molecular reorganization of the internal or surface nanostructure of the ultrathin layers, both of which can be readily detected with AFM. This reorganization is responsible for controlling and altering the physical properties important in applications of colloid stabilization, drug delivery and biomimetic materials, chemical gates, protein adsorption and tuning nanotribological properties for tailored surfaces having self-cleaning, self-healing, and self-refreshing abilities, and reversible superhydrophobic behavior [11–22].

13.1
Surface Brush Layers

Polymer brush layers composed of flexible polymer chains tethered to a solid substrate via one terminal group are currently a subject of intensive theoretical and experimental investigations (Figure 13.1) [23–27]. In these brush layers, physically or chemically adsorbed polymer chains are in some version of a

Scanning Probe Microscopy of Soft Matter: Fundamentals and Practices, First Edition.
Vladimir V. Tsukruk and Srikanth Singamaneni.

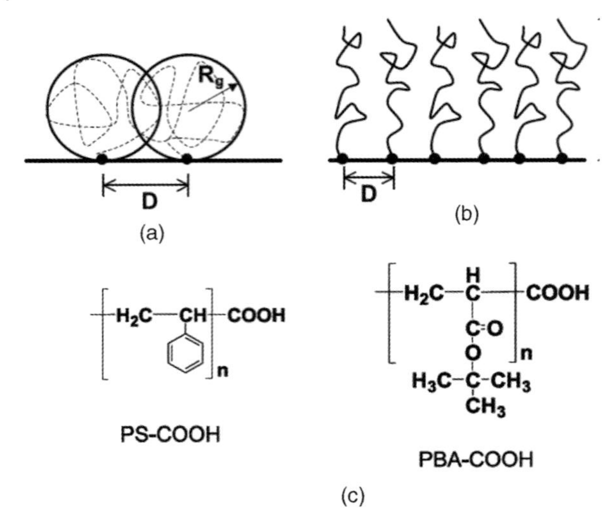

Figure 13.1 Neighboring grafted chains in random coil (a) and (b) brush conformations. (c) Chemical formulas of the polymers studied: PS–COOH and PBA–COOH. Reprinted from Ref. [23].

stretched conformation in contrast to the conventional coil-like conformation of physically adsorbed chains.

Numerous studies are dedicated to ultrathin polymer brush films and their synthesis and behavior [23, 28–32]. There are two conventional methods for the fabrication of polymer brush layers: the *grafting to* approach and the *grafting from* approach. The *grafting from* approach relies on growing polymer chains from precatalyzed surfaces. The *grafting to* approach involves preformed, end-functionalized polymers reacting with a suitable surface under appropriate conditions to form a tethered polymer brush [33–35]. To facilitate the strong chemical attachment, the substrate is usually modified with a reactive precursor that acts as a coupling agent [36–42].

Recently, sophisticated synthetic approaches have resulted in the preparation of surface polymer layers with unique surface morphologies and properties. As a result of the high grafting density and uniformity in composition and chain height throughout the brush, the polymer layers have been tailored to respond communally to very subtle changes in the surrounding environment such as pH, temperature, and solvent quality [1, 5, 9, 43, 44]. In each case, the polymer brush structure was responsible for the observed physical properties and was capable of tuning lubrication, friction, adhesion, and wettability of tailored surfaces [19, 21].

In the case of mixed polymer brush layers (usually with two-component or binary composition), the variety of possible surface morphologies greatly increases owing to the variable chemical composition and microphase separation scenarios. Although it is known that sufficiently random, irreversible grafting of incompatible polymers prevents lateral macrophase separation, the ripple profile describes a system in which the two components are predominantly laterally segregated, with the dimensions of the lateral structures on the order of the radius of the free coiled macromolecular chains.

In this section, we will consider several selected examples of the most popular types of polymer brush surface layers from homopolymers, block copolymers, and mixed covalently grafted polymer brush layers as studied with various SPM modes.

13.1.1
Homopolymer Brush Layers

Homopolymer brush layers are composed of a variety of one or two end-functionalized polymer chains covalently or physically grafted to a solid surface and stretched in the vertical direction, forming relatively uniform nanoscale layers in contrast to random coiled chains in conventional physical films (see Figure 13.1 for some most common molecular structures). Traditionally, AFM is widely explored to visualize the surface morphology of brush layers with high resolution, but more studies are focusing on probing surface properties of polymer brush layers, such as elasticity, molecular characteristics, adhesion and friction properties, viscoelastic response, vertical and lateral segregation, or thermal transitions. Selected results will be discussed here.

Careful consideration of AFM-based probing of polymer brush layers should be implemented due to the possibility of the strong effect of the AFM tip on the surface layer properties during scanning, particularly in contact mode. Indeed, computer simulations have demonstrated that probing with low normal loads and under low elastic deformation results in more accurate information about brush layer morphology, roughness, and thickness. However, significant penetration of the AFM tip under a modest normal load can lead to brush reorganization that compromises the measurement process [45]. Also, "insisting" on harsh probing conditions might result in irreversible plastic deformation, scanning instabilities, and partial or complete removal of the polymer brush layer, all phenomena frequently observed in various AFM experiments.

In one of the early studies, carboxylic acid and anhydride-terminated PS chains of different molecular weights were "grafted to" from the melt state onto silicon substrates modified with an epoxysilane monolayer by Luzinov *et al.* [34]. In the dry state, polymers are in a bad "solvent" (air) and tethered to the surface, and the chains are collapsed to a dense layer and do not dewet from surface. AFM images showed that tethered polymer brush layers with a thickness of several tens of nanometers (as determined from scratch AFM testing) were relatively smooth, uniform, and robust for high grafting densities. However, as was observed for short deposition times, the low and high molecular weight polymer brush layers display very different, nonhomogeneous surface morphologies. The lower molecular weight polymers form densely packed nanometer-scale clusters alternating with unmodified substrate distributed homogeneously throughout the entire substrate. When the brush layers are formed from high molecular weight polymers at short deposition times, the larger clusters are irregularly distributed on the surface displaying the microscopic pattern of a partially dewetted film. However, for all polymers with very long grafting times, the brush layer homogeneously covers the substrate and possesses a very fine domain surface texture.

In another study, relatively thick (50–90 nm) polymer brush layers from two representative glassy and rubbery functionalized polymers, poly(styrene-*co*-2,3,4,5,6-pentafluorostyrene) (PSF) and polymethylacrylate (PMA), have been studied by LeMieux *et al.* [46]. As discussed previously, these polymer brush layers, characterized by a high density of grafting, are synthesized according to a "grafting from" approach on a silicon surface modified with a reactive SAM. In the dry state, both glassy and rubbery brush layers are found to be homogeneous with no signs of significant local lateral chain segregation or developed domain texture (Figure 13.2).

Phase images for these brush layers show a very uniform surface distribution of material. Random variations of layer thickness on a nanoscale were very modest with the resulting microroughness within 1 μm^2 of surface area not exceeding 0.3–0.6 nm, which is common for dense brush layers. The latter value is close to the cross-sectional dimensions of a single polymer chain and is a characteristic of molecularly smooth surfaces. This observation is in contrast to that of polymer brush layers with low to moderate grafting densities for which nanoscale lateral inhomogeneities and domain microstructure are frequently observed and result in much higher surface roughness [28, 47].

The thermal, mechanical, and thermoelastic properties of PSF and PMA polymer brush layers were studied in this work by applying the surface probing approaches discussed in detail in Chapter 5. The authors observed that at room temperature, brush layer mechanical properties were virtually identical to that for unconfined bulk polymer films with nanoscale thickness obtained concurrently via bulk polymerization.

SFS nanomechanical measurements revealed a much lower adhesion of the PSF brush layer as can be seen from FDCs (Figure 13.3). The PSF brush layer is glassy at room temperature and contains fluorine-enriched segments with low surface energy. The difference between the approaching and retracting curves at the point the tip "snaps" out of the physical contact with the sample gives an indication of the total energy needed to fully withdraw the AFM tip, and this difference is clearly much larger with PMA than with PSF (Figure 13.3). In contrast, the easily deformable and sticky rubbery PMA brush layer containing polar segments showed large deformation, significant hysteresis, a long stretchable region, and high adhesion (Figure 13.3).

SFS micromapping of these brush layers with lateral resolutions of about 15 nm showed uniform elastic responses and adhesive forces throughout the probed surface area, except for occasional surface defects, inhomogeneities, and contaminants (Figure 13.3). The analysis of the SFS micromapping data allowed the calculation of corresponding histograms of the surface distribution of elastic moduli and adhesive forces (Figure 13.4). These histograms of surface distributions provided quantitative characterization of the elastic properties of representative glassy and rubbery brush layers.

The PSF brush layer shows a compression elastic modulus of approximately 1.1 GPa, typical for ultrathin glassy polymer layers. Surface distribution shows modest deviations from the average value over microscopic surface areas (within

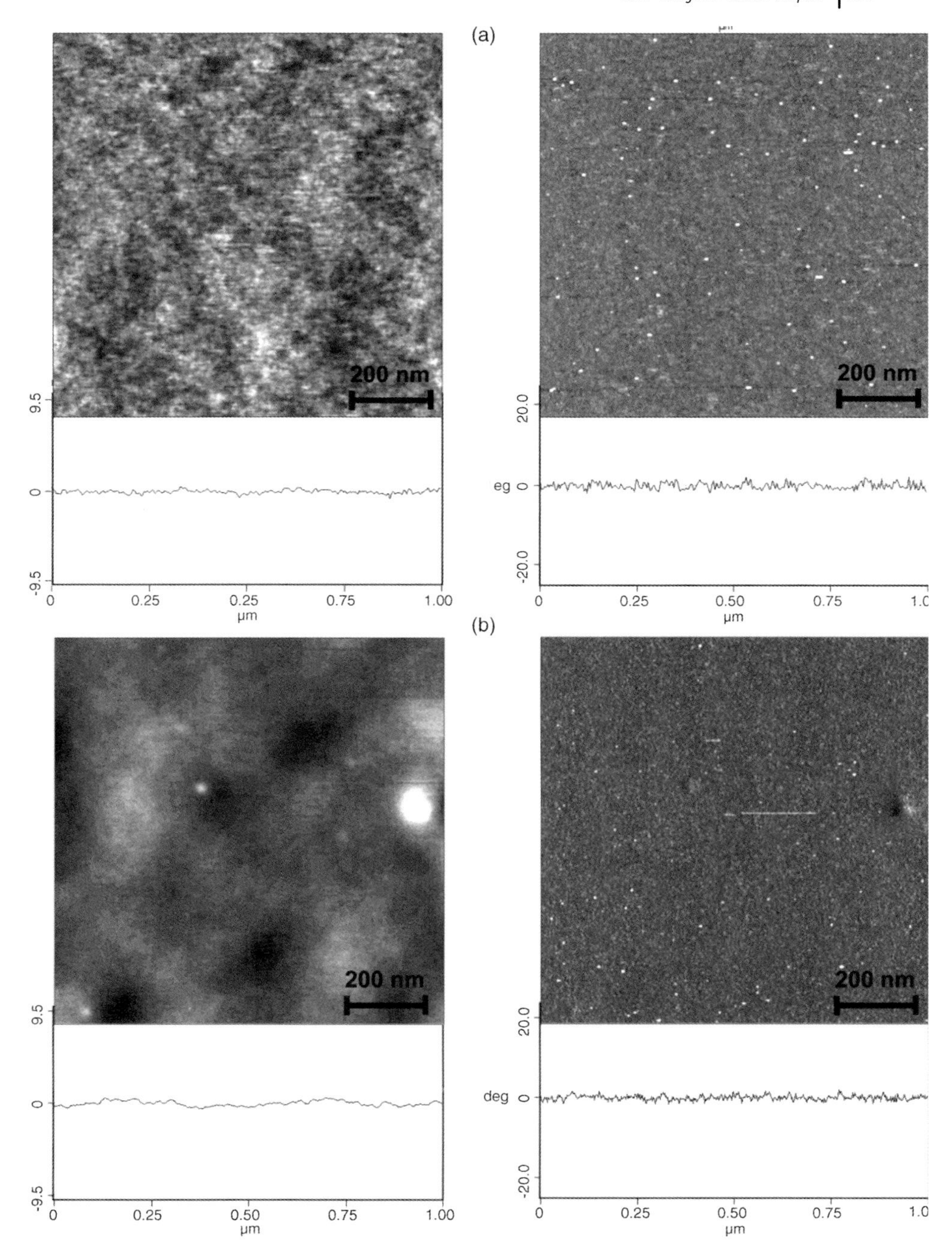

Figure 13.2 AFM images (left, topography; right, phase) of the PSF (a) and PMA (b) layers at 1 μm^2. Corresponding cross sections are included for topography and phase to illustrate the uniformity in topography and chemical composition. Reprinted from Ref. [46].

(a)

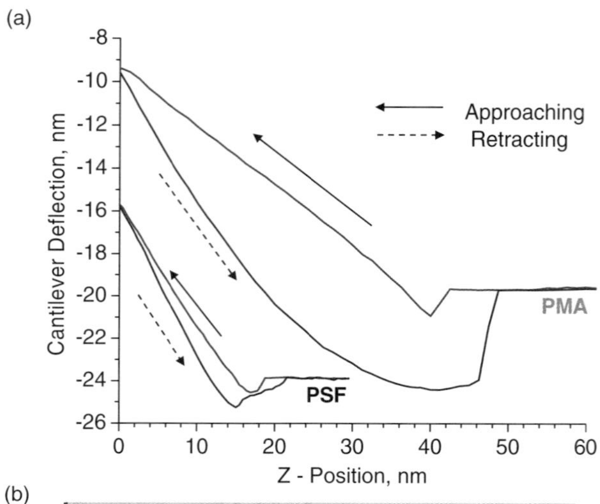
Approaching
Retracting
Cantilever Deflection, nm
Z - Position, nm
PMA
PSF
-8
-10
-12
-14
-16
-18
-20
-22
-24
-26
0
10
20
30
40
50
60

(b)

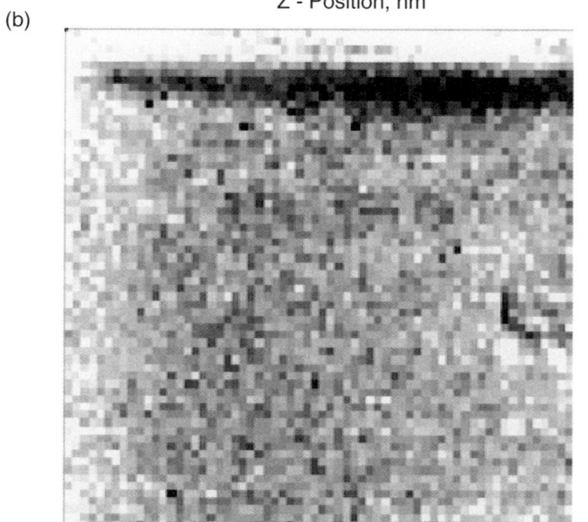

(c)

±0.1 GPa) (Figure 13.4) [48]. In contrast, the rubbery PMA brush layer possesses a much lower elastic modulus of 55 MPa (Figure 13.4). This value is typical for the rubbery state, and a very narrow variation (within ±5 MPa) demonstrates the uniformity of brush elastic properties.

Moreover, the values for both PSF and PMA brush layers are fairly close (slightly below but still within experimental uncertainty) to the experimental values determined for relatively thick spin-coated films of corresponding polymers obtained via bulk polymerization, as well as those for well-known glassy polymers such as polystyrene. Surface histograms demonstrate the narrow distribution of the nanomechanical response for uniform layers, with random deviations of elastic moduli and reduced adhesive forces not exceeding 16 and 11%, respectively, for the entire surface areas tested. These small deviations are characteristic of excellent uniformity in chemical composition, local ordering, surface morphology, and molecular microstructure of various homopolymer brush layers.

The reduced adhesive force, defined as the pull-off force from FDC normalized to the tip radius R or $\Delta F/R$, is essentially a measure of the adhesive energy required to separate the AFM silicon tip from the polymer surface [33, 34]. Analysis of SFS micromapping reveals that PSF and PMA brush layers have very different surface adhesive property distributions, supporting the conclusions made based on single FDC analysis. This difference is caused by the physical distinction, as well as differences in chemical composition, at the surfaces of the glassy and rubbery brush layers as mentioned above. With a high concentration of fluoro-containing groups, PSF has a substantially lower surface energy than PMA brush layers where polar groups contribute to high surface energy in the rubbery material [49, 50]. In addition, as is known for AFM experiments, the pull-off force is affected by a difference in the mechanical contact area between the AFM tip and the layer surface, which is much higher for the more compliant rubbery PMA layer [34, 51]. The larger contact area imposed by the AFM tip in approach mode during the retracing cycle affects the total force required to separate the tip from the surface.

Conversion of the FDCs to indentation (penetration)–loading curves further underlines the different micromechanical responses of glassy and rubbery homobrush polymer layers (Figure 13.5). Under an identical normal load, the rubbery PMA layer apparently undergoes 5–10 times more compressive deformation than the glassy PSF brush layer at all thicknesses and deformations probed. However, the PMA layer cannot support as high a normal load as the PSF layer.

Direct calculations of the depth profile of the elastic modulus from FDCs showed much higher absolute values of the elastic modulus for the PSF brush layer for all

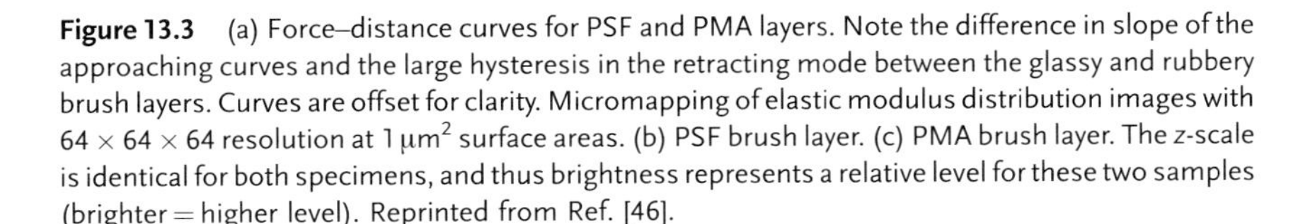

Figure 13.3 (a) Force–distance curves for PSF and PMA layers. Note the difference in slope of the approaching curves and the large hysteresis in the retracting mode between the glassy and rubbery brush layers. Curves are offset for clarity. Micromapping of elastic modulus distribution images with 64 × 64 × 64 resolution at 1 μm^2 surface areas. (b) PSF brush layer. (c) PMA brush layer. The z-scale is identical for both specimens, and thus brightness represents a relative level for these two samples (brighter = higher level). Reprinted from Ref. [46].

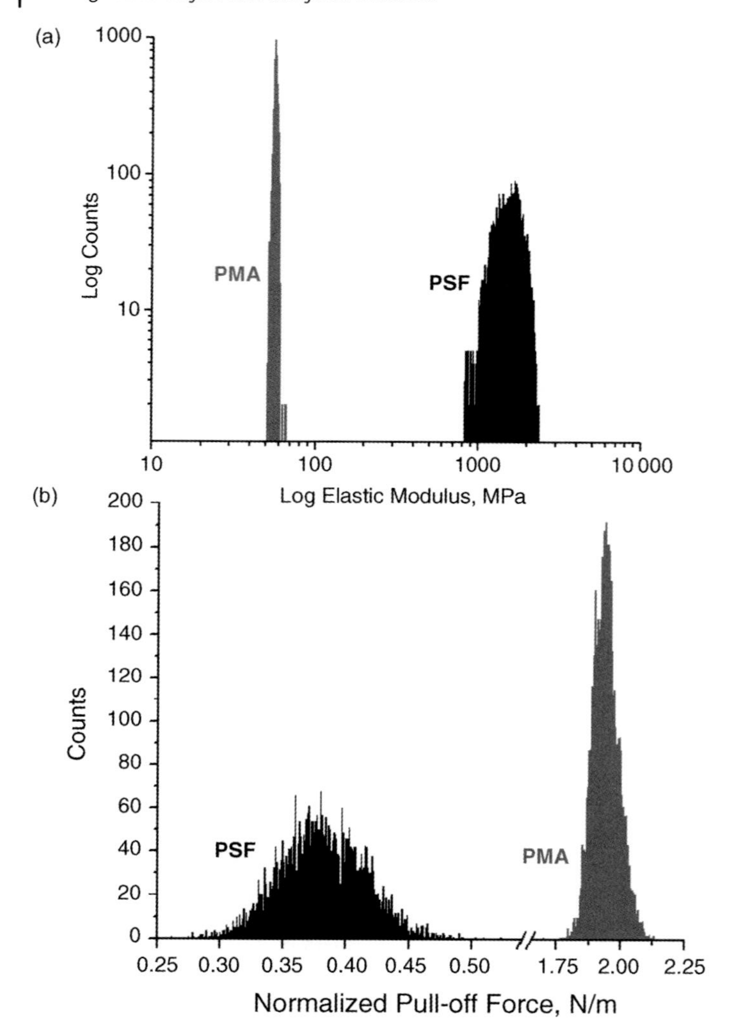

Figure 13.4 Surface histograms comparing the micromechanical mapping data for PMA and PSF layers: elastic moduli (a) and adhesive forces (b). All histograms are taken from $64 \times 64 \times 64$ force volume scans, giving a total of 4096 surface locations tested. The modulus represented here is the average modulus for full penetration of the AFM tip into the layer. Reprinted from Ref. [46].

indentation depths (Figure 13.5). Furthermore, the absolute values of the elastic modulus for the PSF brush layer are close to that for physisorbed spin-coated PSF and PS films with slightly higher values for the topmost surface. Typically, under given experimental conditions, the maximum indentation depth under the normal load of 60 nN was within 25–30 nm for the rubbery PMA brush layer but only 6–8 nm for the PSF brush layer (Figure 13.5).

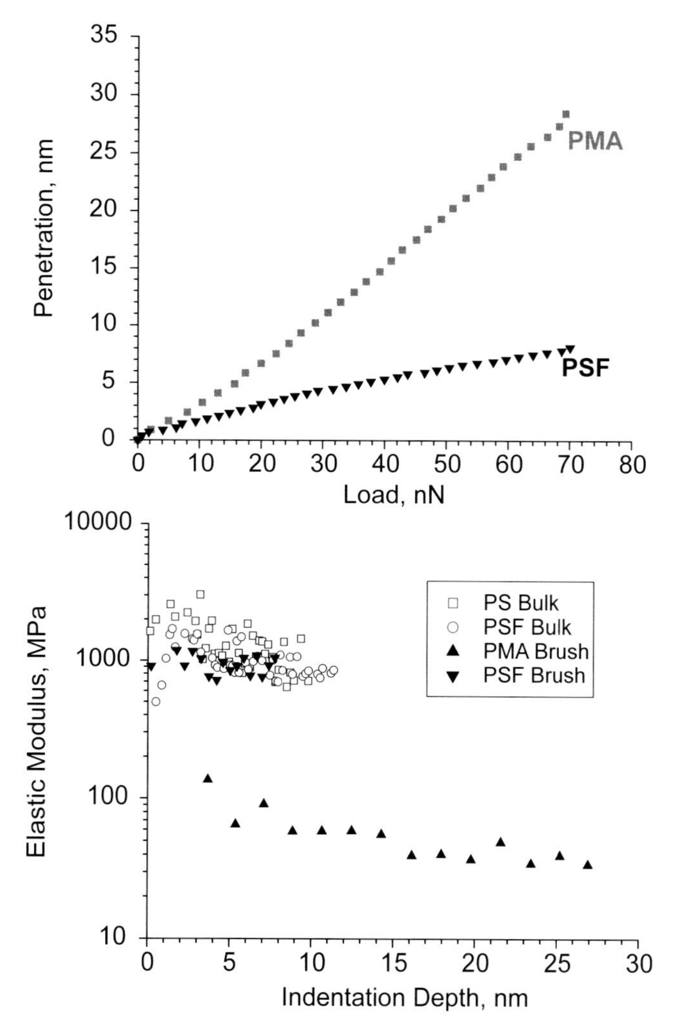

Figure 13.5 (a) Indentation curves for PMA and PSF brush layers. At equal normal loads, a significantly larger penetration took place into the more compliant PMA layer. (b) Elastic modulus depth profiles for PSF and PMA layers. Results are also shown for PS and PSF bulk samples, which are spin-coated films. The thickness of the spin-coated samples is the same as that for the respective grafted brush layers. Reprinted from Ref. [46].

SFS thermomechanical studies of glassy brush layers in a dry state demonstrated that the maximum indentation depth achievable under a constant normal load was virtually constant at different temperatures and stayed within 5–10 nm for temperatures below 60 °C (Figure 13.6). However, at higher temperatures, the compliance of the PSF brush layer increased gradually (Figure 13.6). Furthermore,

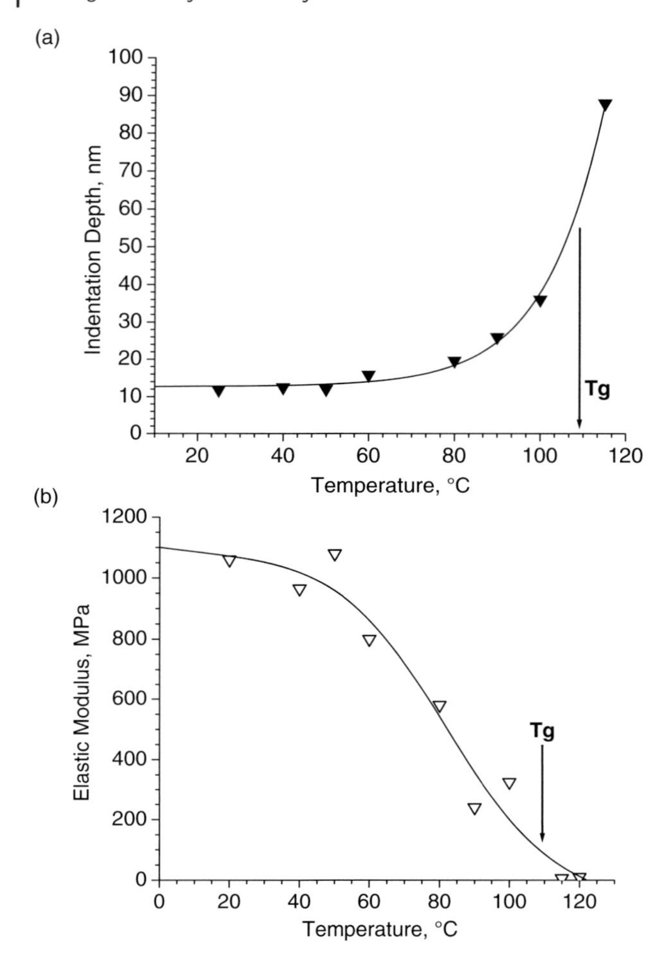

Figure 13.6 (a) Thermomechanical SFS measurements of the PSF brush layer: penetration depth at different temperatures with identical normal load at each temperature. Above the glass transition, full penetration through the 90 nm thick layer was achieved. (b) Temperature variation of the elastic modulus for the PSF brush layer derived from FDCs. Reprinted from Ref. [46].

a sharp rise in the indentation depth up to 90 nm (almost full brush thickness) was observed for temperatures in the region of glass transition (above 90 °C). This indicates a dramatic increase in polymer compliance associated with the unfreezing of chain segments and increased chain mobility during glass transition typical for amorphous polymers.

Indeed, the elastic modulus estimated from FDC data by applying Sneddon's model remains virtually unchanged for temperatures below 60 °C but then drops dramatically (50 times) from 1.1 GPa to 15 MPa for temperatures above 100 °C (Figure 13.6). This type of thermomechanical behavior corresponds to that expected for bulk PSF with a glass transition temperature of 110 °C reported in the literature. The glass transition is somewhat wide, ranging from 60 to 110 °C, but this is not unusual for high molecular weight, amorphous glassy polymers with a broad molecular weight distribution and variable grafting density [52].

In another study, a careful analysis of the FDCs for PMMA homobrush layers grown by the "grafting from" approach was conducted by Yamamoto *et al.* for brush layers in toluene solvent [53]. The authors analyzed the compressibility of brush layers with different molecular weights and directly measured the thickness of the swollen layer using the scratch test. This analysis demonstrated that the compressibility of the brush layer cannot be described by a traditional uniform deformation model. The higher than expected mechanical resistances of these swollen brush layers can be related to their unusually high grafting density that constrains chain stretching and compressive deformation under the AFM tip. Also, pulling of polymer chains from brush layers with the AFM tip was done in this study to characterize the molecular weight distribution. It has been demonstrated that this is an efficient, although time-consuming, sophisticated, and unique technique for the direct measurement of the molecular weight characteristics of grafted polymer chains that cannot be routinely obtained with typical analytical approaches.

Recently, glassy PS and PMMA homopolymer brush layers with large thicknesses grown by the "grafting from" approach have been studied from the perspective of their microtribological properties with the colloidal probe AFM technique by Tsujii *et al.* [54]. High grafting density of the brush layers caused an almost completely stretched chain conformation leading to their unusual microtribological properties, as observed and discussed in this study. In fact, FDC analysis showed that the high grafting density resulted in a much higher mechanical resistance to the AFM tip compression. This high mechanical resistance results in excellent shear properties that, in combination with low adhesion, provides an extremely low friction coefficient (well below 10^{-3}).

The authors demonstrated that the friction behavior can be easily controlled by the quality of solvent, or, for example, by using mixed solvents, as was investigated for several systems. By replacing conventional brush layers with layers grown from PNIPAAM, a known thermoresponsive polymer, unique thermoresponsive lubrication behaviors can be achieved, as demonstrated in this study. FFM studies showed that collapse of PNIPAAM chains at temperatures above 30 °C caused dramatic (three orders of magnitude) increases in the friction coefficient of these thermally responsive brush layers to values usually observed for sticky, mating polymers.

In another recent study, PEG homopolymer brush layers and corresponding copolymer surface layers were studied by Kessel *et al.* [55]. The authors considered changing conformation of these brush layers at variable temperatures or solvent quality. The authors carefully studied the temperature-controlled transition of the PEG copolymer brush layers with an AFM probe directly in liquid. Reversible, temperature-controlled variations of the adhesive forces were revealed with much higher attractions observed for elevated temperatures (above 37 °C, which is above the known LCST). Temporal variation of adhesion was also observed and attributed to the slow kinetics of the brush transformations during their thermal transition.

Protein-modified colloidal AFM probes were also explored by Lim and Deng for probing reversible brush layer properties [56]. The authors focused on the possible role of proteins in temperature-induced cell reactions. Interesting ring-like polymer brush structures were exploited to demonstrate how reversible conformation transformations in response to solvent treatments can be utilized to create an open–close porous brush structure with a diameter of several hundred nanometers [56]. The direct visualization with high-resolution AFM allows the monitoring of thermal transition phenomena in polymer brush layers, which can be contrasted by a variation of the forces applied to the AFM tip.

13.1.2
Grafted Diblock Copolymers

Grafted diblock copolymer may also form brush layers with responsive properties. Diblock-shaped copolymers that combine the two dissimilar polymer chains PS and poly(*t*-butyl acrylate) (PBA) attached to a single focal point (Y-shaped brushes) capable of chemical grafting to a silicon surface have been studied by Julthongpiput *et al.* (Figure 13.7) [57–59]. Postgrafting hydrolysis of the PBA arms resulted in amphiphilic grafted diblock molecules with different volume ratios of PS and highly hydrophilic PAA arms, as well as different lengths of rigid stems (see molecules Y1 and Y2 in Figure 13.7 with different grafting segments). Thus, these brush layers possess both hydrophobic and hydrophilic chains with different lengths and compositional ratios that are confined to a single grafting site. These arms are capable of

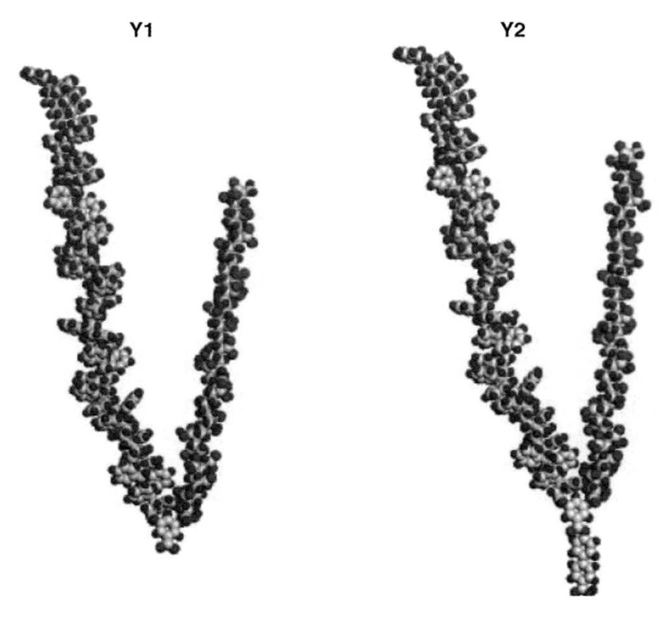

Figure 13.7 Molecular models and chemical structures of Y-shaped diblock copolymers. Reprinted from Ref. [59].

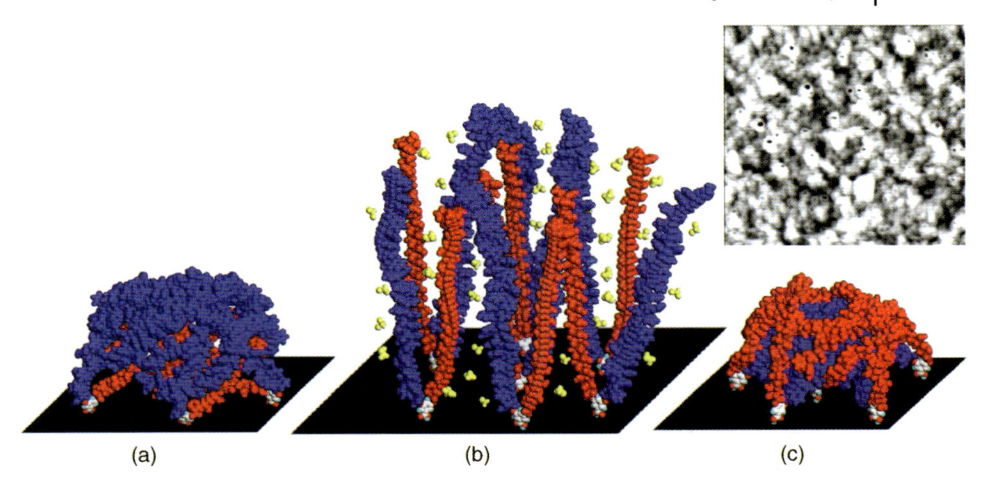

Figure 13.8 Molecular models of the proposed structural rearrangement in Y-shaped brushes. Internally segregated pinned micelle composed of seven grafted PS–PAA molecules spaced 3.5 nm apart. (a) Upon treatment with toluene and following drying, the PS arms (blue) form corona covering the micelle's core consisting of seven PAA arms (red). (b) Representation of the same seven molecules swelled in a nonselective solvent (yellow). (c) Top open, crater-like structure containing seven collapsed PS arms partially covered by seven PAA chains. AFM image corresponds to the right model (c) of pinned structures (300 x 300 nm). Reprinted from Ref. [59].

reversible collapse/swell transitions under treatment with different solvents as explored in this study to control surface morphology (Figure 13.8).

High-resolution AFM images of grafted brush layers demonstrate that Y-shaped molecules exhibit lateral inhomogeneities expected for chemically connected and incompatible blocks. The surface is covered with densely packed islands with lateral dimension close to 10 nm, as estimated after correction for the tip dilation (Figure 13.8). The height of these islands varies from 1.8 nm for molecule Y1 to 2.5 nm for molecule Y2, as estimated from the effective layer thickness obtained from ellipsometry and multiple cross sections of topographical AFM images. This morphology corresponds to internally segregated "pinned micelles" theoretically predicted for densely grafted brush layers of Y-shaped diblock copolymers containing incompatible long-chain arms (see the molecular models in Figure 13.8).

After toluene treatment (a good solvent for PS chains) of these brush layers, the topmost surface layer is predominantly composed of PS arms, which form "coronas" of pinned micelles, while collapsed PAA arms constitute their "cores." Treatment of the brush layers with water, a selective solvent for PAA arms, causes dramatic surface reorganization of the initial morphology. Remarkably, an array of crater-shaped small features is observed at the surfaces of both short- and long-stem Y-shaped diblock molecules (Figure 13.8). The craters represent discrete objects whose rims are slightly elevated with respect to their surrounding due to different collapsing/swelling behaviors of dissimilar arms.

In situ AFM imaging of brush layers directly immersed in toluene (selective solvent for PS arms), water (selective solvent for PAA arms), and mixed solvents revealed

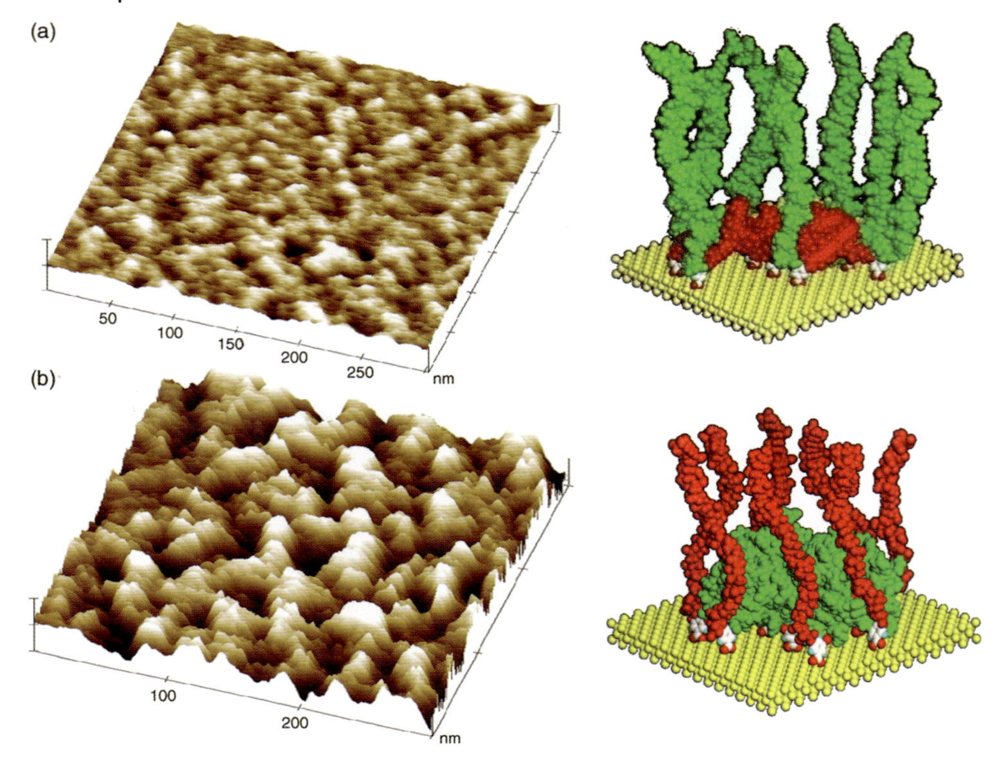

Figure 13.9 AFM images and corresponding models of Y-shaped brushes (a) in water and (b) in mixed solvent. Reprinted from Ref. [60].

variable surfaces with a relatively grainy nanoscale morphology that was controlled by the solvent quality [60]. The average thickness of the brush layers determined from a scratch test directly in fluid was close to 6 nm in different solvents. This value was much higher than the effective thickness of 1.4 and 1.9 nm measured for collapsed brush layers in a dry state (air). This significant (fourfold) increase in thickness and the corresponding increase in microroughness directly in selective and mixed solvents indicate the swelling of arms and the formation of a truly brush-like structure with stretched, "solvent friendly" chains of proper composition (Figure 13.9).

The surface microroughness reached 0.7 nm in water and 0.44 nm in toluene for the Y1 brush layer, as well as 1.6 nm in water and 0.26 nm in toluene for Y2 brush layer These values are much higher than the initial microroughness for a functionalized silicon surface (below 0.2 nm). The surface microroughness of polymer brushes in various solvents was consistently much higher than in the dry state for all brush layers studied in this work. This significant increase (more than twofold) reflects a nonuniform molecular topography after the placement of the brush layers into a selective or mixed solvent. In water, this phenomenon is caused by a vertical extension of PAA arms swollen in a good solvent with PS chains remaining in a collapsed state, as demonstrated in the molecular models in Figure 13.9 (and vice versa in toluene). The highest surface microroughness of diblock polymer brush layers was observed in the mixed solvent, which reflects concurrent swelling of both longer-chain PS arms and shorter-chain PAA arms (Figure 13.9).

The observed conformational rearrangements of different blocks should dramatically affect the mechanical properties of the brush layers. Direct in-liquid SFS measurement of diblock polymer brush layers in the presence of a solvent gives a unique opportunity to assess these characteristics as a function of solvent–solute interactions and chemical composition of the diblock molecules (Figure 13.7).

Indeed, comparative SFS measurements conducted for the diblock brush layers showed that no detectable adhesive forces were observed for both brush layers placed in toluene (Figure 13.10). This phenomenon, along with the repulsive nature of the FDCs in the approaching cycle, is typical for a surface of hydrophobic PS chains that should dominate the topmost surface layer under these conditions. In contrast, for

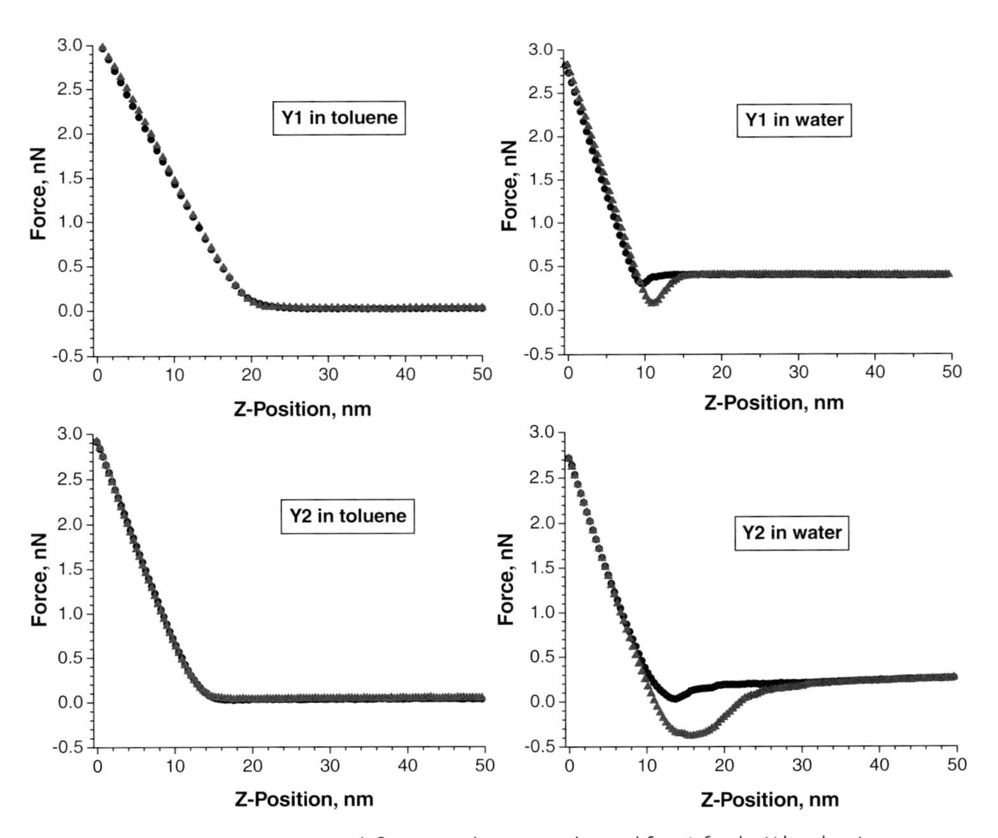

Figure 13.10 Representative FDCs (deflection scale converted to real force) for the Y brushes in selective solvents. The circles represent the approaching curve and the triangles are the retrace curve for the FDCs. Reprinted from Ref. [60].

both brush layers in water, there was a noticeable and consistent pull-off force visible on the retraction portion of the FDC (Figure 13.10). Typical values of this adhesion force under these conditions ranged from 200 to 700 pN, which is a manifold higher than that observed in water.

This relatively high level of adhesion of a brush layer in water is expected because the hydrophilic PAA arms are extending from the surface and thus can directly interact with the hydrophilic silicon oxide surface of the AFM tip. For the Y2 brush layer, the level of adhesion in water was higher, with values reaching the 500–700 pN range. This increasing adhesion indicates strong tip–surface interactions as more swollen PAA chains are able to participate in the physical contact with the hydroxyl-terminated tip [61]. It has been suggested that much stronger adhesion of the AFM tip with the Y2 brush layer in water is caused by the higher effective concentration of the PAA chains on the surface due to higher grafting density.

In toluene, a good solvent for PS chains, the penetration of the AFM tip through the swollen brush layer requires very little force (below 2 nN) since PS chains are highly swollen under these conditions, as is clear from the corresponding loading curves calculated from FDC data (Figure 13.11). Loading curves for both Y1 and Y2 in toluene deviate from conventional Hertzian behavior by leveling off beyond some

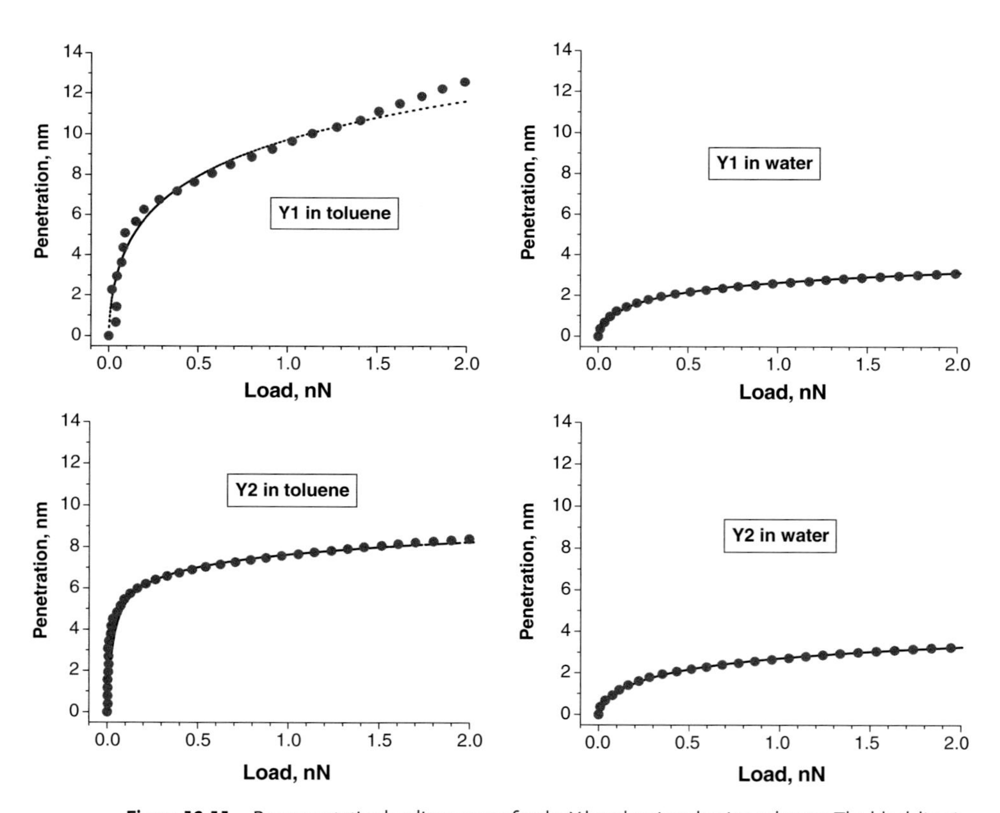

Figure 13.11 Representative loading curves for the Y brushes in selective solvents. The black line in the load–penetration curves is a simulated double-layer model fit. Reprinted from Ref. [60].

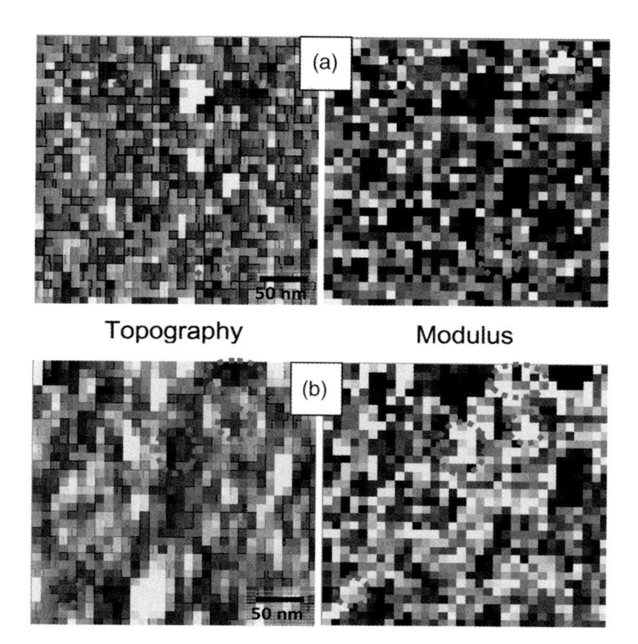

Figure 13.12 Simultaneous topography and force mapping over a 300 nm^2 area with 32 × 32 pixel resolution for FDCs for Y1 (a) and Y2 (b). For topography and modulus maps, the brighter the color, the higher the heights and the relative value of the modulus, respectively. Circles in topography indicate low points (or pits) in topography that correspond to the collapsed PS, and they correlate to the high points in the modulus map indicative of the hard, glassy PS chains in a poor solvent (water). The topography and modulus maps for Y2 also exhibit the same behavior, resulting in a bimodal distribution of the elastic modulus. Reprinted from Ref. [60].

indentation depth due to the presence of the underlying stiff silicon substrate. Apparent maximum brush layer indentation reaches 8 nm for Y2 brushes and is slightly higher for the Y1 brush layer because of the longer PS and PAA arms and lower grafting density. Placing brush layers in water resulted in dramatic decreases in the total effective thickness due to the collapse of longer PS chains. The solid lines in Figure 13.10 are the best fit from a double-layered deformational model with the inclusion of a silicon substrate (see Chapter 5), allowing a quantitative evaluation of the elastic modulus and the brush layer thickness.

The topography and elastic modulus distribution maps for diblock brush layers presented in Figure 13.12 result from taking 1024 individual curves over a 300 nm^2 area directly in water with an effective lateral resolution below 10 nm. Close analysis of the topography and modulus maps of brush layers allows correlation of the low points in the topography (correspond to PS pits) with high points in the elastic modulus. Furthermore, the high points in the topography (corresponding to PAA chains) match very closely to the low elastic modulus values in the modulus map. This force mapping confirms the presence of both chain arms on the brush layer surface in water, as suggested by the molecular models (Figure 13.8).

The integrated analysis of the SFS micromapping results for Y1 and Y2 brush layers in water showed well-defined bimodal surface distribution of the elastic

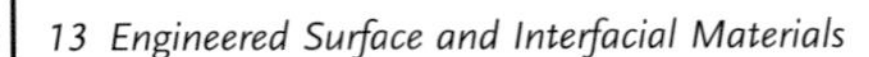

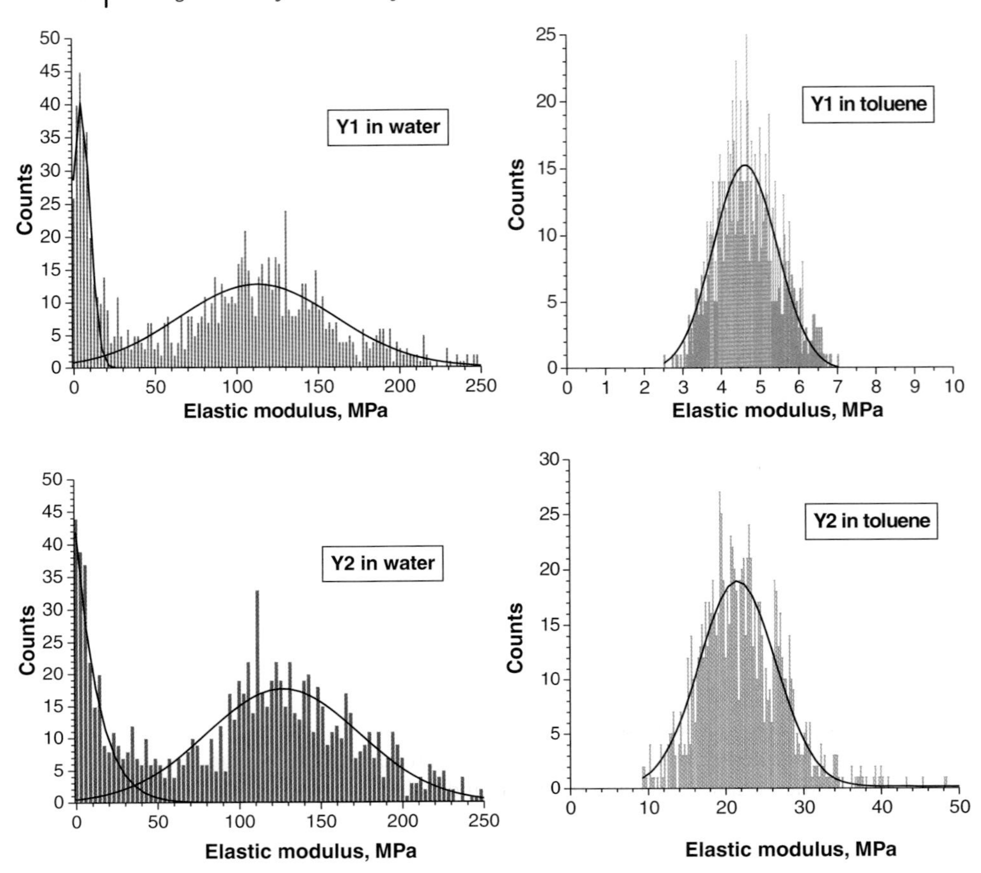

Figure 13.13 Surface histograms of elastic modulus and adhesive forces for Y1 and Y2 brush layers in toluene and water. Reprinted from Ref. [60].

response (Figure 13.13). Bimodal distributions of the elastic modulus for both brush layers obtained in water show a narrow and a strong peak at around 5 MPa coming from the swollen PAA chains. On the other hand, another much broader peak is located at around 110 MPa that corresponds to the apparent elastic modulus of the collapsed, but still partially swollen, PS blocks.

In contrast, the surface distribution of the elastic modulus for Y1 and Y2 brush layers in toluene is unimodal, with much lower values of the apparent elastic moduli (Figure 13.13). An interesting observation is the fivefold increase in the elastic modulus for Y2 (22 MPa) brush layers compared to Y1 (4.6 MPa) brush layers. The decreased compliance of Y2 brush layers can be attributed to shorter arms and higher grafting densities that result in larger space constraints for arms during compression by the AFM tip. The homogeneous character of brush layers that are swollen in toluene indicates that the PS chains almost completely screen the PAA chains and form a continuous, smooth, and swollen PS surface layer over collapsed PAA chains.

13.1.3
Mixed Brush Layers

Quite a few studies consider physically adsorbed diblock copolymers with different affinities to the surface and utilize sophisticated AFM measurements, usually in combination with X-ray reflectivity and ellipsometry (as a recent example, see PS–PAA block copolymers by Muller *et al.* [62] and references therein). However, here we will focus on robust brush layers with high grafting densities of polymer chains, covalently tethered to a solid substrate.

Recently, Vyas *et al.* have studied surface morphology and tribological properties of mixed binary PS–PAA amphiphilic brush layers in comparison with corresponding homobrush layers [63]. In this study, the wear behavior was also studied with a sliding AFM tip to characterize the surface stability of the grafted brush layers in conjunction with other variable molecular parameters. The main difference in the wearing mode was observed for mixed brush layers treated with different solvents: PS-enriched surfaces showed regular rippling morphology and high shear resistance in contrast to brush layers with PAA-enriched surfaces, which could be completely removed under shear stresses. Corresponding AFM studies of single-component brush layers showed similar wearing behavior for individual PS and PAA brush layers.

In another study of different mixed binary brush layers, Vyas *et al.* fabricated brush layers from PS–P2VP (poly(2-vinylpyridine)) compounds to show how reversible phase segregation under variable pH conditions can be utilized to tailor the lubrication properties of brush-modified surfaces [64]. AFM imaging was combined with SFS measurements with different AFM tips to reveal the role of surface topography in the variation of their frictional and adhesive properties. Polymer-modified colloidal probes were also utilized for mixed brush layers in dry and swollen states. Very significant changes in both adhesion and friction were observed after treatment with different selective solvents when AFM measurements were conducted with a PAA-modified colloidal probe. These variations of surface properties were suggested to be caused by increasing polar interactions and by the reduction in topological contributions related to the reduced surface microroughness. The friction coefficient was observed to increase significantly (up to 50%) for brush layers with the increasing content of the polar component, P2VP, as well as with the increasing grafting density of both components.

Combining AFM imaging with layer-by-layer plasma etching of surface layers with controlled increments (the so-called AFM nanotomography) has been exploited by Usov *et al.* to restore the 3D nanoscale morphology of mixed brush layers of PS–PMMA [65]. By applying this labor-intensive approach, which also requires precise localization of a selected surface area, the authors observed significant changes under solvent treatment during switching, resulting in the formation of ripple or dimple segregated morphologies. Careful AFM experiments were conducted on the same selected surface areas after each etching step in both dry and swollen states. These measurements showed that the ripple surface morphology consists of a depressed PS-rich phase with elevated PMMA-rich, elongated domains with a PMMA phase covering all surface features. The authors suggested that very minor variations in grafting point locations are critical for the formation of highly heterogeneous morphology with nanoscale lateral and vertical phase separation of different brush layer components.

Binary brush layers from PDMS and P2VP chains grafted directly on electrodes were probed with AFM to study their ability to gate the electrochemical properties of the electrodes as a result of switching their morphologies by Motornov *et al.* [66]. Along with the direct observation of a highly heterogeneous, phase-separated surface morphology of mixed brush layers, the authors utilized SFS probing to measure reversible changes in their surface properties with changes in pH of solution in conjunction with observed significant changes in electrical resistance. The authors concluded that the molecular reorganizations in mixed binary brushes visualized with AFM can be exploited as switchable chemical gates for biocatalytic electrodes.

In another study, LeMieux *et al.* studied mixed brush layers that were composed of carboxylic acid-terminated PS and PBA chains (Figure 13.1) covalently attached to a silicon surface that was modified with an epoxysilane SAM by a grafting to routine [67]. Polymer surface layers consisted of phase-separated dissimilar chains grafted to a functionalized silicon surface with a total layer thickness of only 1–3 nm in the dry state (Figure 13.14). Under these loose grafting conditions, one-step

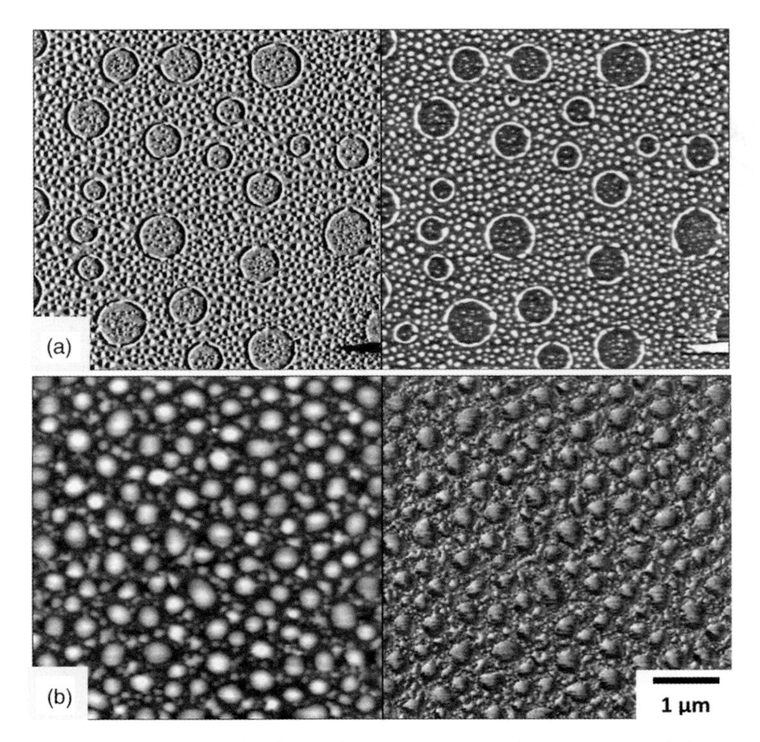

Figure 13.14 Topographical and phase AFM images of (a) one-step and (b) two-step grafting samples of PS and PBA components showing topography (left) and phase (right). Reprinted from Ref. [67].

grafting from a polymer melt resulted in large submicron phase-separated regions in mixed brush layers. In contrast, a refined, two-step sequential grafting procedure for mixed brush layers resulted in extremely small spatial dimensions of phase-separated PS and PBA domains, as observed with both topographical and phase imaging (Figure 13.14). By adjusting the grafting parameters, such as the concentration of each phase and corresponding molecular weight, very finely structured surfaces can be fabricated with domains as small as 10 nm and with less than 0.5 nm overall microroughness. Postgrafting hydrolysis of these mixed brush layers converted PBA to PAA chains to amplify their switching ability of surface wettability and tribological properties as also probed with FFM mode.

In another study by LeMieux *et al.*, two vastly different morphologies appeared when the much thicker mixed brush layer was composed of two grafted from components: PSF and PMA chains [68]. These mixed brush layers were exposed to toluene and acetone, with acetone being the selective solvent for PMA and toluene being the selective solvent for PSF component. AFM imaging of these binary brushes after different solvent treatments reveals the enormous height difference between the two states of mixed brush layers with preferred surface location of PSF or PMA (see AFM images and cross sections in Figure 13.15).

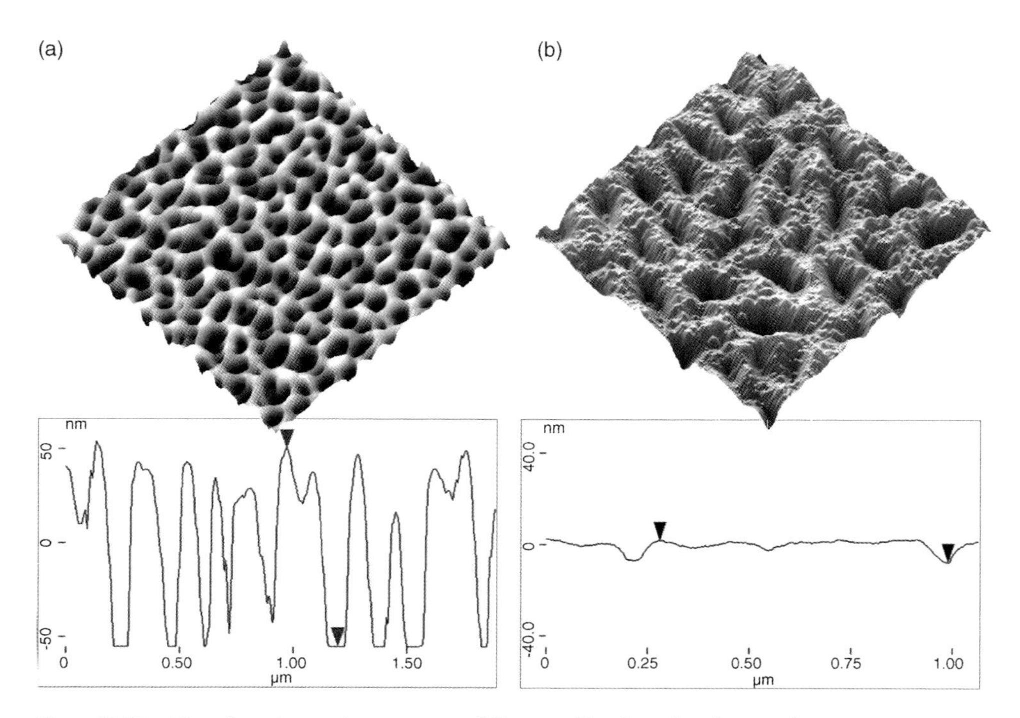

Figure 13.15 3D surface plots and corresponding profile analysis of the PSF–PMA mixed brush layer in the (a) rubbery (height increment is 50 nm) and (b) glassy states (height increment is 40 nm) showing the immense difference of height and surface roughness between the two states. The profile for the glassy state shows that two types of depressions in the PSF layer are found: large (8–10 nm deep) and small (1–7 nm deep). Reprinted from Ref. [68].

Surface microroughness (measured within a 1 μm^2 surface area) of these mixed brush layers increased dramatically from 2.2 nm in the glassy state to 28 nm in the rubbery state. Topographical AFM images confirmed that the lateral phase segregation did not occur for dimensions significantly larger than the polymer coil dimension, meaning that the grafting sites for each component are sufficiently uncorrelated in the course of a brush layer growing from two components. In the rubbery state, the PMA phase seemingly swells to an extremely high degree, forming a web-like cellular layer over the collapsed PSF component (Figure 13.15).

It has been suggested that in contrast to the rubbery state, grafted PSF chains in the glassy state form an ultrathin layer over the collapsed PMA chains. The pronounced vertical segregation (layering) suggested in this study is in agreement with previous experiments and theoretical calculations, although it is unclear from AFM topographical images alone whether clusters (dimples) are forming underneath the swollen component. Therefore, SFS micromapping of the PSF–PMA binary brush layers after exposure to different solvents has been conducted to provide additional information in terms of surface elasticity and vertical distribution of the components.

Indeed, SFS micromapping with a high lateral resolution of 10–30 nm was critical in the identification of different phases with good correlation observed among topography, elasticity, and adhesion associated with both the lateral and the vertical layering of different components (Figure 13.16). SFS measurements of the glassy state of mixed brush layers showed a mechanically heterogeneous surface in contrast to a homogeneous surface of the same brush layer in the rubbery state. For the glassy state, the holes correspond to areas of low elastic modulus and increased adhesion, indicating the presence of a more compliant and sticky rubbery PMA material inside.

The surface histogram derived from SFS micromapping for these mixed brush layers shows a bimodal distribution of the elastic modulus that is expected for a surface with microphase-separated glassy and rubbery regions (Figure 13.17). The value of the main maximum of 900 MPa is close to that measured independently for PSF homopolymer brush layers (1.1 GPa) and is typical for brush layers of glassy polymers. A minor surface contribution with an elastic modulus of about 480 MPa originates from holes and indicates a slightly softer, but still relatively stiff, surface inside. Depth profiling confirms a two-layered structure of selected areas of binary brushes with a rubbery PMA component located beneath the topmost glassy layer, as discussed in detail in the original publication [67].

On the other hand, for the rubbery state, the elastic modulus was within a narrow range of values (50–100 MPa), typical for a rubbery polymer and indicating a more complete and homogeneous soft topmost layer composed of the PMA component. This lower-modulus part of the elastic modulus histogram was also obtained independently using a softer tip with higher sensitivity (Figure 3.17). These results are consistent with independent SFS data for PMA homobrush layers and AFM scratch test and indicate rubbery PMA material grafted to the silicon substrate. The adhesive forces are homogeneous in both glassy and rubbery states, with much broader distribution and higher adhesion observed for the state with the PMA component on the top due to much rougher surface morphology and thus variable contributions composed of slopes, ridges, and pots (Figures 3.15 and 3.17).

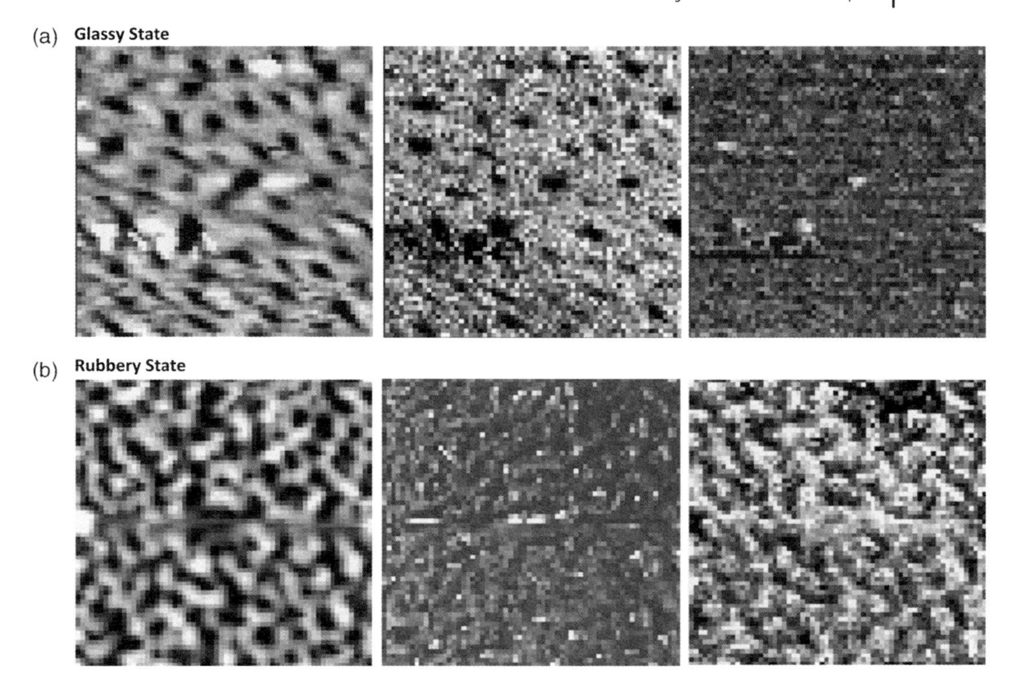

Figure 13.16 Force volume images at 64 × 64 resolution at 1 μm^2 for the glassy state (a) and 2 μm^2 mapping for the rubbery state (b) of PSF–PMA mixed brush layer. Topography (left column), elastic modulus (middle column), and adhesive force (right column) distributions are presented. Bright areas correspond to higher topography, elastic modulus, and adhesion. Reprinted from Ref. [68].

13.2
Self-Assembled Monolayers

Nanometer-thick SAMs composed of functionalized organic molecules covalently attached to a solid substrate have been introduced and extensively studied in the 1980s and 1990s as an interesting class of simple, robust, and facile species for effective surface modification, finding diverse applications in fields such as biosciences and tribology [69]. Generally, conventional organic SAMs are composed of short organic molecules that can be covalently attached to the solid substrate, with strong, high-density grafting, vertical or tilted orientation of densely packed aliphatic long and short tails, and a uniform distribution of the surface-located terminal groups of different types (Figure 13.18) [70].

Initially, a range of traditional surface characterization techniques, such as X-ray reflectivity, SEM, contact angle, and ellipsometry, were employed to study these monolayers, but it was quickly realized that AFM imaging is the most efficient tool to collect quantitative information not only about surface morphology, but also about properties unique to these surface layers. After initial attempts to apply STM to study SAMs, which generated a flux of artifacts caused by the dielectric nature of organic

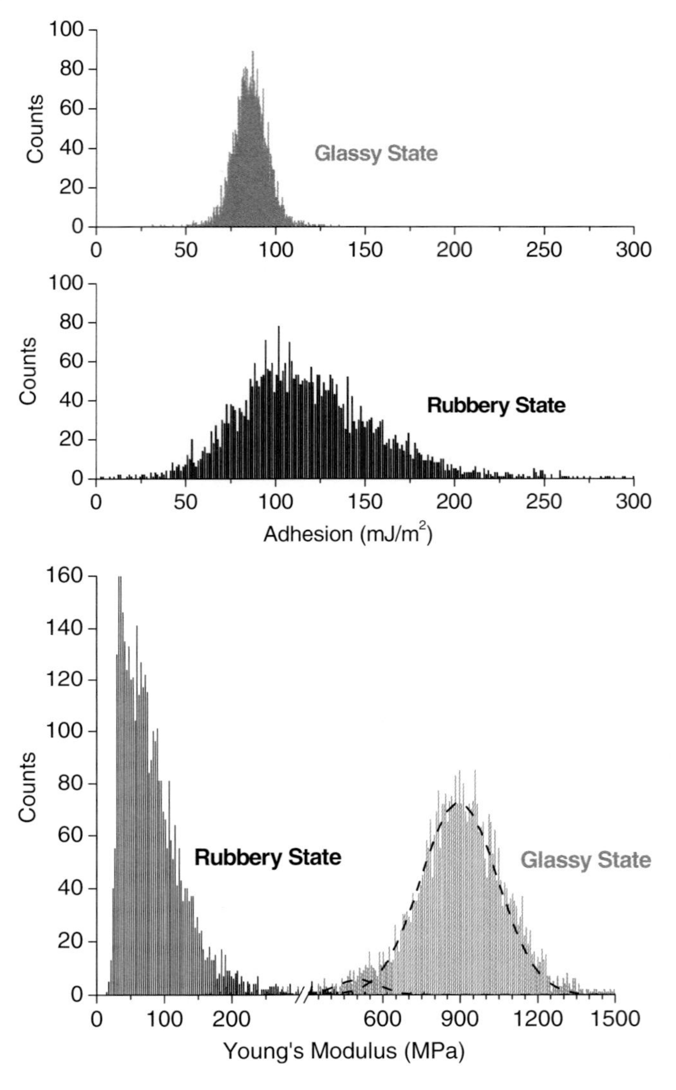

Figure 13.17 Surface histogram distributions from micromapping of PSF–PMA mixed brush layer demonstrating the surface properties such as adhesion (a and b) and elastic modulus (c) for the binary brush layer in two states. Histograms are taken from 64 × 64 force volume scans for a total of 4096 surface locations. The elastic modulus is the average value for each data point over the entire indentation range. The histogram representing the modulus in the glassy state shows a bimodal distribution fitted with two Lorentzian functions. Reprinted from Ref. [68].

SAMs and unfulfilled promises, AFM probing rapidly became the tool of choice for the fine characterization of SAMs under ambient conditions. In the vast majority of studies, AFM serves as a "quality control" tool, with featureless AFM images serving to confirm smooth, uniform, chemically homogeneous surface morphologies with

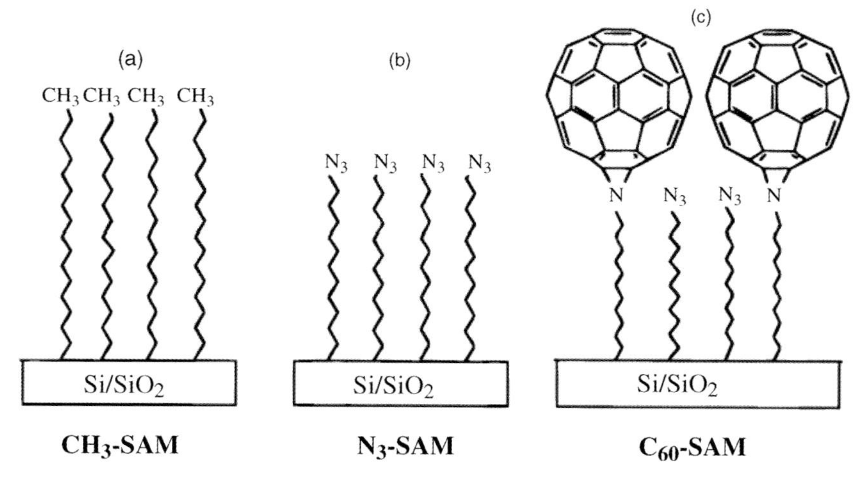

Figure 13.18 Schematics of the methyl-terminated (a), amine-terminated (b), and fullerene grafted (c) alkylsilane SAMs. Reprinted from Ref. [70].

minute microroughness usually well below 0.5 nm and usually fairly close to the underlying substrate roughness.

In this section, we will present selected examples of initial and recent AFM studies of common SAM types beginning from some earlier results on surface morphologies and defects to recent exploration of a range of new SAMs and their fine nanoscale properties critical for modern applications.

13.2.1 Growth Modes of SAMs

The focus of earlier AFM studies beyond routine observation of surface morphologies was on the understanding of modes of adsorption, nucleation, and kinetics of growth of organic SAMs on different substrates by applying high-resolution AFM imaging directly in a liquid cell or to dried SAMs in a stop-and-go approach. Both thiol- and silane-based SAMs at different stages of their formation have been observed not only in the dry state after the interruption of growth at different stages but also directly in solution in the actual process of their growth. Excellent reviews that summarize these earlier studies have been published by Schwartz and Schreiber and only a brief summary of these results will be presented here [71, 72]. In later sections devoted to specific types of SAMs, the recent results on the growth of these surface layers will be highlighted as appropriate.

To date, numerous AFM studies conducted mostly on simple, long-chain thiol and silane SAMs have revealed different modes of organic monolayer growth, depending on the concentration of solution, the type of deposition (vapor or wet), and the nature of substrates. AFM imaging was utilized to monitor the initial nucleation and formation of liquid and solid islands, their thickness, surface coverage, kinetics of coalescence, and point/linear defect formation and healing.

High-resolution AFM imaging revealed that for most SAMs, the initial island formation was followed by the gradual increase in their lateral dimensions. However, the intermediate surface morphologies observed with AFM at different growth stages also include a number of complex scenarios, such as the presence of a disordered organic phase, the formation of second organic layer, partial local dewetting, or the development of fractal-like boundaries.

Usually, a two-step growth process is observed by AFM with a fast first stage, followed by slow "annealing" of these surface monolayers. At this "slow" stage, monolayer islands coalesce with local defects migrating and healing. In many cases, the initial stages of SAM formation were shown to be described by usual Langmuir kinetics as directly determined from the surface coverage obtained from image analysis. However, some modifications of this initial growth mechanism occur when the diffusion-limited growth of organic monolayer islands becomes a predominant mode of growth, as was confirmed on many occasions by direct AFM visualization summarized in seminal comprehensive reviews [71, 72].

Virtually complete thiol SAMs were viewed by AFM as having overall smooth surface morphology in the later stages as a result of the coalescence of the initial individual islands. Occasional pinholes, patches, and linear defects observed with AFM were caused by corresponding defects in the substrates and collapsed and merged island boundaries [71]. It is worth noting that the quenching/drying of SAMs during the stop-and-go approach frequently utilized at earlier stages of AFM measurements could result in the modification of growth kinetics. Therefore, only *in situ* monitoring of the SAMs growth with liquid-cell AFM mode should be considered as a reliable method and should be implemented in all cases if technically possible.

In fact, the complex mode of SAM growth and changes in molecular conformation of molecules involved were confirmed by *in situ* time-dependent AFM imaging. For example, it has been observed that in the course of thiol SAM growth the mode of growth is significantly affected by the interaction of various surface domains with full *trans* conformation or a mixed *trans/gauche* conformation of aliphatic chains. Similar growth modes and surface morphologies were observed for silane SAMs, although a stronger contribution from a continuous disordered layer, which grows around crystalline islands under certain deposition conditions, was pointed out. The expansion of organic islands has also been observed as a mode of lateral growth of silane SAMs by Pillai and Pai [73].

13.2.2
Thiol SAMs

Thiols are capable of strong (although not covalent) bonding with a gold lattice and, as a result, might form uniform, smooth SAMs that can expand over large microscopic surface areas [74]. These smooth and relatively stable thiol monolayers with tailored surface functionalities are widely utilized for various bio-related and electrochemical applications, and their compositions and surface morphologies are critical factors in their performance.

A series of alkylthiol SAMs with the number of carbon groups in the aliphatic chain, n, ranging from 1 to 17 grown on (111) gold surfaces was studied by Alves *et al.* [75]. This study is an example of excellent quality and high-resolution AFM measurements that generated precise information about chain packing. Alves *et al.* revealed that for long aliphatic tails ($n > 4$), SAMs show very smooth surface morphology, with microroughness below 0.3 nm. Long-range crystalline order was also detected with high-resolution AFM on the surface of long-chain SAMs. Such imaging revealed an ordered lattice of the terminal methyl groups that formed a periodic pattern with hexagonal symmetry and 0.52 nm characteristic spacing in most cases. The inability of the AFM to image regular lattices for short-chain alkylthiols was also noted. This difference in molecular ordering was related to both high local deformation of the less densely packed organic monolayers and a possible disordered state of alkylthiol molecules with shorter chains. Furthermore, AFM imaging of some SAMs revealed a coexistence of local surface areas with different ordering and significantly smaller intermolecular spacing that can be associated with either different local packing of distorted chains or with the influence of the underlying gold crystal lattice.

Ordered SAMs serve as an excellent sample for pushing the AFM resolution [71]. The "sensitivity" of the AFM imaging to differences in the repulsive interactions at higher forces or attractive interactions at lower forces and the ability to measure true molecular order are still debated. The molecular-related interpretation of the contribution from local topography created by the alternating terminal groups and gaps between them at fine spatial scales comparable to molecular dimensions is also a controversial and open issue that should be resolved on a case by case basis. One particular problem that should always be considered for imaging alkylthiols on gold at the highest resolution is the inherently grainy surface of most regular gold coatings (except some special cases of single-crystal gold substrates), which limits local high-resolution AFM scans to 100 nm surface areas.

In more recent studies, AFM topographical imaging in conjunction with other complementary imaging and measuring modes is frequently employed for simple alkylthiol SAMs, as well as their exotic counterparts with unique properties. A recent example includes AFM and STM studies of SAMs from thiol-tethered porphyrins as potential candidates for electronic and optical nanoscale coatings conducted by Chan *et al.* [76]. The authors observed local clustering of organic molecules with 2–6 nm lateral dimensions in mixed SAMs with different porphyrin terminal groups. From imaging and measuring conductive properties, the planar arrangement of porphyrin molecules on SAMs has been suggested. It has been suggested that these molecules are predominantly located in the vicinity of defects and this phenomenon limits the tunneling ability of the SAMs – a common problem with organic conductive SAMs, even with high concentration of conjugated molecules and fragments.

Charge transport through dense organic SAMs fabricated from conjugated thiol compounds has been studied with conductive mode SPM and shear force measurements by Fan *et al.* [77]. The authors collected local I–V curves for different SAMs and concluded that the conductive mechanism of these SAMs is partially based on the resonance tunneling phenomenon. They observed the effect of partial charge storing within organic SAMs and related this phenomenon to the presence of some compositional elements and conjugated segments of specially designed SAMs.

Measurements of different surface properties were conducted for SAMs composed of organic molecules with different lengths of aliphatic chains and different types of terminal groups on larger spatial scales. Experimental routines for these measurements are firmly established and well known. These AFM measurements employ such popular modes of operation as friction and chemical force microscopies usually complemented by force spectroscopies and surface micromapping, as will be discussed in detail in following sections with several examples.

Indeed, even the initial attempts of FFM imaging of microprinted, binary alkylthiol SAMs showed significant contrast between different microscopic surface regions caused by differences in chemical composition, converted to dramatic variations of tip–surface interactions [78]. These differences can be easily utilized to visualize and characterize bicomponent SAMs with alternating hydrophobic–hydrophilic microscopic surface areas. Sharp boundaries between SAMs with different surface functionalities were observed with AFM even at intermediate magnifications, and it was shown that the well-defined grainy structure of the underlying gold substrate does not affect the phase separation of different thiol molecules initially imprinted by soft-lithography methods.

Similar differentiation of different organic phases can be obtained as mixed alkylthiol SAMs with random regions of phase separation of different thiol molecules. For instance, Ichii *et al.* demonstrated that submicron phase separation between monothiol and dithiol molecules within SAMs can be detected by conventional AFM topographical imaging and surface potential measurements [79]. The controlled growth of different thiol molecules on selected surface areas "shaved" with an AFM tip has been suggested as a means to conduct comparative studies of the friction properties of thiol SAMs with different terminal groups and the length of aliphatic chains [80]. In this study, Riet *et al.* explored with silicon AFM tip SAMs, with terminal groups ranging from mildly adhesive, such as CF_3, to highly adhesive, such as NH_2, groups. The authors confirmed again that different levels of friction response can be used for unambiguous identification of SAMs with different functionalities [80]. Moreover, AFM observations revealed that the nanografting of additional molecules within the preformed microscopic areas results in less defective monolayer surfaces.

13.2.3
Alkylsilane SAMs

Alkylsilane SAMs are another popular type of SAM that was introduced for an effective surface modification approach long ago [81]. However, unlike thiol molecules, alkylsilanes with mono-, di-, and tri-functionalized groups are capable of forming films grafted with firm covalent bonding to the surface hydroxyl groups of silicon oxides as well as many other oxides and mica (Figure 13.18). These organic SAMs are extremely stable at elevated temperature and under high mechanical stresses. Their high robustness facilitates a wide utilization in more demanding stress-generating applications such as microelectromechanical devices [10].

The surfaces of completely and densely grafted alkylsilane SAMs with methyl terminal groups on silicon wafers are usually very smooth (below 0.3 nm microroughness within a 1 μm^2 surface area), with only occasional holes and bumps observed at larger scales [39]. FFM images show homogeneous distributions of the friction response over surface areas several microns across, confirming extremely uniform monolayer surface composition. In contrast to thiol-based SAMs, alkylsilane SAMs are very stable and cannot be damaged under direct contact mode scanning with modest normal forces.

SAM thickness is routinely determined by a scratch test and is usually confirmed by ellipsometry and X-ray reflectivity. For instance, 2.0–2.2 nm thickness was determined for classical octadecyltrichlorosilane (OTS) SAMs, close to the expected thickness of monolayers estimated from molecular models with slightly tilted alkyl tails [82]. The thicknesses of SAMs were confirmed by spectroscopic ellipsometry, a common confirmation procedure conducted routinely for SAMs. Alkylsilane SAMs can also be fabricated with spin-casting technique and were confirmed to show extremely smooth and uniform surface morphology with well-packed crystalline domains and few defects [83].

In contrast, a light grainy surface topography is observed for alkylsilane SAMs with functionalized terminal groups. For these SAMs, local microroughness increases and can reach 0.3–0.7 nm, which is two to three times higher than that observed for regular SAMs but still low compared to conventional polymer brush layers [82]. Such microroughness can be comparable to SAM thickness and indicates substantial lateral phase separation that can occur in such monolayers due to additional strong interactions of terminal groups and mismatch of cross-sectional areas. It is especially visible for short-chain functionalized SAMs with thicknesses within the range 0.5–0.9 nm (see example for NH_2 and SO_3H SAMs in Ref. [82]).

A combined AFM, XPS, and ellipsometry study of important highly reactive functionalized alkylsilane SAMs was reported by Luzinov and coworkers [40, 84]. In this study, different epoxysilane monolayer films were prepared by adsorption from toluene solutions with 0.1–1% concentration of epoxysilane compound. For the extended deposition, the thickness of the epoxysilane SAM stabilizes and reaches 0.75–1.1 nm, which is close to the extended length of epoxysilane molecules estimated from molecular models (0.95 nm) indicating virtually vertical arrangement of molecules.

Topographical AFM images of these epoxysilane SAM films formed at various concentrations of epoxysilane solution and deposition times revealed that the films obtained from 0.1 and 1% solutions have very different surface morphologies, with SAMs from 1% solution showing multilayer island morphologies. The microroughness of the true epoxysilane monolayer formed from 0.1% solution is virtually constant after the first minute of deposition, reaching 0.3 nm within a 1 μm^2 surface area (Figure 13.19). The SAM microroughness is close but slightly higher than the roughness of the supporting silicon substrates (0.1 nm). This indicates that the grainy surface topography in Figure 13.19 is composed of areas with elevation fluctuations on the order of one bond in the molecular backbone, suggesting that the epoxysilanes may form uniform layers and possess molecularly smooth surfaces beyond the first

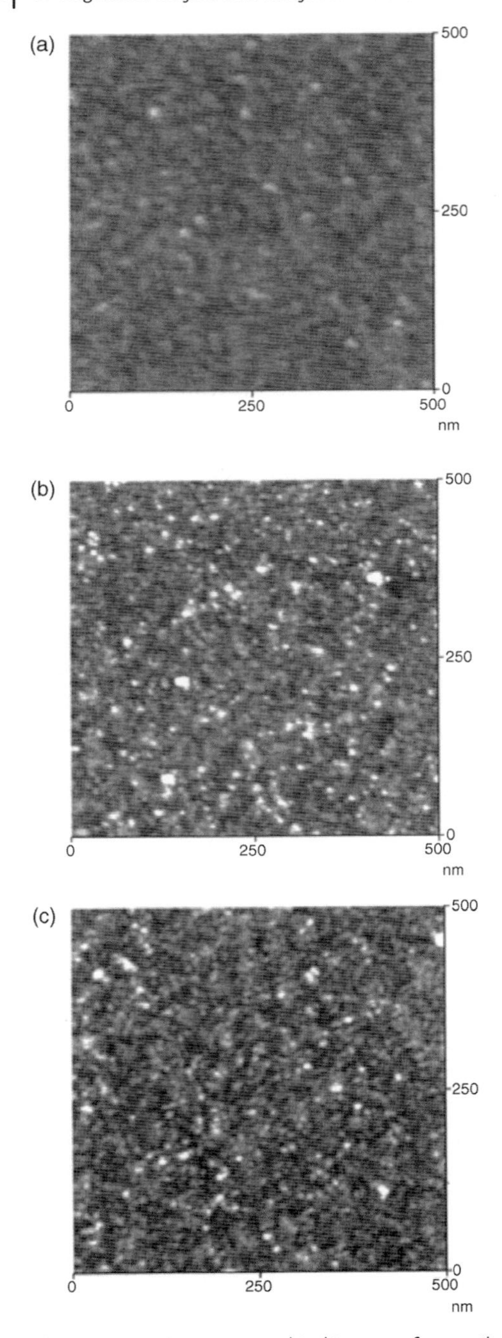

Figure 13.19 AFM topographical images of epoxysilane SAM after 1 min (a), 10 min (b), and 24 h (c) deposition times from 0.1% epoxysilane solution. Vertical scale is 4.0 nm. Reprinted from Ref. [70].

several minutes of formation. Experimental data for these deposition times were interpreted in terms of the later stages of coalescence and "healing" of the initial island microstructure. Surface aggregates can be also observed at certain deposition conditions, making these films heterogeneous and indicating different mechanisms of SAM formation that include concurrent solution and surface-located hydrolyzation and polymerization.

13.2.4
Nanotribological Studies

As mentioned previously, SAMs are considered to be an intriguing version of a robust molecular lubricant that can facilitate dramatic reduction in the frictional and adhesive forces, critical for controlling the interaction of mating surfaces. AFM and FFM are excellent tools for studying the interaction of a single asperity with carefully controlled normal and lateral forces and in a controlled environment, if needed. Along with earlier AFM reports on nanotribological properties of SAMs, it is important to mention studies of the utilization of friction and adhesive contrast of patterned SAMs as a base for CFM chemical force microscopy (CFM) by Noy *et al.* [85], studies of monotonic and nonmonotonic velocity- and humidity-dependent friction forces for a series of SAMs by Liu *et al.* [86], high-resolution studies of chain-dependent frictional properties of alkylthiols and alkylsilanes by Lio *et al.* [87], and systematic studies of SAMs with fluorinated terminal groups with higher friction forces caused by mismatch in geometric dimensions by Kim *et al.* [88, 89], among others.

Among recent studies, we would like to mention two reports on nanoscale adhesion and friction properties of different SAMs. Gojzewski *et al.* exploited dynamic force spectroscopy and other methods to study the role of humidity in nanoscale adhesion of thiol-based SAMs [90]. The authors combined STM, XRD, contact angle measurements, and AFM to separate contributions of different factors in SAM response. STM on robust SAM specimens (with longer alkyl chains) revealed characteristic pinhole defective surface areas and a highly ordered molecular-scale lattice, which was interpreted as a visualization of the terminal methyl groups and was utilized to estimate surface areas occupied by molecules within the SAMs. Besides random site-to-site and temporal fluctuations, a symmetric distribution of adhesive forces was recorded for different loading rates with very little dependence on the absolute value of the mechanical load. Two different regimes in unloading behavior have been observed and associated with the coexistence of two energy barriers for separation of interacting surfaces.

The first regime has been assigned to conventional van der Waals interactions and the second regime was speculated to be controlled by terminal group-specific responses of SAMs. The authors observed that the presence of hydroxyl surface groups results in a dramatic increase in adhesive forces due to the expected formation of hydrogen bonds as well as growing contributions of capillary forces caused by increasing humidity. It is interesting to note that the maximum increase in adhesive forces was observed for intermediate humidities (40–80%). The authors suggested that the decreasing adhesion at higher humidity can be related to the interplay between surface roughness and the small radius of curvature of the AFM tip.

Microtribological properties of SAMs and SAM-based surface layers have recently been studied with combined AFM and XPS analyses by Mo *et al.* [91]. The authors observed that initially nonhomogeneous surfaces of 3-aminopropyltriethoxysilane (APTS) (due to localized polymerization processes as discussed previously) SAMs become more uniform and smooth after grafting additional polymer layers. The presence of the amine surface groups with strong affinity to hydroxyl-terminated silicon oxide groups of the AFM tip resulted in some decreases in the frictional forces with further significant decreases observed for polymer–SAM surface layers. Composite SAMs showed excellent wear resistance after multiple AFM scans with high normal forces, which is much higher than that observed for regular SAMs.

In fact, the increasing wear resistance of the composite SAMs was revealed not only in AFM studies but also under conditions of the mesoscale shearing contact in microtribological measurements [92]. These tests showed decreasing frictional forces after a silicon oxide surface was modified with alkylsilane SAMs and further decreases were observed after additional grafting with polymerized surface layers of different types. In another study, Ahn *et al.* demonstrated that under high loading conditions, the wear resistance is controlled by the ability of the SAM surface to self-heal and restore itself after load release rather than by direct elastic resistance of the surface [93].

Tribological properties of molecular films composed of a fullerene monolayer chemically attached to the functional surface of SAMs (C_{60}-SAM) were studied with friction force microscopy by Tsukruk *et al.* (Figure 13.18) [70]. The authors observed a very high wear stability of the composite fullerene films and highly ordered packing of tethered fullerene molecules. The friction coefficient for these films varies widely from 0.04 at high loads and 0.06 at the highest velocities tested to 0.15 at intermediate velocities and low loads (see corresponding variation of friction forces in Figure 13.20). This nonmonotonic velocity behavior is a striking feature of fullerene-terminated SAM films compared to the steadily rising friction forces for conventional alkylsilane CH_3-terminated SAMs and may be related to the dissipation of the mechanical energy during structural rearrangements of the topmost fullerene molecular layer under high shear forces.

The loading and unloading curves for fullerene-terminated SAMs coincide within the experimental uncertainties, as demonstrated in Figure 13.20. No hysteresis behavior or signs of wear were observed for CH_3-SAMs up to the highest loads and after hundreds of sliding cycles, indicating that the chemically attached SAMs are very robust and stable and display only elastic deformation under high shear forces. The friction forces are higher for the SAMs with terminal azide groups than that for conventional methyl-terminated SAMs, as can be derived from the slope of the loading curves (Figure 13.20). Indeed, the friction coefficient evaluated from loading curves by a linear approximation was 0.018 and 0.04 for CH_3-SAM and N_3-SAMs, respectively, which are typical values for regular and functionalized alkylsilane SAMs.

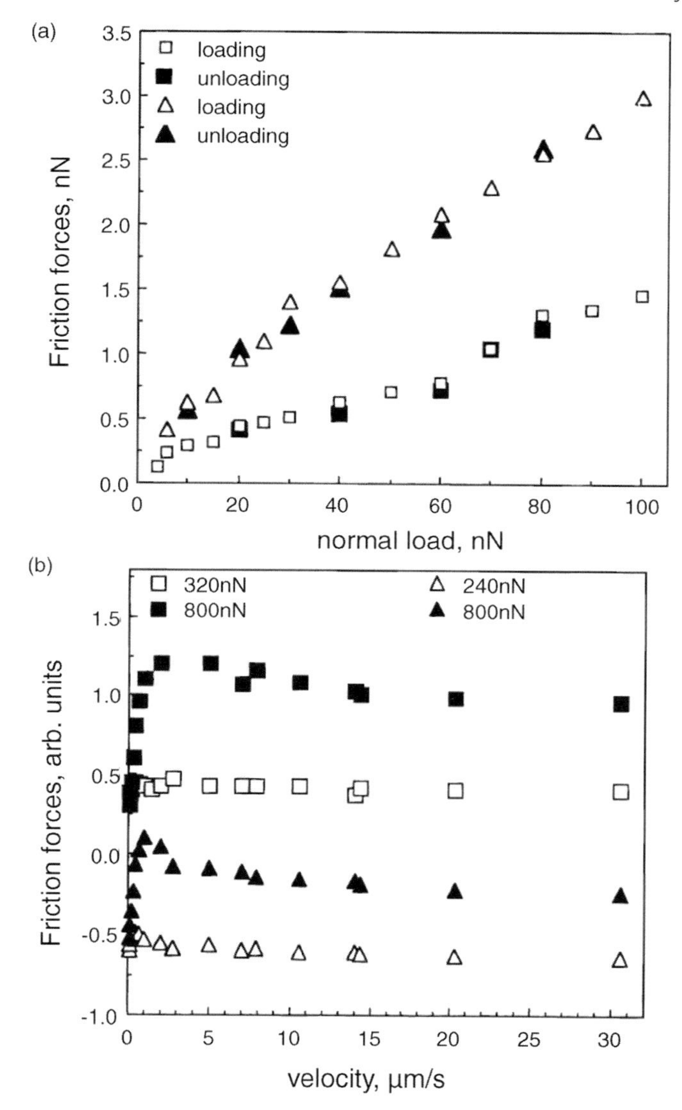

Figure 13.20 (a) Loading and unloading curves for CH_3-SAM (squares) and N_3-SAM (triangles) layers. (b) Friction forces versus the sliding velocity for C_{60}-SAM films (squares and triangles correspond to two independently prepared films) at two different loads. Reprinted from Ref. [70].

13.2.5 Adsorption Control with Surface Modifications

The unique ability of SAMs to dramatically alter adhesive, friction, and elastic properties of selected substrates has been widely utilized to control selective adsorption of synthetic and biological molecules. Here, we will present several recent examples of corresponding studies in which critical information has been obtained by applying AFM.

Uniform and patterned SAMs and active grafted surface layers are widely used to control protein and synthetic macromolecule adsorption, which can be monitored with AFM tip [94]. For instance, biotin and streptavidin adsorption on mixed and PEG-modified SAMs was monitored with AFM and reported by Anderson *et al.* and Kim *et al.* [95, 96]. Significant reduction in the nonspecific adsorption of proteins on PEG surfaces was confirmed with AFM imaging, where occasional 5 nm bumps were assigned mostly to individual biomolecules and surface domains of mixed SAMs [95].

Servoli *et al.* analyzed adsorption of albumin, fibrinogen, and fibronectin on hydrophobic and hydrophilic SAMs by applying antibody-modified AFM tips to prevent adsorption of biomolecules, a common problem in AFM studies of biomolecules [97]. Another important aspect of this study combined phase-contrast AFM imaging with quantitative image analysis and surface plasmon resonance (SPR) measurements, allowing both real-time monitoring of the adsorption process and a direct study of the competitive protein adsorption on SAMs. SPR measurements helped to quantify AFM data and revealed complex relationships between protein adsorption on mixed SAMs and nonlinear adsorption dynamics from multicomponent solutions.

Another example of an AFM study of modified gold surfaces with different functionalized alkylthiol SAMs followed by adsorption of protein A to immobilize immunoglobulin was reported by Briand *et al.* [98]. For these relatively uniform and smooth SAMs, homogeneous adsorption of protein A was observed when one-component homogeneous SAMs were utilized. In contrast, significant aggregation of protein was detected on mixed SAMs.

The analysis of adhesive properties of mixed SAMs in the course of protein adsorption was used to monitor the adsorption process and its intensity in conjunction with the nature of the modified surfaces [95]. Functionalized alkylthiol SAMs were prepared on different substrates to control neuronal cell growth with AFM, and complementary vibrational spectroscopy was utilized to monitor adsorption and binding [99]. Amino-terminated SAMs were observed to be the most efficient surfaces for cell attachment. It has been concluded that the surface microroughness of the underlying metal substrates, which approaches 2 nm, did not play a critical role in the protein adsorption.

The structure of human serum albumin protein adsorption onto well-characterized SAM surfaces at several protein concentrations was studied with combined AFM and X-ray reflectivity by Sheller *et al.* [100]. The duration of deposition was also varied to investigate the influence of the density of grafted hydrocarbon chains in the SAM on protein binding tenacity. A concurrent study of the adsorption to bare silicon wafers with a native oxide surface provided a comparison with a hydrophilic surface similar to widely studied glass and quartz substrates.

In this study, both AFM and X-ray reflectivity measurements showed that after adsorption, rinsing, and drying, the surfaces of all SAM substrates were covered with no more than a single layer of adsorbed protein with incomplete, fractal-like

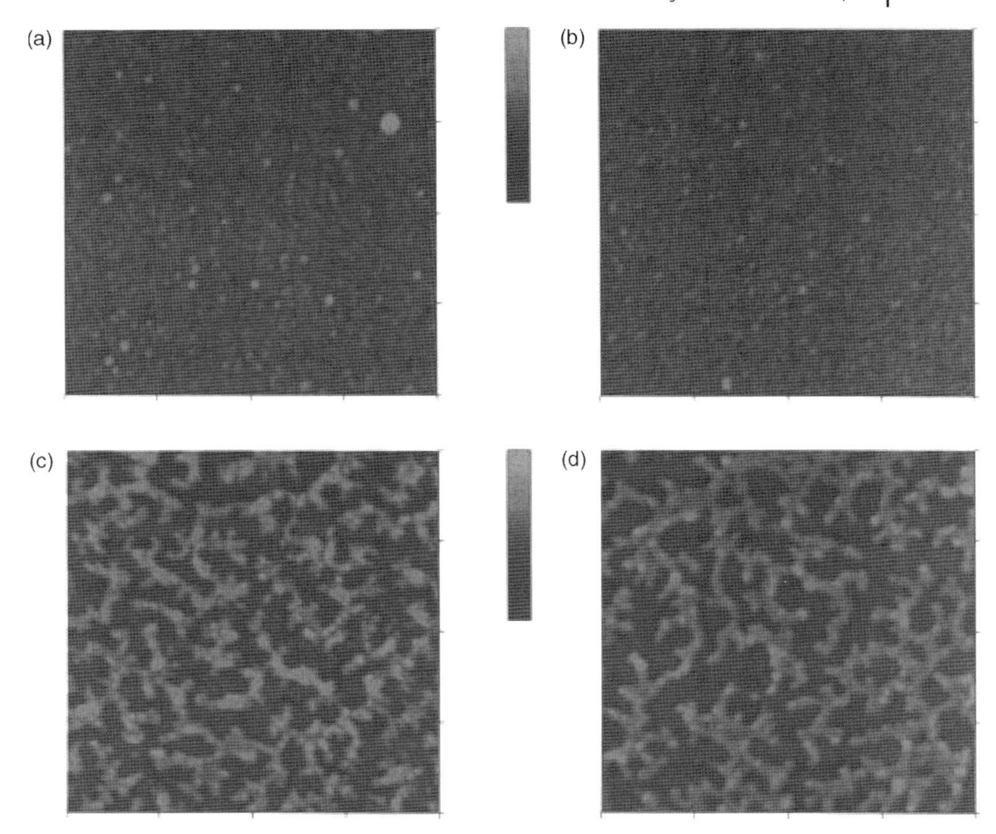

Figure 13.21 AFM images (1 x 1 μm)of substrates covered with HTS SAMs with albumin layers adsorbed at different bulk protein concentrations: (a) 0.5 mg/ml; (b) 0.1 mg/ml; (c) 0.05 mg/ml; (d) 0.01 mg/ml. Reprinted from Ref. [100].

morphology of adsorbed proteins (Figure 13.21). Ultrathin, dense protein layers were seen for the substrates exposed to protein concentrations of 0.1 and 0.5 mg/ml. For higher protein concentrations, the AFM measurements revealed uniform protein layers without pinhole defects. The layer thicknesses were essentially the same for different higher concentrations (1.6 nm thickness with microroughnesses of 0.3 nm), with a good correlation between AFM and X-ray reflectivity data.

AFM images provided a means to directly measure the local thicknesses of the albumin layers (about 3.6 nm). The average thicknesses derived from the X-ray reflectivity data were close to those obtained by AFM imaging. In this study, the tenacity of the protein adsorption on different substrates was also tested by eluting the adsorbed protein with a 1% solution of sodium dodecyl sulfate (SDS) surfactant. This treatment removed almost all protein from the bare silicon surface and from the fully formed, dense SAMs. The AFM image shows that only a few HSA molecules remain on the surface of silicon wafers after this treatment, while comparably large fraction of HSA remains on the surface of the incomplete SAM.

13.3
Adsorbed Macromolecules on Different Substrates

Visualization of long-chain molecules at different surfaces has been an important application of SPM since the first astonishing STM images of alkanes on graphite revealed atomistic details of chain conformations and configurations, molecular ordering including chain end packing and lateral correlations, static and dynamic defects, and anisotropic chain mobilities. All these results were presented and summarized in several fundamental publications, and these STM studies will be not discussed here [101, 102].

In further developments, high-resolution AFM imaging has also been demonstrated to be capable of producing near-molecular resolution images of short- and long-chain molecules on different (not just conductive) substrates, extending the applicability of this technique to a wide variety of surface-specific adsorbed macromolecules [103]. In this section, we briefly refer to a select few AFM studies of classical alkanes. We mostly concentrate on recent studies of adsorbed macromolecules with emphasis on the surface-controlled conformation, determination of molecular weight characteristics, and observation of local and global dynamic behavior as monitored by real-time AFM imaging.

Adsorbed biological macromolecules monitored in the dry state, under wet conditions, and directly in solution have also been widely imaged with AFM for the purposes of revealing their overall conformation and configuration, domain structures, fibrillar morphology, or shape persistence or collapse under variable external conditions but will be not discussed here (see summaries in Refs [104, 105] and an example in the previous section). More examples of adsorbed macromolecules with specific configurations (stars, dendrimers, or block copolymers) can be found in chapters devoted to different classes of macromolecular materials.

13.3.1
Short-Chain Linear Molecules

Numerous examples of AFM applications to study modestly long-chain alkane and related molecules with different chain lengths and configurations obtained under different adsorption conditions have been provided in a seminal book authored by Magonov and Whangbo and a review by Sheiko and Moller [106, 107]. An example of how careful analysis of the geometrical features should be applied to avoid conventional alkane film imaging artifacts can be found in a recent paper published by Bai *et al.* [108].

In another recent study, Magonov and Yerina reported unique high-temperature AFM studies of $C_{60}H_{122}$ alkane chains forming ordered films on graphite surface [109]. The ultrathin films were prepared by spin casting onto graphite. The surface morphology of the formed films and the ordering of first epitaxially adsorbed monolayer were studied at temperatures ranging from room temperature to greater than 150 °C.

On a large scale, AFM images at room temperature showed large flat crystals with multiple steps of 3–5 nm height related to regions where a different number of alkane monolayers stacked during crystallization. High-resolution AFM imaging on some selected areas of these alkane films revealed well-organized stripes with spacing of 7.5 nm, corresponding to the extended length of long-chain alkane molecules in normal packing. The borders between stripes are formed by terminal methyl groups with different heights. Orientation of these molecular stripes varied in different regions with discrete steps of 60°, implying that epitaxial crystallization is controlled by the underlying graphite lattice with threefold symmetry.

Significant changes in surface morphology were observed during gradual heating. Partial melting of surface crystals was observed for temperatures exceeding the nominal bulk melting temperature, with only a true monolayer (0.5 nm thick) still remaining epitaxially ordered at higher temperatures. This epitaxial layer is much more stable than the bulk crystals and remains ordered at temperatures up to 140 °C. At slightly lower temperatures (130–140 °C), occasional reorientations of domains with uniform molecular stripe orientation within domains were observed due to the high mobility of the molecules.

As the chemical compositions of long-chain molecules moves toward higher complexities, their ability to form a wide variety of surface morphologies is greatly enhanced, and a number of unique molecular structures can be observed. Some morphologies resemble those reported for conventional block copolymers (see Chapter 12). For instance, Mourran *et al.* reported AFM images of perfluoroalkyl alkanes ($F(CF_2)_{14}(CH_2)_{20}H$) that formed ribbons, spiral, or toroidal surface structures with an effective width of several tens of nanometers [110]. The type of surface structures spontaneously formed on different surfaces depends on the type of selective solvent used to fabricate these surface films. The height of ordered domains have been measured with AFM and verified by X-ray reflectivity as being much smaller than the extended length of the molecules, leading the authors to suggest that fluorinated chains are oriented normal to the surface with alkyl chains oriented at a certain angle. The authors further suggest that such a two-level arrangement allows dense packing of alkane backbones with smaller cross-sectional areas. Therefore, the overall mismatch compensation mechanism was suggested to be responsible for uniform lateral dimensions of different morphological features observed in the dual segmented molecules.

13.3.2 Long-Chain Macromolecules

Long-chain, flexible macromolecules adsorbed on different surfaces have been widely studied with AFM, given that their interaction with the underlying substrate is strong enough to avoid significant surface diffusion during imaging. Resistance to shear stresses and possible displacements with the AFM tip is also important for stable imaging of adsorbed molecules. Specifically, polyelectrolytes with strong multiple interactions of molecular segments with proper surface functionalities and easily observable conformational changes were the subject of numerous AFM studies, as reviewed recently by Minko and Roiter [111]. Such an approach facilitates direct measurements of molecular weight characteristics and molecular defects.

The high molecular weight of long-chain macromolecules causes their surface mobility to be dramatically lower, thus enabling AFM imaging with high resolution required to monitor individual molecules under different conditions with molecular resolution. Stretching and aligning of macromolecules by spin casting and coadsorption of various surface-active molecules is considered an efficient way to tune molecule–surface interactions and allow higher contrast of molecular chains during AFM scanning [112].

In another approach, adding a thin water layer caused by increasing humidity was also observed to be an efficient method of controlling surface mobility of adsorbed macromolecules and obtaining images of single linear polymer backbones. In this way, the reptation motion of poly(methyl methacrylate) (PMMA) macromolecules on solid substrates was initiated and monitored by Kumaki *et al.* [113]. The authors suggested that both lateral and vertical partial folding of flexible chains caused by such a snake-like motion provides a means for larger-scale chain motion detectable with AFM. Considering that the AFM tip can affect the molecular motion observed, the authors conducted additional tests and suggested that the tip contribution in their study was minimal due to strong molecular interactions with the surface.

Recently, AFM imaging has been employed for revealing the conformation of single polymer chains directly in fluid. For instance, by using light tapping mode ($r_{sp} = 0.98$) imaging in fluid under controlled pH, Roiter and Minko observed the conformation changes in poly(2-vinyl pyridine) (P2VP) chains adsorbed on atomically flat mica substrates [114]. P2VP chains exhibited a sharp globule to coil transition in a narrow pH range (from 4.04 to 3.8). Analysis of the AFM images revealed that the protonation of the P2VP chains with change in pH dramatically altered the end-to-end distance and the radius of gyration of adsorbed chains.

13.3.3 Brush-Like Macromolecules

Visualizations of flexible cylindrical brush macromolecules with different compositions and architectures adsorbed on different surfaces have been extensively conducted by Sheiko and collaborators, as summarized in their reviews [107, 115].

The relatively large cross section of these macromolecules and low surface mobility due to high molecular weight and brush structure made their direct imaging achievable under ambient conditions in contrast to imaging of regular flexible linear chains (Figure 13.22). Not only can dense monolayers be visualized, but also backbones of individual molecules can be tracked within these films and inside individual domains as demonstrated by Sheiko *et al.* (Figure 13.22). Not only the overall shape of macromolecules as individual objects can be tracked in densely packed films, but also their local radii of curvature, state and conformation of side chains, overall molecular weight, and the presence of chemical defects can be readily evaluated and quantified with high-resolution AFM imaging under ambient conditions.

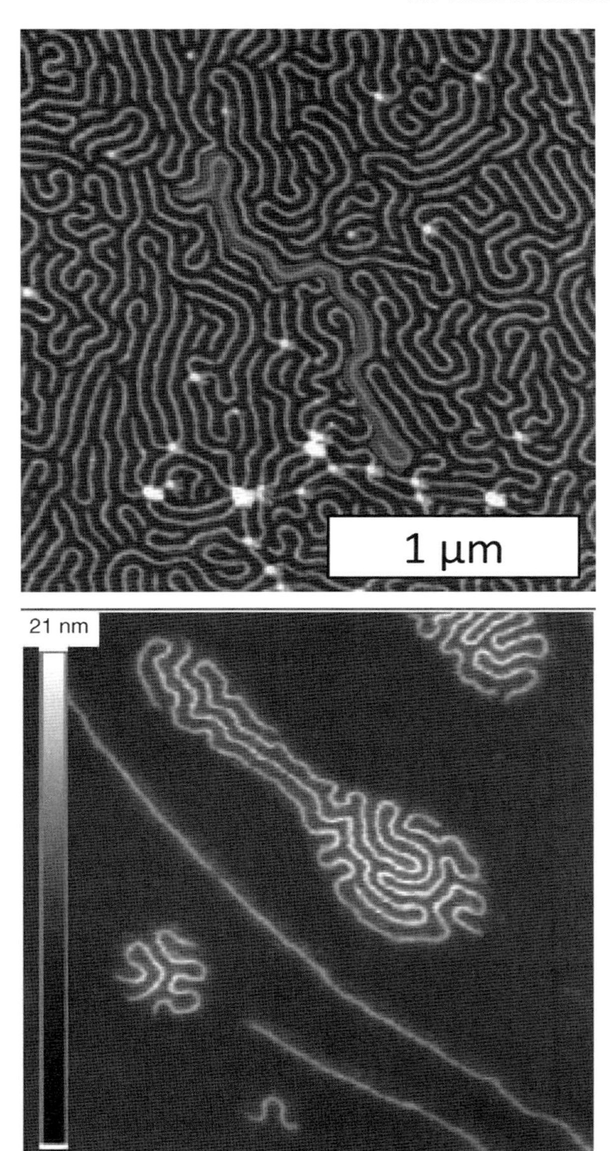

Figure 13.22 Brush-like macromolecules with PBA side chains are densely grafted to a polymethacrylate backbone. Backbone degree of polymerization $N = 3600$, a side chain degree of polymerization and $n = 58$, PDI $= M_w/M_n = 1.6$: dense monolayer and single molecules and clusters on mica. Courtesy of S. Sheiko, K. Matyjaszewski, A. Nese, N. Erina, and C. Su.

More specifically, the role of confinement of brush macromolecules to a flat surface on the spontaneous curvature of backbones, direct methods of measuring the absolute molecular weight, and the accurate visualization of individual and densely packed, multiarmed brush molecules have been studied with high-resolution AFM imaging [116–118].

High-resolution AFM imaging within a relatively short time frame (seconds to minutes) also opened the possibility of addressing questions of the dynamic behavior of these macromolecules in a low mobility situation. True real-time AFM imaging was exploited for observation of conformational transitions of macromolecules by Gallyamov *et al.* [119]. Transitions from an extended worm-like conformation to a globular conformation and vice versa have been observed as a result of changing conditions for adsorption. Coadsorption of small molecules was shown to be critical for controlling this transition, observed to occur via reptation-like motion.

As noted by the authors, although the role of the AFM tip was significant in both dilation of the backbone shape and the tip-induced directional molecular motion by added shear forces, careful scanning with low forces can be conducted largely without disturbing the original conformational state. The authors emphasized a key role of the proper choice of parameters for unambiguous visualization, such as high molecular weight macromolecules for adsorption and careful choice of proper "sticky" substrates [120].

In another advanced study, static images and also flow-induced molecular motion of brush molecules were directly observed with real-time AFM by Xu *et al.* [121]. Material flow observed with AFM was induced by spreading a drying solution droplet on graphite and mica surfaces with thickness reduced down to a monolayer with constant monitoring of the progressing front with AFM (Figure 13.23). Tracing the flow front development at a large scale allows the monitoring of the overall front

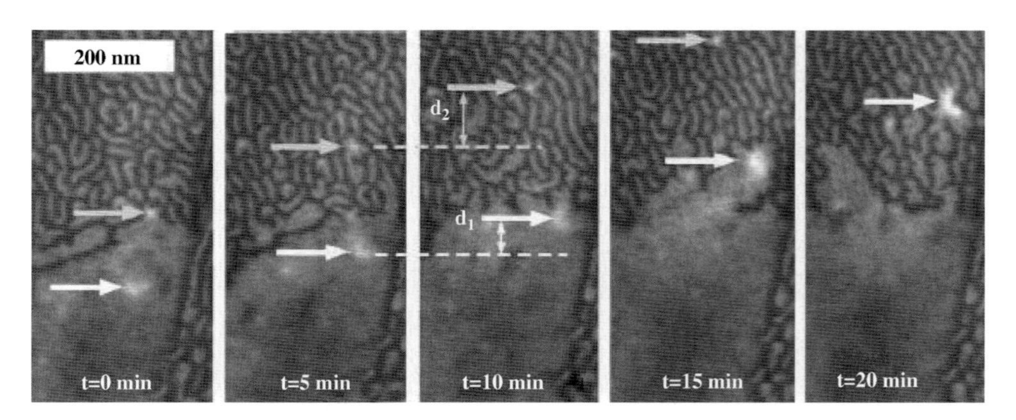

Figure 13.23 AFM monitoring the development of the fingering instability on the molecular length scale with a temporal resolution of 5 min. The height micrographs display one of the most silent features of the fingering instability; that is, molecules in phase 1 (higher pressure) switch their conformation while leaving phase 1 to phase 2 (lower pressure). The light elevated spots due to intercrossed molecules facilitate the observation. Reprinted from Ref. [120].

shape and spreading layer thickness. On the other hand, concurrent real-time imaging of selected surface areas on a molecular level revealed specific molecular landscapes around individual fingering instabilities.

Continuous imaging demonstrated that flow instabilities are triggered by conformational transitions of the molecules in the vicinity of the moving front, as can be tracked in reference to characteristic molecular features shown in Figure 13.23. Interfacial pressure gradients combined with abrupt changes in the molecular friction in the vicinity of the surface defects were suggested to be responsible for destabilizing phase boundary and promoting flow of material. Such unique, high-resolution, real-time AFM measurements of polymer dynamics concurrently on multiple length scales allows the elucidation of mesoscopic material behavior and underlying mechanisms of the molecular motion at different length scales that facilitate this behavior.

References

1 Israëls, R., Leermakers, F.A.M., Fleer, G.J., and Zhulina, E.B. (1994) Charged polymeric brushes: structure and scaling relations. *Macromolecules*, **27** (12), 3249–3261.

2 Lee, M., Kim, J.-W., Yoo, Y.-S., Peleshanko, S., Larson, K., Vaknin, D., Markutsya, S., and Tsukruk, V.V. (2002) Amphiphilic hairy disks with branched hydrophilic tails and a hexa-*peri*-hexabenzocoronene core. *J. Am. Chem. Soc.*, **124** (31), 9121–9128.

3 Takei, Y.G., Aoki, T., Sanui, K., Ogata, N., Sakurai, Y., and Okanao, T. (1994) Dynamic contact angle measurement of temperature-responsive surface properties for poly(*N*-isopropylacrylamide) grafted surfaces. *Macromolecules*, **27** (21), 6163–6166.

4 Hall, R., Hara, M., and Knoll, W. (1996) Isomerization and acid–base behavior in polyion complex Langmuir–Blodgett films. *Langmuir*, **12** (10), 2551–2555.

5 Dante, S., Advincula, R., Frank, C.W., and Stroeve, P. (1999) Photoisomerization of polyionic layer-by-layer films containing azobenzene. *Langmuir*, **15** (1), 193–201.

6 Ichimura, K., Oh, S.-K., and Nakagawa, M. (2000) Light-driven motion of liquids on a photoresponsive surface. *Science*, **288** (5471), 1624–1626.

7 Siewierski, L.M., Brittain, W.J., Petras, S., and Foster, M.D. (1996) Photoresponsive monolayers containing in-chain azobenzene. *Langmuir*, **12** (24), 5838–5844.

8 Auroy, P., Auvray, L., and Léger, L. (1991) Characterization of the brush regime for grafter polymer layers at the solid–liquid interface. *Phys. Rev. Lett.*, **66** (6), 719–722.

9 Raviv, U., Tadmor, R., and Klein, J. (2001) Shear and frictional interactions between adsorbed polymer layers in a good solvent. *J. Phys. Chem. B*, **105** (34), 8125–8134.

10 Tsukruk, V.V. (2001) Molecular lubricants and glues for micro- and nanodevices. *Adv. Mater.*, **13** (2), 95–108.

11 Pincus, P. (1991) Colloid stabilization with grafted polyelectrolytes. *Macromolecules*, **24** (10), 2912–2919.

12 Galaev, I.Y. and Mattiasson, B. (1999) 'Smart' polymers and what they could do in biotechnology. *Trends Biotechnol.*, **17** (8), 335–340.

13 Aksay, I.A., Trau, M., Manne, S., Honma, I., Yao, N., Zhou, L., Fenter, P., Eisenberger, P.M., and Gruner, S.M. (1996) Biomimetic pathways for assembling inorganic thin films. *Science*, **273** (5277), 892–898.

14 Ito, Y., Ochiai, Y., Park, Y.S., and Imanishi, Y. (1997) pH-sensitive gating by conformational change of a polypeptide brush grafted onto a porous polymer membrane. *J. Am. Chem. Soc.*, **119** (7), 1619–1623.

15 Ornatska, M., Jones, S.E., Naik, R.R., Stone, M., and Tsukruk, V.V. (2003) Biomolecular stress-sensitive gauges: surface-mediated immobilization of

mechanosensitive membrane protein. *J. Am. Chem. Soc.*, **125** (42), 12722–12723.

16 Blossey, R. (2003) Self-cleaning surfaces – virtual realities. *Nat. Mater.*, **2** (5), 301–306.

17 Lafuma, A. and Quere, D. (2003) Superhydrophobic states. *Nat. Mater.*, **2** (7), 457–460.

18 Léger, L., Raphaël, E., and Hervert, H. (1999) Surface-anchored polymer chains: their role in adhesion and friction. *Adv. Polym. Sci.*, **138**, 185–225.

19 Bliznyuk, V.N., Everson, M.P., and Tsukruk, V.V. (1998) Nanotribological properties of organic boundary lubricants: Langmuir films versus self-assembled monolayers. *J. Tribol.*, **120** (3), 489–495.

20 Ruths, M., Johannsmann, D., Rühe, J., and Knoll, W. (2000) Repulsive forces and relaxation on the compression of entangled, polydisperse polystyrene brushes. *Macromolecules*, **33** (10), 3860–3870.

21 Mansky, P., Liu, Y., Huang, E., Russell, T.P., and Hawker, C. (1997) Controlling polymer–surface interactions with random copolymer brushes. *Science*, **275** (5305), 1458–1460.

22 Belge, G., Beyerlein, D., Betsch, C., Eichhorn, K., Gauglitz, G., Grundke, K., and Voit, B. (2002) Suitability of hyperbranched polyester for sensoric applications – investigation with reflectometric interference spectroscopy. *Anal. Bioanal. Chem.*, **374** (3), 403–411.

23 de Gennes, P.G. (1980) Conformations of polymers attached to an interface. *Macromolecules*, **13** (5), 1069–1075.

24 Halperin, A., Tirrell, M., and Lodge, T.P. (1992) Tethered chains in polymer microstructures. *Adv. Polym. Sci.*, **100** (1), 31–71.

25 Singh, C., Pickett, G.T., Zhulina, B., and Balazs, A.C. (1997) Modeling the interactions between polymer-coated surfaces. *J. Phys. Chem. B*, **101** (50), 10614–10621.

26 Singh, C., Pickett, G.T., and Balazs, A.C. (1996) Interactions between polymer-coated surfaces in poor solvents. 1. Surfaces coated with A and B homopolymers. *Macromolecules*, **29** (23), 7559–7567.

27 Klein, J. (1996) Shear, friction, and lubrication forces between polymer-bearing surfaces. *Annu. Rev. Mater. Sci.*, **26**, 581–612.

28 Karim, A., Tsukruk, V.V., Douglas, J.F., Satija, S.K., Fetters, L.J., Reneker, D.H., and Foster, M.D. (1995) Self-organization of polymer brush layers in a poor solvent. *J. Phys. II*, **5** (10), 1441–1456.

29 Zhao, B. and Brittain, W.J. (2000) Polymer brushes: surface-immobilized macromolecules. *Prog. Polym. Sci.*, **25** (5), 677–710.

30 Zhao, B. and Brittain, W.J. (1999) Synthesis of tethered polystyrene-*block*-poly(methyl methacrylate) monolayer on a silicate substrate by sequential carbocationic polymerization and atom transfer radical polymerization. *J. Am. Chem. Soc.*, **121** (15), 3557–3558.

31 Zhao, B., Brittain, W.J., Zhou, W., and Cheng, S.Z.D. (2000) Nanopattern formation from tethered PS-*b*-PMMA brushes upon treatment with selective solvents. *J. Am. Chem. Soc.*, **122** (10), 2407–2408.

32 Fredrikson, G.H., Ajdari, A., Leibler, L., and Carton, J.P. (1992) Surface modes and deformation energy of a molten polymer brush. *Macromolecules*, **25** (11), 2882–3288.

33 Minko, S., Patil, S., Datsyuk, V., Simon, F., Eichorn, K.J., Motornov, M., Usov, D., Tokarev, I., and Stamm, M. (2002) Synthesis of adaptive polymer brushes via "grafting to" approach from melt. *Langmuir*, **18** (1), 289–296.

34 Luzinov, I., Julthongpiput, D., Malz, H., Pionteck, J., and Tsukruk, V.V. (2000) Polystyrene layers grafted to epoxy-modified silicon surfaces. *Macromolecules*, **33** (3), 1043–1048.

35 Luzinov, I., Julthongpiput, D., and Tsukruk, V.V. (2000) Thermoplastic elastomer monolayers grafted to a functionalized silicon surface. *Macromolecules*, **33** (20), 7629–7638.

36 Folkers, J.P., Gorman, C.B., Laibinis, P.E., Buchholz, S., and Whitesides, G.M. (1995) *Langmuir*, **11** (3), 813–824;Lee, T.R., Carey, R.I., Biebuyck, H.A., and Whitesides, G.M. (1994) The wetting of monolayer films exposing ionizable acids and bases. *Langmuir*, **10** (3), 741–749.

37 Biebuyck, H.A., Bain, C.D., and Whitesides, G.M. (1994) Comparison of organic monolayers on polycrystalline gold spontaneously assembled from solutions containing dialkyl disulfides or alkanethiols. *Langmuir*, **10** (6), 1825–1831.

38 Ulman, A. (1996) Formation and structure of self-assembled monolayers. *Chem. Rev.*, **96** (4), 1533–1554.

39 Xue, G., Koenig, J.L., Ishida, H., and Wheeler, D.D. (1991) Chemical reactions of resorcinol-formaldehyde latex and coupling agent. *Rubber Chem. Technol.*, **64** (2), 172–180.

40 Tsukruk, V.V., Lander, L.M., and Brittain, W.J. (1994) Atomic force microscopy of C_{60} tethered to a self-assembled monolayer. *Langmuir*, **10** (4), 996–999.

41 Tsukruk, V.V., Luzinov, I., and Julthongpiput, D. (1999) Sticky molecular surfaces: epoxysilane self-assembled monolayers. *Langmuir*, **15** (9), 3029–3032.

42 Luzinov, I., Julthongpiput, D., Liebmann-Vinson, A., Cregger, T., Foster, M.D., and Tsukruk, V.V. (2000) Epoxy-terminated self-assembled monolayers: molecular glues for polymer layers. *Langmuir*, **16** (2), 504–516.

43 Takei, Y.G., Aoki, T., Sanui, K., Ogata, N., Sakurai, Y., and Okanao, T. (1994) Dynamic contact angle measurement of temperature-responsive surface properties for poly(*N*-isopropylacrylamide) grafted surfaces. *Macromolecules*, **27** (21), 6163–6166.

44 Grest, G.S. and Murat, M. (1993) Structure of grafted polymeric brushes in solvents of varying quality: a molecular dynamics study. *Macromolecules*, **26** (12), 3108–3117.

45 Patra, M. and Linse, P. (2006) Reorganization of nanopatterned polymer brushes by the AFM measurement process. *Macromolecules*, **39** (13), 4540–4546.

46 Lemieux, M., Minko, S., Usov, D., Stamm, M., and Tsukruk, V.V. (2003) Direct measurement of thermo-elastic properties of glassy and rubbery polymer brushes grown by grafting from approach. *Langmuir*, **19**, 6126.

47 Kelley, T.W., Schorr, P.A., Johnson, K.D., Tirrell, M., and Frisbie, C.D. (1998) Direct force measurements at polymer brush surfaces by atomic force microscopy. *Macromolecules*, **31** (13), 4297–4300.

48 Chizhik, S.A., Huang, Z., Gorbunov, V.V., Myshkin, N.K., and Tsukruk, V.V. (1998) Micromechanical properties of elastic polymeric materials as probed by scanning force microscopy. *Langmuir*, **14** (10), 2606–2626.

49 Mason, R., Jalbert, C.A., O'Rourke Muisener, P.A.V., Koberstein, J.T., Elman, J.F., Long, T.E., and Gunesin, B.Z. (2001) Surface energy and surface composition of end-fluorinated polystyrene. *Adv. Colloid Interface Sci*, **94** (1–3), 1–19.

50 Israelachvili, J. (1992) *Intermolecular and Surface Forces*, Academic Press, San Diego.

51 Johnson, K.L. (1985) *Contact Mechanics*, Cambridge University Press, Cambridge.

52 Van Krevelen, D.W. and te Nijenhuis, K. (2009) *Properties of Polymers*, 4th edn, Elsevier, Amsterdam.

53 Yamamoto, S., Ejaz, M., Tsujii, Y., Matsumoto, M., and Fukuda, T. (2000) Surface interaction forces of well-defined, high-density polymer brushes studied by atomic force microscopy. 1. Effect of chain length. *Macromolecules*, **33** (15), 5602–5607.

54 Tsujii, Y., Nomura, A., Okayasu, K., Goa, W., Ohno, K., and Fukuda, T. (2009) AFM studies on microtribology of concentrated polymer brushes in solvents. *J. Phys. Conf.*, **184** (1), 012031.

55 Kessel, S., Schmidt, S., Müller, R., Wischerhoff, E., Laschewsky, A., Lutz, J.-F., Uhlig, K., Lankenau, A., Duschl, C., and Fery, A. (2010) Thermoresponsive PEG-based polymer layers: surface characterization with AFM force measurements. *Langmuir*, **26** (5), 3462–3467.

56 Lim, R.Y.H. and Deng, J. (2009) Interaction forces and reversible collapse of a polymer brush-gated nanopore. *ACS Nano*, **3** (10), 2911–2918.

57 Teng, J. and Zubarev, E.R. (2003) Synthesis and self-assembly of a heteroarm star amphiphile with 12

alternating arms and a well-defined core. *J. Am. Chem. Soc.*, **125** (39), 11840–11841.

58 Julthongpiput, D., Lin, Y.-H., Teng, J., Zubarev, E.R., and Tsukruk, V.V. (2003) Y-shaped polymer brushes: nanoscale switchable surfaces. *Langmuir*, **19** (19), 7832–7836.

59 Julthongpiput, D., Lin, Y.-H., Teng, J., Zubarev, E.R., and Tsukruk, V.V. (2003) Y-shaped amphiphilic brushes with switchable micellar surface structures. *J. Am. Chem. Soc.*, **125** (51), 15912–15921.

60 Lin, Y.-H., Teng, J., Zubarev, E.R., Shulha, H., and Tsukruk, V.V. (2005) In-situ observation of switchable nanoscale topography for Y-shaped binary brushes in fluids. *Nano Lett.*, **5** (3), 491–495.

61 Zhang, W. and Zhang, X. (2003) Single molecule mechanochemistry of macromolecules. *Prog. Polym. Sci.*, **28** (8), 1271–1295.

62 Muller, P., Sudre, G., and Théodoly, O. (2008) Wetting transition on hydrophobic surfaces covered by polyelectrolyte brushes. *Langmuir*, **24** (17), 9541–9550.

63 Vyas, M.K., Nandan, B., Schneider, K., and Stamm, M. (2008) Nanowear studies in reversibly switchable polystyrene–poly (acrylic acid) mixed brushes. *J. Colloid Interface Sci.*, **328** (1), 58–66.

64 Vyas, M., Schneider, K., Nandan, B., and Stamm, M. (2008) Switching of friction by binary polymer brushes. *Soft Matter*, **4** (5), 1024–1032.

65 Usov, D., Gruzdev, V., Nitschke, M., Stamm, M., Hoy, O., Luzinov, I., Tokarev, I., and Minko, S. (2007) Three-dimensional analysis of switching mechanism of mixed polymer brushes. *Macromolecules*, **40** (24), 8774–8783.

66 Motornov, M., Sheparovych, R., Katz, E., and Minko, S. (2008) Chemical gating with nanostructured responsive polymer brushes: mixed brush *versus* homopolymer brush. *ACS Nano*, **2** (1), 41–52.

67 Lemieux, M.C., Julthongpiput, D., Duc Cuong, P., Ahn, H.-S., Lin, Y.-H., and Tsukruk, V.V. (2004) Ultrathin binary grafted polymer layers with switchable morphology. *Langmuir*, **20** (23), 10046–10054.

68 Lemieux, M., Usov, D., Minko, S., Stamm, M., Shulha, H., and Tsukruk, V.V. (2003) Reorganization of binary polymer brushes: switching surface microstructures and nanomechanical properties. *Macromolecules*, **36** (19), 7244–7255.

69 Kumar, A., Biebuyck, H.A., and Whitesides, G.M. (1994) Patterning self-assembled monolayers: applications in materials science. *Langmuir*, **10** (5), 1498–1511.

70 Tsukruk, V.V., Everson, M.P., Lander, L.M., and Brittain, W.J. (1996) Nanotribological properties of composite molecular films: C_{60} anchored to a self-assembled monolayer. *Langmuir*, **12** (16), 3905–3910.

71 Schwartz, D.K. (2001) Mechanisms and kinetics of self-assembled monolayer formation. *Annu. Rev. Phys. Chem.*, **52**, 107–137.

72 Schreiber, F. (2000) Structure and growth of self-assembling monolayers. *Prog. Surf. Sci.*, **65** (5–8), 151–257.

73 Pillai, S. and Pai, R.K. (2009) Controlled growth and formation of SAMs investigated by atomic force microscopy. *Ultramicroscopy*, **109** (2), 161–166.

74 Nuzzo, R.G. and Allara, D.L. (1983) Adsorption of bifunctional organic disulfides on gold surfaces. *J. Am. Chem. Soc*, **105** (13), 4481.

75 Alves, C.A., Smith, E.L., and Porter, M.D. (1992) Atomic scale imaging of alkanethiolate monolayers at gold surfaces with atomic force microscopy. *J. Am. Chem. Soc.*, **114** (4), 1222–1227.

76 Chan, Y.-H., Schuckman, A.E., Pérez, L.M., Vinodu, M., Drain, C.M., and Batteas, J.M. (2008) Synthesis and characterization of a thiol-tethered tripyridyl porphyrin on Au(111). *J. Phys. Chem. C*, **112** (15), 6110–6118.

77 Fan, F.-R.F., Yang, J., Cai, L., Price, D.W., Dirk, S.M., Kosynkin, D.V., Yao, Y., Rawlett, A.M., Tour, J.M., and Bard, A.J. (2002) Charge transport through self-assembled monolayers of compounds of interest in molecular electronics. *J. Am. Chem. Soc.*, **124** (19), 5550–5560.

78 Wilbur, J.L., Biebuyck, H.A., MacDonald, J.C., and Whitesides, G.M. (1995) Scanning force microscopies can image patterned self-assembled monolayers. *Langmuir*, **11** (3), 825–831.

79 Ichii, T., Fukama, T., Kobayashi, K., Yamada, H., and Matsushige, K. (2003) Phase-separated alkanethiol self-assembled monolayers investigated by non-contact AFM. *Appl. Surf. Sci.*, **210** (1–2), 99–104.

80 Te Riet, J., Smit, T., Gerritsen, J.W., Cambi, A., Elemans, J.A.A.W., Figdor, C.G., and Speller, S. (2010) Molecular friction as a tool to identify functionalized alkanethiols. *Langmuir*, **26** (9), 6357–6366.

81 Ulman, A. (1991) *Introduction to Ultrathin Organic Films*, Academic Press, Boston.

82 Tsukruk, V.V. and Bliznyuk, V.N. (1998) Adhesive and friction forces between chemically modified silicon and silicon nitride surfaces. *Langmuir*, **14** (2), 446–450.

83 Ito, Y., Virkar, A.A., Mannsfeld, S., Oh, J.H., Toney, M., Locklin, J., and Bao, Z. (2009) Crystalline ultrasmooth self-assembled monolayers of alkylsilanes for organic field-effect transistors. *J. Am. Chem. Soc.*, **131** (26), 9396–9404.

84 Tsukruk, V.V., Luzinov, I., and Julthongpiput, D. (1999) Sticky molecular surfaces: epoxysilane self-assembled monolayers. *Langmuir*, **15** (9), 3029–3032.

85 Noy, I.A., Vezenov, D.V., and Lieber, C.M. (1997) Chemical force microscopy. *Annu. Rev. Mater. Sci.*, **27**, 381–421.

86 Liu, Y., Evans, D.F., Song, Q., and Grainger, D.W. (1996) Structure and frictional properties of self-assembled surfactant monolayers. *Langmuir*, **12** (5), 1235–1244.

87 Lio, A., Charych, D.H., and Salmeron, M. (1997) Comparative atomic force microscopy study of the chain length dependence of frictional properties of alkanethiols on gold and alkylsilanes on mica. *J. Phys. Chem. B*, **101** (19), 3800–3805.

88 Kim, H.I., Kioni, T., Lee, T.R., and Perry, S.S. (1997) Systematic studies of the frictional properties of fluorinated monolayers with atomic force microscopy: comparison of CF_3-, and CH_3-terminated films. *Langmuir*, **13** (26), 7192–7196.

89 Kim, H.I., Graupe, M., Oloba, O., Kioni, T., Imaduddin, S., Lee, T.R., and Perry, S.S. (1999) Molecularly specific studies of the frictional properties of monolayer films: a systematic comparison of CF_3-, $(CH_3)_2CH$-, and CH_3-terminated films. *Langmuir*, **15** (9), 3179–3185.

90 Gojzewski, H., Kappl, M., Ptak, A., and Butt, H.-J. (2010) Effect of humidity on nanoscale adhesion on self-assembled thiol monolayers studied by dynamic force spectroscopy. *Langmuir*, **26** (3), 1837–1847.

91 Mo, Y., Zhu, M., and Bai, M. (2008) Preparation of nano/microtribological properties of perfluorododecanoic acid (PFDA)–3-aminopropyltriethoxysilane (APS) self-assembled dual-layer film deposited on silicon. *Colloid Surf. A*, **322** (1–3), 170–176.

92 Tsukruk, V.V., Ahn, H.-S., Sidorenko, A., and Kim, D. (2002) Triplex molecular layers with nonlinear nanomechanical response. *Appl. Phys. Lett.*, **80** (25), 4825–4827.

93 Ahn, H., Julthongpiput, D., Kim, D.-I., and Tsukruk, V.V. (2003) Dramatic enhancement of the tribological behavior of oil-enriched polymer gel nanolayers. *Wear*, **255** (7–12), 801–807.

94 Pallandre, A., Glinel, K., Jonas, A.M., and Nysten, B. (2004) Binary nanopatterned surfaces prepared from silane monolayers. *Nano Lett.*, **4** (2), 365–371.

95 Anderson, A.S., Dattelbaum, A.M., Montaño, G.A., Price, D.N., Schmidt, J.G., Martinez, J.S., Grace, W.K., Grace, K.M., and Swanson, B.I. (2008) Functional PEG-modified thin films for biological detection. *Langmuir*, **24** (5), 2240–2247.

96 Kim, H., Noh, J., Hara, M., and Lee, H. (2008) Characterization of mixed self-assembled monolayers for immobilization of streptavidin using chemical force microscopy. *Ultramicroscopy*, **108** (10), 1140–1143.

97 Servoli, E., Maniglio, D., Aguilar, M.R., Motta, A., Roman, J.S., Belfiore, L.A., and Migliaresi, C. (2008) Quantitative analysis of protein adsorption via atomic force microscopy and surface plasmon resonance. *Macromol. Biosci.*, **8** (12), 1126–1134.

98 Briand, E., Gu, C., Boujday, S., Salmain, M., Herry, J.M., and Pradier, C.M. (2007)

Functionalisation of gold surfaces with thiolate SAMs: topography/bioactivity relationship – a combined FT-RAIRS, AFM, and QCM investigation. *Surf. Sci.*, **601** (18), 3850–3855.

99 Palyvoda, O., Bordenyuk, A.N., Yatawara, A.K., McCullen, E., Chen, C.-C., Benderskii, A.V., and Auner, G.W. (2008) Molecular organization in SAMs used for neuronal cell growth. *Langmuir*, **24** (8), 4097–4106.

100 Sheller, N.B., Petrash, S., Foster, M.D., and Tsukruk, V.V. (1998) AFM and X-ray reflectivity studies of albumin adsorbed onto self-assembled monolayers. *Langmuir*, **14** (16), 4535–4544.

101 Sleator, T. and Tycko, R. (1988) Observation of individual organic molecules at a crystal surface with use of a scanning tunneling microscope. *Phys. Rev. Lett.*, **60** (14), 1418–1421.

102 Wawkushewski, A., Cantow, H.-J., Magonov, S.N., Möller, M., Liang, W., and Whangbo, M.-H. (1993) STM study of molecular order and defects in the layers of cycloalkanes $(CH_2)_{48}$ and $(CH_2)_{72}$ adsorbed on graphite. *Adv. Mater.*, **5** (11), 821–826.

103 Crämer, K., Pfannemüller, B., Magonov, S.N., Kreutz, W., and Tuzov, I. (1995) Molecular structure of self-organized layers of *N*-octly-D-guconamide. *Adv. Mater.*, **7** (7), 656–659.

104 Müller, D.J., Amrein, M., and Engel, A. (1997) Adsorption of biological molecules to a solid support for scanning probe microscopy. *J. Struct. Biol.*, **119** (2), 172–188.

105 Shulha, H., Foo, C.W.P., Kaplan, D.L., and Tsukruk, V.V. (2006) Unfolding the multi-length scale domain structure of silk fibroin protein. *Polymer*, **47** (16), 5821–5830.

106 Magonov, S.N. and Whangbo, M.-H. (1996) *Surface Analysis with STM and AFM*, Wiley-VCH Verlag GmbH, Weinheim.

107 Sheiko, S.S. and Möller, M. (2001) Visualization of macromolecules – a first step to manipulation and controlled response. *Chem. Rev.*, **101** (12), 4099–4124.

108 Bai, M., Trogisch, S., Magonov, S., and Taub, H. (2008) Explanation and correction of false step heights in amplitude modulation atomic force microscopy measurements on alkane films. *Ultramicroscopy*, **108** (9), 946–952.

109 Magonov, S.N. and Yerina, N.A. (2003) High-temperature atomic force microscopy of normal alkane C_{60} films on graphite. *Langmuir*, **19** (3), 500–504.

110 Mourran, A., Tartsch, B., Gallyamov, M., Magonov, S., Lambreva, D., Ostrovskii, B.I., Dolbnya, I.P., de Jue, W.H., and Moeller, M. (2005) Self-assembly of the perfluoroalkyl-alkane $F_{14}H_{20}$ in ultrathin films. *Langmuir*, **21** (6), 2308–2316.

111 Minko, S. and Roiter, Y. (2005) AFM single molecule studies of adsorbed polyelectrolytes. *Curr. Opin. Colloid Interface Sci.*, **10** (1–2), 9–15.

112 Bocharova, V., Kiriy, A., Stamm, M., Stoffelbach, F., Jérôme, R., and Detrembleur, C. (2006) Simple method for the stretching and alignment of single adsorbed synthetic polycations. *Small*, **2** (7), 910–916.

113 Kumaki, J., Kawauchi, T., and Yashima, E. (2006) "Reptational" movements of single synthetic polymer chains on substrate observed by in-situ atomic force microscopy. *Macromolecules*, **39** (3), 1209–1215.

114 Roiter, Y. and Minko, S. (2005) AFM single molecule experiments at the solid–liquid interface: *in situ* conformation of adsorbed flexible polyelectrolyte chains. *J. Am. Chem. Soc.*, **127** (45), 15688–15689.

115 Sheiko, S.S., Sumerlin, B.S., and Matyjaszweski, K. (2008) Cylindrical molecular brushes: synthesis, characterization, and properties. *Prog. Polym. Sci.*, **33** (7), 759–785.

116 Potemkin, I.I., Khokhlov, A.R., Prokhorova, S., Sheiko, S.S., Möller, M., Beers, K.L., and Matyjaszweski, K. (2004) Spontaneous curvature of comblike polymers at a flat interface. *Macromolecules*, **37** (10), 3918–3923.

117 Sheiko, S.S., da Silva, M., Shirvaniants, D., LaRue, I., Prokhorova, S., Moeller, M., Beers, K., and Matyjaszewski, K. (2003) Measuring molecular weight by atomic force microscopy. *J. Am. Chem. Soc.*, **125** (22), 6725–6728.

118 Boyce, R., Shirvanyants, D., Sheiko, S.S., Ivanov, D.A., Qin, S., Börner, H., and Matyjaszewski, K. (2004) Multiarm molecular brushes: effect of the number of arms on the molecular weight polydispersity and surface ordering. *Langmuir*, **20** (14), 6005–6011.

119 Gallyamov, M.O., Tartsch, B., Khokhlov, A.R., Sheiko, S.S., Börner, H.G., Matyjaszewski, K., and Möller, M. (2004) Real-time scanning force microscopy of macromolecular conformational transitions. *Macromol. Rapid Commun.*, **25** (19), 1703–1707.

120 Xu, H., Shirvanyants, D., Beers, K.L., Matyjaszewski, K., Rubinstein, M., and Sheiko, S.S. (2004) Molecular motion in a spreading precursor film. *Phys. Rev. Lett.*, **93** (20), 206103.

121 Xu, H., Shirvanyants, D., Beers, K.L., Matyjaszewski, K., Dobrynin, A.V., Rubinstein, M., and Sheiko, S.S. (2005) Molecular visualization of conformation-triggered flow instability. *Phys. Rev. Lett.*, **94** (23), 237801.

14
Langmuir–Blodgett and Layer-by-Layer Structures

Nanostructured polymeric materials in the form of thin, ultrathin, and monolayer films with precisely controlled thickness, inner composition, microstructure, and surface morphology can be assembled by a variety of fabrication routines with the most common being spin casting and drop casting. Layer-by-layer (LbL) assembly and Langmuir–Blodgett (LB) deposition of well-ordered nanoscale polymer films are the most popular and versatile approaches widely explored in current studies. Organized LB films introduced in the 1930s are popular for studying interfacial and surface behavior of complex amphiphilic molecules ranging from traditional fatty acids to block copolymers and biomolecules [1].

LbL assembly, introduced and largely promoted by Decher, Lvov, and Möhwald in the 1990s, is a popular technique for the bottom-up fabrication of organized, multilayer, organic, polymeric, and organic–inorganic films based on alternating adsorption of oppositely charged materials in most cases [2–10]. LbL multilayer films, with their well controlled inner microstructure and surface morphology, have received continuous attention in a number of areas such as surface modification and coating, drug delivery, electrochemical devices, fuel cells, chemical sensors, nanomechanical sensors, and nano-, chemi-, and bioreactors [11–15]. LbL assembly, which was initially based on the alternating electrostatic adsorption of polyelectrolytes, has been successfully extended to encompass other driving forces, such as hydrogen bonding, covalent bonding, and other weak intermolecular interactions [16–18]. It is widely accepted that LbL assembly is a powerful practical approach for assembling functional building blocks into ultrathin films with controlled thickness and molecular structures on planar as well as curved and patterned solid substrates. These films can also be made into ultrathin nanocomposites with alternating inorganic nanoparticles and polyelectrolyte layers simply deposited on substrates or as freestanding nanostructures integrated into microfabricated devices [19–24, 29, 46].

In this chapter, we will discuss some interesting recent results regarding surface morphologies, microstructures, and local mechanical (elastic and adhesive) and physical properties of both LbL and LB films as probed with various AFM techniques.

Scanning Probe Microscopy of Soft Matter: Fundamentals and Practices, First Edition.
Vladimir V. Tsukruk and Srikanth Singamaneni.

14.1 LbL films

14.1.1 Conventional LbL Films

A general schematic of the fabrication of LbL films via alternating assembly of oppositely charged polyelectrolytes suggested in a seminal review from Decher is presented in Figure 14.1 [2]. The poly(allylamine hydrochloride) (PAH) and poly (sodium 4-styrene sulphonate) (PSS) polymers presented here are a classical pair of polyelectrolytes still widely exploited for LbL construction (Figure 14.1).

Numerous studies of LbL films with AFM since the introduction of LbL assembly are mostly focused on routine verification of the surface morphology at ambient conditions to assure its uniformity and the integrity of the LbL layers; monitoring of polyelectrolyte or nanoparticle adsorption, aggregation, dewetting behavior; estimation of surface microroughness for LbL films with different numbers of deposited layers and at different spatial scales; and measuring the thickness of the films (usually in combination with ellipsometry and X-ray reflectivity) – all important pieces of information critical for identifying optimal assembly conditions [25–28].

Generally, LbL films fabricated via conventional LbL and spin-assisted LbL assembly show very uniform surface morphology with microroughness (usually measured within 1 μm^2 surface areas) below 0.5 nm, which is the typical diameter of polymeric backbones [29–31]. Contrarily, various advanced films such as hydrogen-bonded LbL films, nanocomposite LbL films, porous LbL films, exponentially grown LbL films, and LbL films with cross-linked components usually show much higher surface microroughness and larger-scale irregularities, exceeding 4–10 nm in some cases. This is due to the localized nonhomogeneous distribution of components, significant intermixing or demixing, and weak interactions involved.

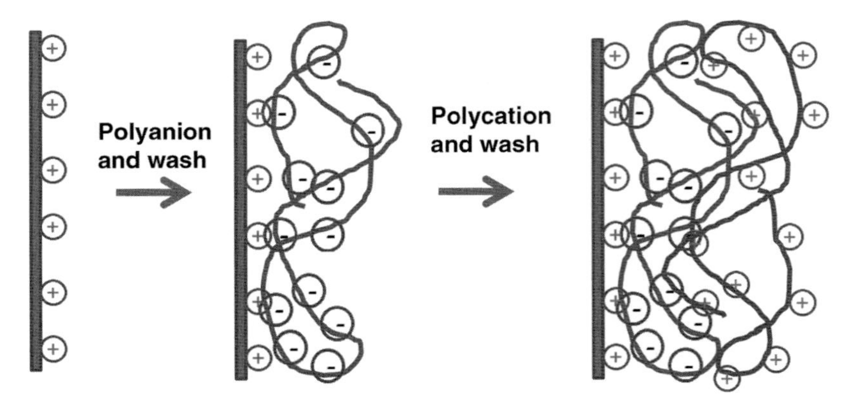

Figure 14.1 LbL assembly of polyelectrolytes, chemical formulas of PSS and PAH are shown as examples. Reprinted from Ref. [2].

More recently, AFM was employed as an imaging tool for *ex situ* and *in situ* monitoring of the LbL assembly processes under variable environments; characterizing the unique morphology of nanocomposite, multicomponent, and porous films; monitoring the initial stages of polyelectrolyte adsorption, mixing, and dewetting; measuring adhesive and nanomechanical properties; and characterizing the fine lateral and vertical distribution of nanostructured components. These aspects of AFM applications to LbL films will be addressed in this section.

In earlier studies, Tsukruk *et al.* explored AFM for the investigation of monolayers and bilayers of polyionic materials deposited by electrostatic deposition on charged SAMs [9]. The formation of SAMs was utilized to control the surface charges and thus initiate polyelectrolyte adsorption. *Ex situ* AFM was used to monitor both PSS adsorption on the charged surfaces of amine-terminated SAMs and PAH adsorption on a complete PSS monolayer. Observations of PSS monolayers at various stages of electrostatic deposition revealed highly inhomogeneous adsorption and chain assembly at the earliest stages of deposition (Figure 14.2). The surfaces of bare glass and commercial silanized glass slides were determined to be too rough to deduce quantitative data for deposited polyelectrolyte monolayers of several nanometers thickness, and thus all quantitative AFM measurements (in this study and the majority of following and current studies) were conducted on silicon wafers with atomically flat surfaces.

As observed in this study, during the first several minutes of polyelectrolyte deposition, negatively charged PSS macromolecules tend to adsorb on selected surface defect sites of positively charged SAMs (such as scratches and edges) (Figure 14.2). Occasionally, the formation of islands composed of a PSS phase around microparticulate contaminants is observed. At this stage, electrostatically

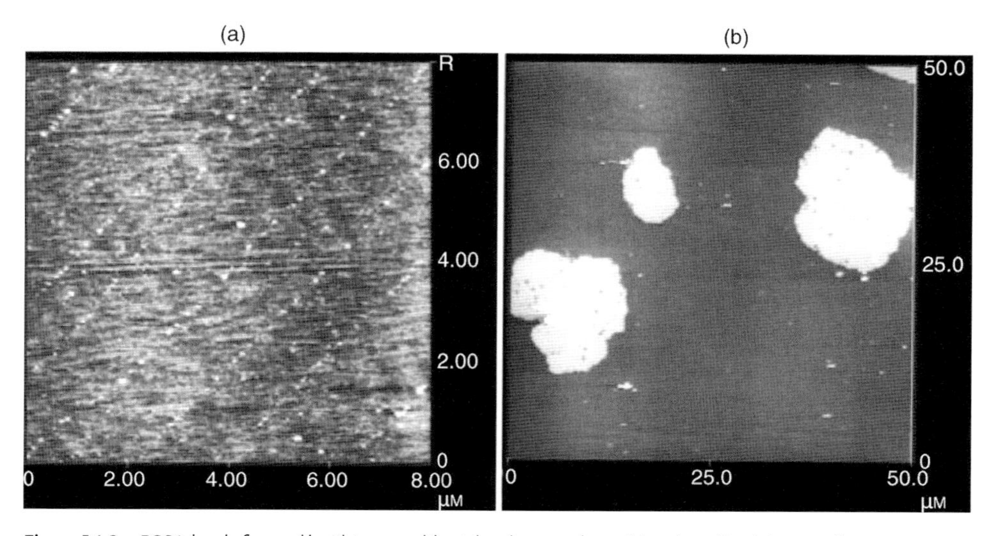

Figure 14.2 PSS islands formed by LbL assembly at the shortest deposition time (1 min) on surface defects of a positively charged SAM (a) along natural scratches and (b) around microparticles. Reprinted from Ref. [9].

driven fast adsorption of PSS chains is predominant, and the equilibration of the surface monolayer structures is not achieved due to the slow surface diffusion of macromolecules. A strong tendency is observed for polyelectrolyte chains to group in selected places of microscopic surface defects such as straight rows of islands located along submicron scratches and the edges of atomic planes occasionally observed for the underlying silicon wafer surface. Occasionally, pieces of PSS monolayers up to several micrometers across can be observed growing around large-scale defects, such as dust particles, which serve as nucleation centers (Figure 14.2). The height of these very smooth pieces of PSS monolayers can be as large as 3.0–4.0 nm and lateral dimensions can reach tens of micrometers.

For longer deposition times (from 2 to 5 min), the authors observed an increase in the surface coverage by randomly distributed nanoscale islands of the PSS component (Figure 14.3). Local coverage increases significantly in the areas of submicron defects, such as the edges of atomic planes. Only longer deposition times ($>$10 min) result in an equilibration of polymer monolayers and the formation of a homogeneous ultrathin PSS monolayer composed of highly flattened and densely packed macromolecular chains all over the surface. At this stage, the PSS monolayer thickness in the dry state is between 1.0 and 1.5 nm, with a low microroughness of about 0.2 nm, which is common for uniform LbL monolayers on flat substrates.

FFM images of these LbL films showed that frictional properties of underlying surface (amine-terminated SAM) and polyelectrolyte islands at initial stages of formation of PSS monolayers were very different and allow fast material composition identification. The friction signal was substantially higher on top of the PSS islands than on surrounding SAM areas. The significant difference in friction responses at identical probing conditions observed for these incomplete monolayers (beyond local surface fluctuations and easily recognizable "geometrical effects") indicates the presence of *two different materials* at the surfaces studied: PSS islands and an amine SAM. Finally, AFM imaging showed that the assembly of a second PAH monolayer on top of a preformed PSS monolayer follows similar tendencies to those discussed above. A two-step assembly process resulted in the formation of homogeneous PAH/PSS bilayers with an overall thickness of 1.7–2.5 nm and a low surface microroughness after prolonged annealing.

In contrast to regular *ex situ* studies of LbL films in a stop-and-go process with sequential washing and drying cycles, *in situ* monitoring of LbL assembly in liquid cell at different pH values has been reported by Menchaca *et al.* [32]. Contact mode imaging exploited in this study revealed a domain surface morphology of initially deposited polyelectrolyte monolayers in the course of formation of PSS–PAH films on PEI-modified substrates. The formation of this domain morphology was related to the initial formation of localized polyelectrolyte complexes with parameters depending on deposition conditions. Size and overall microroughness of the LbL films were monitored as a function of the number of assembled layers, and the in-liquid AFM results were compared with the surface morphology of resulting LbL films observed after their washing and drying. It has been observed that dried LbL films were characterized by surface texture with much finer domain and greater smoothness compared to LbL films observed directly in liquid.

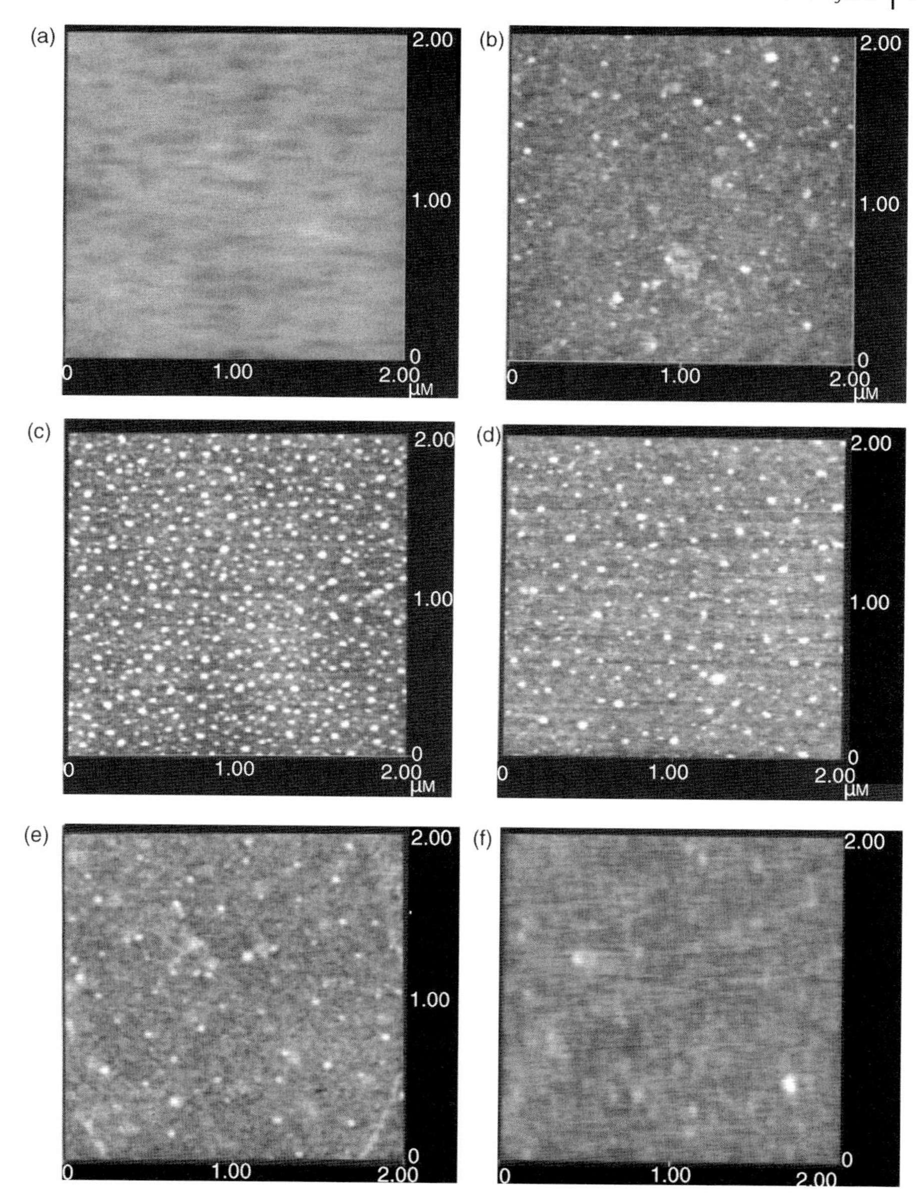

Figure 14.3 AFM images of a PSS layer formation at different adsorption times: (a) 0 s (pristine amine-terminated SAM), (b) 45 s, (c) 2 min, (d) 5 min, (e) 10 min, and (f) 64 min. Reprinted from Ref. [9].

Similar domain morphology formed by localized polyelectrolyte complexation and the critical role of pH in this process were pointed out in a recent AFM study of PEI–PSS LbL films by Elzbieciak *et al.* [33]. Gradual annealing of the initial domain surface morphology in the course of LbL film treatment with salt solutions of different concentrations was reported by McAloney *et al.* [34]. The authors noted that the final, well-annealed, uniform surface of LbL films possessed featureless surface morphologies with minimal surface distortions as a result of salt ion exchange processes during long annealing times in salt solution.

14.1.2
Composite LbL Films

Assembly of composite LbL films composed of different components is usually conducted by addition of inorganic nanostructures such as microparticles, nanoparticles, nanowires, nanorods, and nanoplatelets to the polyelectrolyte film. This is a popular way for the creation of robust organized nanocomposite materials with enhanced mechanical, optical, and conductive properties, which are frequently tested with both AFM imaging and SFS probing technique [35]. One of the most intriguing examples, provided by Kotov and coworkers, involved the fabrication of ultrastrong macroscopic films via LbL assembly of clay nanoplatelets – a layered silicate material with a high inherent stiffness – into a polyelectrolyte film [36]. AFM imaging of individual clay nanoplatelets resolved polyelectrolyte chains spread on the surface, providing maximized interfacial interactions in the resulting nanocomposites (for discussion and images, see Chapter 12). Introducing additional covalent cross-linking into micrometer-thick clay–polyelectrolyte composites facilitated a dramatic increase in Young's modulus (up to around 100 GPa) [37–39].

AFM imaging of composite LbL films is frequently utilized to characterize the distribution of inorganic components, their aggregation state, and mixing of components under different deposition conditions. Several well-known examples of such studies include AFM visualization of well-defined nanodomain surface textures for LbL films that contain silica nanoparticles of polyhedral oligomeric silsesquioxanes (POSS), observation of nanodomain morphology combined with layered composition caused by planar and distorted packing of individual and overlapped clay nanoplatelets in clay-containing PEI- or PDDA-based LbL films, monitoring polyelectrolyte chain adsorption on clay nanoplatelets for assembly of robust and multilayered LbL nanocomposites, and observation of morphological reorganization caused by the incorporation of semiconducting polymers into composite LbL films onto different substrates and in the freestanding state [40–43]. Some selected examples from the recent studies will be discussed in detail.

Reinforced silk–clay LbL films were obtained by integrating a silk fibroin matrix with functional inorganic nanoplatelets via spin-assisted LbL assembly by Kharlampieva *et al.* [44]. The authors demonstrated that organized assembly of this silk protein with clay (montmorillonite) nanosheets resulted in highly transparent and robust ultrathin films with significantly enhanced mechanical properties, including higher strength, increased toughness, and enhanced elastic modulus – all values manyfold higher than that for pristine silk nanomaterials.

In the course of this study, the authors observed that the LbL assembly of silk material with clay-enriched layers resulted in consistent linear growth of the LbL film starting from the fifth bilayer (Figure 14.4). Individual thicknesses of 5 nm and

1.3 nm were measured from AFM scratch tests for silk and clay layers, respectively. The thickness of the clay component added to LbL films corresponds to the thickness of an individual aluminosilicate layer of montmorillonite measured independently and indicates monolayer formation upon adsorption. AFM images indicate full integration of clay nanoplatelets into the silk matrix with a smooth surface (microroughness of about 2.7 nm, comparable to that of a pure silk film). The presence of clay nanoplatelets inside of LbL films is barely visible in phase images due to coverage by polyelectrolyte layers (Figure 14.4). TEM analysis was employed to confirm AFM results and the presence and distribution of clay nanoplatelets. This complementary analysis is especially critical if clay nanoplatelets are buried under polyelectrolyte multilayers with a significant number of layers. In fact, this TEM study, which complemented AFM imaging, shows clay nanoplatelets of a few hundred nanometers embedded in the composite LbL films (Figure 14.4).

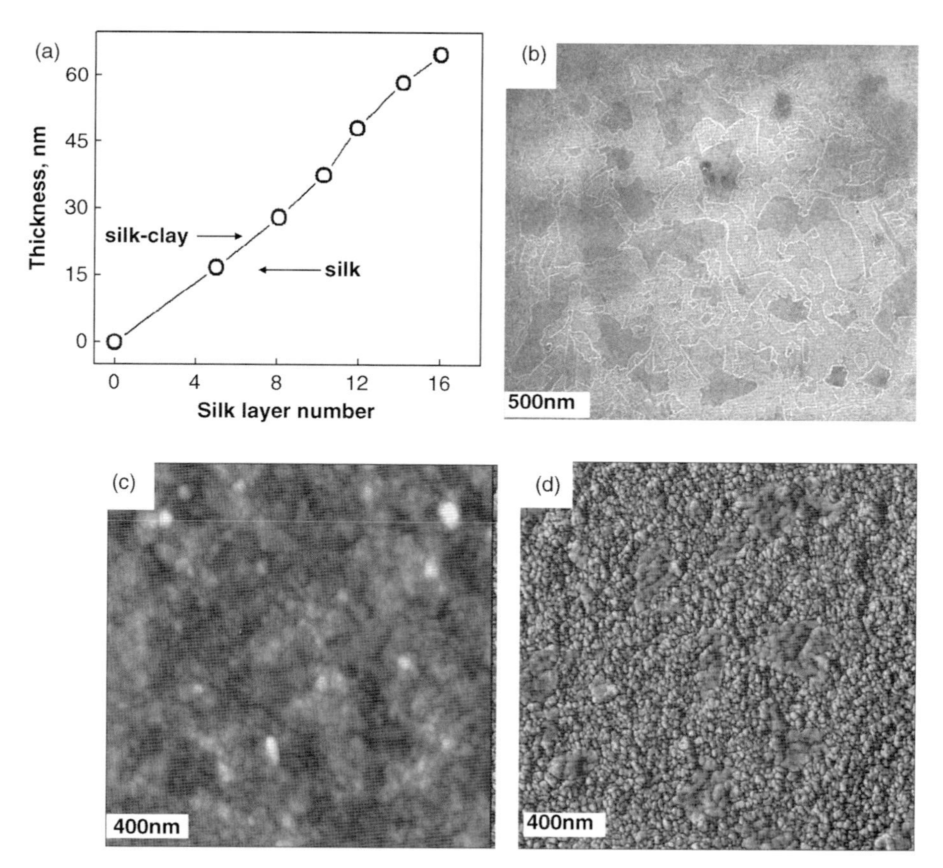

Figure 14.4 (a) Thickness increase of $(silk)_n$ and $(silk\text{–}MMT)_n$ films with increasing number of deposited layers. (b) TEM of a $(silk\text{–}MMT)_{12}$ film. (c) AFM topographical and (d) phase images of the $(silk\text{–}MMT)_{12}$ film. Height is 40 nm (c) and z-scale is 10° (d). Reprinted from Ref. [44].

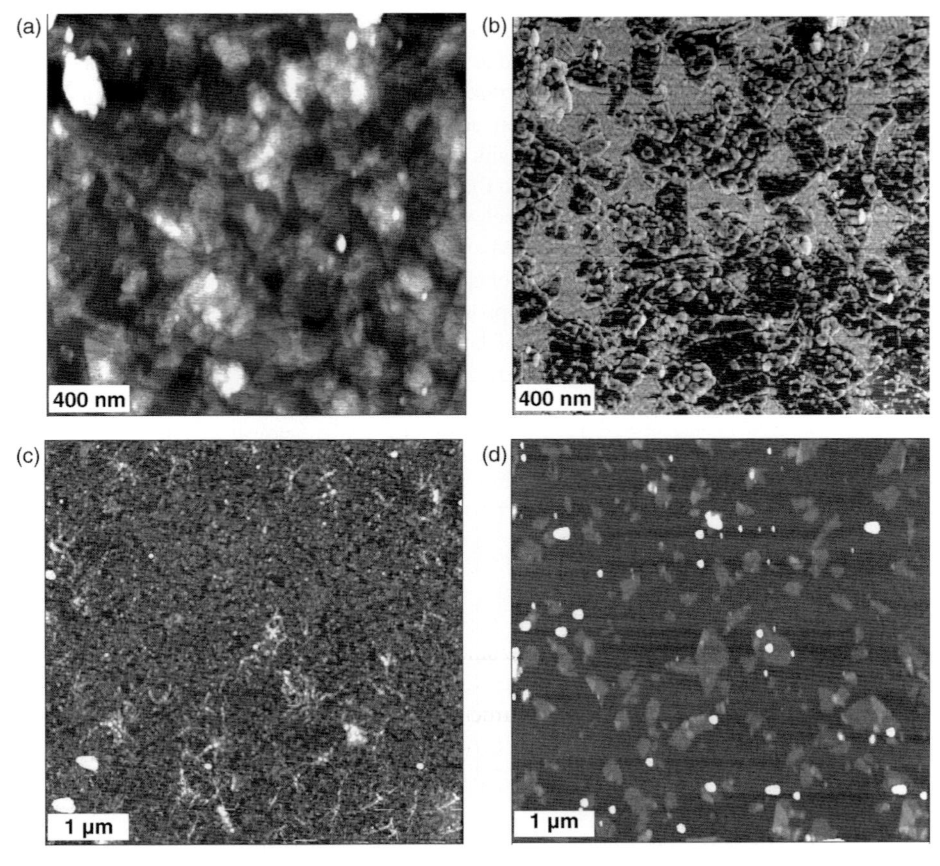

Figure 14.5 (a) AFM topographical and (b) phase images of a $(silk–MMT)_{12}$ film after treatment at 500 °C. Height is 10 nm (a) and *z*-scale is 50° (b). AFM topographical images of a $(silk–MMT)_1$ film (c) before and (d) after treatment at 500 °C. Height is 20 nm (c) and *z*-scale is 10° (d). Reprinted from Ref. [44].

In another approach, applicable to any inorganic–organic LbL films, to independently confirm the presence of inorganic nanoparticles inside these films, the polymeric LbL matrix was further treated at 500 °C to remove the organic component. In this study, this temperature treatment uncovered MMT nanoparticles with average lateral dimensions of 200 nm distributed over the substrate in a predominantly horizontal orientation, clearly shown by AFM images of the residual films (Figure 14.5). The result reveals that MMT adsorbed from 0.05% solution covers only ~8% of the surface area (Figure 14.5). The low surface coverage on silk surface compared to ~80% coverage of that on synthetic PAH-terminated films reflects the lower charge density of the dry silk surface and its much higher hydrophobicity at neutral pH compared to highly charged PAH-coated surfaces with pH ~ 4 [45].

In another study of nanocomposite LbL films, surface morphology of gold nanoparticle-containing LbL membranes assembled with spin-assisted LbL was

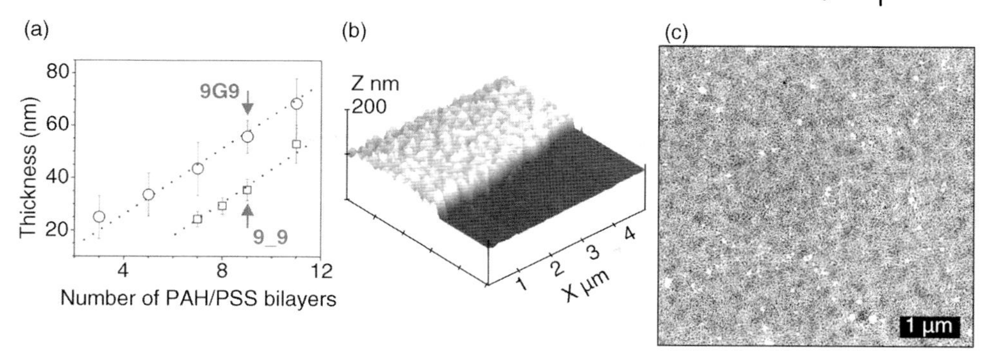

Figure 14.6 Freestanding nanocomposite LbL films containing gold nanoparticles. (a) Thickness variation of LbL films with a different number of polymer bilayers with (circles) and without (squares) gold nanoparticles. (b) 3D AFM image of LbL membrane edge on a silicon wafer surface in the course of scratch testing. (c) Large area TEM micrograph showing distribution of gold nanoparticles inside LbL films. Reprinted from Ref. [20].

studied by Jiang *et al.* [20, 46]. The thickness of these LbL films versus the number of layers was measured from AFM scratch testing and was observed to be much higher than that corresponding to purely polymeric LbL films (Figure 14.6). The AFM image of the edge of the LbL film deposited on a silicon wafer was analyzed with a height histogram to deduce the average film thickness (Figure 14.6a). LbL films were uniform but possessed a relatively high microroughness of about 8 nm (within a 1 μm^2 surface area) (Figure 14.6). This microroughness is much higher than the microroughness usually obtained for purely polymeric LbL films (0.3–0.5 nm) due to conformal coverage of gold nanoparticles with polyelectrolyte layers and the presence of multiple protrusions (Figure 14.6).

Overall, these SPM studies demonstrated that the LbL membranes fabricated without gold nanoparticles showed a much lower microroughness and were thinner due to the absence of the gold nanoparticle layer. The packing density of gold nanoparticles in the central portion of LbL film varied from very low (<2%) to the highest density of 25%. TEM images of the LbL membrane that complemented AFM studies showed a uniform, large-scale distribution of gold nanoparticles with the formation of localized chain-like aggregates (Figure 14.6). In addition, AFM phase images demonstrated that the gold nanoparticles were completely covered by the top polymer layers. No signs of any additional polymer or nanoparticle aggregation (e.g., crystallites and clusters) were found for these nanocomposite LbL membranes.

To evaluate the micromechanical properties of these LbL films, freestanding membranes were suspended over a copper TEM grid with a single cell dimension of 85 μm^2 in studies by Markutsya *et al.*, as shown in Figure 14.7 [47]. The nanoscale deformation behavior of these suspended LbL films was tested with a colloidal AFM probe in the usual manner discussed in Chapter 5. Force–distance curves collected at different surface locations showed significant pull-off forces due to strong interaction of a silica microbead and charged surfaces of LbL films and a near-linear elastic deformation portion at direct physical contact.

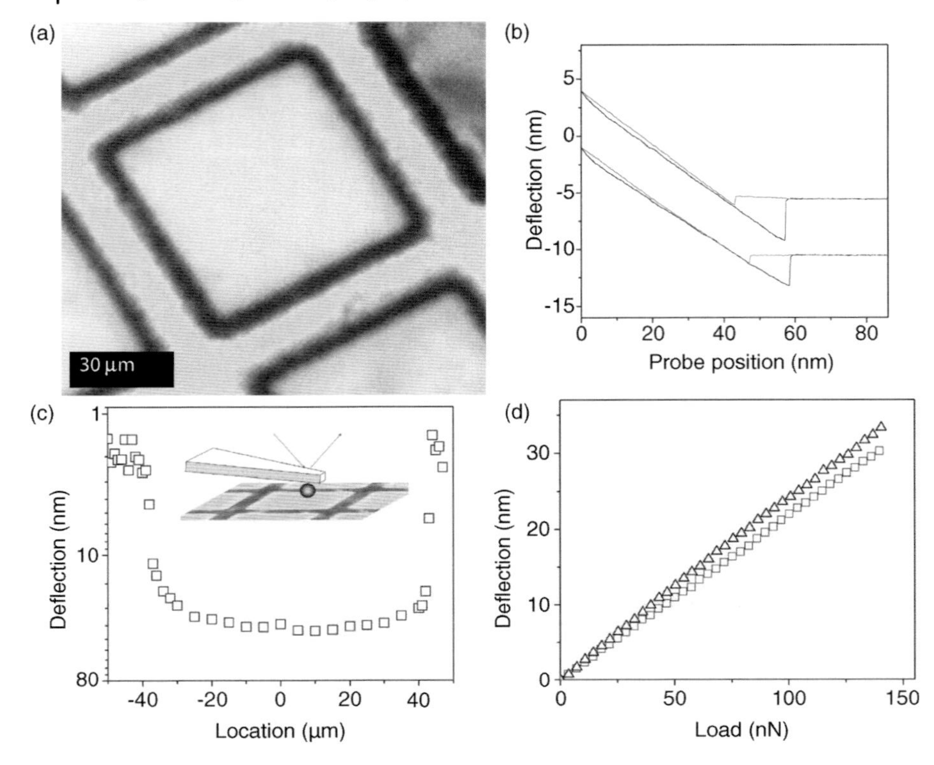

Figure 14.7 (a) SFS mechanical testing of LbL membranes suspended at TEM grid; (b) FDC data for two surface locations; (c) LbL membrane deflection across the TEM grid cells derived from AFM probing; a cartoon of the experimental setup (inset); and (d) the nanoscale deflection of LbL membrane at two different locations. Reprinted from Ref. [20].

As concluded in this study from the analysis of loading curves, the reversible LbL membrane deflection under a point load of 150 nN reached 30 nm and was uniform over tested cells far from the supporting edges (Figure 14.7). An analysis of the loading behavior (deflection versus normal load) far from the edges (>10 μm) revealed a uniform response across the membrane surface (compare data for two randomly selected locations in Figure 14.7) [48]. The linear loading deformational behavior indicated an elastic bending mechanism under low forces. The minimum detected membrane deflection, about 2 nm, was observed under a normal load of only 4 nN. The minimum deflection of a few nanometers, which was much smaller than the total thickness of the LbL membrane, can be detected with this colloidal probe SFS approach. The bending stiffness of the freely suspended LbL membrane was estimated from this experimental data to be about 2 N/m for these polyelectrolyte–gold nanoparticles LbL films.

In another study, Gunawidjaja *et al.* fabricated freestanding nanocomposite LbL films with encapsulated silver nanowires at controlled volume fractions, $\Phi = 2.5$–22.5% [49]. In the course of assembly, silver nanowires were sandwiched between $(PAH/PSS)_{10}$ films, resulting in nanocomposite structures with a general formula $(PAH/PSS)_{10}$-Ag-$(PAH/PSS)_{10}$ (see sketch in Figure 14.8). The silver

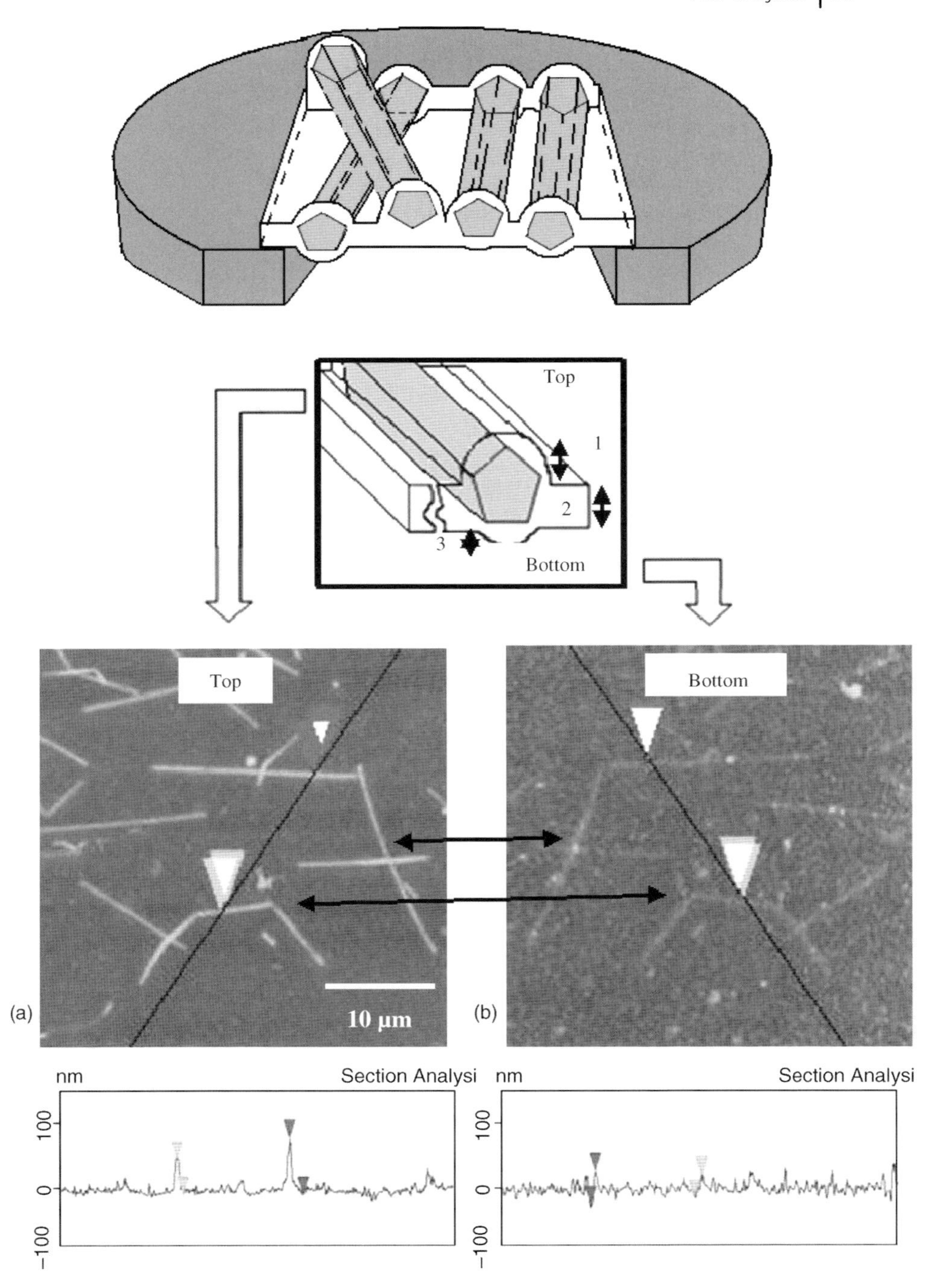

Figure 14.8 Freely suspended LbL films with incorporated silver nanowires (top) and AFM images of the selected surface areas from top (a) and bottom (b) used to restore fine details of film morphology. Reprinted from Ref. [49].

nanowire diameter (73 nm) was much higher than the thickness of the polyelectrolyte LbL films (about 50 nm total). This difference resulted in a peculiar, modulated film morphology with silver nanowires protruding from the planar LbL film on both sides. The local morphology of such contoured sandwich morphologies was revealed by AFM scanning of the same surface area on the top and bottom sides of the LbL film. Cross sections of these morphologies allowed measuring independently diameter and elevations of nanowires, as shown in Figure 14.8.

In an alternative approach, AFM analysis is utilized to monitor the direct formation of gold nanoparticles onto/inside LbL films made from silk by Kharlampieva *et al.* [50]. AFM images of the original, nanoparticle-free LbL films showed that the pristine silk template has a light-grainy morphology with a modestly high surface microroughness of 2–3 nm within 1 μm^2 due to localized crystallization and the appearance of nanoscale domains of the ordered silk II form (Figure 14.9). The thickness of the individual silk layer obtained by spin-assisted LbL assembly was

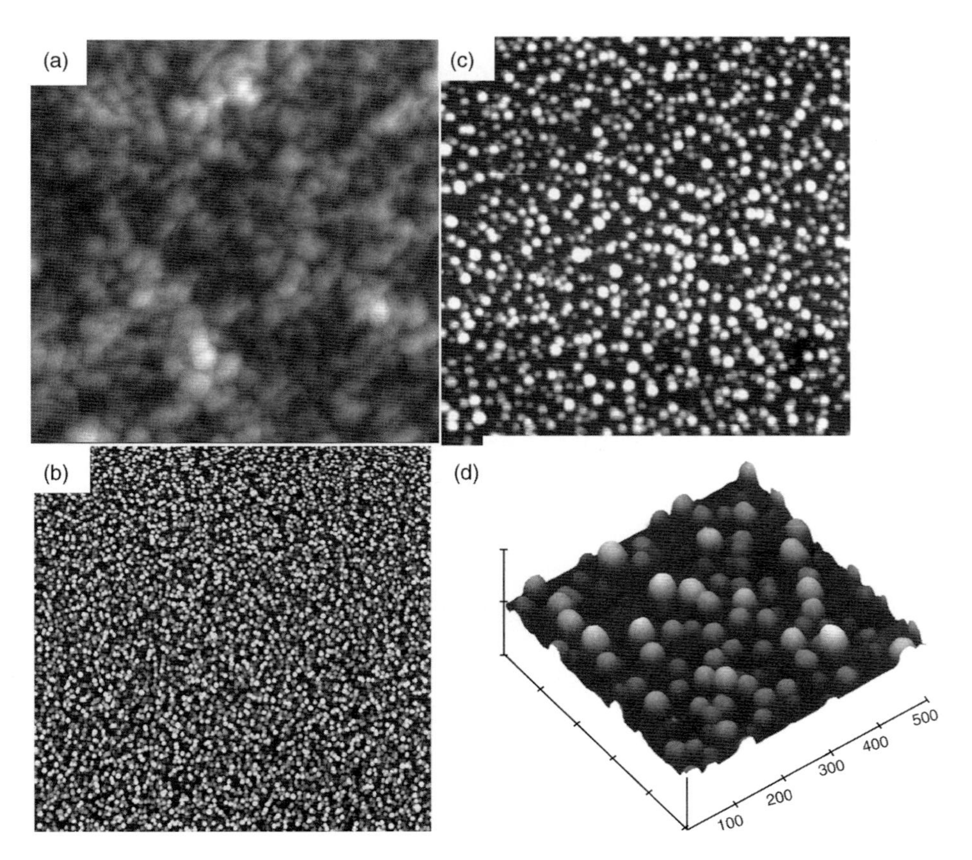

Figure 14.9 AFM images of a $(PAH–PSS)_2PAH$–silk LbL film (a) before immersing into $HAuCl_4$ followed by treatment at 500 °C (b–d). Silk layer was obtained by spin-assisted deposition. *z*-scales are 40 nm (a and d) and 10 nm (b and c). Scan sizes are 1 (a and c), 5 (b), and 500 μm^2 (d). Reprinted from Ref. [50].

5 nm, as measured with ellipsometry and confirmed with AFM scratch testing, and the overall thickness of silk films was between 30 and 60 nm.

To perform biomineralization on these templates, the silk film was exposed to a $HAuCl_4$ solution at pH 10. The exposure resulted in the nucleation and growth of uniformly distributed fine gold nanoparticles on/inside the silk matrix. The formation of nanoparticles caused a modest increase in surface microroughness to 3–4 nm, which is common for nanocomposite LbL films (Figure 14.9). To confirm the presence of gold nanoparticles, the organic LbL matrix, composed of both silk and polyelectrolyte layers, was burned out at 500 °C similarly to that discussed previously for clay LbL films [51]. As a result, gold nanoparticles with diameters of 7.0 nm appeared evenly distributed over the substrate with no large microscale aggregates present on the surface, as clearly seen from AFM images of the thermally treated LbL films (Figure 14.9). The diameter of gold nanoparticles and their core–shell structure suggested from AFM imaging were also confirmed by high-resolution TEM studies, a typical practice of comparative AFM and TEM studies in the investigation of LbL nanocomposites.

By fabricating a different class of soft LbL nanocomposites, Kozlovskaya *et al.* have observed ultrathin, pH-responsive plasmonic membranes of [poly(methacrylic acid)-gold nanorods]$_{20}$ ((PMAA-AuNRs)$_{20}$) with gold nanorods embedded into swollen cross-linked LbL hydrogels [52]. Figure 14.10 demonstrates surface morphology and cross-sectional analysis of these hydrogen-bonded LbL films formed at pH 2.5 before and after covalent cross-linking. A high (3.6 nm) initial microroughness for the initially deposited LbL film, much higher than those observed for traditional ionic one-component LbL films, was observed. However, the initially rough surface became much smoother after chemical treatment. The authors suggested that the overall smoothing of the LbL surface after chemical cross-linking occurred due to dissolution of the previously formed hydrogen-bonded complexes between poly(carboxylic acid) and PVPON. Also, this phenomenon was related to overall polymer matrix swelling due to increased hydrophilicity of the polyacid chains at different pH values [53].

Furthermore, an even larger decrease in the LbL hydrogel surface microroughness was observed when these LbL (PMAA)$_{20}$ films were fully dried. The surface microroughness measured for a 1 μm^2 area on a 20-layer cross-linked (PMAA) LbL film at pretreated silicon wafer surfaces dried from pH 8 and pH 3 was very low, about 0.8 nm, which is comparable to conventional LbL films. Lower microroughness of the cross-linked films at pH 8 and pH 3 also reflected additional swelling of the polyacid LbL matrix caused by electrostatic ionization under these pH conditions. The appearance of larger-scale wrinkles on the hydrogel surfaces in the dried state and chemically cross-linked LbL films was attributed to the restricted and highly localized swelling of the tethered hydrogel LbL films, which causes the development of high residual compressive stresses.

In this study, Kozlovskaya *et al.* have also demonstrated the possibility to incorporate relatively large gold nanorods into the swollen (PMAA)$_{20}$ LbL hydrogel film (Figure 14.11) [52]. AFM imaging of the films loaded with gold nanorods showed a number of flat lying nanorods included in the (PMAA)$_{20}$ film with different concentrations depending on pH state (see different AFM images in Figure 14.11).

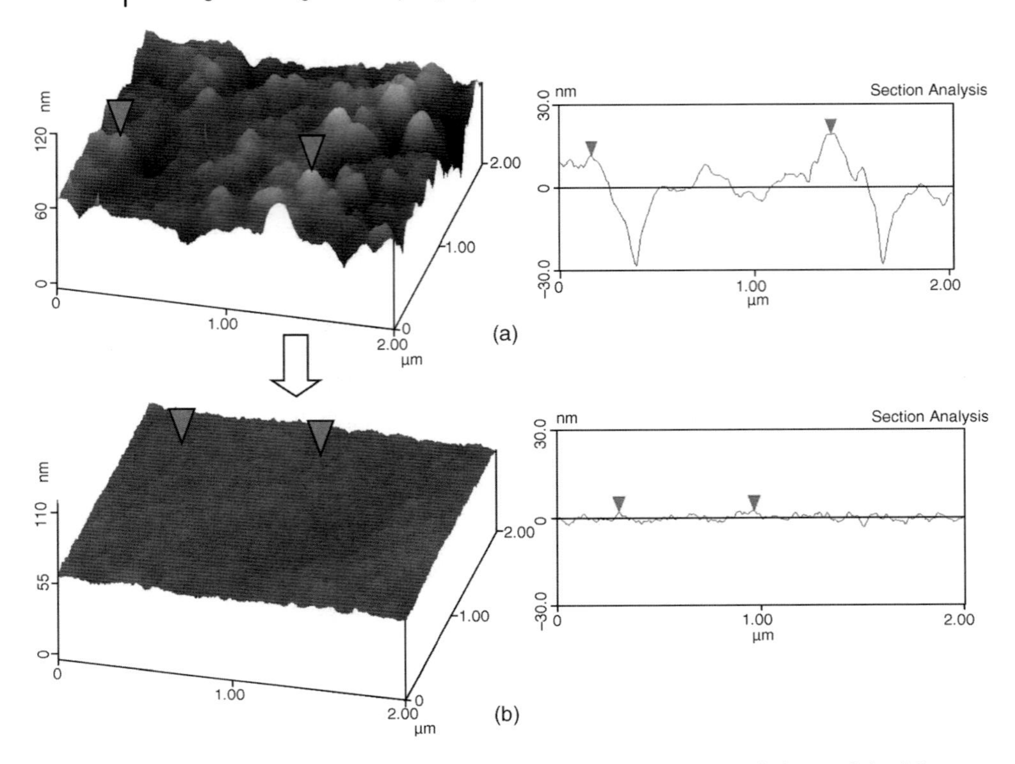

Figure 14.10 AFM images and cross sections of a hydrogen-bonded $(PVPON/PMAA)_{20}$ film deposited at pH 2.5 via spin-assisted LbL assembly (a) and of a hydrogel $(PMAA)_{20}$ film with chemical cross-linking and dried from pH 5 (b). *z*-scale is 30 nm for both images. Markers show the areas along which the cross sections were made. Reprinted from Ref. [52].

Cross-sectional analysis of the surface topography reveals that the gold nanorods, which are seen on the hydrogel surface, are partially embedded within the LbL films and are oriented predominantly in the surface plane (Figure 14.11). In fact, the average height of the gold nanorods obtained from cross-sectional analysis of the $(PMAA\text{-}Au)_{20}$ film was 15 nm, much smaller than the nanorod diameter (20–25 nm), indicating at least partial embedding. The low contrast between free hydrogel surface and surfaces of mostly gold nanorods on corresponding phase images obtained in light tapping mode indicates that the nanorods are mostly covered by the LbL hydrogel material. Thus, these nanorods are completely encapsulated in the hydrogel matrix and not just tethered on the LbL surface [54]. This conclusion was also confirmed by direct comparison of AFM images and TEM results for the same surface areas of nanocomposite LbL films. TEM visualizes all nanorods within ultrathin LbL films, shows a true "volume" content of gold nanorods, and allows quantitative estimation of the ratio of volume to the surface fraction of the nanorods.

Core–shell CdSe/ZnS quantum dots prepared according to the known procedure and functionalized with tri-*n*-octylphosphine oxide (TOPO) were incorporated

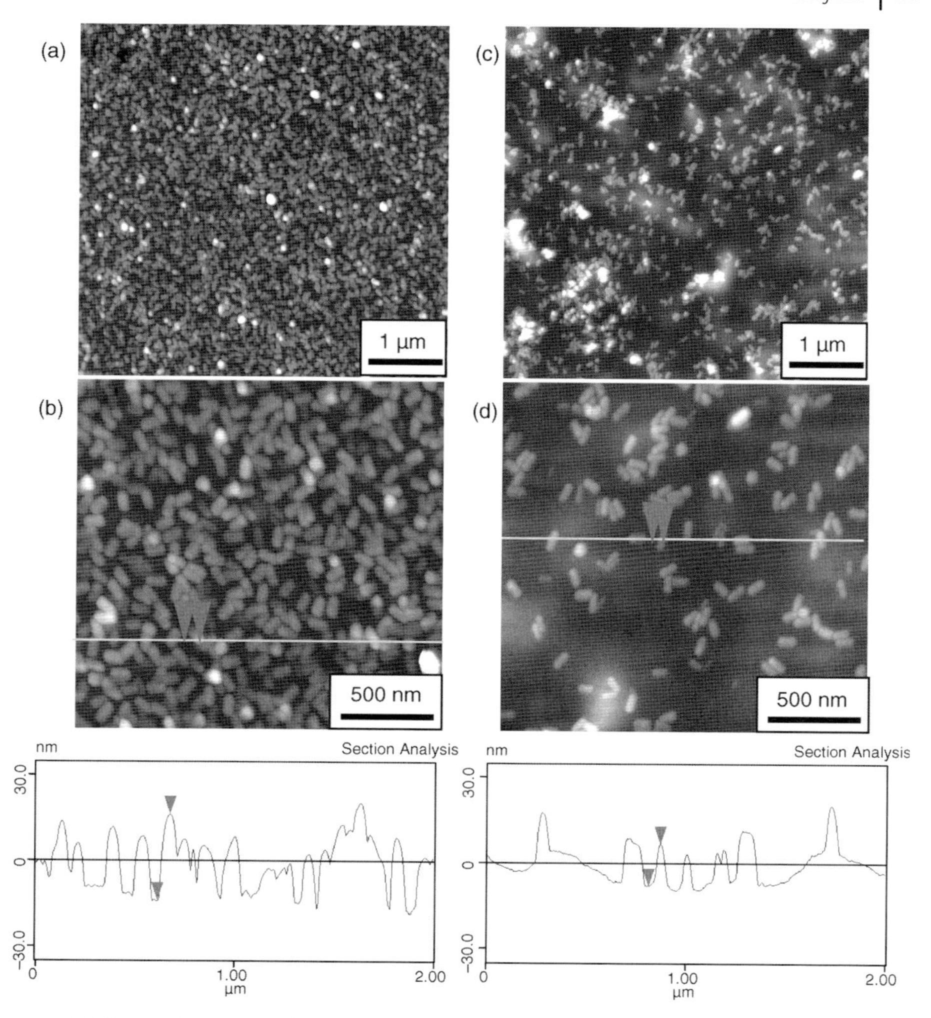

Figure 14.11 AFM images of the $(PAH/PMAA)_3(PMAA)_{20}$ hydrogel films loaded with Au nanorods at pH 3 (a and b) and at pH 8 (c and d) for 10 h and corresponding cross sections. Reprinted from Ref. [52].

into LbL films and investigated with AFM in a recent study by Zimnitsky *et al.* (Figure 14.12) [55–58]. UV–vis absorption shows an absorption maximum at 590 nm and a less pronounced peak at 490 nm, characteristic optical response of CdSe/ZnS quantum dots (Figure 14.12) [59]. These quantum dots showed strong photoluminescence with a pronounced and narrow peak at 633 nm, common for CdSe quantum dots of about 4.5 nm diameter [60]. AFM imaging of these quantum dots adsorbed from solution onto a silicon wafer precovered with a monolayer of PAH showed individual nanoparticles uniformly distributed over the entire PAH-modified surface (Figure 14.12). For AFM estimation of the nanoparticle diameter, deposition of quantum dots was conducted from a dilute (0.01%) solution to visualize individual particles, avoid their aggregation, and estimate nanoparticle size distribution from cross-sectional analysis. The optical image of the freestanding LbL-quantum dot film picked up on a copper plate with a microscopic hole demonstrates its integrity despite high residual stresses developed in the course of this transfer and drying (Figure 14.12).

Although lateral dimensions of nanoparticles usually appear to be larger on AFM images than they actually are due to the dilation effect as discussed in previous chapters, the height of spherical nanoparticles is a reliable proof of their true size. The height histogram calculated from the AFM image shows a relatively narrow size distribution with an average height of 4.9 nm (Figure 14.12). This value is close to that obtained from concurrent TEM data (4.5 nm) used to independently confirm nanoparticle diameter. The minor differences in the apparent diameters estimated from UV–vis spectroscopy, TEM, and AFM can be accounted for by the additional contribution of the ZnS capping layer (Figure 14.12).

14.1.3 Porous LbL Films

Micro- and nanoporous LbL materials can be prepared by different methods, including acidic treatment of initially solid LbL films, removal of embedded nanoparticles, and inducing partial dewetting of polyelectrolyte components. These highly

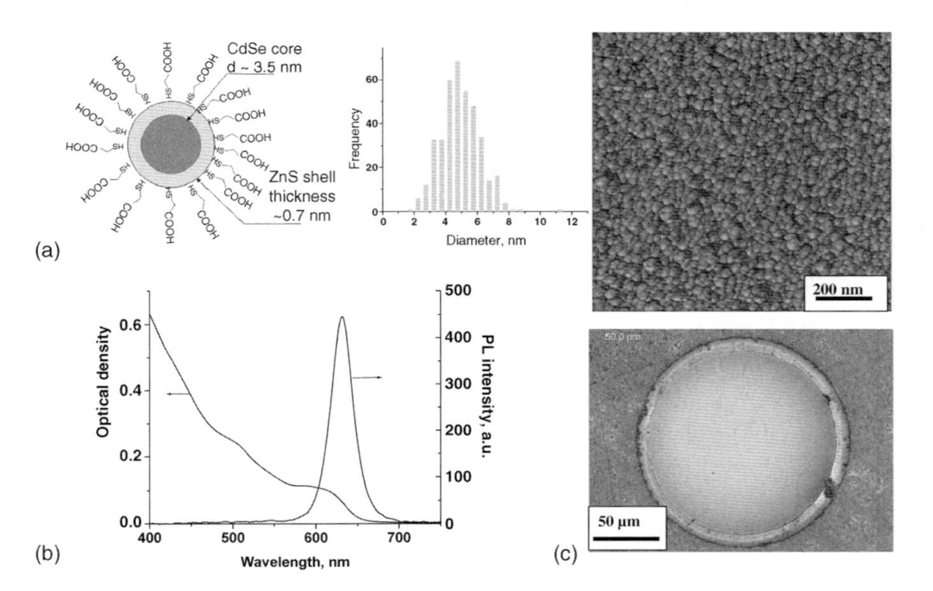

Figure 14.12 Composition of quantum dots with corresponding height distribution from AFM image (a) and corresponding UV absorption and photoluminescent spectra (b). (c) AFM image of adsorbed quantum dots and LbL films with quantum dots freely suspended across opening. Reprinted from Ref. [55].

heterogeneous films are usually carefully examined with AFM, SEM, and TEM imaging to obtain quantitative estimation of their porous morphology. For instance, Mendelsohn *et al.* applied combined SEM and AFM studies to demonstrate the spinodal-like surface morphology of highly porous LbL films obtained by additional solution treatment. The authors estimated average pore dimensions and the level of connectivity in the random network morphology [61]. Next, we present one recent example of a detailed AFM and TEM study of porous LbL films conducted in one of the authors' lab.

Robust, ultrathin, perforated, and freely suspended LbL membranes with uniform fine nanopores in the range of tens of nanometers have been fabricated using a fast and simple method of spin-assisted LbL assembly on hydrophobic substrates by Zimnitsky *et al.* (Figure 14.13) [62]. Nanoporous LbL membranes with thicknesses down to 20 nm were strong enough to be released from the sacrificial substrates, transferred onto various surfaces, and suspended over microscopic openings for further testings. The nanopore size can be controlled by tuning the number of polyelectrolyte bilayers, spinning speed, and properly selecting hydrophobic substrates. The authors demonstrated that the formation of nanopores is caused by partial dewetting of polyelectrolyte layers in the course of their deposition on the underlying hydrophobic surfaces. Nanoscale thickness of perforated membranes with relatively uniform size and high concentration of through nanopores might provide higher rates of transport through freely suspended LbL membranes.

In the freely suspended state, complete 3D morphology of the porous LbL membranes can be studied by scanning both sides of the LbL film with light AFM

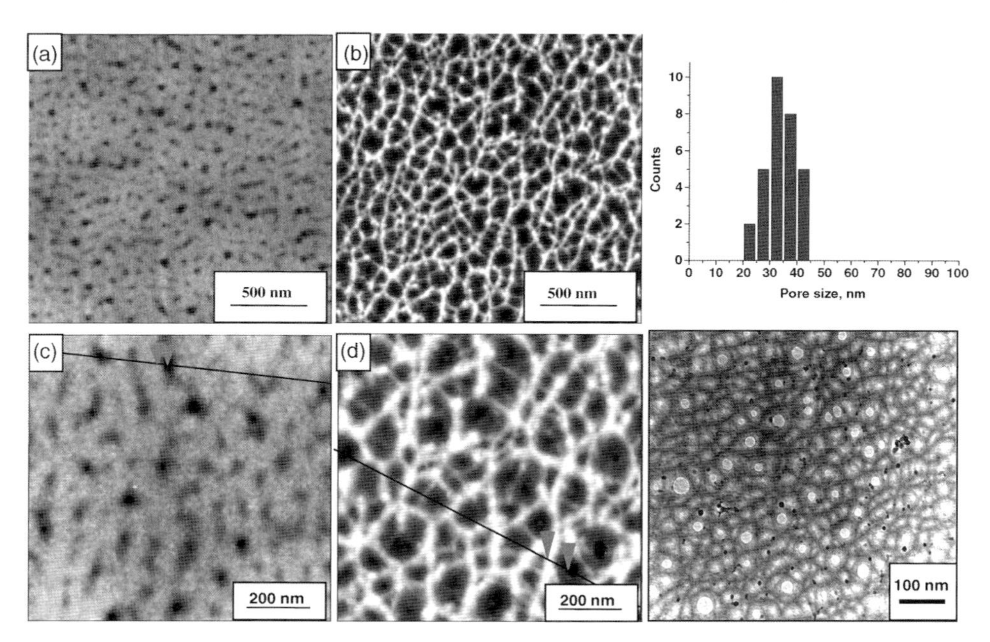

Figure 14.13 AFM images at different magnifications and TEM image of the freely suspended LbL films (top (a and c) and bottom (b and d) sides of the LbL films) and corresponding pore distribution derived from AFM images. Reprinted from Ref. [62].

tapping mode. The AFM image of the top side is similar to the one that was acquired by direct scanning of the LbL film on the substrate. This is an indication of the fact that the bulk morphology of the films does not change much during dissolution of sacrificial layers and handling. The lateral size of the pores varies depending on the number of layers and the nature of the hydrophobic substrates as observed on corresponding AFM images. Dimensions under different conditions vary from as low as 15 nm to as high as 80 nm. Surface distribution of the pore dimensions of any given LbL film is relatively narrow, with a standard deviation usually not exceeding 30% (Figure 14.13).

Comparison of AFM images for both sides of LbL films demonstrated that the surface morphology of the bottom of the LbL membrane that faced the underlying hydrophobic substrate is different from that for the top. The topmost surface shows porous morphology with a modest concentration of partially filled pores. Nanoscopic through holes located in some of the cells are barely detectable with the AFM images from the top of the film. In contrast, the bottom side displays well-developed cellular morphology with cell dimensions of 100–200 nm and well-defined networks of compact polyelectrolyte fibrils formed during the dewetting processes in the course of LbL assembly (Figure 14.13). The section analysis for LbL films with different numbers of layers finally reveals a cone-like shape of porous features as opposed to the common cylindrical shape of traditional porous materials.

In complementary TEM studies of the LbL membrane consisting of 15 polymeric bilayers, the presence of the cell-like morphology and the perforated nature of the LbL film has been confirmed (Figure 14.13). Holes, which correspond to the bright spots on the TEM image, distributed throughout the LbL membrane have a perfectly round shape and are rather monodisperse in size. The average size of the nanopores appears to be bigger on TEM images (40 nm) than on the AFM images (25 nm) due to the limitations of the tip access to the pore interior [63]. The spatial distribution of the cellular network, their sizes, and their geometry are in agreement with the models of dewetting processes on surfaces due to confined microphase separation known as Voronoi figures [64, 65]. The geometrical parameters of such cell-like morphologies and the resulting surface morphologies and physical properties are controlled by the dewetting conditions and the dynamics of solution spreading on surfaces as has been discussed in this study for a variety of different substrates.

14.2
Langmuir–Blodgett Films

Traditional LB films are obtained by sequential deposition of Langmuir monolayers of amphiphilic molecules from the air–water interface to a solid substrate (Figure 14.14) [1]. These highly ordered monolayer and multilayered LB films are frequently utilized for the formation of molecularly ordered organic solids as model systems for studying the surface behavior of a variety of amphiphilic synthetic and biological molecules and macromolecules. Owing to their high versatility, smooth morphology, easy preparation, and the ready control of chemical composition and

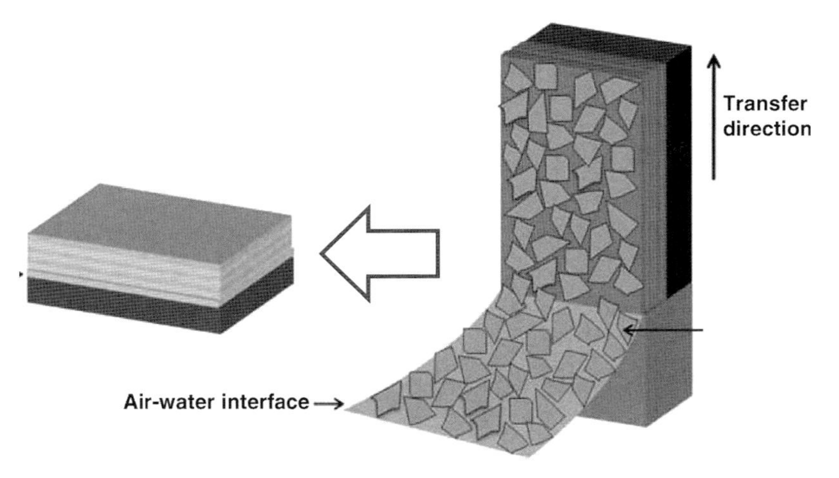

Figure 14.14 Fabrication of composite LB films.

surface molecular area, LB films have been a popular subject of investigation with AFM since the early 1990s.

14.2.1 Molecular Order and Defects

A classic example of the surface behavior of amphiphilic molecules can be represented by stearic acid (STA), an amphiphilic compound that has been widely studied for many years. A pressure–area isotherm for a Langmuir STA monolayer with designated regions of various physical states is shown in Figure 14.15 along with sketches of molecular ordering within monolayers at different surface pressures. These cartoons represent the classical understanding of different phase states within Langmuir monolayers prior to the monolayer collapse: disordered structure of the "gas" state, short-range ordered tilted chains for the "liquid" monolayer, and ordered closely packed chains in the "solid" monolayer [66].

In numerous studies, AFM images of LB monolayers of STA molecules under modest compression demonstrated the two-phase morphology of these Langmuir monolayers. The monolayer is composed of aggregated nanoscale domains and islands, but significant surface areas on the underlying silicon oxide substrate are uncovered with amphiphilic molecules (Figure 14.15). The thickness of individual STA domains of 1.9 nm indicated significant tilting (about 35°) of the alkyl chains within these LB monolayers with relatively large surface areas available for a single molecule.

At a higher lateral compression, when the surface pressure reaches more than 20 mN/m, the STA monolayer is in the "compressed solid" state with crystal packing of tilted alkyl chains expanded over large micrometer regions as was confirmed with AFM. In contrast to poorly ordered monolayers, the solid STA monolayers obtained at high pressure possessed smooth surface morphology with a microroughness in the

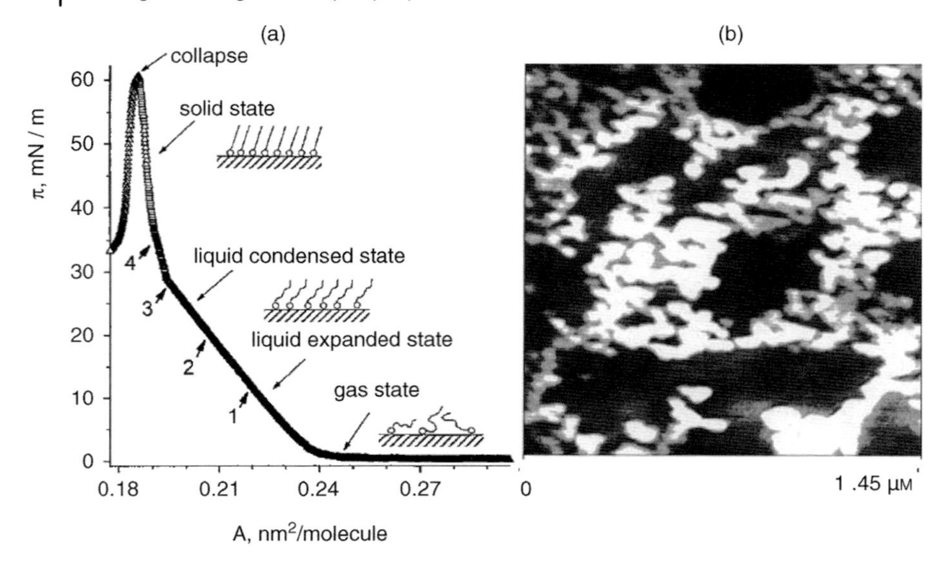

Figure 14.15 (a) Pressure–area diagram for Langmuir monolayer of STA molecules with designated areas of different physical states and sketches of molecular organization. (b) AFM topographical image of STA monolayer at intermediate surface pressures. Reprinted from Ref. [66].

range of 0.4–0.8 nm, common for LB monolayers from a wide variety of amphiphilic molecules. Monolayer defects, such as holes and ridges, were present but rarely observed over microscopic surface areas scanned with AFM. In dense monolayers, the thickness of the solid monolayer was in the range of 2.2–2.5 nm as measured from the depth of the pits that was close to the total length of the STA molecules (about 2.5–2.6 nm in an extended conformation), indicating near-perfect vertical orientation of the aliphatic chains.

Initial studies of LB films were mostly focused on the simple visualization of the molecular order of terminal groups of the vertically aligned alkyl chains of various representative model amphiphilic compounds, such as fatty acids, their salts, and lipids with contact AFM [67–70]. Molecular lattices with different crystal symmetries were revealed for densely packed LB monolayers from amphiphilic molecules with spatial parameters controlled by the chain length, nature of the polar heads, and the molecular density. Reproducible lattice patterns and lattice parameters were confirmed by many AFM studies in which common instrumental artifacts and surface damage could be avoided by applying very low normal forces, proper scanning conditions, and elimination of the usual scanning instabilities.

The application of high-resolution AFM imaging limited and eventually almost eliminated the use of the STM method that was very popular in earlier studies of LB films and generated a number of prominent artifacts and misleading hypes. Scanning in buffer solutions was exploited to minimize forces to avoid monolayer damage and reveal crystal lattices corresponding to those known from X-ray studies [71]. The role of the underlying substrates on the chain packing was demonstrated by imaging LB films, with a growing number of layers with thicker films showing smoother molecular surfaces less resembling the underlying topography [72].

AFM imaging of individual molecular defects within LB films was achieved by high-resolution contact mode at modest normal forces. Naturally occurring pinhole defects were found for fatty acid salts with through hole diameters as small as 10 nm [73]. On the other hand, applying high normal forces was demonstrated to be effective in producing new holes with different controlled sizes. Edge defects in the form of extra bilayer formation and the occurrence of paired dislocations were reported for solid LB monolayers by Bourdieu *et al.* [74]. Long-range molecular ordering for solid LB monolayers with surface modulations similar to that expected for liquid crystalline phases and occasionally occurring dislocations were observed by Peltonen *et al.* [75].

Well-defined domain morphology in lipid-based and mixed Langmuir monolayers similar to those observed with a fluorescent microscope for Langmuir films was characterized in great detail with AFM imaging by Chi *et al.* [76]. Micron-sized surface domains of local ordered molecular regions with irregular and faceted shapes at intermediate compressions were revealed in this study and the following studies from the same group. The authors demonstrated that the dimensions and the symmetry of domain morphology are controlled by the compression rates and deposition conditions. Different stiffness of substrates and ordered domains was suggested to be critically responsible for the high contrast observed in AFM images. Some other intriguing phenomena such as regular rippling effect caused by internal instabilities during LB deposition were also discovered in these studies [77].

Unique LB films from non-traditional amphiphilic organic compounds such as those with columnar ordering of discotic molecules have been widely studied with AFM as well. Densely packed disks within columns and hexagonally arranged columns were reported by Josefowicz *et al.* for LB films from triphenylene compounds [78]. The authors observed uniform and long-range orientation of straight columns along the dipping direction and correlated their findings with X-ray scattering results. Columnar LB films from amphiphilic discotic LC polymeric compounds have been studied in detail by Tsukruk and Janietz [79]. An AFM image of the surface area at the edge of the LB film with a clearly visible separated first and second bilayers of discotic molecules is shown in Figure 14.16. The surface of the top of the discotic LB film is very smooth, contains very few holes and other defects, and shows small deviations from planarity. The microroughness of the films averaged over a 5 μm^2 area was in the range 0.5–0.6 nm, which is very smooth compared to low molecular weight compounds and comparable to that of the underlying silicon substrate (surface microroughness of 0.2 nm for 1 μm^2 surface areas).

It was usually observed that the LB films from the discotic LC polymer are much smoother than traditional LB films made from low molecular weight amphiphilic compounds that usually display holes, edges, domain morphology, and other microscopic and nanoscale surface inhomogeneities. This perfect, low-defect surface morphology is caused by the formation of long-range intralayer ordering of a mesomorphic type with a delicate balance of core-to-core spatial arrangement of discotic groups arranged in long-range ordered columns lying parallel to the

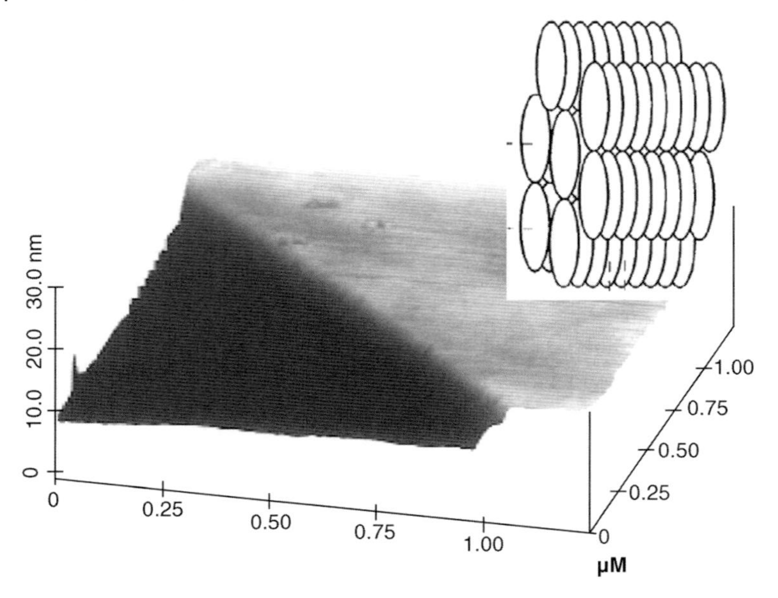

Figure 14.16 The edge of the LB film from discotic LC polymer and corresponding schematics of columnar orientation. Reprinted from Ref. [79].

supporting substrate (Figure 14.16). From the height histograms of the film edge, the total thickness of the first discotic bilayer of 6.8 nm was evaluated to be very different from the 5.3 nm thickness of the second bilayer. The observed increase in the thicknesses of discotic molecular layers in the vicinity of a silicon substrate was suggested to be caused by rearrangements of the internal molecular structures. The relaxation of molecular ordering from the surface driven to the regular "bulk" columnar structure was suggested to occur for molecular layers further from the silicon surface.

Other nontraditional amphiphilic compounds studied with AFM including four generations of monodendrons with multiple dodecyl alkyl tails from 1 to 8, an azobenzene spacer group capable of light-initiated photoisomerization, and a carboxylic acid polar head have been thoroughly investigated at the air–water and air–solid interfaces using AFM, X-ray reflectivity, and UV–vis spectrometry by Genson *et al.* [80]. From X-ray reflectivity data, the authors concluded that one- and two-tail monodendron molecules formed orthorhombic lateral packing with long-range intramonolayer ordering and a vertical orientation of molecules. However, the monodendrons of higher generations formed a kinked structure with the alkyl tails oriented perpendicular to the surface and the azobenzene group tilted at a large degree toward the surface. In contrast to lower generations, the four- and eight-tail dendron molecules formed uniform, smooth monolayers with very limited regions of short lamellae-like structures resolved by high-resolution AFM imaging. The photoisomerization of the azobenzene cores of lower generation monodendrons was studied with X-ray reflectivity at the air–water interface to determine the mobility of the molecules in loosely packed, two-dimensional monolayer films. Correspondingly, AFM measurements of Langmuir monolayers before and after light illumination confirmed the variable thickness of the monolayers during photoisomerization caused by folding central cores.

Evidence of the long-range in-plane ordering of the low-generation monodendron molecules was observed in AFM imaging of solid-supported Langmuir monolayers (Figure 14.17) [81]. At the lowest surface pressure studied, these molecules formed large, leaf-like domains with overall lengths of several micrometers composed of vertically oriented monodendron molecules. Upon increasing the surface pressure to 30 mN/m, which corresponds to the solid Langmuir monolayer state, the ordered domains decreased in overall size to 1 μm and became more densely packed.

Higher resolution AFM imaging of the Langmuir monolayers from monodendrons deposited at higher surface pressure revealed that the domain morphology is composed of a lamellae-like molecular structure formed by highly tilted bilayered monodendron molecules. AFM imaging of individual domains shows the random orientation of the lamellae structure in different domains (more clearly observed in the phase image), with the shift in the directions inside the domain caused by local defects. The spacing of these lamellar structures was calculated from 2D FFT analysis of AFM images to be 5.7 nm, indicating that bilayer packing of monodendron molecules with partially interdigitated alkyl tails is similar to that observed for other monodendrons in the bulk LC state.

Similar layering was observed for various bicomponent amphiphilic molecules with rigid cores and flexible branched tails [82, 83]. Domains with different orientation of planar bilayered molecular packing were observed with high-resolution AFM. Squared local packing and highly uniformly oriented lamellar structures were also observed for different chemical compositions of molecules. The periodicity of these ordered structures observed with AFM was defined by the molecular dimensions and was within 4–7 nm for different molecules. This study also demonstrated that practical resolution of below 1 nm can be easily achieved for molecular films at ambient conditions.

In another study, Gunawidjaja *et al.* demonstrated that highly branched thiophene-containing molecules with COOH-terminal groups formed stable, uniform, and robust Langmuir monolayers with thicknesses of 2–3 nm at the air–water interface at modest surface pressures (<10 mN/m) [84]. These stable Langmuir monolayers from highly photoluminescent conjugated dendritic molecules can be easily transferred to a solid substrate. However, AFM studies showed that the fabrication of ultrathin organized films (tens to 100 nm thicknesses) via multilayer LB deposition or spin casting is limited by the intense dewetting phenomenon. Additional complications came from the formation of globular surface aggregates on top of the first Langmuir monolayer due to strong intermolecular interactions. These thicker surface LB films showed a significant redshift of UV absorption, confirming a dense molecular packing with strong π–π interactions that prevented uniform distribution of these compounds on solid support and buildup of the traditional multilayered LB films, as observed by AFM imaging.

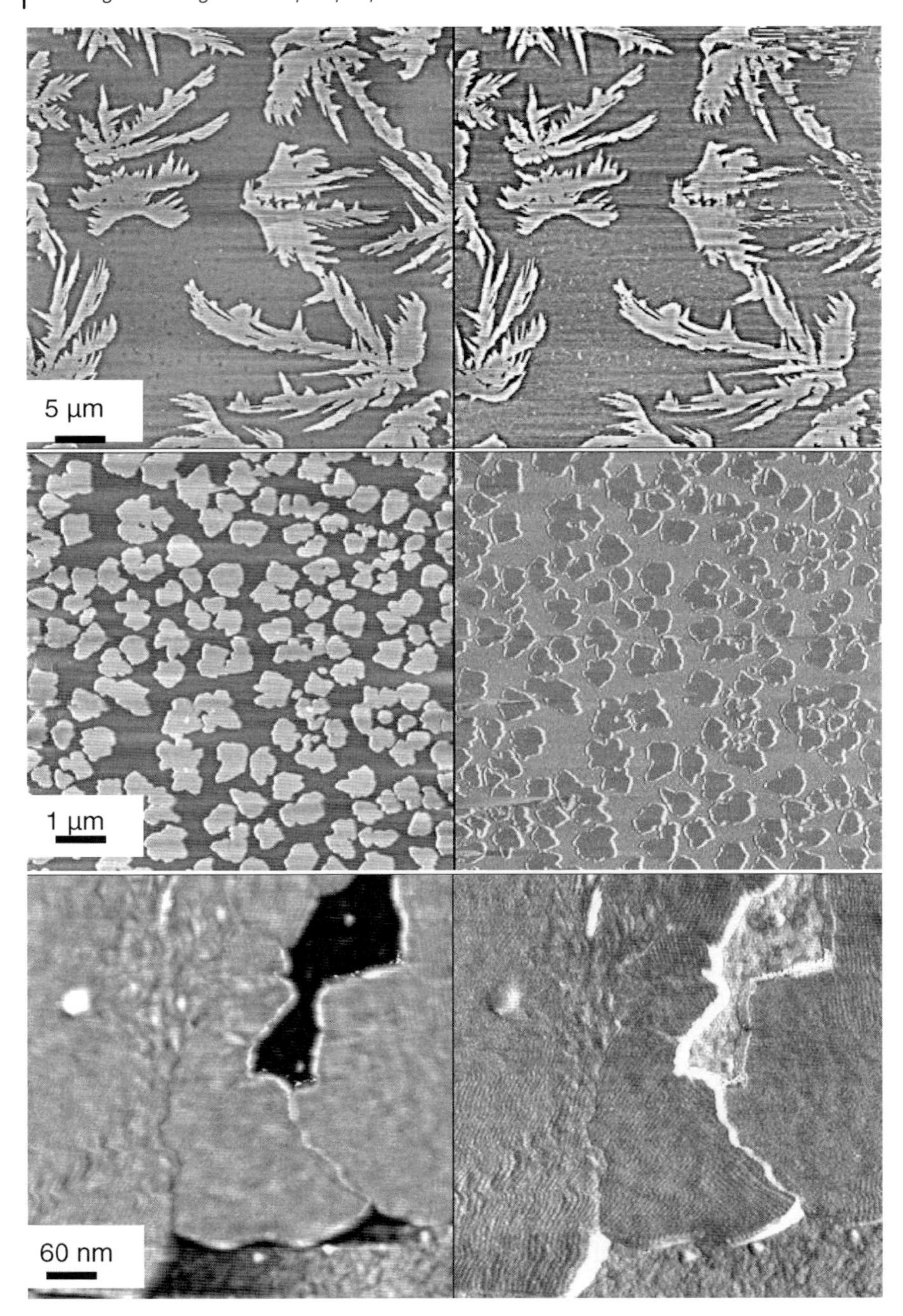

Figure 14.17 AFM images (topography (left) and phase (right)) of a Langmuir monolayer from two-tail monodendron molecules with irregularly shaped domains deposited at 3 mN/m (top) and at 20 mN/m (middle). High-resolution AFM image demonstrating lamellar structure observed within the domains (bottom). Reprinted from Ref. [80].

14.2.2 Mixed and Composite LB Films

AFM is an excellent tool for imaging microphase-separated morphologies within Langmuir monolayers composed of several different components due to high spatial resolution and easy observation of minute height and adhesion differences of different organic phases. The presence of different inorganic phases (metal nanoparticles or clay nanoplatelets) can be easily revealed with conventional AFM tapping imaging even at modest resolution.

Mixed Langmuir monolayers of amphiphilic acids from alkyl- and fluoro-terminated molecules were prepared and imaged by Imae *et al.* [85]. Height differences of various surface domains (usually clearly documented down to a few tenths of a nanometer) were utilized to identify their chemical composition and boundaries of phase stabilities. On the basis of these AFM observations, the authors suggested several molecular models of lateral and vertical ordering of different molecules in a single phase or in separate microscopic surface domains. The same principle of phase identification via the AFM observed difference in the domain heights was utilized to monitor microscopic phase separation in mixed fatty acid films by Ekelund *et al.* [86].

Similar observations were reported by Qaqish *et al.* on mixed Langmuir monolayers [87]. AFM imaging was combined with XPS and fluorescent microscopy experiments to firmly identify different compositions even if the difference in domain height was small to be detected with AFM. The authors concluded that phase-separated domains are controlled by compression conditions and an almost merged distinct morphology can be observed at high surface pressures. In another study, even more distinct rippled surface morphologies were observed for mixed Langmuir monolayers by Qaqish *et al.* [88]. Highly periodic microscopic stripes were observed for transferred Langmuir monolayers and related to the gradient phase separation of different components. The gradient chemical composition within these monolayers has been probed with continuous micromapping of adhesive forces.

Another interesting phenomenon of the nanoscale pattern formation in mixed Langmuir monolayers was observed with AFM by Gamboa *et al.* in the case of fluorinated alkanes added to amphiphilic diblock copolymers [89]. A nanoscale honeycomb lattice was formed due to the spontaneous segregation of the fluorinated component and lateral relaxation of morphology after the monolayer transfer to a solid substrate. In a different study, various patterned mixed Langmuir monolayers were exploited to assemble gold nanoparticles in an organized manner by Watanabe *et al.* [90]. Comprehensive AFM image analysis was conducted to characterize initial phase-separated surface morphologies. Then the location, aggregation, and distribution of small nanoparticles were easily identified by high-resolution AFM observations.

In another study, Peleshanko *et al.* applied AFM to study the surface behavior of the asymmetric amphiphilic heteroarm PEO/PS star polymer at the air–water interface and on solid substrates [91]. These star polymers differ in both architecture (four- and three-arm molecules, PEO-b-PS_3 and PEO-b-PS_2) and in the length of PS chains (molecular weight varied from about 10 000 up to 24 000 g/mol). The authors observed that for a given chemical composition with a predominant content of hydrophobic blocks, the compression behavior of the PS domain structure ultimately controls the surface behavior and the final morphology of the monolayers deposited at different surface pressures. The final surface morphology observed with AFM shows circular micelles with a peculiar vertical segregation of different arms – underlying PEO arms and topmost located PS chains.

In another study of star block copolymers at interfaces, Peleshanko and coworkers focused on the role of functional terminal group combinations for four-arm $(X\text{-PEO})_2\text{-}(PS\text{-}Y)_2$ heteroarm star copolymers in their surface morphology (Figure 14.18) [92]. The authors synthesized a series of star block copolymers with combinations of bromine, amine, hydroxyl, and carboxylic terminal groups and showed with AFM that all circular, cylindrical, and lamellar surface morphologies can be formed depending on the chemical composition and deposition conditions.

The AFM study of a wide variety of morphologies concluded that hydrophilic functional groups attached to hydrophobic chains and hydrophobic functional groups attached to hydrophilic chains resulted in the stabilization of the spherical surface domain morphology, rather than the cylindrical morphology predicted for the given chemical composition in the bulk state of block copolymers. The replacement of functional terminal groups of only hydrophobic polymer chains was found with AFM imaging to be even more effective in creating stable and fine circular domain morphology (Figure 14.18). The authors observed that the ionization of carboxylic terminal groups at higher pH led to a greater solubility of PEO chains in the water subphase that, along with deionization of amine terminal groups, prevented the lateral aggregation of PS-enriched domains and further promoted the formation of the nanoscale circular morphology.

The characteristic mixed small–medium circular and circular-cylindrical domain morphology was observed with AFM for all star block copolymers with Br-terminated PS arms (Figure 14.18). Lateral domain sizes were widely distributed from 10 to 100 nm. Effective monolayer thicknesses measured using spectroscopic ellipsometry were within 2.3–3.8 nm for all Langmuir monolayers studied at low pressures and increased to 3.7–4.5 nm for higher compressions. These thickness values closely resemble those measured by AFM scratch testing, thus confirming the predicted dense micelle conformation on solid substrates. The domain heights measured independently with AFM were much more uniform and stayed within 4.3–6.9 nm for all surface pressures applied. The domain heights increased slightly with rising surface pressure for both star copolymers with hydrophobic terminal groups of PEO chains but remained unchanged for star copolymers with OH- and COOH-terminated PEO chains. The surface coverage with hydrophobic PS domains was within 28–48%, as estimated from the AFM images after taking tip dilation into account.

The replacement of the hydrophobic end groups of PEO chains of these star block copolymers with hydrophilic hydroxyl groups resulted in decreased stability of the circular microstructure and the appearance of mixed circular and cylindrical domains with a wide distribution of lateral sizes from 10 to 200 nm (Figure 14.18). On the other hand, the replacement of the bromine terminal groups with NH_2 groups

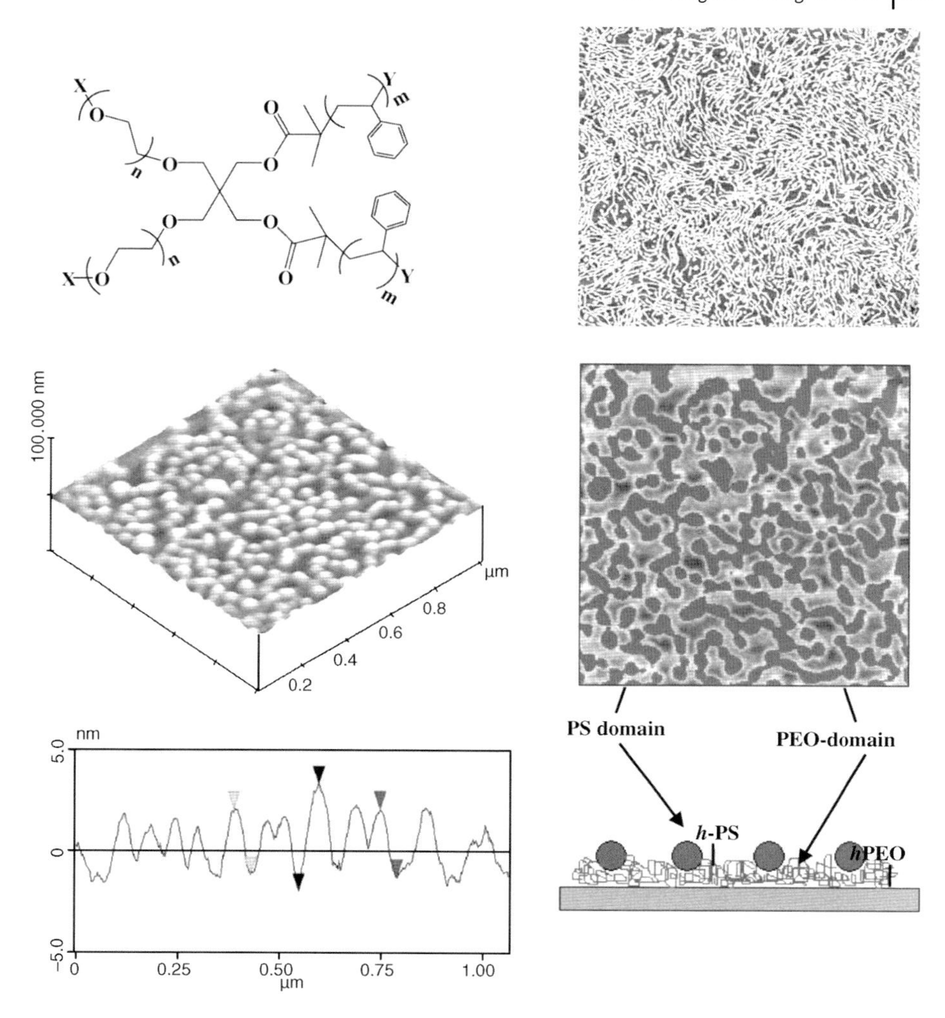

Figure 14.18 AFM topography (1 μm^2) of star block copolymers within Langmuir monolayers showing cylindrical, circular, and bicontinuous morphologies; 3D AFM image of circular domains; corresponding cross section of circular domains; and a model of deposited PEO–PS copolymer on the silicon wafer substrate. Reprinted from Ref. [92].

altered the morphological pattern to fine circular domains with an average diameter below 50 nm, still easily detectable with high-resolution AFM. This indicates that the hydrophobic–hydrophilic combination enhances the microphase separation, leading to stable circular domain morphology. Moreover, microphase separation driven by the dissimilar nature of polymer chains and end functional groups is more effective in controlling aggregation of the hydrophobic PS chains compared to the hydrophilic PEO chains. Finally, considering that the supporting substrate is hydrophilic, the general model suggested that the topmost individual domains are PS and the underlying layer is composed mainly of PEO chains (Figure 14.18) [93]. This general picture of molecular ordering was developed into specific models for star block copolymers with different functionalized terminal groups and confirmed by AFM observations.

Nanofibrillar micellar structures formed by the amphiphilic hyperbranched molecules within Langmuir monolayers were studied by Rybak *et al.* [94]. The authors utilized these monolayers as an organized template for silver nanoparticle formation directly on the metal ion-containing water subphase and applied AFM and *in situ* X-ray reflectivity to monitor these transformations. They observed that the silver nanoparticles were indeed formed within the multifunctional, amphiphilic, hyperbranched molecules with diameters of nanoparticles within 2–4 nm, controlled by the core dimensions and the interfibrillar free surface area (Figure 14.19). Suggested mechanisms of the nanoparticle formation derived from both X-ray and AFM observations involve oxidation of primary amino groups by silver catalysis facilitated by the "caging" of silver ions within surface areas dominated by the multibranched cores.

AFM observations of these Langmuir monolayers on solid substrates showed that, upon addition of potassium nitrate into the subphase, the Langmuir monolayer templated the nanoparticle formation along the nanofibrillar structures (Figure 14.19). Morphology of the silver-containing LB monolayer was affected by the amount of $AgNO_3$ in the subphase. After 7 and 24 h of exposure to a subphase of 5 mM $AgNO_3$, larger aggregates and clusters of silver nanoparticles up to 100 nm across were observed. When the concentration was decreased to 1 mM $AgNO_3$, the silver nanoparticle diameter was reduced. The histogram of nanoparticle heights calculated from AFM images showed the average height to be 2.6 nm for a wide range of conditions.

Typical nanofibrillar morphology of the monolayer was clearly visible for any deposition condition, but in general little correlation was observed between nanofibrillar surface morphology and the resulting nanoparticle surface distribution. Concurrent TEM images of these samples showed a low density of silver nanoparticles adsorbed under these conditions, and the distribution of nanoparticle diameters was similar to the height distribution determined by AFM imaging. The density of silver nanoparticles, which is much lower in AFM images than in TEM images, suggested that a significant portion of silver nanoparticles was screened by the polymeric monolayer having an overall thickness comparable to the nanoparticle diameter. Further, *in situ* X-ray scattering and reflectivity measurements conducted directly on the LB trough during the monolayer exposure to metal ions confirmed the process of the gradual transformation from silver salt to silver nanocrystals.

In an early study, Bliznyuk *et al.* discussed the surface morphology and molecular ordering of LB films from complexes of rigid-backbone poly(naphthoylene benzimidazole) precursor and STA molecules [95]. The anisotropic shape of the rigid, ladder-like polymers with highly conjugated backbones promotes high in-plane ordering of the macromolecules during the formation of LB films, and this ordering was also preserved after thermally induced polymerization and STA desorption. AFM observations complemented by X-ray measurements were used for the characterization of molecular ordering and surface morphology of LB films on a wide range of

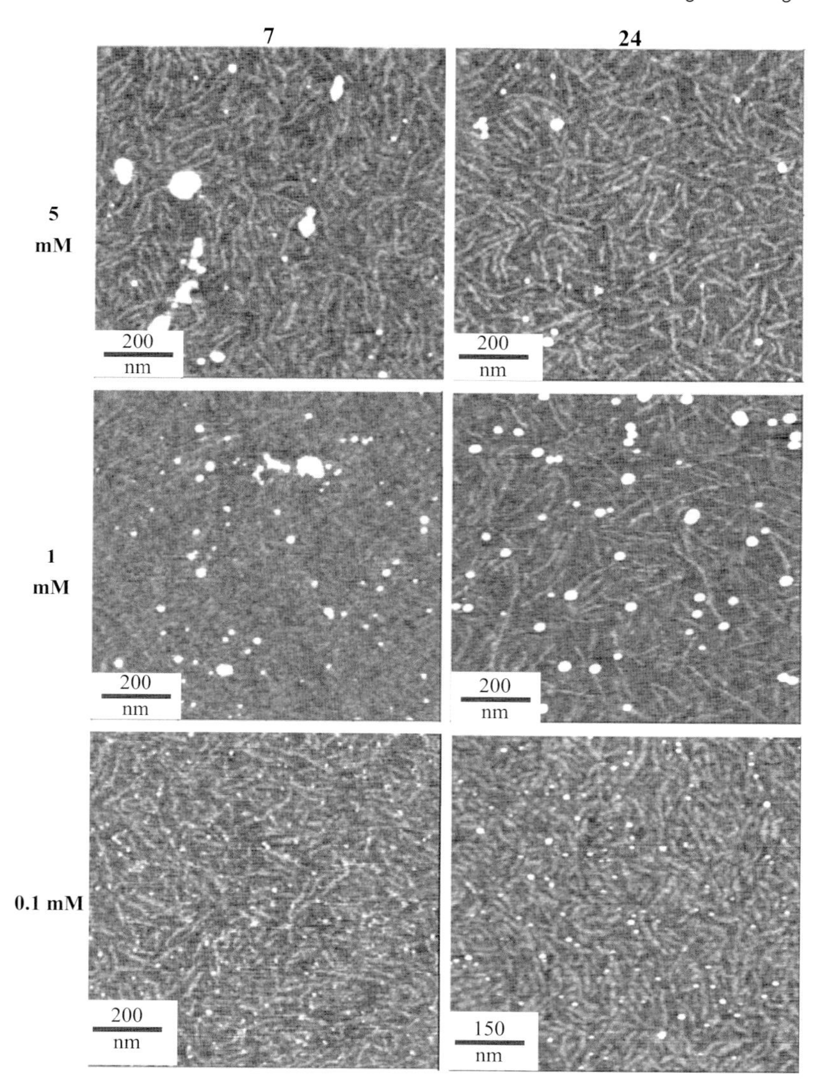

Figure 14.19 AFM topographical images of HBP–nanoparticle LB monolayers as both time of growth and subphase solution concentration are varied. Height scale of all images is 10 nm. Reprinted from Ref. [94].

scales. AFM showed that the surfaces of these composite LB films right after deposition are smooth, with a microroughness of 0.3–0.6 nm and wavy molecular ridges aligned along the dipping direction.

In recent work by Kulkarni *et al.*, negatively charged graphene oxide layers were incorporated into multilayered nanocomposites via LB deposition and studied with AFM [96]. These LbL–LB graphene oxide nanocomposite films were released as robust, freestanding membranes with large lateral dimensions (cm) and thicknesses of around 50 nm. Sectional analysis of the AFM image revealed planar flakes up to few tens of micrometers across having modest polydispersity in the thickness and confirming that the exfoliated flakes are single layers and bilayers (see Chapter 12 for more discussion) [97–99]. By controlling the surface pressure, the coverage of graphene oxide sheets was manipulated to give a uniform deposition with only occasional wrinkles and overlaps observed. The microroughness of graphene oxide sheets of 0.38 nm indicates that the atomic smoothness was largely unaffected by the further coating with polyelectrolyte layers. High contrast in phase images showed a large difference in surface properties of LbL substrates and graphene oxide sheets caused by their very different surface functionalities and stiffness.

A stable Langmuir monolayer of PVP-modified silver nanoplates was formed at the air–water interface and transferred onto a silk substrate in a recent study by Kharlampieva *et al.* (Figure 14.20) [44]. Optical imagine of the Langmuir monolayer showed the highly reflective properties of these composite silver-PVP monolayers with a clear mirror image of the balance visible. Such mirror-like properties indicate the dense packing of horizontal plates with high planarity responsible for excellent reflective properties. Indeed, an AFM image of the Langmuir film deposited onto a silk support showed densely packed triangular silver nanoplatelets with polydisperse lateral dimensions but high surface coverage (Figure 14.20). AFM images show the near-perfect planar orientation with no vertically oriented or tilted platelets and a modest overall surface microroughness of these nanocomposite LB films of only 6 nm, high but comparable to regular nanocomposite LB films.

A study by Gunawidjaja *et al.* dealt with the surface morphology of two series of organically functionalized core–shell silsesquioxane (POSS-M) derivatives with various hydrophobic–hydrophilic terminal group compositions at the air–water interface and on solid surfaces (Figure 14.21) [100]. POSS-M refers to mixed silsesquioxane cores composed of polyhedra, incompletely condensed polyhedra, ladder-type structures, open structures, and linear structures. The surface morphologies obtained with AFM imaging for the POSS-based Langmuir monolayers with various hydrophobic–hydrophilic combinations at low surface pressures (0.5 mN/m) are similar to those observed for the classical amphiphilic star polymer with fine domain morphology, frequent pinholes, and other surface defects.

However, at higher surface pressure (close to 5 mN/m), the POSS-M compounds with a lower content of hydrophilic groups formed more homogeneous Langmuir monolayers with smooth and uniform surface morphologies and multiple nanoscale islands (Figure 14.21). AFM imaging revealed that the molecules can form three distinct surface morphologies depending on the nature of the branches with one-dimensional, circular, and planar aggregates. The lengths of these one-dimensional aggregates and the diameters of the circular aggregates are in the micrometer range. Almost all POSS compounds eventually formed a uniform Langmuir monolayer at higher surface pressures. Furthermore, at the highest surface pressure, the overall surface morphology is eventually dominated by the crowded POSS cores, with the alkyl tails forming the topmost surface layer.

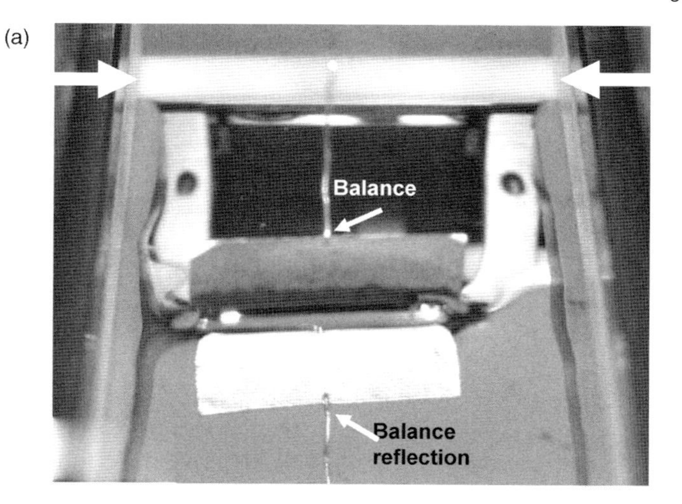

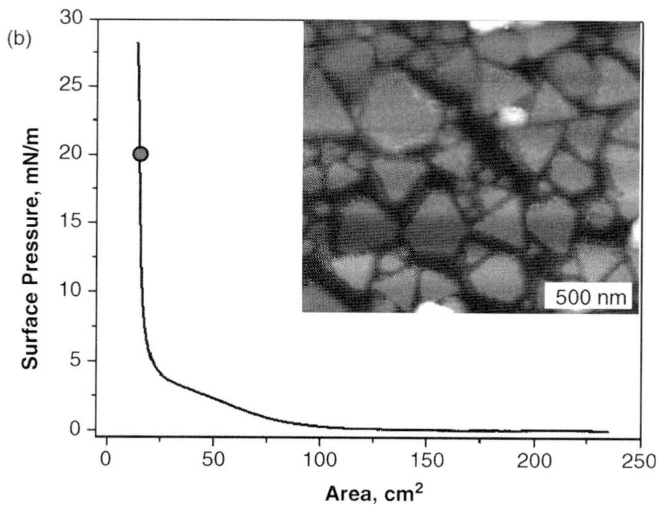

Figure 14.20 (a) LB trough with Langmuir monolayer containing silver nanoplatelets on the water surface after compression (shown with arrows). The clear reflection of the balance illustrates the highly reflective surface of the monolayer. (b) Langmuir isotherm of this monolayer. The dot represents the surface pressure (20 mN/m) at which a monolayer was transferred onto a solid substrate. Inset shows an AFM image of the monolayer transferred onto solid support. Reprinted from Ref. [44].

14.2.3 Mechanical and Tribological Properties

LB films are frequently considered to be excellent, versatile, and well-defined molecular models of boundary lubrication and associated tribological phenomena. Numerous AFM studies, mostly in contact and lateral (FFM) modes, were devoted to

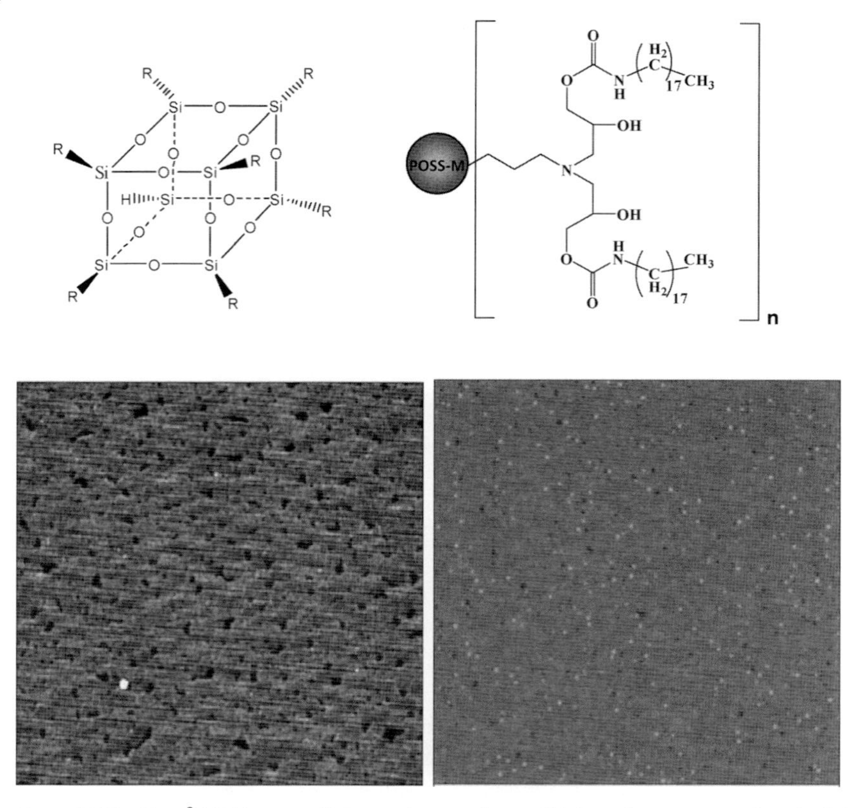

Figure 14.21 5 μm^2 AFM images of LB monolayers at low and high surface pressure of amphiphilic POSS-M compounds along with general schematics, *z*-scale is 10 nm. Reprinted from Ref. [100].

understanding the mechanical, adhesive, and frictional properties of Langmuir monolayers since the introduction of AFM [101]. Apparent significant differences in adhesive and shear properties of different amphiphilic molecules and tethered polymer chains have been widely exploited for the visualization of surface domains with different chemical compositions and even different locally correlated orientations of molecules. Here, we will illustrate common efforts undertaken in this direction with several recent studies of LB films included for illustration.

Friction and wear of organic LB films were studied with FFM by Meyer *et al.* [102]. Significant reduction in the friction response was observed for LB domains compared to the bare substrate. This difference was explained by much lower shear modulus of the organic films (usually with shear modulus well below 10 GPa) and much lower adhesive forces. Some "hidden" defects of the LB surface invisible in regular topographical imaging mode became visible only in shear-detected mode, and thus FFM mode became a very popular approach for the monitoring of the uniformity of the LB films. However, higher shear forces exerted by the AFM tip may result in wearing of the pristine LB surface, rounding domain boundaries, and even lateral motion of microscopic domains and islands, all observed by the authors. The latest results indicated weak interfacial bounding of Langmuir monolayers to a solid substrate and strong cohesive interactions of crystallizable alkyl chains within monolayer islands.

AFM-based nanoindentation of organic and polymeric Langmuir films is widely exploited to measure their mechanical properties. In one of the pioneering studies, Langmuir films of fatty acids were explored to directly measure the elastic modulus and the mechanical contact areas by Weihs *et al.* [103]. The AFM measured value of 9 GPa was reported to be similar to that identified independently from bulk material studies. The contact diameter was estimated to be around 2 nm under regular normal loads. The authors concluded that a large contact area limits the practical AFM resolution at high normal loads. These estimations indicate that only very sharp tips (radius below 10 nm) and scanning at very low forces (e.g., in liquid) might provide proper conditions for near-molecular resolution under regular conditions of AFM scanning.

In recent developments, Oncins *et al.* applied all major AFM, FFM, and SFS modes to classical Langmuir films from fatty acids to unambiguously determine their tribological and mechanical properties [104]. The authors observed that the forces required to puncture the monolayers strongly depend on the orientation of alkyl chains, increasing significantly for more vertical orientation that can be induced by significant lateral compression of the monolayers. The FFM measurements conducted concurrently revealed three distinct shearing regimes with different properties; very low friction of elastically deformed monolayers was observed at low normal load, but it was replaced by high friction at the onset of plastic deformation of organic monolayers at higher loads, followed by their complete removal ("nanoshaving") at even higher shear forces.

In another study, Oncins *et al.* studied mechanical behavior of lipid Langmuir monolayers over a range of temperatures [105]. AFM scanning in liquid was conducted and revealed phase transitions from crystalline to liquid state at elevated temperatures, as was expected from independent measurements. This thermal transition resulted in lower friction forces due to the dramatic reduction in the elastic resistance of the disordered and partially tilted alkyl chains. Nonuniform mechanical properties were revealed during SFS micromapping of LB films containing DNA complexes by Nayak and Suresh [106]. The elastic moduli of DNA-containing Langmuir monolayers estimated from these SFS measurements were much higher for the regions with dominating complex presence, but the dissipation of the energy in these regions was observed to be much smaller.

LB films of a polyglutamate (PG) statistical copolymer have been investigated using both AFM and X-ray reflectivity [107]. AFM showed that PG LB films on silicon surfaces possess an "expanded" thickness of 4.12 nm (larger than in the bulk state) and an average microroughness of 0.7 nm, a characteristic of smooth molecular surfaces. Using the AFM tip, several surface modification modes were demonstrated: the fabrication of large (several hundreds of nanometers) and small (several nanometers) through holes, "writing" of grooves of predictable geometry, the fabrication of pinholes of single layer and bilayer depth, the cleaning of the surface, and the fabrication of holes inside this new area. Formation of a localized "abrasion" surface morphology was related to a plastic deformation caused by stick-slip movement of the AFM tip under selected scanning conditions (Figure 14.22).

The anisotropic mechanical properties of the PG bilayers revealed with AFM "rubbing" were also revealed and attributed to the preferred orientation of the PG backbones along the dipping direction. To observe the difference in the mechanical response, the selected surface area was scanned in the parallel direction and compared with surface areas that were scanned transverse to the dipping direction. Instead of a highly oriented abrasion morphology observed during transversal scanning and caused by periodic stick-slip tip motion, random surface deviations in the 0.4–0.6 nm range were observed in the central region while scanning in the direction of the PG backbones (Figure 14.22). The local microroughness in this area increases almost steadily for the first three scans from 0.4 to 0.9 nm. Succeeding scans with increasing sizes can produce sequential square patterns at the PG surface (Figure 14.22).

AFM observations of LB monolayers of a classic boundary lubricant, STA, in fluid state (at intermediate surface pressures) at higher forces revealed long-range wear and reconstruction of monomolecular films under the localized shear forces caused by the sliding AFM tip [108]. The STA monolayer subjected to shear forces in a fluid state displays a flow of material from the worn area due to weak interaction with solid substrate. As a result of this directional flow induced by shear forces generated by the AFM tip in the contact mode and the material redistribution, a multilayer buildup was observed within the microscopic surface areas far from the AFM affected area. Enhanced surface diffusion of mobile organic material under local shear forces is responsible for the observed long-range effects of the local shear stresses produced within the localized contact area.

It has been also observed that solid and fluid LB monolayers show very different velocity dependencies of friction forces (Figure 14.23). For solid STA Langmuir monolayers, a monotonic increase in the friction forces with velocity rising from 0.02 to 1000 μm/s was observed. In contrast, for the fluid STA monolayers, the friction forces behave nonmonotonically with a pronounced peak achieved at intermediate velocities (Figure 14.23). This nonmonotonic velocity behavior was related to the matching of the characteristic molecular relaxation time and the frequency of interaction of the SFM tip with a characteristic molecular dimension of the fluid monolayers. The friction coefficient for these monolayers remained low and did not exceed 0.05–0.06 at the highest shear velocities, as measured with calibrated lateral tip deflection. This value is close to the friction coefficients for the STA monolayers reported earlier (in the range of 0.07–0.15) for various conditions (sliding pairs, pressures, and velocities).

On the other hand, different deformational behaviors were observed for these LB monolayers in solid state by observing the thickness variation under different normal and shear forces induced by the AFM tip. At low normal loads (<120 nN), very small compressions (below 3% of the initial thickness) were observed for solid STA monolayer islands. In contrast, significant compressions of the STA monolayers (reaching 35% of the initial thickness) were observed before the irreversible damaging

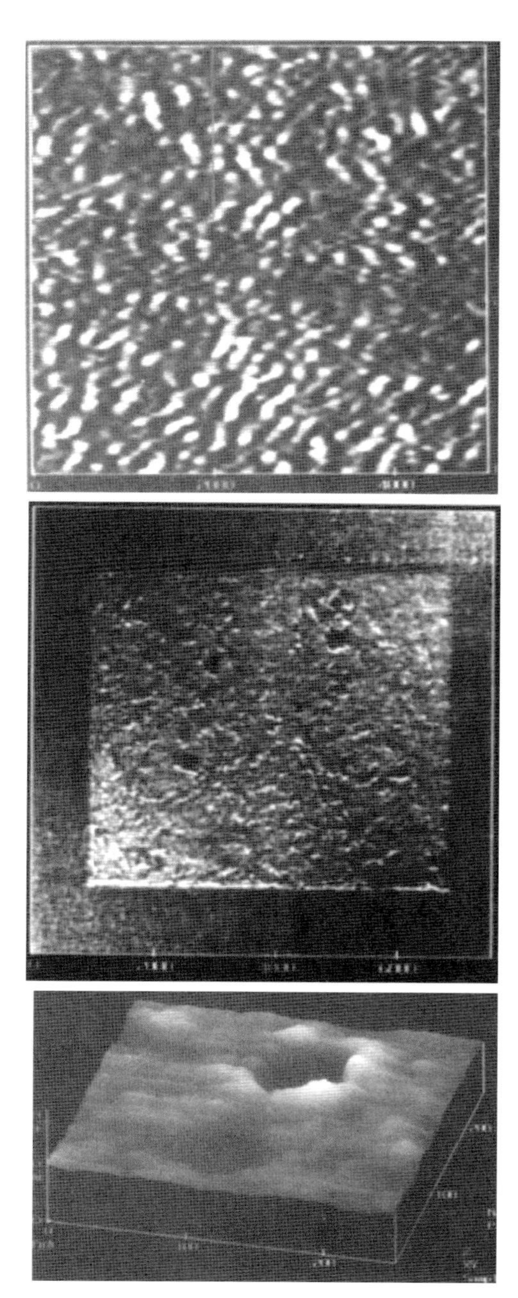

Figure 14.22 AFM image (8 μm^2) of a square region with periodic ripples (top) and randomly distorted surface features (middle) produced on the PG bilayer surface during scanning with high force and a through anisotropic hole “dug” in PG films by the AFM tip (bottom). Reprinted from Ref. [107].

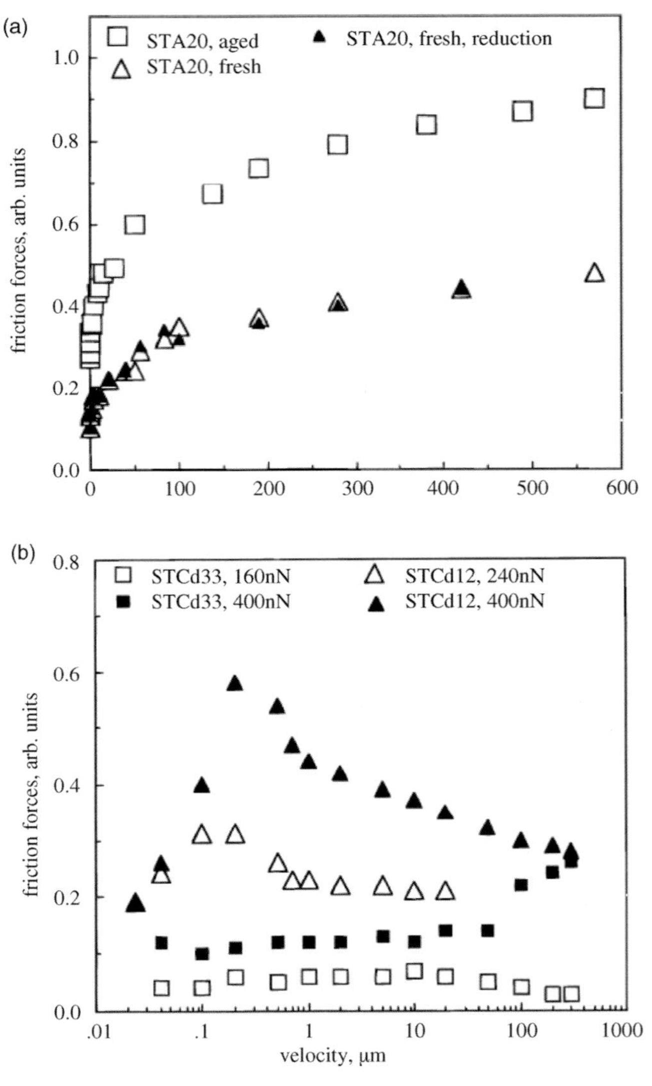

Figure 14.23 Velocity dependence of the friction forces for (a) solid and (b) fluid STA monolayers under different normal loads. Reprinted from Ref. [108].

was observed. The observed compression was related to the collective tilting of the molecules under normal loads due to the formation of gauche conformers in alkyl chains. The estimated Young's modulus from the compression rate was in the range of 0.2–0.7 GPa for very small elastic deformations. However, the elastic modulus decreased sharply at higher deformations when the alkyl chains were highly tilted to accommodate the high indentation depth of the AFM tip. The estimated elastic modulus of STA Langmuir monolayers under these deformational conditions lies within the interval from 20 to 40 MPa, which is characteristic of disordered alkyl chains.

References

1 Roberts, G.G. (ed.) (1990) *Langmuir–Blodgett Films*, Plenum Press, New York.
2 Decher, G. (1997) Fuzzy nanoassemblies: toward layered polymeric multicomposites. *Science*, **277** (5330), 1232–1237.
3 Decher, G. and Schlenoff, J.B. (eds) (2003) *Multilayer Thin Films: Sequential Assembly of Nanocomposite Materials*, Wiley-VCH Verlag GmbH, Weinheim.
4 Lvov, Y. and Möhwald, H. (eds) (2000) *Protein Architecture: Interfacing Molecular Assemblies and Immobilization Biotechnology*, Marcel Dekker, New York.
5 Lvov, Y., Decher, G., and Möhwald, H. (1993) Assembly, structural characterization, and thermal behavior of layer-by-layer deposited ultrathin films of poly(vinyl sulfate) and poly(allylamine). *Langmuir*, **9** (2), 481–486.
6 Arys, X., Jonas, A.M., Laschewsky, A., Legras, R., and Mallwitz, F. (2005) Supramolecular polyelectrolyte assemblies, in *Supramolecular Polymers*, 2nd edn (ed. A. Ciferri), CRC Press, Boca Raton, FL, pp. 651–710.
7 Tsukruk, V.V., Rinderspacher, F., and Bliznyuk, V.N. (1997) Self-assembled multilayer films from dendrimers. *Langmuir*, **13** (8), 2171–2176.
8 Tsukruk, V.V. (1998) Dendritic macromolecules at interfaces. *Adv. Mater.*, **10** (3), 253–257.
9 Tsukruk, V.V., Bliznyuk, V.N., Visser, D., Campbell, A.L., Bunning, T.J., and Adams, W.W. (1997) Electrostatic deposition of polyionic monolayers on charged surfaces. *Macromolecules*, **30** (21), 6615–6625.
10 Jiang, H., Su, W., Hazel, J., Grant, J.T., Tsukruk, V.V., Cooper, T.M., and Bunning, T.J. (2000) Electrostatic self-assembly of sulfated C_{60}–porphyrin complexes on chitosan thin films. *Thin Solid Films*, **372** (1–2), 85–93.
11 Zhai, L., Cebeci, F.C., Cohen, R.E., and Rubner, M.F. (2004) Stable superhydrophobic coatings from polyelectrolyte multilayers. *Nano Lett.*, **4** (7), 1349–1353.
12 Zhai, L., Nolte, A.J., Cohen, R.E., and Rubner, M.F. (2004) pH-gated porosity transitions of polyelectrolyte multilayers in confined geometries and their application as tunable Bragg reflectors. *Macromolecules*, **37** (16), 6113–6123.
13 Shi, F., Wang, Z., Zhao, N., and Zhang, X. (2005) Patterned polyelectrolyte multilayer: surface modification for enhancing selective adsorption. *Langmuir*, **21** (4), 1599–1602.
14 DeLongchamp, D.M. and Hammond, P.T. (2004) High-contrast electrochromism and controllable dissolution of assembled Prussian blue/polymer nanocomposites. *Adv. Funct. Mater.*, **14** (3), 224–232.
15 Farhat, T.R. and Hammond, P.T. (2005) Designing a new generation of proton-exchange membranes using layer-by-layer deposition of polyelectrolytes. *Adv. Funct. Mater.*, **15** (6), 945–954.
16 Quinn, J.F. and Caruso, F. (2004) Facile tailoring of film morphology and release properties using layer-by-layer assembly of thermoresponsive materials. *Langmuir*, **20** (1), 20–22.
17 Zhang, Y., Yang, S., Guan, Y., Cao, W., and Xu, J. (2003) Fabrication of stable hollow capsules by covalent layer-by-layer self-assembly. *Macromolecules*, **36** (11), 4238–4240.
18 Fu, Y., Xu, H., Bai, S., Qiu, D., Sun, J., Wang, Z., and Zhang, X. (2002) Fabrication of a stable polyelectrolyte/Au nanoparticles molecular film. *Macromol. Rapid Commun.*, **23** (4), 256–259.
19 Hammond, P.T. (2004) Form and function in multilayer assembly: new applications at the nanoscale. *Adv. Mater.*, **16** (15), 1271–1293.
20 Jiang, C., Markutsya, S., Pikus, Y., and Tsukruk, V.V. (2004) Freely suspended nanocomposite membranes as highly sensitive sensors. *Nat. Mater.*, **3** (10), 721–728.
21 Mamedov, A.A., Kotov, N.A., Prato, M., Guldi, D.M., Wicksted, J.P., and Hirsch, A. (2002) Molecular design of strong single-wall carbon nanotube/polyelectrolyte multilayer composites. *Nat. Mater.*, **1** (3), 190–194.

22 Chu, Y.-C., Wang, C.-C., and Chen, C.-Y. (2005) A new approach to hybrid CdS nanoparticles in poly(BA-*co*-GMA-*co*-GMA-IDA) copolymer membranes. *J. Membr. Sci.*, **247** (1–2), 201–209.

23 Lu, Y., Liu, G.L., and Lee, L.P. (2005) High-density silver nanoparticle film with temperature-controllable interparticle spacing for a tunable surface enhanced Raman scattering substrate. *Nano Lett.*, **5** (1), 5–9.

24 Ko, H., Jiang, C., and Tsukruk, V.V. (2005) Encapsulating nanoparticle arrays into layer-by-layer multilayers by capillary transfer lithography. *Chem. Mater.*, **17** (22), 5489–5497.

25 Tsukruk, V.V. (1997) Assembly of supramolecular polymers in ultrathin films. *Prog. Polym. Sci.*, **22** (2), 247–311.

26 Lvov, Y., Ariga, K., Onda, M., Ichinose, I., and Kunitake, T. (1997) Alternate assembly of ordered multilayers of SiO_2 and other nanoparticles and polyions. *Langmuir*, **13** (23), 6195–6203.

27 Lvov, Y., Ariga, K., Onda, M., Ichinose, I., and Kunitake, T. (1999) A careful examination of the adsorption step in the alternate layer-by-layer assembly of linear polyanion and polycation. *Colloid Surf. A*, **146** (1–3), 337–346.

28 Lobo, R.F.M., Pereira-da-Silva, M.A., Raposo, M., Faria, R.M., and Oliveira, O.N. (2003) The morphology of layer-by-layer films of polymer/polyelectrolyte studied by atomic force microscopy. *Nanotechnology*, **14** (1), 101–108.

29 Jiang, C. and Tsukruk, V.V. (2006) Freestanding nanostructures via layer-by-layer assembly. *Adv. Mater.*, **18** (7), 829–840.

30 Jiang, C. and Tsukruk, V.V. (2005) Organized arrays of nanostructures in freely suspended nanomembranes. *Soft Matter*, **1** (5), 334–337.

31 Jiang, C., Markutsya, S., and Tsukruk, V.V. (2004) Compliant, robust, and truly nanoscale free-standing multilayer films fabricated using spin-assisted layer-by-layer assembly. *Adv. Mater.*, **16** (2), 157–161.

32 Menchaca, J.-L., Jachimska, B., Cuisinier, F., and Perez, E. (2003) *In situ* surface structure study of polyelectrolyte multilayers by liquid-cell AFM. *Colloid Surf. A*, **222** (1–3), 185–194.

33 Elzbieciak, M., Zapotoczny, S., Nowak, P., Krastev, R., Nowakowska, M., and Warszynski, P. (2009) Influence of pH on the structure of multilayer films composed of strong and weak polyelectrolytes. *Langmuir*, **25** (5), 3255–3259.

34 McAloney, R.A., Dudnik, V., and Goh, M.C. (2003) Kinetics of salt-induced annealing of a polyelectrolyte multilayer film morphology. *Langmuir*, **19** (9), 3947–3952.

35 Drummy, L.F., Koerner, H., Farmer, K., Tan, A., Farmer, B.L., and Vaia, R.A. (2005) High-resolution electron microscopy of montmorillonite and montmorillonite/epoxy nanocomposites. *J. Phys. Chem. B*, **109** (38), 17868–17878.

36 Tang, Z., Kotov, N.A., Magonov, S., and Ozturk, B. (2003) Nanostructured artificial nacre. *Nat. Mater.*, **2** (6), 413–418.

37 Podsiadlo, P., Liu, Z., Paterson, D., Messersmith, P.B., and Kotov, N.A. (2007) Fusion of seashell nacre and marine bioadhesive analogs: high-strength nanocomposite by layer-by-layer assembly of clay and L-3,4-dihydroxyphenylalanine polymer. *Adv. Mater.*, **19** (7), 949–955.

38 Podsiadlo, P., Kaushik, A.K., Shim, B.S., Agarwal, A., Tang, Z., Waas, A.M., Arruda, E.M., and Kotov, N.A. (2008) Can nature's design be improved upon? High strength, transparent nacre-like nanocomposites with double network of sacrificial cross links. *J. Phys. Chem. B*, **112** (46), 14359–14363.

39 Podsiadlo, P., Kaushik, A.K., Arruda, E.M., Waas, A.M., Shim, B.S., Xu, J., Nandivada, H., Pumplin, B.G., Lahann, J., Ramamoorthy, A., and Kotov, N.A. (2007) Ultrastrong and stiff layered polymer nanocomposites. *Science*, **318** (5847), 80–83.

40 Wu, G. and Su, Z. (2006) Polyhedral oligomeric silsesquioxane nanocomposite thin films via layer-by-layer electrostatic self-assembly. *Chem. Mater.*, **18** (16), 3726–3732.

41 Van Duffel, B., Schoonheydt, R.A., Grim, C.P.M., and De Schryver, F.C. (1999) Multilayered clay films: atomic force microscopy study and modeling. *Langmuir*, **15** (22), 7520–7529.

42 Podsiadlo, P., Tang, Z., Shim, B.S., and Kotov, N.A. (2007) Counterintuitive effect of molecular strength and role of molecular rigidity on mechanical properties of layer-by-layer assembled nanocomposites. *Nano Lett.*, **7** (5), 1224–1231.

43 Lin, Y.-H., Jiang, C., Xu, J., Lin, Z., and Tsukruk, V.V. (2007) Robust, fluorescent, and nanoscale freestanding conjugated films. *Soft Matter*, **3** (4), 432–436.

44 Kharlampieva, E., Kozlovskaya, V., Gunawidjaja, R., Shevchenko, V.V., Vaia, R., Naik, R.R., Kaplan, D.L., and Tsukruk, V.V. (2010) Flexible silk-inorganic nanocomposites: from transparent to highly reflective. *Adv. Funct. Mater.*, **20** (5), 840–846.

45 Kim, U.-J., Park, J., Li, C., Jin, H.-J., Valluzzi, R., and Kaplan, D.L. (2004) Structure and properties of silk hydrogels. *Biomacromolecules*, **5** (3), 786–792.

46 Jiang, C., Markutsya, S., Shulha, H., and Tsukruk, V.V. (2005) Freely suspended gold nanoparticle arrays. *Adv. Mater.*, **17** (13), 1669–1673.

47 Markutsya, S., Jiang, Ch., Pikus, Y., and Tsukruk, V.V. (2005) Freely suspended layer-by-layer nanomembranes: testing micromechanical properties. *Adv. Funct. Mater.*, **15** (5), 771–780.

48 Vlassak, J.J. and Nix, W.D. (1992) A new bulge test technique for the determination of Young's modulus and Poisson's ratio of thin films. *J. Mater. Res.*, **7** (12), 3242–3249.

49 Gunawidjaja, R., Jiang, C., Peleshanko, S., Ornatska, M., Singamaneni, S., and Tsukruk, V.V. (2006) Flexible and robust 2D arrays of silver nanowires encapsulated within freestanding layer-by-layer films. *Adv. Funct. Mater.*, **16** (15), 2024–2034.

50 Kharlampieva, E., Zimnitsky, D., Gupta, M., Bergman, K.N., Kaplan, D.L., Naik, R.R., and Tsukruk, V.V. (2009) Redox-active ultrathin template of silk fibroin: effect of secondary structure on gold nanoparticle reduction. *Chem. Mater.*, **21** (13), 2696–2704.

51 Kharlampieva, E., Slocik, J.M., Tsukruk, T., Naik, R.R., and Tsukruk, V.V. (2008) Polyaminoacid-induced growth of metal nanoparticles on layer-by-layer templates. *Chem. Mater.*, **20** (18), 5822–5831.

52 Kozlovskaya, V., Kharlampieva, E., Khanal, B.P., Manna, P., Zubarev, E.R., and Tsukruk, V.V. (2008) Ultrathin layer-by-layer hydrogels with incorporated gold nanorods as pH-sensitive optical materials. *Chem. Mater.*, **20** (24), 7474–7485.

53 Kozlovskaya, V. and Sukhishvili, S.A. (2006) pH-controlled permeability of layered hydrogen-bonded polymer capsules. *Macromolecules*, **39** (16), 5569–5572.

54 Luzinov, I., Julthongpiput, D., and Tsukruk, V.V. (2000) Thermoplastic elastomer monolayers grafted to a functionalized silicon surface. *Macromolecules*, **33** (20), 7629–7638.

55 Peng, X., Manna, U., Yang, W., Wickham, J., Scher, E., Kadavanich, A., and Allvisatos, A.P. (2000) Shape control of CdSe nanocrystals. *Nature*, **404** (6773), 59–61.

56 Xu, J., Xia, J., Wang, J., Shinar, J., and Lin, Z. (2006) Quantum dots confined in nanoporous alumina membranes. *Appl. Phys. Lett.*, **89** (13), 133110.

57 Zimnitsky, D., Jiang, C., Xu, J., Lin, Z., and Tsukruk, V.V. (2007) Substrate- and time-dependent photoluminescence of quantum dots inside the ultrathin polymer LbL film. *Langmuir*, **23** (8), 4509–4515.

58 Zimnitsky, D., Jiang, C., Xu, J., Lin, Z., Zhang, L., and Tsukruk, V.V. (2007) Photoluminescence of a freely suspended monolayer of quantum dots encapsulated into layer-by-layer films. *Langmuir*, **23** (20), 10176–10183.

59 Andreev, A.D., Datsiev, R.M., and Seisyan, R.P. (1999) Absorption spectra of ZnSe/CdSe-based QDs. *Physica Status Solidi B*, **215** (1), 325–330.

60 Yu, W.W., Qu, L., Guo, W., and Peng, X. (2003) Experimental determination of the extinction coefficient of CdTe, CdSe, and CdS nanocrystals. *Chem. Mater.*, **15** (14), 2854–2860.

61 Mendelsohn, J.D., Barrett, C.J., Chan, V.V., Pal, A.J., Mayes, A.M., and Rubner, M.F. (2000) Fabrication of microporous thin films from polyelectrolyte multilayers. *Langmuir*, **16** (11), 5017–5023.

62 Zimnitsky, D., Shevchenko, V.V., and Tsukruk, V.V. (2008) Perforated, freely suspended layer-by-layer nanoscale membranes. *Langmuir*, **24** (12), 5996–6006.

63 Wong, C., West, P.E., Olson, K.S., Mecartney, M.L., and Starostina, N. (2007) Tip dilation and AFM capabilities in the characterization of nanoparticles. *J. Miner. Met. Mater. Soc.*, **59** (1), 12–16.

64 Okabe, A., Boots, B., Sugihara, K., and Chiu, S.N. (eds) (2000) *Spatial Tessellations – Concepts and Applications of Voronoi Diagrams*, 2nd edn, John Wiley & Sons, Ltd, Chichester.

65 Ogawa, H., Kanaya, T., Nishida, K., and Matsuba, G. (2007) Phase separation and dewetting in polymer blend thin films. *Eur. Phys. J. Special Topics*, **141** (1), 189–192.

66 Tsukruk, V.V., Bliznyuk, V.N., Hazel, J., Visser, D., and Everson, M.P. (1996) Organic molecular films under shear forces: fluid and solid Langmuir monolayers. *Langmuir*, **12** (20), 4840–4849.

67 Schwartz, D.K., Garnaes, J., Viswanathan, R., and Zasadzinski, J.A.N. (1992) Surface order and stability of Langmuir–Blodgett films. *Science*, **257** (5069), 508–511.

68 Zasadzinski, J.A., Viswanathan, R., Madsen, L., Garnaes, J., and Schwartz, D.K. (1994) Langmuir–Blodgett films. *Science*, **263** (5154), 1726–1733.

69 Meyer, E., Howald, L., Overney, R.M., Heinzelmann, H., Frommer, J., Guentherodt, H.J., Wagner, T., Schier, H., and Roth, S. (1991) Molecular-resolution images of Langmuir–Blodgett films using atomic force microscopy. *Nature*, **349** (6308), 398–400.

70 Radmacher, M., Tillmann, R.W., Fritz, M., and Gaub, H.E. (1992) From molecules to cells: imaging soft samples with the atomic force microscope. *Science*, **257** (5078), 1900–1905.

71 Weisenhorn, A.L., Egger, M., Ohnesorge, F., Gould, S.A.C., Heyn, S.P., Hansma, H.G., Sinsheimer, R.L., Gaub, H.E., and Hansma, P.K. (1991) Molecular-resolution images of Langmuir–Blodgett films and DNA by atomic force microscopy. *Langmuir*, **7** (1), 8–12.

72 Schwartz, D.K., Viswanathan, R., Garnaes, J., and Zasadzinski, J.A. (1993) Influence of cations, alkane chain length, and substrate on molecular order of Langmuir–Blodgett films. *J. Am. Chem. Soc.*, **115** (16), 7374–7380.

73 Hansma, H.G., Gould, S.A.C., Hansma, P.K., Gaub, H.E., Longo, M.L., and Zasadzinski, J.A.N. (1991) Imaging nanometer scale defects in Langmuir–Blodgett films with the atomic force microscope. *Langmuir*, **7** (6), 1051–1054.

74 Bourdieu, L., Silberzan, P., and Chatenay, D. (1991) Langmuir–Blodgett films: from micron to angstrom. *Phys. Rev. Lett.*, **67** (15), 2029–2032.

75 Peltonen, J.P.K., He, P., and Rosenholm, J.B. (1992) Order and defects of Langmuir–Blodgett films detected with the atomic force microscope. *J. Am. Chem. Soc.*, **114** (20), 7637–7642.

76 Chi, L.F., Anders, M., Fuchs, H., Johnston, R.R., and Ringsdorf, H. (1993) Domain structures in Langmuir–Blodgett films investigated by atomic force microscopy. *Science*, **259** (5092), 213–216.

77 Koepf, M.H., Gurevich, S.V., Friedrich, R., and Chi, L. (2010) Pattern formation in monolayer transfer systems with substrate-mediated condensation. *Langmuir*, **26** (13), 10444–10447.

78 Josefowicz, J.Y., Maliszewskyj, N.C., Idziak, S.H., Heiney, P.A., Mccauley, J.P., and Smith, A. (1993) Structure of Langmuir–Blodgett films of disk-shaped molecules determined by atomic force microscopy. *Science*, **260** (5106), 323–326.

79 Tsukruk, V.V. and Janietz, D. (1996) Cross-interfacial gradient of molecular organization in a discotic polymer molecular film. *Langmuir*, **12** (11), 2825–2829.

80 Genson, K.L., Holzmuller, J., Villacencio, O.F., McGrath, D.V., Vaknin, D., and Tsukruk, V.V. (2005) Langmuir and grafted monolayers of photochromic amphiphilic monodendrons of low generations. *J. Phys. Chem. B*, **109** (43), 20393–20402.

81 Larson, K., Vaknin, D., Villavicencio, O., McGrath, D., and Tsukruk, V.V. (2002) Molecular packing of amphiphiles with crown polar heads at the air–water interface. *J. Phys. Chem. B* **106** 7246–7251.

82 Holzmueller, J., Genson, K.L., Park, Y., Yoo, Y., Park, M.H., Lee, M.S., and Tsukruk, V.V. (2005) Amphiphilic tree-like rods at interfaces: layered stems and circular aggregation. *Langmuir*, **21**, 6392–6398.

83 Tsukruk, V.V., Genson, K.L., Peleshanko, S., Markutsya, S., Greco, A., Lee, M.S., and Yoo, Y. (2003) Molecular reorganizations of rod–coil molecules on a solid surface. *Langmuir*, **19**, 495–502.

84 Gunawidjaja, R., Luponosov, Y.N., Huang, F., Ponomarenko, S.A., Muzafarov, A.M., and Tsukruk, V.V. (2009) Structure and properties of functionalized bithiophenesilane monodendrons. *Langmuir*, **19** (16), 9270–9284.

85 Imae, T., Takeshita, T., and Kato, M. (2000) Phase separation in hybrid Langmuir–Blodgett films of perfluorinated and hydrogenated amphiphiles. Examination by atomic force microscopy. *Langmuir*, **16** (2), 612–621.

86 Ekelund, K., Sparr, E., Engblom, J., Wennerström, H., and Engström, S. (1999) An AFM study of lipid monolayers. 1. Pressure-induced phase behavior of single and mixed fatty acids. *Langmuir*, **15** (20), 6946–6949.

87 Qaqish, S.E., Urquhart, S.G., Lanke, U., Brunet, S.M.K., and Paige, M.F. (2009) Phase separation of palmitic acid and perfluorooctadecanoic acid in mixed Langmuir–Blodgett monolayer films. *Langmuir*, **25** (13), 7401–7409.

88 Qaqish, S.E. and Paige, M.F. (2008) Rippled domain formation in phase-separated mixed Langmuir–Blodgett films. *Langmuir*, **24** (12), 6146–6153.

89 Simões Gamboa, A.L., Filipe, E.J.M., and Brogueira, P. (2002) Nanoscale pattern formation in Langmuir–Blodgett films of a semifluorinated alkane and a polystyrene–poly(ethylene oxide) diblock copolymer. *Nano Lett.*, **2** (10), 1083–1086.

90 Watanabe, S., Shibata, H., Sakamoto, F., Azumi, R., Sakai, H., Abe, M., and Matsumoto, M. (2009) Directed self-assembly of gold nanoparticles and gold thin films on micro- and nanopatterned templates fabricated from mixed phase-separated Langmuir–Blodgett films. *J. Mater. Chem.*, **19** (37), 6796–6803.

91 Peleshanko, S., Jeong, J., Gunawidjaja, R., and Tsukruk, V.V. (2004) Amphiphilic heteroarm PEO-b-PSm star polymers at the air–water interface: aggregation and surface morphology. *Macromolecules*, **37** (17), 6511–6522.

92 Gunawidjaja, R., Peleshanko, S., and Tsukruk, V.V. (2005) Functionalized $(X\text{-}PEO)_2\text{-}(PS\text{-}Y)_2$ star copolymers at the interfaces: role of terminal groups in surface behavior and morphology. *Macromolecules*, **38** (21), 8765–8774.

93 Gonçalves da Silva, A.M., Filipe, E.J.M., d'Oliveira, J.M.R., and Martinho, J.M.G. (1996) Interfacial behavior of poly (styrene)–poly(ethylene oxide) diblock copolymer monolayers at the air–water interface. Hydrophilic block chain length and temperature influence. *Langmuir*, **12** (26), 6547–6553.

94 Rybak, B.M., Bergman, K.N., Ornatska, M., Genson, K.L., and Tsukruk, V.V. (2006) The formation of silver nanoparticles at the air–water interface mediated by the monolayer of functionalized hyperbranched molecules. *Langmuir*, **22** (3), 1027–1037.

95 Bliznyuk, V.N., Neher, D., Ponomarev, I.I., and Tsukruk, V.V. (1996) Structure-fluorescence properties of some naphthoylene-benzimidazole-based LB films. *Thin Solid Films*, **287** (1–2), 232–236.

96 Kulkarni, D., Choi, I., Singamaneni, S., and Tsukruk, V.V. (2010) Large and nano-scale free-standing graphene oxide-polyelectrolyte membranes. *ACS Nano*, **8**, 4667–4676.

97 Medhekar, N.V., Ramasubramaniam, A., Ruoff, R.S., and Shenoy, V.B. (2010) Hydrogen bond networks in graphene oxide composite paper: structure and mechanical properties. *ACS Nano*, **4** (4), 2300–2306.

98 Nakajima, T., Mabuchi, A., and Hagiwara, R. (1988) A new structural model of graphite oxide. *Carbon*, **26** (3), 357–361.

99 Mkhoyan, A.K., Contryman, A.W., Silcox, J., Stewart, D.A., Eda, G., Mattevi, C., Miller, S., and Chhowala, M. (2009) Atomic and electronic structure of graphene-oxide. *Nano Lett.*, **9** (3), 1058–1063.

100 Gunawidjaja, R., Huang, F., Gumenna, M., Klimenko, N., Nunnery, G.A., Shevchenko, V., Tannenbaum, R., and Tsukruk, V.V. (2009) Bulk and surface assembly of branched amphiphilic polyhedral silsesquioxane POSS-M compounds. *Langmuir*, **25** (2), 1196–1209.

101 Tsukruk, V.V. (2001) Molecular lubricants and glues for micro- and nanodevices. *Adv. Mater.*, **13** (2), 95–108.

102 Meyer, E., Overney, R., Brodbeck, D., Howald, L., Lüthi, R., Frommer, J., and Güntherodt, H.-J. (1992) Friction and wear of Langmuir–Blodgett films observed by friction force microscopy. *Phys. Rev. Lett*, **69** (12), 1777–1780.

103 Weihs, T.P., Nawaz, Z., Jarvis, S.P., and Pethica, J.B. (1991) Limits of imaging resolution for atomic force microscopy of molecules. *Appl. Phys. Lett.*, **59** (27), 3536–3538.

104 Oncins, G., Torrent-Burgués, J., and Sanz, F. (2008) Nanomechanical properties of arachidic acid Langmuir–Blodgett films. *J. Phys. Chem. C*, **112** (6), 1967–1974.

105 Oncins, G., Picas, L., Hernández-Borrell, J., Garcia-Manyes, S., and Sanz, F. (2007) Thermal response of Langmuir–Blodgett films of dipalmitoylphosphatidylcholine studied by atomic force microscopy and force spectroscopy. *Biophys. J.*, **93** (8), 2713–2725.

106 Nayak, A. and Suresh, K.A. (2009) Mechanical properties of Langmuir–Blodgett films of a discogen–DNA complex by atomic force microscopy. *J. Phys. Chem. B*, **113** (12), 3669–3675.

107 Tsukruk, V.V., Foster, M.D., Reneker, D.H., Wu, H., Schmidt, A., and Knoll, W. (1994) Stability and modification of polyglutamate Langmuir–Blodgett bilayer films. *Macromolecules*, **27** (5), 1274–1280.

108 Tsukruk, V.V., Bliznyuk, V.N., Hazel, J., Visser, D., and Everson, M.P. (1996) Organic molecular films under shear forces: fluid and solid Langmuir monolayers. *Langmuir*, **12** (20), 4840–4849.

15
Colloids and Microcapsules

A variety of individual colloidal microparticles (spherical and anisotropic, solid and hollow, synthetic and biological, fabricated and assembled) and their arrays on surfaces have been a popular subject for AFM studies since the earliest days. AFM imaging in both dry and wet states allows one to address such fundamental issues as the accurate measurement of colloidal microparticle dimensions and their size distribution, surface quality and the presence, distribution, and density of adsorbed molecules, surface coverage of adsorbed microparticles, mechanical and adhesive properties of microparticles, microparticle adhesion to the supporting surfaces, shell thickness of microcapsules and their deformational properties, and the adhesion between microparticles. These topics will be considered in this chapter.

Because the characteristic dimensions of these colloidal structures are usually on the microscopic scale (from several micrometers to tens of micrometers), adequate AFM visualization to obtain the fine details of their morphologies can be done unambiguously with conventional midresolution AFM imaging in tapping mode. However, if their sizes are pushed to the lower limit, usually below 50 nm, a severe tip dilation effect is observed that might completely distort the topography of colloids. In AFM images, still commonly observed, all such nanoparticles look alike, representing an inverse AFM tip shape that disregards the original shape and dimensions of the nanoparticles. For such small objects, a characteristic average dimension and its distribution can still be obtained from the measurements of the vertical dimensions (which, in the case of spherical particles, represent a true diameter) and a true surface coverage can be estimated from direct counting of the number of particles per specific surface area.

Several examples of AFM studies of such colloidal polymer microparticles and nanoparticles of different types will be discussed in this chapter. These colloidal structures include various microparticles and nanoparticles ranging from simple spherical objects to those with complex shapes, from synthetic colloids to complex biological objects, and from solid microparticles to various thin shell microcapsules (Figure 15.1) [1].

However, we are not going to consider in this chapter such a special and important case as microscopic vesicles from surfactants and corresponding AFM studies of synthetic membranes and biological membranes. Some relevant materials of this

Scanning Probe Microscopy of Soft Matter: Fundamentals and Practices, First Edition.
Vladimir V. Tsukruk and Srikanth Singamaneni.

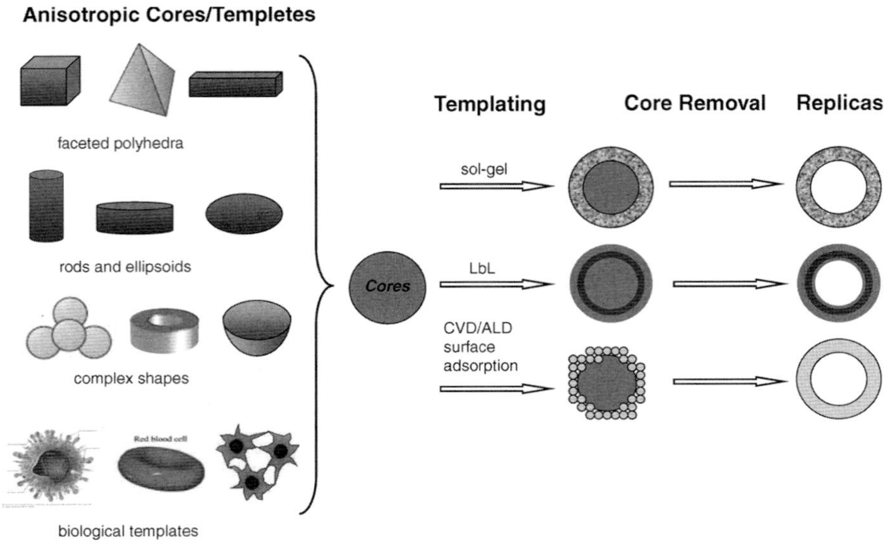

Figure 15.1 Anisotropic cores/templates and possible replication pathways for particle replication. Reprinted from Ref. [1].

class were partially discussed in the section on ultrathin lipid films in Chapter 14. Interested readers are referred to some reviews, discussions, and relevant results on vesicles and membranes [2–4]. Some important aspects of the studies of membrane, micellar, and vesicle structures addressed with AFM technique are briefly listed here with proper references provided for interested readers. In this short list, we mention an excellent review on observing conformational changes of membrane proteins in their native environment, AFM studies on stress-mediated changes of mechanosensitive membrane proteins embedded in SAMs, AFM characterization of supported lipid films in real-time and in a liquid environment, conductance mapping of ion exchange polymeric membranes, and observation of supported lipid bilayers from vesicles [5–9].

15.1 Colloids and Latexes

Colloidal microparticles and nanoparticles obtained from solid polymeric materials with relatively narrow size distribution are a very popular type of microparticle for a broad range of applications ranging from cosmetics and paints to prospective photonic and phononic elements and structures. The common dimension of colloidal microparticles for these applications ranges from several tens of micrometers to several tens of nanometers. Thus, SEM was always an excellent choice for the characterization of their dimension, morphologies, composition, and assemblies [10]. However, in the past decade, AFM became a tool of choice when the dimensions of the colloidal particles are decreased below several hundred nanometers This is especially true if imaging under ambient conditions is required and fine details of their surfaces are critical for understanding.

15.1.1
Individual and Aggregated Solid Microparticles

Individual and aggregated core–shell, rubbery–glassy polymer latexes from poly (butyl acrylate)/poly(methyl methacrylate) (PMMA) materials with different chemical compositions were observed and analyzed in detail in one of the earlier AFM studies by Sommer *et al.* [11]. The authors observed that contact mode AFM imaging can easily damage the soft polymer latex microparticles, while light tapping mode AFM can generate high-quality microparticle images without material damage. Height histograms were generated for these microparticles with a diameter around 300 nm and the average diameter was deduced from the analysis of individual microparticles. The authors considered common cross-sectional artifacts caused by lateral interferences of adjacent microparticles.

Surface microroughness of colloidal microparticles was demonstrated to be very low (0.3 nm within a 100 nm^2 surface area – selected in this study to avoid the significant contribution of the curved shape) for one-component PBA latex microparticles. However, it increased significantly to 1.2 nm for corresponding composite core–shell microparticles. The role of different surfactants in the surface aggregation and their AFM appearance in the AFM images of polymer latex microparticles have also been discussed. Tip dilation was noted and discussed in this and many other AFM studies. To its full extent, this phenomenon was utilized for AFM tip characterization by using smaller diameter nanoparticles in another study [12].

AFM scanning was utilized to follow the polymerization process of cationic copolymer latexes by Duracher *et al.* [13]. In this study, the authors demonstrated some common artifacts in the imaging of latex microparticles, such as instabilities at the edges of particles and the dilation effect for individual and packed arrays of microparticles. The annealing process, which results in the gradual flattening of aggregated latexes and film formation, was monitored with AFM by Perez and Lang [14]. In this study, PBMA latexes were annealed above the glass transition temperature and the surface microroughness was exploited as a parameter that can be used to monitor the flattened particle aggregates at different thermal treatment conditions to compare with theoretical models of latex softening above the glass transition. The ability to visualize the surface of latex microparticles with high resolution was explored by Pfau *et al.* to monitor single molecule adsorption of poly(ethylene imine) polyelectrolytes on charged polystyrene (PS) latex particles [15]. Lateral dimensions of observed surface features obtained with AFM were correlated with the molecular weight of adsorbed PEI macromolecules, which was related to the surface charge density, and compared with the molecular adsorption of a reference planar surface.

PVA latexes and their aggregation into thin, uniform films have been studied by Budhlall *et al.* [16]. The authors measured surface microroughness as a function of composition, polymerization, and drying conditions and analyzed AFM phase images to understand the chemical composition and then suggested the process of the fabrication of uniform films from microparticles. An interesting example of a peculiar long-range pattern formation process via selective extraction of solid nanoparticles to form a honeycomb pattern from latex films was studied by Han *et al.* with both AFM and SEM [17].

A study of mechanical properties of hollow colloidal particles in solution and at ambient air after drying with nanoindentation method was presented by Zoldesi *et al.* [18]. In this study, the authors observed an initial linear response on FDC data for PDMS colloidal particles probed at indentation depths comparable to shell thickness. The elastic modulus measured with force spectroscopy was close to 200 MPa, which is comparable to many other polymer microcapsules (see discussion below). Furthermore, it was observed that this elastic deformation regime was followed by a highly nonlinear response at even larger deformations. The authors concluded that this regime corresponds to the compliant shell's sudden buckling, which can be reversible but becomes permanent after multiple indentations with the AFM tip.

Poly(lactic acid) (PLA), a prospective component for biodegradable polymer films and assemblies, is biocompatible and biodegradable and can provide mechanical strength to a thin film in the absence of degrading conditions [19, 20]. PLA particles, fibers, and films have been demonstrated to be relevant for the biomedical field. PLA and its copolymers with poly(glycolic acid) (PGA) have been widely used for nanoparticle synthesis aimed at controlled drug delivery or controlled administration of different therapeutic agents [21–23]. For example, PLA nanoparticles with avidin-modified surfaces have been reported for diagnostic and therapeutic applications [24]. Films of PLA copolymers or composites cast from various organic solvents have been used as substrates for cell culturing [25, 26].

In a recent AFM and SEM study by Orozco *et al.*, biodegradable PLA nanoparticles for LbL assembly were prepared using acetone as a solvent through the nanoprecipitation method [27]. Two types of PLA nanoparticles, bare PLA and PEI-terminated PLA nanoparticles (PEI–PLA), were prepared in this work by acetone removal via evaporation, and thus PLA nanoparticles formed an aqueous colloidal dispersion. Gold nanoparticles were grown within the PLA nanoparticle assemblies through UV irradiation or under mild reducing conditions to create biodegradable nanocomposites with a distinct optical response that allows the monitoring of the biodegradation of the films with enzymatic treatment by α-chymotrypsin.

Figure 15.2 shows AFM topographical and phase images of the bare PLA nanoparticles and PEI-modified PLA nanoparticles prepared in this work. As seen from this very common AFM image, the degree of aggregation for bare PLA nanoparticles is much larger than that of the PEI-modified PLA nanoparticles due to the increased role of hydrophobic interactions during surface adsorption. Dimensions of PLA nanoparticles and their distribution in the range of 40–50 nm were further accessed from these images and associated with the formation procedure.

Figure 15.3 presents a cross-sectional analysis of selected surface areas of the two types of PLA nanoparticles from this study. The resulting size distributions for the PLA nanoparticles and PEI-modified PLA nanoparticles are also presented in this figure. The average size for bare PLA nanoparticles determined from AFM images is

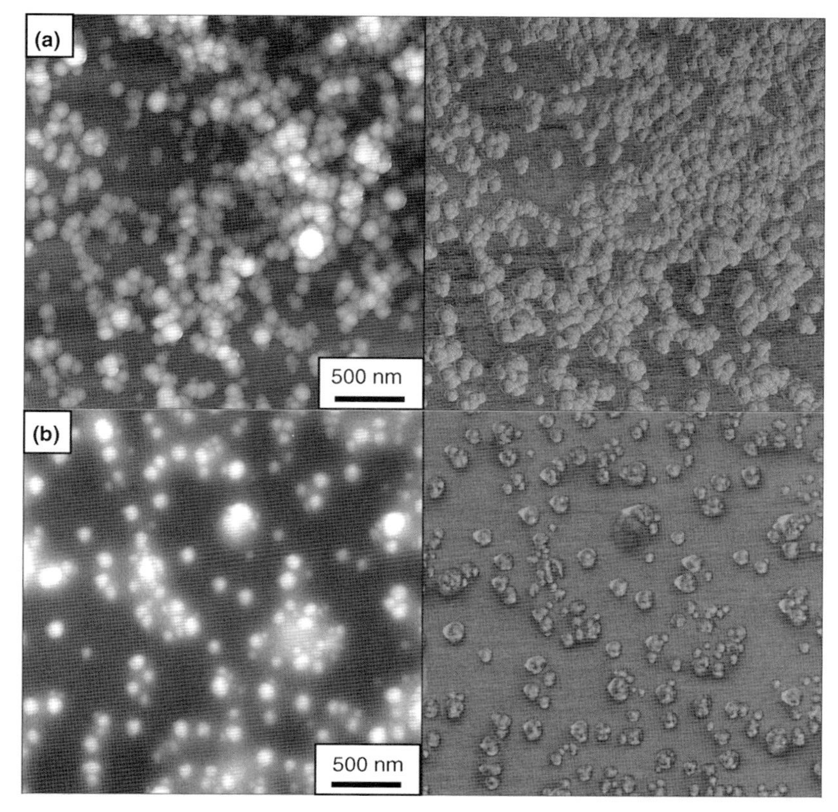

Figure 15.2 AFM topography (left) and phase (right) images of (a) bare PLA and (b) PEI-modified PLA nanoparticles drop-cast from aqueous dispersions onto the surface of silicon wafers. Reprinted from Ref. [23].

slightly larger than that for PEI–PLA nanoparticles (50 versus 45 nm, which is within 1 standard deviation) and the size distribution is narrower for the PEI–PLA nanoparticles. The authors concluded that the PEI coating in this case works as a surfactant that hinders the coalescence of droplets and thus eventually allows the formation of smaller PEI–PLA nanoparticles with a narrower diameter distribution. Composite colloidal microparticles such as PEI–PLA discussed above allow better control of particle size, size distribution, interparticle interactions, and their surface assembly.

15.1.2
Composite Microparticles

Among numerous studies of this type of colloidal systems, we will note only a few recent studies due to their substantial use of AFM imaging to generate relevant conclusions.

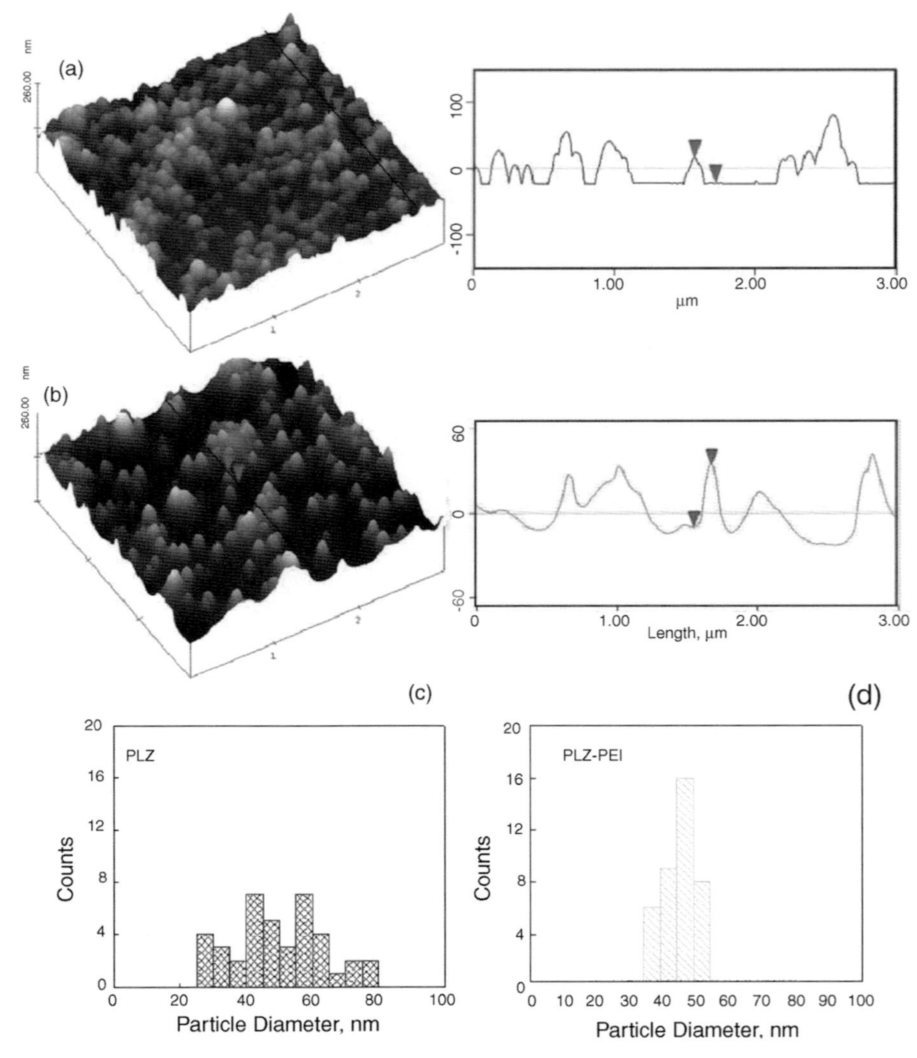

Figure 15.3 3D AFM views and cross section analysis of (a) bare PLA and (b) PEI-modified PLA nanoparticles with their size histograms (c and d, respectively) derived from multiple cross sections. Reprinted from Ref. [23].

In continuation of their studies, Orozco *et al.* demonstrated composite microparticles based on silica microparticles coated with PLA nanoparticles or with LbL shells, $(PLA\text{–}PEI/PAA)_4$, as well as corresponding microcapsules produced after silica core etching (Figure 15.4) [27]. The cationic PLA nanoparticles synthesized in the presence of 0.2 mg/ml PEI possess raspberry-like morphology. AFM height analysis on flat regions of $(PLA\text{–}PEI/PAA)_4$ shells dried on silicon wafers revealed a single-wall thickness of 85 nm. Gold nanoparticles were grown within the PLA

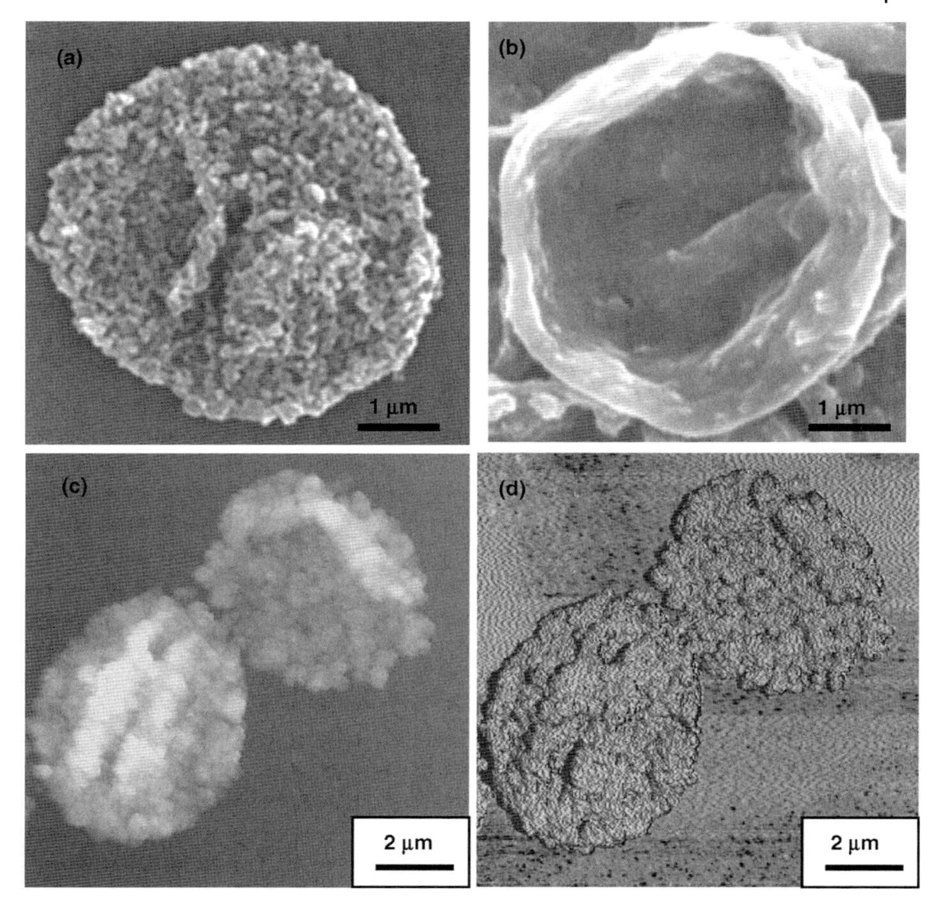

Figure 15.4 SEM images of hollow (PLA–PEI/PAA)$_4$ shells assembled from PLA–PEI nanoparticles at different concentrations (a and b). AFM topographical (c) and phase (d) images of (PLA–PEI–Au/PAA)$_4$ shells made from PLA–PEI nanoparticles. Reprinted from Ref. [23].

nanoparticle films by reducing $[AuCl_4]^-$ ions that were allowed to penetrate inside the films and form ionic pairs with protonated amine groups.

The presence of gold nanoparticles in LbL shells was confirmed by TEM imaging and by SEM and AFM studies (Figure 15.4). From these data, it is clear that the produced nanoshells possess a loosely packed nanoparticle morphology. The silica core removal in turn resulted in the formation of microcapsules with thin shells from PLA-LbL nanoparticles reinforced by gold nanoparticles (Figure 15.4). The authors suggested that the loosely packed nanoparticle arrangement within these polymeric shells is due to the voids between the adsorbed PEI–PLA nanoparticles that are, in turn, formed because of the stabilizing PEI "shell" around the cationic PLA particles, preventing them from close arrangements.

In one of the earlier studies, coating of PS latex particles with multicomponent LbL nanoparticle-containing shells has been exploited to create composite microparticles by Caruso and Mohwald [28]. Silica nanoparticles were deposited from aqueous solution in sequential adsorption with polyelectrolytes on an initially smooth PS latex surface. AFM imaging revealed a uniform coverage of PS colloids with silica nanoparticles without their significant aggregation on the latex surface. On the other hand, the authors noted that the AFM measured diameter for PS latexes in the dry state adsorbed on a solid surface was close to that measured directly in solution with light scattering, thus indicating the insignificant influence of the drying conditions.

In a more recent study, model colloids with/without encapsulated magnetic particles were studied by Rasa *et al.* [29]. AFM imaging was used for routine characterization of their dimensions and shape with an extended discussion of tip dilation effects for individual and aggregated particles provided as well. Magnetic force microscopy has also been applied and the limits of resolution achievable in different media and for nanoparticles larger than 15 nm in diameter were discussed. Even more complex, bimetallic colloids were obtained from PS latexes by Chen *et al.* [30]. These composite metal–polymer colloids were formed by *in situ* growth of gold and platinum nanoparticles in the course of polymerization in corresponding mixed salt solutions. TEM showed a cross-sectional distribution of uniform metal nanoparticles onto metallized PS microspheres, but AFM demonstrated drastic changes in surface morphology in metallized latexes after metallization with a predominant surface location of the grown nanoparticles.

Janus microparticles with biphasic chemical composition and therefore anisotropic surface properties present a new type of composite microparticles introduced relatively recently [31, 32]. Janus microparticles display interesting assembling behavior since they possess both physical and chemical properties of each phase [33, 34]. Assemblies of Janus particles result in minimized interfacial energy and spontaneous aggregation of particles into well-defined clusters [35]. An impressive number of methods have been developed to fabricate Janus particles, the most popular of which is based on toposelective surface modification and template-directed assembly [36, 37]. One example describes Janus particles fabricated by toposelective polymerization with plasma-enhanced chemical vapor deposition (PECVD) on microspheres that are partially embedded in a sacrificial polymer layer [38].

SEM imaging of individual microparticles shows selective, one-face deposition on their surfaces with half of the particle clearly having an additional plasma-polymerized polymer coating (Figure 15.5). SEM images of these PECVD-fabricated 3 μm Janus silica spheres show a plasma-polymerized polymer coating covering approximately 50% of the total surface area of the particle with a well-defined boundary. On both types of particles explored in this study, the plasma-polymerized coating is seen to be conformal to the surface with a sharp delineation between the plasma-polymerized region and the bare silica/titania region that was masked. Estimation of the thickness of the plasma-polymerized layer on the microparticles from the SEM images reports a value within 200–300 nm.

To elucidate the fine morphology of these particles, AFM topography scans of a 1 μm^2 area were collected for polymerized regions of microparticles directly on microparticles embedded in a sacrificial polymer layer (Figure 15.5). All AFM images show a uniform surface morphology of different plasma-polymerized coatings with a

Figure 15.5 SEM images of Janus particles with (a) plasma-polymerized acrylonitrile and (b) ferrocene on silica microspheres. AFM topography (c) and phase (d) of plasma coatings on Janus microparticles. (c) Ferrocene (z-scale = 200 nm, 50°); (e) acrylonitrile (z-scale = 100 nm, 50°). Reprinted from Ref. [38].

microroughness of a few nanometers and the characteristic grainy surface morphology with grain dimensions below 50–100 nm for different coatings. Corresponding phase images clearly demonstrate sharp intergrain boundaries and uniform grain compositions common for plasma-polymerized coatings. Overall, AFM results indicated the presence of a complete, uniform coating free of defects that conformally coats the designated surface area of inorganic microparticles.

15.2 Thin Shell Microcapsules

To fabricate thin shell microcapsules, the particulate sacrificial template is coated with a thin layer of selected material to form a core–shell structure, followed by removal of the template core via their selective etching in a solvent or thermal decomposition in air (Figure 15.1). This selective core removal with various methods results in the formation of hollow microcapsules utilized for the shell-fabricating step, including grafting techniques, chemical vapor deposition, atomic layer deposition, surface adsorption, deposition of preformed colloids, and traditional LbL approach [39–44]. Spherical and anisotropic polymer and inorganic microparticles obtained via a variety of colloidal and nucleation routines have been the focus of long-standing research. LbL assembly of microcapsules, the most popular fabrication approach, will be presented and discussed next.

15.2.1 LbL Microcapsules

LbL assembly is one of the prominent and robust methods used to build controlled ultrathin polymer microshells and nanoshells with nanometer-level control over composition, thickness, and tunable chemical functionality [45–48].

LbL-modified colloidal particles have been extensively exploited to fabricate thin shell spherical microcapsules after the subsequent dissolution of the sacrificial core and AFM was employed to follow various stages of microcapsule fabrication and utilization for different applications [49]. Hollow spherical or complex-shaped LbL microcapsules can be used for encapsulation/release of delicate biomolecules such as DNA or enzymes and drug delivery. For example, Zhu *et al.* achieved stimuli-controlled drug release from hollow silica particles with mesoporous walls by capping the pores with polyelectrolyte multilayers [50]. Sokolova *et al.* reported the fabrication of multi-shell calcium phosphate particles loaded with DNA for delivering genetic materials to cells [51]. Chen *et al.* reported the use of gold nanocages based on microcapsules as an optical imaging contrast agent for optical coherence tomography [52].

The LbL-assembled microcapsules have been employed as microreactors for *in situ* synthesis of nickel-metallized (PAH/PSS) LbL shells by Schukin *et al.* [53]. Indeed, *in situ* nanoparticle synthesis within the multilayer polymer films is an elegant way to obtain metal nanoparticles and corresponding hybrid microparticles (organic–inorganic) due to the confined environment of the polyelectrolyte multilayers, allowing a good control over nanoparticle shapes, morphologies, location, and sizes [54, 55]. Various antibodies and proteins can be readily conjugated to gold nanoparticles through thiol surface chemistry or electrostatic binding, which makes them important for biochemical sensing and detection, as well as for various therapeutic applications [56, 57]. It has been demonstrated that drastic temperature change in gold nanoparticles being in resonance with an excitation light source can be effectively exploited for the delivery of the inner contents of microcapsules to exterior environment [58].

The thermal behavior of LbL microcapsules with different numbers of layers was investigated with confocal fluorescent microscopy, SEM, TEM, and AFM in both the dry and wet states by Köhler *et al.* [59]. In this study, the authors observed that the number of LbL layers, and thus the terminal surface functionality, are critical in defining how the size of microcapsules changes upon heating due to the softening of polyelectrolyte shells. AFM measurements of the collapsed microcapsules after different thermal treatments allowed the authors to conclude how the thermally induced change in the LbL wall thickness correlates with the overall changes in the diameter of the microcapsules. From these results, the authors suggested a routine for the control of microcapsule thermal stabilities and their permeability.

In another study from the same group, Mauser *et al.* exploited AFM to elucidate stimuli-responsive properties of LbL microcapsules assembled from weak polyelectrolytes, PAH and PMA [60]. The hollow LbL microstructures resulting from the removal of silica core showed a common, randomly buckled surface morphology in the dry state and even some residual core material at certain pH conditions as observed with AFM. The thickness of the LbL shells measured from AFM cross sections was shown to be proportional to the number of polyelectrolyte layers with increments of about 3.4 nm per bilayer, which slightly varies for different pH values. These pH dependencies of the microcapsule diameters and shell walls were related to the variable balance of ionic interactions within LbL shells and suggested to be important for tuning the chemically responsive behavior of LbL microcapsules.

Stable LbL microcapsules with photosensitive components have been fabricated by Pastoriza-Santos *et al.* [61]. The authors demonstrated that these microcapsules possess much higher stability to external mechanical stresses due to UV-induced photopolymerization of LbL multilayers. TEM and AFM studies on these structures showed that in the dry state, microcapsulates collapsed and folded on themselves. The microstructure and the thickness of shells were determined from corresponding cross sections of AFM images. The surface of microcapsules was observed to be rougher than that of traditional LbL shells, with characteristic grainy textures caused by nonuniform polymerization of intrashell components. In another study from the same group, Ochs *et al.* assembled biodegradable LbL microcapsules from poly (L-lysine) (PLL) and poly(L-glutamic acid) (PGA) as well as PEG-functionalized microcapsules and studied their behavior under different environmental conditions [62]. In this study, AFM was exploited to study swelling/deswelling behavior of these microcapsules, as well as changes in their surface morphology after PEG surface functionalization, which was observed to bring low-fouling properties important for biomedical applications.

In a recent study, environmentally sensitive LbL microcapsules from thermally responsive PNIPAAm and pH-sensitive alginate acid were made by Wang *et al.* by utilizing different inorganic cores and processing conditions [63]. The authors concluded from AFM measurements that a very low thickness of layers, down to 0.4 nm, resulted in low microcapsule stability and thus special measures should be taken to stabilize these microcapsules to prevent their disintegration. The microroughness of thicker LbL shells estimated from AFM images was unusually high, reaching 85 nm. An inhomogeneous ion distribution was suggested as a cause for this dramatic nonuniformity of surface morphology. Temperature-induced drug release processes from these microcapsules were studied by confocal fluorescent microscopy. The morphology of these microcapsules after drug release was monitored by AFM imaging to focus on fine details. In a related study, further AFM imaging was conducted to determine how thermal transitions affect the morphology and assembly of PAA–PNIPAAm microcapsules [64].

In another study, Kozlovskaya *et al.* demonstrated *in situ* growth of gold nanoparticles in poly(methacrylic acid) (PMAA) hydrogel microcapsules produced from hydrogen-bonded (PMAA/poly-*N*-vinylpyrrolidone) multilayer precursors through cross-linking with ethylenediamine (EDA) (Figure 15.6) [65]. Uniform shapes of collapsed microparticles with diameters of 3 μm were imaged with AFM technique in the dry state and showed the uniform collapse of thin shells without random buckling (Figure 15.6).

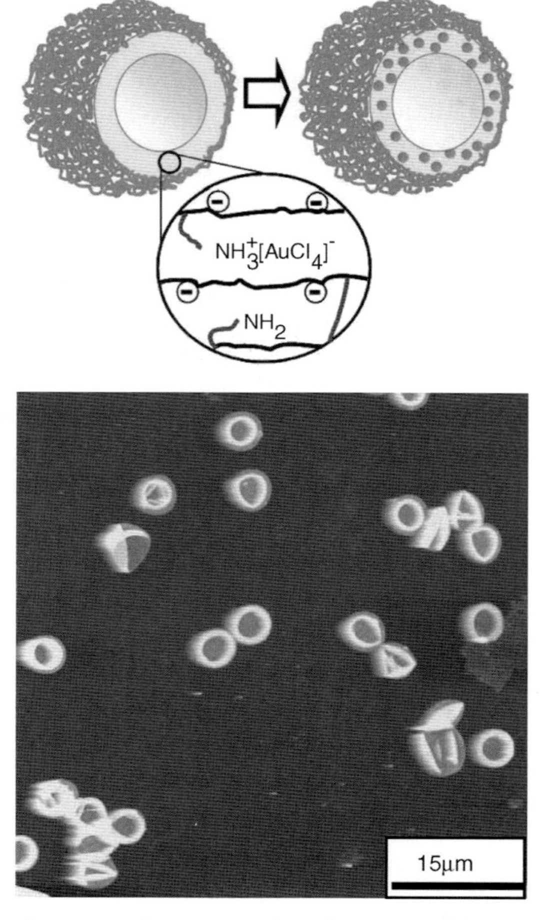

Figure 15.6 LbL microcapsules with grown-in gold nanoparticles and AFM image of dried EDA-$(PMAA)_{15}$ hollow microcapsules; the *z*-scale of the image is 400 nm. Reprinted from Ref. [65].

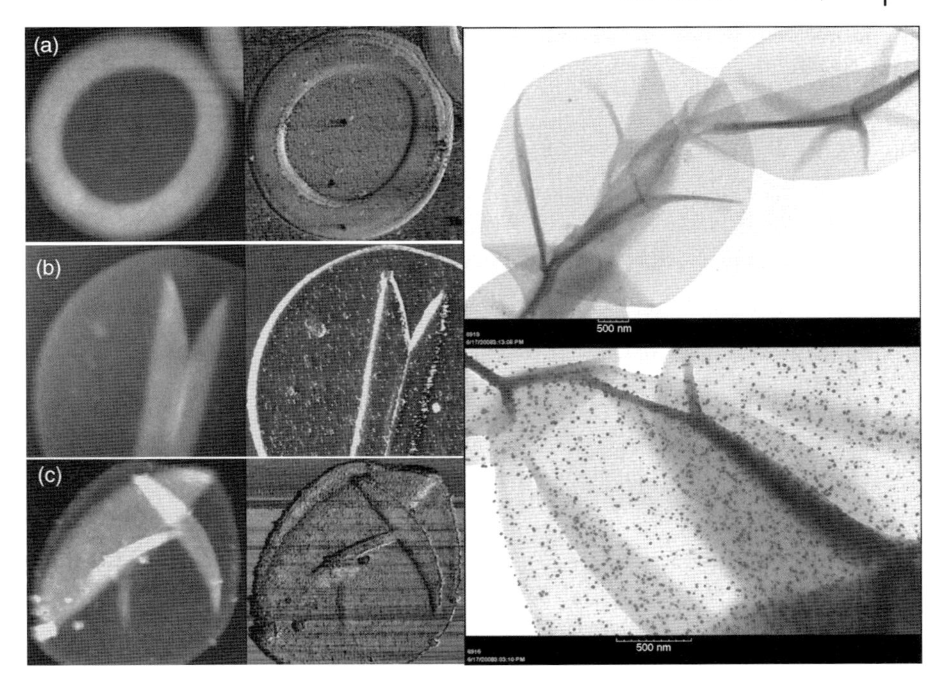

Figure 15.7 *Left*: AFM images (left, height; right, phase) of the (A) collapsed EDA-(PMAA)$_{15}$ capsules without Au nanoparticles and collapsed EDA-(PMAA)$_{15}$ capsules reinforced with Au nanoparticles reduced in borate buffers at (B) pH 10 and (C) pH 5. The scan size is 5 μm and the *z*-scale is 400 nm for all images (height). *Right*: TEM images of original microcapsules and microcapsules with gold nanoparticles. Reprinted from Ref. [65].

The gold nanoparticle growth in LbL shells was controlled by changing the pH-dependent balance of amine/ammonium groups in the ionic cross-links within the LbL multilayers. TEM analysis of dry microcapsules confirmed the presence of monodisperse and uniformly distributed grown-in gold nanoparticles with a relatively narrow size distribution and an average size of 16 nm (Figure 15.7).

The representative AFM single-capsule images of the pristine and reinforced hydrogel EDA-(PMAA)$_{15}$ capsules showed their different collapsed shapes caused by a difference in the stability of the purely polymeric or reinforced polymeric shells (Figure 15.7). Indeed, the purely polymeric hydrogel capsules demonstrate the particular folding behavior upon drying from an aqueous solution. Owing to a very soft wall at pH 7, the capsule collapses uniformly without any wrinkles and with a smooth circular rim as mentioned previously.

In contrast, when the hydrogel capsule walls are reinforced with *in situ* grown gold nanoparticles as confirmed with TEM, the dry capsule morphology changes dramatically. The dried capsules start forming distinctive sharp folds caused by shell buckling, indicating changes in the mechanical properties of the composite hydrogel shells. Folding morphology is clearly evident from the AFM phase images (Figure 15.7).

Multiple wrinkles are formed in central and peripheral regions of the collapsed capsules, indicating the buckling instabilities of the stiffer capsule walls with gold nanoparticles [66].

15.2.2
Hollow Biomolecular and Biotemplated Microcapsules

Various biological structures with multiple length scale shapes and sizes have been utilized as templates for producing complex bio-enabled replicas. LbL assembly and related assembly techniques have been widely explored for construction of hollow microcapsules and planar films either containing biomolecules (e.g., DNA or proteins) in their shells or encapsulating biomolecules in their interior [67]. In an interesting development in this field, a number of studies have been conducted on LbL microcapsules fabricated by using biological cells as templates. As one example, we note the protein filling of LbL microcapsules studied with SEM and AFM by Shenoy and Sukhorukov [68]. In this study, AFM imaging was exploited in a traditional manner to characterize general microstructure, surface morphology, and the thickness of LbL shells of dried microcapsules.

Hybrid LbL multilayers of silica nanoparticles and poly(diallyldimethylammonium chloride) (SiO_2/PDDA) were used to replicate the glutaraldehyde-fixed red blood cells, as reported by Caruso [69]. Corresponding hollow microcapsules obtained after the cell core was destroyed by a deproteinizer mimicked the cell shape including the fine secondary structure, such as sharp nanoscale spikes of the original red blood cell template as revealed by SEM and AFM images.

In another study by Neu *et al.*, red blood cells of discoid or echinocyte shapes were coated with traditional $(PAH/PSS)_5$ LbL shells [70]. After cell core removal by the NaCl–NaOCl solution, the distinctive shape of the original templates was still preserved. AFM imaging in different modes was used to estimate the uniformity of LbL shells, measure the thickness of LbL shells, and prove the presence of the morphological features characteristic of the original cell templates. In addition, *E. coli* bacteria chemically stabilized with a glutaraldehyde cross-linker were used as templates for the fabrication of the $(PSS/PAH)_5$-based hollow cell replicas in the same study. In the case of *E. coli* templates, AFM images showed near-cylindrical hollow LbL microcapsules with characteristic dimensions.

Despite examples of successful applications of the LbL assembly for the replication of living cells, strong cytotoxicity of the traditional polycations used in the conventional LbL assembly posed severe limitations on the applicability of the approach for cell surface engineering if the preservation of the living cell is desired. It has been reported that the cytotoxic effect after LbL encapsulation depends on polycation concentration and exposure time [71, 72]. It has been suggested that the overall toxic effect of polyelectrolytes stems from the positive charge of the polycations inducing pores within the cell membrane [73, 74]. For instance, four-bilayer coatings of PLL/PSS, PEI/PSS, and PAH/PSS resulted in a significant decrease in viability of mammalian cells [75]. $(PLL/alginate)_2$ and PAH/PSS/PAH multilayers have been found to possess extreme toxicity for human pancreatic islets, resulting in their almost instant death [76].

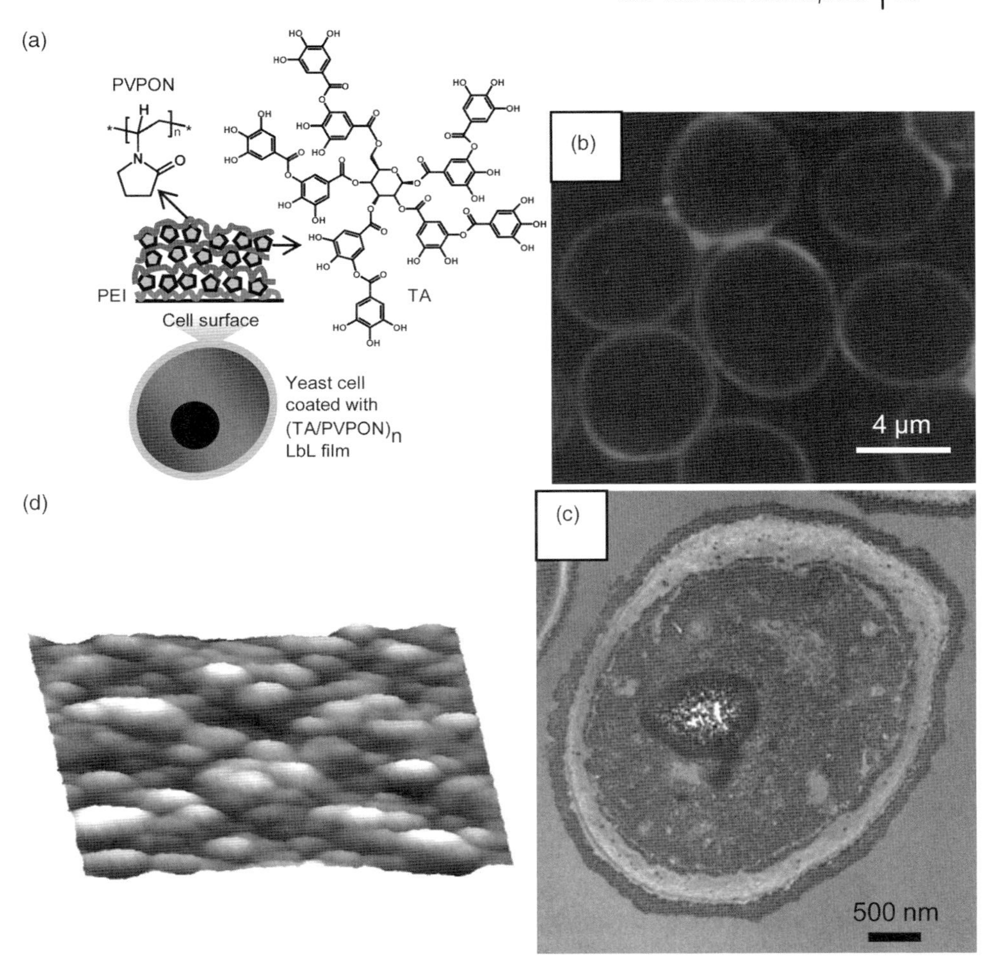

Figure 15.8 (a) The formation of hydrogen-bonded LbL shell on yeast cell surfaces. (b) Confocal image of encapsulated yeast cells with labeled shells. (c) TEM images of freeze-dried encapsulated yeast cells. (d) 3D topography AFM image of collapsed LbL shell; the scale is 500 nm × 500 nm × 50 nm. Reprinted from Ref. [65].

In a very recent study, Kozlovskaya *et al.* demonstrated that the *Saccharomyces cerevisiae* (YPH501) yeast cells can be sequentially coated with hydrogen-bonded (TA/PVPON)$_n$ LbL shells through multiple hydrogen bonding of the hydroxyl groups on tannic acid (TA) and carbonyl groups of PVPON (Figure 15.8) [77]. To ensure the controlled multilayer growth, cell surfaces were primed with a monolayer of PEI, which was possible due to the negative charge on the cell wall surface (−30 mV). The resulting LbL encapsulated cells were capable of undisrupted growth and normal functioning as confirmed via GFP-generated fluorescence. All imaging approaches, namely, fluorescent microscopy for cells coated with labeled polymers, TEM of cell cross sections, and AFM on collapsed shells, were utilized to demonstrate the presence of these shells and characterize their morphology.

The authors observed that the permeability through the hydrogen-bonded LbL shell is dramatically higher (4–5 times) than that of traditional PSS/PAH ionic shells. The observed difference for diffusion coefficients was attributed to loose, nanoporous morphology of the TA/PVPON multilayers characteristic of hydrogen-bonded LbL systems [78]. AFM analysis of the $(TA/PVPON)_n$ shells confirmed grainy and nanoporous surface morphology with the microroughness of about 3 nm (within 1 μm^2 surface areas), which exceeds the common microroughness of 0.5 nm for traditional solid PSS/PAH coatings (Figure 15.8) [79]. These results support the suggestion that the highly permeable structure of porous hydrogen-bonded shells that provides the unrestricted transport of nutrients and proteins is critical for preservation of functioning cells and preventing cell death after coating with nonionic LbL shells.

15.2.3
AFM Testing of Mechanical Properties of LbL Microcapsules

Stability and robustness of LbL microcapsules under variable external conditions such as osmotic pressure, analyte permeation, or solvent quality are critical for the design of these microcapsules as drug delivery vehicles. Therefore, significant efforts have been devoted to developing experimental procedures and conducting measurements for a variety of LbL microcapsules with traditional and colloidal force spectroscopies, which are usually combined with confocal and interference microscopies, and applying sophisticated models of elastic and plastic deformations.

Here, we will present several representative examples of micromechanical testing of microcapsules and microparticles but we will not be discussing experimental details, which can be found in several fundamental studies (see Chapter 5, Refs [80, 81] and references therein). These studies contain detailed discussions of the regimes of deformations, permeability through the walls, elastic and plastic behavior, buckling and wrinkles, combining optical and colloidal force spectroscopy methods, and theoretical approaches employed to analyze experimental data to derive elastic modulus, bending stiffness, ultimate strength, and compressive strains.

In numerous micromechanical studies of LbL microcapsules, CFS was applied to single cells to derive the elastic modulus and bending constants for cell walls under different conditions. For instance, Lulevich *et al.* applied this approach to study T cells and demonstrated a very low elastic modulus (below 7 kPa) of the living cells tested in this study [82]. The CFS probing revealed that the elastic modulus dramatically increased to 200 kPa after chemical fixation. In another approach, tipless AFM microcantilevers were used to measure the deformational behavior of hollow biocompatible PLA microcapsules by Glynos *et al.* [83]. In this study, the authors demonstrated measurements of both the stiffness of microcapsules and limits of elasticity by inducing reversible and irreversible deformation of their shells in a manner similar to that discussed for LbL microcapsules (see above).

Comparisons of linear elastic deformation at small strains, behavior in the viscoelastic regime, and large deformations resulting in buckling have been discussed in detail by Fery and Weinkamer [81]. In this work, thin shell microcapsule deformation was analyzed with experimental results directly compared with theories developed for the deformation of empty and filled microcapsules. The authors also considered the results of the direct simulation of large-scale deformations, including buckling, which were conducted with finite element analysis. Time-dependent deformation was demonstrated to be useful to derive information on the viscoelastic behavior of microcapsules. A combination of CFS with confocal and interference microscopies was demonstrated to be critical for independent evaluation of deformational behavior important for verification of the applicability of theoretical model. Similar combined CFS testing approaches to cells and biological capsules were also suggested and discussed.

Finally, in this study, the stiffness and the elastic modulus of LbL shells were measured and analyzed for several common LbL microcapsules. For classical microcapsules from PAH/PSS polyelectrolytes in pure water, the elastic modulus was found to be within a few hundred of MPa, which is close to that measured for swollen planar polyelectrolyte films. On the other hand, the measured value is much lower than that reported for solid and dry PSS/PAH LbL films (usually within 1–5 GPa, see above). The value of 400 MPa was the highest elastic modulus found for traditional LbL shells in wet state. The elastic modulus, however, decreased significantly for higher ionic strength, as reported in another CFS study from the same group [84].

Different regimes of mechanical deformation for LbL microcapsules have been discussed by Vinogradova [80]. The author distinguished three fundamentally different deformational regimes for these microcapsules. In the elastic regime, completely reversible deformation was observed at low strains (below 20%), with a gradual increase in resistivity observed (Figure 15.9). At higher strains,

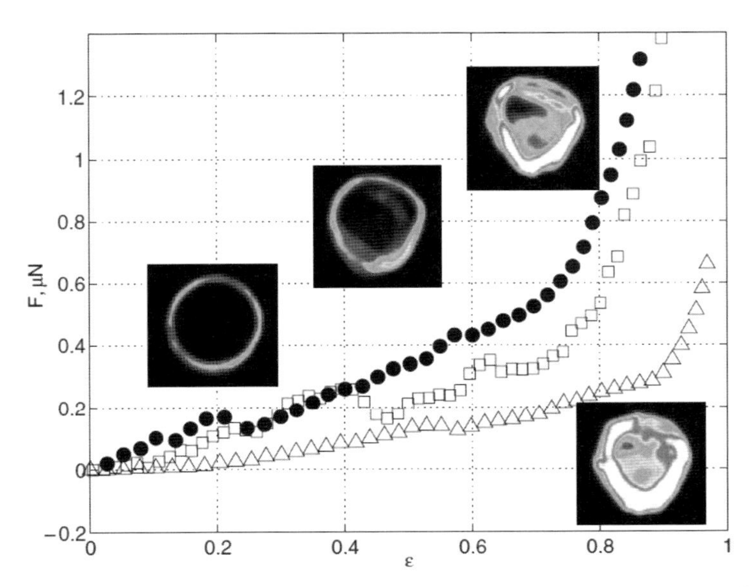

Figure 15.9 Typical force versus strain curves and confocal images obtained at different stages of microcapsule deformation for different microcapsules. Reprinted from Ref. [80].

elastoplastic deformation was observed, which is still partially reversible but with large hysteresis and strong rate-dependent behavior. This deformational behavior is finally replaced by irreversible plastic deformation with dramatic increases in resisting forces for higher strains, above 70%, which is associated with the complete collapse of microcapsules and the deformation of folded LbL walls (Figure 15.9).

From CFS measurements of different microcapsules conducted in this study, it has been concluded that the overall deformational behavior highly depends on type of microcapsules studied with PLA microcapsules showing most compliant behavior. The measured elastic moduli varied widely, ranging from as low as 30 MPa, common for compliant rubbery materials, to as high as 200 MPa, which are characteristic values for hard elastomeric materials. The actual modulus values varied for different diameter microcapsules, LbL microcapsules in different solvents, for different molecular weights of polyelectrolytes, and for different initial cores. The author noted that the modulus values obtained from CFS experiments were close to those obtained concurrently from osmotic measurements that employed confocal fluorescence microscopy and labeled LbL shells. The mechanical deformation of hollow microcapsules was also compared to that for filled microcapsules, which showed richer mechanical behavior due to various contributing mechanisms of permeation, swelling, and buckling.

In another study from the same group, Vinogradova *et al.* demonstrated the effect of adding DNA into LbL microcapsules on their mechanical behavior and stability [85]. AFM imaging showed a very different surface morphology for DNA-containing LbL microcapsules when compared to traditional LbL microcapsules, with the former showing much higher surface microroughness and exceeding porosity. For the DNA-containing outermost layer, large domains of oriented DNA loops were observed on the microcapsule surface, while for microcapsules with a PAH outer surface, a highly porous morphology was observed instead. CFS measurements of the DNA/PAH microcapsules confirmed higher permeability of their walls caused by the porous morphology and showed a much higher compliance, with the elastic modulus being two times lower than that for PSS/PAH microcapsules fabricated and measured under identical conditions. It has been demonstrated that the introduction of polyelectrolyte dendrimers into LbL multilayer walls might also lead to somewhat softened microcapsules, with the elastic modulus decreasing to values below 150 MPa [86].

Micromechanical studies with CFS have also been conducted on hollow LbL tubes with a diameter as large as several micrometers as well as for that below 200 nm to derive elastic modulus and test their permeability [87, 88]. The stiffness for various PSS, PAH, and PEI LbL microtubes was determined by Mueller *et al.* to be in the range 0.01–0.03 N/m, with increasing stiffness detected for smaller microtubes [87]. The elastic modulus of the PSS/PAH tubes was measured to be around 200 MPa, which is close to the value reported for spherical LbL microcapsules. LbL nanotubes from various polyelectrolytes placed across an opening were measured by monitoring the resonant frequencies of the AFM tip resting on the nanotubes by Guenot *et al.* [88]. The resulting elastic modulus of 115 MPa for nanotubes in water was in the expected range (100–300 MPa) measured for similar spherical microcapsules in a swollen state.

15.3
Replicas and Anisotropic Template Structures

A variety of replication techniques have been developed for polymeric structures to mirror unique shapes and morphologies of various inorganic microstructures in addition to the mostly spherical microcapsules considered above (Figure 15.1). Hollow replicas of cores with various complex shapes are of great interest because of their unique characteristics such as low specific density, high specific surface area, variable and high curvature, potentially enhanced catalytic and binding activities, efficient ion storage, complex refractive index, and complex plasmonic and photonic coupling [89–93]. Hollow anisotropic replicas from various polymeric materials have been utilized as microcontainers for encapsulating materials such as therapeutics, fluorescent markers, and field-responsive agents applicable for controlled drug delivery and biomedical imaging [94].

15.3.1
Anisotropic Replicas

Recent studies have shown that transport, lifetime, and local interactions of particles with cells are strongly dependent on their shape and spatially specific intermolecular interactions [95]. For instance, the most important step in drug delivery, ingestion of foreign materials by white blood cells, is frequently dictated by shape [96]. Anisotropic microparticles and nanoparticles have been demonstrated to have the ability to serve as building blocks for assembly into hierarchical synthetic structures with useful properties [97, 98]. Preparation of anisotropic hollow replicas on sacrificial cores/templates introduces significant technical challenges, such as the formation of a uniform conformal coating around surfaces with large variations in curvature. Another challenge in the replication approach is the preservation of the shape of the replicates after core removal and even after drying. Assembly of such anisotropic structures is governed by the balance of attractive and repulsive forces [99].

A variety of easily dissolvable and highly anisotropic inorganic colloids of different shapes, compositions, and functionalities have been introduced as prospective sacrificial templates for replication, including cubes, peanuts, dimers, rods, and pyramids [100–104]. These microparticles can be fabricated by using various approaches and are usually characterized by combined SEM and AFM imaging with AFM utilized not only to visualize their shapes, but also to observe the surface composition of these crystals and measure properties of the resulting microcapsules [105]. For instance, 3 μm^3 calcium carbonate particles can be prepared by the precipitation reaction of sodium carbonate with calcium chloride and cubic-shaped LbL microcapsules can be formed [106]. The preservation of the cubic shape in the wet state was confirmed by confocal microscopy and AFM showed the collapsed cubic shape in the dry state.

Peanut-shaped hematite colloids have been prepared by several groups and were used recently by Liddell and coworkers as sacrificial templates to create hollow shells [100, 107, 108]. To synthesize the core hematite particles, a condensed ferric hydroxide gel was aged under hydrothermal conditions. A silica shell was then grown onto the hematite surface using modified sol–gel chemistry, followed by core removal through selective etching using concentrated hydrochloric acid. The resulting silica peanut-shaped replicas were organized under confinement into an aperiodic crystal structure. A combined confocal microscopy, SEM, and AFM visualized a triangular lattice of peanut microparticles.

There are few examples where anisotropic microcapsules have been constructed with LbL assembly [109, 110]. One of the earliest examples of such an attempt was the alternate deposition of PSS and PAH polyelectrolyte layers on the surfaces of cubic enzyme crystals [111]. The formation of the PSS/PAH LbL shell around the core particle was achieved under conditions when the enzyme crystals were insoluble. The shape of the template crystals was retained during the LbL coating as confirmed with SEM. However, after the enzyme core was solubilized and washed away, the resultant hollow multilayer shells did not show any shape memory, leaving behind almost spherical polymer shells as confirmed with AFM [112]. This transformation is due to the high osmotic pressure built up within the LbL microcapsule because of the high concentration of dissolved enzyme.

LbL multilayer microcapsules based on hydrogen bonding were demonstrated by Kozlovskaya *et al.* to have sufficient structural strength and integrity to maintain the original form of a template particle in solution after core removal [113]. Cubic-shaped microcapsules were fabricated by the LbL assembly onto 10 μm cubic $CdCO_3$ particles. AFM has been explored to directly measure the thickness of the LbL shells on collapsed microcapsules to establish incremental changes with adding polymer bilayers. AFM imaging of these shells showed a highly grainy and porous morphology in the dry state.

In another study, Kozlovskaya *et al.* exploited inorganic $CdCO_3$ and SnS anisotropic microcrystals and nanocrystals as sacrificial cores for the preparation of anisotropic hollow capsules by hydrogen-bonded LbL assembly of TA and PVPON, which provided high compliance combined with chemical stability (Figure 15.10) [114]. PEI-$(TA/PVPON)_3$ hollow microcapsules replicated the cubic shape of the $CdCO_3$ crystals, as can be seen from confocal fluorescent images obtained directly in the wet state (Figure 15.10). The hollow microcapsules preserved the cubic shape in solution after core removal, revealing straight edges, sharp corners, and an overall cubic shape. In addition, for the surface morphology of the collapsed hollow capsule, AFM showed the characteristic TA/PVPON domain surface texture with high porosity previously found for hydrogen-bonded LbL assemblies of tannic acid [115]. From the AFM cross section, the thickness of the three-bilayer TA/PVPON shell was found to be about 9 nm. Remarkably, the original shape of the template particles was demonstrated to be perfectly preserved including well-defined edges despite the ultrathin walls of the resultant capsule, as can be clearly seen in the AFM image in Figure 15.10.

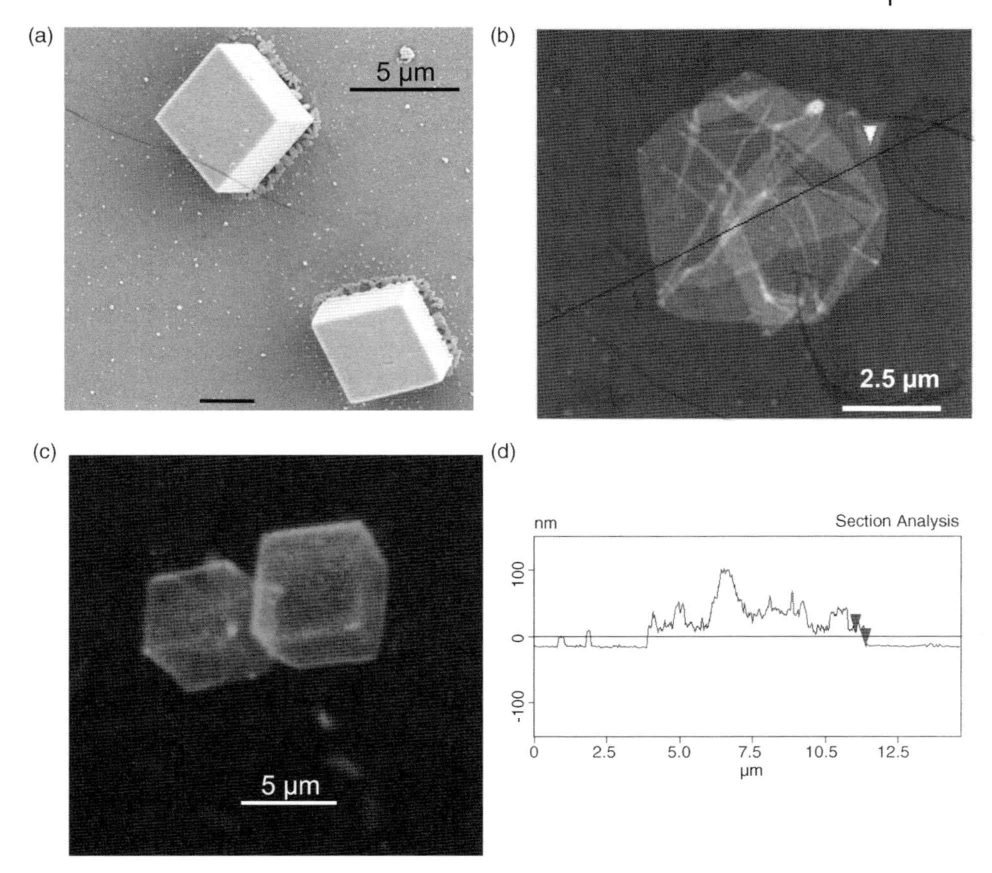

Figure 15.10 (a) SEM images of $CdCO_3$ cores, (b) AFM image of collapsed PEI(TA/PVPON)$_3$ cubic microcapsule with cross section of a collapsed microcapsule (d), and (c) confocal microscopy image of hollow PEI(TA/PVPON)$_3$ capsules in the wet state. Reprinted from Ref. [114].

Another type of anisotropic microcapsules of a rectangular shape with the capsule walls made via the LbL approach has been recently demonstrated by Paunov and coworkers [116]. Nano-cotton fibers derived from sulfuric acid hydrolysis of cellulose carried a negative surface charge and were combined with positively charged PAH onto needle-like aragonite and rhombohedral calcium carbonate crystals. The geometry of the original template was well reproduced in the polymer shell after the crystal template dissolution.

15.3.2
Colloidal Templated Crystals

Colloidal crystals formed by assembled microscopic spherical particles are well known and frequently characterized with AFM usually in a combination with SEM and optical microscopy. Inverse opals can serve as suitable molds to create spherical particles with uniform, predetermined diameters. For example, the inverse opal method was used for the fabrication of well-ordered carbon microspheres by infiltration of phenolic resin into a silica inverse opal structure and subsequent carbonization [117]. This approach has also been used to synthesize spherical colloids of various oxides, metals, silicon, polymers, liquid crystals, and wax [118–120].

Colloidal crystals from nonspherical particles, such as those discussed previously, have the potential for producing inverse opals that have higher complexity, asymmetric unit cells, and anisotropic properties that cannot be achieved with traditional spherical particles. A few examples of such colloidal crystals include asymmetric dimer colloids and mushroom-like colloids fabricated and visualized with AFM and SEM [121, 122]. For example, Colvin and coworkers have shown that polymer inverse templates can be uniformly deformed to produce ellipsoidal inverse opals that can be used to generate anisotropic particle arrays [123].

One of the widely studied types of complex 3D replicas of microparticles is the inverse replicate of the face-centered cubic (*fcc*) opaline templates prepared by colloidal crystal templating [124, 125]. To prepare such structures, void spaces between spherical particles (e.g., PS latexes or silica colloids) are infiltrated by fluid followed by its solidification. Removal of these particles produces the highly ordered porous solids or the inverse opal structure [126]. Silica particles are removed by etching with hydrofluoric acid and polymer spheres by dissolution in the appropriate solvent or by calcination or pyrolysis. The resulting inverse replicas observed with SEM and AFM show long-range ordered structure with uniform spherical voids interconnected through windows at the points where the original spheres touched [127].

15.4
Interfacial Adhesion between Particles and Surfaces

Measuring interfacial strength for nanocomposite materials composed of microparticles and nanoparticles is critically important for understanding the mechanisms of reinforcement, failure, fracturing, and toughening. As pointed out in a recent publication by Kokuoz *et al.*, the precise measurement of local adhesive properties of engineered interfaces under different conditions is critical for controlling a load transfer and the overall strength of the nanocomposites [128, 129].

For microscale and nanoscale engineered interfaces, the fundamental mechanism of adhesion can be probed by applying different AFM-based measurements. In previous chapters, we discussed how adhesive properties can be evaluated by using surface force spectroscopy modes for the direct detachment of microparticles and measuring pull-off forces for modified colloidal and regular AFM tips. The quantification of surface adhesive properties can be conducted by applying some contact mechanics models discussed in Chapter 5. Here, we present some selected recent examples of applying novel alternative AFM-based approaches for more complex

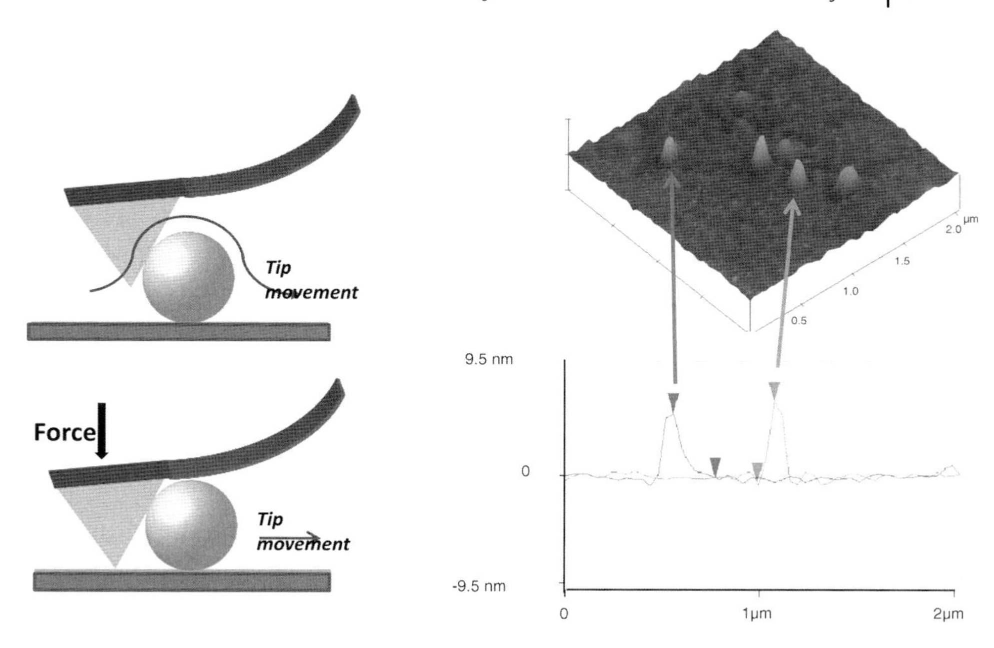

Figure 15.11 Two modes of lateral motion of trapped nanoparticles and AFM image and cross section of residual humps on PVP layer after nanoparticle removal. Image courtesy of I. Luzinov.

measurements of nanoparticle–surface interactions. In addition, we present combined optical and AFM methods for the evaluation of colloidal particle interactions with modified surfaces.

The first method discussed here was successfully employed for the detachment of complex objects such as cells and cellular structures and the estimation of the interfacial strength [130, 131]. Kokuoz *et al.* expanded this method by exploiting the entrapment of nanoparticles under an AFM tip to measure the strength of the nanoparticle–substrate adhesion during AFM tip sliding (Figure 15.11) [128]. For this purpose, silica nanoparticles with a diameter of about 130 nm were trapped in the proximity of a thermally treated AFM tip with an increased radius of curvature while the sliding tip was brought into direct contact with a substrate with randomly adsorbed nanoparticles (Figure 15.11). The authors measured the adhesive forces between the silica nanoparticles and the specifically prepared PVP surface layers with thicknesses below 100 nm. The strength of the adhesive bond was probed in a tearing contact mode, meaning that the nanoparticles were completely removed from the polymer surface by applying a tangential force parallel to the substrate surface. The authors considered two competitive modes of nanoparticle movement – rolling and sliding – to evaluate the phenomenon in quantitative terms.

By analyzing the surface morphology of the residual PVP layer after silica nanoparticle removal, the authors considered two major modes of failure for the adhesive bond. They considered adhesive failure that should result in the clean delamination of the particle from the PVP surface layer. This mode was compared with possible cohesive failure that should include partial removal of the PVP materials strongly adhered to the silica nanoparticles. In AFM studies of the "cleaned" areas, significant plastic deformation was observed on the PVP surface layer with nanoscale humps appearing in the specific locations where the removed nanoparticles had been, as confirmed by AFM observations of the same selected area before and after the removal of nanoparticles (Figure 15.11).

As a result of these observations, the authors suggested that during the silica nanoparticle removal with shear stresses, PVP polymeric chains firmly attached to silica surface elongate to break and then leave behind nanoscale debris that form these humps. It has been suggested that the nanoparticles are removed by destroying the cohesive contact in the course of sliding and shearing the supporting polymer layer. The authors observed that the PVP layer thickness and PVP molecular weight had a pronounced effect on the shearing force needed to disjoint interfacial contact.

Direct visualization of the contact area during CFS experiments is critical for accurate evaluation of adhesion and mechanical properties of colloidal particles and soft films. As discussed in Chapter 5, contact mechanics models are usually applied to analyze experimental data assuming *a priori* certain deformational mechanisms because no direct information on the load-dependent variable contact area with a diameter of 2–20 nm can be unambiguously obtained. Fortunately, for relatively large (micrometer) soft colloids and thick (hundreds of nanometers) polymer surface layers, combining SFS and CFM measurements with concurrent optical observations allows direct measurements (not just estimations) of the contact area and, thus, unambiguous application of the contact mechanics theories for quantitative evaluations of the nanomechanical properties [132]. Corresponding technical aspects of such a combination of different techniques are discussed in Chapter 8 and we will add only a brief note here.

As suggested by Erath and coworkers, replacing stiff colloidal probes (like common silica particles) as probes with elastomeric microparticles allows the direct determination of the contact area at variable normal loads via the application of micro-interferometric measurements [133]. In these measurements, a PDMS microparticle was utilized as an elastomeric colloidal probe to measure the adhesive forces by applying the JKR model to analyze the contact area variation. This approach can also be easily expanded to other polymeric colloidal microparticles. The force spectroscopy data were collected for PDMS particles having a diameter around 10 μm with a small force increment by pressing these compliant particles against hydrophobic and hydrophilic SAM-modified surfaces.

Within the JKR approach, a very low surface energy close to the limit of detection (below 2 mJ/m^2) was found in this study for hydrophilic surfaces in water. However, surface energy increased dramatically, 32 mJ/m^2, for hydrophobic surfaces, with additional contributions emerging from both long-range capillary forces and different surface microroughnesses. The elastic modulus of about 1 MPa was unambiguously measured for elastomeric PDMS colloidal probes, which was close to the expected value. The technique demonstrated the overall robustness of the experimental approach developed in this study.

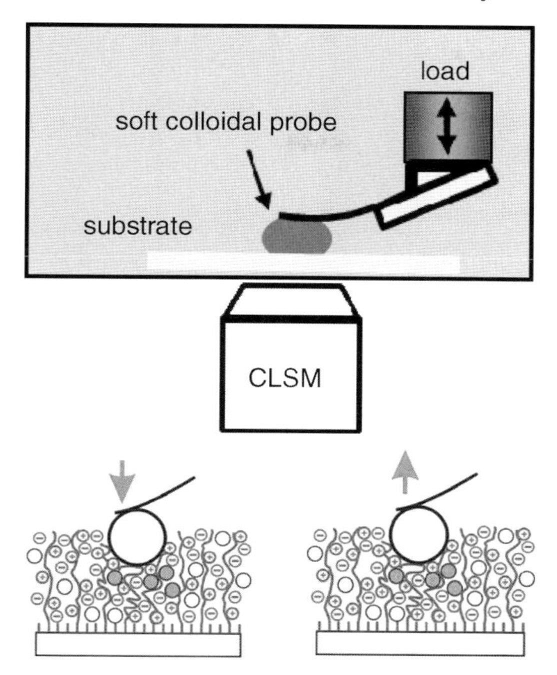

Figure 15.12 Combined colloidal force spectroscopy and fluorescent microscopy measurements of polymer brush layer with elastomeric colloidal probe. Courtesy of J. Erath, J. Bünsow, W. Huck, and A. Fery.

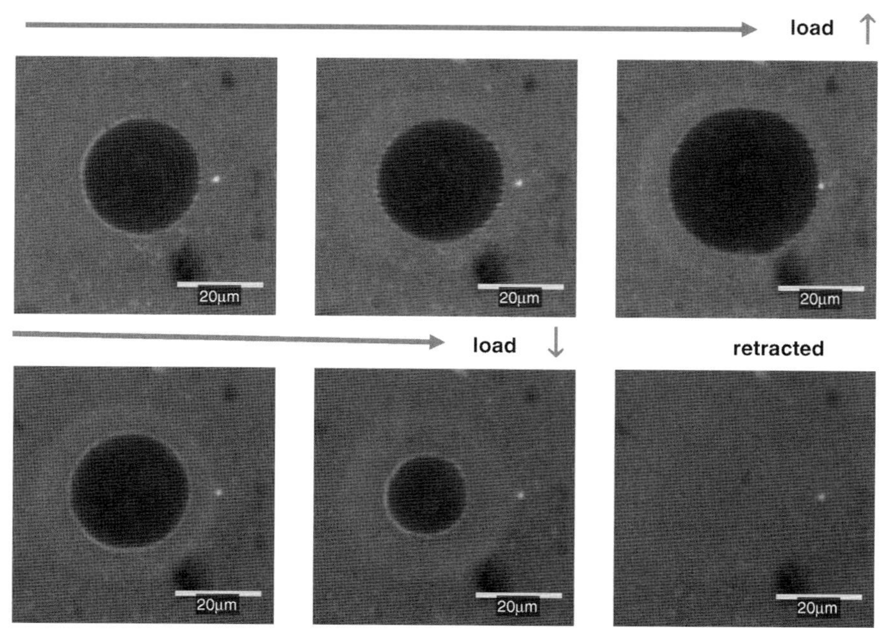

Figure 15.13 Fluorescence microscopy of brush polymer layer under increasing and decreasing normal loads. Courtesy of J. Erath, J. Bünsow, W. Huck, and A. Fery.

In another development of this approach, soft colloidal probes were applied to study polyelectrolyte brushes by Fery and coworkers [134]. Pressure-sensitive fluorescence phenomenon was utilized to directly observe the load-dependent contact area and measure the applied force at the same time (Figure 15.12). A pressure-sensitive brush layer designed by Bünsow *et al.* was labeled with a fluorescent dye whose fluorescence intensity is a function of the charge density inside the brush layer [135]. On increasing the applied load, the brush is compressed, which affects the charge density. As a result, the fluorescence intensity in the contact area decreases.

Indeed, fluorescent microscopy images of the loaded contact area showed decreased fluorescence in the contact area that changes proportionally to the normal load. Increasing load (upper sequence of images) results in a proportional increase in the contact area of the soft colloidal probe and the brush-modified substrate (in the figure, the black spot is due to compression) (Figure 15.13). In contrast, retracting the colloidal probe results in a proportional decrease in the contact area visible in fluorescence microscopy. Such directly visualized contact area variation enables the use of fluorescent microscopy instead of interferometry to accurately monitor the elastic deformation of the polymer brush layer with a thickness below 200 nm and the elastomeric colloidal probe with microscopic dimensions.

References

1 Shchepelina, O., Kozlovskaya, V., Singamaneni, S., Kharlampieva, E., and Tsukruk, V.V. (2010) Synthetic replicas of anisotropic particulate and complex continuous templates. *J. Mater. Chem.*, **20** (32), 6587–6603.

2 Uzun, O., Sanyal, A., and Jeong, Y. (2010) Molecular recognition induced self-assembly of diblock copolymers: microspheres to vesicles. *Macromol. Biosci.*, **10** (5), 481–487.

3 Decuzzi, P. and Ferrari, M. (2008) The receptor-mediated endocytosis of nonspherical particles. *Biophys. J.*, **94** (10), 3790–3797.

4 Simons, K. and Vaz, W.L. (2004) Model systems, lipid rafts, and cell membranes. *Annu. Rev. Biophys. Biomol. Struct.*, **33**, 269–295.

5 Ornatska, M., Jones, S.E., Naik, R.R., Stone, M., and Tsukruk, V.V. (2003) Biomolecular stress-sensitive gauges: surface-mediated immobilization of mechanosensitive membrane protein. *J. Am. Chem. Soc.*, **125** (42), 12722–12723.

6 Tsukruk, V.V., Ornatska, M., and Sidorenko, A. (2003) Synthetic and bio-hybrid nanoscale layers with tailored surface functionalities. *Prog. Org. Coat.*, **47** (3–4), 288–291.

7 Dufrêne, Y. and Lee, G.U. (2000) Advances in the characterization of supported lipid films with the atomic force microscope. *Biochem. Biophys. Acta*, **1509** (1–2), 14–41.

8 Kang, Y., Kwon, O., Xie, X., and Zhu, D.-M. (2009) Conductance mapping of proton exchange membranes by current sensing atomic force microscopy. *J. Phys. Chem.*, **113** (45), 15040–15046.

9 Reviakine, I. and Brisson, A. (2000) Formation of supported phospholipid bilayers from unilamellar vesicles investigated by atomic force microscopy. *Langmuir*, **16** (4), 1806–1815.

10 Keddie, J.L. (1997) Film formation of latex. *Mater Sci. Eng.*, **21** (3), 101–170.

11 Sommer, F., Duc, T.M., Pirri, R., Meunier, G., and Quet, C. (1995) Surface morphology of poly(butyl acrylate)/(polymethyl methacrylate) core–shell

latex by atomic force microscopy. *Langmuir*, **11** (2), 440–448.

12 Ramirez-Aguilar, K.A. and Rowlen, K.L. (1998) Tip characterization from AFM images of nanometric spherical particles. *Langmuir*, **14** (9), 2562–2566.

13 Duracher, D., Sauzedde, F., Elaissari, A., Perrin, A., and Pichot, C. (1998) Cationic amino-containing *N*-isopropyl-acrylamide–styrene copolymer latex particles: 1. Particle size and morphology vs. polymerization process. *Colloid Polym. Sci.*, **276** (3), 219–231.

14 Perez, E. and Lang, J. (1999) Flattening of latex film surface: theory and experiments by atomic force microscopy. *Macromolecules*, **32** (5), 1626–1636.

15 Pfau, A., Schrepp, W., and Horn, D. (1999) Detection of a single molecule adsorption structure of poly (ethylenimine) macromolecules by AFM. *Langmuir*, **15** (9), 3219–3225.

16 Budhlall, B.M., Shaffer, O.L., Sudol, E.D., Dimonie, V.L., and El-Aasser, M.S. (2003) Atomic force microscopy studies of the film surface characteristics of poly(vinyl acetate) latexes prepared with poly(vinyl alcohol). *Langmuir*, **19** (23), 9968–9972.

17 Han, S., Briseno, A.L., Shi, X., Mah, D.A., and Zhou, F. (2002) Polyelectrolyte-coated nanosphere lithographic patterning of surfaces: fabrication and characterization of electropolymerized thin polyaniline honeycomb films. *J. Phys. Chem. B*, **106** (25), 6465–6472.

18 Zoldesi, C.I., Ivanovska, I.L., Quilliet, C., Wuite, G.J.L., and Imhof, A. (2008) Elastic properties of hollow colloidal particles. *Phys. Rev. E*, **78** (5), 051401.

19 Wei, X., Gong, C., Gou, M., Fu, S., Guo, Q., Shi, S., Luo, F., Guo, G., Qiu, L., and Qian, Z. (2009) Biodegradable poly (ε-caprolactone)–poly(ethylene glycol) copolymers as drug delivery system. *Int. J. Pharm.*, **381** (1), 1–18.

20 Nouvel, C., Raynaud, J., Marie, E., Dellacherie, E., Six, J.L., and Durand, A.J. (2009) Biodegradable nanoparticles made from polylactide-grafted dextran copolymers. *J. Colloid Interface Sci.*, **330** (2), 337–343.

21 Wong, S.Y., Pelet, J.M., and Putnam, D. (2007) Polymer systems for gene delivery – past, present, and future. *Prog. Polym. Sci.*, **32** (8–9), 799–837.

22 Reis, C.P., Neufeld, R.J., Ribeiro, A.J., and Veiga, F. (2006) Nanoencapsulation I. Methods for preparation of drug-loaded polymeric nanoparticles. *Nanomed. Nanotechnol. Biol. Med.*, **2** (1), 8–21.

23 Orozco, V.H., Witold, B., Chonkaew, W., and López, B.L. (2009) Preparation and characterization of poly(lactic acid)-g-maleic anhydride + starch blends. *Macromol. Symp.*, **277** (1), 69–80.

24 Nobs, L., Buchegger, F., Gurny, R., and Allémann, E. (2004) Poly(lactic acid) nanoparticles labeled with biologically active Neutravidin™ for active targeting. *Eur. J. Pharm. Biopharm.*, **58** (3), 483–490.

25 Lim, J.Y., Hansen, J.C., Siedlecki, C.A., Hengstebeck, R.W., Cheng, J., Winograd, N., and Donahue, H.J. (2005) Osteoblast adhesion on poly(L-lactic acid)/polystyrene demixed thin film blends: effect of nanotopography, surface chemistry, and wettability. *Biomacromolecules*, **6** (6), 3319–3327.

26 Ding, A.G., Shenderova, A., and Schwendeman, S.P. (2006) Prediction of microclimate pH in poly(lactic-*co*-glycolic acid) films. *J. Am. Chem. Soc.*, **128** (16), 5384–5390.

27 Orozco, V.H., Kozlovskaya, V., López, B.L., and Tsukruk, V.V. (2010) Biodegradable self-reporting nanocomposite films of poly (lactic acid) nanoparticles engineered by layer-by-layer assembly. *Polymer*, **51** (18), 4127–4139.

28 Caruso, F. and Mohwald, H. (1999) Preparation and characterization of ordered nanoparticle and polymer composite multilayers on colloids. *Langmuir*, **15** (23), 8276–8281.

29 Rasa, M., Kuipers, B.W.M., and Philipse, A.P. (2002) Atomic force microscopy and magnetic force microscopy study of model colloids. *J. Colloid Interface Sci.*, **250** (2), 303–315.

30 Chen, C.-W., Serizawa, T., and Akashi, M. (2002) *In situ* formation of Au/Pt bimetallic colloids on polystyrene microspheres: control of particle growth and morphology. *Chem. Mater.*, **14** (5), 2232–2239.

31 De Gennes, P.G. (1992) Soft matter. *Rev. Mod. Phys.*, **64** (3), 645–648.

32 Shchukin, D.G., Kommireddy, D.S., Zhao, Y., Cui, T., Sukhorukov, G.B., and Lvov, Y.M. (2004) Polyelectrolyte micropatterning using a laminar-flow microfluidic device. *Adv. Mater.*, **16** (5), 389–393.

33 Berger, S., Synytska, A., Ionov, L., Eichhorn, K.J., and Stamm, M. (2008) Stimuli-responsive bicomponent polymer Janus particles by "grafting from"/"grafting to" approaches. *Macromolecules*, **41** (24), 9669–9676.

34 Hong, L., Cacciuto, A., Luijten, E., and Granick, S. (2008) Clusters of amphiphilic colloidal spheres. *Langmuir*, **24** (3), 621–625.

35 Jiang, S. and Granick, S. (2007) Janus balance of amphiphilic colloidal particles. *J. Chem. Phys.*, **127** (16), 161102–161104.

36 Hong, L., Jiang, S., and Granick, S. (2006) Simple method to produce Janus colloidal particles in large quantity. *Langmuir*, **22** (23), 9495–9499.

37 Glaser, N., Adams, D.J., Böker, A., and Krausch, G. (2006) Janus particles at liquid–liquid interfaces. *Langmuir*, **22** (12), 5227–5229.

38 Anderson, K.D., Luo, M., Jakubiak, R., Naik, R.R., Bunning, T.J., and Tsukruk, V.V. (2010) Robust plasma polymerized-titania/silica Janus microparticles. *Chem. Mater.*, **22** (10), 3259–3264.

39 Ali, A.M.I. and Mayes, A.G. (2010) Preparation of polymeric core–shell and multilayer nanoparticles: surface-initiated polymerization using *in situ* synthesized photoiniferters. *Macromolecules*, **43** (2), 837–844.

40 Anderson, K.D., Slocik, J.M., McConney, M.E., Enlow, J.O., Jakubiak, R., Bunning, T.J., Naik, R.R., and Tsukruk, V.V. (2009) Facile plasma-enhanced deposition of ultrathin crosslinked amino acid films for conformal biometallization. *Small*, **5** (6), 741–749.

41 Kemell, M., Harkonen, E., Pore, V., Ritala, M., and Leskela, M. (2010) Ta_2O_5- and TiO_2-based nanostructures made by atomic layer deposition. *Nanotechnology*, **21** (3), 035301.

42 Bai, M.-Y., Cheng, Y.-J., Wickline, S.S., and Xia, Y. (2009) Colloidal hollow spheres of conducting polymers with smooth surface and uniform, controllable sizes. *Small*, **5** (15), 1747–1752.

43 Huang, L., Wang, Z., Sun, J., Miao, L., Li, Q., Yan, Y., and Zhao, D. (2000) Fabrication of ordered porous structures by self-assembly of zeolite nanocrystals. *J. Am. Chem. Soc.*, **122** (14), 3530–3531.

44 Caruso, F. (2001) Nanoengineering of particle surfaces. *Adv. Mater.*, **13** (1), 11–22.

45 Caruso, F., Caruso, R.A., and Möhwald, H. (1998) Nanoengineering of inorganic and hybrid hollow spheres by colloidal templating. *Science*, **282** (5391), 1111–1114.

46 Donath, E., Sukhorukov, G.B., Caruso, F., Davis, S.A., and Möhwald, H. (1998) Novel hollow polymer shells by colloid-templated assembly of polyelectrolytes. *Angew. Chem., Int. Ed.*, **37** (16), 2201–2205.

47 Angelatos, A.S., Katagiri, K., and Caruso, F. (2006) Bioinspired colloidal systems via layer-by-layer assembly. *Soft Matter*, **2** (1), 18–23.

48 Jiang, C. and Tsukruk, V.V. (2006) Freestanding nanostructures via layer-by-layer assembly. *Adv. Mater.*, **18** (7), 829–840.

49 Johnston, A.P.R., Cortez, C., Angelatos, A.S., and Caruso, F. (2006) Layer-by-layer engineered capsules and their applications. *Curr. Opin. Colloid Interface Sci.*, **11** (4), 203–209.

50 Zhu, Y.F., Shi, J.L., Shen, W.H., Dong, X.P., Feng, J.W., Ruan, M.L., and Li, Y.S. (2005) Stimuli-responsive controlled drug release from a hollow mesoporous silica sphere/polyelectrolyte multilayer core–shell structure. *Angew. Chem., Int. Ed.*, **44** (32), 5083–5087.

51 Sokolova, V.V., Radtke, I., Heumann, R., and Epple, M. (2006) Effective transfection of cells with multi-shell calcium phosphate–DNA nanoparticles. *Biomaterials*, **27** (16), 3147–3153.

52 Chen, J., Saeki, F., Wiley, B.J., Cang, H., Cobb, M.J., Li, Z.Y., Au, L., Zhang, H., Kimmey, M.B., Li, X.D., and Xia, Y. (2005) Gold nanocages: bioconjugation and their potential use as optical imaging contrast agents. *Nano Lett.*, **5** (3), 473–477.

53 Schukin, D.G., Ustinovich, E.A., Sukhorukov, G.B., Möhwald, H., and Sviridov, D.V. (2005) Metallized polyelectrolyte microcapsules. *Adv. Mater.*, **17** (4), 468–472.

54 Wang, T.C., Rubner, M.F., and Cohen, R.E. (2002) Polyelectrolyte multilayer nanoreactors for preparing silver nanoparticle composites: controlling metal concentration and nanoparticle size. *Langmuir*, **18** (8), 3370–3375.

55 Lee, D., Rubner, M.F., and Cohen, R.E. (2005) Formation of nanoparticle-loaded microcapsules based on hydrogen-bonded multilayers. *Chem. Mater.*, **17** (5), 1099–1105.

56 Riboh, J.C., Haes, A.J., McFarland, A.D., Ranjit, C., and Van Duyne, R.P. (2003) A nanoscale optical biosensor: real-time immunoassay in physiological buffer enabled by improved nanoparticle adhesion. *J. Phys. Chem. B*, **107** (8), 1772–1780.

57 El-Sayed, I.H., Huang, X., and El-Sayed, M.A. (2006) Selective laser photo-thermal therapy of epithelial carcinoma using anti-EGFR antibody conjugated gold nanoparticles. *Cancer Lett.*, **239** (1), 129–135.

58 Petrova, H., Hu, M., and Hartland, G.V. (2007) Photothermal properties of gold nanoparticles. *Z. Phys. Chem.*, **221** (3), 361–376.

59 Köhler, K., Shchukin, D.G., Möhwald, H., and Sukhorukov, G.B. (2005) Thermal behavior of polyelectrolyte multilayer microcapsules. 1. The effect of odd and even layer number. *J. Phys. Chem. B*, **109** (39), 18250–18259.

60 Mauser, T., Déjugnat, C., Möhwald, H., and Sukhorukov, G.B. (2006) Microcapsules made of weak polyelectrolytes: templating and stimuli-responsive properties. *Langmuir*, **22** (13), 5888–5893.

61 Pastoriza-Santos, I., Schöler, B., and Caruso, F. (2001) Core–shell colloids and hollow polyelectrolyte capsules based on diazoresins. *Adv. Funct. Mater.*, **11** (2), 122–128.

62 Ochs, C.J., Such, G.K., Städler, B., and Caruso, F. (2008) Low-fouling, biofunctionalized, and biodegradable click capsules. *Biomacromolecules*, **9** (12), 3389–3396.

63 Wang, A., Tao, C., Cui, Y., Duan, L., Yang, Y., and Li, J. (2009) Assembly of environmental sensitive microcapsules of PNIPAAm and alginate acid and their application in drug release. *J. Colloid Interface Sci.*, **332** (2), 271–279.

64 Wang, F., Zhu, Y., and Gao, C. (2009) Fabrication of complex microcapsules containing poly(allylamine)-g-poly(*N*-isopropylacrylamide) and their thermal responsivity. *Colloid Surf. A*, **349** (1–3), 55–60.

65 Kozlovskaya, V., Kharlampieva, E., Chang, S., Muhlbauer, R., and Tsukruk, V.V. (2009) pH-responsive layered hydrogel microcapsules as gold nanoreactors. *Chem. Mater.*, **21** (10), 2158–2167.

66 Jiang, C., Singamaneni, S., Merrick, E., and Tsukruk, V.V. (2006) Complex buckling instability patterns of nanomembranes with encapsulated gold nanoparticle arrays. *Nano Lett.*, **6** (10), 2254–2259.

67 Jewell, C.M., Zhang, J., Fredin, N.J., and Lynn, D.M. (2005) Multilayered polyelectrolyte films promote the direct and localized delivery of DNA to cells. *J. Control Release*, **106** (1–2), 214–223.

68 Shenoy, D.B. and Sukhorukov, G.B. (2005) Microgel-based engineered nanostructures and their applicability with template-directed layer-by-layer polyelectrolyte assembly in protein encapsulation. *Macromol. Biosci.*, **5** (5), 451–458.

69 Caruso, F. (2000) Hollow capsule processing through colloidal templating and self-assembly. *Chem. Eur. J.*, **6** (3), 413–419.

70 Neu, B., Voigt, A., Mitloehner, R., Leporatti, S., Gao, C.Y., Donath, E.,

Kiesewetter, H., Möhwald, H., Meiselman, H.J., and Bäumler., H. (2001) Biological cells as templates for hollow microcapsules. *J. Microencapsul.*, **18** (3), 385–395.

71 De Koker, S., De Geest, B.G., Cuvelier, C., Ferdinande, L., Deckers, W., Hennink, W.E., De Smedt, S., and Mertens, N. (2007) *In vivo* cellular uptake, degradation, and biocompatibility of polyelectrolyte microcapsules. *Adv. Funct. Mater.*, **17** (18), 3754–3763.

72 Städler, B., Chandrawati, R., Price, A.D., Chong, S.F., Breheney, K., Postma, A., Connal, L.A., Zelikin, A.N., and Caruso, F. (2009) A microreactor with thousands of subcompartments: enzyme-loaded liposomes within polymer capsules. *Angew. Chem., Int. Ed.*, **48** (24), 4359–4362.

73 Bieber, T., Meissner, W., Kostin, S., Niemann, A., and Elsasser, H.-P. (2002) Intracellular route and transcriptional competence of polyethylenimine–DNA complexes. *J. Control Release*, **82** (2–3), 441–454.

74 Godbey, W.T., Wu, K.K., and Mikos, A.G. (1999) Size matters: molecular weight affects the efficiency of poly (ethylenimine) as a gene delivery vehicle. *J. Biomed. Mater. Res.*, **45** (3), 286–275.

75 Germain, M., Balaguer, P., Nicolas, J.-C., Lopez, F., Esteve, J.-P., Sukhorukov, G.B., Winterhalter, M., Richard-Foy, H., and Fournier, D. (2006) Protection of mammalian cell used in biosensors by coating with a polyelectrolyte shell. *Biosens. Bioelectron.*, **21** (8), 1566–1573.

76 Wilson, J.T., Cui, W., and Chaikof, E.L. (2008) Layer-by-layer assembly of a conformal nanothin PEG coating for intraportal islet transplantation. *Nano Lett.*, **8** (7), 1940–1948.

77 Kozlovskaya, V., Harbaugh, S., Drachuk, I., Shchepelina, O., Kelley-Loughnane, N., Stone, M., and Tsukruk, V.V. (2011) Hydrogen-bonded LbL shells for living cell surface engineering. Soft Matt., 7(6), 2364–2372.

78 Kharlampieva, E. and Sukhishvili, S.A. (2006) Hydrogen-bonded layer-by-layer polymer films. *J. Macromol. Sci. C*, **46** (4), 377–395.

79 Jiang, C., Markutsya, S., and Tsukruk, V.V. (2004) Compliant, robust, and truly nanoscale free-standing multilayer films fabricated using spin-assisted layer-by-layer assembly. *Adv. Mater.*, **16** (2), 157–166.

80 Vinogradova, O. (2004) Mechanical properties of polyelectrolyte multilayer microcapsules. *J. Phys. Condens Matter*, **16** (32), R1105.

81 Fery, A. and Weinkamer, R. (2007) Mechanical properties of micro- and nanocapsules: single-capsule measurements. *Polymer*, **48** (25), 7221–7235.

82 Lulevich, V., Zink, T., Chen, H.-Y., Liu, F.-T., and Liu, G. (2006) Cell mechanics using atomic force microscopy-based single-cell compression. *Langmuir*, **22** (19), 8151–8155.

83 Glynos, E., Koutsos, V., McDicken, W.N., Moran, C.M., Pye, S.D., Ross, J.A., and Sboros, V. (2009) Nanomechanics of biocompatible hollow thin-shell polymer microspheres. *Langmuir*, **25** (13), 7514–7522.

84 Mueller, R., Köhler, K., Weinkamer, R., Sukhorukov, G., and Fery, A. (2005) Melting of PDADMAC/PSS capsules investigated with AFM force spectroscopy. *Macromolecules*, **38** (23), 9766–9771.

85 Vinogradova, O.I., Lebedeva, O.V., Vasilev, K., Gong, H., Garcia-Turiel, J., and Kim, B.-S. (2005) Multilayer DNA/poly(allylamine hydrochloride) microcapsules: assembly and mechanical properties. *Biomacromolecules*, **6** (3), 1495–1502.

86 Kim, B.S., Lebedeva, O.V., Kim, D.H., Caminade, A.M., Majoral, J.P., Knoll, W., and Vinogradova, O.I. (2005) Assembly and mechanical properties of phosphorus dendrimer/polyelectrolyte multilayer microcapsules. *Langmuir*, **38** (16), 7200–7206.

87 Mueller, R., Daehne, L., and Fery, A. (2007) Preparation and mechanical characterization of artificial hollow tubes. *Polymer*, **48** (9), 2520–2525.

88 Cuenot, S., Alem, H., Louarn, G., Demoustir-Champagne, S., and Jonas,

A.M. (2008) Mechanical properties of nanotubes of polyelectrolyte multilayers. *Eur. Phys. J. E*, **25**, 343–348.

89 Lou, X.W., Archer, L.A., and Yang, Z. (2008) Hollow micro-/nanostructures: synthesis and applications. *Adv. Mater.*, **20** (21), 3987–4019.

90 Martinez, C.J., Hockey, B., Montgomery, C.B., and Semancik, S. (2005) Porous tin oxide nanostructured microspheres for sensor applications. *Langmuir*, **21** (17), 7937–7944.

91 Lou, X.W., Wang, Y., Yuan, C., Lee, J.Y., and Archer, L.A. (2006) Template-free synthesis of SnO_2 hollow nanostructures with high lithium storage capacity. *Adv. Mater.*, **18** (17), 2325–2329.

92 Ma, H., Cheng, F.Y., Chen, J., Zhao, J.Z., Li, C.S., Tao, Z.L., and Liang, J. (2007) Nest-like silicon nanospheres for high-capacity lithium storage. *Adv. Mater.*, **19** (22), 4067–4070.

93 Martinez, C.J., Hockey, B., Montgomery, C.B., and Semancik, S. (2005) Porous tin oxide nanostructured microspheres for sensor applications. *Langmuir*, **21** (17), 7937–7944.

94 Chen, J.F., Ding, H.M., Wang, J.X., and Shao, L. (2004) Preparation and characterization of porous hollow silica nanoparticles for drug delivery application. *Biomaterials*, **25** (4), 723–727.

95 Gratton, S.E.A., Ropp, P.A., Pohlhaus, P.D., Luft, J.C., Madden, V.J., Napier, M.E., and DeSimone, J.M. (2008) The effect of particle design on cellular internalization pathways. *Proc. Natl. Acad. Sci. USA*, **105** (33), 11613–11618.

96 Champion, J.A., Katare, Y.K., and Mitragotri, S. (2007) Making polymeric micro- and nanoparticles of complex shapes. *Proc. Natl. Acad. Sci. USA*, **104** (29), 11901–11904.

97 Glotzer, S.C. and Solomon, M.L. (2007) Anisotropy of building blocks and their assembly into complex structures. *Nat. Mater.*, **6**, 557–562.

98 Ruiz-Hitzky, E., Darder, M., Aranda, P., and Ariga, K. (2010) Advances in biomimetic and nanostructured biohybrid materials. *Adv. Mater.*, **22** (3), 323–336.

99 Bishop, K.J.M., Wilmer, C.E., Soh, S., and Grzybowski, B.A. (2009) Nanoscale forces and their uses in self-assembly. *Small*, **5** (14), 1600–1630.

100 Matijevic, E. (1993) Preparation and properties of uniform size colloids. *Chem. Mater.*, **5** (4), 412–426.

101 Hao, L.-Y., Zhu, C.-L., Jiang, W.-Q., Chen, C.-N., Hu, Y., and Chen, Z.-Y. (2004) Sandwich Fe_2O_3@SiO_2@PPy ellipsoidal spheres and four types of hollow capsules by hematite olivary particles. *J. Mater. Chem.*, **14** (19), 2929–2934.

102 Liddell, C.M. and Summers, C.J. (2003) Monodispersed ZnS dimers, trimers, and tetramers for lower symmetry photonic crystal lattices. *Adv. Mater.*, **15** (20), 1715–1719.

103 Sugimoto, T., Itoh, H., and Mochida, T. (1998) Shape control of monodisperse hematite particles by organic additives in the gel–sol system. *J. Colloid Interface Sci.*, **205** (1), 42–52.

104 Greyson, E.C., Barton, J.E., and Odom, T.W. (2006) Tetrahedral zinc blende tin sulfide nano- and microcrystals. *Small*, **2** (3), 368–371.

105 Mokari, T., Zhang, M., and Yang, P. (2007) Shape, size, and assembly control of PbTe nanocrystals. *J. Am. Chem. Soc.*, **129** (32), 9864–9865.

106 Bei, C., Ming, L., Jiaguo, Y., and Xiujian, Z. (2004) Preparation of monodispersed cubic calcium carbonate particles via precipitation reaction. *Mater. Lett.*, **58** (10), 1565–1570.

107 Liu, Q. and Osseo-Asare, K. (2000) Synthesis of monodisperse Al-substituted hematite particles from highly condensed metal hydroxide gels. *J. Colloid. Interface. Sci.*, **231** (2), 401–403.

108 Lee, S.H., Gerbode, S.J., John, B.S., Wolfgang, A.K., Escobedo, F.A., Cohen, I., and Liddell, C.M. (2008) Synthesis and assembly of nonspherical hollow silica colloids under confinement. *J. Mater. Chem.*, **18** (41), 4912–4916.

109 Hua, A., Steven, J.A., and Lvov, Y.M. (2003) Biomedical applications of electrostatic layer-by-layer nano-assembly of polymers, enzymes, and nanoparticles. *Cell Biochem. Biophys.*, **39** (1), 23–43.

110 Decher, G. and Schlenoff, J.B. (eds) (2003) *Multilayer Thin Films: Sequential Assembly of Nanocomposite Materials*, Wiley-VCH Verlag GmbH, Weinheim.

111 Silvano, D., Krol, S., Diaspro, A., Cavalleri, O., and Gliozzi, A. (2002) Confocal laser scanning microscopy to study formation and properties of polyelectrolyte nanocapsules derived from $CdCO_3$ templates. *Microsc. Res. Tech.*, **59** (6), 536–541.

112 Caruso, F., Trau, D., Möhwald, H., and Renneberg, R. (2000) Enzyme encapsulation in layer-by-layer engineered polymer multilayer capsules. *Langmuir*, **16** (4), 1485–1488.

113 Kozlovskaya, V., Yakovlev, S., Libera, M., and Sukhishvili, S.A. (2005) Surface priming and the self-assembly of hydrogen-bonded multilayer capsules and films. *Macromolecules*, **38** (11), 4828–4836.

114 Shchepelina, O., Kozlovskaya, V., Kharlampieva, E., Mao, W., Alexeev, A., and Tsukruk, V.V. (2010) Anisotropic micro- and nano-capsules. *Macromol. Rapid Commun.*, **31** (23), 2041–2046.

115 Kozlovskaya, V., Kharlampieva, E., Drachuk, I., Cheng, D., and Tsukruk, V.V. (2010) Responsive microcapsule reactors based on hydrogen-bonded tannic acid layer-by-layer assemblies. *Soft Matter*, **6**, 3596–3608.

116 Holt, B., Lam, R., Meldrum, F.C., Stoyanov, S.D., and Paunov, V.N. (2007) Anisotropic nano-papier mache microcapsules. *Soft Matter*, **3** (2), 188–190.

117 Guan, G., Kusakabe, K., Ozono, H., Taneda, M., Uehara, M., and Maeda, H. (2007) Preparation of carbon microparticle assemblies from phenolic resin using an inverse opal templating method. *J. Mater. Sci.*, **42** (24), 10196–10202.

118 Wang, H., Yu, J.-S., Li, X.-D., and Kim, D.-P. (2004) Inorganic polymer-derived hollow SiC and filled SiCN sphere assemblies from a 3DOM carbon template. *Chem. Commun.*, (20), 2352–2353.

119 Cong, H. and Cao, W. (2006) Preparation of ordered porous NaCl and KCl crystals. *Solid State Sci.*, **8** (9), 1056–1060.

120 Rong, J., Liu, S., and Liu, Y. (2006) Template synthesis of structured titania using inverse opal gels. *Polymer*, **47** (8), 2677–2682.

121 Hosein, D. and Liddell, C.M. (2007) Convectively assembled asymmetric dimer-based colloidal crystals. *Langmuir*, **23** (21), 10479–10485.

122 Hosein, D. and Liddell, C.M. (2007) Convectively assembled nonspherical mushroom cap-based colloidal crystals. *Langmuir*, **23** (17), 8810–8814.

123 Jiang, P., Bertone, J.F., and Colvin, V.L. (2001) A lost-wax approach to monodisperse colloids and their crystals. *Science*, **291** (5503), 453–457.

124 Wong, S., Kitaev, V., and Ozin, G.A. (2003) Colloidal crystal films: advances in universality and perfection. *J. Am. Chem. Soc.*, **125** (50), 15589–15598.

125 Stein, A., Li, F., and Denny, N.R. (2008) Morphological control in colloidal crystal templating of inverse opals, hierarchical structures, and shaped particles. *Chem. Mater.*, **20** (3), 649–666.

126 Xia, Y., Gates, B., and Li, Z.-Y. (2001) Self-assembly approaches to three-dimensional photonic crystals. *Adv. Mater.*, **13** (6), 409–413.

127 Lodahl, P., van Driel, A.F., Nikolaev, I.S., Irman, A., Overgaag, K., Vanmaekelbergh, D.V., and Vos, W.L. (2004) Controlling the dynamics of spontaneous emission from quantum dots by photonic crystals. *Nature*, **430**, 654–657.

128 Kokuoz, B., Kornev, K.G., and Luzinov, I. (2009) Gluing nanoparticles with a polymer bonding layer: the strength of an adhesive bond. *ACS Appl. Mater. Interfaces*, **1** (3), 575–583.

129 Baldan, A. (2004) Adhesively-bonded joints in metallic alloys, polymers and composite materials: mechanical and environmental durability performance. *J. Mater. Sci.*, **39** (15), 4729–4797.

130 Boyd, R.D., Verran, J., Jones, M.V., and Bhakoo, M. (2002) Use of the atomic force microscope to determine the effect of substratum surface topography on bacterial adhesion. *Langmuir*, **18** (6), 2343–2346.

131 Senechal, A., Carrigan, S.D., and Tabrizian, M. (2004) Probing surface adhesion forces of *Enterococcus faecalis* to medical-grade polymers using atomic force microscopy. *Langmuir*, **20** (10), 4172–4177.

132 Butt, H.-J. (1991) Measuring electrostatic, van der Waals, and hydration forces in electrolyte solutions with an atomic force microscope. *Biophys. J.*, **60** (6), 1438–1444.

133 Erath, J., Schmidt, S., and Fery, A. (2010) Characterization of adhesion phenomena and contact of surfaces by soft colloidal probe AFM. *Soft Matter*, **6**, 1432–1437.

134 Erath, J., Bünsow, J., Huck, W., and Fery, A. (2010) Investigation of response of soft matter systems on confinement by soft colloidal probe AFM. Proceedings of International Soft Matter Conference 2010, Granada, Spain, July 5–8, 2010.

135 Bünsow, J., Kelby, T.S., and Huck, W.T.S. (2010) Polymer brushes: routes toward mechanosensitive surfaces. *Acc. Chem. Res.*, **43** (3), 466–474.

16
Biomaterials and Biological Structures

According to Richard Feynman, "What biology needs is not more math, but to see better at the atomic level" [1]. Indeed, conventional microscopy techniques widely used at that time such as optical and electron microscopies either offered limited resolution or imposed foreign and harsh environmental conditions (high vacuum, conductive coatings, and fluorescent labeling), severely limiting our ability for non-damaging imaging and our understanding of biological systems. The field was completely turned upside down in the 1980s due to the invention of scanning force microscopies. Owing to its high resolution, ability to image under native conditions (hydrated state, controlled pH, and temperature), and relatively simple sample preparation, AFM, after its introduction, was immediately embraced by the biophysical and biochemical research communities as an invaluable tool for imaging and probing biological materials.

In particular, surface force measurements of various forms have proven to be crucial in probing native properties of biological materials under *in vivo* conditions [2–5]. In addition to accurate AFM imaging, surface force spectroscopy provided a deep understanding and novel insights into the mechanics of biological structures at different length scales (from tissues to individual protein molecules). In this chapter, using some selected examples, we briefly review the recent results of AFM imaging a wide variety of biomolecules, probing their mechanical properties down to single biomacromolecule, and measuring essential specific biomolecular interactions.

16.1
Imaging Adsorbed Biomacromolecules

16.1.1
General Approaches and Selected Examples

AFM has been extensively used to image a wide variety of biomolecules such as DNA, proteins, and peptides adsorbed onto a solid substrates [6]. We have briefly discussed the imaging of adsorbed DNA molecules onto solid substrates in the context of force control during imaging soft structures in Section 3.4.2 and hence will not consider

Scanning Probe Microscopy of Soft Matter: Fundamentals and Practices, First Edition.
Vladimir V. Tsukruk and Srikanth Singamaneni.

them any further here. Instead, we focus primarily on AFM imaging various kinds of proteins immobilized on surfaces under controlled environmental conditions. Some additional examples of adsorption of biomolecules on functionalized surfaces can also be found in Chapter 13.

The influence of substrate surface properties on protein adsorption processes has been extensively investigated using AFM. The general approach employed in these types of studies involves the adsorption of biomolecules (under various pH and salt conditions) on model substrates with different and controlled surface properties (e.g., functionality, charge, and roughness) designed to mimic selected biological environments or selected biomaterials and, from practical viewpoint, firmly tether biomolecules.

For instance, Denis and coworkers have monitored the adsorbed amount and supramolecular organization of collagen, an important biological component of tissue, on model substrates exhibiting defined topography and surface chemistry [7]. Substrates with controlled surface roughness were prepared by depositing gold on smooth substrates. Substrates with nanoscale protrusions were achieved by colloidal lithography, which involves the adsorption of colloidal microparticles on smooth substrates followed by deposition of gold and polymer layers to form a polymer surface with a well-defined nanoscale topography [8]. Different surface chemistries were achieved by the self-assembly of alkanethiols with CH_3 (hydrophobic) and OH (hydrophilic) terminal groups. AFM imaging of adsorbed collagen molecules was performed under water before and after collagen adsorption from a dilute buffered solution.

The authors observed that on smooth hydroxyl-terminated substrates, rod-like helical collagen molecules formed a ~6 nm thick homogeneous surface layer with a low microroughness. In contrast, a ~20 nm thick surface film exhibiting elongated, aggregated structures was found on smooth but methyl-terminated surfaces. The AFM images (also supported by XPS studies) clearly revealed that a larger amount of collagen proteins were adsorbed onto methyl-terminated surfaces compared to hydroxyl-terminated ones. For rougher surfaces, the adsorbed amount of the protein molecules was similar to that found on smooth substrates. However, the protein molecules no longer exhibited aggregated structures on the hydrophobic surfaces. From these observations, the authors concluded that the adsorbed amount of the protein was mainly affected by the surface functionality, but the supramolecular organization of the adsorbed protein molecules was controlled by both surface chemistry and topography.

Marchin and coworker also employed AFM to investigate the adsorption of the plasma protein fibrinogen on graphite and mica substrates, which again served as model hydrophobic and hydrophilic surfaces, respectively [9]. The tapping mode AFM images revealed that the morphology of submonolayer surface films imaged under ambient conditions was dramatically different on the mica and graphite substrates. While the protein molecules exhibited a tendency to aggregate on the graphite substrate, they were found to adsorb as isolated single molecules in the case of mica substrates. Furthermore, the protein molecules were found to be globular on the mica substrate, while on graphite they exhibited a characteristic trinodular structure.

From AFM images, the average height of the fibrinogen molecules was found to be 1.7 nm on mica and 1.05 nm on the graphite substrate. The lateral dimension (average length) was found to be 31 nm on mica and 63 nm on graphite. Although the dimensions have not been corrected to account for the tip convolution effects, they represent a general trend of the length of the molecules on both substrates. The authors suggested that the differences in the vertical and lateral dimensions of the proteins indicated different conformations of the protein following adsorption on these two surfaces with different surface properties. Their results demonstrate that AFM is an excellent tool to resolve conformational changes in biomacromolecules both due to their interactions with surfaces and due to their ability to discern submolecular structures and elements, which is critically important for understanding the biomaterial–protein interactions.

In a different study, conformational changes in fibronectin (Fn) biomolecules, which were also adsorbed onto surfaces with different properties (from hydrophilic (silica) to hydrophobic (methylated silica surface by modifying with dichlorodimethylsilane)) were investigated by Bergkvist *et al.* using the tapping mode AFM imaging of the protein molecules adsorbed onto these surfaces [10]. The adsorption conditions were controlled to obtain well-separated protein molecules (15–40 Fn molecules/μm^2) evenly distributed over the surface area, allowing reliable single-molecule imaging. In the initial stages, the authors performed the AFM imaging in both liquid and ambient environments. By comparing the AFM images obtained under liquid and ambient conditions, the authors concluded that the morphology of the adsorbed proteins was essentially identical. Further experiments were performed under ambient conditions to simplify the imaging process.

In this study, the authors concluded that on hydrophilic silica surfaces, 70% of the protein molecules exhibited an elongated structure with only partial intramolecular chain interactions (which stabilize the quaternary structure of this protein). On the other hand, protein molecules adsorbed onto hydrophobic, methylated surfaces, predominantly (70%) exhibited a compact structure. The authors reasoned that the observed difference in morphology is due to the negatively charged groups on hydrophilic surfaces, which might interfere with the stabilizing ionic interactions between the two chains in the protein, extending the molecule upon adsorption. The change in the conformation of the protein following adsorption onto the hydrophilic surface is expected to have dramatic influence on the functionality of Fn such as its cell binding ability.

In a different study, Ornatska *et al.* reported on the design of protein–organic surface nanostructures containing the stress-sensitive membrane protein, MscL, from *Salmonella typhymurium* [11]. The authors designed supported organic monolayers (alkylsilane SAMs) and lipid Langmuir–Blodgett monolayers with different packing densities and surface tensions, which can be directly used for membrane protein immobilization from solution. Light tapping mode AFM scanning under ambient humid conditions was utilized for the characterization of adsorbed MscL proteins. All samples were scanned immediately after adsorption to avoid drying of the sample. Dramatic changes in the membrane protein conformation were observed depending upon the packing density of the alkyl

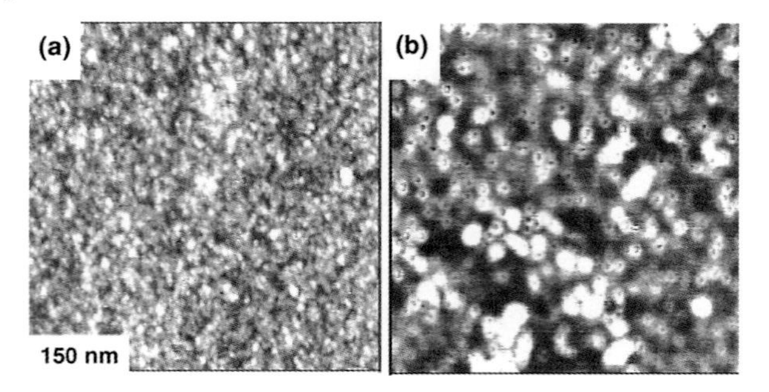

Figure 16.1 AFM images (500 x 500 nm) of immobilized protein molecules in closed (a) and open (b) states (z-scale: 5 nm). Reproduced from Ref. [11].

chains within the supporting organic alkyl monolayer as can be seen in the general observation AFM images (Figure 16.1).

High-resolution AFM images of these protein molecules adsorbed onto modestly hydrophobic monolayers (surface tension within 22–30 mJ/m^2) showed the presence of individual MscL protein molecules embedded in the lipid monolayers as nanoscale elevated features (Figure 16.1). The overall vertical dimension of the molecules was measured to be 4.5 nm, which was smaller than the molecular dimensions in an unperturbed conformation (about 8 nm in a compact conformation with stem) (Figure 16.1). The authors suggested that this height difference was due to the compressive stress induced by the AFM tip, which is usually observed for compliant macromolecules and the partially embedded state of proteins within vertically oriented alkyl chains. The upright protein orientation is expected due to the predominantly hydrophobic nature of the bottom part of the molecule, which favors interactions with nonpolar alkyl chains. No indications of pore structure were found for the protein under these conditions. The authors concluded that under low surface tension and tight alkyl chain packing, the adsorbed and embedded MscL molecules retained their closed pore conformation and prolate shape.

A very different structure was observed when MscL proteins were embedded in less-ordered, loosely packed SAMs with fluid-like, loose packing of alkyl chains and surface tension between 30 and 60 mJ/m^2. Under these conditions, the adsorbed protein molecules consistently showed a donut shape with a large central opening on both low-density SAMs and lipid Langmuir monolayers (Figure 16.1). The height of the protein molecules in this state was within 2.2–2.7 nm, indicating that there was significant flattening of protein molecules.

A large opening was consistently observed at the highest magnification of embedded proteins (Figure 16.2). The apparent width of this opening, calculated as the distance between the edges of the rim, reached 11–15 nm, whereas the overall "apparent" width of the molecules increased to 20–30 nm. These values were greater than those expected for the open pore molecular models of the protein in an upright orientation even considering the tip dilation (Figure 16.3). The authors noted that the proteins can be partially embedded in loosely packed alkyl chains with a higher level of free volume. This high-resolution AFM study clearly revealed that the variable

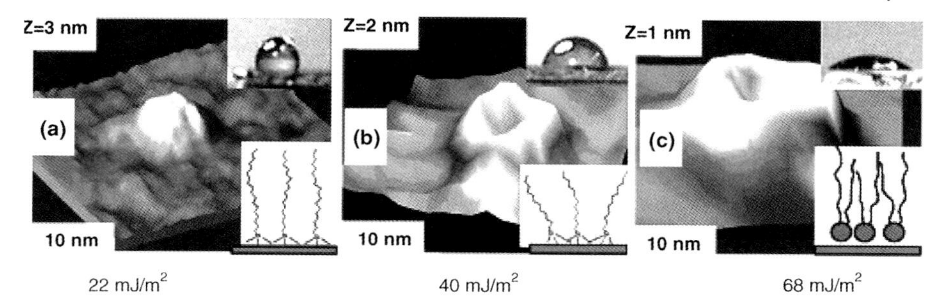

Figure 16.2 AFM images of individual MscL protein molecules embedded in different organic monolayers in (a) closed and (b,c) two open states; insets show microstructure (*right*, bottom corner) of supporting alkylsilane SAMs and lipid monolayers; corresponding water drop shape (*right*, top corner) illustrates increasing hydrophilicity of monolayers with different packing density from left to right. Reproduced from Ref. [11].

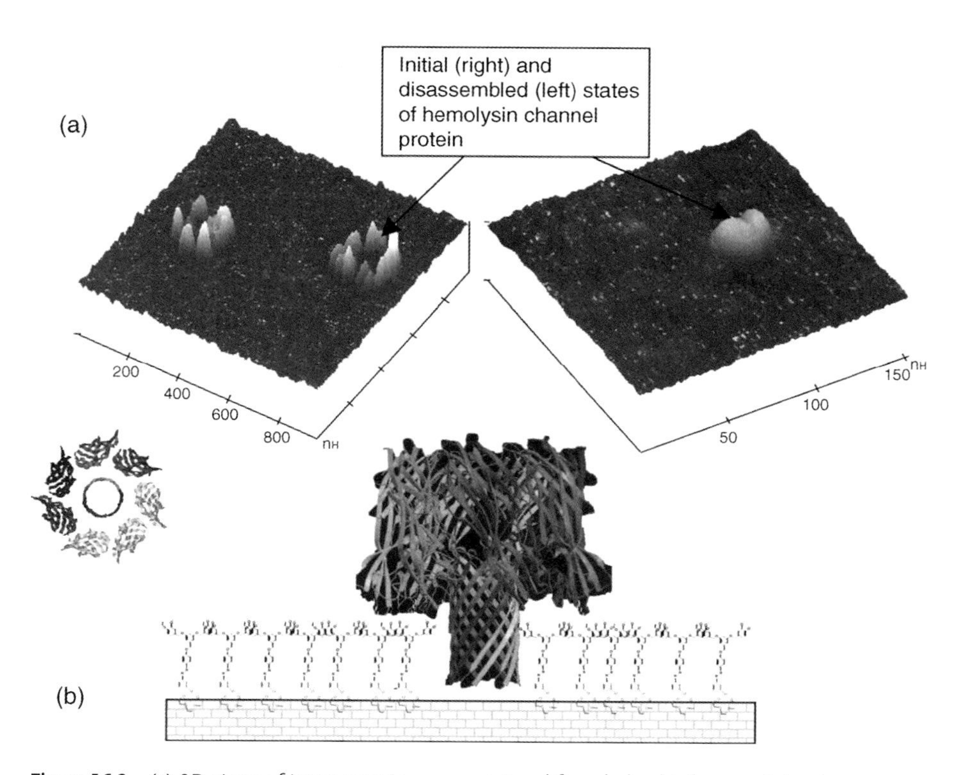

Figure 16.3 (a) 3D views of intact protein (*right*) and disassembled protein (*left*) after photochromic isomerization of dendritic monolayer (amphiphilic monodendron containing one benzyl-15-crown-5 polar focal point and four dodecyl tails as peripheral groups. (b) A scheme of α-hemolysin immobilization in a LB monolayer from photosensitive monodendron molecules. Reproduced from Ref. [12].

surface tension of organic alkylsilane and lipid monolayers used for the immobilization of MscL molecules is instrumental in mediating protein conformation. Therefore, AFM imaging confirmed the transition from a prolate shape of MscL molecules at a low surface tension to a flattened, oblate shape with a wide central opening at a higher surface tension.

In a related work, Tsukruk *et al.* studied the different conformation states of the same membrane protein (MscL) adsorbed onto solid surfaces with various surface functionalities [12]. When the protein was adsorbed onto hydrophilic hydroxyl-terminated silicon surface, AFM imaging did not confirm the presence of open/closed pore structures of individual protein molecules. Protein macromolecules on bare silicon are in a collapsed, compressed shape and structural characteristics are different from the intact conformation. The authors also demonstrated the successful immobilization (preserving the closed pore conformation) of this membrane protein in a Langmuir monolayer of photochromic dendrimer molecules by adsorption from a water subphase directly in LB trough (Figure 16.3).

In the Langmuir monolayer deposited on a solid substrate, the protein preserved its initial conformational state. However, photoisomerization of the supporting photosensitive dendrimer molecules in the monolayer initiated by UV illumination at 365 nm caused a photochromic conversion within the monolayers that, in turn, changes the intralayer pressure and triggers a sudden "disassembling" of the protein in stand-alone subunits (Figure 16.3). Clearly seen in the AFM images are seven subunits grouped similar to their initial arrangement in an assembled protein as can be seen from a comparison with the corresponding molecular model (Figure 16.3). Thus, AFM imaging provides a visual representation of the transformation of the conformational state (between open/closed pore states) of the protein, which could be controlled by the nature of the supporting monolayer and its structural organization and packing density.

In a recent work, Yu *et al.* employed tapping mode AFM to probe the effects of pH, protein concentration, and contact time of the adsorption on the morphology and kinetics of the adsorption of protein A on a hydrophobic PDMS surface [13]. AFM imaging was performed in light tapping mode under ambient conditions to avoid protein damage. AFM images revealed that the adsorption of protein A was significantly affected by pH and the adsorbed surface concentration.

The effects of solution pH on protein A adsorption were investigated by keeping the concentration (1 μg/ml) and adsorption time (30 min) constant. Figure 16.4 shows the AFM images of the pristine PDMS and protein-adsorbed sample under various solution pH conditions. While the PDMS surface exhibits a smooth, featureless morphology in the pristine state, different surface textures of the adsorbed protein A were observed after adsorption in solutions with different pH. The adsorbed protein A displayed a tortuous line-like morphology at pH 2.0 and a tortuous line and dendrites with two or three tails at pH 3.0. At pH 5.0, the authors observed dendrite structures with three or four tails. At higher pH (6.0 and 8.0), the surface morphology reverted to a tortuous line-like morphology from the dendritic structure at pH 5.0.

Corresponding AFM cross sections shown in Figure 16.4 revealed that the average height of the protein A domains adsorbed from pH 5.0 solution was 7.66 nm, which

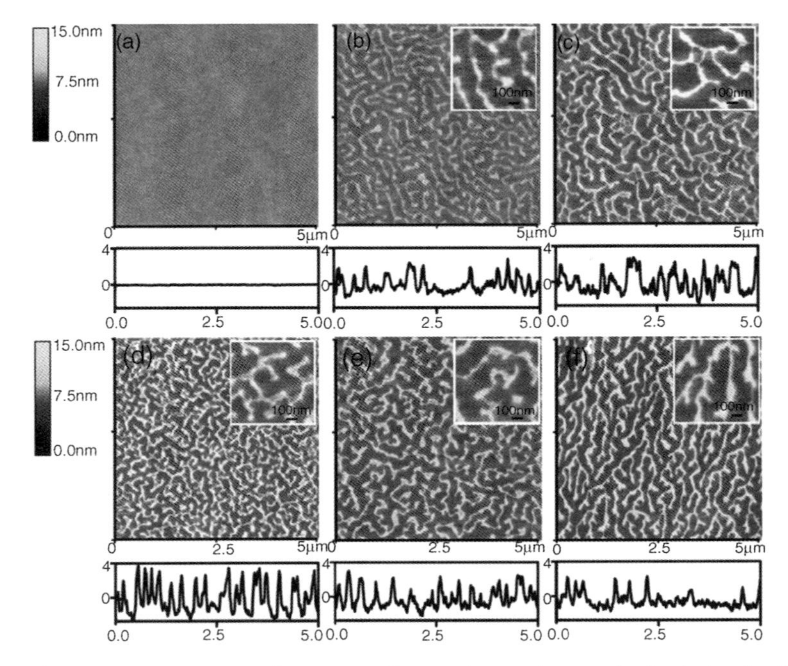

Figure 16.4 AFM images of (a) pristine PDMS and 1 μg/ml protein A adsorbed under different pH conditions at (b) pH 2.0, (c) pH 3.0, (d) pH 5.0, (e) pH 6.0, and (f) pH 8.0. AFM cross sections included with image shows the thickness of the adsorbed protein layer. Reproduced from Ref. [13].

was a much higher value than that of a single protein A molecule (3 nm), indicating the aggregation of protein A molecules at pH 5.0. The authors attributed the significantly different morphology of the adsorbed protein layer at pH 5.0 to the possible hydrophobic integration between the various domains of the protein molecules due to the neutral state of the molecule at pH 5.0, which is the isoelectric point of protein A.

In a study of another well-known fiber-forming biomolecule, Shulha *et al.* monitored the adsorption of silk protein on a hydrophilic silicon substrate [14]. For their studies, they employed silk fibroin from the domesticated silkworm, *Bombyx mori*. The structural portion of the fibers spun by the silkworm in order to form its cocoon consists of two proteins, a heavy chain fibroin (~390 kDa) and light chain fibroin (~25 kDa), linked by a single disulfide bond. The heavy chain fibroin is dominated by large stretches of hydrophobic amino acids, particularly glycine-alanine repeats that form β-sheet crystallites in the spun fibers. The adsorbed silk exhibited uniform surface morphology with occasionally occurring patches of uncovered surface areas (Figure 16.5).

Surface morphology of adsorbed silk protein displayed both individual globular shapes and aggregated nanofibrils composed of several dozen densely packed, near-circular domains. The surface morphology immediately after adsorption was dominated by individual circular features with fibrillar aggregates observed mainly along

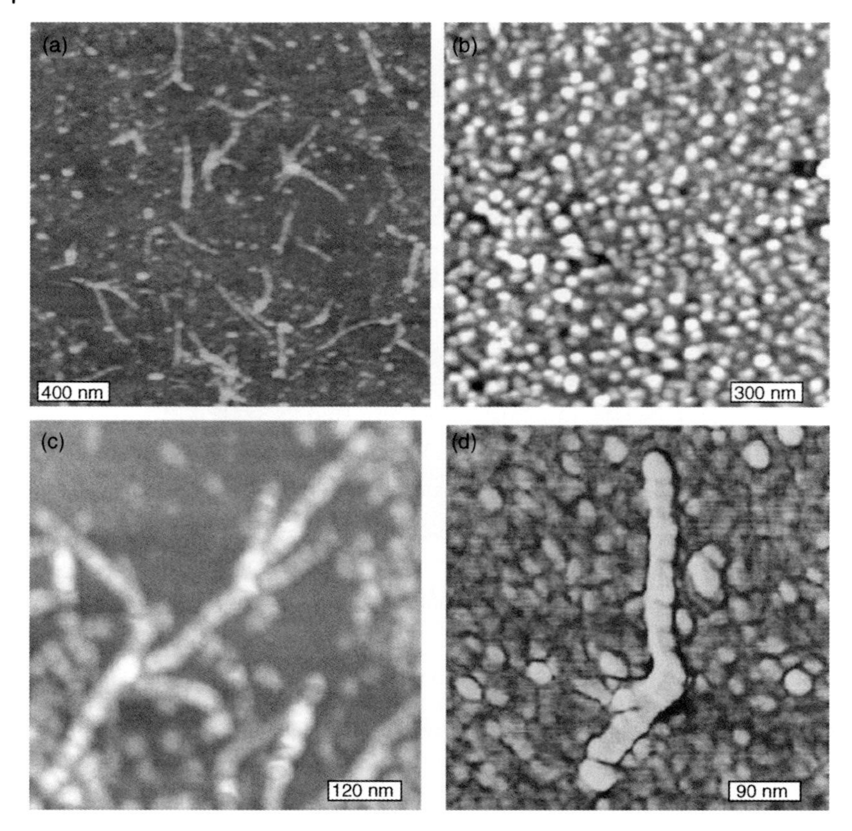

Figure 16.5 Topographical AFM images of silk fibroin protein: (a) near the edge of the protein layer showing globular and fibrillar aggregates; (b) high-resolution images of densely packed globular aggregates; (c) fibrillar structures; and (d) an individual fibril with domain structure. Reproduced from Ref. [14].

the boundaries between completely covered and bare surface areas. The height of these surface nanostructures was 3.1 nm for nanofibrils and 1.6 nm for individual circular features. The average diameter of the individual features (5.4 nm) was estimated from the AFM images accounting for the tip dilation and using a semi-spherical approximation for the tip with a known radius of curvature as discussed in Section 3.3.

The authors justly pointed out that the estimation of molecular dimensions from the tapping AFM images in ambient air should be done very carefully considering several artifacts common for imaging of biomolecules. First, partial dehydration of protein molecules in humid air could result in a change in the overall shape of the adsorbed protein molecules. Applying fast drying by a stream of dry nitrogen should "freeze" the original global shape even if individual domains will be compacted. Second, such dense, compact packing of molecular segments should affect their molecular dimensions as compared to the initial state in solution. Thus, the experimental molecular dimensions and volume (after correction for the tip dilation) should be comparable to the theoretical molecular volume calculated from the molecular weight and packing density that is typical for loosely packed bulk material.

Finally, tapping AFM imaging in air could result in the compression of a soft protein adsorbed onto a solid substrate. To this end, the authors applied very light tapping with a set point ratio (r_{sp}) higher than 0.95. Under these scanning conditions, normal forces applied to soft surfaces do not exceed 0.1 nN, which could result in their very modest (below 10%) deformation, thus introducing insignificant error. Therefore, the true volume of an individual circular feature estimated with all possible precautions was close to 40 nm^3. This value is many-fold smaller than the estimated molecular volume of a single-protein molecule with a molecular weight of 390 kDa and a predominantly α-helical backbone compacted into a globular shape based on prior models (180 nm^3). This relatively small volume observed in the AFM images was reasoned to be caused by the spreading of the hydrophilic parts (thus maximizing the interaction with the hydrophilic silicon substrate) of the protein and the segregation of the hydrophobic regions from one or more molecules resulted in the globules observed in AFM images.

In a recent report, AFM has been employed to probe the functional activity (not just the structure and morphology) of adsorbed protein molecules on mica surfaces. Time-dependent functional changes of adsorbed protein (fibrinogen in this study) were measured by Soman *et al.* using antigen–antibody debonding forces [15]. This measurement was performed by using AFM probes functionalized with monoclonal antibodies that can recognize peptide γ392-411 of fibrinogen. The probability of the antigen recognition was calculated from the surface force spectroscopy measured between the antibody on the functionalized tip and the fibrinogen on mica substrates. Statistical analysis revealed that the probability of antibody–antigen recognition was a maximum at ~45 min after adsorption followed by a decrease in the probability with increasing residence time. The nanoscale surface force spectroscopic measurements exhibited excellent agreement with the macroscale platelet adhesion measurements on these mica substrates that also had a maximum at ~45 min after adsorption. The remarkable correlation between the nano- and the macroscale measurements of the functional activity of the fibrinogen adsorbed onto the surface clearly indicates that AFM can be an extremely important tool to gain insight into the functionality of biomolecules with molecular resolution.

Among important biological molecules, IgG is the most abundant immunoglobulin in the blood and is produced in large quantities during secondary immune responses [16]. The binding of the Fc region of IgG, which coats microorganisms in the blood, and the Fc receptors of macrophages and neutrophils allows these phagocytic cells to bind, ingest, and destroy invading bacteria [17]. It is a ~180 kDa biomolecule with a characteristic Y-shaped structure. It is 15 nm across as determined by single-crystal X-ray diffraction measurements. Since the molecules are on the order of the radius of the tips (10–20 nm) routinely employed for AFM imaging, tip convolution significantly affects the apparent shape of the molecules visualized using conventional imaging techniques. Only under the best imaging conditions, with a very sharp tip and low forces, a heart-shaped structure and Y-shape have been

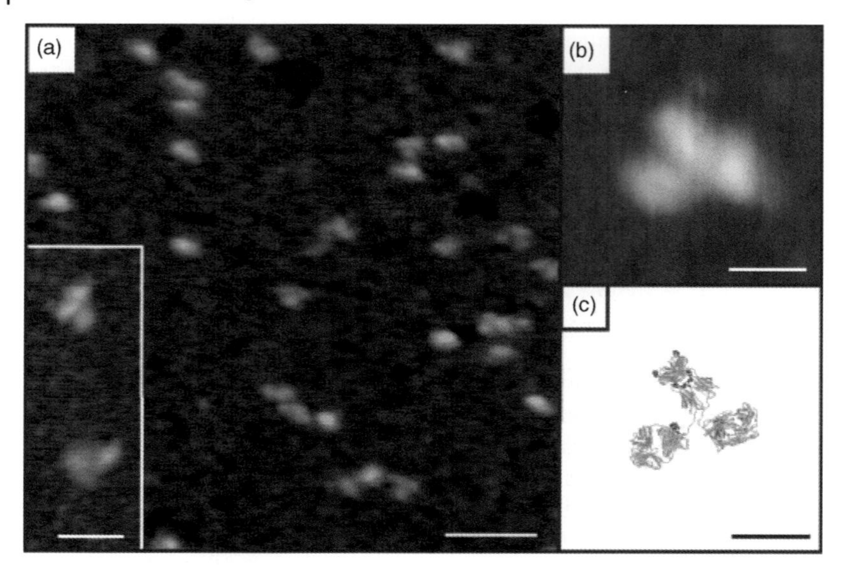

Figure 16.6 AFM images showing the IgG molecules on solid substrates imaged using carbon nanotube tip. (a) Relatively large scan area showing several molecules that exhibit the characteristic Y-shape (scale bar: 50 nm); (b) high-resolution images of IgG showing small tip-induced broadening (scale bar: 20 nm). Reproduced from Ref. [18].

observed. In fact, Cheung *et al.* performed high-resolution AFM imaging using carbon nanotube tips clearly showing the Y-shaped structure of IgG deposited on a solid substrate (Figure 16.6) [18]. Furthermore, the dimensions of the IgG closely agreed with the well-known dimensions (from X-ray diffraction measurements) of the biomolecule, indicating minimal tip-induced broadening of the features.

16.1.2 Peptides

Peptides and short oligomers of amino acids are similar to proteins in terms of chemistry, but much shorter in length. Owing to their recognition capabilities, peptides are a topic of great interest in the field of biochemistry, medicinal chemistry, pharmacology, and more recently in materials science [19]. Among numerous AFM studies of adsorbed peptides, So *et al.* investigated the adsorption of GBP1, a gold binding peptide, with an amino acid sequence MHGKTQATSGTIQS, repeated three times to enhance its binding affinity [20]. The solution 3rGBP1 is intrinsically disordered consisting of a repetitive random coil-extended structure conformation ((MHGKTQA) random coil and (TSGTIQS) extended). AFM was employed to monitor the morphology of 3rGBP1 surface films prepared over large grained atomically flat surfaces of Au {111}.

Figure 16.7 shows AFM topography images of the resulting peptide films adsorbed from a 1 μg/mL peptide solution. Two different rinsing procedures were investigated,

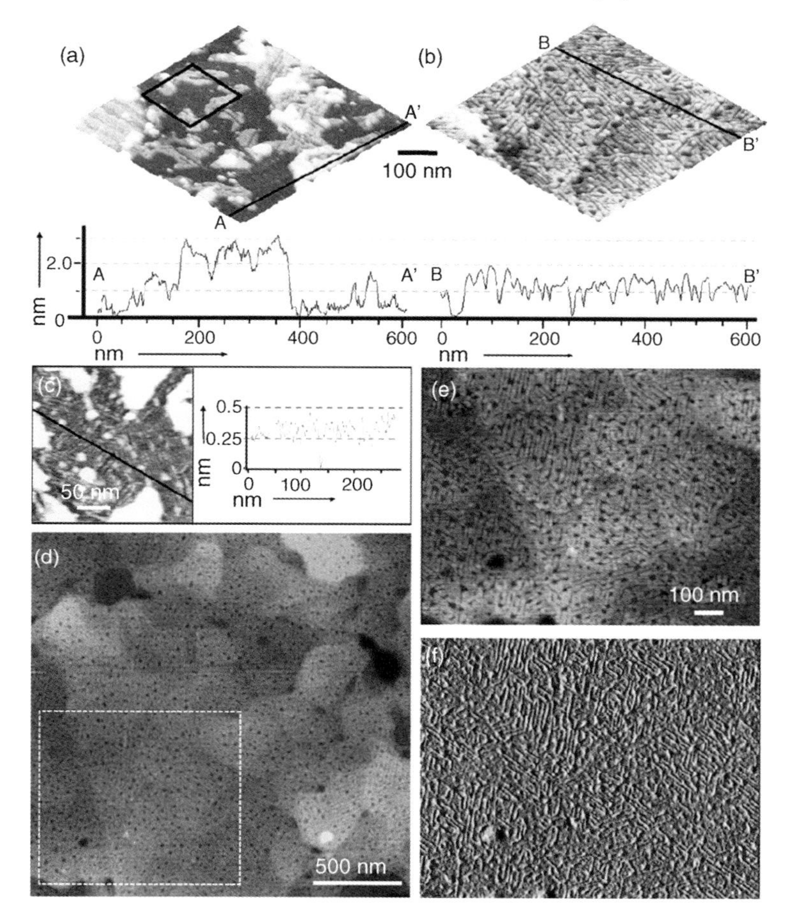

Figure 16.7 AFM images of the 3rGBP1 peptides adsorbed onto Au {111} surface showing the supramolecular assembly of (a) partial (harshly rinsed) and (b) confluent (minimally rinsed) structured layers. The cross sections corresponding to the AFM images show step heights of a harshly rinsed surface and the minimally rinsed surfaces. (c) Higher-resolution AFM image showing the finely corrugated sixfold symmetrical structures. (d–f) AFM scan over an area of 4 μm^2 showing the long-range ordering of the peptide film on several ⟨111⟩ oriented gold grains, in both (d,e) height and (f) amplitude signals. Reproduced from Ref. [20].

aggressive and gentle rinsing, from the perspective of how they affect peptide adsorption behavior. Figure 16.7a shows the AFM image following aggressive rinsing, which reveals step heights in the films corresponding to single and dual increments of ~1.5 nm from the gold surface. These vertical dimensions indicate that the peptide assembled with their longest axis parallel to the surface. This particular orientation allows the random coil and extended segments to maximize the side-chain interactions with the Au {111} surface. Higher-resolution images revealed that the peptide film covers a finely corrugated close-packed structure with a sixfold symmetry (see Figure 16.7c). These corrugations were found to have a height of 0.2–0.3 nm and a periodicity of 6 nm. Gentle rinsing resulted in a confluent, structured surface film with extended sixfold domains (Figure 16.7b,d,f). These high-resolution AFM observations, which revealed the sixfold symmetry of the adsorbed peptides, suggest a possible lattice correlation mechanism considering the sixfold symmetry of the gold surface lattice beneath.

16.2 Probing Specific Biomolecular Interactions

16.2.1 General Approaches to Nanoprobing

Specific interactions, which are the basis for the recognition of various biomolecules, are central to many aspects of life. Understanding such interactions is extremely important in order to gain insight into cell signaling, cell adhesion, genome replication and transcription, and immune responses. Thermodynamic and kinetic aspects of the specific recognition processes, which rely on a collection of non-covalent interactions, have been probed by classical biochemical methods based on labeling approaches, which are limited to the bulk and by modest spatial resolution. Owing to the nature of these methods, aspects such as individual events, conformational changes, and heterogeneity of responses remain masked. Single molecule methods have become extremely important to further our understanding of the specific interactions between complementary biological species. Owing to its ability to precisely measure forces much smaller than a single covalent bonding and comparable to various weak interactions and its ability to interrogate biomolecules in a native environment, AFM serves as a unique tool to probe the specific interactions of biomolecules.

Generally, probing specific interactions with AFM requires one of the biomolecules to be firmly bound at the apex of the AFM tip, while the other species is immobilized on the surface of a substrate. Numerous functionalization chemistries have been developed each with its own set of advantages and shortcomings to these immobilizations. The readers may referred to several reviews, which provide an excellent summary of various chemical routes for the covalent binding of biomolecules to the AFM tip [21, 22].

One particularly promising approach for binding biomolecules to an AFM tip, which has consistently yielded reliable results, employs polyethylene glycol (PEG) as a spacer, which is terminated by *N*-hydroxy succinimide (NHS) on the one end and a maleimide group on the other end. Michael addition between the maleimide group and a thiol group (either naturally present on the biomolecule or introduced by mutagenesis) of the biomolecule is used to bind the biomolecule to the tip. A recent report by Zimmermann *et al.* provides a detailed protocol for this immobilization procedure [23]. The PEG spacer chains (typically a few tens of nanometers long), which are commonly employed for immobilizing biomolecules, spaces the biomolecules from the surface of the tip. This arrangement preserves the natural conformation of the immobilized biomolecule, significantly reduces nonspecific adsorption onto the tip, provides necessary conformation freedom, and more importantly serves as an internal control for force measurements. Extension of the PEG (unraveling the polymer chain) during the retraction of the AFM tip results in a characteristic distance in the FDC indicating force curves with specific interaction events among other unwanted nonspecific interactions.

As pointed out by Bizzarri and Cannistraro, the force curves collected for biomacromolecules exhibit a sort of "zoology" of different shapes and trends [21]. Before proceeding to specific examples in which surface force spectroscopy has been employed to probe specific interactions between various biomolecules, we will briefly discuss the diversity of force curves obtained during these measurements, which are in stark contrast to those obtained while probing conventional polymer samples. The interpretation of these force curves is often challenging and requires careful consideration of the unbinding length, which should be equal to the spacer chain (PEG molecules as mentioned above) and the quantized unbinding forces, which indicate the unbinding of multiple pairs simultaneously.

Figure 16.8 shows several examples of the FDC typically observed during the probing of specific interactions of biomolecules [21]. While curve 1 does not show any detectable binding, the identical slope of the extension and the retraction portions of curve 2 shows a nonspecific adhesion event. Curves 3 and 4, on the other hand, show a trend of the change in the slope of the retraction curve indicating the relaxation of the cantilever, with further retraction causing the unraveling of the flexible linker (unbinding length) molecule followed by a specific unbinding event. Since the length of the linker molecule is known, the unbinding length in such force curves can form an internal control to confirm the validity of the specific unbinding event. Finally, curves 5 and 6 exhibit multiple unbinding events representing partial stretching of the molecules, nonspecific interactions, and the like. When such multiple jumps are observed, the last unbinding event is considered as the specific unbinding event if the starting and ending of the event is at zero deflection (as in curve 5) and if not the curve is discarded (as in curve 6).

16.2.2
Examples of Biomolecular Interactions

Returning to specific examples, Sulchek *et al.* employed force spectroscopy to measure the binding forces between a Mucin1 (MUC1) peptide and a single-chain variable fragment (scFv) antibody [24]. As discussed above, to differentiate between the specific and the nonspecific interactions, the authors tethered biomolecules to the tip and sample surface with flexible PEG spacer molecules. The spacer PEG chains had a length of 30 nm, which is significantly higher than the persistence length of these molecules, providing the required flexibility and conformational freedom. The authors analyzed the unbinding events at single and double linker (PEG) lengths.

Figure 16.9 shows the histogram of the unbinding forces at the two lengths revealed in this study. While a broad distribution of forces was observed for the single

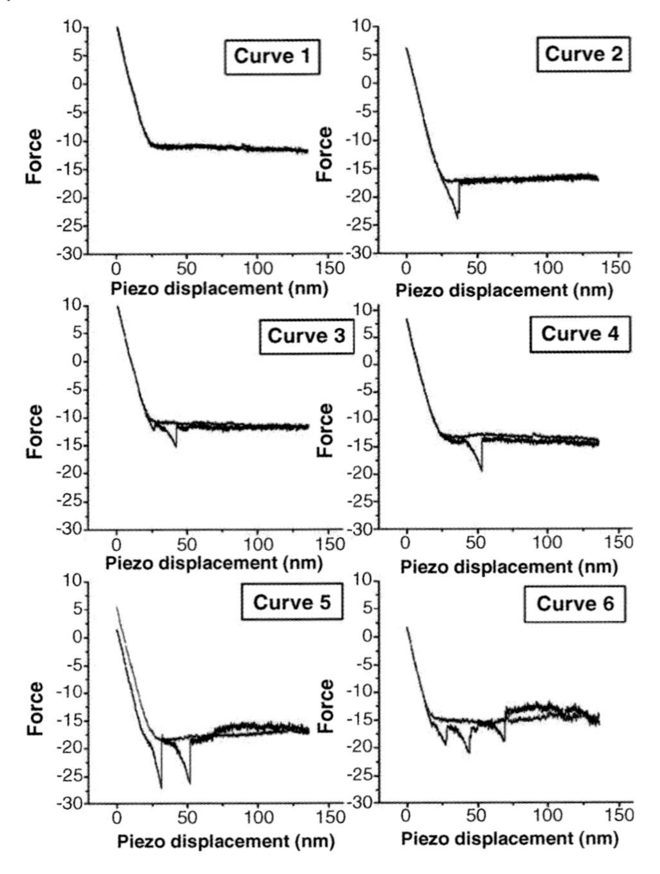

Figure 16.8 Experimental force–distance curves obtained while measuring specific interaction between biomolecules. Curve 1: no event; curve 2: an example of nonspecific binding event; curves 3 and 4: specific unbinding events; curves 5 and 6: multiple events. Reproduced from Ref. [21].

tether length, a sharp peak at 150 pN was noted for the unbinding events (jumps in the force curve) at the double tether length. The broad distribution of the force at the single tether length signifies the nonspecific nature of these events. On other hand, at the double tether length (extension of PEG chains at both tip and substrate), the sharp peak at 150 pN corresponds to the specific interaction between the MUC1 peptide and the scFv antibody. This example clearly demonstrates that the rupture force magnitude and elastic characteristics of the spacers allow the identification of specific rupture events.

As a last example for this section, we would like to point out a recent study by Kirat *et al.* in which AFM-based force spectroscopy was employed to probe the specific interactions between a biomacromolecule and a molecular imprinted synthetic receptor [25]. In fact, this is the first and so far the only study to have employed AFM to probe such interactions. Molecular imprinting involves the polymerization and cross-linking of a functional monomer around a template molecule. Following

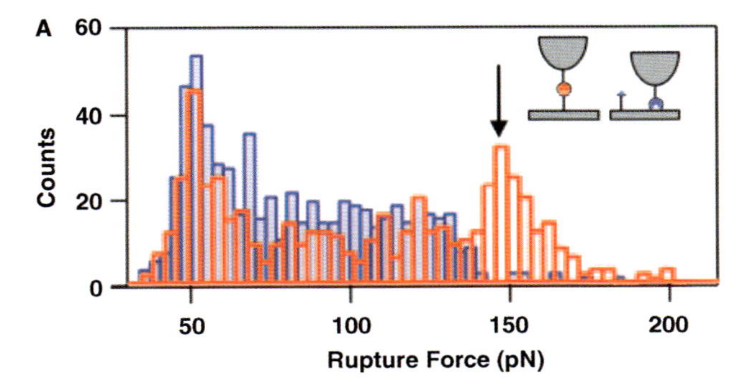

Figure 16.9 Spectrum of specific interactions between MUC1 peptide and single-chain variable fragment antibody. (a) Histograms of the rupture forces measured in the one-linker length (blue filled bars) and two-linker length (red unfilled bars) rupture regions. The arrow indicates the peak corresponding to the specific MUC1–antibody interactions. Reproduced from Ref. [24].

the polymerization and cross-linking process, the template molecules are removed to leave an artificial receptor with a complementary shape and functionality (noncovalent interactions such as van der Waals and hydrogen bonding) to the template molecules. While AFM has been extensively employed for imaging molecular imprints (especially via surface imprinting), application of SFS for probing specific interactions provides a promising avenue to gain insight into the structure and recognition capabilities of the molecular imprints.

In this study, Kirat *et al.* employed copolymers of acrylamide and different acrylic acid-based cross-linkers as a functional matrix and cytochrome C (Cyt C) as a template for imprinting and targeted analyte at later stages [25]. Force spectroscopy measurements were performed using AFM cantilevers with covalently attached Cyt C molecules and without Cyt C molecules (control tip). Figure 16.10 shows the adhesion histograms obtained using AFM tips with and without Cyt C molecules on imprinted and flat surfaces. It can be clearly seen from the histograms that the adhesion forces observed for measurements performed using Cyt C-functionalized tips on an imprinted surface shows a distinct peak at $\sim$90 pN, while the other cases exhibit a broader distribution. The narrow distribution of forces signifies the specific interaction between the Cyt C molecules and the artificial receptors, while the boarder distribution of forces ranging from 50–350 pN indicates the nonspecific interaction in these control samples.

16.3 Mechanics of Individual Biomacromolecules

16.3.1 Stretching and Pulling of Long-Chain Molecules

Owing to its remarkable sensitivity in monitoring extremely small forces, force spectroscopy has been employed to monitor the mechanics of single biomacromo-

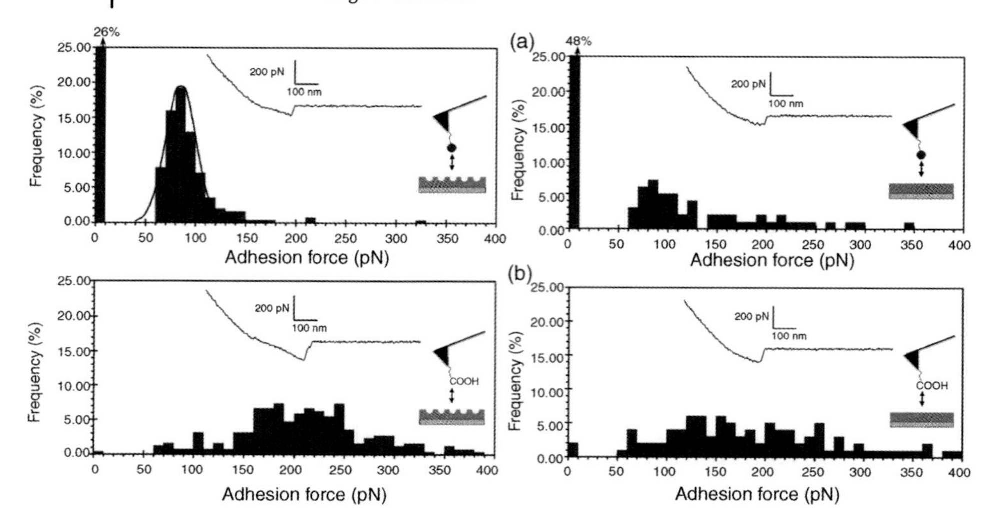

Figure 16.10 Representative force–distance curves and the adhesion force histograms obtained from force spectroscopy of the (*top*) 2D molecular imprinted polymer layer (*bottom*) nonimprinted polymer layer. (a) A Cyt C-tip or (b) a control tip. For the Cyt C-tip on imprint film, the distribution of adhesion forces was fitted with a Gaussian function centered at 85 pN. Reproduced from Ref. [25].

lecules, namely, stretching and unfolding. The technique often termed "protein pulling" involves the adsorption of the biomolecules onto the surface of the substrate followed by bringing the AFM tip in contact with the biomolecule to assure the formation of a firm joint. After a short amount of time is allowed to enable the adsorption of the biomolecule onto the tip, the AFM probe is retracted to stretch and unfold the protein, thus gaining quantitative insight into the mechanics of the individual biomacromolecules [26–28].

The SFS technique exploits the ability of the AFM tip to tether to strongly interacting macromolecular ends and pull a single macromolecular chain (Figure 16.11) [29]. During these experiments, one chain end is attached to the AFM tip and another to the surface while data are collected as force–distance curves (Figure 16.11) [30]. To describe FDC behavior observed in this study, several models were developed that can predict the dependence of the force applied on stretching the biomolecule. In the simplest homopolymer case, the chain can be considered as a random Gaussian coil. As the next step, the model of the freely joined chain (FJC) can be applied [31, 32]. This model, in which the chain is considered as a set of N rigid elements, each of them having length l (Kuhn length) and connecting through flexible joints, provides the relationship between the stretching distance and the force in the form [33, 34]:

$$R_z = Nl\left[\coth\left(\frac{Fl}{k_B T}\right) - \frac{k_B T}{Fl}\right] \tag{16.1}$$

where R_z is chain's end-to-end distance, T is temperature, and F is stretching force. A modified FJC model is used to describe the FDC behavior to account for the elastic

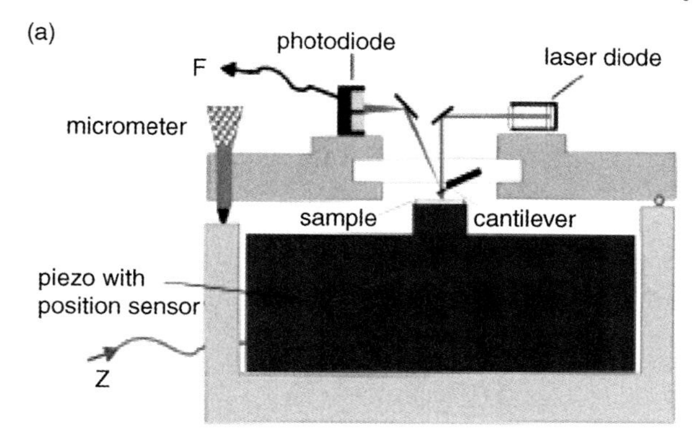

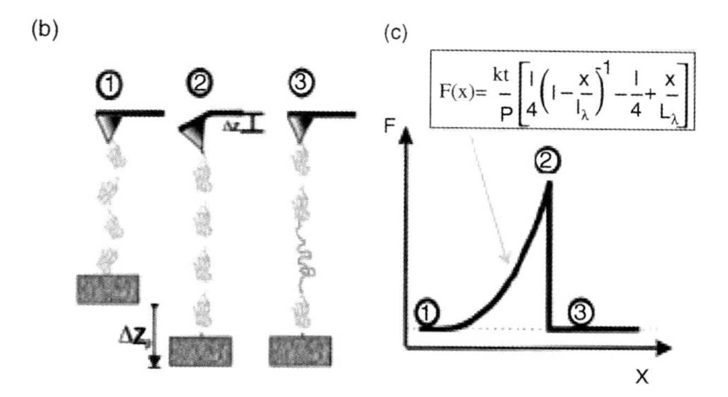

Figure 16.11 SFS principles of single protein stretching. (a) The position of the head is determined with 0.1 nm accuracy. (b) The process of unfolding. When a tip goes up, a restoring force bends a cantilever (from stage 1 to 2). After domain unfolding, a cantilever returns to the previous position (stage 3). (c) Theoretical model predicts the gradual increase in force with the tip extension. Reproduced from Ref. [30].

deformations of bonds and enthalpy effects by using the additional fitting parameter K, segment elasticity.

Another widely used model for analyzing unfolding dynamics is a worm-like chain (WLC) model [35, 36]. In this model, the chain is considered as being continuously curved with the curvature at every point being random. The WLC model accounts for local entropic stiffness in the chain by persistence length, lp. The extension of the chain in this model is limited to the contour length. Under small stretching forces and long-chain conditions, the persistence length equals half the Kuhn segment length from the FJC model. The WLC behavior is described by the equation

$$F=\frac{k_{\mathrm{B}}T}{l_{\mathrm{p}}}\left[\frac{R_{\mathrm{g}}}{L}-\frac{1}{4}+\frac{1}{4(1-R_{\mathrm{g}}/L)^{2}}\right] \tag{16.2}$$

where L is the total chain length and R_g is the radius of gyration. By using these three mechanistic models, it is possible to estimate the elastic properties of a single flexible chain at different length scales [37, 38].

In all tedious and labor-intensive studies of biomacromolecule unfolding, SFS is conducted in an aqueous environment that allows us to avoid capillary forces and recreate a fluidic environment close to the natural environment of biomolecules. The most widespread method is when the sample is deposited on the substrate with a tip picking it up and stretching [39]. In some cases, organic molecules or polymer chains are attached to the tip to mediate stretching (AFM tip modification discussed earlier in Section 2.5). This method is required when the interactions of the same chain with different surfaces are the focus of the research [40]. Appropriate tip modification is conducted to expose chemical groups with affinity for the end groups of biomolecules [41–43].

In the majority of cases of complex macromolecules, stretching FDCs show the presence of several peaks as was noted above and demonstrated in Figure 16.12 [30]. The forces applied and analyzed are usually in the pico-Newton range, which covers all intramolecular and intersegmental weak interactions contributing to the process. Due to the complexity of the interactions, the origin of peaks is frequently a matter of discussion and is a key point for the interpretation of the unfolding behavior of

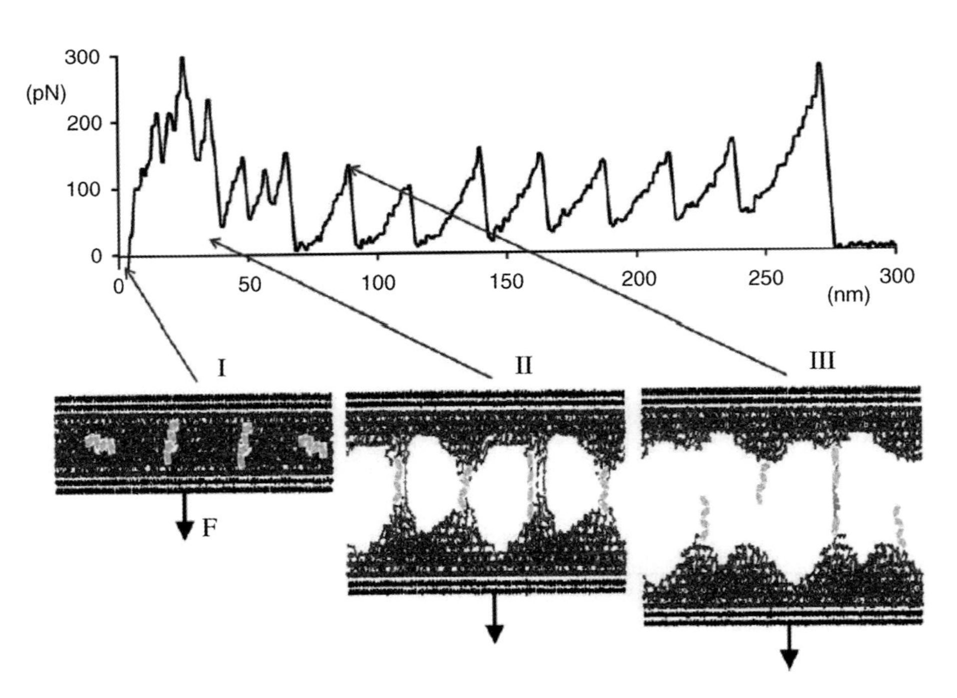

Figure 16.12 A typical FDC for unfolded polyprotein with size of domains of 40 nm. Initial ~70 nm represent nonspecific interactions and multiple chain attachments to the tip. After initial contact, molecules detach from the tip until only one remains and unfolding of a single molecule occurs. Reproduced from Ref. [30].

proteins. Even the same protein can have a different set of peaks after modification [44]. It was found that the unfolding of each domain occurs at a certain force.

A special case for analysis is the unfolding of a single protein molecule [45]. An FDC histogram of this event shows clear quantization that corresponds to the stretching of one, two, three, or more molecules. It is observed that during protein molecule unfolding, the forces on FDC are set in an ordered fashion: first, the weakest domain is unfolded and the last event is the rupture of the tip–chain bond. It was shown that if several parts of one chain are randomly attached to the substrate, the FDC will have a random distribution of peaks (Figure 16.12) [46, 47]. The changes in the shape of the FDC, force required for unfolding, and significant rate dependence of the observed force are used for accurate assignment of different unfolding processes [48, 49].

SFS analysis of proteins is considered the most promising method for the identification of the intramolecular domain structure of proteins in a natural fluidic environment despite several key problems remaining because of their complex pattern of unfolding. This complexity includes the time–space overlapping of several concurrent processes: multiple chain pulling, simultaneous unfolding of several domains, detachment from the tip, detachment from the substrate, a rate-dependent sequence of unfolding, secondary structure element deformation, trapped local conformations, and the contribution of lateral interactions. Separation is frequently ambiguous, but has been demonstrated for biomacromolecules with well-known and uniform domain structure. However, the interpretation of unfolding scenarios for proteins with multidomain structures adsorbed and in unknown states is challenging [30, 50].

16.3.2 Unfolding of Different Biomacromolecules

In the first example, fibronectin was investigated with SFS on a gold-coated substrate under water [51]. The stretching range for fibronectin molecules was selected to be around 250 nm. The experiment was repeated under different speeds ranging from 0.01 to 10 nm/ms to investigate the rate dependence of the unfolding events. Sawtooth-like FDCs were observed with a rupture force of 145 pN and a contour length of 285 nm. Analysis of the unfolding events for fibronectin domain showed that this length corresponds to a domain length composed of 90 amino acids. Proteins with different fibronectin domains and polyproteins composed of individual domains were investigated as well in order to confirm the suggestions (Figure 16.13) [52]. It was found that multidomain molecules showed several histogram peaks for the force and length in good agreement with theoretical models. The WLC analysis produced the persistent length of 0.4–2.5 nm depending on the domain sequence.

Meadows *et al.* conducted SFS for the same proteins deposited on a mica or glass surface and dried [53]. The experiment was done in water or phosphate-buffered saline (PBS) environment with cantilevers having spring constants from 30 to 70 mN/nm. The *z*-range was from 60 to 300 nm and the pulling speed was 1 nm/ms. The authors found that the isolated molecules have a tendency to denature, and

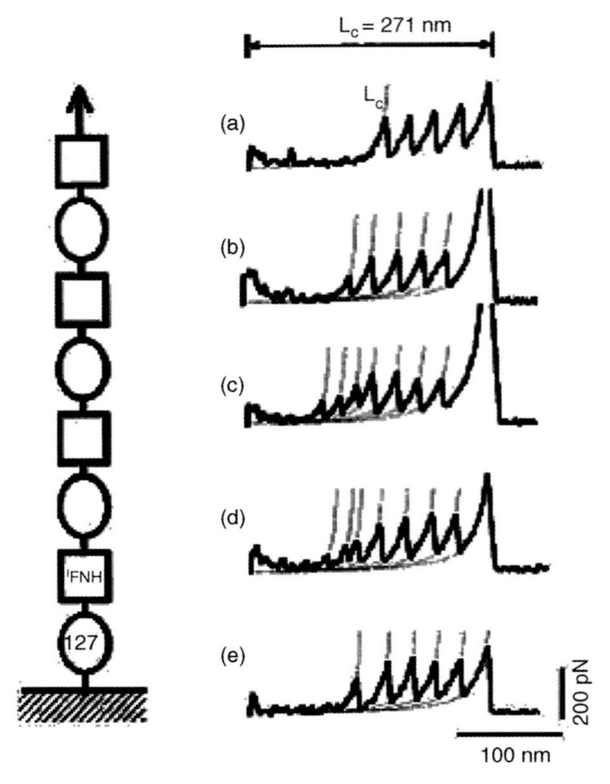

Figure 16.13 The FNIII (fibronectin) FDC, domain model, and fitting (thin lines) by WLC model. All contour lengths are around 280 nm that is close to the expected value. Reproduced from Ref. [52].

the contour length tends to be 12 nm. Aggregation of proteins in densely packed films on some surface areas did not affect their unfolding behavior.

Gutsmann *et al.* conducted extensive SFS experiments for collagen molecules [54]. They used AFM cantilevers with a spring constant of ~25 mN/m and different pulling speeds of 7 and 14 nm/ms. It was found that the distribution of rupture events (contour lengths) possesses two peaks. The authors explained them as the presence of two different domains and it was noticed that AFM imaging shows a domain with a size close to the size of a single peak on the FDC data.

In a comprehensive study, Shulha *et al.* employed SFS to measure the unfolding properties and the internal domain structure of *B. mori* silk fibroin [55]. As discussed in Section 16.1, the surface morphology of the silk protein adsorbed onto hydrophilic silicon substrates displayed both individual globular shapes and aggregated nanofibrils composed of several dozen densely packed near-circular domains (see Figure 16.5). Considering the complex multidomain *B. mori* silk protein composed of 12 hydrophilic and 12 hydrophobic domains, which comprise a wide range of possible molecular lengths, three independent types of SFS measurements were exploited to include different strategies of stretching with appropriate forces, ranges, and rates.

The first method involved stretching with relatively high forces and extensions reaching 3 μm, which exceeds the completely extended length of the molecules (estimated to be about 1.8 μm). This probing was conducted for surface areas modified with "bulk" protein (occasional thicker surface patches). In the second approach, the stretching was limited to a distance below 1.5 μm combined with high resolution in the Z-direction to collect data on sequential unfolding events for a "normal" surface layer of protein (Figure 16.14). In this case, a detailed sawtooth shape analysis was applied for the extension range within 1000 nm. Finally, gentle, low-rate, limited stretching with low stretching distances (below 300 nm) was performed to focus on sequences of weak unfolding events related to the hydrophilic domains (Figure 16.14). It is worth noting that due to instrumentation limitations (vertical resolution, drifts, and temporal limits), it is impossible to obtain high-resolution data that reflect all particular events in a single pulling experiment.

The overall interpretation of possible stretching scenarios has been made on the basis of the analysis of individual statistics for spacings and forces derived from

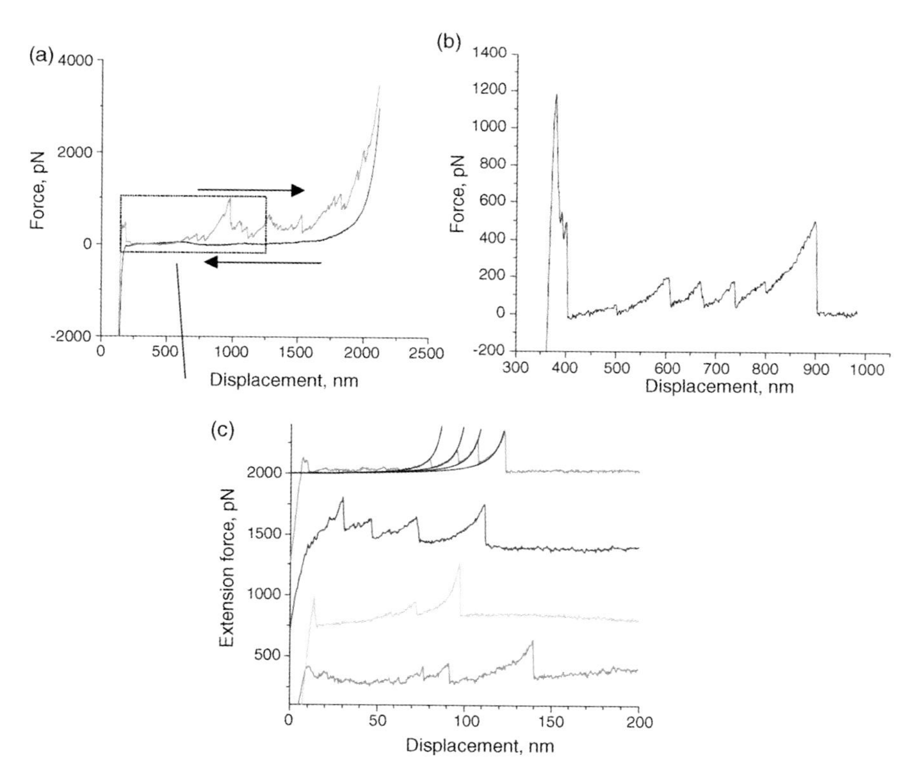

Figure 16.14 Force–displacement curves for silk fibroin. (a) Largest extension (>2.5 μm, stretching rate is 3000 nm/s), both stretching and relaxation are shown; (b) extension curve for midextension (<1 μm, stretching rate is 1000 nm/s). (c) Examples of several force–displacement curves on a small extension scale (below 300 nm) and an example of corresponding WLC fitting curves used to determine extension domain lengths and Kuhn segment. Reproduced from Ref. [55].

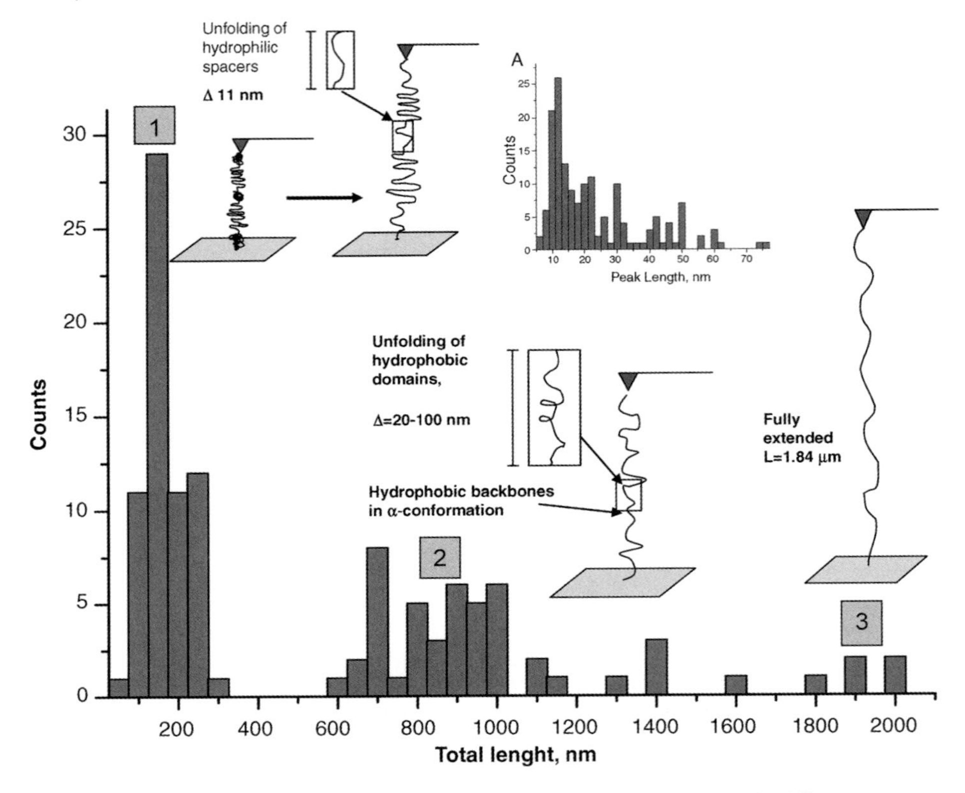

Figure 16.15 Statistics for maximum stretched length (3) and the total length of the events (1 and 2) obtained in different extension scenarios and theoretical estimations (boxes) for total extended lengths for different scenarios (cartoons). Inset shows unfolding events at small extensions. Reproduced from Ref. [55].

the model of block sizes and distributions in this protein [56]. The unfolding pattern for silk molecules suggested a complex character of fibroin stretching, which includes weak–intermediate–strong events of unfolding for the combination of 12 hydrophilic and 12 hydrophobic domains and the ultimate stretching of the backbones. For small and intermediate stretching distances, more organized events were observed, which were associated with regular unfolding in the protein backbone (Figure 16.15).

For low stretching, the well-defined spacing of 11.3 nm dominates the unfolding pattern (see inset in Figure 16.15). Overall, 11.3, 22.5, and 32 nm spacings with a ratio close to 1 : 2 : 3 were consistently detected. The main periodicity coincided precisely with the fully extended length of the hydrophilic domains calculated from their amino acid composition (11.2 nm). Forces for a single unfolding event were close to 60 pN, which is common for this type of backbone segment unfolding. The analysis of individual domain unfolding in terms of the Hookean spring model suggested a spring constant of about 0.01 N/m, which is expected for stretching of flexible backbones governed by an entropic mechanism. The persistence length determined from SFS fitting with the worm-like chain model was 0.24 nm, which is a characteristic of very flexible chains. Other consistent lengths can be assigned to double and triple unfolding events. The extended length of the series of events (estimated as the total length of all peaks) spread from 100 to 230 nm and are centered at 160 nm (Figure 16.15). These lengths correspond to the length of backbones with folded α-helical hydrophobic and stretched hydrophilic domains (including two terminal hydrophilic domains). The terminal domains can be stretched to the contour length of 53 and 18 nm, respectively.

Very different statistics observed for the intermediate stretching regime hinted a different cause for the unfolding events observed in this range (Figure 16.15). It is worth noting that small-extension events are "hiding" in the initial portion of the curves and are not well resolved when optimized for mid-range stretching conditions. This statistic shows a wide distribution around 800–1000 nm without a well-defined single preferential length, which reflects the irregular distribution of unfolding lengths observed in the SFS experiment (Figure 16.15). The unfolding pattern observed correlated well with the multidomain hydrophilic–hydrophobic model suggested before on the basis of the amino acid composition of the protein backbone with all major parameters. Finally, some of the longest unfolding events corresponded to the total length of protein molecule.

Another important protein, spectrin, is a cytoskeletal protein composed of two subunits, α- and β-chain, which form laterally associated heterodimers. Spectrin is on the intracellular side of the plasma membrane of many cell types playing an important role in the maintenance of the mechanical integrity of the plasma membrane and cytoskeletal structure. Repeat units of this protein fold into triple helical coiled coils comprised of 106 amino acid residues. The SFS approach was employed by Rief *et al.* to mechanically unfold these repeats and the force required to unfold was found to be 25 to 35 pN [57]. The stretching of the protein was performed at two different speeds: 0.08 μm/s and 0.8 μm/s. Conceivably, it was observed that protein unfolding is a rate-dependent process. Pulling at a higher speed (0.8 μm/s) required higher unfolding forces compared to a lower speed (0.08 μm/s). Monte Carlo simulations of unfolding events exhibited excellent quantitative agreement with the observed force–displacement curves. This example demonstrates that AFM force measurements can also provide insight into the dynamic nature of the mechanical behavior of the protein molecules.

16.4 Single-Cell Elasticity

Over the past decade, AFM-based nanomechanical measurements have been extensively applied for measuring the mechanical, vibrational, and surface properties of live cells. As was concluded, the first issue that should be considered for either imaging or applying surface force spectroscopic or any other AFM technique on cell surfaces is an efficient strategy to immobilize the cells on a surface. Poor immobilization of the cells leads to detachment of the cells from the surface during the probing process.

There are several techniques that have been investigated for the immobilization process, each with its own advantages and disadvantages and one more suitable than the other for specific cell types. Approaches such as air drying and chemical fixation (covalent binding using silanized substrates), although relatively simple, cause denaturing of the biomolecules [58–60]. Another technique is to use polycation-modified surfaces, which electrostatically bind the cells with negative charges on the surface [61]. Yet another approach is to use mechanical means by fixing the cells in agar gels or immobilizing them on a porous polymeric template with a pore size comparable to the cell size [62, 63]. However, as mentioned above, the choice of technique depends on the type of the cell under investigation and often requires the trial of several approaches to identify the best method for each particular case.

In a related study, Touhami *et al.* employed nanomechanical measurements to reveal the high elastic modulus of chitin-rich regions compared to the surrounding regions during yeast cell division [64]. Under normal circumstances, chitin is not present in the cell walls of yeast cells. However, during cell division (budding), chitin accumulates into localized regions, called bud scar regions, where it causes the stiffening of the cell wall. Force–distance curves were obtained in the bud scar regions and the surrounding regions. Using the Hertz model, the authors calculated the elastic modulus in these regions. The elastic modulus in the bud scar regions (6.1 MPa), where the chitin accumulates, was found to be nearly an order of magnitude higher compared to the surrounding regions without chitin (as low as 0.6 MPa). This is an excellent example where the unprecedented lateral resolution of AFM was employed to unveil these unique aspects of yeast cell growth.

Gaboriaud *et al.* have performed *in situ* AFM nanomechanical measurements on *Shewanella putrefaciens* (gram-negative bacteria) at pH 4 and pH 10 [65]. Force–-distance curves obtained under these pH conditions exhibited both linear and nonlinear regimes. The authors attributed the nonlinear part of the force–distance curves to the progressive indentation of the AFM tip in the bacterial cell wall, which included the priori polymeric fringe, and the linear part to the compression of the plasma membrane. The indentation observed at pH 10 was much higher than that observed at pH 4 for the same applied load. This difference in the mechanical properties was ascribed to the swelling of the polymeric fringes, which were found in several *Shewanella* species, with a thickness ranging from 20 to 130 nm.

Yet another exciting facet of the application of AFM in the context of living cells is the ability to monitor the metabolic process, which results in nanomechanical motion of these cellular structures in real time. Pelling *et al.* employed AFM to monitor the nanoscale mechanical vibrations of living yeast cell (*Saccharomyces cerevisiae*) [66]. As was observed, the vibrations of the cell walls exhibited a local temperature-dependent motion at characteristic frequencies. For monitoring the nanoscale vibrations, the authors employed extremely soft microcantilevers (with a spring constant about 0.05 N/m). The cell wall vibrations were monitored by engaging the AFM tip on the relatively stiff cell wall of the yeast cell followed by monitoring the microcantilever deflection under variable conditions.

Knowing the local stiffness of the cell and using a cantilever with a known spring constant, the authors of this study employed the spring-on-spring approach to translate the deflection of the cantilever into the motion of the cell wall. Periodic motions with a frequency of 0.8 to 1.6 kHz and amplitude of ~3 nm were recorded in these experiments. Exposure of the cells to a metabolic inhibitor (sodium azide) resulted in a complete halt of the periodic motion. The metabolically driven nanomechanical periodic motion of the cell wall observed in these AFM probing was believed to be caused by the concerted action of numerous molecular motors (proteins such as kinesin, dynein, and myosin).

Single cell operations observed with nanoscale resolution have been performed using an AFM tip as a sharp indenter to penetrate through the cell membrane [67–71]. Obataya and coworkers modified the conventional micromachined AFM tip using a focused ion beam to form sharp nanoneedles with a well-defined geometry (nanocones and nanocylinders) [69]. Figure 16.16 shows the force–distance curves obtained during the penetration of the conical and the cylindrical tips through melanocyte cell membranes.

For the needle with a cone-shaped tip, the authors observed a quadratic force increase after it came in contact with the cell surface followed by a linear increase in the force. The point of the change in the force–distance function signified the point of the geometrical change from a cone to a cylinder experienced by the cell membrane. On the other hand, the force increase exhibited only a linear behavior for the cylindrical needle. The authors analyzed these results in the framework of the Hertz model for the corresponding shape of the indenter, which suggested the linear dependence of the force with indentation depth for a cylindrical geometry and

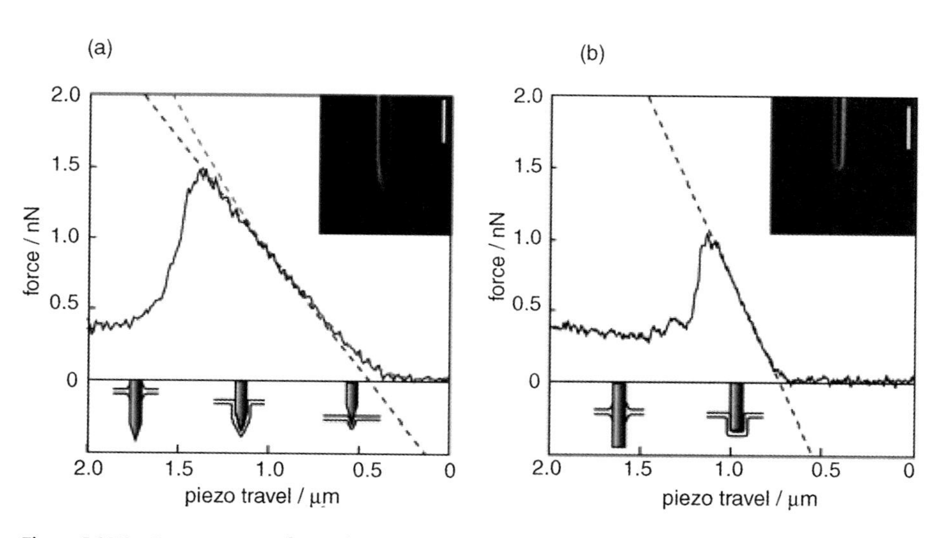

Figure 16.16 Representative force–distance curves for a melanocyte using nanoneedles having (a) cone-like shapes and (b) cylindrical shapes at the edges. Insets show magnified SEM images around the tip of the needle. Insets also include a schematic of the proposed situations of the nanoneedle tip and the cell surface. Reproduced from Ref. [69].

quadratic dependence for a conical geometry. The modeling analysis was in good agreement with the experimental observations. Using Hertzian analysis, the authors calculated the elastic moduli from the force curves and found them to be within 2–10 KPa. Furthermore, the authors employed confocal fluorescence and phase contrast microscopies to directly monitor the indentation of the nanoneedle into the cell membrane on a larger scale.

In an alternative approach, Kouklin *et al.* and Vakarelski *et al.* employed carbon nanotubes as nanoneedles for performing the single-cell nanoprobing operation [70, 71]. One of the distinct advantages of using much finer carbon nanotube tips as opposed to the micromachined silicon cantilevers is that the nanoindentation experiment can be performed in a minimally invasive manner. These studies demonstrated that carbon nanotubes can be effective in carrying nanoparticulate payloads and penetrating the plasma membrane of living pleural mesothelial cells at small indentation depths (100–200 nm as opposed to the 1–2 μm when conventional micromachined tips are employed) and small penetration forces (100–200 pN).

In several recent studies, researchers have employed SFS to probe the nanomechanical properties of normal and metastatic cancer cells. For example, Cross and coworkers measured the nanomechanical properties of metastatic adenocarcinoma cells and benign mesothelial cells [72]. The tumor and benign cells were preliminarily identified on the basis of the shape changes between these cells. The tumor cells displayed anchorage-independent growth patterns, such as rounding of cells, whereas normal mesothelial cells exhibited a large, flat morphology.

These assumptions employed for the AFM measurements were further confirmed by the immunofluorescence analysis using biomarkers specific to the tumor and healthy cells. Within the same sample, the authors found that the metastatic cancer cells were nearly 70% softer (with a small standard deviation) compared to the benign cells. Considering the large diversity of the biochemical nature of the cells, the nanomechanical measurements revealed rather universal nanomechanical behavior for different cancer types and patient effusions, suggesting that AFM-based nanomechanical measurements could be used as a complementary technique for identifying tumor cells.

16.5
Lipid Bilayers as Cell Membrane Mimics

From the very beginning, AFM has been considered as a key microscopic technique for the study of cell phospholipid membranes. Numerous physical attributes such as bilayer thickness and its spacing, viscosity, formation of two-dimensional arrays of phospholipids, phase separation within lipids, dimensions of membrane domains, and their shape are routinely investigated using AFM imaging (see related discussion on LB films in Chapter 13). In an earlier study, Xie *et al.* performed *in situ* AFM measurements to monitor the phase transition of bilayers formed from a zwitterionic phospholipid (1,2-dimyristoyl-*sn*-glycero-3-phosphocholine (DMPC)) on mica substrates [73]. The AFM images revealed that the fluid-to-gel phase transition process is

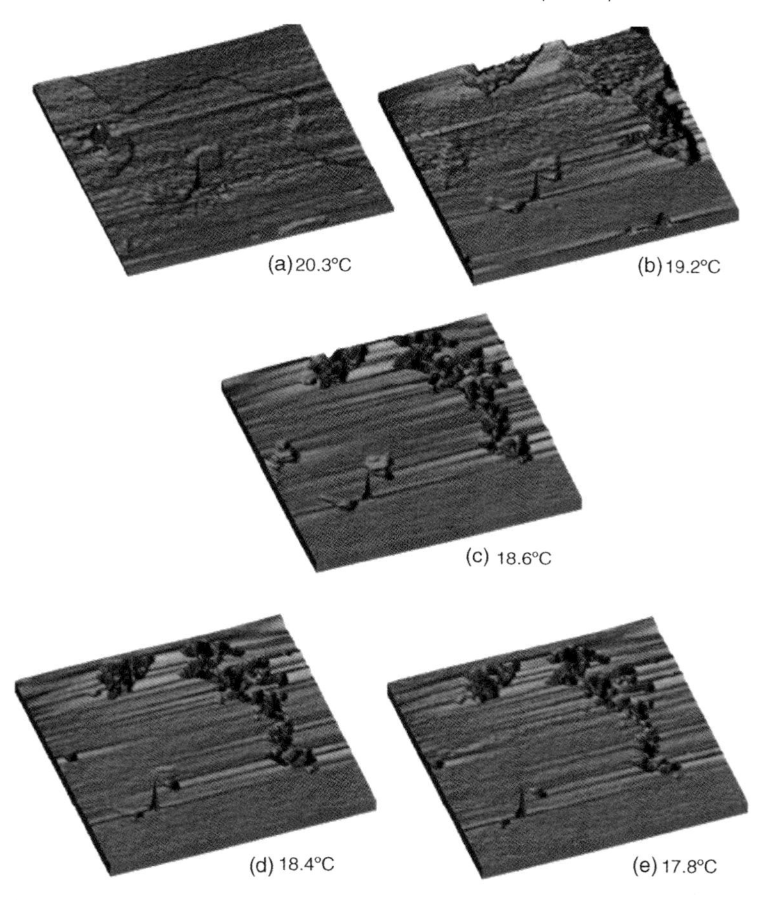

Figure 16.17 3D AFM images (10 μm × 10 μm) of DMPC bilayer captured at (a) 20. 3 °C, (b) 19.2 °C, (c) 18.6 °C, (d) 18.4 °C, and (e) 17.8 °C as the temperature was decreased from 26 °C to 17 °C. Reproduced from Ref. [73].

associated with substantial tearing of the bilayer due to the density difference between the two phases.

Figure 16.17 shows the AFM images of the progressive phase transition of the lipid layer as the temperature is decreased below the phase transition temperature. The height difference between the liquid and the gel phases was used to distinguish the two lipid phases in the AFM images. AFM observations reveal that as the gel phase continued to form (with lowering temperature), large defects developed (see the upper right portion of the images). The fluid bilayer was found to rupture upon cooling, possibly due to the different densities of the fluid and gel phases. The AFM images also revealed that growth from the fluid nuclei showed strong anisotropy with elongated fluid areas possibly due to the strain in the gel phase, causing the film to rupture. This study shows that AFM is an excellent tool for conducting *in situ* investigation to reveal nanoscale details of these rather complex lipid systems.

Giant unilamellar vesicles (GUVs) are an interesting class of cell membrane mimicking systems with a broad range of size from 1 to 100 μm. Owing to their large size, confocal fluorescence microscopy has been employed to monitor the structure and phase behavior of GUVs. To obtain higher spatial resolution information, Tokumasu *et al.* employed AFM to study the phase properties of domains in GUVs formed from three component mixed lipid bilayers (1,2-dipalmitoyl-*sn*-glycero-3-phosphocholine (DPPC), 1,2-dilauroyl-*sn*-glycero-3-phosphocholine (DLPC), and cholesterol (chol)) [74]. Figure 16.18 shows the AFM and the corresponding confocal

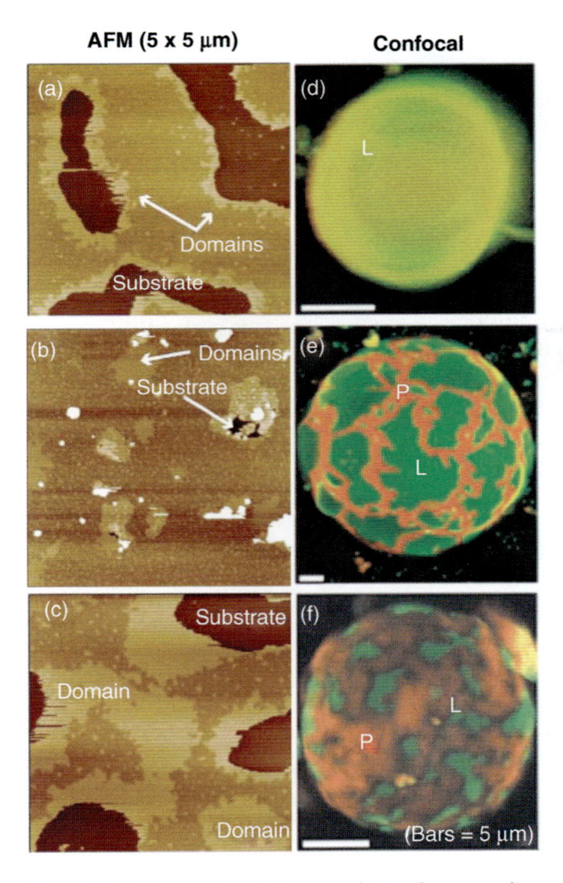

Figure 16.18 GUV domain images obtained using AFM (*left*) and CFM (*right*) (scale bars correspond to 5 μm) with varying composition of DPPC and DLPC lipids with no cholesterol. (a,d) At 20% DPPC: 80% of DLPC, microscopic domains formed only at the edges of the membrane; (b,e) membrane containing 50% DPPC showed a slight increase in domain area; (c,f) membrane with 80% DPPC, the area occupied by DPPC-rich domains markedly increased. Reproduced from Ref. [74].

fluorescence microscopy images of GUVs with different ratios of DPPC and DPLC lipids.

As was observed, for a GUV with DPPC:DPLC of 1 : 4, the confocal fluorescence images do not show any domain morphology, while the AFM images show distinct domains at the boundaries. When the content of DPPC was increased to a DPPC: DPLC ratio of 1 : 1, the confocal fluorescence images exhibited a domain structure (with DLPC-rich phases indicated as L in the images), while the AFM image showed a slight increase in the lipid domain as pointed out in the AFM image. Finally, when the DPPC : DPLC ratio was 4 : 1, the DPPC-rich domains dramatically increased as seen in the AFM image. Thus, the AFM imaging revealed nanoscale details of the phase behavior of this important class of vesicles, which were inaccessible to conventional optical microscopic techniques. However, most of the studies related to GUVs are performed on supported membranes possibly due to the difficulty in immobilizing the GUVs on substrates and the large vibrations (few hundred nanometers) of these structures.

In a recent study, Roiter *et al.* investigated the structural integrity of lipid (L-α-dimyristoyl phosphatidylcholine) membranes as they are formed on polar nanoparticles of different diameters [75]. For monitoring the formation of the lipid

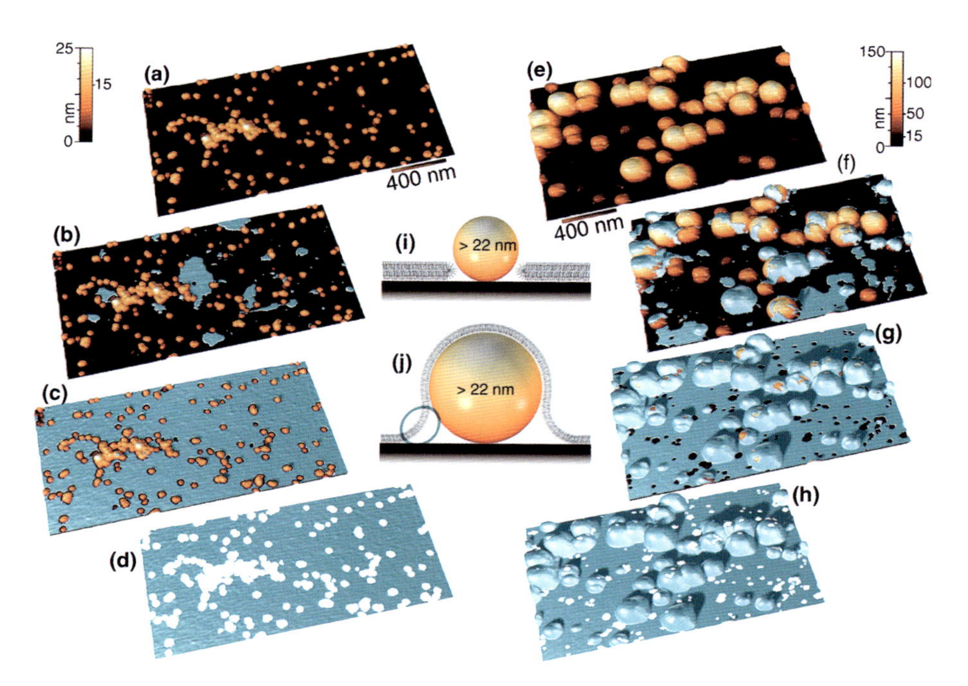

Figure 16.19 AFM images of the lipid bilayer formation in the presence of particles larger than lipid bilayer thickness (*left*) of lipid bilayer formation over a surface with 5–20 nm silica nanoparticles (*right*) of lipid bilayer formation over the surface with mixed 5–140 nm silica particles. Schematic at the center shows the lipid bilayer forming a pore around particles smaller than 22 nm and how enveloping around the larger particles. Reproduced from Ref. [75].

membranes, a series of AFM images were obtained, starting with the silica nanoparticles on a substrate (i.e., without lipid) and finishing with the surface covered by the lipid bilayer. In order to monitor the formation of the lipid membranes on the nanoparticles, successive AFM images were aligned precisely by matching surface features and then subtracted to reveal the location of the formed bilayer at each particular stage.

As was observed for relatively small nanoparticles (5–20 nm), the addition of lipid into the medium leads to the formation of lipid bilayer islands with all the nanoparticles remaining uncoated, both at partial and final sample coverage. Figure 16.19 (*left*) shows the series of AFM images of the partial and full coverage of the lipid showing the uncovered nanoparticles. Especially, Figure 16.19d shows the processed image after the subtraction of the nanoparticles where the substrate clearly shows the uncovered lipid membranes only. The series of AFM images on the right of Figure 16.19 shows the partial and full coverage of the lipid membrane on nanoparticles with a broad size distribution (5–140 nm).

Overall, AFM images show that the lipid membrane covers the larger particles (>22 nm), while the smaller nanoparticles remain uncovered by the membrane under the given conditions. Furthermore, a closer observation revealed tiny ruptures in the lipid membranes even on larger particles when they encountered surface features with high curvature on these particles. This detailed study with high-resolution AFM images revealed that the lipid membranes were stable when deposited on extremely small nanoparticles (less than 1.2 nm) and on nanoparticles with a diameter larger than 22 nm. When deposited on nanoparticles within particular diameters (from 1.2 to 22 nm), the lipid membranes ruptured.

References

1 Feynman, R.P. (1999) *The Pleasure of Finding Things Out: The Best Short Works of Richard P. Feynman*, Perseus Books, New York.

2 Kienberger, F., Ebner, A., Gruber, H.J., and Hinterdorfer, P. (2006) Molecular recognition imaging and force spectroscopy of single biomolecules. *Acc. Chem. Res.*, **39** (1), 29–36.

3 Engel, A. and Muller, D.J. (2000) Observing single biomolecules at work with the atomic force microscope. *Nat. Struct. Biol.*, **7**, 715–718.

4 Hansma, H.G., Kim, K.J., and Laney, D.E. (1997) Properties of biomolecules measured from atomic force microscope images: a review. *J. Struct. Biol.*, **119** (2), 99–108.

5 Fotiadis, D., Scheuring, S., Muller, S.A., Engel, A., and Muller, D.J. (2002) Imaging and manipulation of biological structures with the AFM. *Micron*, **33** (4), 385–397.

6 Fritz, J. and Anselmetti, D. (1997) Probing single biomolecules with atomic force microscopy. *J. Struct. Biol.*, **119** (2), 165–171.

7 Denis, F.A., Hanarp, P., Sutherland, D.S., Gold, J., Mustin, C., Rouxhet, P.G., and Dufrêne, Y.F. (2002) Protein adsorption on model surfaces with controlled nanotopography and chemistry. *Langmuir*, **18** (3), 819–828.

8 Denis, F.A., Hanarp, P., Sutherland, D.S., and Dufrêne, Y.F. (2002) Fabrication of nanostructured polymer surfaces using colloidal lithography and spin-coating. *Nano Lett.*, **2** (12), 1419–1425.

9 Marchin, K.L. and Berrie, C.L. (2003) Conformational changes in the plasma

protein fibrinogen upon adsorption to graphite and mica investigated by atomic force microscopy. *Langmuir*, **19** (23), 9883–9888.

10 Bergkvist, M., Carlsson, J., and Oscarsson, S. (2003) Surface-dependent conformations of human plasma fibronectin adsorbed to silica, mica, and hydrophobic surfaces, studied with use of atomic force microscopy. *J. Biomed. Mater. Res.*, **64A** (2), 349–356.

11 Ornatska, M., Jones, S.E., Naik, R.R., Stone, M., and Tsukruk, V.V. (2003) Biomolecular stress-sensitive gauges: surface-mediated immobilization of mechanosensitive membrane protein. *J. Am. Chem. Soc.*, **125** (42), 12722–12723.

12 Tsukruk, V.V., Ornatska, M., and Sidorenko, A. (2003) Synthetic and bio-hybrid nanoscale layers with tailored surface functionalities. *Prog. Organic Coatings*, **47** (3), 288–291.

13 Yu, L., Lu, Z., Gan, Y., Liu, Y., and Li, C.M. (2009) AFM study of adsorption of protein A on a poly(dimethylsiloxane) surface. *Nanotechnology*, **20** (28), 285101.

14 Shulha, H., Wong, C., Kaplan, D.L., and Tsukruk, V.V. (2006) Unfolding the multi-length scale domain structure of silk fibroin protein. *Polymer*, **47** (16), 5821–5830.

15 Soman, P., Rice, Z., and Siedlecki, C.A. (2008) Measuring the time-dependent functional activity of adsorbed fibrinogen by atomic force microscopy. *Langmuir*, **24** (16), 8801–8806.

16 Campbell, N.A., Reece, J.B., and Mitchell, L.G. (1999) *Biology*, 5th edn, Benjamin/Cummings, Menlo Park, p. 853.

17 Alberts, B., Johnson, A., Lewis, J., Raff, M., Roberts, K., and Walter, P. (2003) *Molecular Biology of the Cell*, 4th edn, Garland Science, New York, p. 1376.

18 Cheung, C.-L., Hafner, J.H., and Lieber, C.M. (2000) Carbon nanotube atomic force microscopy tips: direct growth by chemical vapor deposition and application to high-resolution imaging. *Proc. Natl. Acad. Sci.*, **97** (8), 3809–3813.

19 Sewald, N. and Jakubke, H.-D. (2009) *Peptides: Chemistry and Biology*, 2nd edn, John Wiley & Sons, Inc., New York.

20 So, C.R., Kulp, J.L., III, Oren, M.M., Zareie, H., Tamerler, C., Evans, J.S., and Sarikaya, M. (2009) Molecular recognition and supramolecular self-assembly of a genetically engineered gold binding peptide on Au{111}. *ACS Nano*, **3** (6), 1525–1531.

21 Bizzarri, A.R. and Cannistraro, S. (2010) The application of atomic force spectroscopy to the study of biological complexes undergoing a biorecognition process. *Chem Soc. Rev.*, **39**, 734–749.

22 Bizzarri, A.R. and Cannistraro, S. (2009) Atomic force spectroscopy in biological complex formation: strategies and perspectives. *J. Phys. Chem. B*, **113** (52), 16449–16464.

23 Zimmermann, J.L., Nicolaus, T., Neuert, G., and Blank, K. (2010) Thiol-based, site-specific and covalent immobilization of biomolecules for single-molecule experiments. *Nat. Protocol*, **5**, 975–985.

24 Sulchek, T.A., Friddle, R.W., Langry, K., Lau, E.Y., Albrecht, H., Ratto, T.V., DeNardo, S.J., Colvin, M.E., and Noy, A. (2005) Dynamic force spectroscopy of parallel individual Mucin1–antibody bonds. *PNAS*, **102** (46), 16638–16643.

25 Kirata, K.E., Bartkowskib, M., and Haupt, K. (2009) Probing the recognition specificity of a protein molecularly imprinted polymer using force spectroscopy. *Biosens. Bioelectron.*, **24** (8), 2618–2624.

26 Rief, M., Gautel, M., Oesterhelt, F., Fernandez, J.M., and Gaub, H.E. (1997) Reversible unfolding of individual titin immunoglobulin domains by AFM. *Science*, **276** (5315), 1109–1112.

27 Rief, M., Pascual, J., Saraste, M., and Gaub, H.E. (1999) Single molecule force spectroscopy of spectrin repeats: low unfolding forces in helix bundles. *J. Mol. Biol.*, **286** (2), 553–561.

28 Brockwell, D.J., Paci, E., Zinober, R.C., Beddard, G.S., Olmsted, P.D., Smith, D.A., Perham, R.N., and Radford, S.E. (2003) Pulling geometry defines the mechanical resistance of a bold beta-sheet protein. *Nat. Struct. Biol.*, **10** (9), 731–737.

29 Carrion-Vazquez, M., Oberhauser, A., Fisher, T., Marszalek, P., Li, H., and

Fernandez, J. (2000) Mechanical design of proteins studied by single-molecule force spectroscopy and protein engineering. *Prog. Biophys. Mol. Biol.*, **74** (1–2), 63–91.

30 Ludwig, M., Rief, M., Schmidt, H., Li, H., Oesterhelt, F., Gautel, M., and Gaub, H.E. (1999) AFM, a tool for single-molecule experiments. *Appl. Phys. A*, **68** (2), 173–176.

31 Li, H., Zhang, W., Zhang, X., Shen, J., Liu, B., Gao, C., and Zao, G. (1998) Single molecule force spectroscopy on poly(vinyl alcohol) by AFM. *Macromol. Rapid Commun.*, **19** (12), 609–611.

32 Senden, T., di Meglio, J., and Auroy, P. (1998) Anomalous adhesion in adsorbed polymer layers. *Eur. Phys. J.*, **3** (2), 211–216.

33 Smith, S.B., Cui, Y., and Bustamante, C. (1996) Overstretching B-DNA: the elastic response of individual double-stranded and single-stranded DNA molecules. *Science*, **271** (5250), 795–799.

34 Rief, M., Oesterhelt, F., Heymann, B., and Gaub, H.E. (1997) Single molecule force spectroscopy on polysaccharides by atomic force microscopy. *Science*, **275** (5304), 1295–1297.

35 Porod, G. (1949) Zusammenhang zwischen mittlerem Endpunktsabstand und Kettenlänge bei Fadenmolekülen. *Monatsh. Chem.*, **80** (2), 251–255.

36 Rubenstein, M. and Colby, R.H. (2003) *Polymer Physics*, Oxford University Press, New York.

37 Hugel, T. and Seitz, M. (2001) The study of molecular interactions by AFM force spectroscopy. *Macromol. Rapid Commun.*, **22** (13), 989–1016.

38 Fisher, T., Marszalek, P., Oberhauser, A., Carrion-Vazquez, M., and Fernandez, J. (1999) The micro-mechanics of single molecules studied with atomic force microscopy. *J. Physiology*, **520** (1), 5–14.

39 Li, H., Rief, M., Oesterhelt, F., and Gaub, H. (1998) Single-molecule force spectroscopy on xanthan by AFM. *Adv. Mater.*, **10** (4), 316–319.

40 Friedsam, C., Del Campo Becares, A., Jonas, U., Seitz, M., and Gaub, H. (2004) Adsorption of polyacrylic acid on self-assembled monolayers investigated by single-molecule force spectroscopy. *New J. Physics*, **6** (1), 9.

41 Friedsam, C., Del Campo Becares, A., Jonas, U., Gaub, H., and Seitz, M. (2004) Polymer functionalized AFM tips for long-term measurements in single-molecule force spectroscopy. *Chem. Phys. Chem.*, **5** (3), 388–393.

42 Xu, Q., Zou, S., Zhang, W., and Zhang, X. (2001) Single-molecule force spectroscopy on carrageenan by means of AFM. *Macromol. Rapid Commun.*, **22** (14), 1163–1167.

43 Zou, S., Zhang, W., Zhang, X., and Jiang, B. (2001) Study on polymer micelles of hydrophobically modified ethyl hydroxyethyl cellulose using single-molecule force spectroscopy. *Langmuir*, **17** (16), 4799–4808.

44 Hertadi, R., Gruswitz, F., Silver, L., Koide, A., Koide, S., Arakawa, H., and Ikai, A. (2003) Unfolding mechanics of multiple OspA substructures investigated with single molecule force spectroscopy. *J. Mol. Biol.*, **333** (5), 993–1002.

45 Rief, M. and Grubmuller, H. (2002) Force spectroscopy of single biomolecules. *Chem. Phys. Chem.*, **3** (3), 255–261.

46 Zhang, W. and Zhang, X. (2003) Single molecule mechanochemistry of macromolecules. *Prog. Polym. Sci.*, **28** (8), 1271–1295.

47 Haupt, B., Ennis, J., and Sevick, E. (1999) The detachment of a polymer chain from a weakly adsorbing surface using an AFM tip. *Langmuir*, **15** (11), 3886–3892.

48 Schlierf, M., Li, H., and Fernandez, J. (2004) The unfolding kinetics of ubiquitin captured with single-molecule force-clamp techniques. *PNAS*, **101** (19), 7299–7304.

49 Oberhauser, A., Hansma, P., Carrion-Vazquez, M., and Fernandez, J. (2001) Stepwise unfolding of titin under force-clamp atomic force microscopy. *PNAS*, **98** (2), 468–472.

50 Carrion-Vazquez, M., Obrhauser, A., Fowler, S., Marszalek, P., Broedel, S., Clarke, J., and Fernandez, J. (1999) Mechanical and chemical unfolding of a single protein: a comparison. *PNAS*, **96** (7), 3694–3699.

51 Li, H., Oberhauser, A., Redick, S., Carrion-Vazquez, M., Erickson, H., and Fernandez, J. (2001) Multiple conformations of PEVK proteins detected by single-molecule techniques. *PNAS*, **98** (19), 10682–10686.

52 Oberhauser, A.F., Badilla-Fernandez, C., Carrion-Vazquez, M., and Fernandez, J.M. (2002) The mechanical hierarchies of fibronectin observed with single-molecule AFM. *J. Mol. Biol.*, **319** (2), 433–447.

53 Meadows, P., Bemis, J., and Walker, G. (2003) Single-molecule force spectroscopy of isolated and aggregated fibronectin proteins on negatively charged surfaces in aqueous liquids. *Langmuir*, **19** (23), 9566–9572.

54 Gutsmann, T., Fantner, G., Kindt, J., Venturoni, M., Danielsen, S., and Hansma, P. (2004) Force spectroscopy of collagen fibers to investigate their mechanical properties and structural organization. *Biophys. J.*, **86** (5), 3186–3193.

55 Shulha, H., Wong, C., Kaplan, D.L., and Tsukruk, V.V. (2006) Unfolding the multi-length scale domain structure of silk fibroin protein. *Polymer*, **47** (16), 5821–5830.

56 Omenetto, F.G. and Kaplan, D.L. (2010) New opportunities for an ancient material. *Science*, **329** (5991), 528–531.

57 Rief, M., Pascual, J., Saraste, M., and Gaub, H.E. (1999) Single molecule force spectroscopy of spectrin repeats: low unfolding forces in helix bundles. *J. Mol. Biol.*, **286** (12), 553–561.

58 Camesano, T.A. and Logan, B.E. (2000) Probing electrostatic interactions using atomic force microscopy. *Environ. Sci. Technol.*, **34** (16), 3354–3362.

59 Camesano, T.A., Natan, M.J., and Logan, B.E. (2000) Observation of changes in bacterial cell morphology using tapping mode atomic force microscopy. *Langmuir*, **16** (10), 4563–4572.

60 Lister, T.E. and Pinhero, P.J. (2001) *In vivo* atomic force microscopy of surface proteins on *Deinococcus radiodurans*. *Langmuir*, **17** (9), 2624–2628.

61 Schaer-Zammaretti, P. and Ubbink, J. (2003) Imaging of lactic acid bacteria with AFM: elasticity and adhesion maps and their relationship to biological and structural data. *Ultramicroscopy*, **97**, 199–208.

62 Kasas, S. and Ikai, A. (1995) A method for anchoring round shaped cells for atomic force microscope imaging. *Biophys. J.*, **68** (5), 1678–1680.

63 Dufrene, Y.F., Boonaert, C.J.P., Gerin, P.A., Asther, M., and Rouxhet, P.G. (1999) Direct probing of the surface ultrastructure and molecular interactions of dormant and germinating spores of *Phanerochaete chrysosporium*. *J. Bacteriol.*, **181** (17), 5350–5354.

64 Touhami, A., Nysten, B., and Dufrêne, Y.F. (2003) Nanoscale mapping of the elasticity of microbial cells by atomic force microscopy. *Langmuir*, **19** (11), 4539–4543.

65 Gaboriaud, F., Bailet, S., Dague, E., and Jorand, F. (2005) Surface structure and nanomechanical properties of *Shewanella putrefaciens* bacteria at two pH values (4 and 10) determined by atomic force microscopy. *J. Bacteriol.*, **187** (11), 3864–3868.

66 Pelling, A.E., Sehati, S., Gralla., E.B., Valentine, J.S., and Gimzewski, J.K. (2004) Local nanomechanical motion of the cell wall of *Saccharomyces cerevisiae*. *Science*, **305** (5687), 1147–1150.

67 Han, S., Nakamura, C., Obataya, I., Nakamura, N., and Miyake, J. (2005) Gene expression using an ultrathin needle enabling accurate displacement and low invasiveness. *Biochem. Biophys. Res. Commun.*, **332** (3), 633–639.

68 Han, S.W., Nakamura, C., Obataya, I., Nakamura, N., and Miyake, J. (2005) A molecular delivery system by using AFM and nanoneedle. *Biosens. Bioelectron.*, **20** (10), 2120–2125.

69 Obataya, I., Nakamura, C., Han, S., Nakamura, N., and Miyake, J. (2005) Nanoscale operation of a living cell using an atomic force microscope with a nanoneedle. *Nano Lett.*, **5** (1), 27–30.

70 Vakarelski, I.U., Brown, S.C., Higashitani, K., and Moudgil, B.M. (2007) Penetration of living cell membranes with fortified carbon nanotube tips. *Langmuir*, **23** (22), 10893–10896.

71 Kouklin, N.A., Kim, W.E., Lazareck, A.D., and Xu, J.M. (2005) Carbon nanotube

probes for single-cell experimentation and assays. *Appl. Phys. Lett.*, **87** (17), 173901/1–173901/3.

72 Cross, S.E., Jin, Y.-S., Rao, J., and Gimzewski, J.K. (2007) Nanomechanical analysis of cells from cancer patients. *Nat. Nanotechnol.*, **2**, 780–783.

73 Xie, A.F., Yamada, R., Gewirth, A.A., and Granick, S. (2002) Materials science of the gel to fluid phase transition in a supported phospholipid bilayer. *Phys. Rev. Lett.*, **89** (24), 246103/1–246103/4.

74 Tokumasu, F., Jin, A.J., Feigenson, G.W., and Dvorak, J.A. (2003) Nanoscopic lipid domain dynamics revealed by atomic force microscopy. *Biophys. J.*, **84** (4), 2609–2618.

75 Roiter, Y., Ornatska, M., Rammohan, A.R., Balakrishnan, J., Heine, D.R., and Minko, S. (2008) Interaction of nanoparticles with lipid membrane. *Nano Lett.*, **8** (3), 941–944.

Part Four
Nanomanipulation, Patterning, and Sensing

Scanning Probe Microscopy of Soft Matter: Fundamentals and Practices, First Edition.
Vladimir V. Tsukruk and Srikanth Singamaneni.

17
Scanning Probe Microscopy on Practical Devices

As previously discussed in several chapters, the ability to probe soft samples under ambient conditions with little to no sample preparation distinguishes AFM from electron microscopy. Furthermore, in at least some commercial designs, the headspace of the AFM is sufficient to probe reasonably large samples (e.g., large wafers, devices, and live specimens). The powerful combination of nanoscale resolution of both surface structures and properties and ambient probing makes AFM uniquely suited for probing practical and active micro- and nanodevices in their natural environments (parts of circuitry) and under working conditions, which provides new insights into their operation. In particular, AFM enables the correlation of the structure and structural homogeneities with the properties and performance of the active components of the devices during their actual operation. Such knowledge at the device level is extremely valuable in the design of novel materials and processing conditions to improve device performance. In this chapter, using several specific practical examples, we briefly discuss some of the recent progress in the application of different scanning probe microscopy techniques to probe active micro- and nanodevices under operational conditions.

17.1
Electrical SPM of Active Electronic and Optoelectronic Devices

In Chapter 4, we discussed the basics of electrical scanning probe microscopy with specific examples involving their application to organic conductors and semiconducting materials. In this chapter, we primarily focus on the application of these techniques (conductive AFM, scanning Kelvin probe microscopy, electrostatic force microscopy) for active device measurements. The ability to probe the electrical properties (conductivity, surface potential, and work function) with nanoscale resolution provides extremely valuable information for understanding the structure–property–performance relationships and nanoscale heterogeneity of active surface layers of organic devices such as organic field effect transistors, light emitting diodes, and photovoltaic devices.

Over the past decade, AFM and related techniques have emerged as powerful tools for the characterization of nano- to micrometer-scale heterogeneities of various

Scanning Probe Microscopy of Soft Matter: Fundamentals and Practices, First Edition.
Vladimir V. Tsukruk and Srikanth Singamaneni.

electrical properties of active and passive device components [1–9]. Extensive efforts have been dedicated in imparting device-probing capabilities to the conventional approaches such as C-AFM, EFM, KPFM, which involved hardware modification of the existing (commercially available) AFM hardware and/or addition of a new signal channel to tap into other concurrently collected information. One of the very good and representative examples in this regard is the photoconductive AFM mode, which involves surface mapping of the local photocurrent generated in an active organic photovoltaic layer with excellent spatial resolution, as discussed below.

The spatial distribution of electrostatic potential is an important factor of the device performance, and cross-sectional mapping of the potential in an active device provides a better understanding of the performance of organic thin film transistors (OTFTs). In a related study, Ikeda *et al.* employed KPFM to reveal the cross-sectional potential map of an active OTFT [1]. The bottom-contact-type OTFT with a copper-phthalocyanine (CuPc) channel was cleaved and the internal potential distribution of its channel region was monitored. It was revealed that the potential distribution on the cross section varied with the applied drain and gate voltage. The KPFM-imaged area demonstrated in this study includes Si, SiO_2, CuPc, air, and also a part of the source and drain electrodes.

In the potential images, various changes were identified with the changes in gate voltage and drain voltage. In particular, the contour maps clearly revealed the spatial distribution of the potential in the CuPc film. In the contour map obtained under the conditions of $V_D = -2$ V and $V_G = 0$ V, the negative potential region was seen only near the drain electrode of the CuPc channel. On the other hand, in the map obtained with $V_D = -2$ V and $V_G = -5$ V, the negative potential region was found to extend to the source electrode from the drain electrode. When $+5$ V was applied to the gate, the negative potential region was found to be localized near the drain electrode and did not extend toward the source electrode. The vertical potential distribution from the bottom (gate) to the top (CuPc film) portion exhibited a potential peak along the semiconductor/insulator interface when a negative voltage was applied to the gate. The authors suggested two possible reasons for the observed electrostatic potential peak: (i) it indicates the amount of induced charges and (ii) it originates from the combination of two kinds of charges – the charge induced at the CuPc/SiO_2 interface and a small opposite charge induced at the grain boundaries.

Time-resolved electrostatic force microscopy (trEFM) has been employed to probe the photogenerated charge in organic photovoltaic devices [3, 10, 11]. As discussed in Chapter 4 in detail, conventional electrostatic force microscopy involves monitoring the distribution of the surface potential or work function of the surface under investigation using the long-range electrostatic forces. Typically, the frequency shift (phase signal) of the oscillatory cantilever with biased tip, which is proportional to local capacitive gradient and the potential difference between the tip and the sample, is monitored to map the surface charge distribution. However, conventional EFM is limited to static and quasi-static measurements and thus is not suited to probe the efficiency of a solar cell as a function charge generation and charge injection, which determine the efficiency and are dynamic processes. Recently, Coffey and Ginger have introduced trEFM, which enables the probing of transient electrostatic force gradients and thus enables dynamic monitoring of the transient phenomena mentioned above [10].

In another study, Jaquith *et al.* have employed trEFM to directly and locally probe the kinetics of charge trap formations in polycrystalline pentacene TFTs [3]. For mapping the charge trapping in the active layer in the TFT device, holes were introduced into the pentacene film by applying a negative gate voltage, which resulted in a planar charge density (σ) of injected holes equal to $C_g(V_g - V_T)$, where C_g is the gate capacitance per unit area and V_T is the transistor's threshold voltage. The gate was held at this voltage for specific time (50–1000 ms), during which time some of the mobile charge is converted into a trapped charge. The gate was then returned to zero voltage in order to drive the mobile charge from the channel, followed by the trEFM method.

In these experiments, it was found that the trapping rate depends strongly on the initial concentration of free holes and that trapped charge is highly localized. From the observed dependence of the trapping rate on the hole chemical potential, the authors suggest that the trapping process should not be viewed as a filling of midgap energy levels, but instead as a process in which the very creation of trapped states requires the presence of free holes. The trEFM results suggested that the grain boundary trapping in polycrystalline pentacene is not as critical as one would expect. Instead, charge trapping via an activated process, such as a chemical reaction, assisted by a localized structural defect is more critical.

In a more recent study, Yu *et al.* applied SKPM to back-gated graphene-based devices to demonstrate that the work function of graphene can be controlled by electric field effect-induced modulation of carrier concentration [12]. SKPM was employed to probe the surface potential variation across the device at different gate voltages. The carrier density, and hence the work function of the graphene device, is controlled by the applied gate voltage. On biased graphene devices, SKPM also allows for accurate measurements of graphene/metal contact resistances by mapping the surface potential of a device.

Utilizing multi-terminal device geometry, the authors compared the surface potential and transport measurement on the same device as a function of V_g (see Figure 17.1) [12]. Using transport measurements, the charge neutral gate voltage of the particular device probed was determined to be 48 V. In the surface potential map within the electrode and channel, no significant spatial variation was observed. On the other hand, a stepwise increase in V_{CPD} (contact potential difference) at the junctions between the graphene and the electrode regions was observed with a fixed gate voltage. Similar features were also observed at different gate voltages, where the overall line profiles of V_{CPD} in the two regions shift upward with a common background as gate voltage increases. The authors ascribed the background signal observed in the surface potential maps to the unscreened long-range electrostatic interactions between the conducting cantilever probe and the back gate. In this design, the background is insensitive to small spatial position changes of the SKPM tip.

Furthermore, as the gate voltage should not influence the local surface potential in the metallic electrode, the authors employed the background observed on the metal

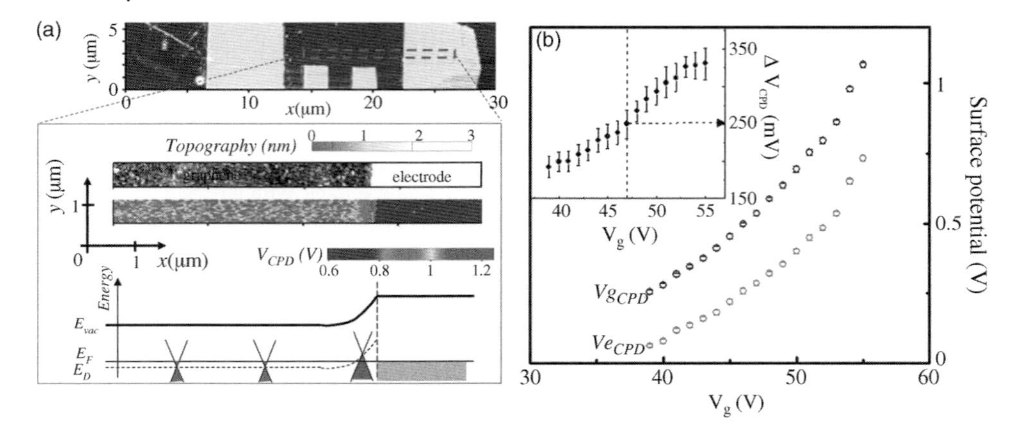

Figure 17.1 (a) AFM topography of the device and surface potential map of a selected area shown in the topography image. A schematic energy alignment diagram for the graphene sample and the metallic electrode is also shown. (b) The CPD of graphene and electrode at different gate voltages, obtained from the average surface potential. Reproduced from Ref. [12].

electrode surface to separate the spatially constant background from the relative change in the local surface potential. In particular, the electric field effect-induced local surface potential change in graphene was obtained from the following relationship: $\Delta V_{\mathrm{CPD}} = V^{\mathrm{g}}_{\mathrm{CPD}} - V^{\mathrm{e}}_{\mathrm{CPD}}$, where $V^{\mathrm{e}}_{\mathrm{CPD}}$ and $V^{\mathrm{g}}_{\mathrm{CPD}}$ are the average V_{CPD} in the electrode and the graphene, respectively (see Figure 17.1). A sudden change in ΔV_{CPD} can be clearly observed at the charge neutrality point, as indicated by the vertical dashed line. Such sudden change in the ΔV_{CPD} was consistently observed for single-layer graphene devices, while in the case of bilayer graphene devices, no such sudden change in ΔV_{CPD} was observed, indicating the difference in the electronic structures and properties of these samples.

Lee *et al.* have employed a multifunctional AFM cantilever for applying highly localized thermal and electric fields to interrogate transport in single-wall carbon nanotube field effect transistors (carbon nanotube FETs) [13]. The probe could be operated in contact mode, intermittent contact mode, or as a Kelvin probe, and enabled independent control of the electric field, mechanical force, and temperature applied to the surface.

Of the numerous applications such a multifunctional cantilever can be used for, we describe one interesting experiment performed in this study [13]. The cantilever was brought into contact with a carbon nanotube, which formed the channel of a field effect transistor and was resistively heated to very high temperatures (above 1000 °C). The contact force was kept below 12 nN using the feedback control in the AFM mode. It is interesting to note that in the initial stages of the experiment, at room temperature, the contact force was fixed constant at 12 nN and then subsequently decreased as the cantilever was heated and became softer. Figure 17.2 shows the current (I_{DS}) in the carbon nanotube FET as the power supplied to the resistive

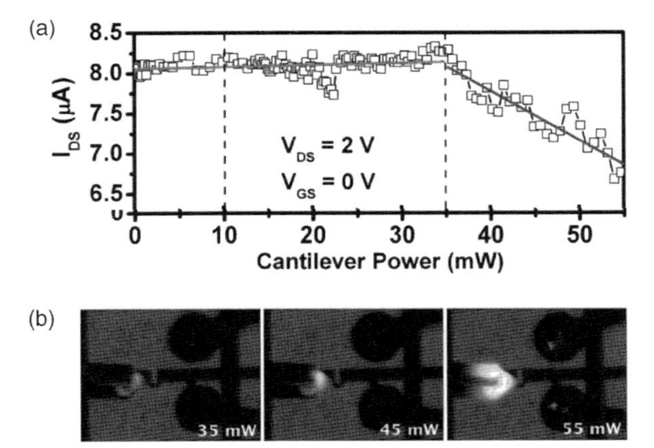

Figure 17.2 (a) Current (I_{DS}) in the carbon nanotube FET as a function of the power supplied to the cantilever resistive element, while the tip makes contact with the carbon nanotube at the midpoint between source and drain. (b) Optical images of the setup at cantilever power of 35, 45, and 55 mW, showing the glowing hot tip. Reproduced from Ref. [13].

element of the cantilever was increased stepwise. From Figure 17.2, it can be observed that the current through carbon nanotubes was mostly insensitive to local heating from the cantilever when the cantilever power was below 35 mW, which was suggested to be due to the presence of water and organic layers on the nanotube and the silica surface at relatively low temperatures.

A linear decrease in the current was observed as the cantilever power was increased above 35 mW. Figure 17.2 also shows optical images of the cantilever when the power supplied to the resistive element of the cantilever was 35, 45, and 55 mW. It can be seen that the cantilever is glowing and emitting light from its free end with the glowing area and intensity of light increasing with the power supplied to cantilever. The temperature of the cantilever associated with the 35, 45, and 55 mW powers were estimated to be 865, 990, and 1170 °C, respectively. The observed resistance changes in the carbon nanotube were employed to estimate the thermal resistance between the AFM tip and the carbon nanotube, which was concluded to be 1.6×10^7 and 9.6×10^6 K/W at 45 and 55 mW cantilever power, respectively.

Lau *et al.* employed a modified version of conductive atomic force microscopy to understand the electrical switching mechanism of metal–molecule–metal junctions [14]. In particular, the authors studied the electrical switching characteristics of Ti–stearic acid–Pt sandwich structure. Electrical characterization of the molecular device was performed by applying a bias voltage V_b relative to the Ti electrodes. The *I–V* curves of the pristine samples were nonlinear, with zero-bias resistance typically higher than 10^5 Ω. The devices could be switched reversibly and repeatedly between ON and OFF conductance states by applying sufficiently large voltage $V_b < 0$ or $V_b > 0$, respectively. To gain an insight into the switching mechanism, the authors employed an AFM tip to apply high point load (0.8 μN) with simultaneous recording

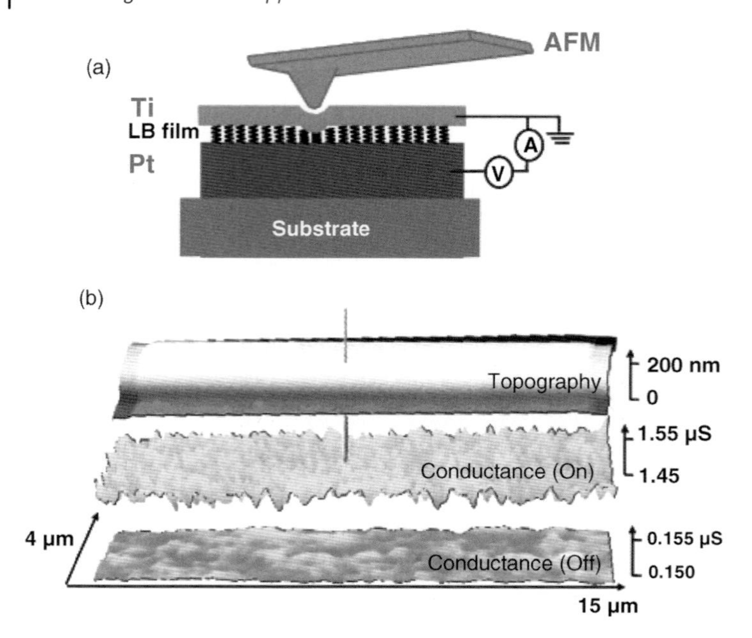

Figure 17.3 (a) Schematic of the experimental setup, (b) topographical (*top*), and conductance images (*middle* and *bottom*) of a junction. The conductance is relatively uniform across the junction when the junction is in the "off" state (bottom image), while a peak in conductance appears (red spike) when the junction is switched "on" (middle image). Reproduced from Ref. [14].

of the current under constant bias. It is important to note that the AFM tip in this experiment just applied a local load while the electrical bias and current were recorded using macroscopic Pt and Ti electrodes. The local load applied by the AFM tip on the electrodes causes the compression of molecular structures sandwiched between the electrodes.

Initially, the authors acquired a conductance image before any switching was performed (Figure 17.3). The conductance distribution was found to be relatively uniform across the scan region, including both the electrodes and the adjacent substrate areas. This result indicated that the junction conductance did not exhibit any significant response to the local AFM tip perturbation. After the junction was switched to the "on" state, a highly localized increase in conductance appeared when the tip pressurized a particular surface area. The middle image in Figure 17.3b shows the sharp conductance peak that spikes through the above image as the AFM tip applies a load to the particular spatial location. The selected spot was highly localized with a full-width at half-maximum of around 40 nm, indicating that the switching center or the mechanically sensitive region is less than 40 nm in size.

Conversion of mechanical energy into electrical energy using piezoelectric nanowires has been extensively investigated. In particular, owing to their combined piezoelectric and semiconducting nature, various forms of ZnO nanostructures have been employed to fabricate nanogenerators [15]. Wang and coworker have demon-

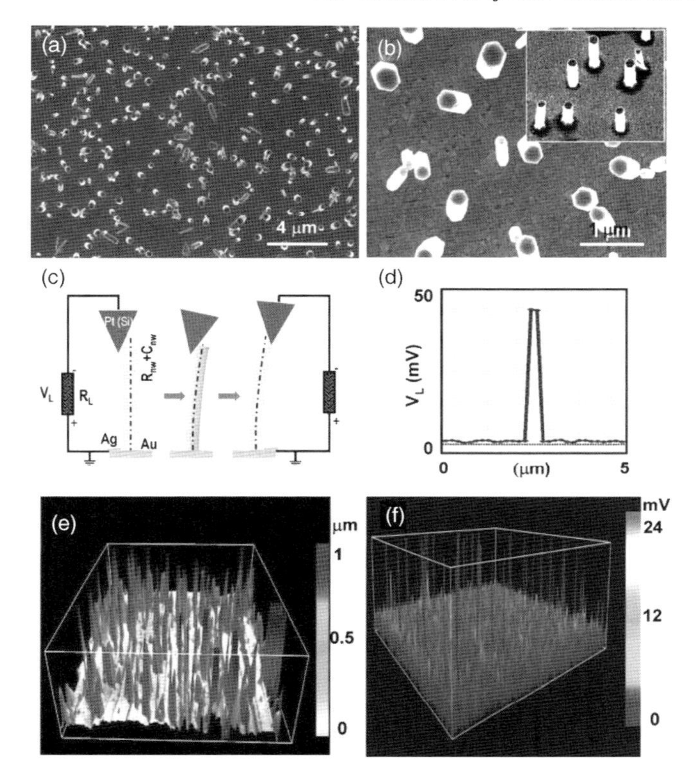

Figure 17.4 (a) A low-magnification and (b) a higher-resolution SEM image showing a sparsely grown ZnO NW array on a plastic substrate. (c) Schematic showing the experimental setup for the conductive AFM measurement of the NW array. (d) Output voltage profile for the AFM tip scanning over a single NW on the plastic substrate. (e) A 3D plot of the AFM topography image and (f) output voltage profile obtained by scanning the AFM tip over a 40 µm × 40 µm area of aligned ZnO NWs on a plastic substrate [17].

strated a nanogenerator using vertically aligned ZnO nanowires on rigid silicon and flexible plastic substrates [16, 17]. They used a conductive AFM to deform a single, vertically aligned nanowire and probe the output voltage. The conducting AFM provided a detailed insight into the nature of the Schottky barrier between the ZnO nanowire and the metal AFM tip. The AFM tip traversing across the vertically aligned nanowire bends the nanowire, causing one side of the nanowire to be compressed while the other side is stretched, resulting in the polarization of the ionic charges in the ZnO lattice and an associated electric field within the nanowire.

Figure 17.4 shows SEM images of the vertically aligned ZnO nanorod array grown on plastic substrates and output voltage images that were recorded simultaneously in this study. While scanning in contact mode, the AFM tip deformed the ZnO nanowires and the bending distance acquired from the images was employed to estimate the mechanical properties of the nanowires. Under these scanning conditions, many sharp peaks were observed in the output voltage image, which were usually 4 to 50 times higher than the noise level observed in the image. However, under these imaging conditions the discharge peaks corresponded only to two or three pixels, which was insufficient to reveal the discharge profile.

As suggested by the authors, the chopped tops of the voltage discharge peaks are possibly due to the discrepancy between the dwell time of the AFM tip and the lifetime of the discharge pulse (Figure 17.4). From the discharge profile, it can be observed that no voltage output was observed in the initial stages of the deformation of the nanowire. Output voltage was detected only when the deflection of the nanowire approached its maximum and finally when the nanowire was released by the AFM tip, output voltage dropped to zero. The AFM probing clearly revealed that the output of piezoelectricity was detected toward the end of the AFM scan over the nanowire. This observation was explained by the Schottky nature of the contact and the positive bias condition as the tip made contact at the trailing edge of the nanowire. *In situ* AFM measurements performed in this study provided a direct insight into the nanogenerator energy harvesting mechanism, guiding advanced and real-world designs for energy harvesting.

As is known, indium tin oxide (ITO) is an extremely important electrode-related material in organic photovoltaic and light-emitting devices due to its high optical transparency and electrical conductivity [18, 19]. It is routinely employed as window electrodes in photovoltaic devices and understanding its structure and electrical properties with nanoscale resolution is extremely important. Brumbach *et al.* have employed a combination of C-AFM and X-ray photoelectron spectroscopy to understand the electrical activity and near-surface chemical structure of ITO films activated by plasma cleaning and haloacid treatment [20].

Tapping mode AFM revealed that commercial ITO thin films had a highly textured surface morphology because of grain formation during the sputtering process. The surface microroughness was found to be extremely variable even within a single commercial batch of ITO. Small subgrains, with diameters of 50–100 nm, are typically grouped to form larger grains with diameters of 0.3–0.5 μm. To obtain current images using C-AFM in the case of detergent/solvent-cleaned ITO samples, a bias of −2 V between the tip and the sample was necessary to detect currents on the 0 to 1 nA scale. Even at this relatively high bias voltage, only 10–20% of the geometric area in $2 \times 2\,\mu m^2$ scans exhibited current in excess of 0.5 nA in regions with an average diameter of 50–200 nm.

For plasma-cleaned or acid-etched ITO samples, a significantly lower bias voltage (−0.02 V) was sufficient to obtain currents in the ranges of 0.1–0.3 and 1–10 nA. The oxygen-plasma-etched ITO surface exhibited the most uniform distribution of current with an average current of 1–2 nA in regions of nearly 50–200 nm diameter. In the case of HI-etched and HCl-etched samples, nearly 90% of the ITO surface gave almost equivalent tip–sample currents (10 nA) over wide surface areas. The authors concluded that the acid activation steps formed a clean ITO surface for a few milliseconds, but exposure to water and atmosphere quickly reforms a hydroxylated surface. As suggested by the authors, it is worth noting that the overall solution electrochemical activity of the ITO surfaces can be significantly improved by these activation steps.

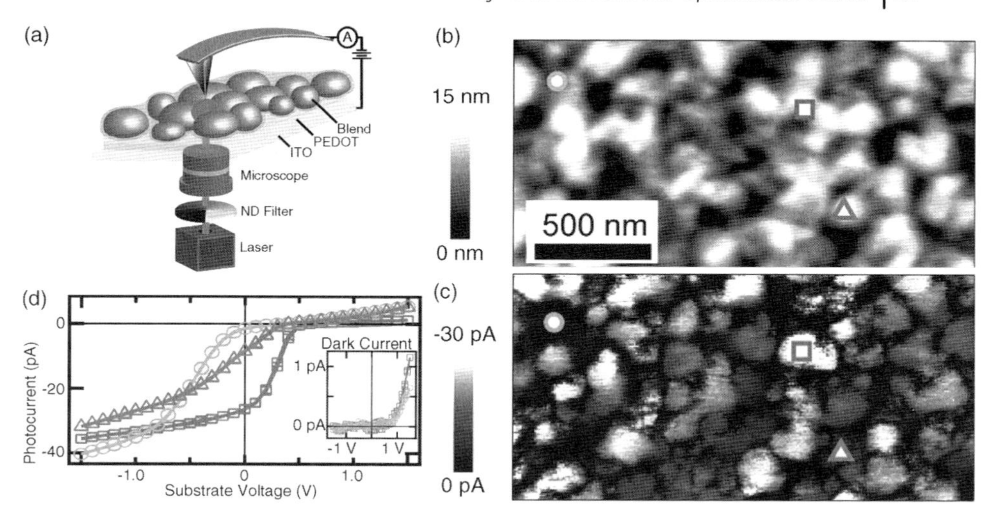

Figure 17.5 (a) Schematic of the experimental setup showing the laser being focused through a transparent electrode onto a photovoltaic blend film. Current is collected with a metal-coated AFM tip. (b) AFM topography image of an MDMO-PPV : PCBM 20 : 80 blend film. (c) Photocurrent map measured with zero external bias. (d) Local *I*–*V* data acquired at the three locations indicated in (b) and (c). Reproduced from Ref. [21].

In another study, Coffey *et al.* applied photoconductive AFM (pcAFM) to monitor the local photocurrent collection in poly(2-methoxy-5-(3′,7′-dimethyloctyl-oxy)-1,4-phenylene vinylene) (MDMO-PPV)/[6,6]-phenyl-C_{61}-butyric acid methyl ester (PCBM) blends [21]. The technique pioneered by this group involves the surface mapping of the topography and conductivity while simultaneously shining light (via a diffraction-limited laser spot) under the photoactive film (Figure 17.5). For this purpose, the AFM is combined with an inverted optical microscope, facilitating the illumination of the sample from the bottom. The photoconductive AFM enables one to locally map photocurrents with spatial resolutions approaching the exciton diffusion length.

Their measurements revealed significant heterogeneity in the photocurrents of blends processed from different solvents. Figure 17.5 shows the topography, spatially resolved photocurrent generated by a 532 nm laser, and representative (positions indicated in topography and photocurrent map) *J*–*V* curves obtained on a blend film spin cast from xylene, as measured by pcAFM. As can be seen from the topography image collected in this study, the surface of the film clearly reveals features of around 100 nm in size, which correspond to PCBM crystallites suspended within the blend matrix, as suggested by the authors. Figure 17.5 also shows the photocurrent collected with the pcAFM, which varied by more than an order of magnitude over the area examined. Furthermore, the photocurrent also shows surface features of 20–100 nm in size within which the photocurrent is largely uniform.

Although it is suggested from the topography and the corresponding photocurrent images that the photocurrent features are associated with the PCBM crystallites, the authors pointed out that it is important to note that not all crystallites affect the current in the same way. The authors attributed this variation to differences in vertical features of surface morphology, suggesting that the elevated location of the PCBM crystallites can significantly influence the local performance in addition to laterally inhomogeneous morphology. The open-circuit voltages were found to vary from 0.35 to 0.53 V and fill factors (ratio of maximum obtainable power to maximum theoretical power) varied from 0.42 to 0.58.

This study clearly revealed that the nanoscale heterogeneity in the film morphology causes significant local variations in electrical performance. In light of these observations, the authors suggested that the bulk heterojunction devices should be thought of as being composed of many nanoscale devices wired in parallel. Each of these nanoscale devices possesses distinct morphological and, hence, electrical properties that result in their complex integrated network-based properties. The authors also concluded that the macroscopic device performance should be limited by the presence of the poorly performing domains. This example shows that the ability to probe morphology and nanoelectrical properties in the device state can be exploited to gain such invaluable insight into the structure–property–performance relationship of the organic photovoltaic devices.

trEFM was employed by Coffey and Ginger to explore the photoinduced charging behavior of an organic photovoltaic cell comprised of a polymer blend of poly(9,9′-dioctylfluorene-*co*-benzothiadiazole) (F8BT) and poly(9,9′-dioctylfluorene-*co*-bis-*N*,*N*′-phenyl-1,3-phenylenediamine) (PFB) [10]. trEFM was used to follow the fast trapping and detrapping kinetics of charge carriers on submicrosecond timescales. They were able to record a charging curve and determine the local charging rate in the material by fitting the curve with an exponential decay function. The process is repeated pixel by pixel to generate a spatial image of the charging rate. The simultaneous topography and charging rate images provided the structure–property relationship (charging behavior and local PFB:F8BT film composition) with 100 nm resolution.

An excellent correlation between the spatially averaged local charging rate and the measured external quantum efficiency for a wide range of blend compositions was demonstrated in this study. Clearly, this is an important result in that a trEFM image of a polymer blend can be used to accurately predict the efficiency of the polymer solar cell that can be obtained from a particular conductive film. In other words, SPM imaging not only predicts the properties of the film obtained under the specific processing conditions but also immediately makes an excellent prediction of the efficiency of the entire device, thus acting as an extremely useful diagnostic tool. Furthermore, the authors noted that it is possible to use trEFM to monitor other quantities of interest, such as spatially correlated charge trapping and detrapping processes.

Application of KPFM to active light-emitting electrochemical cell has significantly enhanced the understanding of the operation of such devices. One of the important questions in the performance of such devices is the localization of the region where the potential drop occurs in the device. There are several models that differ in explaining the fundamental operation of the devices that need to be verified. One model suggests the formation of an organic p–i–n junction in which most of the potential drop occurs across an intrinsic region in the center of the device. A different model predicts that most of the potential drop occurs at the thin layers of ions that accumulate very close to cathode and anode, while the rest of the device does not witness any large electric field.

In studying this phenomenon, Pingree *et al.* measured the potential profiles of both dynamic and fixed junction planar light-emitting electrochemical cells using KPFM and compared the results against models of light-emitting electrochemical cell operation [22]. The authors found that in conventional dynamic junction light-emitting electrochemical cells fabricated using lithium trifluoromethanesulfonate (LiTf), poly(ethylene oxide) (PEO), and the soluble alkoxy-PPV derivative poly-[2-methoxy-5-(3′,7′-dimethyl-octyloxy)-*p*-phenylenevinylene (MDMO-PPV), the bulk (>90%) of the potential dropped near the cathode. Contrarily, little potential drop was observed across either the film or the anode/polymer interface. Figure 17.6 shows the surface potential of an MDMO-PPV/PEO/LiTf device immediately after applying a +4 V bias, as obtained in this study.

Figure 17.6 also shows the corresponding potential profile (averaged over 70 scan lines), electric field profiles (obtained by numerical differentiation of the potential profile), and net elemental charge profile (obtained by numerical differentiation of the voltage profile) perpendicular to the electrode edge. The SKPM images corresponding line traces show a nearly constant potential across the device, with over 90% of the applied voltage dropping at the cathode, and almost no noticeable potential drop at the polymer/anode interface. This experimental observation is clearly in contradiction with the theories discussed above, providing new unexpected insights into the possible mechanisms of the operation of LEC devices. In the case of fixed junction LECs where the LiTf was replaced with [2-(methacryloyloxy) ethyl]

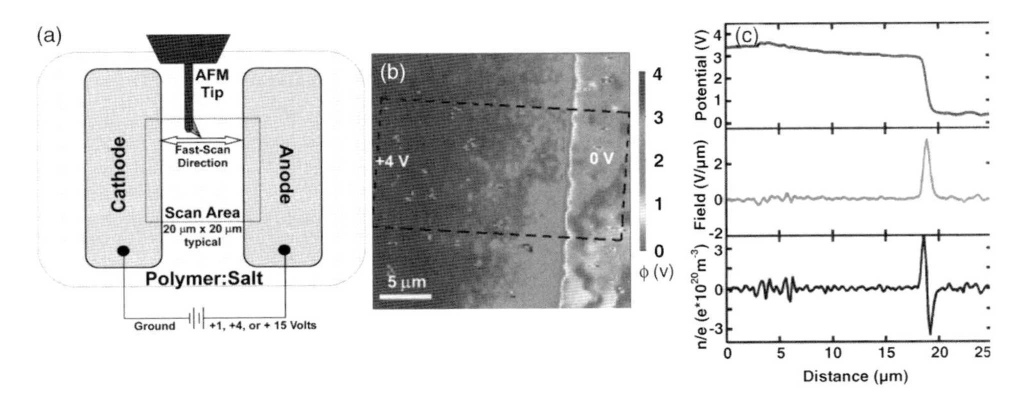

Figure 17.6 (a) Schematic showing the active probing of the light-emitting electrochemical cell; (b) KPFM images of LiTf-doped MDMO-PPV light-emitting electrochemical cells under a 4 V bias immediately after application of the bias; (c) corresponding line-averaged 1D potential profile, electric field profile, and net elemental charge density profile perpendicular to the electrode. Reproduced from Ref. [22].

trimethylammonium 2-(methacryloyloxy)ethane-sulfonate (METMA/MES), the potential drop occurred at both contacts during the initial poling. The ability to probe the surface potential distribution across a working device with a submicron spatial resolution clearly makes KPFM an important tool for such studies.

17.2
Magnetic Force Microscopy of Storage Devices

Several prominent examples of MFM applications to real memory and electronic devices have been published. In one of these studies, Ato *et al.* employed current-induced magnetic force microscopy to investigate the channel properties of carbon nanotube FETs [23]. The shape of the MFM cantilever employed in these studies was tailored to enhance its response to weak magnetic force. A DC voltage V_{gs} was applied to the back gate of the carbon nanotube FET, and an AC voltage V_{ds}, with some DC offset, was used as the source-drain bias. The frequency of the AC voltage applied between the source and the drain was tuned to the torsional resonant frequency of the cantilever, which was extracted as a measure of the magnetic force.

Figure 17.7 shows line profiles and the representative images of the topography and the amplitude of the MFM signal across a single carbon nanotube channel for $V_{gs} = 0\,\text{V}$ and $V_{ds} = 2\,\text{V}_{p-p}$ in the bipolar method obtained in the same study [23]. The

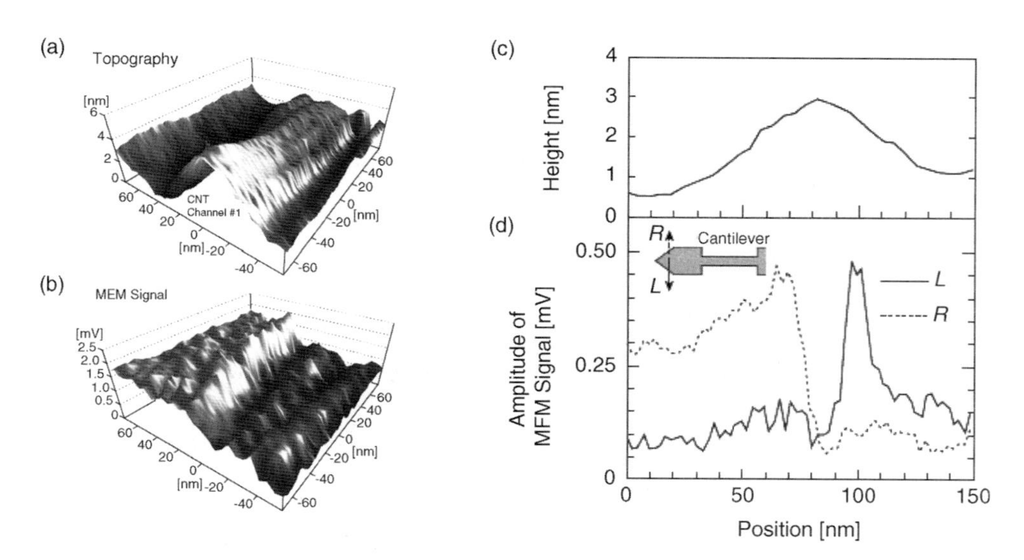

Figure 17.7 (a) Topography and (b) MFM signal images taken by the bipolar method around a single carbon nanotube channel for $V_{gs} = -4\,\text{V}$ and $V_{ds} = 2\,\text{V}_{p-p}$. Line profiles of (c) topography and (b) the amplitude of the MFM signal across a single carbon nanotube channel. Inset in (d) shows the magnetization direction of the MFM tip seen from above. Reproduced from Ref. [23].

line profile of the MFM amplitude was obtained as the MFM tip was magnetized in the L-direction, as indicated in the inset (Figure 17.7d). This line profile and the image clearly show that the highest magnetic force appeared on one side of the carbon nanotube channel where the gradient of the magnetic field was maximum. When the magnetization direction was switched from L to R, the peak position in the MFM signal also switched to the other side of the carbon nanotube (Figure 17.7d). This behavior demonstrates that the detected peak of the MFM signal is a function of the magnetic force rather than the electrostatic force. The technique was employed to obtain magnetic field strength as a function of applied gate voltage. The curves obtained in this study were analogous to the conventional FET transfer characteristics, revealing the threshold voltage and trans-conductance of FET devices. Using this approach, the authors were able to reveal the differences in the threshold gate bias and trans-conductance among different carbon nanotube FET devices and in the asymmetric conductance of a single carbon nanotube channel.

MFM has also been employed to probe stability of the storage bits of a hard disk device. Jiang and Guo have investigated the stability of magnetic domains of a hard disk upon annealing at various temperatures for different times [24]. Figure 17.8 shows the MFM images of the hard disk annealed at different temperatures. The authors noted that the bits remained stable below 200 °C (for 30 min annealing) with no decay in the MFM signal. For higher temperatures, a progressive decay in the magnetization of the bits was observed. As can be seen in Figure 17.8, for annealing at 200 °C, some of the bits become indistinct. Annealing at 250 °C resulted in bridging between the bits caused by the reversal of the magnetic domains and finally at 280 °C

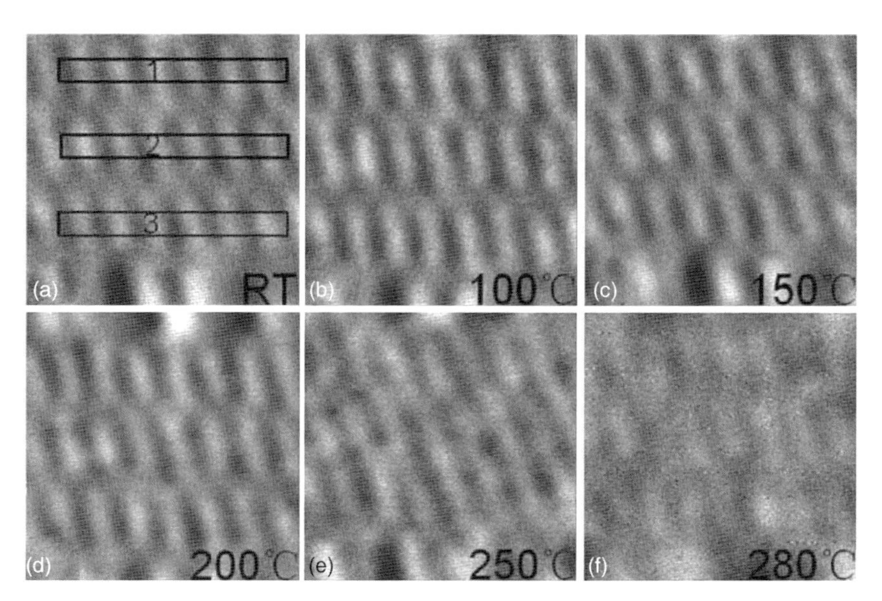

Figure 17.8 MFM images (1.2 μm × 1.2 μm) showing the bit stability of the hard disk under various thermal annealing conditions. Reproduced from Ref. [24].

the profiles are completely blurry with a significant decay in the magnetization. It is interesting to note that the bits remained stable under repeated annealing for shorter duration (10 min) at even 280 °C. Thus, the ability to perform MFM on hard disk device under controlled environmental conditions revealed the thermal stability of the recording bits.

17.3
NSOM of Electrooptical Devices and Nanostructures

Near-field scanning optical microscopy can be applied to understand important mechanisms of the interaction of light and materials at nanoscale, such as plasmon resonances. A surface plasmon involves the collective coherent oscillation of the conductive electrons at the interface of a metal and dielectric material. A broad term, plasmonics phenomenon involves the control of light at the nanoscale using surface plasmons [25, 26]. It encompasses such unique optical phenomena as highly localized enhancement of the electromagnetic (EM) field at nanostructured metal surfaces, high sensitivity of the surface plasmon resonance to external stimuli, extraordinary transmission through subwavelength apertures in thin metal films, and subwavelength waveguiding in metal nanostructures [27–34]. Plasmonics is gaining immense interest due to potential applications of the above-mentioned phenomena in nanooptical components for plasmonic circuits, fabrication of nanoscale structures, photovoltaics, and chemical/biological sensors [35–41]. Understanding the localized surface plasmonic properties of metal nanostructures with nanoscale resolution is critically important for the design and fabrication of nanoplasmonic devices. The routinely employed tools such as UV–vis extinction spectroscopy and dark field scattering spectroscopy provide limited spatial resolution (single nanoparticles at best) in probing the plasmonic properties of these nanostructures.

We discussed NSOM application in Chapter 8 and here we only briefly present some very recent NSOM studies of plasmonic devices. Recently, Rang *et al.* employed interferometric homodyne tip-scattering scanning near-field optical microscopy (*s*-SNOM) to probe the vector field components of the local field distribution of individual silver nanoprisms [42]. The technique's spatial resolution of ~15 nm enabled the mapping of surface plasmon polariton modes and the variation of the same within the nanostructure and with the size of the nanostructure. Figure 17.9 shows the near-field pattern obtained for Ag nanoprisms with an edge length of 120 nm and a height of 35 nm exhibiting a resonant dipole excitation for different excitation and detection polarization combinations.

Figure 17.9 shows the out-of-plane field localization in the z-direction obtained in this study. The image shows the field in the Y–Z plane obtained along the trajectory indicated by the white dashed line in the topography. From the field map, it is clear that the observed electric field is localized within just several tens of nanometers above the nanoprism confirming the evanescent nature of the field detected using the scattering NSOM.

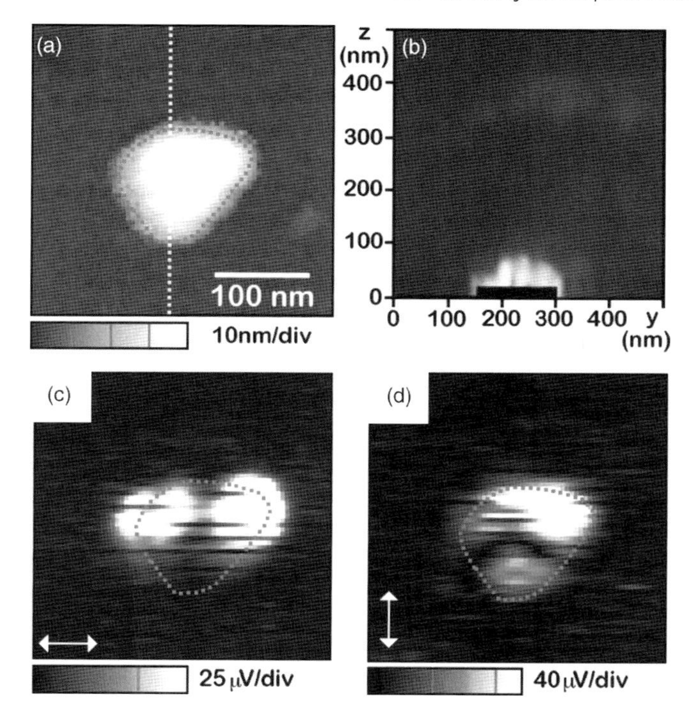

Figure 17.9 (a) Topography and (b) corresponding *s*-SNOM images of Ag nanoprism exhibiting a dipolar excitation for p-polarization excitation along the trajectory indicated by the dashed line in the topography. Homodyne amplification detection in (c) $s_{in}s_{amp}$ polarization configuration of the in-plane field component is probed as compared to (d) $|Ez|$ for $p_{in}p_{amp}$. Reproduced from Ref. [42].

Figure 17.9 also shows scattering-NSOM results of the same particle applying interferometric homodyne signal amplification for s-polarized excitation and s-polarized detection, amplifying the in-plane ($|E_x|$) field distribution. Similarly, for p-polarized excitation and detection, the $|E_z|$ component is probed. It is interesting that in contrast to the field in the *z*-direction, which is predominantly concentrated at the edge regions and confined within the boundaries of the nanoprism, the in-plane field exhibited significant intensity beyond the outer boundary of the nanoprism. The authors also observed higher-order (quadrupole) excitation for larger nanoprisms (edge length of 450 nm) and large spatial variations of the field on length scales as small as 20 nm. The most interesting observation of their study was that the largest field enhancement is not necessarily found to be associated with the tips of the nanoprisms.

In another study, McNeil *et al.* employed the NSOM method to directly map the efficiency of an organic solar cell. These measurements revealed the local spatial variations in efficiency of the two-component organic blend system [43]. The authors employed near-field scanning photocurrent microscopy to map the photocurrent conversion efficiency over the polymer blend samples. The technique involves local

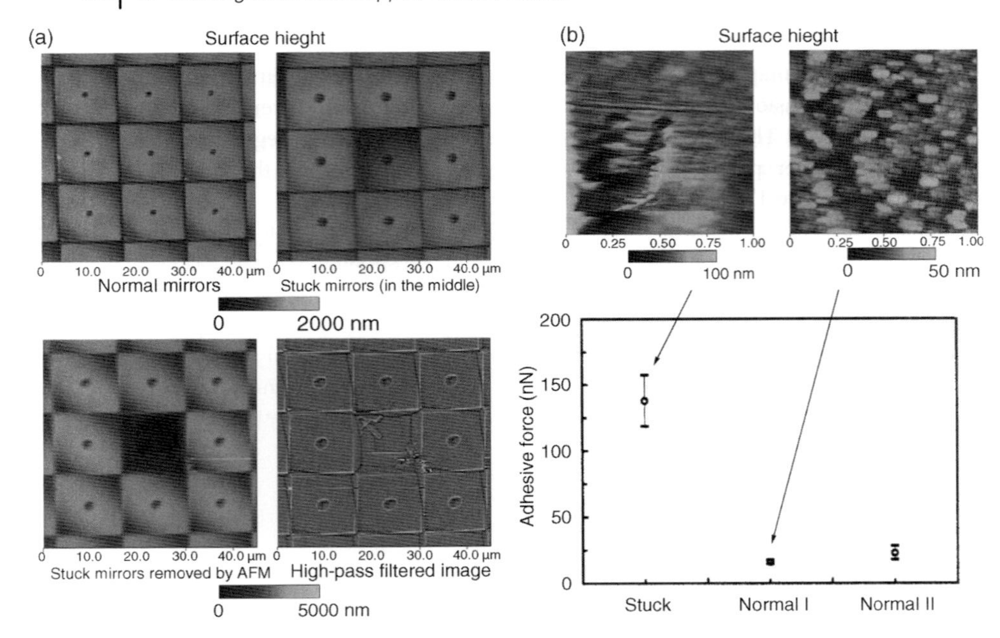

Figure 17.11 (a) The top row shows AFM surface height images of a stuck micromirror surrounded by eight normal micromirrors. Bottom row shows the stuck micromirror that was removed by an AFM tip after repeated scanning at high normal load. Right image in the bottom row shows the high-pass filtered image showing the residual hinge that sits underneath the removed micromirror. (b) AFM topography images and adhesive forces of the landing sites underneath the normal and stuck micromirrors. Reproduced from Ref. [55].

during contact between aluminum alloy spring tips and landing sites, hinge memory effects, hinge fatigue, and vibration failure all need to be carefully considered to ensure reliable operation. One of the common issues with such DMD devices is the stiction between the tip and the landing site. Self-assembled monolayers of perfluorodecanoic acid are employed to overcome the stiction, the quality of which is susceptible to the presence of moisture.

In an attempt to clarify the aforementioned issues, Liu and Bhushan employed AFM to identify a stuck micromirror (as opposed to a freely rotatable micromirror) in the DMD array by probing their nanotribological properties [55]. In these measurements, it is hard to tilt the stuck micromirror back to its normal position by adding a modest normal load at the rotatable corner of the micromirror. Figure 17.11 shows such a stuck mirror amid of an array of normally movable mirrors. Upon finding such a stuck mirror, the surface was repeatedly scanned at a large normal load, up to 300 nN. After several scans, the stuck micromirror can be removed and subsequently the surrounding micromirrors also could be removed by continuous scanning under a large normal load.

The measurements of the adhesive force of the landing site underneath the stuck micromirror and the normal micromirror clearly reveal significantly higher adhesion in the case of stuck mirror compared to the normal movable mirrors. Figure 17.11 shows the topography of landing sites of understuck in comparison with normal micromirrors. The landing site under the stuck micromirror exhibits a U-shaped wear mark, which is surrounded by a smeared area, thus indicating the nature of the fatigue mechanism behind the failed element.

References

1 Ikeda, S., Shimada, T., Kiguchi, M., and Saiki, K. (2007) Visualization of induced charge in an organic thin-film transistor by cross-sectional potential mapping. *J. Appl. Phys.*, **101** (9), 094509.

2 Tal, O., Rosenwaks, Y., Roichman, Y., Preezant, Y., Tessler, N., Chan, C.K., and Kahn, A. (2006) Threshold voltage as a measure of molecular level shift in organic thin film transistors. *Appl. Phys. Lett.*, **88** (4), 043509.

3 Jaquith, M., Muller, E.M., and Marohn, J.A. (2007) Time-resolved electric force microscopy of charge trapping in polycrystalline pentacene. *J. Phys. Chem. B.*, **111** (27), 7711–7714.

4 Muller, E.M. and Marohn, J.A. (2005) Microscopic evidence for spatially homogeneous charge trapping in pentacene. *Adv. Mater.*, **17** (11), 1410–1414.

5 Kemerink, M., Timpanaro, S., de Kok, M.M., Meulenkamp, E.A., and Touwslager, F.J. (2004) Three-dimensional inhomogeneities in PEDOT: PSS films. *J. Phys. Chem. B.*, **108** (49), 18820–18825.

6 Nardes, A.M., Kemerink, M., Janssen, R.A.J., Bastiaansen, J.A.M., Kiggen, N.M.M., Langeveld, B.M.W., van Breemen, A.J.J.M., and de Kok, M.M. (2007) Microscopic understanding of the anisotropic conductivity of PEDOT: PSS thin films. *Adv. Mater.*, **19** (9), 1196–2000.

7 Pingree, L.S.C., Macleod, B.A., and Ginger, D.S. (2008) The changing face of PEDOT:PSS films: substrate, bias, and processing effects on vertical charge transport. *J. Phys. Chem. C*, **112** (21), 7922–7927.

8 Silveira, W.R. and Marohn, J.A. (2004) Microscopic view of charge injection in an organic semiconductor. *Phys. Rev. Lett.*, **93** (11), 116104.

9 Burgi, L., Richards, T.J., Friend, R.H., and Sirringhaus, H. (2003) Close look at charge carrier injection in polymer field-effect transistors. *J. Appl. Phys.*, **94** (9), 6129.

10 Coffey, D.C. and Ginger, D.S. (2006) Time-resolved electrostatic force microscopy of polymer solar cells. *Nat. Mater.*, **5**, 735–740.

11 Coffey, D.C., Reid, O., Pingree, L.C.S., and Ginger, D. (2007) Understanding nanostructured solar cell performance with time-resolved electrostatic force microscopy, in Laser Science, OSA Technical Digest (CD), Optical Society of America, paper LTuL1.

12 Yu, Y.-J., Zhao, Y., Ryu, S., Brus, L.E., Kim, K.S., and Kim, P. (2009) Tuning the graphene work function by electric field effect. *Nano Lett.*, **9** (10), 3430–3434.

13 Lee, J., Liao, A., Pop, E., and King, W.P. (2009) Electrical and thermal coupling to a single-wall carbon nanotube device using an electrothermal nanoprobe. *Nano Lett.*, **9** (4), 1356–1361.

14 Lau, C.N., Stewart, D.R., Williams, R.S., and Bockrath, M. (2004) Direct observation of nanoscale switching centers in metal/molecule/metal structures. *Nano Lett.*, **4** (4), 569–572.

15 Wang, Z.L. (2008) Towards self-powered nanosystems: from nanogenerators to nanopiezotronics. *Adv. Funct. Mater.*, **18** (22), 3553–3567.

16 Wang, A.L. and Song, J. (2006) Piezoelectric nanogenerators based on zinc oxide nanowire arrays. *Science*, **312** (5771), 242–246.

17 Gao, P.G., Song, J.H., Liu, J., and Wang, Z.L. (2007) Nanowire piezoelectric nanogenerators on plastic substrates as

flexible power sources for nanodevices. *Adv. Mater.*, **19** (1), 67–72.

18 Minami, T. (2005) Transparent conducting oxide semiconductors for transparent electrodes. *Semicond. Sci. Tech.*, **20** (4), S35–S44.

19 Gu, G., Bulovic, V., Burrows, P.E., Forrest, S.R., and Thompson, M.E. (1996) Transparent organic light emitting devices. *Appl. Phys. Lett.*, **68** (19), 2606–2608.

20 Brumbach, M., Veneman, P.A., Marrikar, F.S., Schulmeyer, T., Simmonds, A., Xia, W., Lee, P., and Armstrong, N.R. (2007) Surface composition and electrical and electrochemical properties of freshly deposited and acid-etched indium tin oxide electrodes. *Langmuir*, **23** (22), 11089–11099.

21 Coffey, D.C., Reid, O.G., Rodovsky, D.B., Bartholomew, G.P., and Ginger, D.S. (2007) Mapping local photocurrents in polymer/fullerene solar cells with photoconductive atomic force microscopy. *Nano Lett.*, **7** (3), 738–744.

22 Pingree, L.S.C., Rodovsky, D.B., Coffey, D.C., Bartholomev, G.P., and Ginger, D.S. (2007) Scanning Kelvin probe imaging of the potential profiles in fixed and dynamic planar LECs. *J. Am. Chem. Soc.*, **129** (51), 15903–15910.

23 Ato, M., Takahashi, T., Okigawa, Y., and Mizutani, T. (2009) Conductance of individual channels in a carbon nanotube field-effect transistor studied by magnetic force microscopy. *J. Appl. Phys.*, **106** (11), 114315.

24 Jiang, Y. and Guo, W. (2009) Temperature dependence of the stability of written bits in a magnetic hard-disk medium investigated by magnetic force microscopy. *J. Magn. Magn. Mater.*, **321** (18), 2963–2965.

25 Maier, M.S. (2007) *Plasmonics: Fundamentals and Applications*, 1st edn, Springer, New York.

26 Ozbay, E. (2006) Plasmonics: merging photonics and electronics at nanoscale dimensions. *Science*, **311** (5758), 189–193.

27 Hao, E. and Schatz, G.C. (2004) Electromagnetic fields around silver nanoparticles and dimers. *J. Chem. Phys.*, **120** (1), 357–366.

28 Gersten, J. and Nitzan, A. (1980) Electromagnetic theory of enhanced Raman scattering by molecules adsorbed on rough surfaces. *J. Chem. Phys.*, **73** (7), 3023–3037.

29 Willets, K.A. and Van Duyne, R.P. (2007) Localized surface plasmon resonance spectroscopy and sensing. *Annu. Rev. Phys. Chem.*, **58**, 267–297.

30 Barnes, W.L., Dereux, A., and Ebbesen, T.W. (2003) Surface plasmon subwavelength optics. *Nature*, **424**, 824–830.

31 Ebbesen, T.W., Lezec, H.J., Ghaemi, H.F., Thio, T., and Wolff, P.A. (1998) Extraordinary optical transmission through sub-wavelength hole arrays. *Nature*, **391**, 667–669.

32 Ditlbacher, H., Hohenau, A., Wagner, D., Kreibig, U., Rogers, M., Hofer, F., Aussenegg, F.R., and Krenn, J.R. (2005) Silver nanowires as surface plasmon resonators. *Phys. Rev. Lett.*, **95** (25), 257403(4).

33 Pyayt, A.L., Wiley, B., Xia, Y., Chen, A., and Dalton, L. (2008) Integration of photonic and silver nanowire plasmonic waveguides. *Nat. Nanotechnol.*, **3**, 660–665.

34 Lal, S., Link, S., and Halas, N.J. (2007) Nano-optics from sensing to waveguiding. *Nat. Photonics*, **1**, 641–648.

35 Nikolajsen, T., Leosson, K., and Bozhevolnyi, S.I. (2004) Surface plasmon polariton based modulators and switches operating at telecom wavelengths. *Appl. Phys. Lett.*, **85** (24), 5833–5835.

36 Cai, W., White, J.S., and Brongersma, M.L. (2009) Compact, high-speed and power-efficient electrooptic plasmonic modulators. *Nano Lett.*, **9** (12), 4403–4411.

37 Srituravanich, W., Pan, L., Wang, Y., Sun, C., Bogy, D.B., and Zhang, X. (2008) Flying plasmonic lens in the nearfield for high-speed nanolithography. *Nat. Nanotechnol.*, **3**, 733–737.

38 Rontzsch, L., Heinig, K.H., Schuller, J.A., and Brongersma, M.L. (2007) Thin film patterning by surface plasmon induced thermocapillarity. *Appl. Phys. Lett.*, **90** (4), 044105.

39 Stenzel, O., Stendal, A., Voigtsberger, K., and von Borczyskowski, C. (1995) Enhancement of the photovoltaic conversion efficiency of copper phthalocyanine thin film devices by incorporation of metal clusters. *Sol. Energy Mater. Sol. Cells*, **37** (3–4), 337–348.

40 Atwater, H.A. and Polman, A. (2010) Plasmonics for improved photovoltaic devices. *Nat. Mater.*, **9**, 205–213.

41 Anker, J.N., Hall, W.P., Lyandres, O., Shah, N.C., Zhao, J., and Van Duyne, R.P. (2008) Biosensing with plasmonic nanosensors. *Nat. Mater.*, **7**, 442–453.

42 Rang, M., Jones, A.C., Zhou, F., Li, Z.-Y., Wiley, B.J., Xia, Y., and Raschke, M.B. (2008) Optical near-field mapping of plasmonic nanoprisms. *Nano Lett.*, **8** (10), 3357–3363.

43 McNeill, C.R., Frohne, H., Holdsworth, J.L., Furst, J.E., King, B.V., and Dastoor, P.C. (2004) Direct photocurrent mapping of organic solar cells using a near-field scanning optical microscope. *Nano Lett.*, **4** (2), 219–223.

44 Dennler, G., Scharber, M.C., and Brabec, C.J. (2009) Polymer-fullerene bulk-heterojunction solar cells. *Adv. Mater.*, **21** (13), 1323–1338.

45 Luzinov, I., Julthongpiput, D., Malz, H., Pionteck, J., and Tsukruk, V.V. (2000) Polystyrene layers grafted to epoxy-modified silicon surfaces. *Macromolecules*, **33** (3), 1043–1048.

46 Mate, C.M. and Homola, A.M. (1997) *Micro/Nanotribology and Its Applications* (ed. B. Bhushan), Kluwer, Dordrecht, The Netherlands, p. 647.

47 Ma, X., Gui, J., Grannen, K.J., Smoliar, L.A., Marchon, B., Jhon, M.S., and Bauer, C.L. (1999) Spreading of PFPE lubricants on carbon surfaces: effect of hydrogen and nitrogen content. *Tribol. Lett.*, **6** (1), 9–14.

48 Zhao, Z., Bhushan, B., and Kajdas, C. (1999) Tribological performance of PFPE and X-1P lubricants at head-disk interface. Part II. Mechanics. *Tribol. Lett.*, **6** (2), 141–148.

49 Singer, E. and Pollack, H. (eds) (1992) *Fundamentals of Friction*, Kluwer, Dordrecht, The Netherlands.

50 Rabinowicz, E. (1995) *Friction and Wear of Materials*, John Wiley & Sons, Inc., New York, p. 1965.

51 Gellman, A.J. (1998) Lubricants and overcoats for magnetic storage media. *Curr. Opin. Colloid Interface Sci.*, **3** (4), 368–372.

52 Bao, G. and Li, S.F. (1998) Lubricant effect on noncontact-mode atomic force microscopy images of hard-disk surfaces. *Langmuir*, **14** (5), 1263–1271.

53 Tsukruk, V.V. (2001) Molecular lubricants and glues for micro- and nanodevices. *Adv. Mater.*, **13** (2), 95–108.

54 Bhushan, B. (2007) Nanotribology and nanomechanics of MEMS/NEMS and BioMEMS/BioNEMS materials and devices. *Microelectron. Eng.*, **84**, 387–412.

55 Liu, H. and Bhushan, B. (2004) Nanotribological characterization of digital micromirror devices using an atomic force microscope. *Ultramicroscopy*, **100** (3–4), 391–412.

18
Nanolithography with Intrusive AFM Tip

18.1
Introduction to AFM Nanolithography

In this chapter, we provide a brief overview of the most recent and interesting results on SPM-based nanolithography and nanopatterning techniques applicable to soft materials.

As is well known, one of the important sources of artifacts of AFM imaging of soft matter is the physical perturbation and ultimate plastic deformation of the compliant surface being imaged under relatively high forces due to the intrusive nature of the AFM tip, which exerts local stresses (normal and shear). The same intrusive nature of the AFM tip under high shear and/or normal loads (especially in contact mode), which might result in few GPa local pressure, forms the solid basis for the scanning probe nanolithography of soft materials, an important and relatively recent development. The ability to control the applied force precisely and down to a few pN, the high spatial resolution reaching only a few nanometers, and small dimensions of the resulting deformable regions down to a few tens of nanometers, all are critical for the development of numerous nanomanipulation and nanopatterning techniques that are based on various scanning probe modes.

Historically, scanning tunneling microscope was the first to be used to demonstrate the unique ability for highly controlled manipulation of individual atoms and nanocrystals and form predetermined patterns of the same on the surface [1–3]. STM imaging revealed the Xenon atoms reconstructed on Ni {110} surface to form the letters IBM [1]. Scanning tunneling lithography has also been employed as a nanolithographic technique, through local modification of a resist layer using a beam of electrons. One of the important limitations of scanning tunneling nanolithography is the high vacuum requirement to avoid electrical breakdown between the tip and the surface and the extremely low temperature required to reduce atom mobility. Furthermore, due to the extremely small penetration of the low-energy electrons, STM-based lithography imposes additional constraints in the form of thin resist layers. There have been a few studies where STM lithography has been performed at room temperature by significantly lowering the voltage on the tip to avoid breakdown, but this further lowers the maximum thickness of the resist that

Scanning Probe Microscopy of Soft Matter: Fundamentals and Practices, First Edition.
Vladimir V. Tsukruk and Srikanth Singamaneni.

can be employed [4]. We will not consider scanning tunneling lithography here further and the readers may refer to several comprehensive reviews on this subject [5–7].

Here, we primarily focus on the application of AFM-based techniques as lithographic and patterning tools considering the wide applicability of these techniques to soft dielectric matter. There are generally several common methods of AFM-based nanolithography that can be classified as mechanical nanolithography, electrostatic nanolithography, oxidative nanolithography, thermomechanical nanolithography, and dip-pen nanolithography, naming just the most widely exploited modes. These approaches rely on different physical or chemical means for forming nanoscale patterns on a surface with the AFM tip [8–11].

Many of the AFM nanolithography techniques mentioned above involve the application of relatively high loads (which in turn translate to high local pressures, up to hundreds of MPa – several GPa) to cause ultimate mechanical (plastic) deformation of the surface. Other AFM nanolithographical techniques might cause highly localized chemical reactions under the tip to create surface features or local instabilities under an electrical field. For example, mechanical nanolithography, which historically has been introduced first, includes surface deformations by plowing, scribing, scratching, shaving, indentation, and controlled displacement (manipulation) of materials and exploits relatively large mechanical loads to achieve surface patterns. The load or pressure required for mechanical lithography depends on the mechanical properties of the soft material, the surface properties, and the tip–surface interactions. For example, for shaving off alkanethiol self-assembled monolayers on Au {111}, the surface load (pressure) required is approximately 20–50 nN (which corresponds to 0.2–0.5 GPa) [12]. For mechanical patterning of polymer surfaces, the applied local pressure needs to be higher than the yield stress of the polymer, which is usually within the range of several MPa – several hundred MPa [13–15].

On the other hand, oxidative nanolithography relies on the highly localized oxidation of the surface by the anodic reaction enabled by the surface water layer or the capillary formed between the tip and the surface. Unlike mechanical nanolithography, oxidative nanolithography does not directly rely on high mechanical forces. In contrast, electrostatic and thermomechanical nanolithography rely on softening the polymer surface layer by local heating, followed by the application of local mechanical forces (relatively small) to inscribe patterns of deformation into the softened material. Finally, dip-pen nanolithography, which involves the localized deposition of desired chemical species on a surface with a nanoscale resolution using the "dirty" AFM tip as a quill pen, will be discussed in detail in Chapter 19 [16].

18.2
Mechanical Lithography

AFM-based mechanical lithography involves the physical manipulation (plowing, scribing, scratching, indentation, and hammering) of soft matter surfaces to create a designed microscopic pattern with controlled lateral and vertical dimensions with a nanoscale resolution [17–23]. Mechanical nanolithography is primarily performed both in the indentation regime and in the static and dynamic scanning modes. In the simplest indentation mode, the AFM tip is brought into physical contact with the surface and the surface is indented at a constant load above a certain threshold to plastically deform the surface. In the static contact mode, the AFM tip is dragged at a high force to create user defined patterns of plastically deformed regions on the surface. Finally, in the dynamic mode the surface deformation is performed under tapping mode conditions (i.e., with the cantilever oscillating close to the resonance frequency of the cantilever) with high amplitude damping.

The tip–sample forces are minimized during imaging soft matter in the regular tapping mode by maximizing the set point ratio (r_{sp}) and minimizing the amplitude of oscillation, as was discussed in previous chapters. For mechanical lithography, the same parameters are adjusted to maximize the force in order to achieve sufficient local pressure to plastically deform the surface [18]. Similar to imaging, in dynamic mode of surface plowing, pattern instabilities caused by the torsional deflection of the cantilever are minimized compared to the static mode. The resolution of the AFM-generated patterns that can be achieved using this technique is defined by both the pressure applied and the sharpness of the AFM tip employed. The lateral dimensions of the surface features that can be achieved by conventional microfabricated AFM tips are usually on the order of a few tens of nanometers, while the vertical dimensions mostly depend on the normal load applied during the fabrication and can vary from a few nanometers to hundreds of nanometers.

Several clear and common disadvantages of mechanical lithography are the possibility of tip damage (less important for soft materials) and contamination (critical issue for soft materials) due to the direct physical contact and high forces between the tip and the surface of the highly deformed sample. The damage and contamination of the tip causes pattern irregularities due to change in the tip size, shape, and physical properties.

Mechanical nanolithography has been extensively applied to various organic SAMs, where the AFM tip is used to scratch off molecular monolayers without damaging stiff substrates in predetermined regions to create molecular grooves. We will highlight several important examples and interested readers may refer to several comprehensive reviews dedicated to this subject [24–26]. The process often termed as nanoshaving involves the mechanical scratching of the molecules adsorbed onto the surface using relatively high shear forces to break the chemical bond between the molecules and the surface. The force (load) required to perform the scratching is typically determined by a trial and error method, which involves the successive increase of the load while monitoring changes in the local structure (by scanning a slightly larger surface region, i.e., zooming out). While too small forces do not result in removal of the adsorbed molecules on the surface, too large forces can result in plastic deformation or damage of the AFM tip.

The nanoshaving approach has been widely demonstrated for both thiol and silane SAMs [27–29]. As the monolayers of the organic molecules are scratched off, the molecules are removed from the surface and are suspended in the corresponding solvent (typically ethanol) surrounding the AFM tip, thus leaving a clean pit under ideal conditions. However, care has to be taken to ensure clean and sharp patterns of the molecules on the surface by avoiding ripping part of the molecule or merely changing the orientation of the molecule and readsorption of the removed molecules. To avoid readsorption of the molecules in the shaved locations, the adsorbate molecules have to be highly soluble in the liquid medium surrounding the AFM tip. In the case of thiol molecules, the liquid medium is typically an alcohol such as ethanol or 2-butanol, as the thiols have limited solubility in water. On the other hand, controlled nanoshaving of silane-based SAMs can be performed in water owing to the fast hydrolization of these molecules in water.

Sometimes, such molecular cleavage is combined with the deposition of molecular species with different functionalities or physical properties as well as other nanostructures to create complex bicomponent SAM surfaces, called nanografting. The process typically involves performing nanoshaving in the presence of different molecular species in the solvent around the AFM tip. This results in the assembly of the new molecular species in the scratched regions resulting in nanoscale patterns of the molecules [30–32].

In earlier studies, Xu *et al.* have demonstrated the nanografting of alkanethiol molecules of various lengths and functionalities on Au {111} surfaces [31]. The authors employed thiols with various chain lengths from 2 to 37 carbons (C_2SH, C_6SH, $C_{16}SH$, $C_{22}SH$, and $C_{18}OC_{19}SH$) for nanografting. The authors grafted a rectangular $C_{18}S$ pattern ($20 \times 60\,nm^2$) within a $C_{10}S$ matrix by scratching the SAM. Following the initial nanografting, the solution was then replaced with a C_{22}-SH solution, and a second rectangular pattern ($28 \times 70\,nm^2$) of $C_{22}S$ was also grafted in an adjacent location (40 nm apart) (see Figure 18.1) [31]. The $C_{18}S$ and $C_{22}S$ patterns were parallel to each other and were 0.80 and 1.17 nm higher than the surrounding $C_{10}S$ matrix, respectively. This demonstration clearly revealed the ability to produce multiple patterns regiospecifically from different adsorbates. In other work, Amro *et al.* demonstrated AFM complex mechanical nanolithography by combining nanografting and dip-pen nanolithography (DPN) (see Chapter 19 for more discussion on DPN-related lithographical techniques) [33].

Heyde *et al.* have demonstrated mechanical nanolithography in tapping mode AFM using high set points [21]. The authors justly pointed out that mechanical lithography offers a distinct advantage in that the lithographic features could be imaged in a nondestructive manner by just increasing the set point ratio (applying light tapping mode) eliminating the need for a change in the mode of scanning, change in the tip, reidentification of the location, or even change in the instrument. In particular, the authors demonstrated dynamic plowing lithography (DPL) on PMMA glassy films. Figure 18.2 shows the AFM image of the three lines written in the PMMA film with high forces and the corresponding line profile obtained in light tapping. The AFM also reveals that the tip embossed a grooved and carved PMMA surface, accumulating material on the sides of the groove. The width of the grooves were measured to be 40 ± 3 nm, 38 ± 4 nm, and 37 ± 2 nm, which depended on the shape and dimensions of the AFM tip and the local surface properties. On the other hand, the depth of the three lines was found to be around 3.4 ± 0.4 nm for all three

(a)

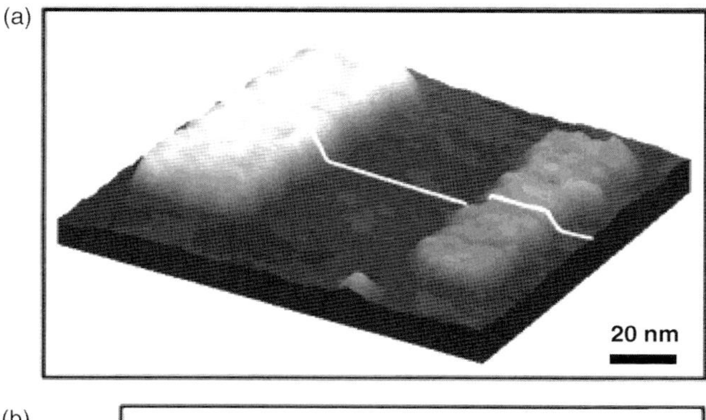

(b)

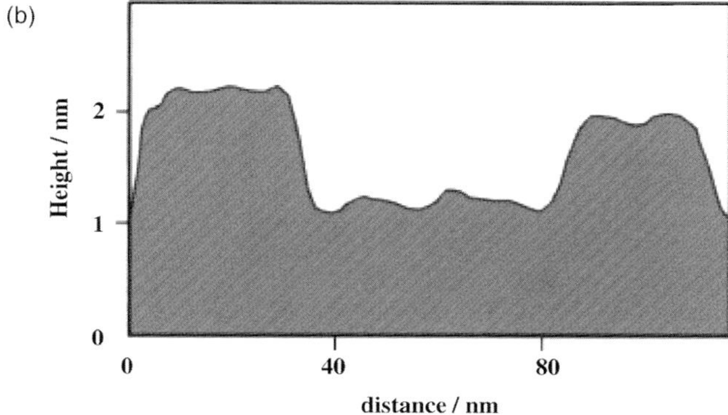

Figure 18.1 (a) Topographic images of a $C_{10}S$ matrix with a $25 \times 60\,nm^2$ $C_{22}S$ and $20 \times 60\,nm^2$ $C_{18}S$ nanoislands patterned using a nanografting approach. (b) A corresponding cursor profile showing that the surfaces of the $C_{22}S$ and $C_{18}S$ islands are 1.17 and 0.85 nm, respectively, higher than the matrix monolayer. The displacement force was 15 nN and the scan rate was 50 μm/s. The vertical origin in the cursor profile is the surface of gold, whose position is determined by displacing the adsorbed thiol layer. Reproduced from Ref. [31].

lines. The depth of the line depends on the amplitude set point used during the plowing process and remains stable for different surface areas. In fact, the depth of the linear indentations was found to increase linearly with the amplitude set point.

In another study, Cappella and coworkers have investigated two different methods of mechanical lithography, namely, dynamic plowing lithography and regular nanoindentation [17, 18]. The authors studied two common amorphous glassy polymers, PMMA and PS. Figure 18.3 shows the plot of the indentation depth as a function of the ratio of the amplitude of oscillation in writing and imaging modes for DPL and with applied force in the case of the nanoindentation mode. The insets in the figure also depict the AFM images showing the corresponding indentations generated in the glassy polymer surface.

While the authors found no fundamental differences in the efficiency of the indentation (depth of the features as a function of loading parameters) between the two methods, they noted that the DPL process is a much faster and more practical

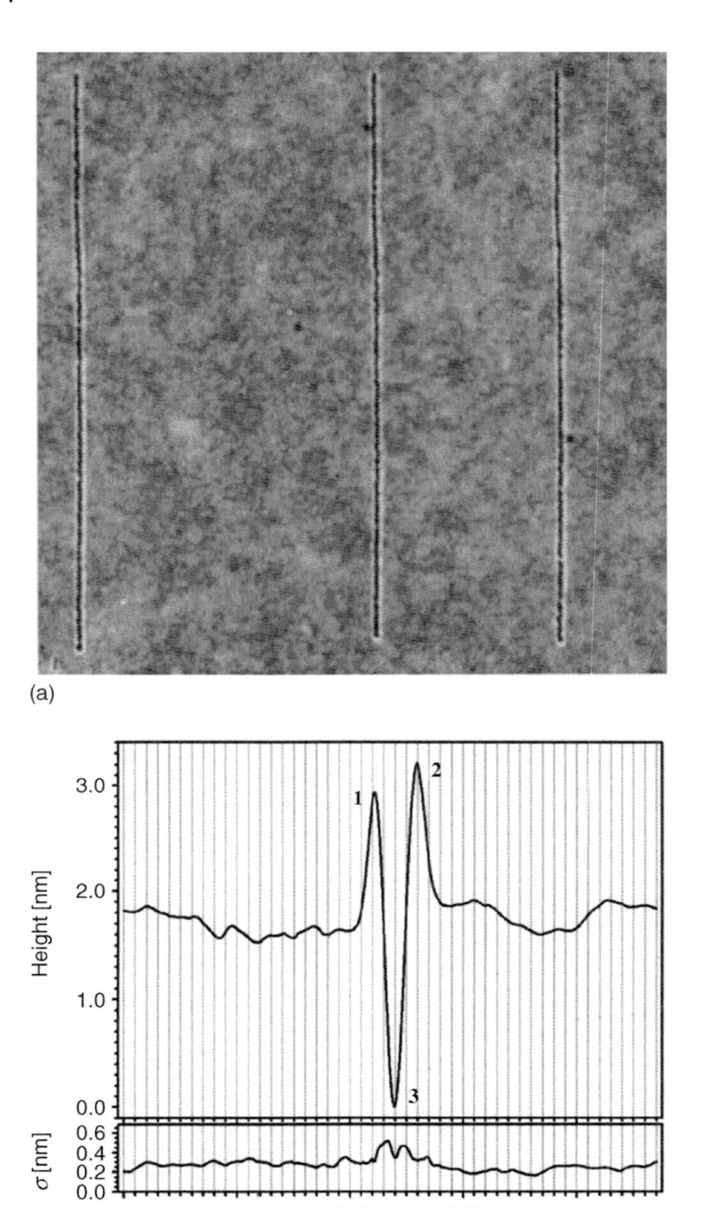

Figure 18.2 AFM height image showing the lines written in PMMA ($3 \times 3\ \mu m^2$, z-scale = 4.8 nm). (b) Cross-section line, which is the average and standard deviation of 100 scan lines. Reproduced from Ref. [21].

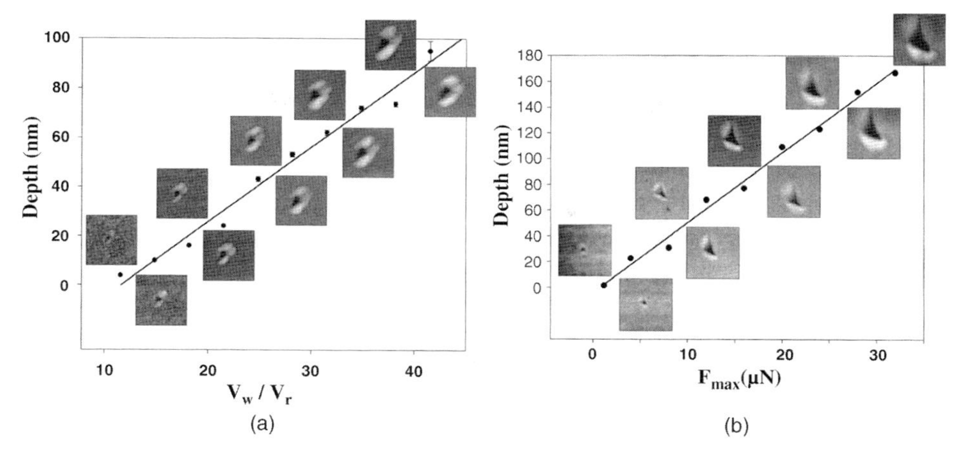

Figure 18.3 (a) Plot of the depth of holes written by means of dynamic plowing lithography versus the ratio of the amplitude of oscillation in writing and imaging modes. The insets show tapping mode topographies of single holes. All insets are 700 × 700 nm. (b) Plot of the depth of holes written by means of nanoindentation versus F_{max}. The insets show tapping mode topographies of single holes. All insets are 700 × 700 nm. Reproduced from Ref. [17].

approach compared to the regular indentation method. Yet another important difference between both methods was that the border walls that surrounded the lithographed structure in the DPL process were found to be much larger than those formed during indentation. Furthermore, the border walls were found to have a larger volume compared to the carved out material. This difference in the volume was ascribed to the much looser polymer structure of the border walls due to the fast plowing process. The energy dissipated into the polymer during plowing was recognized to be sufficient to break covalent bonds and cause plastic deformation of the polymer. Furthermore, from the force–displacement curves, the adhesion properties of the modified and unmodified polymer (hydrophobic) were found to be significantly different owing to the chain scission during the plowing process, and the subsequent oxidation of the materials that converts the highly deformed regions selected to highly hydrophilic areas.

In one of the earlier studies, Tsukruk *et al.* have investigated the morphological transition in LB bilayers from polyglutamate (PG) deposited on a silicon surface and on cadmium arachidate LB multilayers while scanning the surface under different normal loads and at different tip velocities [34]. Interaction of the AFM tip with the bilayer surface generally led to surface roughening for PG bilayers deposited directly on silicon. In contrast, PG bilayers deposited on cadmium arachidate multilayers displayed a greater mechanical stability during scanning with the AFM tip. This roughening of the PG bilayers was found to be more severe when scanning perpendicular to the dipping direction, which is also the direction of the rigid PG backbone alignment. The roughness was also found to vary with the scan rate due to continuous surface damage. The initial roughness increased slightly with the scan rate for tip velocities of 0.2–8 μm/s and then dropped sharply, suggesting the existence of a critical scan velocity below which a stick-slip mechanism is active in damaging the soft surface.

Periodic heterogeneities on the surface such as that observed in this work are very common and could be formed by friction instabilities of the sliding surfaces of the so-called stick-slip type [35]. In this type of movement, changes in the surface sliding mode are connected with periodic transitions between "static" (solid-like) and "kinetic" (fluid-like) states, which are analogous to freezing–melting transitions [35, 36]. Minor variation in shearing force is needed to produce expected jumps from one state to another. In their study on PG layers, several nanoscale features (down to 10 nm across) were produced in the PG bilayers using the AFM tip, each feature produced being sensitive to the anisotropic nature of the LB layers.

Manipulation of various single macromolecules on solid substrates has been demonstrated using SPM-based nanomechanical lithography as well. DNA has been extensively employed as a model molecule for the demonstration of the nanomanipulation using sharp SPM tips [37]. Manipulation procedures included cutting, moving, and bending the DNA to form complex patterns of the same on the surface of the substrate. The typical procedure involves the deposition of the DNA on the surface of a (modified) substrate followed by tapping mode imaging of the surface to locate the target molecules for manipulation. Subsequently, the same region is scanned again and the set point is lowered (causing hard tapping) at selected regions. Several such scans are performed to result in cutting the DNA at localized and predetermined regions. Following the cutting process, the shape of DNA can be manipulated by sweeping these molecules, which are weakly adsorbed to the surface by shear forces.

Hu *et al.* have demonstrated the formation of complex DNA patterns on surface by using a combination of molecular combing and SPM mechanical nanolithography [38]. This combination is a powerful tool due the complementary length scales of the manipulation. The molecular combing process, which involves the alignment of the DNA (or any rigid one-dimensional structures such as carbon nanotubes or metal nanowires) under fluid flow, results in large-area uniform orientation of the molecules [39, 40]. The spacing between the molecules can be controlled by adjusting the concentration of the DNA in the solution. Following such alignment, intricate complex patterns and different shapes of the DNA were achieved using SPM nanolithography as described above. Figure 18.4 shows the tapping mode AFM image of the DNA molecules that were initially aligned by the molecular combing followed by cutting the DNA at desired location to form the isolated rhombohedron patterns on the surface.

SPM-based mechanical nanomanipulation has also been performed on individual carbon nanotubes to show the feasibility of using this approach for building functional electronic devices [41–44]. For example, Roschier *et al.* have fabricated a single-electron transistor by manipulating individual carbon nanotubes on the substrate [45]. The process involved scanning the surface in conventional tapping mode with the feedback loop on, followed by positioning the tip to scan a single line to move the nanotube end with the feedback loop switched off. The tip–surface distance was decreased in small steps (1–10 nm) and the vibration amplitude of the cantilever was monitored to observe the location of the nanotube *in situ*. In the initial stages, the tip was pushed very hard against the surface in order to overcome the nanotube–

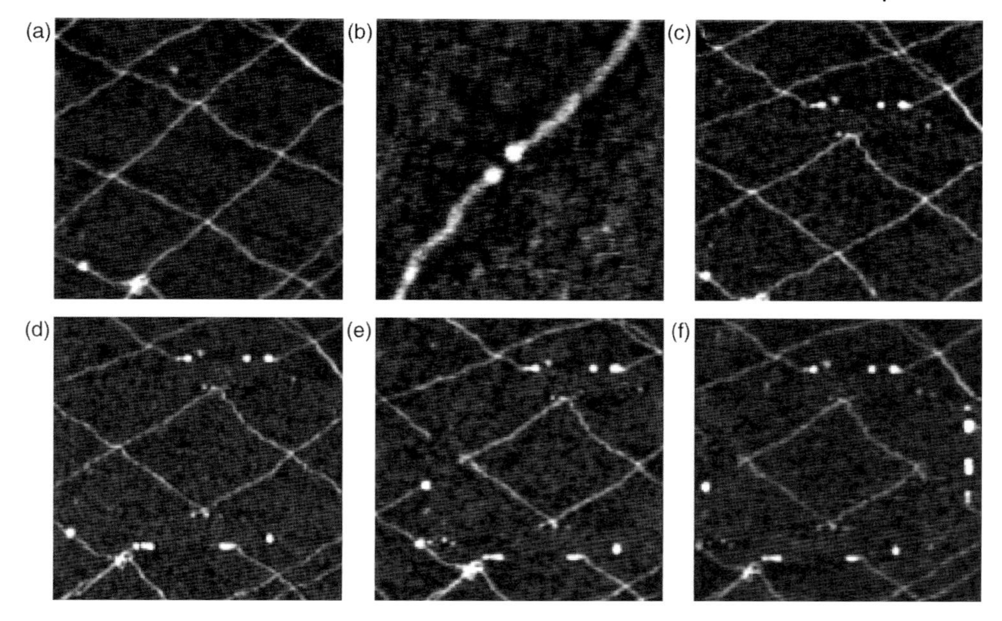

Figure 18.4 Tapping mode AFM image of a DNA network on mica formed by the molecular combing technique followed by SPM-based cutting. Reproduced from Ref. [38].

surface interactions (stiction) followed by moving in small rotations to pivot the nanotube about one end. In other words, the SPM manipulation involved iterative dragging and pushing of the nanotube over small distances (around 1 μm) to bridge it across the predeposited electrodes. The electrical characteristics of the device built with these SPM tools exhibited single-electron charging effect with a bandgap of 15 meV for the semiconducting carbon nanotubes.

18.3 Local Oxidative Lithography

The formation of localized oxide layers for surface patterning and controlling the doping level and metallization is a common step in the conventional photolithography process. The ability to form complex oxide structures using the AFM tip is naturally an important and powerful process in AFM-based nanolithography. The oxide layers formed by the AFM tip on the surface can act as dielectric barriers, masks for selective etching, and serve as nanoscale surface templates for further nanofabrication and patterning. In the local oxide layer forming process, the AFM tip is used as a cathode and the water meniscus formed between the AFM tip and the sample surface serves as anode with the electrolyte forming a nanoelectrolytic cell. The water meniscus is controlled by the electrical field induced between the tip and the sample through the application of voltage or just by the physical contact between the tip and the surface.

Fine control over the meniscus size down to 20 nm is routinely possible with this approach, which enables the formation of sub-10 nm oxide structures. The lateral dimensions of the oxide structures scale linearly with the applied voltage while the height of these dimensions follows power law dependence. The voltage applied during local oxidation is typically between 5–30 V and the time of application of the voltage varies from 5 ms to 1 s. Although oxidation was originally demonstrated for a {111} silicon surface, it has been extended to a wide variety of metal (titanium, tantalum, aluminum, molybdenum, and nickel), semiconductor, and even organic surfaces (as discussed later). It is this versatility of the local oxidation using the AFM probe that makes it such a powerful and widely employed technique. In this section, we limit the discussion to several examples of local oxidation as applied to soft matter (molecular assemblies, polymers, and biomolecules).

Local oxidation of molecular assemblies enables the formation of both positive and negative features by changing the bias applied to the AFM tip. Although a negative tip bias (tip as cathode) is more commonly used to form positive (elevated) features on the surface, Lee *et al.* demonstrated that the negative features can also be achieved by reversing the bias on the AFM tip [46]. By applying + 10 V voltage on the tip, the authors formed grooves in a monolayer of palmitic acid formed on a silicon surface using LB deposition. The measured depth of the grooves was found to be 1.2 nm, which was equal to the length of the palmitic acid molecule. By changing the voltage, the authors demonstrated that the depth of the grooves could be controlled. This control over the groove size suggested the mechanism for the formation of the groves to be a gradual degradation of the organic film under large local electrical field as opposed to other possibilities such as chemical reaction causing the degradation of the molecular layer under ambient conditions.

As another example, Sagiv and coworkers have demonstrated an elegant method for the modification of SAMs using the local oxidation method [47, 48]. Their approach involved the application of voltage to oxidize the terminal methyl groups of OTS SAM into carboxylic groups, followed by forming nonadecenyltrichlorosilane with functional terminal $-CH{=}CH_2$ groups in the modified parts. In the next step, a further chemical treatment with H_2S was performed to convert ethylenic surface groups and form a thiol-terminated SAM layer. The authors demonstrated that gold clusters can be selectively adsorbed onto the local thiol- or amine-terminated regions (Figure 18.5). In fact, very intricate patterns such as a Picaso painting could be generated by raster scanning the tip and applying the voltage (around 8.5 V) point by point. Figure 18.5 shows the AFM image of the gold nanoclusters (visible at higher magnification) patterned along the amine-terminated regions of the silicon surface with the locally functionalized SAM formed in a similar approach.

Yoshinobu *et al.* have demonstrated the patterning of protein (ferritin) using local oxidation of a silicon surface by the AFM tip [49]. Figure 18.6 shows the schematic of the positive and negative patterning of the proteins adapted in this study. In the negative patterning, the silicon surface was modified with γ-APTES, to form amine termination (proteinphilic surface regions). Oxide patterns were formed on the silicon surface by applying controlled voltage between the gold-coated AFM tip and the substrate. Two different patterns (positive and negative) of adsorbed proteins

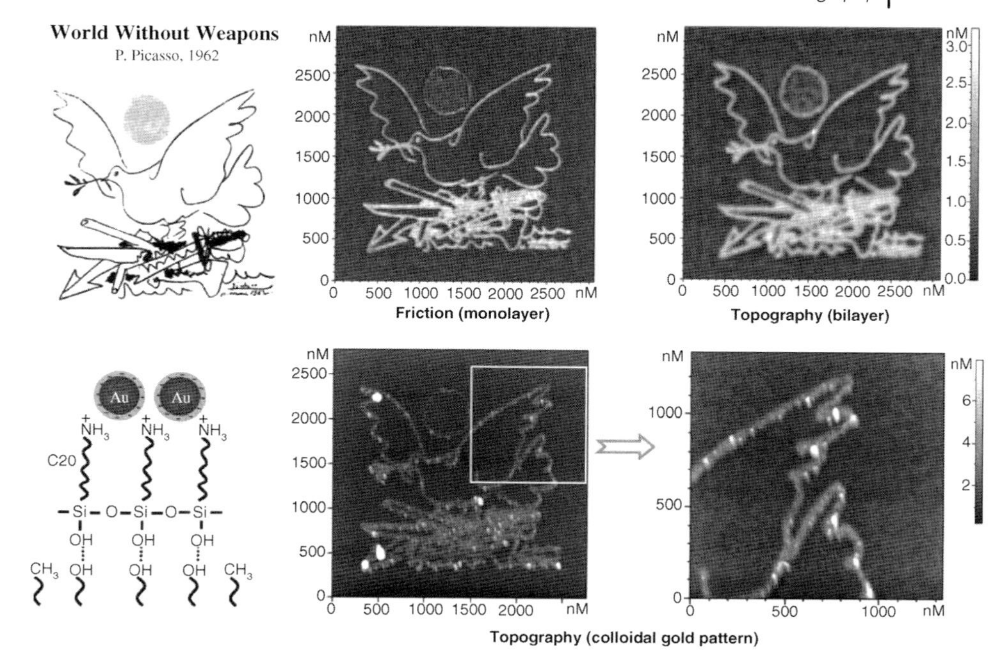

Figure 18.5 Tapping mode AFM images showing an example of complex metal–organic nanoarchitecture fabricated via the template-guided hierarchical self-assembly route. Reproduced from Ref. [48].

were produced using the oxide pattern as a template. During the AFM tip-based local oxidation process, the original amine termination is locally lost. Exposure of the surface to protein solution resulted in preferential adsorption of the protein to amine terminated regions, while the pretreated regions remained pristine.

Figure 18.7 shows the AFM of the protein pattern achieved by the negative patterning approach. In the positive patterning approach, a silicon surface was modified with OTS SAM to form hydrophobic methyl termination. The OTS SAM was locally removed during local oxidation followed by exposure to γ-APTES to result in protein-philic amine termination in these regions. Exposure of the patterned surface to protein solution resulted in preferential adsorption of proteins in amine-terminated regions (see Figure 18.7). This study demonstrated that by employing essentially the same surface chemistry, both positive and negative patterning of various biomolecules with high lateral resolution can be archived.

18.4 Electrostatic Nanolithography

Electrostatic nanolithography involves the application of an electric field using a conductive AFM tip to the polymer surface causing highly localized deformation in

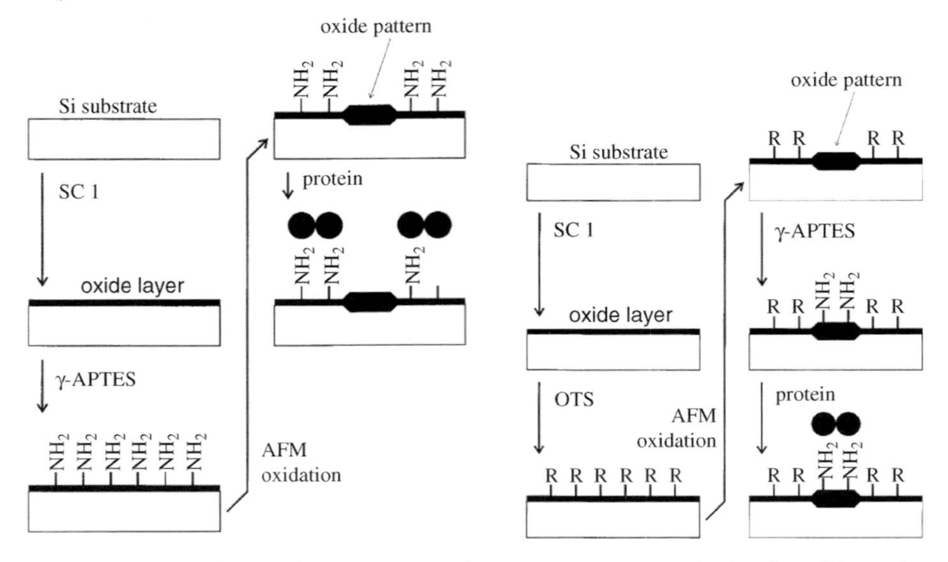

Figure 18.6 Schematic showing positive and negative patterning using local oxidation lithography. Reproduced from Ref. [49].

the polymer surface. In an original study, Lyuksyutov *et al.* demonstrated nanopatterning of the polymer film using localized Joule heating and dielectrophoretic manipulation achieved by passing electrical current between a conductive (tungsten carbide) AFM tip and a conductive substrate [50]. Although the polymer itself was nonconductive, current flow through a thin polymer film caused by the dielectric

Figure 18.7 AFM images showing positive and negative patterned ferritin proteins on silicon surface. Reproduced from Ref. [49].

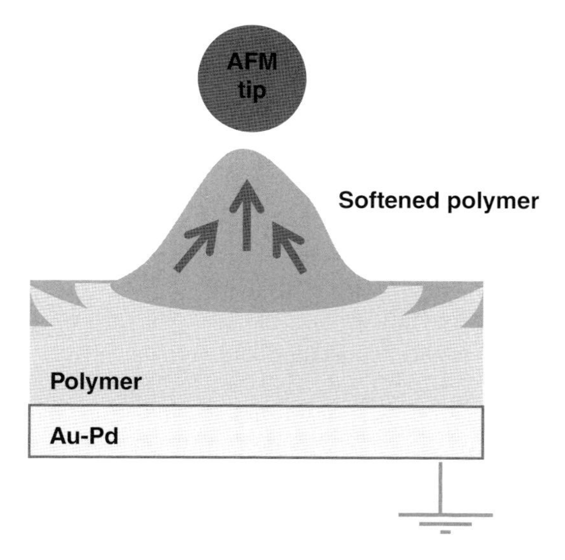

Figure 18.8 Schematic showing the AFM tip and film geometry with the conductive AFM tip resulting in a conductive cylinder in the polymer film due to Joule heating and softening of the polymer. The softened polymer film is deformed under the strong electric field resulting in dots on the surface.

breakdown of the polymer film resulted in highly localized Joule heating (few attoliters). The polymer in this localized region is above the glass transition temperature, significantly softened. The strong non-uniform electric field (10^8–10^9 V/m) causes the polarization and electrostatic attraction of the softened polymer toward the AFM tip resulting in raised surface structures.

Figure 18.8 shows the schematic of the AFM tip interaction with electrical field resulting in local softening of the polymer films followed by the flow of the softened polymer toward the AFM tip (overcoming the Laplace pressure) due to the strong electric field applied. AFM imaging was employed to monitor the periodic array of various nanoscale features (lines, undulating lines, and grooves) inscribed into a PMMA film with this approach.

The process of surface modification is extremely fast and efficient and enables a wide variety of nanostructures to be engraved into the film within a few minutes. For example, polymer dots (raised features) were formed by pausing the AFM tip for 0.2–5 s with a constant bias. On the other hand, lines were created by continuous scanning at tip velocities from 0.1 μm/s to 8 μm/s. The study also demonstrated that electrostatic nanolithography can be employed to form patterns in a broad range of polymers with different physicochemical properties, although the feature sizes produced (1 nm–1 μm) strongly depended on the magnitude of the current through the polymer film.

In a subsequent study, the same group demonstrated amplitude-modulated AFM-assisted electrostatic nanopattering in ultrathin glassy polymer (PMMA, PS) films [51]. The study revealed that the lateral dimensions of the polymer dots could be varied from 10 to 50 nm by controlling the tip–film separation and the bias voltage applied. Furthermore, the aspect ratio of the dots achieved by the amplitude-modulated method was found to be significantly higher compared to that achieved by the earlier contact mode method. The aspect ratio exhibited a linear dependence on the bias voltage. The authors pointed out that during the high frequency oscillation (200–400 kHz) of the cantilever, timescale of the tip–surface contact is extremely small (microsecond range), providing enhanced control over current injection and electronic breakdown of the film and eliminating the lateral forces as in the case of conventional tapping mode AFM scanning. Furthermore, controlling the oscillation frequency provides the ability to control heat generation and subsequently the polymer film deposition.

In a later study, the authors derived an analytical relationship for describing the electric field and the electrostatic pressure associated with the electric field between the charged AFM tip and the dielectric polymer surface [52]. The electric field between the tip and the surface of the polymer is due to the charges accumulated either on the surface of the film or on the AFM tip through triboelectrification or an applied external bias. The authors realized two distinct nanolithography mechanisms that can occur during the electric field-assisted nanolithography. In the first case where electric breakdown occurs in the polymer film, the surface features are formed by the mass transport of the locally heated and softened polymer film. The authors demonstrated that the lateral size of the features, which are typically a few tens of nanometers (10–100 nm) in height, are governed by tip–surface separation and the glass transition of the polymer film.

In the second process, the AFM tip comes extremely close (less than 1 nm) to the polymer surface and the electrostatic pressure, yielding the polymer and creating an irreversible (plastic deformation) feature in the polymer surface. Unlike the earlier process, which is not severely affected by the tip shape, the later approach depends strongly on the AFM tip shape and can be performed only in contact AFM mode. The lateral dimensions of the surface features created are typically between 10 and 30 nm and the z-dimensions (height) are typically between 0.1 and 1 nm. As pointed out by the authors, the second approach is amenable for the formation of surface features in polymer films with arbitrary thickness as the method does not rely on the breakdown of the polymer film and the local Joule heating.

In another study, Juhl *et al.* presented an improved nanolithographical approach, termed z-lift electrostatic lithography, to form nanoscale features on thin PS films (within 10–50 nm) [53]. In this method, the z-position and the z-lift rate of the cantilever were modulated (extended or retracted) during the application of voltage to precisely control the height of the nanostructures. z-lift electrostatic lithography was performed by systematically varying the parameters that included bias voltage (0–40 V), voltage ramp (0.1–0.2 s), and z-lift rate (0.5–2.0 μm/s), resulting in variable-height nanostructures with excellent precision and repeatability. The authors noted that the height of the features was mainly controlled by and is proportional to the z-lift magnitude and only a narrow range of voltages yield stable surface structures for any given film thickness. Similar to the earlier approach, the polymer is not removed or cross-linked during structure formation, making the surface features erasable during annealing.

Chung *et al.* have demonstrated the creation of Taylor cones in a glassy polymer (PMMA surface) using higher tip voltages of 30–60 V [54]. Under these conditions, the strong electric field between the tip and the polymer surface initiates electrohydrodynamic (EHD) instability and nanofluidic motion of the polymer melt. Local EHD instability also results in unstable surface waves formed under electrostatic pressure and the decoupling of the vertical transport from these lateral surface waves, which results in the formation of the conical surface structures. Each surface feature formed this way is comprised of a Taylor cone surrounded by a circular groove. Figure 18.9 shows such a Taylor cone consisting of a central cone with height of 50 nm and full-width at half-maximum of 150 nm and a surrounding circular groove with a diameter of a few microns. While a single central cone was observed for sharp AFM tips, the formation of multiple central cones with one common groove was observed for blunt AFM tips.

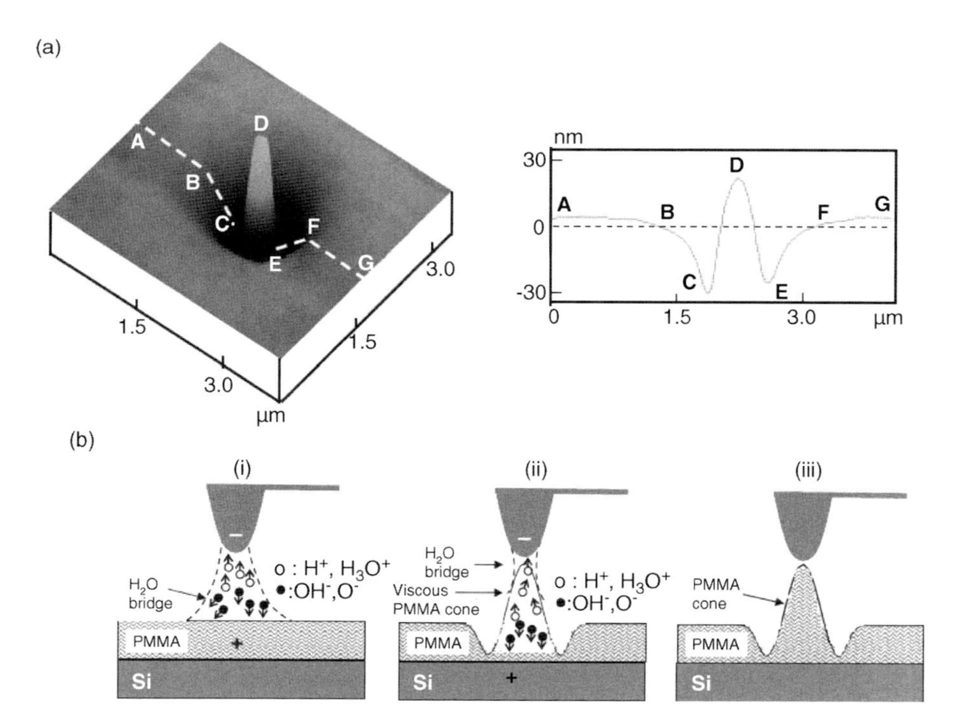

Figure 18.9 (a) AFM image showing the Taylor cone, the circular groove, and the corresponding cross section of the same. (b) Schematics showing the sequential formation of the conical structure on PMMA: (i) ionic dissociation of the water bridge upon the application of a bias on the AFM tip; (ii) ionic conduction in the polymer melts generated by local Joule heating; (iii) formation of a polymeric Taylor cone pattern after deactivating the tip bias. Reproduced from Ref. [54].

In further development, Xie *et al.* have investigated the conduction mechanism by monitoring the current–voltage and current–time curves during the formation of the Taylor cone under different humidity conditions [55]. Their measurements revealed that the charge transport is dominated by water bridge-assisted ionic conduction. With the application of voltage bias to the tip, the water meniscus between the tip apex and the surface dissociated into H^+ and OH^- ions. The ionic current resulted in local Joule heating, in turn leading to polymer melting, and the water bridge forms the media for electrical conduction and polymer mass transport.

The schematic in Figure 18.9 shows the dissociation of the water meniscus and the formation of the Taylor cone suggested in this work. Perceivably, the electrical conduction and the joule heating cease when the bias on the tip is removed. The ionic conductivity and the structure of the Taylor cone formed on the polymer surface were found to be highly sensitive to humidity conditions, which directly affect the water bridge (meniscus) between the AFM tip and the substrate. When the humidity is reduced, the electric conductivity is lower while higher humidity enhances water condensation between the tip and the surface and subsequent water dissociation, leading to higher H^+ and OH^- concentrations, thus resulting in larger conductivity. Higher humidity also helps to enhance carrier mobility in polymer melts, as the polymer viscosity would be lower if it was heated by an initial high ionic current, resulting in higher Taylor cones under high-humidity conditions.

Martin *et al.* have demonstrated electric field-assisted nanopatterns in a polymer film deposited on a silicon substrate [56]. The authors demonstrated that the application of voltages above 28 V resulted in a hole in the 30 nm PMMA layer and concurrent oxidation of the silicon substrate underneath. In fact, the silicon surface underneath was found to be oxidized even in cases when no holes were formed in the PMMA layer. The authors suggested that the holes in the polymer film were formed by the local electrochemical reaction resulting in the transport of OH^- groups through the polymer film.

Figure 18.10 shows the tapping mode AFM images of the line pattern formed in the PMMA film exposing the silicon substrate. Subsequent deposition of the metal layer (aluminum) followed by lift-off of the PMMA layer resulted in metal patterning on the polymer surface. The image shows the incomplete removal of the PMMA film during the lift-off process. In other words, this study demonstrated that electric field-assisted nanolithography can be combined with conventional lithographic approaches such as metal deposition and lift-off processes to realize nanoscale metal features on the surface of the substrate.

An alternative approach to electrostatic nanolithography involves the creation of a "nanoexplosion" under the AFM tip to modify materials beneath. Xie *et al.* demonstrated the highly localized (nanoscale) explosion and shock wave generation in a nanometer-sized air/water media caused by a voltage-biased AFM probe (Figure 18.11) [57]. When the tip was biased and brought in proximity to a dielectric surface, the electric field between the tip and the surface exceeded the dielectric breakdown strength, causing the explosive discharge of the air/water media in the tip–substrate gap. The nanoexplosion was demonstrated to be highly localized in the vicinity of the tip apex and resulted in the formation of a tiny surface feature.

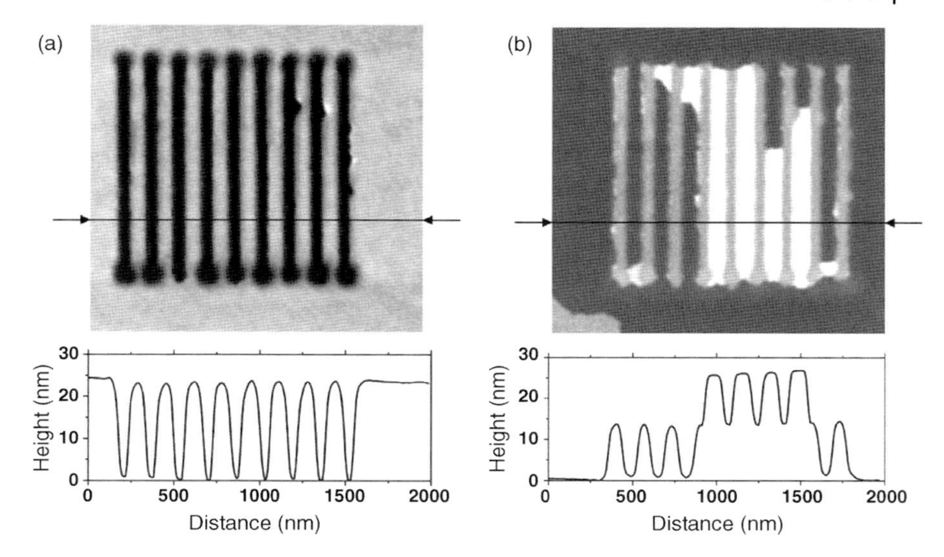

Figure 18.10 AFM nanolithography performed on a 25 nm PMMA layer deposited on silicon. (a) AFM image after the lithography process without any further development step. For drawing the lines, the applied voltage is 30 V and the tip is moved at 0.5 μm/s with the feedback disabled. (b) AFM image of the same pattern after depositing 8 nm of aluminum and a lift-off process. AFM image shows the resist has not been completely eliminated. Reproduced from Ref. [56].

Figure 18.11 shows the general schematic of the nanoexplosion mechanism suggested in this study and a corresponding AFM image of the resulting surface feature formed.

Apart from the tiny central surface feature, which can be formed under the biased tip, the nanoexplosion could also result in a strong transient shock wave that propagates along the surface of the dielectric film, expanding the trajectory of discharged species out of the explosion zone. The shock wave causes the formation of an outer ring surrounding the central structure formed under the AFM tip (Figure 18.11). Nanoexplosion-based nanolithography has been performed on different polymer surfaces such as PS and poly(vinyl carbazole). The authors also employed finite elemental method simulation to show that the nanoexplosion is highly localized and the electric field was confined to a surface area with a radius of 80 nm and the charge distribution was mostly confined to a radius of 25 nm.

18.5 Thermomechanical Nanolithography

Thermomechanical nanolithography is another widely used approach that involves locally heating the polymer surface using an AFM tip as a heat source with

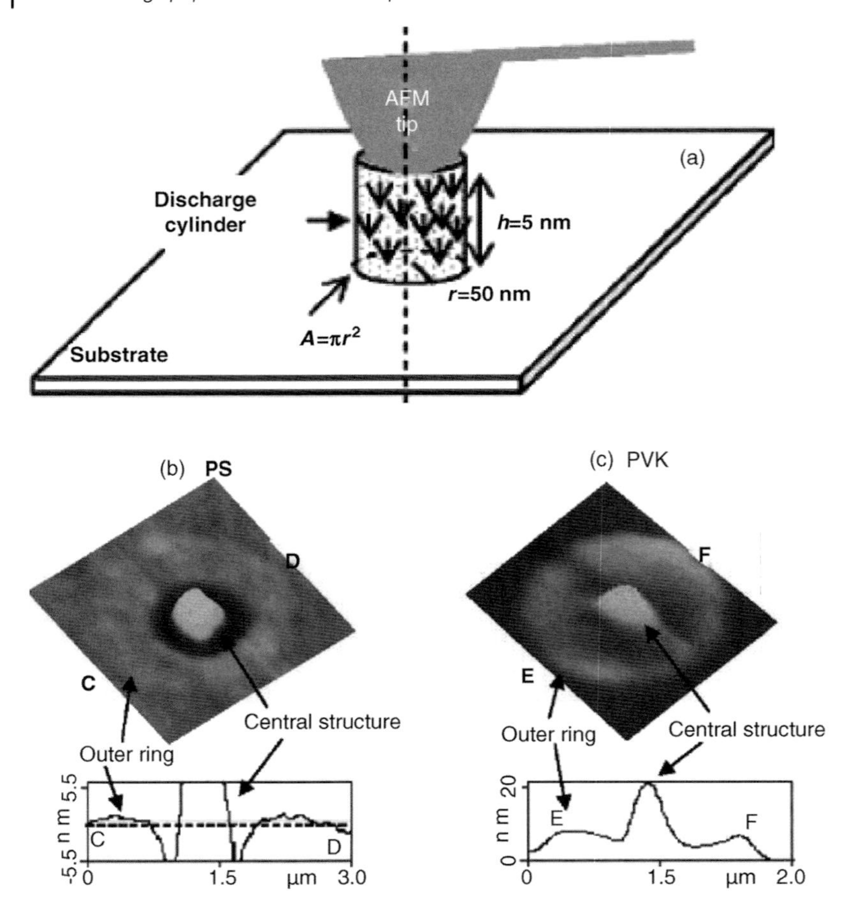

Figure 18.11 (a) Geometry of the discharge device consisting of the AFM tip, substrate, and the discharge cylinder with a height of (h) 5 nm and basal area of $A = \pi r^2$, where (r) = 50 nm. AFM height images showing the nanoexplosion performed on (b) PS with a 35 V bias and a 2 s duration, and (c) poly(vinyl carbazole) with a 10 V bias and a 2 s duration. Reproduced from Ref. [57].

simultaneous nanoindentation into the locally softened polymer surface. This approach utilizes scanning thermal microscopy instrumentation and principles that were discussed in Chapter 6 [58, 59].

A combination of the heat and mechanical force causes the polymer to soften and flow, resulting in patterning of the polymer, but more frequently writing data bits using the polymer film as a storage medium. This technique, pioneered by the IBM research team, has been extensively investigated by them and others [60–63]. This nanolithographical technique involves the use of a special cantilever in which the legs of the cantilever are highly doped silicon to enable high electrical conductivity, while the tip region is poorly doped, resulting in high resistivity in these regions.

The high resistance of the tip causes resistive heating of the tip when electrical current is passed through the cantilever.

As mentioned above, the writing process in this mode involves bringing the preheated tip in direct physical contact with a polymer film deposited on a substrate. The substrate can be a hard silicon substrate, which prevents the tip from indenting beyond the recording medium (thickness of the polymer film) and enables the rapid transfer of heat away from the heated region owing to the inherent high thermal conductivity of the substrate. It is worth noting that one of the issues in using stiff silicon material as the substrate is the exceedingly high wear of the AFM tip due to the repeated direct contact of the tip with the hard substrate under high forces. One way to overcome this issue is to use a thin epoxy layer (SU8) between the silicon and the recording polymer layer. It has been demonstrated that in the initial stages, when the tip–surface contact area is extremely small, a very small fraction (0.2%) of the heat is utilized for softening the polymer material. Most of heat is lost by fast dissipation through the cantilever legs into the chip hosting the cantilever and as radiation through the air gap between the tip and the surface.

After the softening of the polymer surface, the contact area between the tip and the polymer surface starts increasing, thus resulting in higher heat transfer to the polymer material finally reaching significant fraction of about 2%. It has been demonstrated by Vettiger *et al.* that the physical behavior of the polymer follows the well-established time–temperature superposition principle, which states the interdependence of the time and temperature in the thermomechanical behavior of polymeric materials [64]. In particular, the authors showed that the threshold temperature versus heating time in the writing process followed the established time–temperature superposition. Data bits as small as 40 nm have been carved using this technique, which corresponds to a prospective data storage density of 400 Gb/in^2.

The cantilever design, which was originally used for thermomechanical writing can also be used for reading the data bits from the recording medium. For reading the data bits, the tip is also maintained at higher temperature ($\sim$350 °C), although this temperature is much lower than the tip temperature during writing process. The heat flow between the substrate and the heating platform causes changes in the temperature of the resistor and a change in the resistance of the heater, which is monitored for reading of the bits. The reading mechanism is based on the fact that the thermal conductance between the recording substrate and the heater platform critically depends on the distance between them with the medium (typically air) in between them transporting the heat. When the tip is at the nonindented position, the thermal conductance between the heating platform and the substrate is smaller than in the case of an indent position due to the smaller distance between the tip and the substrate in the latter case. The higher thermal conductance between the tip and the substrate results in a decrease in the temperature, hence a decrease in the resistance of the heater. Effectively, the cantilever resistance reaches two distinct values as it traverses over regions with and without indents, that is, “1” and “0,” respectively. Monitoring the resistance is a highly sensitive method for reading the thermomechanical data bits written into the polymer film.

In another related study, Vettiger *et al.* studied the shape of the pile-up regions that are formed during the thermal writing process into the polymer film [64]. One important conclusion from their study was that the pile-up regions originate not only from the simple volume conservation process but also due to the flash heating and rapid cooling of the polymer surface. Considering the relatively large pile-up regions formed during thermomechanical indentation, one of the important questions that would arise is the high spatial resolution that can be achieved using this technique. When the indentations are sufficiently far from each other, the pile-up regions do not interact and the individual indentations can be clearly resolved. As the indentations get closer, the pile-up region of the second indentation overlaps with the pitch of the first indentation, resulting in a significant decrease in the indentation depth. Finally, when the two indentations are even closer, the pile-up regions and the pitches merge, resulting in a single anisotropic indentation.

As discussed above, thermomechanical indentation of the polymers has been considered as a means for data storage technology in which a polymer is used as the recording medium. Although the small size of the indentations provides very high data storage density, the rate of operation of a single AFM tip is nearly three orders of magnitude slower compared to the magnetic storage medium. To overcome this problem, researchers at IBM have introduced the concept of *Millipede*, which involves the use of a 32×32 array of the preheated AFM tips [62, 65–67]. For such an array of cantilevers, 1024 indentations can be formed concurrently.

Either the entire cantilever array or the polymer recording medium can be scanned in the x- and y-directions similar to the conventional AFM mode. The contact between the probe array and the polymer sample is controlled by adjusting the z-piezo using only one feedback to control all the probes involved. To ensure uniform contact between the cantilever tips and the storage medium, additional approaching sensors, which provide the feedback signals, are integrated into the corners of the array chip. During data writing and reading, the chip is raster scanned over the storage field with each cantilever in the array operating (writing and reading) within its own storage field. For a 32×32 array within a 3×3 mm area, the storage field of each cantilever corresponds to 0.9 Gb, assuming the storage density to be 500 Gb/in^2.

Apart from high-density data storage, a heated AFM tip has also been applied to induce local chemical reduction of graphene oxide layers on a substrate to tune the electrical conductivity by nearly four orders of magnitude [68]. In this study, Wei *et al.* demonstrated the reduction of graphene oxide in highly localized regions, with a full-width at half-maximum of 12 nm, by traversing the heated AFM tip at speeds of a few μm/s and a normal load of $\sim$100 nN. The authors noted that the AFM tip-induced reduction did not induce any noticeable tip wear or sample tearing, suggesting a continuous carbon skeleton, which is important for the electrical properties. The degree of reduction of the graphene oxide, as verified by friction force microscopy, could also be controlled by the temperature of the AFM tip.

Following the reduction of the graphene oxide along an arbitrary pattern, the authors verified the local changes in the electrical properties of the layers using conductive AFM (CAFM). CAFM and topography images revealed zigzag reduction pattern performed on the epitaxial graphene oxide layer. The topography image clearly reveals the zigzag pattern with the reduced regions slightly lower (~1 nm) than the graphene oxide regions. This reduction in the thickness is due to the loss of the oxygen-rich functional groups. The CAFM image also revealed the high conductivity of the reduced regions showing high currents under constant tip bias of 2.5 V. Average profiles of the height and the conductivity reveal the narrow (FWHM of 12 nm) reduction patterns that were achieved using an AFM tip.

Finally, we can conclude that the combination of high resolution and versatility make the SPM-based nanofabrication as a powerful technique that is applicable to numerous classes of materials (metals, polymers, semiconductors, biomolecules, and molecular assemblies). However, various reports in literature are seemingly contradictory since the interactions between the AFM tip and the sample dramatically change with even small changes in the conditions (e.g., tip–sample forces, scanning speeds, and ambient humidity) under which the SPM lithography is performed. Fundamental understanding of various processes such as material transfer, mechanical deformation, and nanoscale chemical reactions in SPM-based nanolithography still remains incomplete. Deeper understanding of these local processes is important to enable reliable methods of SPM lithography for industrial applications.

One of the important considerations in AFM-based nanolithography techniques in general is the intense wear on the AFM tip during multiple indentations, which is one of the important sources of irregularities in the patterns. The AFM tip, which comes into intimate contact with the surface and in some cases under high normal and shear stresses, is prone to damage or, even more importantly, contamination resulting in significant changes in the size and shape of the tip. Several cleaning methods have been demonstrated to clean AFM tips by removing external contaminations. Some of these techniques have been discussed in Section 3.3.

In some cases of SPM nanolithography, such as thermomechanical indentation, special care is taken to avoid such damage. A buffer polymer layer is employed to avoid the contact of the silicon cantilever with the silicon surface and thus minimizing the wear on the tip. However, in the case of the mechanical force lithography technique, tip wear is a significant problem. Diamond and diamond-coated tips have been employed to avoid tip wear in mechanical lithography. Using these harder tips, mechanical lithography has been successfully applied to not only organic layers but also soft metals such as gold, copper, nickel, and silver [21, 69].

Yet another factor that limits the application of the SPM-based nanolithography is the serial approach of the technique making it a rather slow process. One of the ways this has been addressed is by replacing a single cantilever with an array of cantilevers and parallel operation of them significantly speeding up the process. This approach has been particularly successful in the case of thermomechanical nanoindentation (known as the Millipede approach) and dip-pen nanothography, which are discussed in Sections 18.5 and 19.3, respectively [70].

While the approaches reported so far clearly demonstrate the applicability of these techniques for laboratory-scale scientific investigations, scaling up these techniques to large-area applications will critically depend on the success of multiprobe approaches and minimizing the irregularities from different sources, some of them discussed above. Some of the issues related to pattern irregularities can be addressed by a rigorous experimental and theoretical investigation of the mechanisms involved in the nanolithography processes. These developments will enable tighter control over these critical variables in industrial nanofabrication.

Overall, SPM-based nanolithography and data storage applications critically depend on making the multicantilever arrays cost-effective and well integrated with the existing scanning technologies. Remarkable progress has been made in using multiple-probe methods and integrating these with other unconventional patterning approaches such as microcontact printing in the case of dip-pen nanolithography, which will be discussed in Chapter 19 [71]. Such progress is yet to happen in the mechanical, electrostatic, and electrochemical approaches to make them a robust tool viable for large-scale applications.

References

1 Eigler, D.M. and Schweizer, E.K. (1990) Positioning single atoms with a scanning tunnelling microscope. *Nature*, **344**, 524–526.

2 Zhang, J., Liu, J., Huang, J.L., Kim, P., and Lieber, C.M. (1996) Creation of nanocrystals through a solid–solid phase transition induced by an STM tip. *Science*, **274** (5288), 757–760.

3 Gerber, C. and Lang, H.P. (2006) How the doors to the nanoworld were opened. *Nat Nanotechnol.*, **1** (1), 3–5.

4 Xu, L.S. and Allee, D.R. (1995) Ambient scanning tunneling lithography of Langmuir–Blodgett and self-assembled monolayers. *J. Vac. Sci. Technol. B*, **13** (6), 2837–2840.

5 Hartwich, J., Sundermann, M., Kleineberg, U., and Heinzmann, U. (1999) STM writing of artificial nanostructures in alkanethiol-type self-assembled monolayers. *Appl. Surf. Sci.*, **144–145**, 538–542.

6 Tao, N.J., Li, C.Z., and He, H.X. (2000) Scanning tunneling microscopy applications in electrochemistry: beyond imaging. *J. Electroanal. Chem.*, **492** (2), 81–93.

7 Walsh, M.A. and Hersam, M.C. (2009) Atomic-scale templates patterned by ultrahigh vacuum scanning tunneling microscopy on silicon. *Annu. Rev. Phys. Chem.*, **60**, 193–216.

8 Wouters, D. and Schubert, U.S. (2004) Nanolithography and nanochemistry: probe-related patterning techniques and chemical modification for nanometer-sized devices. *Angew. Chem. Int. Ed.*, **43** (19), 2480–2495.

9 Garcia, R., Martinez, R.V., and Martinez, J. (2006) Nano-chemistry and scanning probe nanolithographies. *Chem. Soc. Rev.*, **35** (1), 29–38.

10 Sheehan, P.E. and Lieber, C.M. (1996) Nanomachining. manipulation and fabrication by force microscopy. *Nanotechnology*, **7** (3), 236–240.

11 Sheehan, P.E. and Lieber, C.M. (1996) Nanotribology and nanofabrication of MoO_3 structures by force microscopy. *Science*, **272** (5265), 1158–1161.

12 Liu, J.F., Von Her, J.R., Baur, C., Stallcup, R., Randall, J., and Bray, K. (2004) Fabrication of high-density nanostructures with an atomic force microscope. *Appl. Phys. Lett.*, **84** (8), 1359–1361.

13 Lee, W.K. and Sheehan, P.E. (2008) Scanning probe lithography of polymers: tailoring morphology and functionality at the nanometer scale. *Scanning*, **30** (2), 172–183.

14 Kassavetis, S., Mitsakakis, K., and Logothetidis, S. (2007) Nanoscale patterning and deformation of soft matter by scanning probe microscopy. *Mater. Sci. Eng. C*, **27** (5–8), 1456–1460.

15 Ward, I.M. and Sweeney, J. (2004) *An Introduction to the Mechanical Properties of Solid Polymers*, 2nd edn, John Wiley & Sons, Inc., New York.

16 Piner, R.D., Xu, F., Zhu, J., Hong, S., and Mirkin, C.A. (1999) Dip pen nanolithography. *Science*, **283** (5402), 661–663.
17 Cappella, B. and Sturm, H. (2002) Comparison between dynamic plowing lithography and nanoindentation methods. *J. Appl. Phys.*, **91** (1), 506–512.
18 Cappella, B., Sturm, H., and Weidner, S.M. (2002) Breaking polymer chains by dynamic plowing lithography. *Polymer*, **43** (16), 4461–4466.
19 Jung, T.A., Moser, A., Hug, H.J., Brodbeck, D., Hofer, R., Hidber, H.R., and Schwarz, U.D. (1992) The atomic force microscope used as a powerful tool for machining surfaces. *Ultramicroscopy*, **42–44**, 1446–1451.
20 Nie, H.-Y., Motomatsu, M., Mizutani, W., and Tokumoto, H. (1995) Local modification of elastic properties of polystyrene–polyethyleneoxide blend surfaces. *J. Vac. Sci. Technol. B*, **13** (3), 1163–1166.
21 Heyde, M., Rademann, K., Cappella, B., Geuss, M., Sturm, H., Spangenberg, T., and Niehus, H. (2001) Dynamic plowing nanolithography on polymethylmethacrylate using an atomic force microscope. *Rev. Sci. Instr.*, **72** (1), 136–141.
22 Xie, X.N., Chung, H.J., Sow, C.H., and Wee, A.T.S. (2006) Nanoscale materials patterning and engineering by atomic force microscopy nanolithography. *Mater. Sci. Eng. R*, **54** (1–2), 1–48.
23 Wang, Y., Hong, X., Zeng, J., Liu, B., Guo, B., and Yan, H. (2009) AFM tip hammering nanolithography. *Small*, **5** (4), 477–483.
24 Liu, G., Xu, S., and Qian, Y. (2000) Nanofabrication of self-assembled monolayers using scanning probe lithography. *Acc. Chem. Res.*, **33** (7), 457–466.
25 Liang, J., Rosa, L.G., and Scoles, G. (2007) Nanostructuring. Imaging and molecular manipulation of dithiol monolayers on Au (111) surfaces by atomic force microscopy. *J. Phys. Chem. C*, **111** (46), 17275–17284.
26 Rosa, L.G. and Liang, J. (2009) Atomic force microscope nanolithography: dip-pen, nanoshaving, nanografting, tapping mode, electrochemical and thermal nanolithography. *J. Phys. Condens. Matter*, **21**, 483001.
27 Vandamme, N., Snauwaert, J., Janssens, E., Vandeweert, E., Lievens, P., and Haesendonck, C.V. (2004) Visualization of gold clusters deposited on a dithiol self-assembled monolayer by tapping mode atomic force microscopy. *Surf. Sci.*, **558** (1–3), 57–64.
28 Chen, S. (2000) Self-assembling of monolayer-protected gold nanoparticles. *J. Phys. Chem. B*, **104** (4), 663–667.
29 de Boer, B., Frank, M.M., Chabal, Y.J., Jiang, W., Garfunkel, E., and Bao, Z. (2004) Metallic contact formation for molecular electronics: interactions between vapor-deposited metals and self-assembled monolayers of conjugated mono- and dithiols. *Langmuir*, **20** (5), 1539–1542.
30 Xu, S. and Liu, G.-Y. (1997) Nanometer-scale fabrication by simultaneous nanoshaving and molecular self-assembly. *Langmuir*, **13** (2), 127–129.
31 Xu, S., Miller, S., Laibinis, P.E., and Liu, G.-Y. (1999) Fabrication of nanometer scale patterns within self-assembled monolayers by nanografting. *Langmuir*, **15** (21), 7244–7251.
32 Xu, S., Laibinis, P.E., and Liu, G.-Y. (1998) Accelerating the kinetics of thiol self-assembly on GoldsA spatial confinement effect. *J. Am. Chem. Soc.*, **120** (36), 9356–9361.
33 Amro, N.A., Xu, S., and Liu, G.-Y. (2000) Patterning surfaces using tip-directed displacement and self-assembly. *Langmuir*, **16** (7), 3006–3009.
34 Tsukruk, V.V., Foster, M.D., Reneker, D.H., Schmidt, A., Wu, H., and Knoll, W. (1994) Stability and modification of polyglutamate Langmuir–Blodgett bilayer films. *Macromolecules*, **27** (5), 1274–1280.
35 Yoahkawa, H., McGuiggan, P., and Israelachvili, J.N. (1993) Identification of a second dynamic state during stick-slip motion. *Science*, **259** (5099), 1305–1308.
36 Yoshizawa, H., Chen, Y.-L., and Israelachvili, J. (1993) Fundamental mechanisms of interfacial friction. 1. Relation between adhesion and friction. *J. Phys. Chem.*, **97** (16), 4128–4140.

37 Li, M.Q., Hansma, H.G., Vesenka, J., Kelderman, G., and Hansma, P.K.J. (1992) Atomic force microscopy of uncoated plasmid DNA: nanometer resolution with only nanogram amounts of sample. *Biomol. Struct. Dyn.*, **10** (3), 607–610.

38 Hu, J., Zhang, Y., Gao, H., Li, M., and Hartmann, U. (2002) Lambda DNA molecules on a mica surface, mechanically arranged by manipulation with the probe of an atomic force microscope. *Nano Lett.*, **2** (1), 55–57.

39 Ouyang, Z., Hu, J., Chen, S.F., Sun, J.L., and Li, M.Q. (1997) Molecular patterns by manipulating DNA molecules. *J. Vac. Sci. Technol. B*, **15** (4), 1385–1387.

40 Ko, H., Peleshanko, S., and Tsukruk, V.V. (2004) Combing and bending of carbon nanotube arrays with confined microfluidic flow on patterned surfaces. *J. Phys. Chem.*, **108** (14), 4385–4393.

41 Falvo, M.R., Taylor, R.M., Helser, A., Chi, V., Brooks, F.P., Jr., Washburn, S., and Superfine, R. (1999) Nanometre-scale rolling and sliding of carbon nanotubes. *Nature (London)*, **397** (6716), 236–238.

42 Postma, H.W.C., Sellmeijer, A., and Dekker, C. (2000) Manipulation and imaging of individual single-walled carbon nanotubes with an atomic force microscope. *Adv. Mater.*, **12** (17), 1299–1302.

43 Postma, H.W.C., Jonge, M.D., Yao, Z., and Dekker, C. (2000) Electrical transport through carbon nanotube junctions created by mechanical manipulation. *Phys. Rev. B*, **62** (16), 10653–10656.

44 Hertel, T., Martel, R., and Avouris, P. (1998) Manipulation of individual carbon nanotubes and their interaction with surfaces. *J. Phys. Chem. B*, **102** (6), 910–915.

45 Roschier, L., Penttilä, J., Martin, M., Hakonen, P., Paalanen, M., Tapper, U., Kauppinen, E.I., Journet, C., and Bernier, P. (1999) Single-electron transistor made of multiwalled carbon nanotube using scanning probe manipulation. *Appl. Phys. Lett.*, **75** (5), 728–730.

46 Lee, H., Kim, S.A., Ahn, S.J., and Lee, H. (2002) Positive and negative patterning on a palmitic acid Langmuir–Blodgett monolayer on Si surface using bias-dependent atomic force microscopy lithography. *Appl. Phys. Lett.*, **81** (1), 138–140.

47 Maoz, R., Frydman, E., Cohen, S.R., and Sagiv, J. (2000) Constructive nanolithography: inert monolayers as patternable templates for *in-situ* nanofabrication of metal–semiconductor–organic surface structures: a generic approach. *Adv. Mater.*, **12** (10), 725–731.

48 Liu, S.T., Maoz, R., and Sagiv, J. (2004) Planned nanostructures of colloidal gold via self-assembly on hierarchically assembled organic bilayer template patterns with *in-situ* generated terminal amino functionality. *Nano Lett.*, **4** (5), 845–851.

49 Yoshinobu, T., Suzuki, J., Kurooka, H., Moon, W.C., and Iwasaki, H. (2003) Fabrication of oxide patterns and immobilization of biomolecules on Si surface. *Electrochim. Acta*, **48** (20–22), 3131–3135.

50 Lyuksyutov, S.F., Vaia, R.A., Paramonov, P.B., Juhl, S., Waterhouse, L., Ralich, R.M., Sigalov, G., and Sancaktar, E. (2003) Electrostatic nanolithography in polymers using atomic force microscopy. *Nat. Mater.*, **2** (7), 468–472.

51 Lyuksyutov, S.F., Paramonov, P.B., Juhl, S., and Vaia, R.A. (2003) Amplitude-modulated electrostatic nanolithography in polymers based on atomic force microscopy. *Appl. Phys. Lett.*, **83** (21), 4405–4408.

52 Lyuksyutov, S.F. and Paramonov, P.B. (2004) Induced nanoscale deformations in polymers using atomic force microscopy. *Phys. Rev. B*, **70** (17), 174110.

53 Juhl, S., Phillips, D., Vaia, R.A., Lyuksyutov, S.F., and Paramonov, P.B. (2004) Precise formation of nanoscopic dots on polystyrene film using z-lift electrostatic lithography. *Appl. Phys. Lett.*, **85** (17), 3836–3839.

54 Chung, H.J., Xie, X.N., Sow, C.H., Bettiol, A.A., and Wee, A.T.S. (2006) Polymeric conical structure formation by probe-induced electrohydrodynamical nanofluidic motion. *Appl. Phys. Lett.*, **88** (2), 023116.

55 Xie, X.N., Chung, H.J., Sow, C.H., Bettiol, A.A., and Wee, A.T.S. (2005) Water-bridge-assisted ionic conduction in

probe-induced conical polymer pattern formation. *Adv. Mater.*, **17** (11), 1386–1390.

56 Martín, C., Rius, G., Borrisé, X., and Pérez-Murano, F. (2005) Nanolithography on thin layers of PMMA using atomic force microscopy. *Nanotechnology*, **16** (8), 1016–1022.

57 Xie, X.N., Chung, H.J., Sow, C.H., Adamiak, K., and Wee, A.T.S. (2005) Electrical discharge in a nanometer-sized air/water gap observed by atomic force microscopy. *J. Am. Chem. Soc.*, **127** (44), 15562–15567.

58 Gorbunov, V.V., Fuchigami, N., and Tsukruk, V.V. (2000) Microthermal analysis with scanning thermal microscopy. I. Methodology and experimental. *Probe Microscopy*, **2** (1), 53–63.

59 Gorbunov, V.V., Fuchigami, N., and Tsukruk, V.V. (2000) Microthermal analysis with scanning thermal microscopy. II: Calibration, modeling, and interpretation. *Probe Microscopy*, **2** (1), 65–75.

60 King, W.P., Santiago, J.G., Kenny, T.W., and Goodson, K.E. (1999) Modeling and simulation of sub-micrometer heat transfer in AFM thermomechanical data storage. *ASME MEMS*, **1**, 583–588.

61 Binnig, G., Despont, M., Drechsler, U., Häberle, W., Lutwyche, M., Vettiger, P., Mamin, H.J., Chui, B.W., and Kenny, T.W. (1999) Ultrahigh-density atomic force microscopy data storage with erase capability. *Appl. Phys. Lett.*, **74** (9), 1329–1331.

62 Lutwyche, M.I., Despont, M., Drechsler, U., Dürig, U., Häberle, W., Rothuizen, H., Stutz, R., Widmer, R., Binnig, G.K., and Vettiger, P. (2000) Highly parallel data storage system based on scanning probe arrays. *Appl. Phys. Lett.*, **77** (20), 3299–3301.

63 Rowland, H.D. and King, W.P. (2007) Predicting polymer flow during high-temperature atomic force microscope nanoindentation. *Macromolecules*, **40** (22), 8096–8103.

64 Vettiger, P., Cross, G., Despont, M., Drechsler, U., Durig, U., Gotsmann, B., Haberle, W., Lantz, M.A., Rothuizen, H.E., Stutz, R., and Binnig, G.K. (2002) The "millipede": nanotechnology entering data storage. *IEEE Trans. Nanotechnol.*, **1** (1), 39–55.

65 Lutwyche, M., Andreoli, C., Binnig, G., Brugger, J., Drechsler, U., Häberle, W., Rohrer, H., Rothuizen, H., Vettiger, P., Yaralioglu, G., and Quate, C. (1999) 2D AFM cantilever arrays a first step towards a terabit storage device. *Sens. Actuators A Phys.*, **73** (1–2), 89–94.

66 Vettiger, P., Brugger, J., Despont, M., Drechsler, U., Dürig, U., Häberle, W., Lutwyche, M., Rothuizen, H., Stutz, R., Widmer, R., and Binnig, G. (1999) Ultrahigh density, high-data-rate NEMS-based AFM data storage system. *J. Microelectron Eng.*, **46** (1–4), 11–17.

67 Despont, M., Brugger, J., Drechsler, U., Dürig, U., Häberle, W., Lutwyche, M., Rothuizen, H., Stutz, R., Widmer, R., Binnig, G., Rohrer, H., and Vettiger, P. (2000) VLSI-NEMS chip for parallel AFM data storage. *Sens. Actuators A Phys.*, **80** (2), 100–107.

68 Wei, Z., Wang, D., Kim, S., Kim, S.-Y., Hu, Y., Yakes, M.K., Laracuente, A.R., and Dai, Z. (2010) Nanoscale tunable reduction of graphene oxide for graphene electronics. *Science*, **328** (5984), 1373–1376.

69 Klehn, B. and Kunze, U. (1999) Nanolithography with an atomic force microscope by means of vector-scan controlled dynamic plowing. *J. Appl. Phys.*, **85** (7), 3897–3903.

70 Hong, S., Zhu, J., and Mirkin, C.A. (1999) Multiple ink nanolithography: toward a multiple-pen nano-plotter. *Science*, **286** (5439), 523–525.

71 Huo, F.W., Zheng, Z.J., Zheng, G.F., Giam, L.R., Zhang, H., and Mirkin, C.A. (2008) Polymer pen lithography. *Science*, **321** (5896), 1658–1660.

19
Dip-Pen Nanolithography

19.1
Basics of the Ink and Pen Approach

In the previous chapter, we discussed popular SPM-based nanolithography techniques such as mechanical, electrostatic, electrochemical, and thermomechanical, all of which rely on the delivery of some form of energy/forces/stresses to the surface under the SPM tip to cause local physical or chemical changes, which are preserved after releasing the local stimulus. In contrast, dip-pen nanolithography (DPN), which has been briefly mentioned earlier, involves the highly localized delivery or deposition of mobile molecules or materials on the surface as mediated by a properly treated SPM tip. In the DPN method, the SPM tip is used as a microfabricated pen with a nanoscale tip for direct writing of chemically distinct micropatterns on a surface with sub-20 nm resolution under ideal deposition conditions. The ultimate resolution of the pattern depends on the nature of the ink, surface chemistry of the substrate, speed of the writing process, ambient humidity, volume of the meniscus, and temperature as will be discussed in this chapter.

DPN is a constructive technique in which the wetted SPM tip is utilized as a nanoscale quill pen tip to deliver and deposit properly tuned ink (chosen chemical species) with nanoscale resolution using the liquid meniscus naturally formed between the tip and the surface (Figure 19.1) [1]. The molecules delivered to the surface in this manner might physisorb or chemisorb on the properly selected surface areas forming nanoscale patterns behind the continuously moving sharp tip.

Ever since the DPN technique was introduced by Mirkin's group in the late 1990s, a wide variety of organic and inorganic chemical inks have been deposited on various surfaces, proving it to be a versatile patterning technique [2]. The variety of inks that have been successfully deposited include alkylthiols [3–6], alkoxysilanes [7], alkynes [8], ferrocenylthiols [9], silazanes [10], sols [11, 12], proteins [13–16], metals salts [17–19], DNA [20, 21], conjugated polymers [13, 22, 23], and colloidal particles [24, 25], just to name a few of the most popular choices. Table 19.1 summarizes a wide variety of substrate–ink combinations that have been successfully

Scanning Probe Microscopy of Soft Matter: Fundamentals and Practices, First Edition.
Vladimir V. Tsukruk and Srikanth Singamaneni.

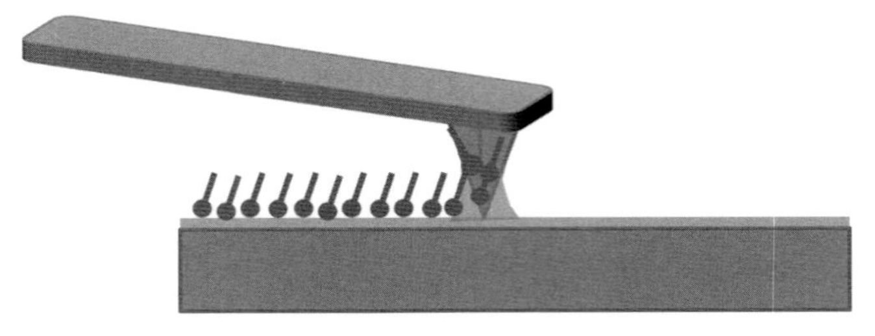

Figure 19.1 Schematic showing the DPN process where the molecules on the tip are deposited on the surface through meniscus-mediated transport.

employed in DPN within the first few years after the introduction of the technique with many more combinations available now [26].

There have been extensive efforts to gain a fundamental understanding of molecular transport from the AFM tip to the surface, which plays a key role in determining the resolution, pattern uniformity, and reliability of this direct writing technique. The molecular transport is often a complex interplay involving numerous physical and chemical factors that depend upon the chemical nature of the ink and substrate, ambient humidity, temperature, size and shape of the tip, and distribution of the ink on the tip. In most of the cases, these various factors are interrelated in a rather complex manner.

For example, the effect of humidity on the rate of deposition depends on the nature of the ink used in the DPN experiment. Indeed, the deposition rate of mercaptohexadecanoic acid (MHA) was found to have a significant dependence on the ambient

Table 19.1 Summary of the various DPN ink–substrate combinations that have been reported in literature.

Ink	Substrate	Notes
Alkylthiols (e.g., ODT and MHA)	Au	15 nm resolution with sharp tips on single crystal surfaces, <50 nm on polycrystalline surfaces
Ferrocenylthiols	Au	Redox-active nanostructures
Silazanes	SiO_x, GaAs	Patterning on oxides
Proteins	Au, SiO_x	Both direct write and indirect assembly
Conjugated polymers	SiO_x	Polymer deposition verified spectroscopically and electrochemically
DNA	Au, SiO_x	Sensitive to humidity and tip silanization conditions
Fluorescent dyes	SiO_x	Luminescent patterns
Sols	SiO_x	Solid-state features
Metal salts	Si, Ge	Electrochemical and electroless deposition
Colloidal particles	SiO_x	Viscous solution patterned from tip
Alkynes	Si	C–Si bond formation
Alkoxysilanes	SiO_x	Humidity control important
ROMP materials	SiO_x	Combinatorial polymer brush arrays

Source: Reproduced from Ref. [26].

humidity with a higher rate of deposition at higher humidity [27]. On the other hand, the deposition rate of octadecanethiol (ODT) was found to exhibit negligible dependence on the ambient humidity [28, 29]. This seemingly contradictory dependence of the deposition rate on the humidity is due to the difference in the solubility of these two molecules in water. It is known that under ambient conditions, an ultrathin adlayer (several molecular layers) of water exists on the surface. This water condenses between the tip and the surface forming a meniscus between them. This water meniscus acts as a medium of transport for the ink molecules and therefore the solubility of the ink in water is an important parameter in the deposition process. Considering the significant dependence of the DPN process on the ambient humidity, DPN systems are typically enclosed in a humidity-controlled chamber or glove box for reliable writing under controlled relative humidity.

On the other hand, diffusion dynamics of the ink transported to the surface of the substrate plays a critically important role in the formation of patterns in the DPN process. To this end, Jang *et al.* investigated the self-assembly process using a two-dimensional diffusion model and assuming the tip as a source of constant ink flux [30]. The authors suggested that the temporal variation in the surface area of the features deposited takes the functional form of

$$A_i = a + b\tau_i \tag{19.1}$$

where A_i is the area of the feature and τ is the contact time at each point. In other words, for dot patterns the diameter of the dot scales with the contact time at any given point is $\tau^{0.5}$. The model suggested to describe the ink deposition process was found to be consistent with numerous experimental DPN observations involving a wide variety of inks such as alkanethiols, oligonucleotides, and conducting polymers and various surfaces [20, 23, 39].

The authors of these studies also noted that the random diffusion nature of the self-assembly process causes the patterns to exhibit significant fluctuations at the peripheries. In a later study, Sheehan and Whitman modeled the deposition process under the assumption that the tip is a constant ink concentration source as opposed to the source of constant ink flux [28]. The authors suggested the feature size to be a function of both time of contact and the distance of the tip from the surface. The more complex model was found to show better agreement with the experimental observation over the range of contact times studied for different inks. Experimental observations clearly suggest that the meniscus formed between the AFM tip and substrate plays a critical role in ink transport kinetics.

While most of the early efforts in understanding the nature (size, shape, and humidity dependence) of the meniscus used analytical theories and computer simulations, a more direct and experimental approach to visualize the meniscus under a controlled environment (humidity) has been introduced at more mature stages of development. Weeks *et al.* have, for the first time, imaged the meniscus formed between a silicon nitride tip and silicon and gold surfaces by using environmental SEM (ESEM) [31]. Figure 19.2 shows a representative ESEM image of the meniscus at three different relative humidities. It can be clearly seen that the height of the meniscus and its width decreases dramatically with decreasing relative humidity.

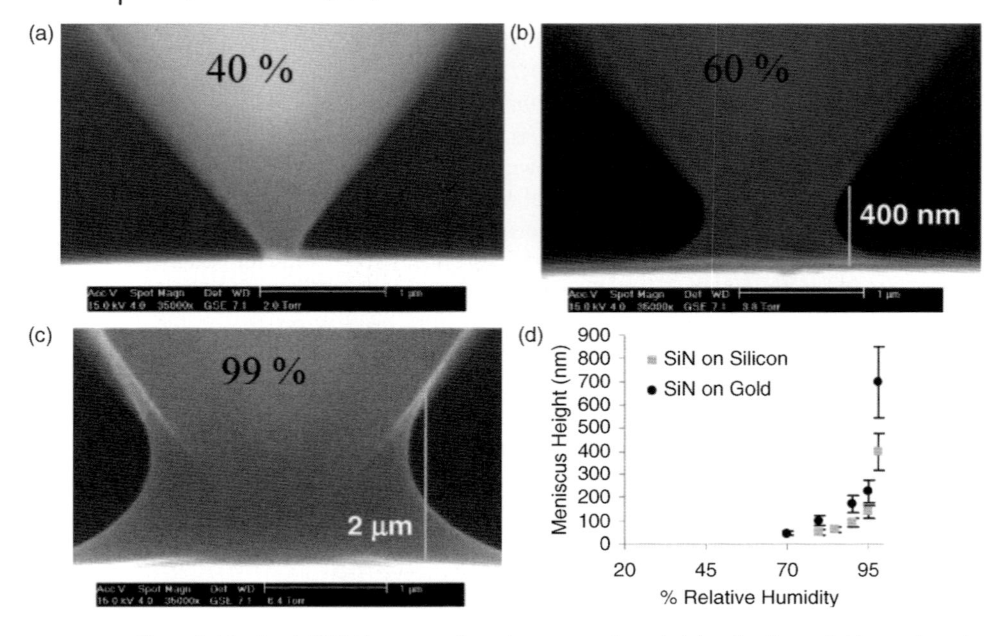

Figure 19.2 (a–c) ESEM images collected at various relative humidities showing the meniscus formed between the tip and the surface. (d) Plot showing the measured meniscus height of a silicon nitride cantilever in contact with both a silicon surface and a gold surface. Reproduced from Ref. [31].

Figure 19.2 also shows the plot of the height of the meniscus at different relative humidities for both silicon and gold surfaces using a silicon nitride tip. In the case of the gold surface, a meniscus was not observed for humidity values below 70%, while the threshold of meniscus formation for a hydrophilic silicon surface was found to be 80%. However, the authors justly pointed out that considering the modest resolution of ESEM (around 50 nm), it remains unclear whether or not a meniscus exists in a relatively dry environment.

Yet another interesting observation of this study is the hysteresis in the height of the meniscus observed between the increasing and decreasing RH values. During decreasing humidity, a stable meniscus was observed even down to 40% RH while the liquid meniscus was not observed until nearly 75% RH during increasing the RH. This study answered one very important question in understanding the ink transport phenomenon: Is the volume of the meniscus large enough to allow large rates of molecular transport by bulk diffusion alone? The direct ESEM confirmed the large size of the meniscus at high RH values, while the situation at low RH remains unclear due to the resolution limitation of ESEM.

In a subsequent study, the same research group employed ESEM to observe and understand the dynamics of the formation of a meniscus [32]. Their most striking observation was that the time taken for the meniscus to reach equilibrium can be on the order of a few minutes. This observation clearly suggested the importance of the dynamics of liquid meniscus formation in understanding some of the experimental observations of the change in the rate of deposition of ink over time (not only contact time but also overall temporal scanning conditions).

There were numerous reports where it was observed that the overall ink transport rate decreases with increasing time [33–35]. The decrease in the transport rate was believed to be caused by the change in the surface energy of the substrate upon adsorption of the ink molecules, depletion of the ink near the tip over time, and the dissolution kinetics of ink molecules in a liquid meniscus. The direct ESEM observation clearly highlighted the importance of the dynamics of meniscus formation and challenged the previous assumption that the meniscus formation and equilibration occurs at a much smaller timescale compared to the timescale of ink dissolution into the meniscus and the contact time.

In conventional dip-pen nanolithography, silicon-based hydrophilic hard materials such as silicon and silicon nitride are employed for the fabrication of cantilevers and the SPM tips. One of the limitations of such an approach is the small range of the feature size that can be achieved using a tip of a given size. In other words, patterns comprised of different feature sizes (i.e., nano- versus microscale) that require tips of different sizes (sharp tips versus colloidal probes) cannot be generated easily. Overcoming this limitation, polymer pen nanolithography, which employs flexible elastomeric pyramidal tips, has recently been recently introduced (see Section 19.3 for detailed discussion).

In another alternative approach, Kramer *et al.* have demonstrated the use of polymer colloids attached to a cantilever as deformable (swellable) tips to write patterns on the surface [36]. In this approach, the polymer elastomeric tip is brought in contact with the surface and retracted, followed by a change in the external conditions (e.g., humidity, pH, etc.). The change in the external ambient causes the polymer colloid to change in dimensions resulting in contact with the surface. The authors demonstrated this form of deposition using PMMA colloids and by changing the relative humidity causing the PMMA sphere to swell and contact the surface, thus transferring the ink molecules on the tip to the surface. The deflection of the cantilever could be controlled with the relative humidity of the ambient conditions. The technique was extended to an array of eight cantilevers and simultaneous deposition of the ink from the eight different cantilevers was demonstrated. The authors noted that by varying the nature of the polymer material used as a colloidal probe (i.e., polymers with varying swelling behavior), it is possible to actuate the cantilevers one at a time by simply choosing the environmental conditions.

19.2 Writing with a Single Pen

DPN has been employed to produce LbL polyelectrolyte multilayer (PEM) films that are patterned with nanoscale resolution. In a typical writing procedure, Lee *et al.* created dot patterns of MHA on a gold surface using a conventional DPN approach

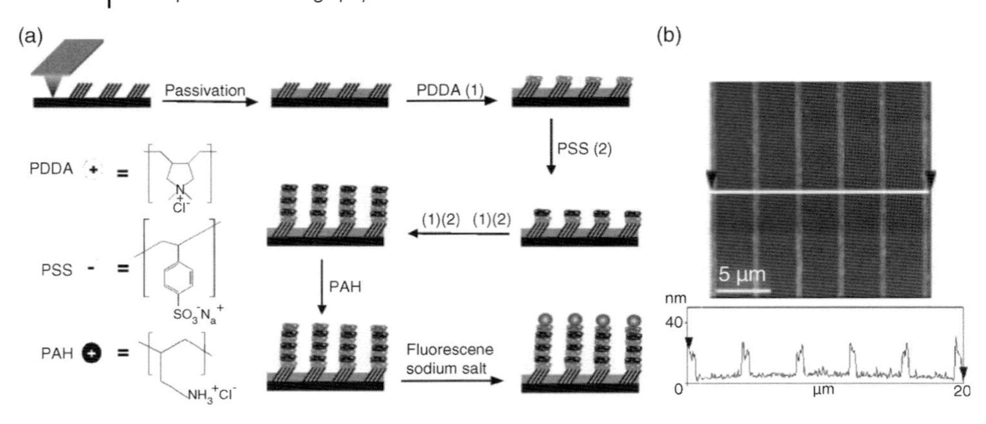

Figure 19.3 (a) Schematic showing the process employed to obtain LbL multilayers on SAMs patterned using DPN. (b) Topographical AFM image of an LbL array with $(PDDA/PSS)_3PAH$ layers and its corresponding height profile. Reproduced from Ref. [37].

followed by exposure to a passivating chemical species to avoid nonspecific adsorption of the polyelectrolytes in the regions without MHA [37]. The authors examined a series of passivating species such as ODT, 16-mercapto-1-hexadecanol, and 11-mercaptoundecyl-tri(ethylene glycol) to form SAMs and found that a PEG-passivated surface had the least nonspecific adsorption. Following the surface passivation, the substrate was exposed to alternating solutions of poly(diallyldimethylammonium chloride) (PDDA) (polycation) and PSS as a traditional polyanion with thorough ultrasonic rinsing in between to remove any loosely bound polyelectrolyte (Figure 19.3). Finally, the LbL film was terminated with a polycationic PAH layer. Similar to conventional LbL, excellent control over the thickness of the patterns was observed with the number of bilayers deposited.

Figure 19.3 also shows the AFM image of the line patterns of $(PDDA/PSS)_3$ PAH PEM films and the corresponding topographical cross section showing excellent uniformity in the thickness of the LbL-assembled lines. The width of the DPN-formed PEM lines was found to be close to 200 nm, which was in excellent agreement with the width of the underlying SAM patterns.

In another study, Amro *et al.* demonstrated AFM mechanical nanolithography by combining nanografting and DPN (see Chapter 18 for discussion on the nanografting method) [38]. This combined lithographical technique termed nanopen reader and writer (NPRW) involved mechanically removing the alkanethiol SAMs in certain predetermined regions using high shear forces in the first stage. Removal of the resist SAM layer in certain regions enables the adsorption of different species in these regions with ink trasport. In the NPRW technique, the AFM tip was inked with the new molecular species, leading to a simultaneous adsorption of the different alkanethiols (ink) during the removal of the resist polymer layer. By using alkanethiols with different chain lengths (10, 12, and 18 carbon atoms), the authors successfully demonstrated the precise lateral replacement of one thiol species with another within the shaved surface regions.

The technique was taken even further by using metal nanoparticles as part of the ink rather than alkenthiols with different lengths. Garno *et al.* have employed AFM-based mechanical lithography to precisely deposit gold nanoparticles in predetermined regions of a surface coated with alkanethiol SAMs [39]. Two different approaches were demonstrated for controlled deposition of the nanoparticles. In the first method, the SAM layer was shaved away by AFM scanning under high load followed by exposing the substrates to a gold nanoparticle solution. The nanoparticles preferentially adsorbed to the shaved regions while the SAM layer resisted nonspecific adsorption and any lateral diffusion of the nanoparticles.

In the second approach explored, the nanoparticles were applied to the AFM tip (the so-called NPRW mode as described above). Imaging of the original and patterned SAM surface was performed under small normal loads (higher set point ratio) during which the nanoparticles remained on the surface. On the other hand, larger normal loads and shear stresses resulted in the removal of the alkanethiol molecules with simultaneous adsorption of the nanoparticles to the surface from surrounding solution. Although both techniques enabled the deposition of nanoparticles with very high lateral resolution, the second approach was quite versatile in that it could be performed in different environments (ambient or in liquid) and the entire patterning process was extremely fast (within 5–6 min).

Assembly of various thiol molecules on gold surfaces forming SAMs of these molecules with extremely high lateral resolution is probably the most extensively investigated and very well-developed technique in the DPN family. For instance, the SAM patterns formed on the various surfaces with DPN have been used to create protein arrays on the surface [16]. In this study, Lee *et al.* have demonstrated the formation of dot (with a diameter as small as 100 nm) arrays of rabit IgG on a surface. For this purpose, the authors first formed dot patterns of MHA on a gold thin film substrate followed by passivation of the surface using 11-mercaptoundecyl-tri(ethylene glycol) to minimize nonspecific adsorption of the proteins to the surrounding surface. The substrate was subsequently exposed to the rabit-IgG solution at pH 7 that promotes the effective protein adsorption to the carboxylic groups of MHA surface.

To demonstrate the specific recognition capability of the biomolecules immobilized on the surface, the protein array was exposed to a solution containing lysozyme, retronectin, goat anti-IgG, and human anti-IgG. AFM imaging revealed no signs (i.e., no change in the height) of protein adsorption to the rabbit IgG. In contrast, exposure to a solution containing lysozyme, goat anti-IgG, human anti-IgG, and rabbit anti-IgG resulted in binding of the rabbit anti-IgG to the rabbit IgG immobilized on the surface. AFM images clearly revealed the increase in the height of the protein dots after such an exposure, suggesting the specific binding of rabbit anti-IgG and rabbit IgG.

A similar strategy was also employed for site-specific and orientation-controlled attachment of tobacco mosaic virus (TMV) on surfaces. Vega *et al.* have demonstrated the fine attachment of individual TMV on surfaces with a high degree of control (perpendicular arrays) over their orientation [40]. Mutually orthogonal (in both *x*- and

y-directions), rectangular patterns of MHA with dimensions of 350 nm × 110 nm were deposited using DPN. The substrate was passivated using thiolated poly (ethylene glycol) to minimize nonspecific adsorption of the virus to the remaining surface. Subsequently, the carboxylic groups of MHA molecules were coordinated with Zn^{2+} ions by exposing the surface to a $Zn(NO_3)_2 \cdot 6H_2O$ solution. The Zn^{2+} ions enabled the anchoring of the carboxylic-rich surface of TMV to the MHA patterns on the surface. This treatment of MHA patterns was found to be extremely important with almost no TMV adsorbing onto the MHA patterns in the absence of such treatment. The authors successfully demonstrated complex patterns of individual virus distributions on the surface. These examples clearly reveal that the high spatial control of surface functionality offered by DPN can be favorably employed to achieve directed assembly of technologically important and complex nanostructures, both synthetic organic and biomolecular.

SAM patterns with desired surface functionality created with DPN have also been extensively used to guide the surface assembly of various nanostructures such as nanoparticles, nanowires, polymer blends, and so on. One particularly impressive example is the assembly of single-walled carbon nanotubes (SWNTs) into arbitrary patterns using patterned functionality of the surface [41]. Wang *et al.* have exploited the preferential affinity of single-walled carbon nanotubes to carboxylic groups to guide the adsorption of SWNTs on SAM-modified surfaces. Patterns of MHA (carboxylic functionality) with strong affinity were created on the surface followed by passivation of the surface with ODT (methyl functionality). Exposure to the SWNT solution caused the nanotubes to strongly adhere to the selected MHA surface regions (Figure 19.4).

When the MHA surface regions were smaller than the total length of the SWNTs, the nanotubes aligned along the shorter interface of the MHA and ODT patterns exhibited the ability of adaptive curving. Figure 19.4 shows the AFM image of these SWNTs assembled in linear, curved, and random line structures, following the MHA patterns defined by the DPN process.

Patterns of alkanethiol molecules chemisorbed on a gold surface have been used to control the phase separation of polymer blends. Coffey *et al.* have formed dots of MHA on the surface of gold followed by exposing the surface to benzenethiol to passivate the surface [42]. Following the preparation of the substrate, a polymer blend solution of conducting P3HT and PS was spin cast onto the surface resulting in spatially organized phase separation of the polymer cylinders of P3HT-rich domains forming in the MHA regions. The preferential segregation of the P3HT-rich domains was confirmed using all fluorescence microscopy, AFM, and conductive atomic force microscopy.

When the polymer blend solution was deposited on the substrate with patterns of MHA, PS preferentially adsorbed on the chemically similar benzenthiol regions. During the spin coating process, as the solvent evaporates, the polymers begin to phase separate. On a PS/benzenethiol surface, the interfacial energy required to form a PS/P3HT interface results in a barrier to lateral domain formation and the film segregates vertically. On the other hand, the MHA/P3HT interfacial energy is lower than the PS/P3HT energy, resulting in lateral phase separation and the nucleation of P3HT-rich domains. Following the nucleation process, these surface-templated

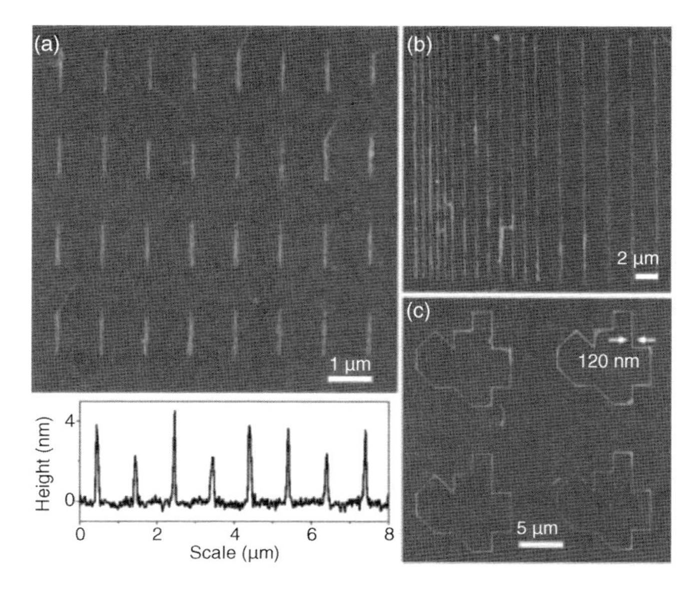

Figure 19.4 AFM tapping mode topographic images of SWNT arrays. (a) Parallel aligned SWNTs with a line density. (b) SWNTs adsorbed along the MHA linear DPN patterns (20 µm × 200 nm) spaced by 2 µm, 1 µm, and 600 nm. (c) Random line structure deposited on the surface of substrate showing the precise positioning, bending, and linking of SWNTs to a MHA DPL template. Reproduced from Ref. [41].

domains are formed and continue to grow until the film vitrifies. This patterning technique has been extended to other polymer pairs such as PVP and PS by using nonpolar DPN templates.

Electrostatic interaction was also employed to deposit patterns of charged conducting polymer on an oppositely charged surface. For this study, Lim and Mirkin have demonstrated the deposition of sulfonated polyaniline (negatively charged) and doped polypyrrole (positively charged) on amine-terminated (positively charged) and piranha-cleaned (negatively charged) silicon surfaces, respectively [23]. Polymer patterns with line widths below 300 nm and dots with diameters as small as 130 nm were achieved using this approach. The authors noted that the polymer deposition could not be performed on an electrically neutral surface or on surfaces with the same charge as the polymer itself. These observations corroborated the electrostatic nature of the adsorption of the polymer chains to the surface. The dimensions of the patterns (diameter of the circular dots) was found to scale with the square root of the contact time, in agreement with the diffusion model assuming the tip as a source of constant ink flux (as discussed in Section 19.1).

In a recent report, Wang *et al.* have demonstrated the direct deposition and precise micropatterning of gold nanoparticles using the DPN approach [43]. Their procedure involved using gold nanoparticles (both positively and negatively charged) as ink and patterning the same on hydrophilic and oppositely charged surfaces using the electrostatic interaction between them. Drop casting and evaporation of the gold nanoparticle solution on the tip was employed for inking the AFM tip as opposed to the conventional approach of dip coating. The schematic of the inking of the AFM tip with gold nanoparticles, which is exploited in this study, is shown in Figure 19.5. However, such an ink formation resulted in a smaller loading of gold nanoparticles on the tip and hence inconsistent results under variable scanning conditions.

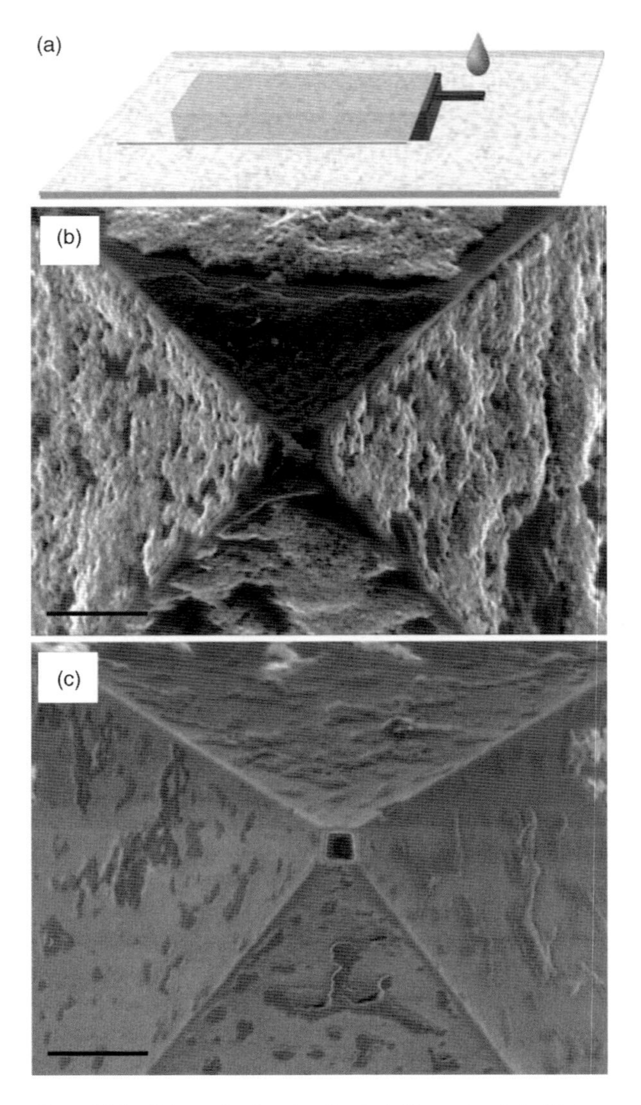

Figure 19.5 Schematic showing the tip inking setup. The droplet of AuNPs is confined around the AFM tip using a parafilm surrounding the probe. SEM images of AFM tip loaded with Au nanoparticles (b) and a depleted AFM tip after patterning (c). Scale bars are 500 nm. Reproduced from Ref. [43].

SEM images of a treated AFM tip show the high coverage of the tip with gold nanoparticles and the depletion of gold nanoparticles before and after patterning, respectively. The authors noted that in this approach, direct patterning of the gold nanoparticles on the surface of the substrate requires a highly hydrophilic substrate. Moreover, patterning was not possible using moderately hydrophilic and hydrophobic substrates. One important difference between the conventional inks (such as alkylthiols) and the nanoparticle ink is the absence of the noticeable lateral diffusion of the ink in the case of the gold nanoparticles. Furthermore, the quality and resolution of writing with gold nanoparticles did not show any significant dependence on the writing speed with the range of the speeds probed (0.01–2 μm/s), which indicated much slower nanoparticle transport in the meniscus bridge.

In contrast to the conventional approaches of anchoring targeted chemical or biological species to the substrate by scanning the inked AFM tip along the predetermined design, an alternative approach involves modifying the existing film on the surface by local delivery of a chemical species. Using this approach, Wang *et al.* have demonstrated the formation of predetermined patterns on a pH responsive polymer (poly (4-vinyl pyridine) (P4VP)) film by delivering acidic buffer (pH 4.0) using an AFM tip [44]. Figure 19.6 shows the schematic of the swelling and corresponding pattern formation in the P4VP film with local acidic ink. The acidic buffer causes the P4VP film to protonate, which results in local swelling of the polymer film. It was found that the acidic ions (i.e., H_3O^+) on the tip were more reliably transferred to the P4VP film in the presence of an applied electric field.

In this study, the authors investigated the effects of patterning parameters, such as applied bias and contact force. The authors observed that both higher bias voltage and higher contact force resulted in a marked increase in the height of the features formed in the P4VP film (Figure 19.6). The images show the patterns generated in a P4VP film under constant load (1 μN) and at three different values of applied bias ($v = 5$, 3, and 0 V). It can be clearly seen that the pattern height increases with the applied voltage under constant load. A similar increase in the pattern height was observed for an increase in the load under constant external voltage. Interestingly, the surface features in the P4VP film formed by local swelling could be erased by applying basic (phosphate buffer at pH 8.3) ink to the selected surface areas. The basic ink deposited with the AFM tip caused deprotonation of the side pyridine units of the backbones resulting in a collapse of the polymer chains and hence the formation of the topographical features.

19.3 Simultaneous Writing with Multiple Pens and Large-Scale DPN

In the discussion in this section so far, we have primarily discussed writing patterns on a surface using a single AFM tip and focused on corresponding processes in a single contact area. Although the technique offers excellent spatial resolution and virtually unlimited pattern complexity, one of the key limitation of this approach is the relatively low throughput (serial nature) of the technique, which is a severe issue for

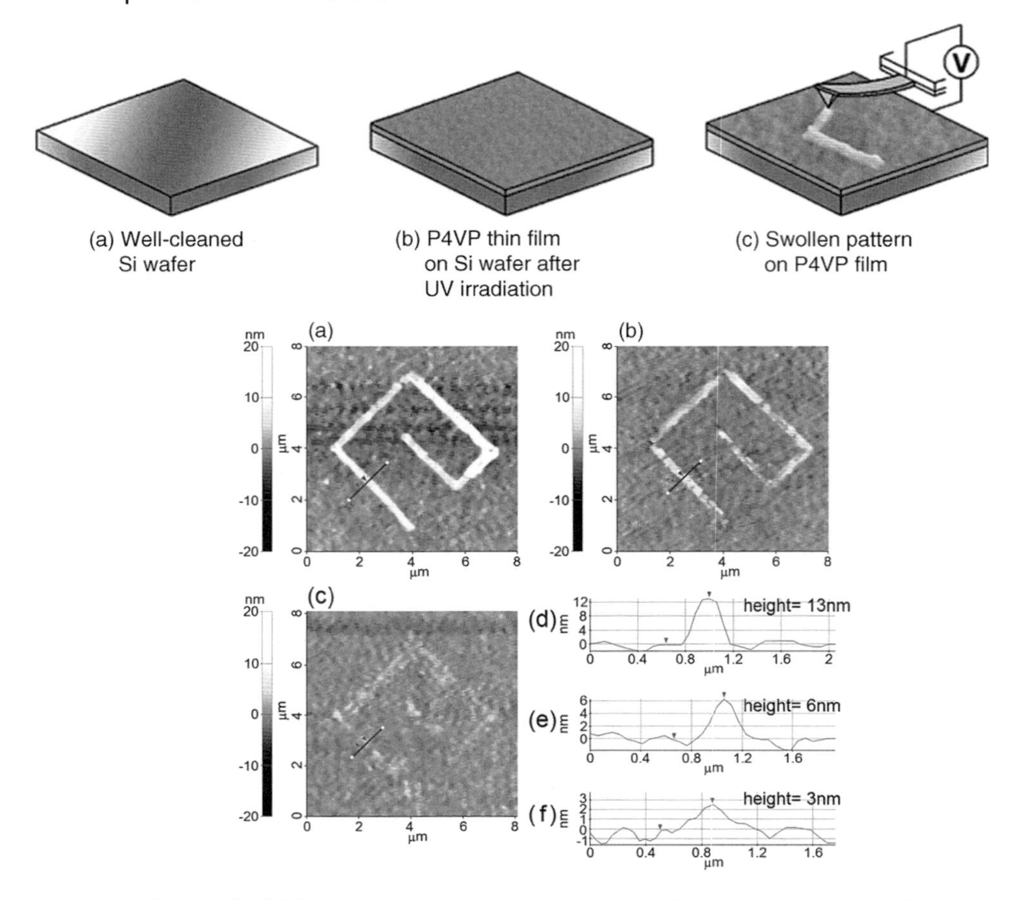

Figure 19.6 (a) Schematic showing the formation of topographic patterns by localized swelling of P4VP by the localized delivery of acidic ions to P4VP film. (b) AFM topographic images of surface features drawn by applying a constant force of 1.0 μN. The applied bias voltages (V) and measured heights of selected lines (nm) are (i) 5, 13; (ii) 3, 6; and (iii) 0, 3. Line profile plots of (iv–v) correspond to the features selected in (i–iii), respectively. Reproduced from Ref. [45].

prospective practical (beyond lab bench) applications. In order to overcome this critical limitation, multiple-pen cantilever arrays have been extensively investigated and are now commercially available, dramatically enhancing the throughput ability of the DPN process and its expansion to large-scale patterning.

The multiple-pen DPN technique has also been introduced in recent studies as will be discussed below. This approach can be broadly classified as having two different designs, namely, passive and active writing with parallel pens. In the passive parallel pen approach, the 1D or 2D array of pens are just duplicates of a single pen, enabling parallel writing of numerous identical patterns. The second approach, which is called the active pen approach, involves the independent actuation of the tips of an array of microcantilevers. In this approach, using different inks on different tips and independent actuations, complex patterns of multiple chemical or biological species can be utilized for complex deposition on the surface. In the following discussion, we present several selected examples from recent literature where multiple pens (in both passive and active modes) have been employed to create large-scale complex patterns on surfaces.

One of the remarkable demonstrations of large-scale DPN involved the creation of an 88 000 000 gold dot array using a 2D array of 55 000 microcantilevers (see Figure 19.7). The microfabricated 2D cantilever array was carefully designed to achieve this unique lithographical tool. The pyramidal tips of the cantilever were sufficiently long (about 7.5 μm) to have sufficient clearance to engage on the surface and the square base of the pyramid had an edge length equal to the width of the cantilever. The microcantilevers were bent by approximately 20° from the supporting base, which was achieved by the differential thermal expansion between the silicon nitride cantilever and the thin metal layers (Au/Ti) deposited on the top of the microcantilever. This bent structure ensures the contact of the tip to the surface

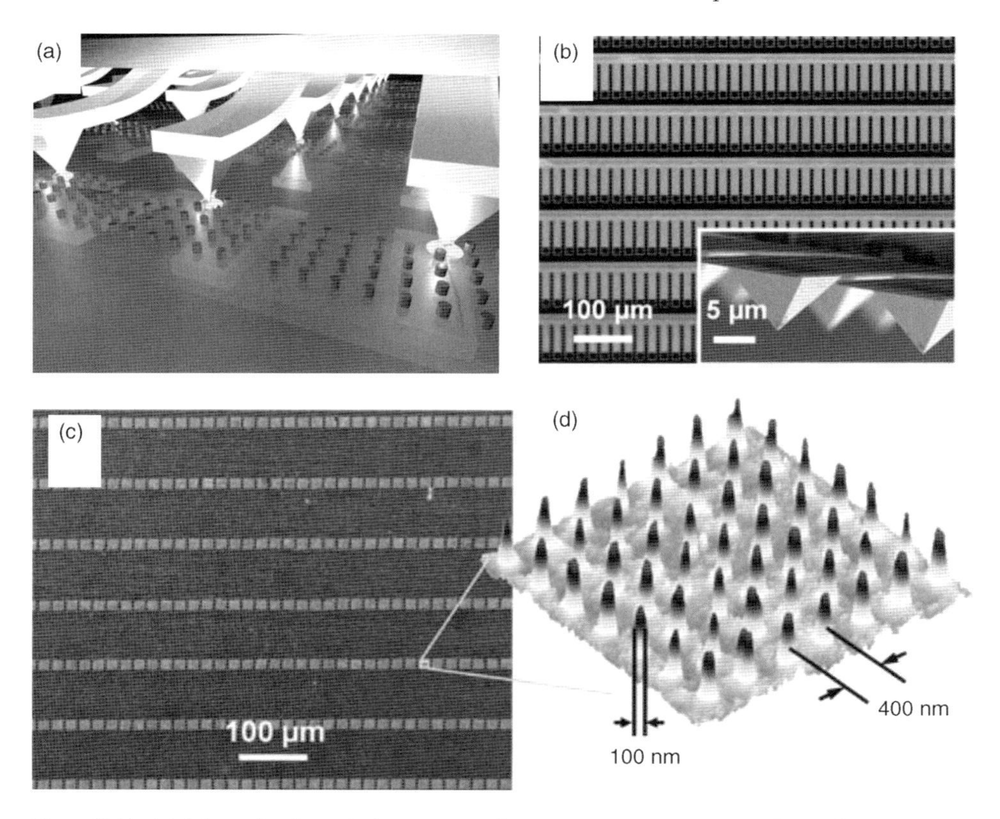

Figure 19.7 (a) Schematic of massively parallel DPN with a passive, wire-free, 2D cantilever array. (b) Optical micrograph of a part of the 2D array of cantilevers. (Inset) SEM image of the cantilever arrays at a different viewing angle. (c) Large-area SEM image of part of an 88 000 000 gold dot array. (d) AFM topographical image of one of the blocks, where the dot-to-dot distance is 400 nm, and the dot diameter is 100 nm. Reproduced from Ref. [45].

overcomes the uneven surface morphology (roughness and thickness variations) of the substrate and the inherent misalignment of the numerous cantilevers. This microcantilever array can also be engaged onto the surface using gravity instead of a conventional feedback system followed by locking the array into position with respect to the piezo scanner head using epoxy resin.

Figure 19.7 shows the SEM image of a part of the large-scale pattern of gold nanodots on the surface and the 88 million dot features generated by using ODT molecules as an ink. Each tip generated 1600 dots in a 40×40 array, where the dot-to-dot distance was 400 nm. It can be seen from the higher magnification image that the dots had a diameter of about 100 nm and a height of 30 nm. The dots were spaced by 20 μm in the x-direction and by 90 μm in the y-direction, which correspond to the distances determined by the array architecture. In the same study, the authors demonstrated multiple duplication of a sophisticated pattern (Thomas Jefferson image from a US nickel coin) 55 000 times by moving the cantilever array in a predetermined manner using the integrated software that controls the precise motion of the microcantilever array.

As mentioned earlier, the second approach of parallel pen or multipen DPN involves active control over the individual pens, a more sophisticated approach. In this case, the array of pens are not mere duplicates of a single pen, instead the user can individually control each pen to create more complex patterns on the surface. The active multipen has been demonstrated using two different methods. In the first approach, thermal actuation of the individual tips is employed to controllably bring the tip in contact with the surface. The thermal actuation of individual cantilevers was achieved by surface stresses induced in the cantilever due to the differential thermal expansion of the bimaterial structure (e.g., gold-coated silicon nitride cantilevers) [46]. The cantilevers were resistively heated by passing electrical current through the resistive heater on the cantilever, which could be individually addressed (see Figure 19.8 for the SEM image of the cantilever array).

One interesting demonstration involved writing microscopic digital numbers from 0 to 9 by a 10-pen array. This impressive patterning of ODT molecules was achieved by moving the array of tips in the form 8 and the individual tips contacted the surface only at specific locations depending on the number assigned. Figure 19.8 shows the lateral force microscopy images (5×5 μm) of the ODT patterns formed in the form of numbers from 0 to 9.

In the second design, electrostatic actuation of the individual cantilevers is exploited as opposed to the thermal actuation discussed above as described by Bullen and Liu [47]. Several advantages to the later technique as summarized by the authors are that it avoids the need for additional heating of the microcantilevers, avoids the thermal cross-talk between the neighboring cantilevers, and results in more overall stable actuation of larger arrays.

In another dramatic improvement of DPN scalability, Huo *et al.* have recently demonstrated polymer pen nanolithography(PPN) [48]. This novel design involves using a very large array of soft and flexible elastomeric pyramidal structures to deliver molecules to the surface across large surface areas. The technique is an interesting combination of conventional microcontact printing, which involves

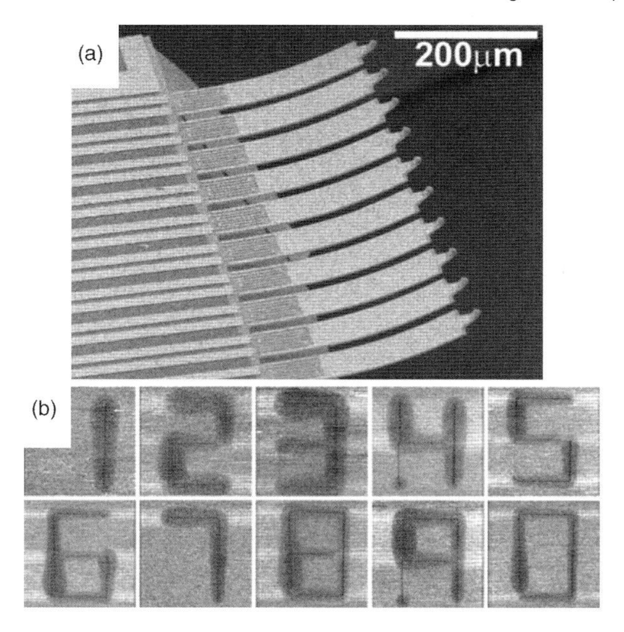

Figure 19.8 (a) SEM image showing the array of 10 thermally actuated DPN probes. It can be seen that each of the probe is 300 μm long, 80 μm wide, and 1.3 μm thick. (b) LFM scans, 8 μm^2, of 10 simultaneously generated ODT patterns on a gold surface. Each numeral is 6 μm tall, 4 μm wide, and was written at 1 μm/s. Reproduced from Ref. [46].

"static" patterning of various chemical species using an elastomeric stamp and conventional dip-pen nanolithography described above.

Tips were made of a highly elastomeric and robust material, PDMS, using a silicon master fabricated using conventional photolithography [48]. The elastomeric substrate with the array of these highly deformable tips is attached to the z-piezoelement of the AFM instrument. The highly compliant tips enable fine control over the surface feature size by controlling the applied pressure by piezoelement extension/contraction and the elastic modulus of the PDMS. In this study, it was found that the feature size of MHA molecules, which were used as the ink, exhibited a linear growth with increasing normal load. With a pyramid tip radius of about 60 nm, the feature sizes were controlled from smallest, several hundred nanometers, to 6 μm by varying the normal load over a wide range.

The fine dimensional control with force applied and the maskless nature of polymer pen nanolithography enabled the parallel fabrication of complex patterns on a large surface obviating the need for new masters as in conventional soft lithographic approaches. Furthermore, polymer pen nanolithography allows even sub-100 nm resolution at very low elastic deformation owing to the extremely precise closed loop scanner and small tip radii.

As an impressive demonstration of this method, the authors used polymer pen nanolithography to generate 15 000 replicas of the 2008 Olympic logo on a gold thin film substrate using MHA as the ink and subsequent wet chemical etching. Each logo fabricated this way was a bit map of 70 × 60 μm generated by taking advantage of the multiscale capabilities of polymer pen nanolithography. In other words, the logo was comprised of individual bits with different physical size, enabled by the control over the feature size offered by the elastomeric tips and precise control of normal load.

In turn, the letters and numbers "Beijing 2008" were formed from nearly 20 000 dots with a 90 nm diameter upon initial tip contact. The picture and Olympic rings were made from nearly 4000 dots with a 600 nm diameter that were formed with higher tip–surface contact forces. These surface structures were created by holding the pen array at each spot for 0.05 s and traveling between spots at a much higher linear speed of 60 μm/s.

The authors demonstrated 15 000 replicas formed across a 1 cm^2 substrate by the polymer-pen method. The yield of such patterning was found to be extremely high reaching 99% and the overall uniformity over a 1 cm^2 substrate surface area is unprecedented for SPM-related nanolithographical techniques. Impressively, the total time required to fabricate these large-scale complex surface structures was very modest for SPM methods, less than 40 min.

Taking polymer-pen lithography further, the same group demonstrated beam-pen lithography (BPL), which exploited the optical transparency of PDMS tips used in polymer pen lithography [49]. The PDMS pens were converted to highly localized light sources by creating apertures at the tip. This was achieved by plasma oxidation of PDMS followed by the deposition of a gold layer on the surface of templates. Subsequently, the PDMS tip array was brought in contact with PMMA deposited on glass slide to remove the gold layer from the tip and form an aperture. The size of the aperture could be controlled (500 nm to 5 μm) by adjusting the external force exerted as the tip meets the PMMA surface. Smaller apertures (around 50 nm) could be achieved using a focused ion beam technique. With the PDMS tip array illuminated from the top, each tip with a nanoscale aperture acted as a near-field light source.

The authors employed the near-field source to pattern features using a conventional positive photoresist, which was exposed in a predetermined manner by piezo-controlled movement of the light source array. The arbitrary nature of the patterns that could be generated using BPL distinguishes it from conventional photolithography or contact microprinting, which only allows one to create and duplicate preformed complex micropatterns.

References

1 Salaita, K., Wang, Y., and Mirkin, C.A. (2007) Applications of dip-pen lithography. *Nat. Nanotechnol.*, **2** (2), 145–155.

2 Piner, R.D., Zhu, J., Xu, F., Hong, S., and Mirkin, C.A. (1999) Dip-pen lithography. *Science*, **283** (5402), 661–663.

3 Hong, S., Zhu, J., and Mirkin, C.A. (1999) Multiple ink nanolithography: toward a multiple-pen nano-plotter. *Science*, **286** (5439), 523–525.

4 Hong, S.H. and Mirkin, C.A. (2000) A nanoplotter with both parallel and serial writing capabilities. *Science*, **288** (5472), 1808–1811.

5 Weinberger, D.A., Hong, S., Mirkin, C.A., Wessels, B.W., and Higgins, T.B. (2000) Combinatorial generation and analysis of

nanometer-and micrometer-scale silicon features via "dip-pen" nanolithography and wet chemical etching. *Adv. Mater.*, **12** (21), 1600–1603.

6 Ivanisevic, A., McCumber, K.V., and Mirkin, C.A. (2002) Site-directed exchange studies with combinatorial libraries of nanostructures. *J. Am. Chem. Soc.*, **124** (40), 11997–12001.

7 Jung, H., Kulkarni, R., and Collier, C.P. (2003) Dip-pen lithography of reactive alkoxysilanes on glass. *J. Am. Chem. Soc.*, **125** (40), 12096–12097.

8 Hurley, P.T., Ribbe, A.E., and Buriak, J.M. (2003) Nanopatterning of alkynes on hydrogen-terminated silicon surfaces by scanning probe-induced cathodic electrografting. *J. Am. Chem. Soc.*, **125** (37), 11334–11339.

9 Ivanisevic, A., Im, J.H., Lee, K.B., Park, S.J., Demers, L.M., Watson, K.J., and Mirkin, C.A. (2001) Redox-controlled orthogonal assembly of charged nanostructures. *J. Am. Chem. Soc.*, **123** (49), 12424–12425.

10 Pena, D.J., Raphael, M.P., and Byers, J.M. (2003) "Dip-Pen" nanolithography in registry with photolithography for biosensor development. *Langmuir*, **19** (21), 9028–9032.

11 Fu, L., Liu, X., Zhang, Y., Dravid, V.P., and Mirkin, C.A. (2003) Nanopatterning of "hard" magnetic nanostructures via dip-pen nanolithography and a sol-based ink. *Nano Lett.*, **3** (6), 757–760.

12 Su, M., Liu, X., Li, S.-Y., Dravid, V.P., and Mirkin, C.A. (2002) Moving beyond molecules: patterning solid-state features via dip-pen nanolithography with sol-based inks. *J. Am. Chem. Soc.*, **124** (8), 1560–1561.

13 Noy, A., Miller, A.E., Klare, J.E., Weeks, B.L., Woods, B.W., and De Yoreo, J.J. (2002) Fabrication of luminescent nanostructures and polymer nanowires using dip-pen nanolithography. *Nano Lett.*, **2** (2), 109–112.

14 Wilson, D.L., Martin, R., Hong, S., Cronin-Golomb, M., Mirkin, C.A., and Kaplan, D.L. (2001) Surface organization and nanopatterning of collagen by dip-pen nanolithography. *Proc. Natl. Acad. Sci. USA*, **98** (24), 13660–13664.

15 Wei, L., Hong, X., Guo, W., Bai, Y.-B., and Li, T.-J. (2002) Direct fabrication of protein arrays using dip-pen nanolithography. *Chem. J. Chin. Univ.*, **23** (7), 1386–1388.

16 Lee, K.B., Park, S.J., Mirkin, C.A., Smith, J.C., and Mrksich, M. (2002) Protein nanoarrays generated by dip-pen nanolithography. *Science*, **295** (5560), 1702–1705.

17 Li, Y., Maynor, B.W., and Liu, J. (2001) Electrochemical AFM "dip-pen" nanolithography. *J. Am. Chem. Soc.*, **123** (9), 2105–2106.

18 Maynor, B.W., Li, Y., and Liu, J. (2001) Au: ink" for AFM "dip-pen" nanolithography. *Langmuir*, **17** (9), 2575–2578.

19 Porter, L.A., Choi, H.C., Schmeltzer, J.M., Ribbe, A.E., Elliott, L.C.C., and Buriak, J.M. (2002) Electroless nanoparticle film deposition compatible with photolithography, microcontact printing, and dip-pen nanolithography. *Nano Lett.*, **2** (12), 1369–1372.

20 Demers, L.M., Ginger, D.S., Park, S.-J., Li, Z., Chung, S.-W., and Mirkin, C.A. (2002) Direct patterning of modified oligonucleotides on metals and insulators by dip-pen nanolithography. *Science*, **296** (5574), 1836–1838.

21 Demers, L.M., Park, S.-J., Taton, T.A., Li, Z., and Mirkin, C.A. (2001) Orthogonal assembly of nanoparticle building blocks on dip-pen nanolithographically generated templates of DNA. *Angew. Chem. Int. Ed.*, **40**, 3071–3073.

22 Maynor, B.W., Filocamo, S.F., Grinstaff, M.W., and Liu, J. (2002) Direct-writing of polymer nanostructures: poly (thiophene) nanowires on semiconducting and insulating surfaces. *J. Am. Chem. Soc.*, **124** (4), 522–523.

23 Lim, J.-H. and Mirkin, C.A. (2002) Electrostatistically driven dip-pen nanolithography of conducting polymers. *Adv. Mater.*, **14** (20), 1474–1477.

24 Ali, M.B., Ondarcuhu, T., Brust, M., and Joachim, C. (2002) Atomic force microscope tip nanoprinting of gold nanoclusters. *Langmuir*, **18** (3), 872–876.

25 Liao, J.-H., Huang, L., and Gu, N. (2002) Fabrication of nanoparticle pattern through atomic force microscopy tip-

induced deposition on modified silicon surfaces. *Chin. Phys. Lett.*, **19** (1), 134–136.

26 Ginger, D.S., Zhang, H., and Mirkin, C.A. (2003) The evolution of dip-pen nanolithography. *Angew. Chem. Int. Ed.*, **43** (1), 30–45.

27 Weeks, B.L., Noy, A., Miller, E., and De Yoreo, J.J. (2002) Effect of dissolution kinetics on feature size in dip-pen nanolithography. *Phys. Rev. Lett.*, **88** (25), 255505.

28 Sheehan, P.E. and Whitman, L.J. (2002) Thiol diffusion and the role of humidity in "dip pen nanolithography". *Phys. Rev. Lett.*, **88** (15), 156104.

29 Rozhok, S., Piner, R., and Mirkin, C.A. (2003) Dip-pen nanolithography: what controls ink transport? *J. Phys. Chem. B*, **107** (3), 751–757.

30 Jang, J., Hong, S., Schatz, G.C., and Ratner, M.A. (2001) Self-assembly of ink molecules in dip-pen nanolithography: a diffusion model. *J. Chem. Phys.*, **115** (6), 2721.

31 Weeks, B.L., Vaughn, M.W., and DeYoreo, J.J. (2005) Direct imaging of meniscus formation in atomic force microscopy using environmental scanning electron microscopy. *Langmuir*, **21** (18), 8096–8098.

32 Weeks, B.L. and Deyoreo, J.J. (2006) Dynamic meniscus growth at a scanning probe tip in contact with a gold substrate. *J. Phys. Chem. B*, **110** (21), 10231–10233.

33 Hampton, J.R., Dameron, A.A., and Weiss, P.S. (2005) Transport rates vary with deposition time in dip-pen nanolithography. *J. Phys. Chem. B*, **109** (49), 23118–23120.

34 Hampton, J.R., Dameron, A.A., and Weiss, P.S. (2006) Double-ink dip-pen nanolithography studies elucidate molecular transport. *J. Am. Chem. Soc.*, **128** (5), 1648–1653.

35 Peterson, E.J., Weeks, B.L., De Yoreo, J.J., and Schwartz, P.V. (2004) Effect of environmental conditions on dip pen nanolithography of mercaptohexadeconic acid. *J. Phys. Chem. B*, **108** (39), 15206–15210.

36 Kramer, M.A., Jaganathan, H., and Ivanisevic, A. (2010) Serial and parallel dip-pen nanolithography using a colloidal probe tip. *J. Am. Chem. Soc.*, **132** (13), 4532–4533.

37 Lee, S.W., Sanderin, R.G., Oh, B.-K., and Mirkin, C.A. (2005) Nanostructured polyelectrolyte multilayer organic thin films generated via parallel dip-pen nanolithography. *Adv. Mater.*, **17** (22), 2749–2753.

38 Amro, N.A., Xu, S., and Liu, G.-Y. (2000) Patterning surfaces using tip-directed displacement and self-assembly. *Langmuir*, **16** (7), 3006–3009.

39 Jayne, C.G., Yang, Y., Amro, N.A., Cruchon-Dupeyrat, S., Chen, S., and Liu, G.-Y. (2003) Precise positioning of nanoparticles on surfaces using scanning probe lithography. *Nano Lett.*, **3** (3), 389–395.

40 Vega, R.A., Maspoch, D., Salaita, K., and Mirkin, C.A. (2005) Nanoarrays of single virus particles. *Angew. Chem. Int. Ed.*, **44** (37), 6013–6015.

41 Wang, Y., Maspoch, D., Zou, S., Schatz, G.C., Smalley, R.E., and Mirkin, C.A. (2006) Controlling the shape, orientation, and linkage of carbon nanotube features with nano affinity templates. *Proc. Natl. Acad. Sci. USA*, **103** (7), 2026–2031.

42 Coffey, D.C. and Ginger, D.S. (2005) Patterning phase separation in polymer films with dip-pen nanolithography. *J. Am. Chem. Soc.*, **127** (13), 4564–4565.

43 Wang, W.M., Stoltenberg, R.M., Liu, S., and Bao, Z. (2008) Direct patterning of gold nanoparticles using dip-pen nanolithography. *ACS Nano*, **2** (10), 2135–2142.

44 Wang, X., Wang, X., Fernandez, R., Ocola, L., Yan, M., and La Rosa, A. (2010) Electric-field-assisted dip-pen nanolithography on poly(4-vinylpyridine) (P4VP) thin films. *ACS Appl. Mater. Interfaces*, **2** (10), 2904–2909.

45 Salaita, K., Wang, Y., Fragala, J., Vega, R.A., Liu, C., and Mirkin, C.A. (2006) Massively parallel dip-pen nanolithography with 55000-pen two-dimensional arrays. *Angew. Chem. Int. Ed.*, **45** (43), 7220–7223.

46 Bullen, D., Chung, S.W., Wang, X.F., Zou, J., Mirkin, C.A., and Liu, C. (2004) Parallel dip-pen nanolithography with arrays of individually addressable cantilevers. *Appl. Phys. Lett.*, **84** (5), 789–791.

47 Bullen, D. and Liu, C. (2006) Electrostatically actuated dip pen nanolithography probe arrays. *Sens. Actuators A*, **125** (2), 504–511.

48 Huo, F., Zheng, Z., Zheng, G., Giam, L.R., Zhang, H., and Mirkin, C.A. (2008) Polymer pen lithography. *Science*, **321** (5896), 1658–1660.

49 Huo, F., Zheng, G., Liao, X., Giam, L.R., Chai, J., Chen, X., Shim, W., and Mirkin, C.A. (2010) Beam pen lithography. *Nat. Nanotechnol.*, **5** (10), 637–640.

20
Microcantilever-Based Sensors

20.1
Basic Modes of Operation

20.1.1
General Introduction

The AFM technique has also revived interest in nanofabrication and a plethora of applications of micromechanical structures [1]. AFM has long relied on microcantilevers as force transducers for its numerous imaging modes including topographical, electric potential, magnetic, and force imaging [2, 3]. As a natural succession to their application as force transducers beyond traditional AFM, microfabricated cantilevers were selected as a new platform for transduction in sensing technology more than a decade ago [4]. Since then, the new technology has emerged to find important prospective applications for microcantilever-based sensors in chemical, biological, and thermal sensing employing these flexible microcantilevers as sensitive force transducers [5, 6]. In this last chapter of the book, we will briefly introduce a different but related field of microcantilever-based sensors for chemical and biological applications, which were originally inspired by AFM technology.

There are a number of cantilever configurations (e.g., with and without intrinsic stress, silicon versus polymer, or "diving boards" versus V-shaped) adapted for various prospective applications such as thermal sensing, IR sensing, chemical sensing, or biosensing. Within each sensing paradigm, different implementation principles have been established. For instance, for chemical sensing, one can monitor induced stress, weight change, reflectance change, or shape change, as represented in Figure 20.1. On the other hand, for example, the change in stress can be measured by piezoresistive elements, piezoelectric elements, thin film transistors, or light beam deflections, to name a few.

The microscopic levers for sensing can be fabricated in various geometries with specific materials and coatings, using a vast range of semiconducting and metallic materials to optimize stiffness, thermal noise, conductivity, reflectance, and Q factor. It is a challenge to cover all existing accomplishments and future trends in a whole subfield of microcantilever-based sensing, and it will not be undertaken in this book.

Scanning Probe Microscopy of Soft Matter: Fundamentals and Practices, First Edition.
Vladimir V. Tsukruk and Srikanth Singamaneni.

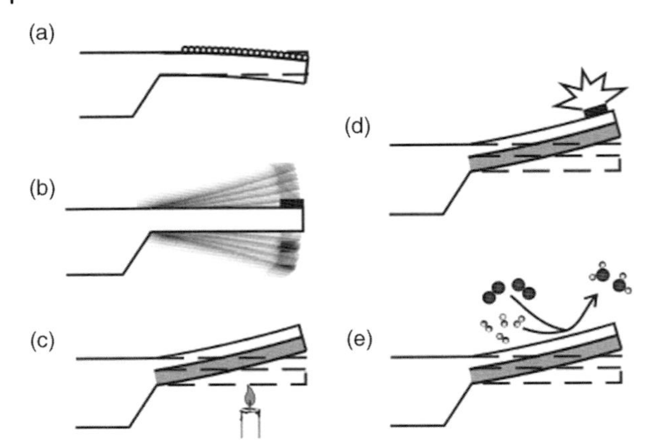

Figure 20.1 Various modes of operation of microcantilevers: (a) surface stress due to absorption of molecules causing static deflection; (b) dynamic resonance frequency shift mode due to change in effective mass; (c) heat sensing mode due to differential thermal expansion; (d) deflagration of explosive on the heated microcantilever surface; and (e) catalytic reaction on the cantilever surface. Reproduced from Ref. [5].

A detailed discussion of the full range of possible cantilever designs is not the subject of this chapter, but rather, we will focus mainly on selected materials' aspects of their design that are critical for ultimate sensing applications. In this section, we briefly introduce the microcantilever-based transduction principles along with their operation modes and detection methods of microcantilever-based sensors (Figure 20.1).

To provide sensing ability to microcantilever beams, their top and/or bottom surfaces must be coated in a chemically well-defined way to provide a functional surface capable of reacting with the target molecules and a passivated surface that will not significantly react with the target molecules, thus creating differential adsorption and stresses. Chemical reaction and deflagration methods of detection are also presented here as practically important approaches (Figure 20.1).

There are two basic modes of operation in microcantilever-based sensing, namely, static (physical deflection of the microcantilever) and dynamic (change in resonance frequency/phase), besides several ways to initiate cantilever reactions such as heat (deflection due to differential thermal expansion) or chemical reaction. In another example, adsorption of molecules onto the surface of the microcantilever causes a bending due to increasing interfacial stress. Each mode differs from others in terms of the principle of transduction, functionalization, and detection mechanisms. Here, we briefly introduce major modes of operation and highlight the design considerations specific to each mode.

20.1.2
Static Deflection Mode

The asymmetry of a functionalized top surface and a passivated bottom surface is especially important for the implementation of static deflection mode. The microcantilever flexural behavior is controlled by the spring constant k of the cantilever, which is defined by material properties and microcantilever geometrical dimensions [7, 8]. For a rectangular microcantilever of length (l), thickness (t), and width (w), the spring constant, k, is calculated as [9]

$$k = \frac{Ewt^3}{4l^3} \tag{20.1}$$

where E is Young's modulus ($E_{Si} = 1.3 \times 10^{11}\ \mathrm{N/m^2}$ for Si(100)).

Typical spring constants for common microcantilevers exploited as sensors (several hundred microns in length with a thickness around 1 μm) fall in the range of 0.001–1 N/m. Actual spring constants can be calculated and measured for various complicated shapes and compositions by using a range of theoretical and experimental approaches, as has been discussed elsewhere [7, 10–12].

Assuming that uniform surface stress, $\Delta\sigma$, over the whole area of the cantilever causes bending, the shape of the bent microcantilever (Figure 20.2) can be approximated as part of a circle with radius R, as given by Stoney's equation:

$$\frac{1}{R} = \frac{6(1-\nu)}{Et^2}\Delta\sigma \tag{20.2}$$

where ν is the Poisson's ratio ($\nu_{Si} = 0.24$) [13, 14]. For a given deflection, the surface stress change (schematically represented in Figure 20.2) can be derived by using

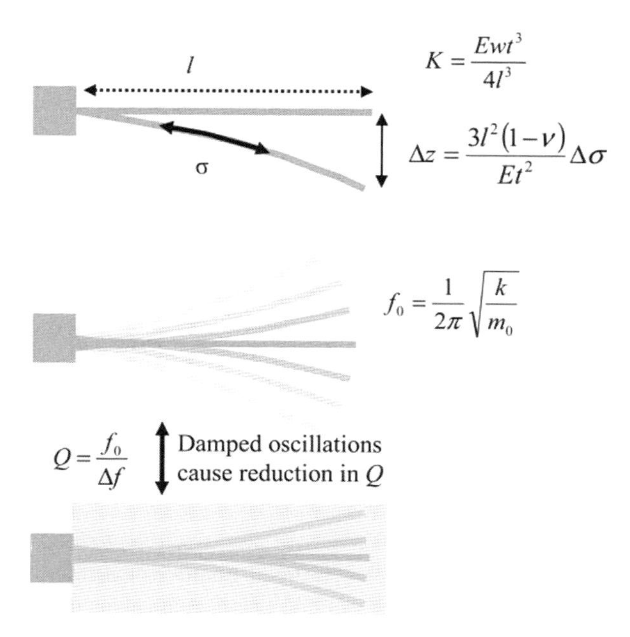

Figure 20.2 Deflection modes: (a) microcantilever deflecting in static mode under surface stress; (b) microcantilever oscillating at fundamental frequency, f_0; (c) viscous damping for underliquid operation along with corresponding parameters. Reproduced from Ref. [5].

Eq. (20.2), which is, however, valid only for a surface layer much thinner than the beam itself (<20%) [13]. There have been several attempts to modify Stoney's equation for thicker surface layers, the accuracy of which has been reviewed in a recent article [15].

Static deflection operation based upon measuring constant deflection at a given constant stress is possible in various environments such as vacuum, ambient, and fluidic. In a gaseous environment, molecules adsorb onto the functionalized sensing surface and form a molecular layer, provided there is affinity for the molecules to adhere to the surface. Static mode operation in liquids, however, usually requires rather specific sensing layers, based on molecular recognition, such as DNA hybridization or antigen–antibody recognition, as will be discussed later.

Polymer sensing layers frequently show a partial selectivity because of the selective swelling phenomenon and the fact that the molecules from the environment diffuse into the polymer layer at different rates mainly depending on the size and affinity of the molecules to the polymer layer. By selecting polymer surface layers expressing a wide range of hydrophilic or hydrophobic ligands, the chemical affinity can be manipulated to bind various molecules through selected and specific intermolecular forces such as ionic or hydrogen bonding, or van der Waals forces.

20.1.3 Dynamic Resonance Frequency Shift Mode

As is known, the resonance frequency, f_0, of an oscillating microcantilever is a structure/material's characteristic constant if its elastic properties remain unchanged during the molecule adsorption/desorption process and damping effects are negligible. However, changing environmental conditions or changing mass alters the primary resonance frequency. Thus, by oscillating a microcantilever at its eigenfrequency, information of adsorption or desorption of mass can be obtained under the prerequisite that the molecules on the surface might be in a dynamic equilibrium with molecules from the environment.

The corresponding mass changes can be determined by tracking the change in the eigenfrequency (Δf_0) of the microcantilever during mass adsorption or desorption (as shown schematically in Figure 20.2). In this dynamic mode, the microcantilever is used as a special microbalance, with added mass on the surface causing the resonance frequency to shift to a lower value. The mass change on a rectangular cantilever during molecular adsorption is related to the resonance frequency shift according to the known equation:

$$\Delta m = \frac{k}{4\pi^2 n} \times \left(\frac{1}{f_0^2} - \frac{1}{f_1^2}\right) \qquad (20.3)$$

where n is a geometric parameter that depends upon the geometry of the microcantilever and equals 0.24 for rectangular cantilevers and f_1 is the eigenfrequency after the mass change [16].

Mass change determination can be combined with controlled temperature variations to facilitate "micromechanical thermogravimetry," as proposed by Berger *et al.* [17]. In the mass balance mode, the sample under investigation is mounted at the apex of the cantilever; however, its mass should not exceed several hundred nanograms in this case. In the case of adsorption or desorption (as well as decomposition processes), mass changes in the low picogram range can then be detected in real time.

20.1.4 Heat Sensing Behavior

Bimaterial microcantilevers comprised of the layers of two very different materials exhibit bending with change in temperature due to thermal expansion difference (bimorph effect) (Figure 20.1). This very well-known phenomenon is frequently referred to as the "bimetallic effect" and corresponding structures are called bimorphs [18, 19]. In reference to the microcantilever-based sensors, this mode of operation is frequently referred to as "heat mode" [4].

Due to the differential thermal expansion, silicon nitride cantilevers, for example, with a thin gold film on one side undergo measurable bending in response to extremely small temperature changes. Minute temperature variation results in significant differential stress in the cantilever generated by dissimilar thermal expansion of the silicon nitride cantilever and the gold coating [15]. Heat flow change may not only be caused by external influences, such as a change in the environmental temperature (thermal detection), but can also occur directly on the surface by a catalytic reaction or initiated by the variable thermal material properties of a sample material attached to the apex of the cantilever (the so-called micromechanical calorimetry) [20, 21].

A number of different phenomena acting concurrently might cause microcantilever static and dynamic responses under variable conditions. For instance, differential thermal expansion of the cantilever substrate and coating layer results in the bending of the bimaterial cantilever that can reach 60 nm per degree (Figure 20.3). However, the differential thermal expansion can be caused by other factors such as radiation-induced background heating or an exothermic reaction caused by analyte adsorption. Alternatively, the changes in the differential surface energy can be caused by the preferential adsorption of the analyte without any reaction or the surface tension stress due to the liquid phase on one side of the cantilever (capillary phenomenon). Mass change due to simple physical adsorption might also result in cantilever deflections.

Here, we will briefly discuss the role and level of contributions in the cantilever bending from different factors. Figure 20.3 summarizes the most common transduction mechanisms with typical attributes such as the stress and deflection of the microcantilever or the resonance frequency shift. The estimation of the linear stresses developed and corresponding typical cantilever deflections have been conducted by using Stoney's equation (20.2) from the available literature data.

In the capillary phenomenon, the liquid droplets adsorbed on the surface of the cantilever apply normal forces on the cantilever due to the vertical component of the surface energy of the liquid–vapor interface [22]. It has been recently experimentally

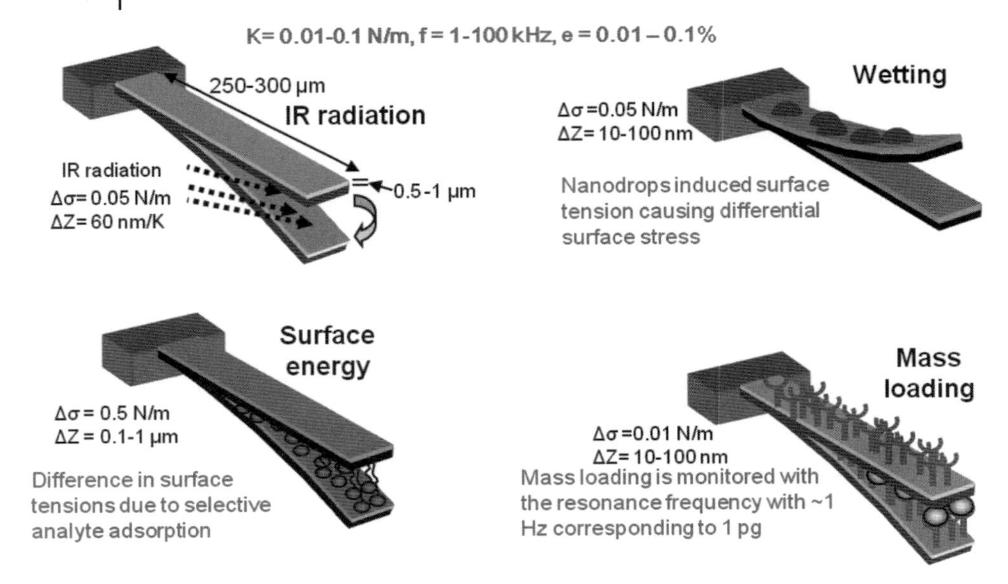

Figure 20.3 Transduction mechanisms of microcantilever-based sensors with the typical range of surface stresses and deflections achieved in each case for typical microcantilever dimensions. Reproduced from Ref. [5].

demonstrated that the vertical component ($\gamma_{LV}\sin\theta$) of the surface energy, which is neglected for immovable surfaces in the Young's equation, can cause a significant deflection of freely suspended structures but, generally, the static deflection caused by this phenomenon does not exceed 100 nm for regular microcantilevers with common spring constants [22].

The differential surface stress due to preferential adsorption of the analyte is the primary transduction mechanism for microcantilevers functionalized with SAMs and metal coatings that have a special affinity to the analyte molecules, while the other surface remains largely insensitive to them. In such a process, the differential stress can be as high as 0.5 N/m, resulting in large static deflections approaching 1000 nm, with both values far exceeding any other secondary contribution (Figure 20.3). Moreover, dimensional changes in sensitive materials (e.g., polymer layers) due to the sorption of analyte molecules termed as swelling or deswelling might result in even larger interfacial stress (as high as a few MPa in some cases, discussed later) in the bimaterial structure causing the bending of the entire structure by many microns, clearly overshadowing other contributions.

On the other hand, adsorption and desorption of the analyte on the surface and bulk of the functional coating result in a change in the effective mass. The overall stress developed for such a process is not very high (usually below 0.01 N/m) and the deflection is modest (around 10 nm in most cases), resulting in the parameters being difficult to detect in both cases (Figure 20.3). Biomolecular interactions (DNA hybridization, protein conformation changes, and antibody–antigen interactions) also cause a shift in resonance frequency or cantilever deflection due to differential surface stress [23]. This deflection originates from osmotic pressure when biomolecules closely bind on one surface of the lever. Practical deflection measurements thus typically rely on a high surface density of receptor molecules and a close packing of bound analytes rather than on just added mass effects. The resonance frequency of the microcantilever decreases as mass is bound to the surface with typical sensitivities approaching 1 Hz/pg. Overall, sensitivity based upon dynamic detection in changes in resonance frequency is concurrent with sensitivity based upon measuring static deflection of the cantilever and frequently exceeds it.

20.2 Thermal and Vapor Sensing

20.2.1 Microcantilever Thermal Sensors

The detection of infrared (IR) radiation, and, in particular, the wavelength regions from 3 to 5 μm and 8 to 14 μm, is important since atmospheric absorption in these regions is especially low [24, 25]. Thermal IR detectors can be based on pyroelectric [26], thermoelectric, thermoresistive (bolometers) [27–31], and micromechanical transducers [32–39].

Bimaterial microcantilevers can also be used to turn heat into a mechanical response and can be referred to as thermomechanical detectors. An important advantage of thermomechanical detectors is that they are essentially free of intrinsic electronic noise and can be combined with a number of different readout techniques with extremely high sensitivities. The bimaterial design can be exploited for IR detection by fabricating microcantilevers whereby bending of a cantilever upon incident radiation results from a mismatch in thermal expansion coefficients (α) of the materials, as was discussed earlier [21, 40].

This approach was pioneered by Barnes and Gimzewski when they coated microcantilevers with a metal (as the sensing active layer) to form a bimorph [41]. Later, Datskos *et al.* made the point that 2D arrays of these heat-sensitive cantilevers can serve as thermal imaging devices [32, 42]. The ideal bimaterial properties of a microcantilever engineered for IR sensing include large mismatch of α and thermal conductivity (λ) between the two materials with one of the materials having extremely low λ. Low residual stresses are useful to reduce nonthermal bending and one of the materials should absorb in the desired IR range. However, LeMieux and coworkers have demonstrated that trapped strong residual stress in the coating layers or at the polymer–inorganic interface can significantly enhance the response of the bimaterial structures, as will be discussed later [43].

In one of the recent designs, Quate and coworkers used silicon cantilevers exotically shaped in a flat spiral with an aluminum coating to complete the efficient bimorph [44]. Datskos and coworkers developed a microcantilever bimorph with silicon as a substrate and a 150 nm gold coating as the high α component that exhibited temperature sensitivity of 0.4 °C [45]. Majumdar and coworkers applied

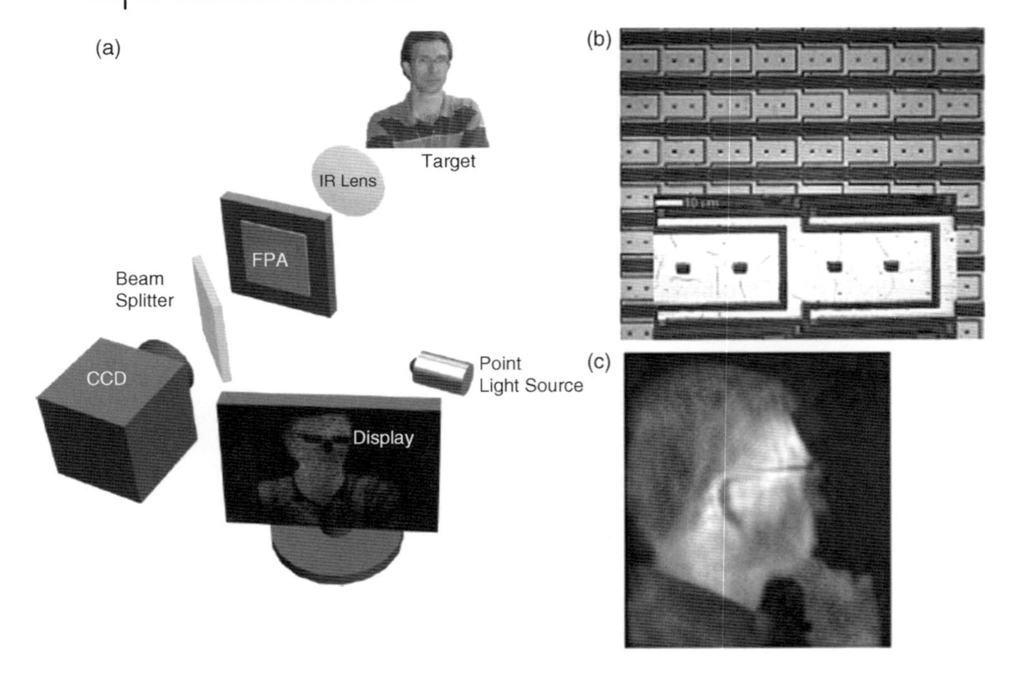

Figure 20.4 (a) The components and the arrangement of the optical readout of a 256 × 256 array of bimaterial microcantilevers. (b) Part of a fabricated 256 × 256 array. The selected geometry is characterized by a high fill factor and relaxed design rules. (c) Example of a thermal image of a human being obtained using an optical readout. Reproduced from Ref. [48].

bimaterial cantilevers of silicon nitride and gold into a complicated comb-like MEMS structure, which resulted in a sensitivity of 3–5 K [46]. Sarcon Microsystems in collaboration with Oak Ridge National Laboratory (ORNL) has developed bimaterial cantilevers with a theoretical sensitivity approaching 5 mK, which is the lowest value reported for uncooled IR detectors based on metal-coated cantilevers (Figure 20.4) [47]. Figure 20.4 displays an example of a thermal imaging system based on an array of 256 × 256 bimaterial microcantilevers with an optical readout [48]. In this design, SiC was the low thermal expansion component, again being combined with aluminum as the high thermal expansion layer. An optical readout was used to simultaneously interrogate all the microcantilevers.

Considering that the difference in thermal expansion coefficients for metal–ceramic bimaterial designs discussed so far is inherently limited ($\Delta\alpha < 20 \times 10^{-6}\,\mathrm{K}^{-1}$), the polymer–ceramic bimaterial cantilevers have been suggested to dramatically enhance thermally induced bending due to much more efficient actuation of readily expandable polymer nanolayers ($\Delta\alpha \geq 2 \times 10^{-4}\,\mathrm{K}^{-1}$) combined with low thermal conductivity [49]. Thus, polymer–nanoparticle composite structures have been introduced with a combination of polymer brush layers, silver nanoparticles, and carbon nanotubes to enhance IR adsorption and reinforce the nanocomposite coating. The application of such a reinforced nanocomposite coating with a high thermal expansion coefficient (α) produces a nearly fourfold improvement (theoretical, noise-limited detection limit of 0.5 mK) in the thermal sensitivity compared to the metal-coated counterparts [49]. Although there was substantial improvement in the sensitivity of the thermal bimorphs, the wet grafting technique employed to modify the microcantilevers was tedious, frequently resulted in cantilever damage, and was not compatible with traditional microfabrication technology.

To overcome these shortcomings, plasma polymerization has been employed to fabricate actuators in microcantilever thermal bimorphs [43]. Plasma polymerized styrene (ppS)-coated microcantilevers exhibited an extremely high deflection sensitivity of nearly 2 nm/mK making the theoretical detection limit to be as low as 0.2 mK (Figure 20.5). It is interesting to note that the response of the plasma polymer-coated microcantilevers is opposite to thermal expansion since the variation in the internal stresses in the polymer layer with temperature dominates the bending event.

In fact, the estimation of the internal plasma polymerized coating stresses gives differential stresses close to 100 MPa, which is a very high value indicating high compression of the stressed polymer layer at room temperature. It is suggested that the high cross-link density of the polymer layer and their chemical grafting at the

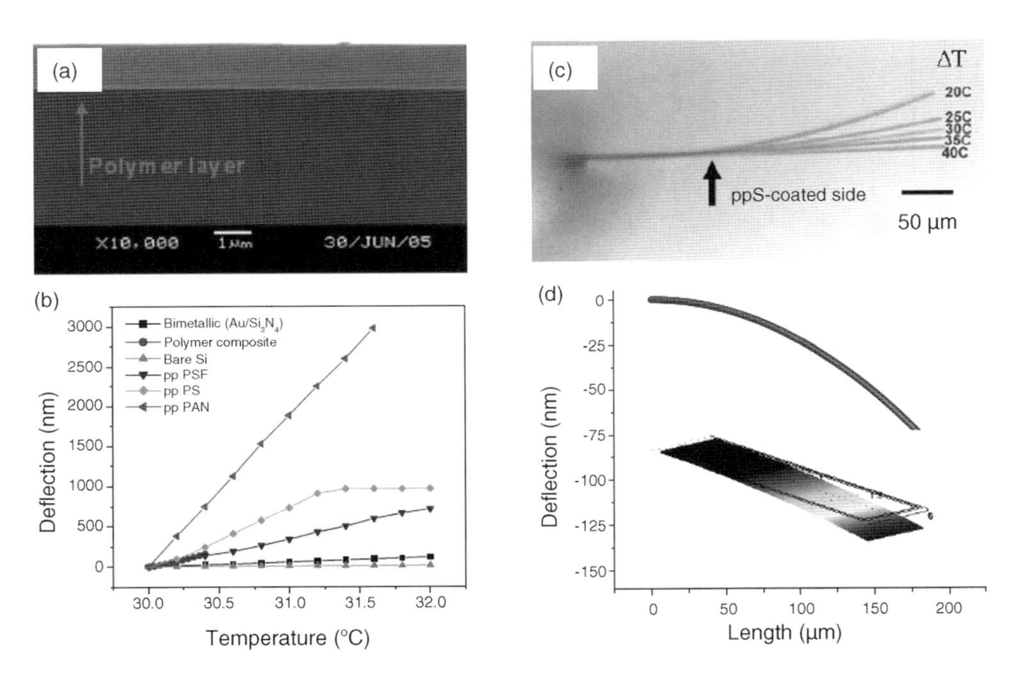

Figure 20.5 (a) SEM image of microcantilever coated on a single side with ppS. (b) Deflection response to temperature for microcantilevers (350 μm × 35 μm × 1 μm) coated with gold, polymer composite (polyacrylonitrile/carbon nanotube), plasma polymerized pentafluorostyrene, styrene, acrylonitrile, and uncoated reference. Reproduced from Ref. [49]. (c) Overlapped optical micrographs of the ppPS-microcantilever at various temperatures. (d) Finite element modeling showing the deflection profile of the microcantilever for 1 K change in temperature. Reproduced from Ref. [43].

interface should provide enhanced mechanical and thermal stability even under such high stress. Calculation of the differential surface stresses with finite element analysis (Figure 20.5) gives a value of 10 N/m, which is much higher than stresses usually generated by grafting polymer layers (usually within 0.3–1 N/m) and resulting from molecular adsorption (<0.2 N/m) [50]. Moreover, it shows that compressive stress within the polymer layer reaches 60 MPa, and is balanced by the tensile stress located exactly at the polymer–silicon interface combined with compressive stress at a bare silicon surface.

20.2.2 Chemical Sensors

Sensors for a reliable detection of solvent vapors are important in chemical processing technology. For example, safe handling during storage and transport of large amounts of solvents in a container, where a simple and fast test for real-time monitoring of chemical vapors may be required [51–53].

The microcantilever bending due to the interaction between the solvent vapor and the polymer with respect to time and magnitude evolution was exploited for vapor monitoring in a number of studies [54–69]. In a typical laboratory test, 0.1 mL of various solvents is placed in vials, and the vapor from the headspace above the liquid was sampled using microcantilever sensor arrays, operated in static deflection mode as a kind of artificial nose (Figure 20.6) [52]. Detection of vapors takes place via diffusion of the analyte molecules into the polymer coating, resulting in a swelling of the polymer, interfacial stresses, and bending of the cantilever. Each cantilever is coated with a different polymer or polymer blend to provide high selectivity (see Figure 20.6).

Examples of cantilever deflection traces upon injection of dichloromethane vapor at 50 s for 10 s are shown in Figure 20.6. The cantilever deflections at the time points t_1 to t_5 describe the time development of the curves in a reduced data set, that is, 40 (8×5) cantilever deflection amplitudes ("fingerprints") that account for a measurement data set (Figure 20.6). This data set is then evaluated using PCA techniques, extracting the most dominant deviations in the responses for the various vapors (Figure 20.6) [52]. The axes refer to the projections of the multidimensional data sets into two dimensions (principal components). Vapor injections involved water, ethanol, dichloromethane, and toluene. The PCA plot shows well-separated clusters of measurements indicating clear identification of vapor samples.

In another example, microcantilevers coated with 300 nm of plasma polymerized methacrylonitrile (ppMAN) exhibited a nearly 3.5 μm static deflection for a 1% change in relative humidity, making the detection limit a very low value of 10 ppb, or differences in 0.00005% in relative humidity (Figure 20.7) [70]. These microcantilevers exhibited a monotonous deflection response from 5% to 70% (total deflection >200 μm) with small hysteresis (<2%). The microcantilevers also exhibited an extremely stable response to water vapor over long storage time (nearly 2 years) with less than 5% variation. The microcantilevers coated with different plasma polymers have been exploited for organic vapor and plastic explosive detection and

(a)

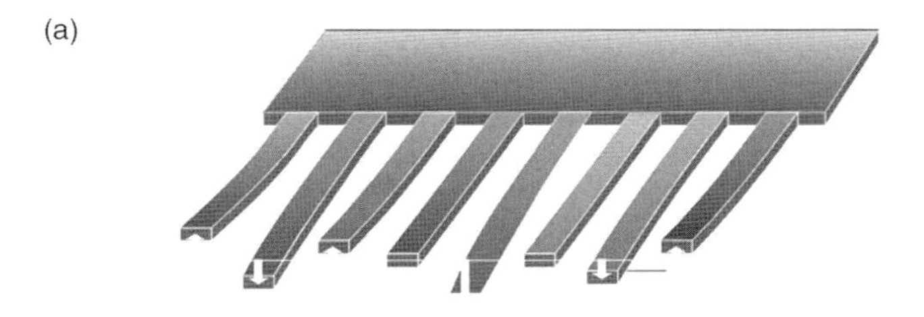

(b)

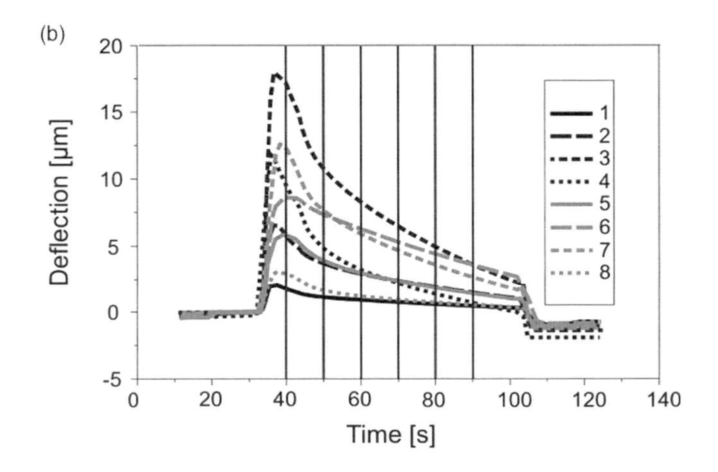

(c)

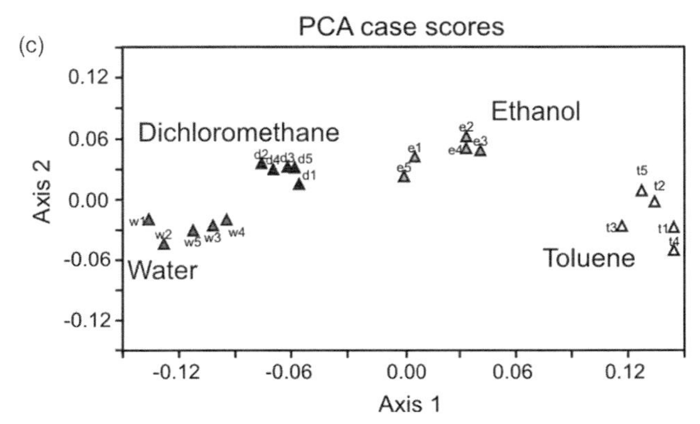

Figure 20.6 (a) A cantilever array functionalized with polymers exhibiting a deflection pattern that can be exploited for artificial nose applications. (b) Cantilever deflection traces during exposure to dichloromethane vapor. The following polymers were used: 1 = PVP, 2 = PVP/PU/PS/PMMA, 3 = PU/PS/PMMA, 4 = PU/PS, 5 = PU, 6 = PS/PMMA, 7 = PS, and 8 = PMMA. PVP = polyvinylpyridine, PU = polyurethane, PS = polystyrene, PMMA = polymethyl methacrylate. (c) PCA plot demonstrating the recognition capability of the cantilever array. Reproduced from Ref. [52].

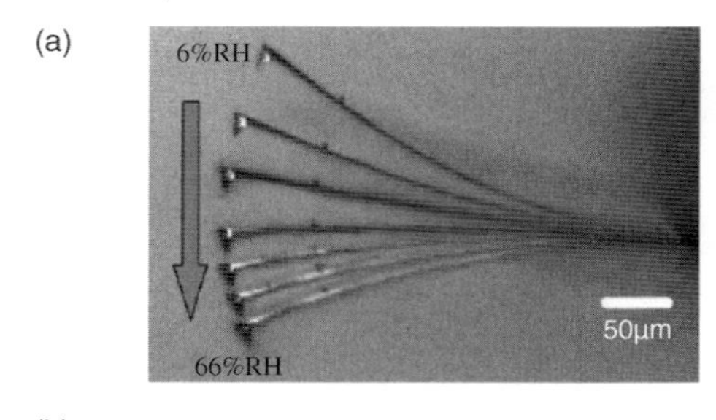

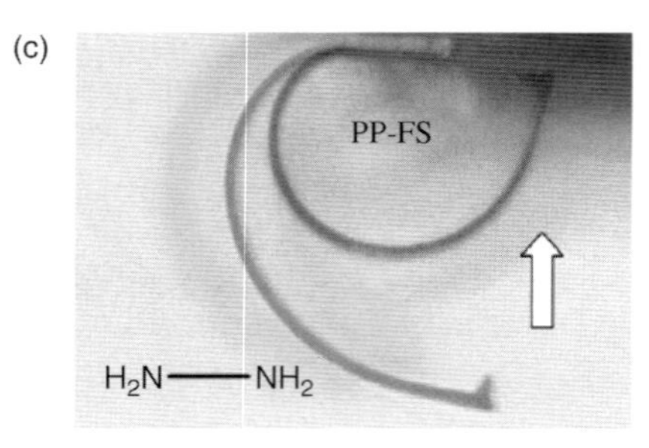

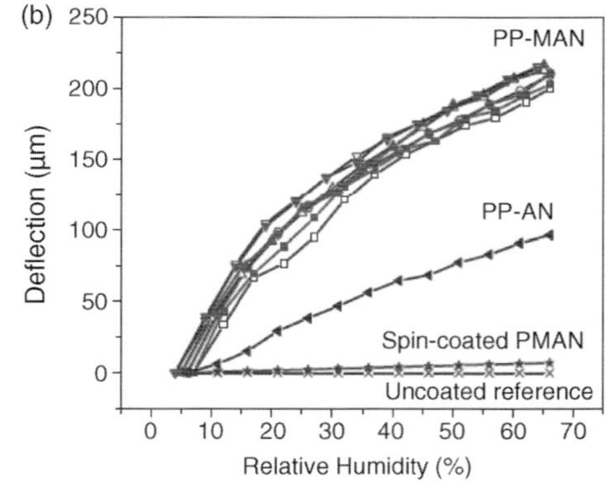

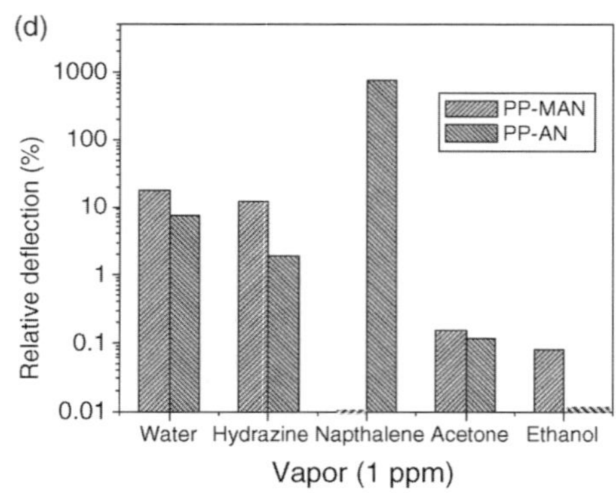

Figure 20.7 (a) Optical images showing the bending of the ppMAN-coated cantilever for humidity changing from 6% to 66% RH at an interval of 10% RH with deflection at 6% taken as a reference point. (b) The deflection versus humidity of cantilevers coated with ppMAN, ppAN, spin-coated PMAN, and a bare silicon cantilever. Empty symbols indicate humidification and corresponding filled symbols indicate desiccation: triangles (5th consecutive cycle), inverted triangles (10th consecutive cycle), squares (4 months after fabrication), and circles (after 18 months). (c) Optical image of a ppSF-coated microcantilever bent nearly 180° due to high residual stress and response to hydrazine vapor. (d) Deflection response of ppAN and ppMAN to 1 ppm concentration of different vapors. Reproduced from Ref. [70].

proved to be highly sensitive, very robust, and extremely selective with the response spanning four orders of magnitude to various analyte molecules (Figure 20.7). Moreover, under certain conditions, the interaction of plasma polymer-coated cantilever with analytes resulted in such strong stress that the microcantilever becomes completely bent (Figure 20.7).

It is suggested that, unlike conventional bimaterial structures, which rely on small differences in the surface tension for active and passive sides, plasma polymerized coating facilitates a mechanism involving large interfacial stresses causing inherently higher bending forces. High stress at the plasma polymer–silicon interface facilitates large micron-scale responses to a minute external stimulus. The exceptional performance of the plasma polymerized ppMAN nanocoating is likely caused by the peculiar nanodomain morphology, the nanoporous structure, and the presence of polar segments and hydrophobic methyl groups in a highly randomized cross-linked chemical topology, all of which facilitate fast uptake and removal of water molecules.

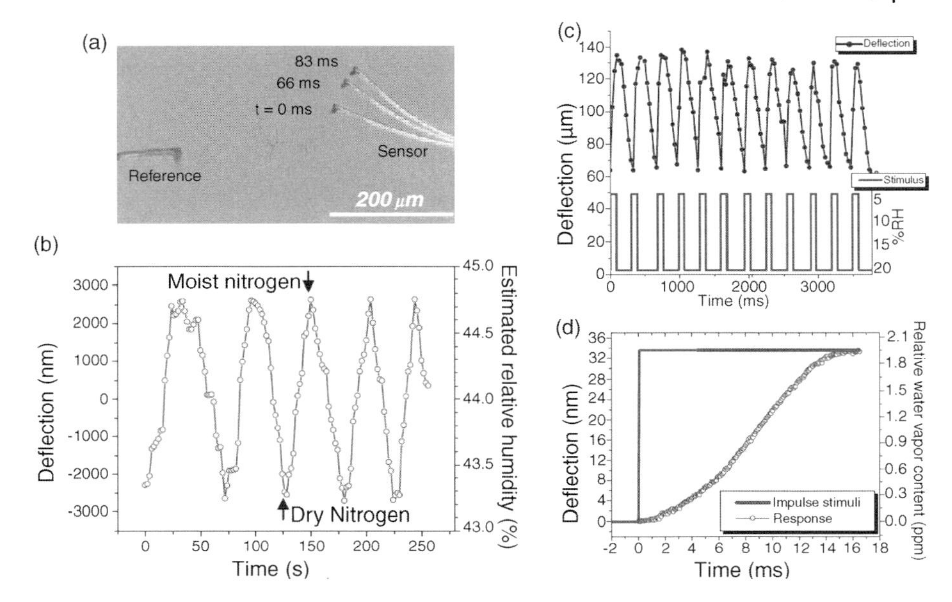

Figure 20.8 Dynamic response of microcantilever coated with ppMAN for repeated cycles of humidification and desiccation: (a) overlaid snapshots of the cantilevers depicting the response to nitrogen pulse; (b) deflection of the cantilever under desiccating nitrogen pulses followed by relaxation to humid state; (c) response to cycles of small variations in humidity; (d) deflection of cantilever to a sudden change in humidity (0.01% step). Reproduced from Ref. [70].

The reported sensitivity of 3500 nm/1% RH for plasma polymerized cantilevers is more than two orders of magnitude better than that achieved with regular coatings, indicating efficient transfer of swelling-induced stress to the polymer–inorganic interface. The most interesting aspect of this type of microcantilever is the very fast response time, which is essentially instantaneous (better than 80 ms) for large humidity changes and close to 10 ms for small changes in water content (Figure 20.8) [70].

Although explosive detection falls into the broad category of chemical sensing, it is unique not only due to the exigency in combating the potential threats but also due to the extremely high sensitivity requirements. Preventive countermeasures require inexpensive, highly selective, and very sensitive small sensors that can be mass-produced and microfabricated. Such low-cost sensors could be arranged as a sensor grid for large area coverage of sensitive infrastructure, such as airports, public buildings, or traffic infrastructure. Microfabricated detectors for explosives will be very useful as compact versions of established monitoring technologies such as ion mobility spectrometry or nuclear quadrupole resonance, which have been developed but are not likely to be miniaturized further [71, 72].

Two approaches have been adapted for the detection of explosives using microcantilevers: (1) static or dynamic mode of operation, in which microcantilevers are functionalized with SAMs or polymer layers to achieve selective binding, and

(2) microdeflagration of the explosives on the microcantilever surface [70, 73–78]. Several approaches to detect dangerous chemicals are described in literature: photomechanical chemical microsensors based on adsorption-induced and photo-induced stress changes due to the presence of diisopropyl methyl phosphonate (DIMP), which is a model compound for phosphorous-containing chemical warfare agents, and trinitrotoluene (TNT) [56]. Other explosives frequently used include pentaerythritol tetranitrate (PETN) and hexahydro-1,3,5-triazine (RDX) [73]. These compounds are very stable but their explosive power is very large, and moreover, the vapor pressures of PETN and RDX are very low, in the range of ppb and ppt, making them very difficult to detect.

Pinnaduwage *et al.* have reported the detection of 10–30 ppt of PETN and RDX using microcantilevers functionalized with a SAM of 4-mercaptobenzoic acid [73]. The authors suggested that the hydrogen bonding between the nitro groups of the explosives molecule and the hydroxyl groups of the MBA is responsible for the reversible adsorption. The same group has demonstrated detection of DNT using SXFA-[poly(1-(4-hydroxy-4-trifluoromethyl-5,5,5-trifluoro)pent-1-enyl)methylsiloxane] polymer-coated microcantilevers [73]. The nitro (NO_2) group on all nitro aromatic explosives is highly electron deficient resulting in a high electron accepting ability for these molecules, which has been exploited for the specific recognition of the nitroaromatic explosives. In a recent study, Singamaneni *et al.* have demonstrated that plasma polymerized benzonitrile-coated microcantilevers exhibited extremely high static deflection with a low detection limit below 10 ppb for hydrazine, another potentially explosive component (Figure 20.7) [70].

The second method involves a microexplosion of the molecules sticking to microcantilever surface by an electrical pulse, which results in an exothermic spike in the static deflection signal of the microcantilever [73]. This spike was found to be related to the heat produced during deflagration. The amount of heat released is proportional to the area versus bending signal plot of the process. The detection of TNT via deflagration was demonstrated by Pinnaduwage *et al.* who used piezoresistive microcantilevers [77]. It is worth noting that the inherent stickiness of the TNT molecules was exploited without any special functionalization as well. For example, TNT was found to readily stick to Si surfaces [79, 80]. Due to the small difference in the affinity between the two surfaces of the microcantilever, finite deflection was observed before the voltage pulse. TNT vapor was observed to adsorb onto its surface resulting in a decrease in the resonance frequency. Application of an electrical pulse (10 V, 10 ms) to the piezoresistive cantilever resulted in deflagration of the TNT causing a bump in the cantilever bending. The deflagration was found to be complete, as the same resonance frequency as before the experiment was observed. The amount of TNT mass detected in this study was determined at 50 pg.

This technique applied to the detection of PETN and RDX displayed a much slower kinetic response [73, 78]. Traces of DNT in TNT have also been used for detection of TNT because it is the major impurity in production grade TNT and is a decomposition product of TNT. The saturation concentration of DNT in air at 20 °C is 25 times higher than that of TNT. DNT was reported to be detected at the 300 ppt level using polysiloxane polymer layers [73]. Microfabrication of electrostatically actuated resonant microcantilevers in complementary metal oxide semiconductor technology for detection of the nerve agent stimulant dimethylmethylphosphonate (DMMP) using polycarbosilane-coated microcantilevers is an important step toward an integrated sensing platform [81, 82].

20.3 Sensing in Liquid Environment

Microcantilever-based sensors can be deployed in liquid environments, which make them applicable to biosensing. Over the past 10 years, microcantilevers have also been applied as biosensors as stand-alone structures and in arrays for detecting protein interactions, DNA binding, and microorganism behavior on surfaces [83–96, 112]. By specifically functionalizing only one microcantilever surface, either the frequency or the deflection mode can be used to detect specific biomolecular binding events.

It is important to note that while dynamic mode works efficiently in the gas phase where the quality factor remains virtually unchanged compared to vacuum (the resonance frequency shifts by a few percentages), in a liquid environment, this approach suffers from substantial damping of the cantilever oscillation. This damping is due to the high viscosity of the surrounding medium increasing drag forces by several orders of magnitude. This fluid damping results in a low-quality factor $Q = f_0/\Delta f$, where Δf is the full-width at half-maximum of the frequency spectrum. The dramatic drop in the quality factor usually observed in liquid is from the typical range of 100–1000 in air to values below 50 [97, 98].

Under these conditions, the resonance frequency shift is difficult to track with high accuracy and thus the overall sensitivity in liquid decreases dramatically. While in air a frequency resolution of below 1 Hz is easily achieved for common cantilevers, resolution values of only about 20 Hz is considered very good for measurements in a liquid environment. Moreover, in the case of damping or changes in the elastic properties of the cantilever during the experiment, the measured resonance frequency will not be exactly the same as the eigenfrequency, and the mass derived from the frequency shift will be inaccurate and require special efforts [99].

A novel design for the interrogation of solutions that eliminates these difficulties has been reported very recently by Burg *et al.* [100]. The authors suggested the fabrication of complex microcantilevers with microfluidic channels embedded into the cantilevers, as shown in Figure 20.9. The fluid continuously flowing through the channel to deliver the analyte species causes a change in the resonant frequency of the suspended microchannel due to binding of the analyte without compromising the cantilever performance (Figure 20.9).

A transient flow of particles through the channel results in temporal dips in the resonance frequency (Figure 20.9) depending on the position of the particles along the channel. An excellent quality factor of 15 000 was reported for a microresonator channel filled with water or air, and the ability to detect single biomolecules and nanoparticles in fluid with the lowest detected mass around 300 attograms was

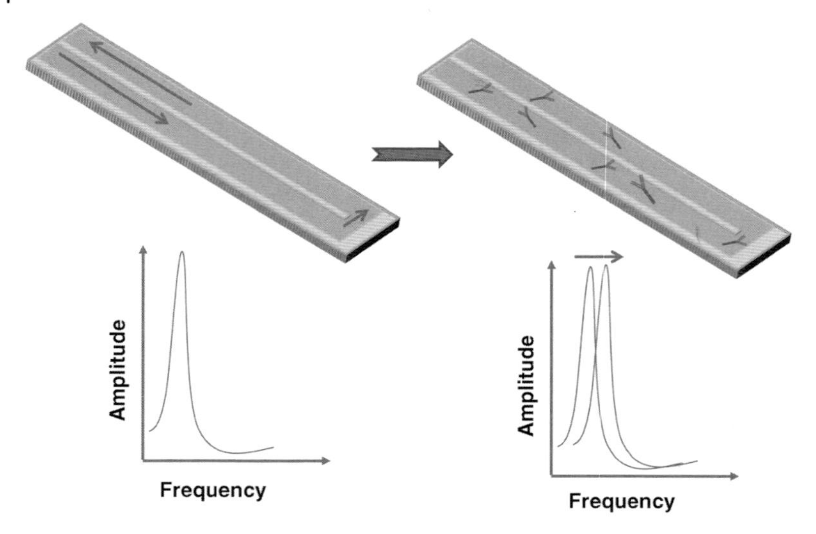

Figure 20.9 Schematic showing the microfluidic resonator structure which exhibits a change in the resonance frequency as the analyte binding to the complementary species.

demonstrated [100]. This new design paradigm has the potential to significantly improve the applications of microcantilevers for fluidic environments.

Commonly used instruments for observing the kinetics of biomolecular interactions on a surface include surface plasmon resonances and quartz crystal microbalance with dissipation monitoring (QCM-D) [101, 102]. Due to the extreme sensitivity of microcantilever transduction (nearly three orders higher mass sensitivity than QCM [103]), it has attracted considerable attention for probing the dynamics of physical and chemical processes occurring at molecular levels. The first report of monitoring a chemical reaction on the surface of a microcantilever was done by Barnes *et al.*, in which they monitored the catalytic conversion of $H_2 + O_2$ to H_2O on a platinum-coated microcantilever, introducing the concept of a microcantilever-based microcalorimeter [21]. The kinetics of chemisorption of alkanethiols on gold-coated microcantilevers has been investigated by various groups [50, 104, 105]. Berger *et al.* have observed a linear increase in the surface stress at the monolayer with the length of the alkyl chain of the molecules. It is important to note that the deflection of microcantilevers during the assembly process was dominated by the differential surface stress, while the thermal and gravimetric effects were negligible [104].

Microcantilever-based sensing has also been employed to probe the swelling of polymer layers, self-assembly of polyelectrolyte monolayers, formation of lipid layers, and conformational changes in proteins [103, 106–111]. It is worth noting that since microcantilevers are sensitive to the slightest environmental fluctuations due to flow and thermal gradients, data acquisition during analyte introduction in liquids often precludes observation of the association kinetics within a given time interval. The time necessary for full equilibration ranges between minutes and hours for deflection sensing [106, 112, 113]. A time delay also exists between molecular binding and generation of sufficient surface stress to initiate measurable cantilever deflections, which also complicates the experimental results.

The selectivity of the microcantilever response is based on the specificity of the capture molecule. For instance, in the case of a glucose sensor, glucose oxidase (GO_x) was immobilized on a single gold-coated rectangular microcantilever [114]. The study showed that microcantilevers had the ability to detect physiologically relevant levels of glucose, but they could not yet conduct selective detection from a complex mixture of proteins and plasma normally present in a blood sample. Selectivity is decreased in this electrochemical setup due to the presence of interfering electroactive species (e.g., ascorbic acid, catechol, or uric acid). Microcantilever technology, however, does not detect any such spectator species, and thus its selectivity is modest in most cases.

Recently, Chen *et al.* employed polymer brushes as actuators for the detection of glucose at physiologically relevant concentrations [115]. The authors formed glucose-responsive poly(*N*-isopropylacrylamide)-copoly(acrylic acid)-(3-aminophenyl-boronic acid) (PNIPAAM-*co*-PAA-PBA) brushes on microcantilevers and compared the response to PNIPAAM-*co*-PAA. In order to probe the swelling response of the polymer brushes, the brushes were also formed on a planar surface and AFM was employed to monitor the swelling of the micropatterned brushes. Upon exposure of PNIPAAM-*co*-PAA-PBA (~140 nm) and PNIPAAM-*co*-PAA (~140 nm) polymer brushes to a buffer at pH 9, significant swelling to 420 and 400 nm, respectively, was observed. When immersed in 50 mM glucose, only PNIPAAM-*co*-PAA-PBA brushes exhibited swelling reaching a thickness of 560 nm, while no additional swelling was observed in the case of PNIPAAM-*co*-PAA brushes.

The swelling of the polymer brush in the glucose causes differential stress at the surface of the cantilever resulting in bending of the cantilever. The authors compared the bending response of the PNIPAAM-*co*-PAA-PBA and PNIPAAM-*co*-PAA brush-functionalized cantilevers at two different pH conditions [115]. The microcantilever functionalized with PNIPAAM-*co*-PAA-PBA exhibited a much higher response to glucose at pH 9 compared to the cantilever functionalized with PNIPAAM-*co*-PAA. The authors suggested that at pH 9 both PAA and PBA are ionized and the repulsive interaction of charges due to the increase in the ion concentration in the polymer chains leads to an increase in swelling and hence a higher response, which was also in agreement with AFM observations.

In another study, Wu *et al.* used microcantilevers to detect PSA from a mixture of blood proteins [112]. Commercially available V-shaped, gold-coated microcantilevers were decorated with a layer of anti-PSA antibody. Significant deflection was observed upon binding of free PSA. When exposed to concentrations spanning the diagnostically relevant range (0.2 ng/mL–60 μg/mL), the microcantilevers showed a distinct deflection in a background of 1 mg/mL BSA. The deflection signal at 60 ng/mL versus 6 ng/mL was also clearly distinguishable even in the presence of 1 mg/mL human serum albumin. Changes in microcantilever geometry resulted in shifted deflections for the same concentration of PSA. The authors found that surface stress, however, was independent of geometry and directly related to PSA concentration.

Nonspecific interactions had very little effect on deflection ability, suggesting that surface stress is a sensitive reporter of specific PSA binding.

The high selectivity of these assays resides in the specificity of the biomolecular reactions occurring on the properly modified microcantilever surface. Unlike electrochemical or labeling detection methods, the free-energy change resulting from molecular recognition binding is directly translated into a mechanical deflection, reducing the effects of sources of interference (e.g., optical, electrochemical, etc.). Smaller cantilevers, increasing ligand surface density, and decreasing surface roughness are suggested as a means of increasing detection sensitivity. The use of differential signals by employing reference, unmodified cantilevers may further increase sensitivity and resolution, while decreasing errors from nonspecific interactions.

Microcantilevers have been used not only to study binding of specific ligands to functionalized sides but also to monitor the conformation of biological macromolecules after adsorption on the surface. Since microcantilever detection does not rely on a particular material substrate, a wide variety of surface functionalities can be presented. Thus, a range of interactions can be screened and monitored relatively easily. In fact, protein interactions, DNA hybridization, protein conformations, and, more recently, lipid bilayer formation have all been observed using microcantilever sensors [83–90, 92, 93, 106, 112].

DNA hybridization was one of the first processes studied using functionalized microcantilevers in liquid. Fritz *et al.* showed that a single base pair mismatch could be detected by observing microcantilever nanoscale deflection [83]. In these experiments, a linear array of eight rectangular microcantilevers, each functionalized with a different oligonucleotide base sequence, was employed. By determining the differential deflections between adjacent cantilevers, confounding effects such as nonspecific interactions were significantly minimized, and thus the sensitivity to a single pairing event has been optimized. Moreover, it was observed that the cantilevers were reusable after cleaving the bound DNA with urea.

Many subsequent studies have expanded this work beyond its initial scope. For example, Braun *et al.* used microcantilevers to monitor the conformational changes in membrane-bound bacteriorhodpsin on a gold surface [116]. Ink jet spotting was used to deposit bacteriorhodpsin on a linear array of eight gold-coated rectangular microcantilevers, where some cantilevers were also left pristine to serve as a reference. Cantilever deflection was then monitored using a standard optical photodiode system. bR was prebleached by removing various degrees of retinal, an internally bound ligand, resulting in different bacteriorhodpsin conformations. Hydrolysis of retinal was also simulated *in situ* by addition of hydroxylamine. It was found that 33% prebleached bacteriorhodpsin resulted in a much lower deflection upon addition of hydroxylamine when compared to unbleached bacteriorhodpsin.

As noted in other studies and mentioned briefly above, the cantilever deflection was unstable during flow, compromising stability, and data were thus recorded before and after bleaching. Since nonspecific adsorption of hydroxylamine was observed on both bR and control cantilevers, the differential deflection was thought to be independent of nonspecific binding, and was caused by conformational changes due to retinal removal. This conformational change in bacteriorhodopsin was postulated to result in the expansion of the membrane patches, further altering the surface stress and affecting cantilever deflection.

Furthermore, vesicle fusion and lipid bilayer formation have been observed using microcantilever deflection in liquid [109]. In these experiments, the deflection of a microcantilever linear array was monitored by individual laser beams. To observe physisorbed bilayer formation on the bottom SiO_2 surface, the top gold surface was coated with 2-mercaptoethanol to minimize vesicle fusion there. After incubating with vesicles and flushing with a buffer, tensile stresses developed due to the bilayer formation led to cantilever deflection on the order of 80 nm, corresponding to a low but detectable surface stress of about 30 mN/m.

In another experiment, chemisorbed bilayer formation was observed on the top gold surface with thiolated lipids [109]. This bilayer formation resulted in the compressive stress causing the cantilever deflection in the hundreds of nanometers indicating 10 times the surface stresses from simply physisorbed bilayers. When mixed with unmodified lipids, the deflection response decreased proportionally. The "pinning" of the bilayer was postulated to increase the surface stress, and the overall increase was similar to that seen for alkylthiol SAMs.

References

1 Binnig, G., Quate, C.F., and Gerber, C. (1986) Atomic force microscope. *Phys. Rev. Lett.*, **56** (9), 930–933.

2 Karrasch, S., Hegerl, R., Hoh, J.H., Baumeister, W., and Engel, A. (1994) Atomic force microscopy produces faithful high-resolution images of protein surfaces in an aqueous environment. *Proc. Natl. Acad. Sci. USA*, **91** (3), 836–838.

3 Muller, D.J., Schabert, F.A., Buldt, G., and Engel, A. (1995) Imaging purple membranes in aqueous solutions at sub-nanometer resolution by atomic force microscopy. *Biophys. J.*, **68** (5), 1681–1686.

4 Gimzewski, J.K., Gerber, Ch., Meyer, E., and Schlittler, R.R. (1994) Observation of a chemical reaction using a micromechanical sensor. *Chem. Phys. Lett.*, **217** (5–6), 589–594.

5 Singamaneni, S., LeMieux, M.C., Lang, H.P., Gerber, Ch., Lam, Y., Zauscher, S., Datskos, P.G., Lavrik, N.V., Jiang, H., Naik, R.R., Bunning, T.J., and Tsukruk, V.V. (2008) Bimaterial microcantilevers as a hybrid sensing platform. *Adv. Mater.*, **20** (4), 653–680.

6 Goeders, K.M., Colton, J.S., and Bottomley, L.A. (2008) Microcantilevers: sensing chemical interactions via mechanical motion. *Chem. Rev.*, **108** (2), 522–542.

7 Hazel, J.L. and Tsukruk, V.V. (1999) Spring constants of composite ceramic/gold cantilevers for scanning probe microscopy. *Thin Solid Films*, **339** (1–2), 249–257.

8 Sader, J.E., Larson, I., Mulvaney, P., and White, L.R. (1995) Method for the calibration of atomic force microscope cantilevers. *Rev. Sci. Instrum.*, **66** (7), 3789–3798.

9 Sader, J.E. (1995) Parallel beam approximation for V-shaped atomic force microscope cantilevers. *Rev. Sci. Instrum.*, **66** (9), 4583–4587.

10 Hazel, J. and Tsukruk, V.V. (1998) Friction force microscopy measurements: normal and torsional spring constants for V-shaped cantilevers. *J. Tribol.*, **120** (4), 814–819.

11 Sader, J.E., Chon, J.W.M., and Mulvaney, P. (1999) Calibration of rectangular atomic force microscope cantilevers. *Rev. Sci. Instrum.*, **70** (10), 3967–3969.

12 Hutter, J.L. and Bechhoefer, J. (1993) Calibration of atomic force microscope tips. *Rev. Sci. Instrum.*, **64** (7), 1868–1873.

13 Stoney, G.G. (1909) The tension of metallic films deposited by electrolysis. *Proc. R. Soc. Lond. A*, **82** (553), 172–175.

14 von Preissig, F.J. (1989) Applicability of the classical curvature–stress relation for thin films on plate substrates. *J. Appl. Phys.*, **66** (9), 4262–4268.

15 Klein, C.A. (2000) How accurate are Stoney's equation and recent modifications. *J. Appl. Phys.*, **88** (7), 5487–5489.

16 Thundat, T., Warmack, R.J., Chen, G.Y., and Allison, D.P. (1994) Thermal and ambient-induced deflections of scanning force microscope cantilevers. *Appl. Phys. Lett.*, **64** (21), 2894–2896.

17 Berger, R., Lang, H.P., Gerber, C., Gimzewski, J.K., Fabian, J.H., Scandella, L., Meyer, E., and Güntherodt, H.-J. (1998) Micromechanical thermogravimetry. *Chem. Phys. Lett.*, **294** (4–5), 363–369.

18 Shaver, P.J. (1969) Bimetal strip hydrogen gas detectors. *Rev. Sci. Instrum.*, **40** (7), 901–905.

19 Timoshenko, S.P. (1925) Analysis of bi-metal thermostats. *J. Opt. Soc. Am.*, **11** (6), 233–255.

20 Moulin, A.M., Stephenson, R.J., and Welland, M.E. (1997) Micromechanical thermal sensors: comparison of experimental results and simulations. *J. Vac. Sci. Technol. B*, **15** (3), 590–596.

21 Barnes, J.R., Stephenson, R.J., Welland, M.E., Gerber, C., and Gimzewski, J.K. (1994) Photothermal spectroscopy with femtojoule sensitivity using a micromechanical device. *Nature*, **372** (6501), 79–81.

22 Jeon, S., Desikan, R., Tian, F., and Thundat, T. (2006) Influence of nanobubbles on the bending of microcantilevers. *Appl. Phys. Lett.*, **88** (10), 103118.

23 Chen, G.Y., Thundat, T., Wachter, E.A., and Warmack, R.J. (1995) Adsorption-induced surface stress and its effects on resonance frequency of microcantilevers. *J. Appl. Phys.*, **77** (8), 3618–3622.

24 Mao, M., Perazzo, T., and Kwon, O. (1999) Direct-view uncooled micro-optomechanical infrared camera. Microelectromechanical Systems, Nashville, Twelfth IEEE International Conference, pp. 100–105.

25 Datskos, P.G., Rajic, S., Senesac, L.R., and Datskou, I. (2001) Fabrication of quantum well microcantilever photon detectors. *Ultramicroscopy*, **86** (1–2), 191–206.

26 Hanson, C. (1993) Uncooled thermal imaging at Texas Instruments. Infrared Technology XXI. *SPIE*, **2020**, 330–339.

27 Skatrud, D.D., Kruse, P.W., Willardson, R.K., and Weber, E.R. (1997) *Uncooled Infrared Imaging Arrays and Systems, Semiconductors and Semimetals*, Academic Press, San Diego.

28 Kruse, P.W. (1995) A comparison of the limits to the performance of thermal and photon detector imaging arrays. *Infrared Phys. Technol.*, **36** (5), 869–882.

29 Butler, N., Blackwell, R., Murphy, R., Silva, R., and Marshall, C. (1995) Low-cost uncooled microbolometer imaging system for dual use. Infrared Technology XXI. *SPIE*, **2552**, 583–591.

30 Wood, R.A. (1993) Uncooled thermal imaging with monolithic silicon focal planes. Infrared Technology XXI. *SPIE*, **2020**, 322–329.

31 Wood, R.A. and Foss, N.A. (1993) Micromachined bolometer arrays achieve low-cost imaging. *Laser Focus World*, June, 101–106.

32 Datskos, P.G., Oden, P.I., Thundat, T., Wachter, E.A., Warmack, R.J., and Hunter, S.R. (1996) Remote infrared radiation detection using piezoresistive microcantilevers. *Appl. Phys. Lett.*, **69** (20), 2986–2988.

33 Oden, P.I., Wachter, E.A., Datskos, P.G., Thundat, T., and Warmack, R.J. (1996) Optical and infrared detection using microcantilevers. Infrared Technology XXII. *SPIE*, **2744**, 345–354.

34 Perazzo, T., Mao, M., Kwon, O., Majumdar, A., Varesi, J.B., and Norton, P. (1999) Infrared vision using uncooled micro-optomechanical camera. *Appl. Phys. Lett.*, **74** (23), 3567–3569.

35 Datskos, P.G., Rajic, S., and Datskou, I. (1998) Photoinduced and thermal stress in silicon microcantilevers. *Appl. Phys. Lett.*, **73** (16), 2319–2321.

36 Amantea, R., Goodman, L.A., and Pantuso, F. (1998) An uncooled IR imager with 5 mK NETD. Infrared Technology and Applications XXIV. *Proc. SPIE*, **3436**, 647–659.

37 Lai, J., Perazzo, T., Shi, Z., and Majumdar, A. (1997) Optimization and performance of high-resolution micro-optomechanical thermal sensors. *Sens. Actuators A*, **58** (2), 113–119.

38 Senesac, L.R., Corbeil, J.L., Rajic, S., Lavrik, N.V., and Datskos, P.G. (2003) IR imaging using uncooled microcantilever detectors. *Ultramicroscopy*, **97** (1–4), 451–458.

39 Wachter, E.A., Thundat, T., Oden, P.I., Warmack, R.J., Datskos, P.G., and Sharp, S.L. (1996) Remote optical detection using microcantilevers. *Rev. Sci. Instrum.*, **67** (10), 3434–3439.

40 Varesi, J., Lai, J., Perazzo, T., Shi, Z., and Majumdar, A. (1997) Photothermal measurements at picowatt resolution using uncooled micro-optomechanical sensors. *Appl. Phys. Lett.*, **71** (3), 306–308.

41 Barnes, J.R., Stephenson, R.J., Woodburn, C.N., O'Shea, S.J., Welland, M.E., Rayment, T., Gimzewski, J.K., and Gerber, Ch. (1994) A femtojoule calorimeter using micromechanical sensors. *Rev. Sci. Instrum.*, **65** (12), 3793–3798.

42 Oden, P.I., Datskos, P.G., Thundat, T., and Warmack, R.J. (1996) Uncooled thermal imaging using a piezoresistive microcantilever. *Appl. Phys. Lett.*, **69** (21), 3277–3279.

43 LeMieux, M.C., McConney, M.E., Lin, Y.H., Singamaneni, S., Jiang, H., Bunning, T.J., and Tsukruk, V.V. (2006) Polymeric nanolayers as actuators for ultrasensitive thermal bimorphs. *Nano Lett.*, **6** (4), 730–734.

44 Manalis, S.R., Minne, S.C., Quate, C.F., Yaralioglu, G.G., and Atalar, A. (1997) Two-dimensional micromechanical bimorph arrays for detection of thermal radiation. *Appl. Phys. Lett.*, **70** (24), 3311–3313.

45 Corbeil, J.L., Lavrik, N.V., Rajic, S., and Datskos, P.G. (2002) "Self-leveling" uncooled microcantilever thermal detector. *Appl. Phys. Lett.*, **81** (7), 1306–1308.

46 Zhao, Y., Mao, M., Horowitz, R., Majumdar, A., Varesi, J., Norton, P., and Kitching, J. (2002) Optomechanical uncooled infrared imaging system: design, microfabrication, and performance. *J. MEMS*, **11** (2), 136–146.

47 Hunter, S.R., Amantea, R.A., Goodman, L.A., Kharas, D.B., Gershtein, S., Matey, J.R., Perna, S.N., Yu, Y., Maley, N., and White, L.K. (2003) High sensitivity uncooled microcantilever infrared imaging arrays. *Proc. SPIE*, **5074**, 469–480.

48 Grbovic, D., Lavrik, N.V., Datskos, P.G., Forrai, D., Nelson, E., Devitt, J., and McIntyre, B. (2006) Uncooled infrared imaging using bimaterial microcantilever arrays. *Appl. Phys. Lett.*, **89** (7), 073118.

49 Lin, Y.H., McConney, M.E., LeMieux., M.C., Peleshanko, S., Jiang, C., Singamaneni, S., and Tsukruk, V.V. (2006) Trilayered ceramic–metal–polymer microcantilevers with dramatically enhanced thermal sensitivity. *Adv. Mater.*, **18** (9), 1157–1161.

50 Godin, M., Williams, P.J., Tabard-Cossa, V., Laroche, O., Beaulieu, L.Y., Lennox, R.B., and Grütter, P. (2004) Surface stress, kinetics, and structure of alkanethiol self-assembled monolayers. *Langmuir*, **20** (17), 7090–7096.

51 Battiston, F.M., Ramseyer, J.-P., Lang, H.P., Baller, M.K., Gerber, Ch., Gimzewski, J.K., Meyer, E., and Guntherodt, H.-J. (2001) Chemical sensor based on a microfabricated cantilever array with simultaneous resonance-frequency and bending readout. *Sens. Actuators B*, **77** (1–2), 122–131.

52 Baller, M.K., Lang, H.P., Fritz, J., Gerber, Ch., Gimzewski, J.K., Drechsler, U., Rothuizen, H., Despont, M., Vettinger, P., Battiston, F.M., Ramseyer, J.P., Fornaro, P., Meyer, E., and Guntherodt, H.-J. (2000) A cantilever array-based artificial nose. *Ultramicroscopy*, **82** (1–4), 1–9.

53 Bumbu, G.-G., Kircher, G., Wolkenhauer, M., Berger, R., and Gutmann, J.S. (2004) Synthesis and characterization of polymer brushes on micromechanical cantilevers. *Macromol. Chem. Phys.*, **205** (13), 1713–1720.

54 Lang, H.-P., Baller, M.K., Berger, R., Gerber, Ch., Gimzewski, J.K., Battiston, F.M., Fornaro, P., Ramseyer, J.P., Meyer, E., and Güntherodt, H.-J. (1999) An artificial nose based on a micromechanical cantilever array. *Anal. Chim. Acta*, **393** (1–3), 59–65.

55 Okuyama, S., Mitobe, Y., Okuyama, K., and Matsushita, K. (2000) Hydrogen gas sensing using a Pd-coated cantilever. *Jpn. J. Appl. Phys.*, **39**, 3584–3590.

56 Datskos, P.G., Sepaniak, M.J., Tipple, C.A., and Lavrik, N. (2001) Photomechanical chemical microsensors. *Sens. Actuators B*, **76** (1–3), 393–402.

57 Datskos, P.G., Rajic, S., Sepaniak, M.J., Lavrik, N., Tipple, C.A., Senesac, L.R., and Datskou, I. (2001) Chemical detection based on adsorption-induced and photoinduced stresses in microelectromechanical systems devices. *J. Vac. Sci. Technol. B*, **19** (4), 1173–1179.

58 Rangelow, I.W., Grabiec, P., Gotszalk, T., and Edinger, K. (2002) Piezoresistive SXM sensors. *Surf. Interf. Anal.*, **33** (2), 59–64.

59 Lange, D., Hagleitner, C., Hierlemann, A., Brand, O., and Baltes, H. (2002) Complementary metal oxide semiconductor cantilever arrays on a single chip: mass-sensitive detection of volatile organic compounds. *Anal. Chem.*, **74** (13), 3084–3095.

60 Henkel, K. and Schmeisser, D. (2002) Erektile dysfunktion nach beckenfrakturen und beckentraumen. *Anal. Bioanal. Chem.*, **374** (6), 329–333.

61 Arutyunov, P.A. and Tolstikhina, A.L. (2002) Atomic force microscopy as a universal means of measuring physical quantities in the mesoscopic length range. *Meas. Techniques*, **45** (7), 714–721.

62 Zhou, J., Li, P., Zhang, S., Long, Y.C., Zhou, F., Huang, Y.P., Yang, P.Y., and Bao, M.H. (2003) Zeolite-modified microcantilever gas sensor for indoor air quality control. *Sens. Actuators B*, **94** (3), 337–342.

63 Gunter, R.L., Delinger, W.G., Manygoats, K., Kooser, A., and Porter, T.L. (2003) Viral detection using an embedded piezoresistive microcantilever sensor. *Sens. Actuators A*, **107** (3), 219–224.

64 Kooser, A., Manygoats, K., Eastman, M.P., and Porter, T.L. (2003) Investigation of the antigen antibody reaction between anti-bovine serum albumin (a-BSA) and bovine serum albumin (BSA) using piezoresistive microcantilever based sensors. *Biosens. Bioelectron.*, **19** (5), 503–508.

65 Kooser, A., Gunter, R.L., Delinger, W.D., Porter, T.L., and Eastman, M.P. (2003) Gas sensing using embedded piezoresistive microcantilever sensors. *Sens. Actuators B*, **99** (2–3), 474–479.

66 Wright, Y.J., Kar, A.K., Kim, Y.W., Scholz, C., and George, M.A. (2005) Study of microcapillary pipette-assisted method to prepare polyethylene glycol-coated microcantilever sensors. *Sens. Actuators B*, **107** (1), 242–251.

67 Dutta, P., Senesac, L.R., Lavrik, N.V., Datskos, P.G., and Sepaniak, M.J. (2004) Response signatures for nanostructured, optically-probed, functionalized microcantilever sensing arrays. *Sens. Lett.*, **2** (3–4), 238–245.

68 Senesac, L.R., Dutta, P., Datskos, P.G., and Sepaniak, M.J. (2006) Analyte species and concentration identification using differentially functionalized microcantilever arrays and artificial neural networks. *Anal. Chim. Acta*, **558** (1–2), 94–101.

69 Archibald, R., Datskos, P., Devault, G., Lamberti, V., Lavrik, N., Noid, D., Sepaniak, M., and Dutta, P. (2007) Independent component analysis of nanomechanical responses of cantilever arrays. *Anal. Chim. Acta*, **584** (1), 101–105.

70 Singamaneni, S., McConney, M., LeMieux, M.C., Jiang, H., Enlow, J., Naik, R., Bunning, T.J., and Tsukruk, V.V. (2007) Polymer–silicon flexible structures for fast chemical vapor detection. *Adv. Mater.*, **19** (23), 4248–4255.

71 Ewing, R.G. and Miller, C.J. (2001) Detection of volatile vapors emitted from explosives with a handheld ion mobility spectrometer. *Field Anal. Chem. Technol.*, **5** (5), 215–221.

72 Garroway, A.N., Buess, M.L., Miller, J.B., Suits, B.H., Hibbs, A.D., Barrall, G.A., Matthews, R., and Burnett, L.J. (2001) Remote sensing by nuclear quadrupole resonance. *IEEE Trans. Geosci. Remote Sens.*, **39** (6), 1108–1118.

73 Pinnaduwage, L.A., Boiadjiev, V., Hawk, J.E., and Thudat, T. (2003) Sensitive detection of plastic explosives with self-assembled monolayer-coated microcantilevers. *Appl. Phys. Lett.*, **83** (7), 1471–1473.

74 Pinnaduwage, L.A., Thundat, T., Hawk, J.E., Hedden, D.L., Britt, P.F., Houser, E.J., Stepnowski, S., McGill, R.A., and Bubb, D. (2004) Detection of 2, 4-dinitrotoluene using microcantilever sensors. *Sens. Actuators B*, **99** (2–3), 223–229.

75 Datskos, P.G., Lavrik, N.V., and Sepaniak, M.J. (2003) Detection of explosive compounds with the use of microcantilevers with nanoporous coatings. *Sens. Lett.*, **1** (1), 25–32.

76 Pinnaduwage, L.A., Gehl, A., Hedden, D.L., Muralidharan, G., Thundat, T., Lareau, R.T., Sulchek, T., Manning, L., Rogers, B., Jones, M., and Adams, J.D. (2003) Explosives: a microsensor for trinitrotoluene vapour. *Nature*, **425**, 474–474.

77 Pinnaduwage, L.A., Wig, A., Hedden, D.L., Gehl, A., Yi, D., and Thundat, T. (2004) Detection of trinitrotoluene via deflagration on a microcantilever. *J. Appl. Phys.*, **95** (10), 5871–5875.

78 Pinnaduwage, L.A., Thundat, T., Gehl, A., Wilson, S.D., Hedden, D.L., and Lareau, R.T. (2004) Desorption characteristics of uncoated silicon microcantilever surfaces for explosive and common nonexplosive vapors. *Ultramicroscopy*, **100** (3–4), 211–216.

79 Muralidharan, G., Wig, A., Pinnaduwage, L.A., Hedden, D., Thundat, T., and Lareau, R.T. (2003) Adsorption–desorption characteristics of explosive vapors investigated with microcantilevers. *Ultramicroscopy*, **97** (1–4), 433–439.

80 Pinnaduwage, L.A., Yi, D., Tian, F., Thundat, T., and Lareau, R.T. (2004) Adsorption of trinitrotoluene on uncoated silicon microcantilever surfaces. *Langmuir*, **20** (7), 2690–2694.

81 Voiculescu, I., Zaghloul, M.E., McGill, R.A., Houser, E.J., and Fedder, G.K. (2005) Electrostatically actuated resonant microcantilever beam in CMOS technology for the detection of chemical weapons. *IEEE Sensors J.*, **5** (4), 641–647.

82 Pinnaduwage, L.A., Ji, H.F., and Thundat, T. (2005) Moore's law in homeland defense: an integrated sensor platform based on silicon microcantilevers. *IEEE Sensors J.*, **5** (4), 774–785.

83 Arntz, Y., Seelig, J.D., Lang, H.P., Zhang, J., Hunziker, P., Ramseyer, J.P., Meyer, E., Hegner, M., and Gerber, Ch. (2003) Label-free protein assay based on a nanomechanical cantilever array. *Nanotechnology*, **14** (1), 86–90.

84 Backmann, N., Zahnd, C., Huber, F., Bietsch, A., Pluckthun, A., Lang, H.-P., Güntherodt, H.-J., Hegner, M., and Gerber, Ch. (2005) A label-free immunosensor array using single-chain antibody fragments. *Proc. Natl. Acad. Sci. USA*, **102** (41), 14587–14592.

85 Lam, Y., Abu-Lail, N.I., Alam, S.M., and Zauscher, S. (2006) Using microcantilever deflection to detect HIV-1 envelope glycoprotein gp120. *Nanomedicine*, **2** (4), 222–229.

86 Milburn, C., Zhou, J., Bravo, O., Kumar, C., and Soboyejo, W.O. (2005) Sensing interactions between vimentin antibodies and antigens for early cancer detection. *J. Biomed. Nanotechnol.*, **1** (1), 30–38.

87 Pinnaduwage, L.A., Hawk, J.E., Boiadjiev, V., Yi, D., and Thundat, T. (2003) Use of

microcantilevers for the monitoring of molecular binding to self-assembled monolayers. *Langmuir*, **19** (19), 7841–7844.

88 Raiteri, R., Nelles, G., Butt, H.-J., Knoll, W., and Skladal, P. (1999) Sensing of biological substances based on the bending of microfabricated cantilevers. *Sens. Actuators B*, **61** (1), 213–217.

89 Savran, C.A., Knudsen, S.M., Ellington, A.D., and Manalis, S.R. (2004) Micromechanical detection of proteins using aptamer-based receptor molecules. *Anal. Chem.*, **76** (11), 3194–3198.

90 Wu, G., Ji, H.-F., Hansen, K.M., Thundat, T., Datar, R., Cote, R., Hagan, M.F., Chakraborty, A.K., and Majumdar, A. (2001) Origin of nanomechanical cantilever motion generated from biomolecular interactions. *Proc. Natl. Acad. Sci. USA*, **98** (4), 1560–1564.

91 Fritz, J., Baller, M.K., Lang, H.-P., Rothuizen, H., Vettiger, P., Meyer, E., Güntherodt, H.-J., Gerber, Ch., and Gimzewski, J.K. (2000) Translating biomolecular recognition into nanomechanics. *Science*, **288** (5464), 316–318.

92 Hansen, K.M., Ji, H.-F., Wu, G., Dater, R., Cote, R., Majumdar, A., and Thundat, T. (2001) Cantilever-based optical deflection assay for discrimination of DNA single-nucleotide mismatches. *Anal. Chem.*, **73** (7), 1567–1571.

93 McKendry, R., Zhang, J., Arntz, Y., Strunz, T., Hegner, M., Lang, H.-P., Baller, M.K., Certa, U., Meyer, E., Guntherodt, H.-J., and Gerber, Ch. (2002) Multiple label-free biodetection and quantitative DNA-binding assays on a nanomechanical cantilever array. *Proc. Natl. Acad. Sci. USA*, **99** (15), 9783–9788.

94 Gfeller, K.Y., Nugaeva, N., and Hegner, M. (2005) Rapid biosensor for detection of antibiotic-selective growth of *Escherichia coli*. *Appl. Environ. Microbiol.*, **71** (5), 2626–2631.

95 Gupta, A., Akin, D., and Bashir, R. (2004) Single virus particle mass detection using microresonators with nanoscale thickness. *Appl. Phys. Lett.*, **84** (11), 1976–1978.

96 Nugaeva, N., Gfeller, K.Y., Backmann, N., Lang, H.-P., Düggelin, M., and Hegner, M. (2005) Micromechanical cantilever array sensors for selective fungal immobilization and fast growth detection. *Biosens. Bioelectron.*, **21** (6), 849–856.

97 Chen, G.Y., Warmack, R.J., Oden, P.I., and Thundat, T. (1996) Transient response of tapping scanning force microscopy in liquids. *J. Vac. Sci. Technol. B.*, **14** (2), 1313–1317.

98 Butt, H.J., Siedle, P., Seifert, K., Fendler, K., Seeger, T., Bamberg, E., Weisenhorn, A.L., Goldie, K., and Engel, A. (1993) Scan speed limit in atomic force microscopy. *J. Microscopy*, **169** (1), 75–84.

99 Braun, T., Barwich, V., Ghatkesar, M.K., Bredekamp, A.H., Gerber, Ch., Hegner, M., and Lang, H.-P. (2005) Micromechanical mass sensors for biomolecular detection in a physiological environment. *Phys. Rev. E*, **72** (3), 031907-1–031907-9.

100 Burg, T.P., Godin, M., Kundsen, S.M., Shen, W., Carlson, G., Foster, J.S., Babcock, K., and Manalis, S.R. (2007) Weighing of biomolecules, single cells and single nanoparticles in fluid. *Nature*, **446** (7139), 1066–1069.

101 Karlsson, R., Michaelsson, A., and Mattsson, L. (1991) Kinetic analysis of monoclonal antibody–antigen interactions with a new biosensor based analytical system. *J. Immunol. Methods*, **145** (1–2), 229–240.

102 Glasmästar, K., Larsson, C., Höök, F., and Kasemo, B. (2002) Protein adsorption on supported phospholipid bilayers. *J. Colloid Interface Sci.*, **246** (1), 40–47.

103 Ganesan, P.G., Wang, X., and Nalamasu, O. (2006) Method for sensing the self-assembly of polyelectrolyte monolayers using scanning probe microscope cantilever. *Appl. Phys. Lett.*, **89** (21), 213107–213113.

104 Berger, R., Delamarche, E., Lang, H.-P., Gerber, C., Gimzewski, J.K., Meyer, E., and Güntherodt, H.-J. (1997) Surface stress in the self-assembly of alkanethiols on gold. *Science*, **276** (5321), 2021–2024.

105 Kohale, S., Molina, S.M., Weeks, B.L., Khare, R., and Hope-Weeks, L.J. (2007) Monitoring the formation of self-assembled monolayers of alkanedithiols using a micromechanical cantilever sensor. *Langmuir*, **23** (3), 1258–1263.

106 Zhang, R., Graf, K., and Berger, R. (2006) Swelling of cross-linked polystyrene spheres in toluene vapor. *Appl. Phys. Lett.*, **89** (22), 223114–223114-3.

107 Bumbu, G.-G., Wolkenhauer, M., Kircher, G., Gutmann, J.S., and Berger, R. (2007) Micromechanical cantilever technique: a tool for investigating the swelling of polymer brushes. *Langmuir*, **23** (4), 2203–2207.

108 Igarashi, S., Itakura, A.N., Toda, M., Kitajima, M., Chu, L., Chifen, A.N., Förch, R., and Berger, R. (2006) Swelling signals of polymer films measured by a combination of micromechanical cantilever sensor and surface plasmon resonance spectroscopy. *Sens. Actuators B*, **117** (1), 43–49.

109 Pera, I. and Fritz, J. (2007) Sensing lipid bilayer formation and expansion with a microfabricated cantilever array. *Langmuir*, **23** (3), 1543–1547.

110 Moulin, A.M., O'Shea, S.J., Bradley, R.A., Doyle, P., and Welland, M.E. (1999) Measuring surface-induced conformational changes in proteins. *Langmuir*, **15** (26), 8776–8779.

111 Mukhopadhyay, R., Sumbayev, V.V., Lorentzen, M., Kjems, J., Andreasen, P.A., and Besenbacher, F. (2005) Cantilever sensor for nanomechanical detection of specific protein conformations. *Nano Lett.*, **5** (12), 2385–2388.

112 Wu, G., Datar, R.H., Hansen, K.M., Thundat, T., Cote, R.J., and Majumdar, A. (2001) Bioassay of prostate-specific antigen (PSA) using microcantilevers. *Nat. Biotechnol.*, **19**, 856–860.

113 Valiaev, A., Abu-Lail, N.I., Lim, D.W., Chilkoti, A., and Zauscher, S. (2007) Microcantilever sensing and actuation with end-grafted stimulus-responsive elastin-like polypeptides. *Langmuir*, **23** (1), 339–344.

114 Pei, J., Tian, F., and Thundat, T. (2004) Glucose biosensor based on the microcantilever. *Anal. Chem.*, **76** (2), 292–297.

115 Chen, T., Chang, D.P., Liu, T., Desikan, R., Datar, R., Thundat, T., Berger, R., and Zauscher, S. (2010) Glucose-responsive polymer brushes for microcantilever sensing. *J. Mater. Chem.*, **20** (17), 3391–3395.

116 Braun, T., Backmann, N., Vogtli, M., Bietsch, A., Engle, A., Lang, H.-P., Gerber, Ch., and Hegner, M. (2006) Conformational change of bacteriorhodopsin quantitatively monitored by microcantilever sensors. *Biophys. J.*, **90** (8), 2970–2977.

Index

Scanning Probe Microscopy of Soft Matter: Fundamentals and Practices, First Edition.
Vladimir V. Tsukruk and Srikanth Singamaneni.

b

d

g

n

t

u

v